MW00843506

THIRD EDITION

# MANUFACTURING FACILITIES

## Location, Planning, and Design

THIRD EDITION

# MANUFACTURING FACILITIES

## Location, Planning, and Design

Dileep R. Sule

CRC Press is an imprint of the
Taylor & Francis Group, an **informa** business

CRC Press
Taylor & Francis Group
6000 Broken Sound Parkway NW, Suite 300
Boca Raton, FL 33487-2742


Printed in the United States of America on acid-free paper
10 9 8 7 6 5 4 3 2 1

International Standard Book Number-13: 978-1-4200-4422-5 (Hardcover)



**Library of Congress Cataloging-in-Publication Data**

Sule, D. R. (Dileep R.)
Manufacturing facilities : location, planning, and design / Dileep R. Sule. -- 3rd ed.
p. cm.
Includes bibliographical references and index.
ISBN 978-1-4200-4422-5
1. Factories--Design and construction. 2. Factories--Location. I. Title.

TS177.S84 2008
658.2--dc22 2008017803

**Visit the Taylor & Francis Web site at**
**http://www.taylorandfrancis.com**

**and the CRC Press Web site at**
**http://www.crcpress.com**

# Contents



## *PART 1 Planning and Design*

## *PART 2 Facility Location*

# Preface

Since the introduction of the previous edition of *Manufacturing Facilities*, worldwide industrial competition for consumer products has grown intense. International trade barriers are disappearing and few manufacturers have established themselves as trend setters while others have captured the mass market with high quality product at low cost. Thus to compete, one must plan, organize, and skillfully produce manufacturing units from concepts to the final product in an equally well planned facility that provides the means to achieve these objectives efficiently without waste.

Facility planners and engineers play an important part in understanding and enhancing all major aspects of manufacturing. A facility planner can also serve as a bridge between different departments within an organization that are trying to implement a comprehensive plan of action to produce economically and effectively. This edition provides the information and analytic tools necessary to convert a product design into a production strategy and then describes in detail all of the planning techniques needed to build and operate an efficient manufacturing facility. The new edition expands on the information a facility planner must have to take full advantage of today's technological advances.

Chapter 2 covers product design and development and includes techniques such as market research, forecasting and quality function deployment. These methods establish if there is a need for a product and important features that should be included in the product that will lead to improved customer satisfaction.

To increase operational efficiency, use of automation in manufacturing is almost always necessary. Chapter 3 covers automation. It includes topics such as bar codes, machine vision, radio frequency identifier and programmable logic controllers.

Chapter 4 describes some tools necessary to measure and improve operational performance. Production charts are used to display present and/or proposed methods of operation. Depending on volume and number of different products produced in the plant, distinct plant layout and operational flexibility are needed. Here, one type of production arrangement may be better than the other. Chapter 4 describes different production systems including job shop, batch processing, assembly line and application of group technology (GT) in developing cellular and flexible manufacturing.

Chapter 5 incorporates methods for selection of labor and equipment. Appropriate staffing and capacity of both manpower and equipment are necessary for keeping the production cost down.

To respond to the need of efficient plant operations, a number of new approaches have been recently developed. They are grouped together under an umbrella of Lean Manufacturing. The techniques includes 5S, Six Sigma, Andon, Takt Time, SMED, Value Stream Mapping, Visual Management, Poka-yoke and even supply chain management. All these techniques are described in Chapter 6.

Some supporting topics such as type and requirements of plant buildings and support facilities, as well as within plant functioning organization and methods of communication, are described in Chapter 7. This background information is useful for looking at the entire picture.

Chapters 8, 9, and 10 are about material handling. Efficient material handling must be able to function without waste. Material handling has been evolving and information included in these chapters covers the basics of material handling, equipment descriptions, operations research techniques in material-handling system designs, as well as the use of robots and automated guided vehicles in material handling. Group technology is being increasingly used to divide product mix into coherent job cells and has become an important tool in developing production strategies and machine layouts. The topic of group technology is discussed in considerable depth in Chapter 10. The chapter also includes methods for efficient machine placement in a job shop and flexible manufacturing systems.

In production, even if we try to achieve Just in Time (JIT) production, some storage and warehousing may be essential. How to develop and operate a storage facility is illustrated in Chapter 11. Storage requires floor space, which can be expensive. MRP (Material Requirement Planning) has significantly reduced the need for excess inventory and discussion on MRP is included in the storage and warehousing section in Chapter 11.

Chapters 12, 13, and 14 discuss another very important topic, that of a plant layout. It is not sufficient to optimize only the layout of the facility. Facility planners are faced with simultaneously developing a plant layout and selecting material-handling equipment and ensuring material-handling systems. Chapter 14 discusses these issues and describes a procedure for concurrently minimizing plant layout and associated material-handling costs.

Every facility planner should be able to estimate the lighting, heating, and air-conditioning requirements for a new or existing facility. Techniques to determine these requirements are introduced in Chapter 15. Also included is a discussion of the Americans with Disabilities Act of 1990 (Chapter 15 and Appendix E), which has considerable influence on facility design.

Chapter 16 introduces computer integrated manufacturing systems (CIMS). It describes basic components of CIMS, benefits and deficiencies of CIMS, and how to plan for incorporating CIMS.

Chapters 17 through 19 focus on rudimentary methods of facilities locations. These heuristic methods are quick and easy to apply and provide good solutions to select facilities location problems. In most cases, these are sufficient to incorporate efficiency into the system.

A *solutions manual* for instructors is also available from the publisher; it contains complete solutions to all the problems in the text. In addition, there are suggested solutions for the two case studies included in Appendix F.

Although all the chapters in this text are useful, it may not be possible to cover them all in a one-semester course. This book provides the instructor with the opportunity to develop a course based on the students' backgrounds and particular areas that he/she feels must be emphasized in their curriculum. Lean manufacturing covered in Chapter 6 is a collection of methods that are presently used extensively in improving plant performance. In addition, an instructor wanting to stress plant layout and material handling may concentrate on Chapters 4, and 8 through 14. Those who are primarily interested in understanding concepts of simultaneous plant layout and material handling may study Chapter 14 example without going through all the detailed calculations involved. To add some flavor of facility location, instructors may add Chapters 17 through 19 into their studies. Curricula that require students to synthesize the information on various topics in a capstone design course will find this book particularly useful because it integrates all the important aspects of industrial engineering as they relate to facilities planning.


## ACKNOWLEDGMENTS

A number of individuals contributed in development of this edition. A significant role was played by Mathhew Chaisson by providing pictures used throughout the book. He also confirmed the concepts of lean manufacturing by illustrating a case study. Manish Chavan and Oner Moral were also Instrumental in acquiring narrative photos and providing help when needed. Kishor Joshi, Ravi Karthik, and Saroj Neupane were very helpful in getting the manuscript in order. To all of them I wish to express my sincere thanks and appreciation.

The quality of the manuscript was improved because of the comments and suggestions from CRC staff. In particular I want to thank Cindy Carelli, the senior acquisition editor, and Jay Margolis, project editor, for their help and comments throughout the project.

Finally, to my wife, Ulka, and children, Sangeeta and Sandeep, my thanks for their support during this project.


**Dileep R. Sule**

# 1 Introduction

Planning is vital for the efficient utilization of available resources. It is a necessary component in various areas of a nation's economy — in both the public and private sectors, manufacturing and service organizations, business, and education.

Manufacturing organizations are especially challenging because they involve the performance of multiple activities; all must be planned, designed, coordinated, and executed through a collective effort. Although each member of the team (designer, production worker, material handler) may be an expert in his or her own area, each person's work must achieve the overall objective of the plant. Only an efficient and productive organization can survive in today's competitive market.

Because of this, managers and supervisors must have an extensive background in all phases of manufacturing. They should be able to communicate with one another and understand problems and their proposed solutions, as well as their effects on the performance of the entire plant. In essence, managers should have an overview of all functions involved in successfully designing and operating a plant.

In this book, we approach manufacturing planning in two phases. Part I covers all phases of manufacturing within a plant, such as designing a product, choosing manufacturing processes, developing production systems and associated plant layout, designing material-handling and storage systems, selecting the essential labor resources to operate the plant, and considering the cost factors that go into the design and operation of the plant. Part II covers basic location/allocation analysis, which determines (1) the most economic location for a plant or plants within a country (or machines within a plant), and (2) which customers should be served by which plant (or machine) if there is more than one manufacturing facility.

## 1.1 NATURE AND SCOPE OF PART I

Designing a manufacturing plant is a bold undertaking that necessitates many complex decisions, each of which significantly affects the profitability of the venture. In broad generalities, we must select a product, design it, produce it, and sell it. Besides the financial consideration of raising capital, which we will not discuss, numerous decisions that affect the time and cost of manufacturing the product must be made in each step. For example,

1. Improper design or layout of a plant can add considerably to the cost of manufacturing. It may increase the throughput time, setup times, and in-process inventories, and may contribute to the overall inefficiency of operations.
2. For repeated high-volume and tedious tasks, use of robots might reduce production cost, but the same robots could actually add to the cost if the

facility must handle many tasks, each with different characteristics and having a low volume of production.

3. Being overinsured or underinsured can add to costs and potential liabilities.
4. A careful design, appropriate processing methods, and a well-laid out plant with well-trained personnel might produce a product that is functionally superior but still not competitive. This could be because the plant is located far from its market, thus increasing the distribution cost; because not enough attention was paid to the long-term performance (reliability) of the product; or perhaps because the packaging was not attractive enough.
5. The choice of a material-handling system and equipment can define the efficiency of operation, and, once selected, the system is generally expensive to change.
6. Use of computers, numerically controlled machines, and robots can add to integration and automation in manufacturing, but it also adds to the initial (installation) cost.
7. Storage and warehouse operations are critical in maintaining smooth production and low inventory costs.

These are just several of the numerous problems that a decision maker must face. The purpose of Part I of this book is to provide the necessary background information so that such decisions are based on an understanding of the alternatives. Successful introduction of products or services in the market must consider some or all of the following factors:

- market research and forecasting of demand
- design and specification of the product
- lean manufacturing and supply chain analysis
- degree of automation
- operation analysis
- make-or-buy decisions
- production processes
- building size and type of construction
- development of organization and determination of support personnel
- material-handling requirements
- development of material-handling systems
- selection of material-handling equipment
- evaluation of robots and automatic guided vehicles in material handling
- selection and design of packaging
- storage decisions and warehousing operations
- dock design and development
- plant layout
- relationship between material handling and plant layout
- plant site selection
- utility selection and specification
- safety practices and policies
- compliance with the Americans with Disabilities Act

- insurance and tax considerations
- computer integration in manufacturing

We discuss each of these topics in some detail in the ensuing chapters. In many of these chapters, we also present computer programs that automate tedious calculations associated with some analyses.

## 1.2 NATURE AND SCOPE OF PART II

One problem area in which planning can improve efficiency and save considerable effort and resources is that of facility location and allocation. The basic questions here are: how to choose the optimum location in which to install the facilities, and how to assign customers to each facility so as to minimize the overall cost of operation. The nature of the problem defines *facility* and *customer*. For example, in determining suitable locations for industrial plants so that they can best serve demands from various regions in the country, the plants are the facilities and the product users are the customers. In determining the market territories to assign to sales personnel, the territories are the customers and the salespeople are the facilities.

**FIGURE 1.1** Manufacturing. Translating ideas into products. Top left: The first step in the process is designing the product. Top right: A prototype is then produced and tested. Bottom left: Production of a product in a modern, well-designed facility can result in fewer mistakes and lower costs. Bottom right: These steps result in a useful and necessary product. (Courtesy of Grob Inc., Bluffton, OH.)

Examples of facility location/allocation problems are numerous. Although the few methods that we illustrate in this book might not address all the applications listed here, it is still useful to note the fields in which the problems relating to facility location may arise:

- Selecting sites for emergency service facilities, such as hospitals and fire stations
- Determining the best locations for tool rooms, machines, water fountains, wash areas, concession areas, and first-aid stations in a manufacturing plant
- Choosing sites for warehouses and distribution centers
- Selecting subcontractors and assigning of appropriate work to each
- Picking vendors and determining items to purchase from each, with or without quantity discounts
- Choosing sites for maintenance departments in plants or, on a larger scale, those for state agencies, such as a highway department or garage
- Selecting sites and capacities of machines to meet expected demands from customers distributed throughout a given area
- Developing a layout for a machine shop or an instrument panel

In this book, we study only the basic models in facility location. Other methods and extensions are illustrated in the vast literature of industrial engineering and management science. Interested readers may refer to Francis and White (1991) for an extensive bibliography.

**FIGURE 1.2** Facilities layout. Integration of people and equipment. A modern manufacturing plant allows for efficient material handling and storage. The machines are well placed, allowing room for storing and moving materials in an effective manner. (Courtesy of Grob Inc.)

## 1.3 THE ROLE OF A FACILITY PLANNER

All the functions we discuss in this book may be performed by a *facility planner*. Who is a facility planner and what is his/her job description? The person responsible for coordinating different departments and achieving efficiency in manufacturing operations — a facility planner — may be designated differently according to the size of facilities. In a small plant, this person could be a supervisor, a manager, or even the president of the company. In a large plant, this person is more likely to be designated as a facility manager or planner. Traditionally, one of the major responsibilities of a facility planner is to develop an efficient plant layout and material-handling systems. This individual may also be called upon to integrate many tasks and operations within manufacturing.

Facility planners may participate in the design and development of a quality product that satisfies a known market demand and can be produced efficiently in the plant. They may help select economic processes to manufacture the product and specify the necessary machines and the qualifications of the labor for operating these machines. They may formulate the production policies and specify the degree and type of automation. They may develop the plant layout, define the direction of material flow, and select the material-handling equipment to accomplish the material transfer. They may develop the locations for in-process storage and determine storage capacities. They may determine the areas needed for storage and warehousing, design their layouts, and formulate their operating policies. They may be responsible for seeing that the safety procedures are followed and that the plant is in compliance with disability regulations. They may also be responsible for keeping the costs of insurance, taxes, and health care within limits. In short, facility planners seek to maximize the production efficiency of the plant.

The list of the planning tasks may seem rather long, but in effect, the facility manager coordinates sales, engineering, and personnel with production and operations activities, such as planning, scheduling, and controlling, and contributes to management of these areas. Although the activities specified here are discussed in a manufacturing context, similar facility planning activities are found in nonmanufacturing or service environments. For example, food services in a hospital must be satisfactory to the patients and staff. In operating a retail store, managers must decide which items to store, the inventory level for each item, the number of cashiers to hire, and the layout of the store. Solutions to these problems can be derived from the subject matter covered in this book.

## 1.4 COMPUTER PROGRAMS

The CD attached to the back cover contains a number of computer programs, a full listing of which is given in Appendix G. It is hoped, however, that the reader will first become familiar with the illustrated techniques and then use a computer to solve large-scale problems and to do sensitivity analysis with the input/output parameters. Like any other prepackaged computer software, the programs should not be regarded as instant providers of answers. Each one is built on certain assumptions and logic, and only by understanding the methods shown in various chapters can one appreciate the results presented by the programs.

The programs concern both location analysis and the planning and design segment of the book. The introduction to a program follows a general pattern within a given chapter. First, a specific procedure is explained; then, a numerical example is solved. At the end of the chapter, input requirements for the computer program associated with the procedure are listed. The programs are written in BASIC, FORTRAN (WATFIV), or C. Both BASIC and FORTRAN are readily available in most computer facilities, so the associated programs are in uncompiled forms. This presentation allows the reader to modify the programs — for example, increase the dimension statements — to accommodate larger data sets. Other programs are written in Better-BASIC or C and are provided in executable forms. Such programs include:

1. Material requirement planning
2. Group technology (one each for small- and large-dimension problems)
3. Computer-aided plant layout (COMLADII)
4. Computer-aided plant layout and material handling (COMLAD3).

## SUGGESTED READINGS

Francis, R.L., and White, J.A., *Facility Layout and Location — An Analytical Approach,* 2nd edition, Prentice Hall, Englewood Cliffs, NJ, 1991.

Phillips, D.T., Ravindran, A., and Solberg, J.J., *Operations Research — Principles and Practice,* 2nd edition, John Wiley, New York, 1987.

Zanakis, S.H., and Evans, J.R., "Heuristic optimization: why, when, and how to use it," *Interfaces,* 5, 84–86, 1981.

# *PART 1*

## *Planning and Design*

# 2 Product Development

In this chapter, we introduce several important steps in product development. *Market research* evaluates the sales potential of a new or a revised product that is being proposed. *Quality function deployment* analyzes the product to understand real customer desires and how to convert those into the design features. *Forecasting* estimates the expected demand for the product. *Design* transfers an idea into engineering drawings that form the basis for developing the necessary manufacturing facilities. *Design for manufacturing* evaluates and modifies the product design so that it can be produced efficiently. *Concurrent engineering* analyzes the existing processes and procedures to improve efficiency during the production phase.

## 2.1 MARKET RESEARCH

In the business world, companies very seldom remain constant in terms of the products they manufacture. New products are often introduced, and existing items are modified to accommodate the changing needs and preferences of consumers. For old and newly emerging corporations alike, information about what people desire, the price they are willing to pay, and the extent of the market potential are of tremendous value.

Market research involves the use of scientific methods, mainly statistical techniques, in collecting and analyzing data to discover consumer needs and desires in relation to the product a company might wish to manufacture and/or introduce into the market. Such research could determine consumers' attitudes (e.g., need for the product as well as quality and packaging requirements) and buying habits (e.g., frequency of purchase, brand loyalty, and acceptable price), and might help develop sales and lower marketing costs when the product is finally introduced.

Product planning plays an important part in sales potential because customers are influenced by such factors as design, price, performance, ergonomics, and aesthetic appeal of a product. These factors, however, are interrelated. A customer must have some reason to purchase the product — perhaps the item performs a useful task or the purchaser has a desire to possess the object. The product's design must meet the visual requirements of the consumer; this often influences the purchasing decision and the price that the consumer is willing to pay. Although failure to provide a satisfactory ergonomic interface for frequently used items may not affect initial sales, it will definitely have an impact on future sales to the same customer, as well as to the customer's associates.

Market surveys can also be used to determine customer-desired characteristics that might influence the design — for example, aesthetic appeal. The public might have preferences in color, shape, handling, and other factors that appeal to their senses. Surveys can further indicate an approximate selling price and how this price might change on the basis of additional design features.

Introducing a new product is especially challenging. The company must ensure that the item will not only compare well against current competitor's, but should also include all the features that will be available during and shortly after its introduction to handle the challenge posed by a competing product selling at approximately the same price. To be assured of sales, the product has to cost less than its competitors or must offer features that are not found in other products. Unless the product is better or cheaper, customers will not choose the item; the objective should not be to match the competitors but to move beyond them.

### 2.1.1 Market Potential

If similar products are already on the market, it is vital to determine the volume produced and sold by other manufacturers to estimate the sales potential. This information can often be obtained from the manufacturers themselves with surveys by letter and telephone, and at times by research at the library or on the Internet in government documents, stockholders' reports, and trade journals. Analysis of past and present data on sales, along with the factors that affect those sales, such as population, income level, frequency of purchase, and changes in attitudes and fashions, may indicate the market potential for the foreseeable future.

Government documents provide vast amounts of data that otherwise would be almost impossible or very expensive to gather. For instance, the U.S. Department of Commerce, Bureau of the Census publishes annual reports about the various industries in the United States. These reports include such information as production, sales, imports, and exports. The government also publishes reviews on the "Revised Monthly Retail Sales and Inventories," which record production and sales figures for each industry. These monthly reports divide the retail industries into two segments: durable and nondurable. Durable goods include commodities such as furniture and automobiles — items that will be used for at least a short duration; on the other hand, nondurable goods include food, gasoline, and other things that are suitable for immediate consumption.

Some other sources of data include:

1. The Conference Board, *Guide to Consumer Market*, New York. It provides data on population, employment, income, expenditures, and related items.
2. *Business Week*. The first issue of each year presents data on expected trends for industries during that year.
3. Dun's *Census of American Business* and The Future Directory published by *Time*. Both provide financial data about American businesses.
4. Market research firms. Many sell their findings; prominent among these are:
   a. A.C. Nielsen Company
   b. Marketing Research Corporation of America
   c. Synovate
   d. National Family Opinion

Many markets are changing rapidly. Not only is sales history not necessarily a good indicator of the future market, but as a result of the introduction of new technology, often no information is available on which to base a forecast. How do you accurately predict the future of multimedia players when you are the first one to enter the market? The use of microchips in appliances, automobiles, and communications has created many new opportunities for companies introducing unique products in the market. In such cases, estimation of the market potential is especially difficult and mainly based on factors that the company thinks might influence sales. The data for some of these factors may be obtained either from the sources mentioned earlier or by the methods described here.

### 2.1.2 Collection of Data

Although data can be gathered in many ways, there are three basic types of surveys that obtain personal views and values, each with its own advantages and disadvantages. The methods are mail survey, telephone survey, and personal interview.

The mail survey is relatively easy to perform and requires minimal labor. A survey form is developed and distributed to people on a mailing list. However, it can be difficult to obtain responses from these individuals. Maintenance of mailing lists of customers who might be interested in the product can also be a difficult task. It may be possible to buy an appropriate mailing list from any of several organizations.

The telephone interview requires more labor than the mail survey. As the term implies, questions are asked over the phone, and responses are tabulated. Again, it is important to converse with the "right type" of customers — people who would be interested in buying the product. The mail survey (with a low response rate) and the telephone survey (with a high response rate) can both be conducted at minimal cost, albeit at the expense of personal contacts. Conveying all the features of the product may at times require displaying the product or describing it with visual aids; these alternatives are impossible in mail and telephone surveys.

The personal interview technique requires polling personnel to contact potential customers at their homes or places of business. This is a slower method, in which the interviewer and the consumer may discuss the questions on the survey form, and the observer can note such things as the consumer's attitude and enthusiasm about the product. However, this form of data collection can be very expensive if the group to be surveyed is spread over a large area.

### 2.1.3 Survey Form

As an example of a survey form, a questionnaire from a prospective manufacturer of a coffeemaker, is illustrated in Figure 2.1. The questions should be asked for both home and office use of coffeemakers, and the form should display the information the company wishes to acquire. Questions 1, 2, and 3 are asked to determine whether there is a cross section of population with special interest in the product; if such a group could be identified, advertising and sales campaigns might be geared to appeal to these people. Questions 4, 5, and 6 are included to learn what percentage of the people might be interested in the company's product. Knowledge of what features to include in the product is important, for which purpose questions 7, 10, 11, 14, 15, and 16 are used. Questions 8 and 9 are designed to determine the extent of competition. Questions 12 and 13 are asked to aid in setting prices. Comments might suggest new design features to be considered for future inclusion.

The questionnaire is generally developed on the basis of the information that is needed and can be used. Only very infrequently should questions that have no

In order to better serve our customers, the ABC Company is conducting a survey, the results of which will be used in developing our latest coffee maker. We value your opinion and would appreciate your assistance in our research. Please take a few minutes to complete the following questionnaire and return it in the enclosed envelope. Thank you very much for your time and cooperation.

(1) Circle appropriate age:
20 or under 21 to 30 31 to 40 41 to 50 Over 50

(2) Sex: Male Female

(3) Occupation:

(4) Are you a coffee drinker? Yes No
How many cups per day? 1 2 3 4 5 Other

(5) Present method of making coffee:
Instant Percolator Drip Other

(6) Preferred method of making coffee:
Instant Percolator Drip Other

(7) What is your preference in color for coffee makers?
White Yellow Brown Black
Other (Please specify)

(8) If you own a drip-type coffee maker, what brand is it?

(9) How long have you owned your drip coffee maker?
Less than 6 moths 6–12 months 1–2 years More than 2 years

(10) How often do you make coffee?
Per day: 0 1 2 3 More than 3
Per week: 0 1–4 5–8 9–15 More than 15

(11) How many cups do you make at a time?
2 4 6 8 10

(12) How much did you pay for your present coffee maker?
Below $30 $30–$50 More than $50

(13) How much would you be willing to pay for your coffee maker today?
Below $30 $30–$50 More than $50

(14) Which do you prefer?
Porcelain Stainless Glass Plastic Other

(15) Would you like a timer on your coffee maker? Yes No

(16) Do you use filters? Yes No Cost?

Comments:

**FIGURE 2.1** Market research questionnaire

informational value be included in this form. Collection and analysis of the data are expensive and time consuming, and there is hardly any room for collection of trivial information. Occasionally, a simple question might be used at the beginning of the survey to help put the respondent at ease.

### 2.1.4 Basic Steps in Market Research

We now formally define the steps required to perform market research. The time and effort required in each step depend on how detailed the analysis must be for each of the factors involved.

1. *Analyze the situation.* Review company records, trade and professional publications, library material, and available past market research reports to understand the markets. Be aware of the company's potentialities and limitations. Know why the firm wishes to enter into this venture and what it stands to lose in the process.
2. *Plan the research.* Determine what information is needed. What is the purpose of this research, and what are the methods for gathering such information? Decide how the study is to be organized.
3. *Collect data.* Which survey, if any, may be used to collect the information from consumers? Can government reports, reports released by private corporations, and/or trade magazines be used to discover present sales trends? What facts should be gathered about conditions affecting the sales?
4. *Analyze the data.* Which statistical methods should be used to analyze the collected data? It is important to review the information and determine the implications. An attempt should be made to find out whether market size would increase if the public were (more) interested in the product, what price could be charged, and the answers to other related questions about design and sales.
5. *Report the findings.* All results and interpretations should be properly documented and reported to the appropriate levels of management.

### 2.1.5 Market Decision

The preceding type of analysis should enable management to decide whether it will be profitable for the company to enter the present market. By using the data on the expected life of a product and anticipated future consumer needs, the market potential for the company can be estimated. Initially, rough calculations of production cost should be made to determine the estimated profitability of the product. The cost may depend on what the company feels its share of the market will be, or on how many units it must produce and sell to make the project worthwhile. If the venture seems profitable, then arrangements can be made to develop manufacturing facilities to produce the targeted amount.

### 2.1.6 An Example of Market Research

An equipment manufacturing company wishes to begin production and sales of steel tables. The plant is located in a large city that has several colleges and two universities in addition to the usual public school system. The company's equipment sales are distributed among many customers: approximately 30% to the government and offices, and approximately 70% to the schools. Management expects that 90% to 100% of the new steel tables will be sold to the schools. New school buildings are seldom constructed, so these will not be a major source of sales; the company will therefore depend heavily on its present customers.

There are currently four major suppliers of school tables. One of these has decided to stop making shipments to the area. That company's plant is located quite a distance from the city, and when the cost for shipment was added to that for production, the price to the consumers no longer made the company competitive. This has opened the market to a player already in the area who can supply the tables.

**TABLE 2.1**
**Physical Characteristics of the Proposed Tables**

| Table Capacity in Persons | | Dimensions (Length × Width × Height) (in.) |
|---|---|---|
| One Side | Total | |
| 2 | 4 | 54 × 36 × 30 |
| 3 | 8 | 84 × 36 × 30 |
| 6 | 14 | 180 × 45 × 40 |

The company plans to produce three sizes of tables. The smallest table seats two students on one side and four if both sides are used. The medium-sized table seats three students on one side and up to eight using both sides and the ends. The largest table is a conference table that seats up to 14 people. Table 2.1 shows the dimensions of the three proposed tables.

The company would first like to survey its present customers to determine the potential demand for the product. A large percentage of these customers are located in the vicinity of the plant; because sales personnel often visit the schools and institutions, the market research teams decide to use the personal interview survey.

Of all the people questioned in the survey, 92% responded, all being current customers of the company. Seventy percent of the customers are schools that are required by law to take bids. Approximately 9% of the customers were under contract with their present supplier. All the other customers, being private schools, are under no particular restriction for buying.

Virtually all the respondents expressed at least some interest in the product. Forty percent of the schools planned to change over from desks to tables as soon as their desks needed replacing. All the other schools would consider the changeover, and even without a general replacement, they would still have use for some tables. Most expected their desks to last from 5 to 10 years, with an average life cycle of 8 years.

The respondents also replied concerning the sizes of tables that they think are, or would be, in demand; their relative demand ratios; and the average price that they are presently paying for these tables. The statistical analysis provided the distribution shown in Table 2.2.

**TABLE 2.2**
**Estimated Relative Demand and Price for Each Type of Table**

| Table Capacity in Persons Per Side | Sales (%) | Price ($) |
|---|---|---|
| 2 | 63 | 63.00 |
| 3 | 30 | 88.00 |
| 6 | 7 | 210.00 |

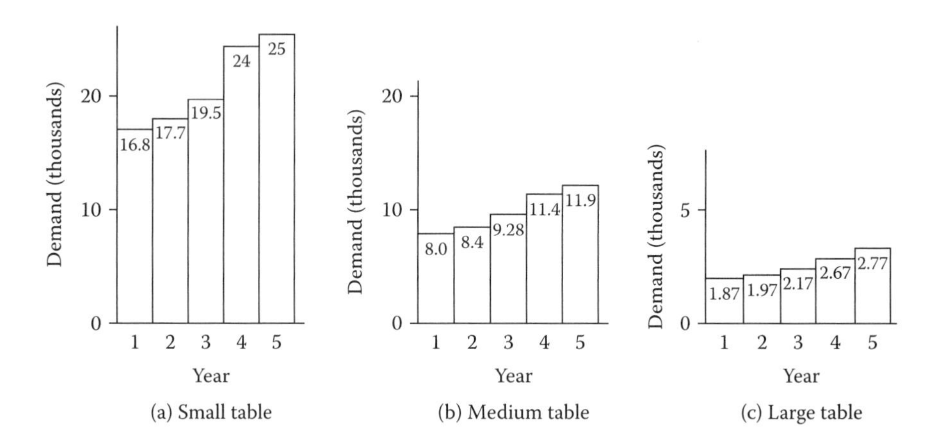


**FIGURE 2.2** Five-year demand forecast

The company plans to distribute its products through both direct sales and sales to independent distributors. All major sales of 25 units or more will be handled directly by the company. Smaller sales will go through independent distributors.

Production plans call for start-up in January of next year. It is estimated that to make the venture profitable, the company will have to obtain 17% of the area market by the end of 5 years. The sales personnel will strive for 8% at the end of the first year, 11% at the second year, and a 2% increase in the market share each year for the following 3 years. These figures were developed after studying past market reports for similar products that the company has manufactured and sold. Demand for school tables in the area has been steadily increasing for the past 5 years. The desks currently in use have an expected life span of 8 years, which means that approximately 12.5% of the desks should be replaced each year. More and more schools are replacing their desks with tables because the latter are found to be more comfortable and lend flexibility to the classroom setting. The forecast demand for the next 5 years is shown in the graphs presented in Figure 2.2. Because only a short period was to be considered, a linear forecast was performed for each type of table.

The company that is planning to cease operations currently accounts for 25% of the market. Using this information and all the other statistics gathered, the team concludes that it is realistic to strive for 17% of the market. The number of units to be produced and sold each year is calculated by multiplying the percentage of the market that the company plans to capture each year by the annual demands. The results are shown in Table 2.3.

At these production rates, is it feasible to produce the product? The answer depends on its cost. A cost estimate of direct material, equipment, and labor to produce the parts must now be made. To this, overhead costs to account for such items as material handling, utilities, and supervision must be added. The overhead cost could be considerable; 150% to 200% of direct labor is not uncommon. A planning sheet, as shown in Figure 2.3, describing all operations and the equipment and tools to be used could help in developing the cost estimate for the product. Chapter 5 describes production charts that are helpful in planning the detailed steps necessary in developing cost estimates.

**TABLE 2.3**
**Projected Sales**

| Year | Percentage of Total Market (%) | Units Produced and Sold | | |
|---|---|---|---|---|
| | | Small | Medium | Large |
| 1 | 8 | 1300 | 640 | 150 |
| 2 | 11 | 1947 | 924 | 217 |
| 3 | 13 | 2535 | 1204 | 282 |
| 4 | 15 | 3160 | 1710 | 401 |
| 5 | 17 | 4000 | 1904 | 444 |

With an estimated efficiency of 90%, the direct cost would be $32.57 = 29.32/0.9. The indirect cost for the plant is 160% of the direct labor, or $20.27 (11.40/0.90 × 1.6), giving a cost for manufacturing of $52.84. If we add $1.20 for administration and packaging/shipping, the net cost to the company is $54.04. Because the unit is expected to sell for $63.00, it will generate a profit of $8.96.

Similar analyses on other types of tables result in a profit of $10.20 for medium tables and $18.15 for large tables. Based on projected sales from the previous chart, the yearly profit is:

| Year | Profit |
|---|---|
| 1 | $21,292.74 |
| 2 | $30,808.47 |
| 3 | $40,143.30 |
| 4 | $56,976.15 |
| 5 | $63,319.40 |

Small Table

| OPERATION | | MACHINE NAME | TOOLS/EQUIP. REQUIREMENTS | DIRECT COST | | |
|---|---|---|---|---|---|---|
| NO. | DESCRIP. | | | LABOR | MATERIAL | EQUIP. |
| 10 | Cut to size (legs) | Radial arm saw | | 5 min. at $5/hr $.42 | Steel tubing 4 × 29" × 1¼" dia. × ⅛" = $2.90 | 5 min. at $6/hr = $.50 |
| 20 | Weld bracket | Welding M/C | Clamps | 10 min. at $6.60/ hr = $1.10 | Rods $.10 Brackets $ 2.20 | 10 min. at $4/hr = $.67 |
| | | | Total | $11.40 | $12.20 | $5.72 |

The direct cost of production is $29.32 (11.40 + 12.20 + 5.72).

**FIGURE 2.3** Sampling planning sheet

This gives a rough estimate of the average yearly profit of $42,508.01. The management expects the need for an additional investment of $125,000 to engage in the planned level of activity. Assuming a 5-year duration for the project, the internal rate of return for this investment is 17%, and, since the company's minimum attractive rate of return before taxes is 15%, the project is accepted.

## 2.2 QUALITY FUNCTION DEPLOYMENT

Along with market research, quality function deployment (QFD) is playing an increasingly important role in product development. QFD was first applied in Japan; the concept is now becoming popular and successful in American industries. QFD calls for a structured process of converting customer requirements into design and manufacturing needs. It is a management planning technique wherein group discussions among concerned personnel are held and a clear consensus on how the customer needs are to be achieved is reached. QFD can be used in any phase of production. For example, for the entire facility, the customer may be the ultimate consumer of the product, while within the plant, the consumer may be the next sequential department to process the part. QFD is also used to compare a product against its competitors in the market.

Development of QFD may start with identifying the customers (internal or external to the organization) and their wants. Questions such as *what, when, where, why,* and *how* the customers will use the product may help in identifying their needs. For example, in developing a coffeemaker, the customer is clearly an external consumer who would make and drink the coffee. The customer could be a restaurant operator or an individual consumer within a family. We shall develop a QFD for an individual consumer. Because we propose to market the product to a large population, we will interview many consumers to determine what they want. Typical answers are shown here.

The product (coffeemaker) will be used for

What: to make coffee in the amount I want, two to eight cups, to keep coffee warm, no burns, easy to clean
When: any time, immediately when I get up, several times during the day
Where: in office, home, car, recreational vehicle, campsite
Why: satisfying beverage, to dampen hunger, to perk up
How: hot, right taste, right concentration, with sugar, cream, or powdered cream, no grounds

The data can also be presented in a fishbone diagram, which is easy to visualize. A fishbone diagram for these data is given in Figure 2.4.

Clearly, the power source for heating the water plays an important part in the decision. Homes and offices use 110-V power supplies, while cars and recreational vehicles work with 12-V batteries. A campsite may lack power, so a wood fire is needed to warm the water. Thus, we must first decide which market to serve. It is possible to develop a power supply (perhaps using a transformer) to serve more than

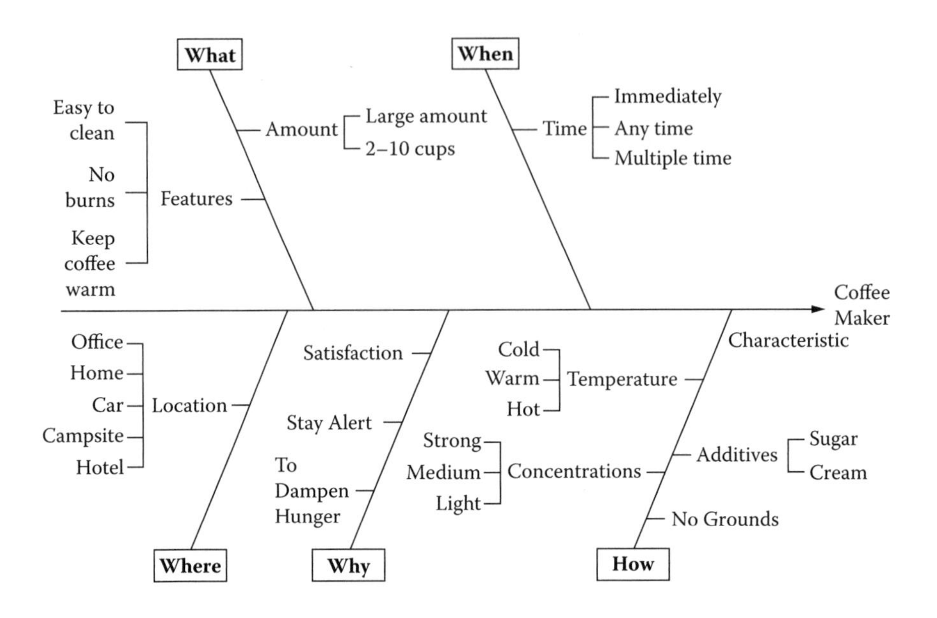


**FIGURE 2.4** Fishbone diagram for a coffeemaker

one type of user, but the resulting cost may be prohibitive. Therefore, we may decide to concentrate on the home and office use of the coffeemaker.

The next step is to translate the customers' comments into design attributes and develop a QFD chart (Figure 2.5). This chart is a cross-matrix chart that shows the correlation between customer needs and how these needs can be achieved by the proposed design features. Before entering the needs into the chart, however, the raw statements should be evaluated and identical needs that are worded differently should be consolidated into one logical statement. This prevents one demand feature from being repeated several times in the chart.

The design attributes are listed after considerable discussions between the team members. Typically, such a group may consist of a design engineer, a manufacturing engineer, and administrative personnel. For example, good taste is a function of the rate of steam flow through coffee grounds. Flow, in turn, is a function of the number of cups of coffee to be made. A variable heater setting that controls the wattage would ensure uniform steam generation and a consistent taste. Thus, the design of the control lever is included in the QFD. The QFD chart not only lists the positive relations between the needs and proposed design features but also notes any negative effects. For example, a double-barrel coffeemaker (a two-filter system, with one filter used to make the first pot and another used to make the second pot without refilling the coffee) would not be easy to clean, indicated by an X.

The stage is now set to determine which design features should be included in the product. Again, a group discussion can lead to a subjective assignment of weights for each need statement from the consumer. The factors that may influence the weight

| Design Features / Customer Wants | Auto Shut Off | Few Parts | Stopper | Timer | Double-barrel Coffee Maker | Shape | Size | Thermostat | Control Lever | Sugar Cream Dispenser | Filter (Disposable) | Filter Cup (Size/Shape) | Weights | Total |
|---|---|---|---|---|---|---|---|---|---|---|---|---|---|---|
| **What** | | | | | | | | | | | | | | |
| 2–10 Cups | — — | — — | | | | Δ 9 | ⊙ 81 | ○ 27 | | — — | | — — | 9 | |
| Should not burn | ⊙ 45 | — — | | | | Δ 5 | — — | — — | | — — | | Δ 5 | 5 | |
| Easy to clean | Δ 6 | ⊙ 54 | | | | ⊙ 54 | ○ 18 | — — | | ○ 8 | | ⊙ 54 | 6 | |
| **When** | | | | | | | | | | | | | | |
| Immediately | | | ⊙ 27 | ⊙ 27 | — — | | | | | | — — | | 3 | |
| At Particular times | | | — — | ⊙ 27 | — — | | | | | | — — | | 3 | |
| Multiple times | | | — — | ⊙ 18 | ⊙ 18 | | | | | | Δ 2 | | 2 | |
| **How** | | | | | | | | | | | | | | |
| Hot | | | | Δ 7 | | | | ⊙ 53 | — — | — — | — — | — — | 7 | |
| Right taste | | | | — — | | ○ 24 | ⊙ 72 | ○ 21 | ⊙ 56 | ○ 21 | ⊙ 53 | ○ 21 | 7 | |
| Cream/sugar | | | | — — | | ⊙ 45 | ○ 45 | — — | — — | ⊙ 36 | — — | — — | 4 | |
| No grounds | | | | — — | | ⊙ 72 | ⊙ 72 | — — | — — | — — | ⊙ 45 | Δ 5 | 5 | |
| **Where** | | | | | | | | | | | | | | |
| At home/office | | | | | — — | | | | | | | | 8 | |
| Tabletop | | | | | Δ 5 | | | | | | | | 5 | |
| Under a counter | | | | | — — | | | | | | | | 8 | |
| **Why** | | | | | | | | | | | | | | |
| To stay alert | | | | | | | | | ○ 3 | Δ 1 | ○ 3 | ○ 3 | 1 | |
| To dampen hunger | | | | | | | | | ○ 3 | Δ 1 | ○ 3 | ○ 3 | 1 | |
| **Score** | 51 | 54 | 27 | 79 | 23 | 209 | 288 | 111 | 62 | 77 | 116 | 91 | | 1188 |
| Percentage | 4.29 | 4.54 | 2.27 | 6.64 | 1.93 | 17.59 | 24.24 | 9.34 | 5.22 | 6.48 | 9.76 | 7.66 | | |

| Symbol | Relationship | Weight |
|---|---|---|
| ⊙ | Strong | 9 |
| ○ | Medium | 3 |
| Δ | Weak | 1 |

**FIGURE 2.5** Quality function deployment chart

may include possible sales potential, features of a competing product, production feasibility, reliability, and quality. The last column in Figure 2.5 shows the relative weights for this example. For instance, under the "What" criteria, "Capacity of 2–10 cups" carries a weight of 9, while "It should be easy to clean" carries a weight of 6. These are high weights compared with the "When" or "Why" categories because of the designers' subjective opinions as to what may sell. Each relationship symbol, such as a small circle, also carries a weight, as shown in Figure 2.5. The product of symbol weight and relative weight is noted for each entry in the figure. For example, the attribute "should not burn" has a strong relationship with "auto shutoff," giving a design weight of 9, and has a relative weight of 5. Thus, the product 45 is displayed beside the entry under "Auto Shutoff." The sum of these products for each column determines the score for that design feature. The relative importance of a design feature is obtained by dividing the score for the feature by the sum of the scores for all design features. For instance, the most important features in our coffeemaker are the size and shape of the body, with relative weights of 24.24 and 17.59, respectively.

Developing a QFD to compare one brand of a product against others requires a similar analysis, except that the symbols are used to compare the relative satisfaction given by each product. For each need, the design features within each brand satisfying that need are compared. For example, brand A may heat water faster than brand B. Therefore, the response for immediate water heating for brand A may be two concentric circles, while that for brand B is one circle, indicating that brand B could be improved by further analysis of this design feature.

## 2.3 FORECASTING

Most people like to have sufficient lead time to plan for future activities. This is especially true for engineers and managers who are responsible for plant installation and operations. They must decide, for example, how large the plant should be, which machines to buy, how many workers to employ and train, and how to arrange for financing. The answers might depend on projected or forecast sales of the products.

Forecasting techniques can be divided into two major categories: quantitative and qualitative. Quantitative techniques are applicable when some quantifiable data about past performance are available and it is assumed that the same pattern will continue in the foreseeable future. Qualitative forecasting, on the other hand, does not require specific information and is based mainly on intuition, judgment, and the opinions of people who are knowledgeable and have experience in a specific activity or product. When past quantitative data are not available, as may be the case in introducing a new product, qualitative methods might be the only way of forecasting. Such techniques could also supplement quantitative methods when the pattern for future activities might change via factors that were not prominent in the past. We will study some of the basic techniques of quantitative forecasting, but we must first characterize the means of identifying what a good forecast is.

### 2.3.1 Error Measuring

The difference between the forecast value, $F$, and the actual demand, $D$, is the error in forecasting, $e$. The objective is to choose a forecasting method and/or associated parameters, such as weight, that will minimize this error.

Mathematically, an error in period $t$ can be expressed as $e_t = F_t - D_t$. If there are $n$ periods for which both forecast and actual demands are known, then we have several ways of recording the total error:

$$\mathrm{ME} = \sum_{t=1}^{n} \frac{e_t}{n}$$

$$\mathrm{MAE} = \sum_{n=1}^{n} \frac{|e_t|}{n}$$

$$\mathrm{MSE} = \sum_{t=1}^{n} \frac{e_t^2}{n}$$

$$\mathrm{SDE} = \sqrt{\sum_{t=1}^{n} \frac{e_t^2}{n-1}}$$

where

ME = mean error
MAE = mean absolute error
MSE = mean squared error
SDE = standard deviation of error.

There are also other error measures that are more suitable under certain conditions [interested readers may refer to the work of Makidakis et al. (1984)], but calculating for the MSE is one of the more popular procedures to follow in evaluating the forecasting methods described herein.

### 2.3.2 Forecasting Methods

#### 2.3.2.1 Moving Average

A simple forecasting method for averaging relevant past data, the moving average method requires us to define the number of observations that will be included in the calculation of the average. As new data become available, the oldest observation is dropped and the latest observation is included in calculating the moving average, the prediction for the next period. If we are predicting demand, for example, mathematically, we have

$$F_t = \frac{\Sigma_{i=1}^{n} D_{t-i}}{n}$$

where $F_t$ is the forecast demand during the period $t$, $D_{t-i}$ is the actual demand in period $t - i$, with $i = 1, 2, \ldots, n$, and $n$ is the number of observations to be included in the moving average.

For example, given that $n = 3$ and the actual demands for the past 6 months are:

| Month | Observed Demand |
|---|---|
| January | 500 |
| February | 515 |
| March | 600 |
| April | 620 |
| May | 595 |
| June | 635 |

the predicted demand for July is

$$F_{JULY} = \frac{(620 + 595 + 635)}{3} = 616.66, \text{ or } 617 \text{ units}$$

If the actual demand for July were found to be 625, then the predicted demand for August would be

$$F_{AUGUST} = \frac{(595 + 635 + 625)}{3} = 618.33, \text{ or } 618 \text{ uni}$$

### 2.3.2.2 Weighted Moving Average

In the preceding method, each observation within the moving average calculation made an equal contribution, or had the same weight, in predicting the forecast value. It may be desirable to give more importance to the more recent observations and less to the more distant past. This is achieved by "weighting" each data point. In the previous example, suppose we assign the weights shown in Table 2.4; then the forecast demand in July would be

$$\frac{1}{8}(620) + \frac{1}{4}(595) + \frac{5}{8}(635) = 623$$

or, in general,

$$F_t = \frac{\Sigma_{t=1}^{n} (W_{t-i} \times D_{t-i})}{n}$$

where $W_{t-i}$ is the weight placed on observation $t - i$.

**TABLE 2.4**
**Relative Weight on the Demands**

| Observation | Weight | Relative Weight |
|---|---|---|
| One month old | 5 | 5/8 |
| Two months old | 2 | 1/4 |
| Three months old | 1 | 1/8 |
| Total | 8 | |

#### 2.3.2.3 Exponential Smoothing

Although it is generally preferred that recent values be given more weight in forecasting than older observations, determining the appropriate weight for each as is required in the weighted moving average method is normally a complex task. Exponential smoothing is another alternative in which the weight assigned to observations decreases with age, specifically in an exponential manner. Furthermore, it is accomplished by knowing only the actual and forecast demands for the last period and the smoothing constant. In practice, this method is mainly used for short-term forecasting.

The method is developed with the following thoughts in mind. The variation between the forecast demand and the actual demand may be caused by two factors: (1) a trend, and (2) random fluctuations called noise. Ideally, the next forecast should follow the trend and not react to the noise. We must decide what percentage (called the smoothing constant) of the difference between the actual and forecast demands is due to the trend. The forecast for the next period is adjusted for the trend.

Mathematically, it can be summarized as

$$F_{t+1} = F_t + \alpha (D_t - F_t)$$

where

$F_{t+1}$ = demand forecast in period $t + 1$

$D_t$ = actual demand in period $t$

$\alpha$ = smoothing constant where $0 < \alpha < 1$.

By rearranging the terms, we can write

$$F_{t+1} = \alpha D_t + (1 - \alpha)F_t.$$

To show how the influence of each observation decreases as it becomes older, substitute for $F_t$ in this expression:

$$\begin{aligned} F_{t+1} &= \alpha D_t + (1-\alpha)(\alpha D_{t-1} + (1-\alpha)F_{t-1}) \\ &= \alpha D_t \alpha(1-\alpha)D_{t-1} + (1-\alpha)^2 F_{t-1} \end{aligned}$$

**TABLE 2.5**
**Demand Forecast for $\alpha = 0.2$**

| Month | Forecast | Actual Demand |
|---|---|---|
| January | 500.0 | 500.0 |
| February | 500.0 | 515.0 |
| March | 503.0 | 600.0 |
| April | 522.4 | 620.0 |
| May | 541.9 | 595.0 |
| June | 552.5 | 635.0 |
| July | 569.0 | |

Continuing the process, we obtain

$$F_{t+1} = \alpha D_t + \alpha(1-\alpha)D_{t-1} + \alpha(1-\alpha)^2 D_{t-2}$$
$$+ \alpha(1-\alpha)^{n-1} D_{t-(n-1)} + (1-\alpha)^n F_{t-(n-1)}$$

as $\alpha < 1$, and the weight for each successive observation is decreasing at the rate of $1 - \alpha$; therefore, the weight of the older observation is decreasing exponentially.

Initially, we must first fix the value of $\alpha$. Suppose in the prior example that $\alpha = 0.2$, then the forecast demands for each period would have been as shown in Table 2.5.

To begin the procedure, the forecast value is assumed to be the same as the actual demand for the first period, January. The forecast for February is the same as that for January, because there is no forecasting error in January. The forecast for March is $500 + 0.2(515 - 500) = 503$, and so on. The forecast values for the last six periods do not compare very well with the actual values. Try another value for $\alpha$, for instance, 0.6, chosen at random for illustration (Table 2.6).

**TABLE 2.6**
**Demand Forecast for $\alpha = 0.6$**

| Month | Forecast | Actual Demand |
|---|---|---|
| January | 500.0 | 500.0 |
| February | 500.0 | 515.0 |
| March | 509.0 | 600.0 |
| April | 563.6 | 620.0 |
| May | 597.4 | 595.0 |
| June | 594.9 | 635.0 |

Now the forecast is improved, but it is desirable to determine the optimum value of $\alpha$. One method is to try different values of $\alpha$, calculate the mean square error associated with each, and select the one that gives the minimum MSE. In this example, the best value for $\alpha$, to one-digit accuracy, is 0.9.

#### 2.3.2.4 Curve Fitting

One way to forecast the demand is simply to plot the curve or calculate the value of an equation from available data and extend it to make predictions. The method allows a number of variables to be independent, the values for which can be either preset, controlled, or measured. The estimate of the expected demand can then be made when these values are known. The coefficients of the equations are evaluated following the least-squares fit method — that is, minimizing the sum of squares of the errors.

Suppose the equation to be fitted is

$$Y = b_0 + b_1 X_{1i} + b_2 X_{2i} + \cdots + b_m X_{mi}$$

where there are $m$ independent variables and $n$ observations, $\hat{Y}_i$ is the estimated value of the demand, and $Y_i$ is the actual observed demand in period $i$. The error in measurement is $e = (Y_i - \hat{Y}_i)$, and the least-squares fit requires

$$\min e^2 = \min \sum_{i=1}^{n} (Y_i - \hat{Y}_i)^2$$

$$= \min \sum_{i=1}^{n} (Y_i - (b_0 + b_1 X_{1i} + b_2 X_{2i} + b_m X_{mi}))^2$$

The minimization is obtained by taking the first derivative with respect to each coefficient, that is, $b_0$, $b_1$, …, $b_m$, equating each to zero, and solving the resulting equations simultaneously to obtain the values of each coefficient. In its simplest form, it is known as simple linear regression, which uses only one independent variable, for example, the period in the equation $Y = b_0 + b_1 X_i$. The curve is a straight line, and the corresponding coefficients are obtained by

$$b_1 = \frac{n\Sigma X_i Y_i - (\Sigma X_i)(\Sigma Y_i)}{n\Sigma X_i^2 - (\Sigma X_i)^2}$$

$$b_0 = \bar{Y} - b_1 \bar{X}$$

where

$$\bar{Y} = \frac{\Sigma Y_i}{n} \quad \text{and} \quad \bar{X} = \frac{X_i}{n}$$

### Example: Demand Prediction by Curve Fitting

Consider a lumber manufacturer that has sawdust as its by-product and is contemplating developing a fire log-producing facility to utilize the sawdust. After contacting the stores within a hundred-mile radius, where it believes it can sell its product, the manufacturer has received the following information on the demand for the fire logs in this area over the last six seasons:

| Season ($X$) | Demand (in Thousands) ($Y$) |
|---|---|
| 1980 | 102 |
| 1981 | 105 |
| 1982 | 110 |
| 1983 | 108 |
| 1984 | 115 |
| 1985 | 113 |

The manufacturer expects that it can capture approximately 10% of the market in the first year and can increase its share by 3% in each of the next 3 years. Determine the expected demand in each of the next 4 years.

#### Solution

The plot of the data points is shown in Figure 2.6. It seems that the relationship between years and demand can be expressed satisfactorily by a linear fit; that is,

$$\hat{Y}_i = b_0 + b_1 X_i$$

When $\hat{Y}_i$ is the predicted demand in year $X_i$, note that the calculations can be made with $X_i$ as perhaps 1980 or 1981, or by coding the years by subtracting 1980 to result in 0, 1, 2, 3, etc. Using $X_i$ as 1980, 1981, …, we get the following results:

$$b_1 = \frac{n\Sigma X_i Y_i - (\Sigma X_i)(\Sigma Y_i)}{n\Sigma X^2{}_i - (\Sigma X_i)^2}$$

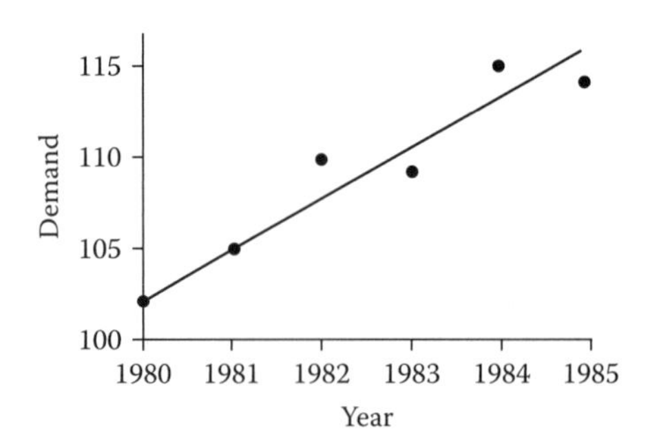


**FIGURE 2.6** Relationship between years and demand

and

$$b_0 = \bar{Y} - b_1 \bar{X}$$

$$\sum X_i = 11{,}895, \quad \sum X^2_i = 23{,}581{,}855,$$

$$\bar{X} = 1982.5$$

$$\sum Y_i = 653, \quad \left(\sum X_i Y_i\right) = 1{,}294{,}614,$$

$$\bar{Y} = 180.83, \quad \left(\sum X_i\right)\left(\sum Y_i\right) = 7{,}767{,}435$$

Subtracting the appropriate values in these expressions, we have

$$b_1 = 2.3714 \quad \text{and} \quad b_0 = -\,4592.47$$

giving the following equation for predicting demand:

$$\hat{Y} = -\,4950.47 + 2.3714\, X_i$$

Now we can forecast the total demand for the next 4 years by extending this graph and by knowing the percentage of the market the manufacturer has predicted for itself. The manufacturer estimates demand as shown in Table 2.7.

Incidentally, had we used the coded values for the years (1980 = 0, 1981 = 1, …), the calculation would have resulted in the following equation:

$$\hat{Y}_i = 102.90 + 2.3714X$$

Using 1986 as 6 would have resulted in the same market predictions as we have observed before.

This solution assumes a continuous linear trend. However, predicting so far into the future is very subjective. For example, the demand might saturate and remain at some constant value. If the manufacturer thinks that this might happen in the planning period, it must change the equation of the curve that will take into consideration

**TABLE 2.7**
**Forecasted Demand**

| $X_i$ | $\hat{Y}$ (in thousands) | Market Share (%) | Predicted Demand (in thousands) |
|---|---|---|---|
| 1986 | 117.13 | 10 | 11.71 |
| 1987 | 119.50 | 13 | 15.54 |
| 1988 | 121.87 | 16 | 19.50 |
| 1989 | 124.24 | 19 | 23.61 |

**TABLE 2.8**
**Domestic Consumption**

| Year | Chrome Faucets | Chrome Showerheads |
|---|---|---|
| 1980 | 579,337 | 566,170 |
| 1981 | 524,958 | 552,970 |
| 1982 | 514,989 | 458,936 |
| 1983 | 519,239 | 551,481 |
| 1984 | 632,444 | 595,291 |

this belief. For instance, the equation might be of the form

$$\hat{Y}_i = boX_i^{bi}$$

### 2.3.3 Demand Forecasting for a New Company

The following two cases indicate how forecasting could be accomplished by using government documents and surveys. Both have old data but illustrate the principles. The first case is for chrome faucets and showerheads. A U.S. government document entitled "Current Industrial Reports of Plumbing Fixtures" shows domestic sales of chrome faucets in the United States between 1980 and 1984 (Table 2.8). The data were obtained by adding U.S. manufacturers' production to imports and subtracting exports from the total.

Simple linear regression analysis results in the following forecasting equations.

For chrome faucets:

$$\text{Demand in current year} = 524{,}004.9 + 10{,}049.5\ (\text{current year} - 1980)$$

For showerheads:

$$\text{Demand in current year} = 527{,}943.7 + 5675.3\ (\text{current year} - 1980)$$

The results from the regression equations could be improved by including factors that affect major sales of the faucets and showerheads, such as new building construction and retail replacement sales. But as an initial estimate, the above equations are fairly good predictors.

The second case consists of a company trying to forecast the demand for a product called "CompuTable." This is a set consisting of a chair and a table to accommodate a desktop computer, its monitor, and printer. The product is to be used mainly by owners of home computers, although it can also be used in small businesses.

The government document does not list such an item; therefore, future demands could not be predicted on the basis of known past sales. Surveys were conducted by telephone and by personal interviews, and the data associated with home computer users were analyzed further. The results indicated that 61.8% of home computer users would like to own a product similar to CompuTable. A search in *Facts on File*

(April 1985, published by Facts on File, Inc., New York, NY) showed that 10% of U.S. households had a home computer as of 1985. Using statistics from the U.S. Bureau of Commerce as a guide to the number of households in 1984, the number of home computer owners is estimated to be 8.5 million. Further research in *Business Week* magazine (June 24, 1985) supported these findings. *Business Week* predicted a leveling off of home computer demand in the near future to approximately 2 million units per year and a slight rise in the sale of portable computers, which can also use CompuTable.

Neglecting minor losses due to death and the "fad factor," a conservative estimate of the potential buying population as of February 1985 would be 61.8% of 8.5 million computer owners, or 5.25 million people. This population would increase annually by 61.8% of 2 million new computer buyers, or by 1.23 million per year. The surveys indicate that approximately 10.9% of this market would purchase a table in any given year. Hence, the expected demands for the 5 years, starting in 1987, which is the first production year for the company, where demand is in thousands, are:

1987: $0.109[(5.25) + 3(1.23)] = 974.46$
(projected from 1985 to 1987)
1988: $0.109\ [0.70\ (8.94) + 1.23] = 1002.23$
1989: $0.109\ [0.891\ (7.49) + 1.23] = 861.491$
1990: $0.109\ [0.891\ (6.473) + 1.23] = 762.72$
1991: $0.109\ [0.891\ (5.76) + 1.23] = 693.475$

where 8.94 was the total prospective population in 1987, 10.9% of which bought a table, leaving 89.1% as prospective buyers.

If the company targets 2%, 4%, 5.5%, 6.5%, and 7.5% of the total market in each consecutive year, its potential demand would be 19,489, 40,089, 47,382, 49,577, and 52,011, respectively.

## 2.4 DESIGN

Once the market analysis is performed and it has been determined that the product has a sufficient probable market, the next step is to develop the detailed design that is suitable for production. Resource requirements and complexity in design are almost directly proportional to the extent to which the prospective product requires an original analysis. If a similar product is already on the market, one can take advantage of the situation by synthesizing its design features. However, if a new product is being introduced, the development and testing phases can be very expensive and time consuming.

Development of design follows basically the same steps as any engineering analysis, namely:

1. Identify the problem and develop preliminary ideas.
2. Refine the ideas.
3. Analyze and select the design. Test the suitability of that design.
4. Implement the decision.

Each step can be further subdivided to analyze its specific properties. For example, refinement may include modification of

- shapes and forms
- weights and volumes
- physical properties, such as strength, elasticity, and/or impact resistance
- scale drawings

Or implementation may include development of

- working drawings and specifications
- models and details
- testing and modification of a prototype

Before we initiate the design of a product, it is helpful to keep a few points in mind.

Close tolerances are important because they define the quality of the product, but expensive machine tools and machine operations are required to produce parts with great accuracy. Figure 2.7 shows the cost-tolerance relationship for a machining operation: the cost of production increases very rapidly as tolerance is reduced.

A product should not be overdesigned. There always is a need to allow for some safety factor, but overdesigning is expensive. Most items need not be the best in the market; however, they must be competitive.

A product is often judged by its appearance and how well it performs its designated function. As is the case with tolerance, however, the cost for the appearance (surface finish, coating, trim) increases rapidly.

A designer must be aware of the estimated volume of production of the finished product. The product designed for mass sales must be adaptable for manufacture on mass production machines with a minimum of different setups.

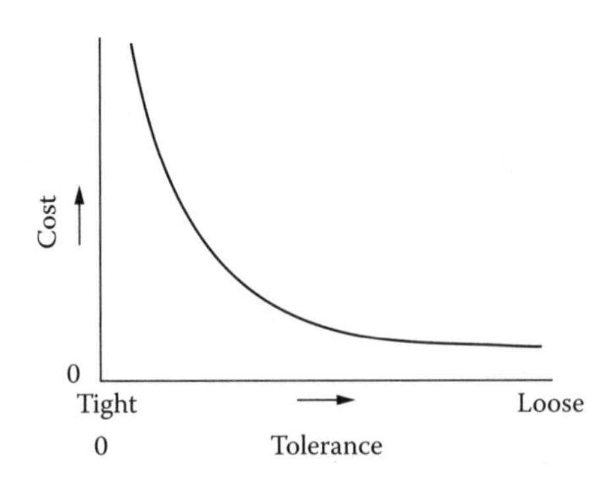


**FIGURE 2.7** Cost-tolerance relationship for machining

## 2.5 DESIGN FOR MANUFACTURE

The primary objective in designing a product is to meet the customer's needs. At the same time, we should be able to transform the design into a workable product. To expend less effort, cost, and time in manufacturing, the designer should take into account the economics of manufacturability (and even maintainability) of the design from its initial stages of development. The term *design for manufacture* (DFM) is synonymous with design for production (production design), design for economic manufacture, design for assembly, design for automated manufacture, etc. The motive behind all these concepts is the same — to cut time and costs in finished-product development and manufacturing by taking into account the issues involved in product manufacturability.

Traditionally, design efforts relate to only 5% of the total product cost but influence 70% of the manufacturing cost. Designing a product that is easy to produce may require additional time and expenditure in the design phase; however, this cost is more than offset in the production phase by a product that is easy to manufacture and assemble, resulting in less reworking. A design with features that are difficult to produce will incur additional costs in production throughout its life cycle. If the problem is noticed after the design is released for mass production, the effort involved in changing the design, redesigning the production processes, tools, jigs, and fixtures, and stopping the production line will be very expensive. It is therefore good practice to consider the manufacturability aspects of the design before it is released for production.

DFM is a set of techniques for efficient production, assembly, and testing of a product. It involves a team approach between designers and manufacturing engineers in the analysis of the product during the design and prototype-testing phases. A database or an expert system provided by the manufacturing department may help provide the manufacturability specification to the designers in early stages. The formation of a team consisting of designers and manufacturing engineers is also helpful for analyzing the product in its journey through all the manufacturing phases.

### 2.5.1 Design Principles in DFM

Steps in DFM may involve a range of activities from analysis of tolerance of the parts to complete redesign of fabrication, machining, and assembly operations with newer and fewer parts. For example, the analysis may include identifying redundant parts and parts or subassemblies that are difficult to assemble. Increasing the percentage of parts common to other products in the plant and improving quality to meet customers' expectations are also major goals.

Some of the design principles in DFM are

1. The product should be designed with as few parts as possible. For example, a test can be given to determine if separate parts are necessary by asking such questions as
   a. Do the parts move relative to each other?
   b. Are the parts required to be of different materials?
   c. Is it necessary for the part to be separate for maintenance purposes?
   d. Is the part necessary for assembling remaining parts?

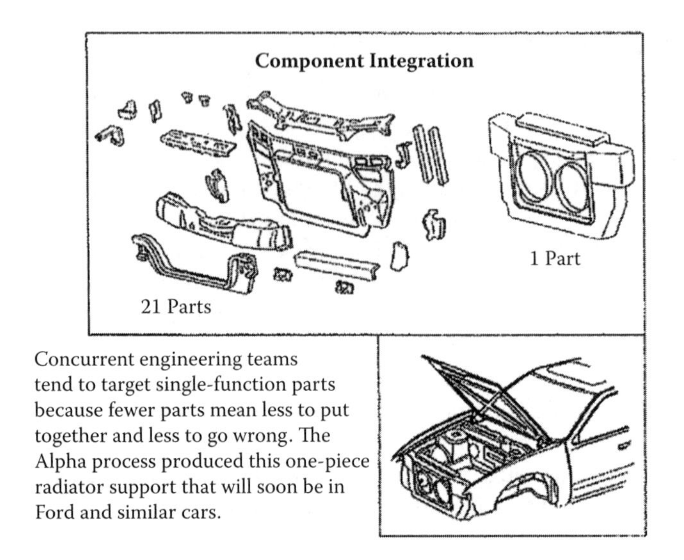


**FIGURE 2.8** Part designed for multiple functions

One popular means of combining parts is to mold the parts together. For example, in a coffeemaker, the base (used for inserting the heating element), the stand, and the top (used for sliding the coffee holder) are all molded together as a body to avoid assembly. In a small appliance involving gears and a shaft, both are molded together as one piece. A molded dashboard in a car is another example of a part designed for multiple functions (Figure 2.8).

2. When assembly is to be performed, the bottom part should act as a base for other parts. A part with a flat base provides a good support for itself without the need for additional fixtures. Other parts can be assembled onto it with unidirectional insertions, primarily vertically downward. Giving chamfers and lead-ins helps the assembly process.
3. Fasteners such as screws, nuts, bolts, and rivets should be eliminated in favor of snap fits, welding, and gluing. Also, the different types and sizes of fasteners should be reduced. Parts with snap fits are quicker and easier to assemble than those joined with screws. Parts assembled with different-size screws require more time, because additional time is spent in initial sorting of the screws.
4. A design in which an already assembled part must be removed to gain access to other parts should be modified. Replacing a spark plug in a car engine should not require removal of its distributor.
5. Parts should be easy to handle (large vs. small) and orient (symmetric vs. nonsymmetric, color- and shape-coded).
6. Tooling analysis should be performed to reduce the variety of tools used in assembly or disassembly operations. There is no reason for requiring, in a single assembly, the use of screwdrivers with Phillips, flat, and star heads.

7. Standard and interchangeable parts should be used.
8. As large a tolerance on the parts as possible should be specified, and the use of wires and cables that are difficult to handle should be limited.

Boothroyd and Dewhurst (1982) developed a method for analyzing a product and determining assembly efficiency. The procedure is similar to the one used for establishing a standard time on a new or existing work using predetermined time methods such as *methods time measurement*. Each motion is examined and analyzed to see if it can be simplified or, even better, removed altogether. Such analysis leads to questioning the value of complex motions and, in all likelihood, their elimination due to redesigning the parts and/or procedures.

### 2.5.2 Concurrent Engineering

Concurrent engineering (CE) is similar to DFM; indeed, DFM and CE are often thought of as one and the same. Some analysts, however, consider CE to additionally involve concurrent design of processes and preplanning manufacturing activities during the DFM analysis. Thus, CE is not only concerned with the development of a new product that is easy to manufacture, but also concentrates on improving, modernizing, and/or developing new processes and new operating policies. For example, it may involve analysis of new high-speed diamond milling in processing, introduction of robots in material handling, and use of fuzzy logic in numerically controlled machines. In a successful CE (or DFM) effort, teamwork plays a significant part. In CE, based on the problem, the team members are drawn from different areas of expertise. They must work together for the betterment of the company and not necessarily for the departments from which they come. They must tolerate criticism of their work and in return be able to evaluate their colleagues' work objectively. They should have a realistic schedule for completion of the job and avoid a tendency to constantly change the product requirements, which in turn will change tooling and manufacturing demands. They should carefully evaluate the cost of automation and preferably simplify and reexamine the product, processes, and operating policies before engaging in high-cost automation. The management should encourage the team effort by appropriate reward for the success of the project.

### 2.5.3 Role of QFD, DFM, and CE in Facility Planning

QFD, DFM, and CE each play an important role in facility planning. QFD ensures that there will be fewer changes down the line as the product matures in the market. A product designed to satisfy customer needs with its manufacturability in mind (with application of DFM and CE) is a well-thought-out product that is efficient to produce. For example, with fewer parts, there is an associated reduction in the number and types of machines needed to produce the parts. Inventory and associated material handling are reduced. Because machines can be less varied, maintenance of machines is simplified and the variety of skills required by the workers is limited. Production planning and control may be less complex, resulting in a corresponding reduction in computer hardware and trained personnel associated with these activities. As we show in subsequent chapters, these and other factors contribute to the area calculations, material-handling considerations, and affinity indexing between the departments. Relationships between the departments and the size of each department play an important part in an efficient plant layout. Type and quantity of material-handling equipment also contribute to the overall efficiency of the operations. In addition, with initial QFD analysis there will be fewer changes in the product design that would necessitate changes in the layout of the plant.

## 2.6 DRAWINGS

The final solution (design) must have a sufficient degree of detail and a complete set of working drawings (or models) that may form the legal basis for an outside contractor to bid on the job, if so desired.

These drawings should include an assembly drawing, detail drawings, a bill of materials or parts list, and perhaps a special, exploded pictorial drawing.

The assembly drawing is necessary to demonstrate how the various parts of the finished product are finally assembled or fit together, and to reveal, as far as is practical, how the finished product might function. Usually, only overall dimensions and those necessary for assembly are given on the drawings, which themselves can take many forms. For example, they may consist of one or more of the following, depending on the designer's desire for clarity and economy:

- orthographic sectional views (cutaways) (Figure 2.9)
- axonometric or oblique pictorials (Figure 2.10)
- perspective drawings (artist sketches)
- exploded axonometric views and perspective sketches

Detail drawings, such as the illustration shown in Figure 2.11, are inevitably orthographic projections of each individual piece (part) of the product and provide a detailed description of the shape and size (dimensions) of the finished piece. Detailed drawings for standard parts such as standard nuts, bolts, and keys need not be made, but such parts are shown in the assembly drawing and are included in the bill of materials or parts list.

The specification for each piece (in addition to the shape and size as shown on the detail drawing) is given in either the bill of materials, which is described in the next section, or the parts list, which is usually included on the face of the assembly drawing. (Exceptions are permitted when clarity is thus enhanced, and in-house standards might vary from company to company.) The parts list includes the following information for each piece required for the final assembly: piece number, piece name, quantity required for the final assembly, material, sometimes the stock size of the raw material, detail drawing numbers, weight, and any other pertinent details. A space is also provided beside each piece for any remarks, such as heat treatment.

### 2.6.1 BILL OF MATERIALS

A bill of materials displays a list of parts and is directly required to make a complete assembly. At a minimum, it should indicate, for each part, its number (or drawing

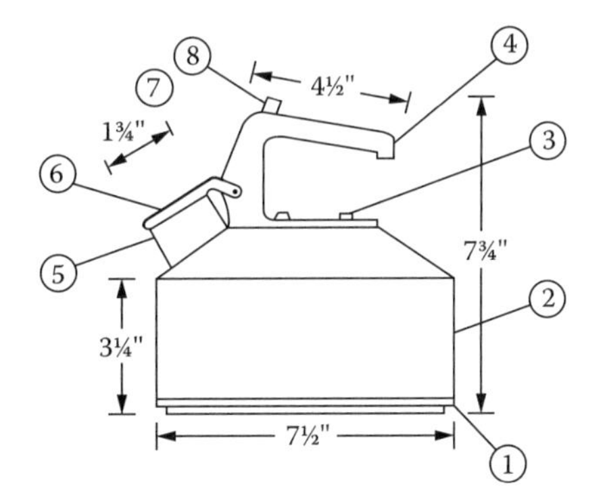


| NO. | PART NAME | REQ'D | MAT'L |
|---|---|---|---|
| 1 | BASE | 1 | ALUM |
| 2 | BODY | 1 | ALUM |
| 3 | SCREW | 2 | STEEL |
| 4 | HANDLE | 1 | BAKELITE |
| 5 | SPOUT | 1 | ALUM |
| 6 | UPPER LID PART | 1 | BAKELITE |
| 7 | ROLL PIN | 1 | STEEL |
| 8 | LID LEVER | 1 | BAKELITE |

Original drawing scale: 1" = 2"

**FIGURE 2.9** Teakettle assembly

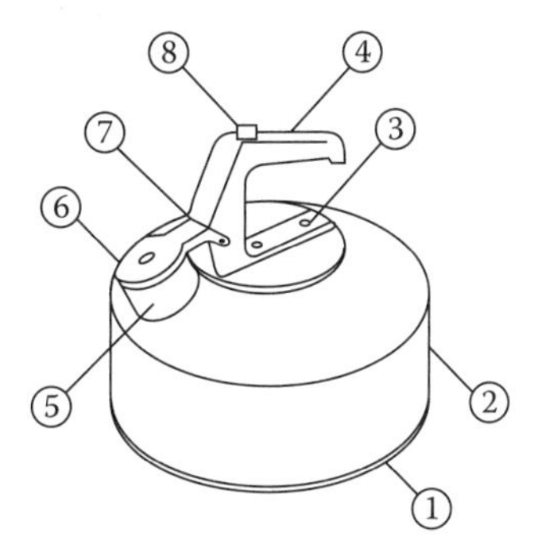


| NO. | PART NAME | REQ'D | MAT'L |
|---|---|---|---|
| 1 | BASE | 1 | ALUM |
| 2 | BODY | 1 | ALUM |
| 3 | SCREW | 2 | STEEL |
| 4 | HANDLE | 1 | BAKELITE |
| 5 | SPOUT | 1 | ALUM |
| 6 | UPPER LID PART | 1 | BAKELITE |
| 7 | ROLL PIN | 1 | STEEL |
| 8 | LID LEVER | 1 | BAKELITE |

Original drawing scale: 1" = 2"

**FIGURE 2.10** Teakettle: isometric view

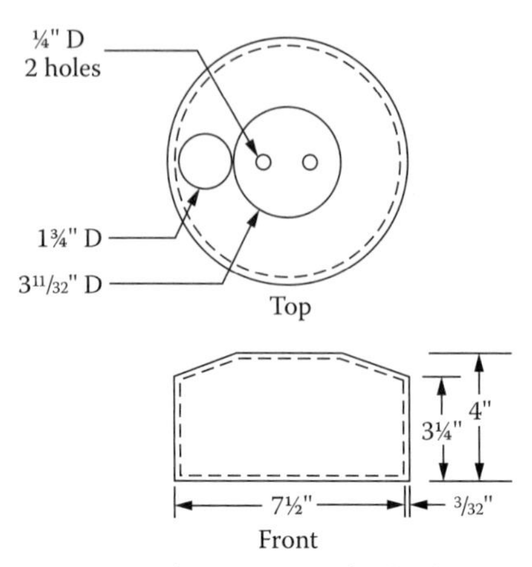


**FIGURE 2.11** Details of a teakettle

number), its description, the quantity necessary in the assembly, and whether the part is to be made within the plant or to be purchased from a supplier. The listing may include additional information such as material description, weight, and unit price. The name and part number or model number of the assembly are also shown at the top of the chart. Because a bill of materials should list only the parts and subassemblies that go directly into that assembly, a separate bill of materials is required for each assembly and then again for each subassembly that is part of the assembly, and so on. Figure 2.12 shows a bill of materials for the teakettle illustrated earlier.

### 2.6.2 Additional Drawings

In addition to the assembly and detail drawings, there may be auxiliary drawings that include special needs of users. In the case of a machine, these might include, for example, plans for the foundation to accommodate the machine, oiling diagrams for use with the preventive maintenance program, and arid wiring diagrams showing electrical connectors.

The next step is to develop route sheets, one for each part that is to be produced in the plant. These show, step by step, how the part is to be made, which machines to use, the operations to perform, and standard time allocations.

A detailed routing sheet is illustrated in Chapter 5, but at this point it should be noted that machines and standard times are required to develop such a sheet. Chapter 3 briefly describes various manufacturing processes that could be used in making a part. The choice of processes influences the selection of machines and thus affects the entries in the routing sheets.

**Product: Teakettle (Final Assembly)**
**Stock Number: 100**

| STOCK # DRAW # | PART NAME | UNITS REQ. | MATERIAL DESCRIP. | WT (LB) | UNIT COST ($) | SOURCE FOR MAT./PART |
|---|---|---|---|---|---|---|
| 1 | Base | 1 | 3/64" sheet aluminum 3003 alloy (AL3003) | .1678 | .150 | Reynolds Aluminum Supply (RAS) Make |
| 2 | Body | 1 | 3/64" (AL3003) | .201 | .18 | RAS Make |
| 3 | Machine screw | 2 | Low-carbon plated steel | .02 | .008 | Lone Star Screw Co. (LSS) Buy |
| 4 | Handle | 1 | Thermo-setting phenolic compound mineral (Bakelite) | .1830 | .10 | Plastic Engr. Co. (PEC) Buy |
| 5 | Spout | 1 | 3/64" (AL3003) | .0368 | .040 | RAS Make |
| 6 | Upper lid part | 1 | Bakelite | .0166 | .010 | PEC Buy |
| 7 | Roll pin | 1 | Stainless steel spring | .010 | .006 | Schinder's Machine Works Buy |
| 8 | Lid lever | 1 | Bakelite | .020 | .008 | LSS Buy |

Total weight = 0.6407 lb
Total material cost = $0.518

**FIGURE 2.12** Bill of materials for a teakettle

## 2.7 ECONOMIC EVALUATION OF PROCESSES

There are many different processes and machines that can be used in manufacturing, and the question may arise: Which method is best suited for the intended operation?

In most instances, the selection of a process is dominated by such design needs as shape, material tolerance, and the required quantity of production. However, whenever alternative methods that will give the same quality product are available, economic evaluation may be performed to select the best alternative. An example illustrates such an approach.

A machine manufacturer requires close tolerance of ±2 miles (0.002 in.) of the cams it produces. The manufacturer can perform machining using one of the following methods:

I. Cut the part on milling machine and then grind to the required tolerance on a grinder.
II. Use wire electrical discharge machining (EDM) that will produce the cam with the required tolerance.

The manufacturer produces 1000 cams of various sizes per month and, on the average, estimates the information in Table 2.9 to be the necessary data. The interest rate is currently 12% per year.

Using the factors in Table 2.9, determine the process to use by calculating the monthly costs of the two alternatives.

The capital recovery costs per month ($i = 1\%$, $n = 120$, $A/P = 0.0143$) for each machine are:

Milling machine: $20{,}000 \times 0.0143 = \$286.00$
Grinder: $15{,}000 \times 0.0143 = \$214.50$
EDM: $90{,}000 \times 0.0143 = \$1287.00$

The numbers of units that must be initiated per month to allow for defective units are

$$\textit{Grinder}: \frac{1000}{0.995} = 1005.0$$

$$\textit{Milling machine}: \frac{1005}{0.97} = 1036.1 \approx 1037$$

**TABLE 2.9**
**Data for Machine Manufacturing Example**

| | Profile Milling Machine | Surface Grinder | Wire Electrical Discharge Machining |
|---|---|---|---|
| Initial cost ($) | 20.00 | 15.00 | 90.00 |
| Time for setup operation and handling per unit (min) | 10 | 8 | 9 |
| Percentage of defective units | 3 | 1/2 | 1/2 |
| Unit cost of rejects ($) | 1.00 | 18.00 | 20.00 |
| Operating expenses, including maintenance and tool replacement per hour ($) | 1.20 | 2.00 | 5.00 |
| Useful life (years) | 10 | 10 | 10 |
| Operator cost per hour ($) | 15.00 | 15.00 | 20.00 |

Therefore, the expected numbers of rejects on each machine are

Grinder: 1005 – 1000 = 5
Milling machine: 1037 – 1005 = 32
EDM: 1005 – 1000 = 5

Assuming an 80% efficiency in operation, the average monthly time required on each process is

$$\text{Milling Machine: } \frac{1037 \times 10}{60 \times 0.8} = 216.04 \text{ hours}$$

$$\text{Grinder: } \frac{1005 \times 8}{60 \times 0.8} = 167.50 \text{ hours}$$

$$\text{EDM: } \frac{1005 \times 9}{60 \times 0.8} = 188.43 \text{ hours}$$

Now we can determine the cost of manufacture on each machine, which includes capital recovery, rejects, labor, operation, and overhead cost (overhead is assumed to be 150% of the direct labor cost).

Thus, the costs for milling are

Capital recovery: $286
Rejects: 32 × $15 = $480
Labor: 216.04 × $15 = $3240.60
Machine cost: 216.04 × $1.50 = $324.06
Overhead: 216.04 × $15 × 1.5 = $4860.94
Total: $9191.60

Similarly, the cost for grinding is

$$214.5 + 5 \times 18 + 167.5\,(15 + 2.0 + 1.5 \times 15) = \$6920.75$$

and the cost for EDM is

$$1287.0 + 5 \times 20 + 188.43\,(20 + 5.0 + 1.5 \times 20) = \$11{,}750.65$$

Therefore, the cost of process I is

$$9191.60 + 6920.75 = \$16{,}112.35$$

and that of process II is $11,750.65. It is obvious that process II, that of using EDM, would be selected.

Further investigation showed that a CNC (computer numerical control) milling machine that would allow an operator to tend to both milling and grinding operations could be used; however, the cycle time would be extended to 11 min (Section 6.3), and the overhead cost would increase to 175% of the direct labor cost. The operator is still paid $15 per hour. The data for the CNC milling machine are

- Initial cost: $35,000
- Percentage of defective units: $\frac{1}{2}$%
- Operating expense, including maintenance tool/hour: 2.00
- Useful life (years)

By following the same procedure as before, the capital recovery cost per month is 35,000 × 0.0143 = \$500.5. The number of bad units produced is (1005/0.995) – 1005 = 1010 – 1005 = 5.

Assuming an 80% efficiency, the monthly time required of CNC milling is $(1010 \times 11)/(60 \times 0.8) = 231.46$ hr. The required time on the grinder would be $(1005 \times 11)/(60 \times 0.8) = 230.31$ hr. Because both operations are to be carried out simultaneously, the longer of the two times — 231.46 h — will be used in our calculations as the cycle time. Therefore, the cost per month for process I is

$$500.5 + 214.5 + 5 \times 15 + 5 \times 18$$
$$+ 231.46(15 + 2 + 2 + 1.75 \times 15) = \$11,353.56$$

This cost is less than that for EDM alone, and the decision will now be to adopt the CNC milling and grinder combination; however, the closeness of the costs of the two alternatives would suggest a need for more careful data collection and analysis.

## 2.8 COMPUTER-AIDED DESIGN

Computers are now a standard tool for design development. Besides performing engineering calculations, computer-aided design (CAD) can be used to improve speed and accuracy in designing. CAD is extremely useful in modifying an existing design or in developing a new design in a family of parts, such as printed circuit boards. The computer permits pretesting of the proposed product by simulating its use under various conditions. It is also possible to use the computer to test many design features without having to actually manufacture the model (Figure 2.13).

The CAD system consists of three major components: the designer/draftsperson, the hardware or a computer of any reasonable size, and the software. In the interactive mode, the computer is used to display data in the form of pictures and symbols. The designer can modify the data and immediately see the effects on the design. The software calculates the results of the changes and displays the corresponding figures on the screen. The designer may similarly change shapes and forms, and the computer will show the effects of those changes on individual parts and on the assembly. The designer may alter design tolerances and observe the effects on mating parts. The designer can also enlarge a design on the screen and look for any irregularities, such as interference between moving parts.

CAD systems are also used extensively in drafting. These systems include several basic drawings such as lines, polygons, circles, arcs, rectangles, and other simple shapes. From these drawings, three-dimensional (3-D) figures can be constructed to form such shapes as cubes, pyramids, cones, cylinders, and spheres. Common shapes such as slots, holes, and pockets can be included in a solid model of an object. They can be in solid colors or other patterns (called hatching). Basic shapes can be modified by filleting to produce round shapes.

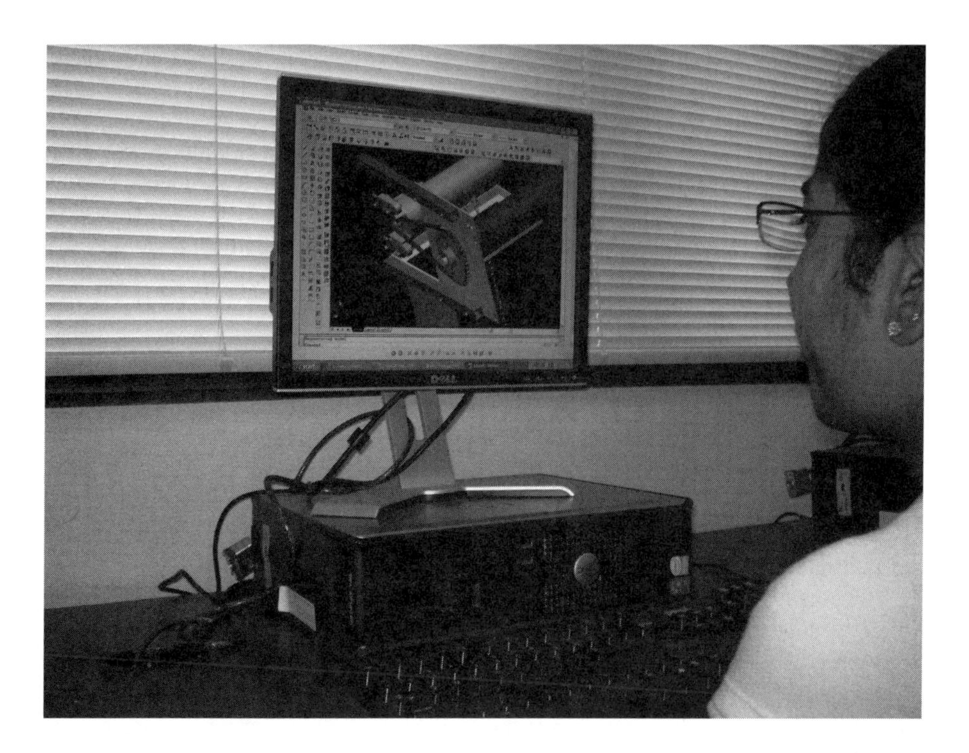

Computer-aided design

Designers construct a skeletal wire form model that is a 3-D representation of the shapes to show all edges and features as lines. A more realistic 3-D mode is called a solid model. Solid modeling involves functions for creating 3-D shapes and combining shapes. Solid modeling also maintains a set of relationships between the components of an object so that changes can be propagated to following constructions.

CAD models can be manipulated and viewed in a wide variety of environments. They can be viewed from any angle, broken apart, or sliced. These shapes can be simulated, rotated, transformed to different locations, or reformed, to analyze for strengths and defects in the design.

Animation can be created by moving these shapes to make virtual reality possible. Such animation can simulate a real-world interaction with the object being designed. For example, if the object is a plant, the virtual reality system may allow visualization of the scene as we walk inside and outside of the plant, giving a real-time view of the plant from a multitude of perspectives.

Revising existing drawings, a task that is often put off, becomes very easy to accomplish because of the system's ability to store and retrieve drawings. Software that is capable of producing 3-D views has proved to be quite effective in automatically determining interference of components. CAD systems have also been developed to produce parts lists and bills of materials from drawing specifications and to check for the validity of specifications.

The advantages of a CAD system are

1. It allows visualization of an item being designed. The computer analyzes and displays the effects of any design changes on the part and/or the subassembly and the entire assembly.
2. It can be used to produce drawings, specifications, and bills of materials.

Even a small organization can probably afford some type of a CAD system. As to which type is most suitable depends on the work being done and the expected growth in the use of the machine. Careful thought during the initial installation phase is necessary, because each manufacturer's system is different and most systems are not compatible with those of other manufacturers.

Although CAD systems can be developed by the user (an expensive undertaking), most are purchased from software vendors. Indeed, it is generally more convenient to have a vendor provide the hardware and the software, but to retain in-house responsibility for all maintenance.

The use of the CAD system may be extended further if its database can be directly connected to computer-aided manufacturing (CAM) machines, such as numerically controlled machines. Figure 2.14 shows functional relationships between CAD-CAM and manufacturing. More discussion on this integration is provided in Chapter 16, by which point the reader will be familiar with more aspects of automation.

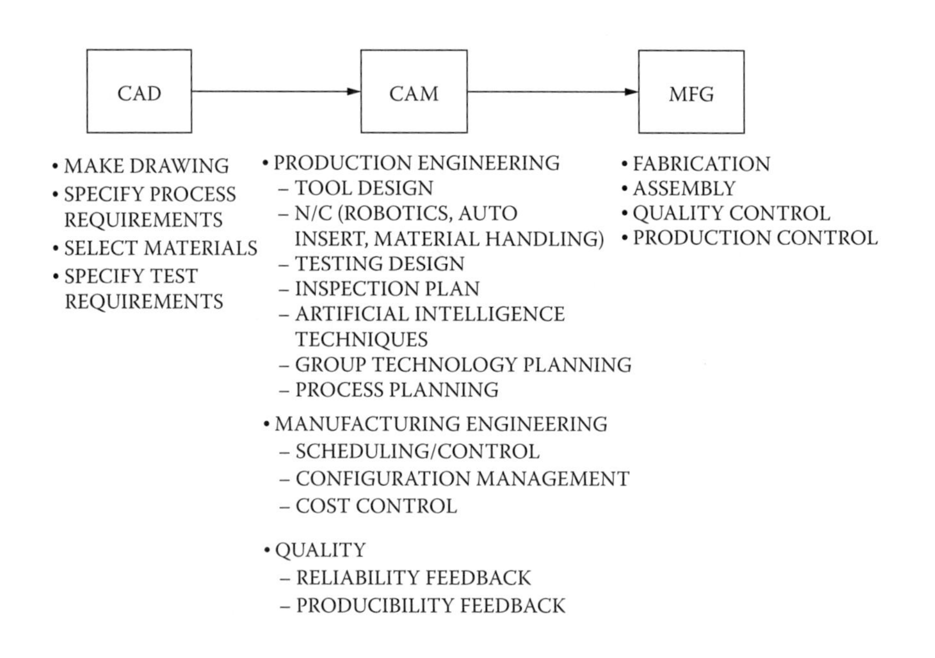


**FIGURE 2.14** Functional tasks

## 2.9 COMPUTER PROGRAM DESCRIPTION

The program calculates the rate of return given the initial investment and yearly income. It is in interactive mode with the information to be supplied based on the questions:

1. *What is the initial investment?* (Enter the dollars invested at year 0.)
2. *How many years will the project operate?* (Enter the estimated life of the project.)
3. *What is the income for year $j$?* (Enter the estimated earnings for year $j$. One entry is required for each year of operation.)

This program outputs the internal rate of return.

### 2.9.1 SUMMARY

To decide whether the introduction of a new product is potentially profitable, market research must be conducted. This can be accomplished by gathering information from well-established marketing firms or by performing market research of our own. Mail surveys, telephone surveys, and personal interviews are three techniques for conducting such research, and each has advantages and disadvantages. An analysis with a planning sheet, along with the data from market research, may provide clues to the profitability of a venture.

Forecasting future markets based on available information plays an important part in estimating customer demands. If the data are insufficient or inconclusive, one must resort to a qualitative prediction based mainly on judgment and experience. When data are available, quantitative methods, such as a moving average, exponential smoothing, and trend analysis, may be used. The question of which method to use is, to a large extent, answered by determining which forecasting method provides the minimum error.

Close attention must be paid to the details of product design during the initial planning phases. A product should be functional and reliable without being overdesigned; however, the method of manufacturing also influences some of the design features.

QFD analysis helps pinpoint who the customers are and what they really want. To be successful, a product must include design features that satisfy customers, which are not necessarily the features that designers think are important. DFM and CE modify and simplify the design so that it can be produced efficiently. Furthermore, the analysis verifies that the design is producible and that the plant has the capacity and capability to produce it. The use of QFD, DFM, and CE also reduces future design and procedural changes, preventing the need for constantly changing equipment and facility layouts.

Mechanical drawings are the link between design engineers and manufacturing personnel. These drawings should include assembly drawings, detail drawings, a bill of materials, parts lists, and even special and exploded views for clarity.

CAD can significantly improve speed and accuracy in design. It is especially valuable if the new product can be built using or modifying existing components. CAD allows one to visualize the items, test them for design changes, simulate their use to understand potential problems, and produce detail drawings and bills of materials for plant use — all without going to the expense of actual production.

In examining market research, forecasting, and product design, this chapter covers the initial planning phases for a product. Whether the product is successful in the market or not is, to a large extent, dependent on the thoroughness of the analysis.

## PROBLEMS

1. What is market research?
2. Where can information about market potential be found?
3. Review the steps for performing market research.
4. A furniture manufacturer is thinking of adding a new line of waterbeds to its existing line of products. The manufacturer believes it must sell at least 750 beds per month to make any profit. With waterbeds being a new product, sales will probably not amount to more than 10% of the market in the first year. Market research showed the sales of waterbeds for the last 5 years, given in the accompanying table. Apply the least-squares method to predict the demand and to determine whether or not the manufacturer should add the new line.

| Year | Total Sales of Waterbeds per Month |
|---|---|
| 1981 | 1000 |
| 1982 | 3000 |
| 1983 | 4430 |
| 1984 | 5129 |
| 1985 | 6538 |

5. BP Manufacturing is trying to enter the fishing equipment market by introducing a sound-emitting device called the Dial-A-Fish, which is scientifically proven to attract fish in laboratories. This new device should capture 10% of the present market according to market research. The facilities needed are 5000 ft$^2$ for production, 1350 ft$^2$ for packaging, and 1500 ft$^2$ for shipping and receiving. The plant will be located in a region where the labor rate is \$5 per hour for assembly line workers and overhead costs are 150% of direct labor costs. Land cost is \$7000 per acre, and the construction cost is \$200 per square foot.

The present market for similar devices is 250,000 units per year, and it is estimated that a total of 0.5 h of labor is required per piece. The unit will sell for $15.00. Over a 5-year period, will BP Manufacturing make a profit if the present interest rate is 10%?

6. Develop a QFD for a table fan. What are the most desired features from the customer's viewpoint? From the designer's viewpoint? Are these the same features?
7. Analyze a product that is at least 5 years old and see if its design could be improved with the application of DFM principles.
8. Develop a fishbone diagram for a table lamp. Have you exposed any new characteristics that otherwise are not obvious?
9. Compare quantitative and qualitative forecasting techniques. Is any one technique preferable to another?
10. What is meant by error in forecasting? What are the best means of measuring such error? How do you know if you have a fairly good forecasting model?
11. Known demand for a product (in hundreds) for the last eight periods is listed here.

| Period | 1 | 2 | 3 | 4 | 5 | 6 | 7 | 8 |
|---|---|---|---|---|---|---|---|---|
| Demand | 5.0 | 8.3 | 13.9 | 16.2 | 15.4 | 18.6 | 16.4 | 17.5 |

    a. Using the 3-month moving average, forecast the demand for period 9. If the actual demand for period 9 is 18.3, forecast the demand for period 10.
    b. Predict the demand for periods 9 and 10 using the exponential smoothing technique. Select the best value of $\alpha$.
    c. Use linear regression to forecast the demand for these two periods.

12. A fiberglass insulation manufacturer is trying to decide how much insulation to produce in August. The manufacturer knows that it will sell about twice as much insulation in the first months of summer and winter (June, July and December, January) than in any other month of the year. The previous 8 months of data that are available are given in the accompanying table. Predict the demands for July and August.

| Month | Number of Rolls |
|---|---|
| September | 820 |
| October | 790 |
| November | 835 |
| December | 1575 |
| January | 1724 |
| February | 783 |
| March | 811 |
| April | 827 |
| May | 845 |
| June | 1710 |

13. An electronic product such as a TV can be made with plug-in modules. A major advantage of a plug-in module is in service, because it can be replaced easily. Suppose a component costing $5 can fail at a rate of 2% per year. It takes, on average, 2 h to replace the component, and the repairperson charges $40 per hour. The same component can be installed on a solid-state module at a cost of $30. Because of the solid state nature, the failure rate drops to 0.05% per year; because of the plug-in design, it can be replaced in 30 min. Determine the minimum required life of the equipment to make a switch from the old design to the plug-in design economically feasible.
14. In problem13, the manufacturer thinks there is considerable loss of "goodwill" if the product fails and has to be replaced again and again. To avoid this, the manufacturer decides to replace the old component with the plug-in module. If the useful life of the equipment (before it becomes obsolete) is 3 years, what is the dollar value placed on the cost of goodwill?
15. Why is it beneficial for a designer to consult with a manufacturing engineer in the plant during the design phase of a product?
16. Review the steps in developing a product design.
17. As a manufacturer, list six considerations, in order of importance, that should be kept in mind in designing an electric toaster and a lawn mower. How would the rankings change when viewed by a consumer?
18. What information should be included in product drawing? In the bill of materials?
19. Using Figure 2.15, develop a bill of materials for an 8-in aluminum frying pan including weight of parts (aluminum weighs 0.098 $lb/in^3$; Bakelite, 0.162 $lb/in^3$), number of each part needed, part name, material, material description and supplier, and unit cost.

    The cost per 1000 lb of aluminum is $1350.00. The cost per 1000 lb of Bakelite is $700.00.
20. Using the bill of materials developed for problem 19, calculate the yearly material cost for 200,000 of these frying pans.
21. Develop a planning sheet for manufacturing finished picture frames. The costs are: labor, $4.80 an hour; wood, $0.50 a foot; nails, $0.02 a piece; and varnish, $1.30 a quart. Estimate equipment usage and cost.
22. A manufacturer of staplers has the choice of two processes to produce the housing of the staplers. Method I produces the part with a punch press and smooths it with an abrasive belt grinder. Method II produces the part by casting and smooths it with an abrasive belt grinder. The manufacturer expects to produce 10,000 units per month. The present interest rate is 12%. The factors on which the manufacturer will base the decision are shown in the accompanying table. Determine the monthly cost of each alternative.

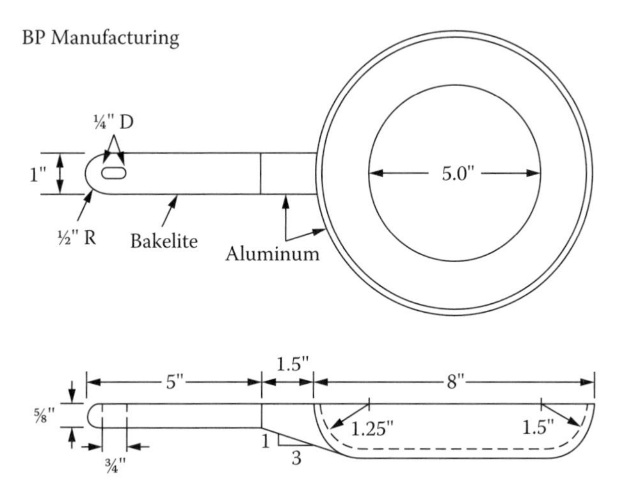


**FIGURE 2.15** BP manufacturing

| | I | | II | |
|---|---|---|---|---|
| | **Punch Press** | **Belt Grinder** | **Casting (Oven Mold Label)** | **Belt Grinder** |
| Initial cost ($) | 20.000 | 20.000 | 7000 | 2500 |
| Time for setup operation and handling per unit (min) | 1 | 0.2 | 1.5 | 1.5 |
| Percentage of defective units | 1/2 | 1/5 | 5 | 1/5 |
| Unit cost of rejects ($) | 0.80 | 0.60 | 0.90 | 0.70 |
| Operating expense, including maintenance and tool replacement per hour ($) | 0.20 | 0.05 | 0.08 | 0.05 |
| Useful life (years) | 8 | 5 | 8 | 5 |
| Operator cost per hour ($) | 12.00 | 8.00 | 11.00 | 8.00 |

23. How can a computer assist in product design? What are the components of such a system? Does CAD make communication between a designer and a manufacturer easier?
24. List important functional tasks associated with CAD, CAM, and manufacturing.

## SUGGESTED READINGS


### Market Research

Cox, W.E., Jr., *Industrial Marketing Research*, John Wiley, New York, 1979.
Ferber, R., ed., *Handbook of Market Research*, McGraw-Hill, New York, 1974.
Freiman, D.J., *The Marketing Path to Global Profits*, AMACOM, New York, 1979.
Williams, R., Jr., *Technical Market Research*, Roger Williams Technical and Economic Services, Inc., Switzerland, 1962.

### Quality Function Deployment

Akao, Yoji, ed., *Quality Function Deployment: Integrating Customer Requirements into Product Design*, Productivity Press, 1990.
Moran, J.W., Marsh, S., Nakui, S., and Hoffherr, G.D., *QFD: Facilitating and Training QFD*, QPC Publishers, Methuen, MA, 1991.
American Supplier Institute, *Quality Function Deployment: Manual for Three Day QFD Workshop*, Dearborn, MI, 1989.

### Forecasting

Abramson, A.G., Operations Forecasting, American Marketing Association, 1967.
Makridakis, S., Wheelwright, S.C., and Hyndman, R.J., *Forecasting Methods and Applications*, 3rd edition, John Wiley, New York, 1998.
Wenzel, C.D., "Look at the foreseeable trends in parts handling offers strategies for system planners," *Industrial Engineering*, 16(3), 46–54, 1984.

### Design and Drawing

Eide, A., Jenison, R., Mashaw, L., Northup, L., and Sanders, C.G., *Engineering Graphics Fundamentals*, 2nd edition, McGraw-Hill, New York, 1995.
*Engineering Design Graphics Journal*, American Association for Engineering Education, Ohio State University, Columbus, OH, 1985.
Giesecke, F., Mitchell, A., Spencer, H., Hill, I., and Loving, R., *Engineering Graphics*, 5th edition, Macmillan, New York, 1993.

### Designing for Manufacturability

Boothroyd, G., and Dewhurst, P., *Design for Assembly: A Designer's Handbook*, University of Massachusetts, Amherst, MA, 1982.
Shina, S.G., *Concurrent Engineering and Design for Manufacture of Electronics Products*, Van Nostrand Reinhold, New York, 1991.
Suh, N.P., *The Principles of Design*, Oxford University Press, Fairlawn, NJ, 1990.

### Computer-Aided Design

Groover, M.P., and Zimmers, E.W., *CAD/CAM Computer-Aided Design and Manufacturing*, Prentice Hall, Englewood Cliffs, NJ, 1984.
Industrial Design Magazine, Design Publications, Inc., New York, 1985.
Pao, Y.C, *Elements of Computer-Aided Design and Manufacturing*, John Wiley, New York, 1984.

# 3 Automation

In the computer age, productivity in a manufacturing organization has been substantially increased by refinements in the methods of operations rather than development of new manufacturing processes. Our ability to acquire data in real time and to analyze and to act on the information either directly or indirectly through the actions of devices has contributed significantly to the efficiency of an operation. Automation in manufacturing is a loosely defined term implying applications of machines, controls, and computers to increase productivity. Factory automation may include many facets. In a broad sense, it may comprise four distinct but closely interrelated parts: (1) the manufacturing system, (2) material handling, (3) sensing equipment, and (4) the control system.

We will study types of manufacturing systems and material handling in later chapters. In this chapter, we introduce the aspects of automation: (1) sensory techniques, (2) bar codes, (3) radio frequency identification systems, (4) machine vision, (5) programmable logic controllers, (6) numerically controlled machines, and (7) robots and their applications. Illustrative examples are given, where feasible, to show a spectrum of applications.

## 3.1 SENSING METHODS

Sensing equipment plays a role similar to that of the physical senses. It observes what is happening and transmits the information to the control unit. Sensing techniques use photoelectric cells, infrared cells, high-frequency electronic devices, and units making use of isotopes, X-rays, ultrasonics, and resonance. Sensing devices offer many advantages. Speeds of operation are many times faster than the maximum sensing that is possible for humans. There is no human-fatigue problem, and absolute accuracy of inspection is assured within machine limits. Observations can also be made in places that are inaccessible or unsafe for human beings.

Sensors are the eyes and ears of any automated process. One cannot operate a numerically controlled (NC) machine without knowing the position of the stock relative to the cutting tool. Feedback is needed to determine when the stock has moved to the proper cutting position or whether the cutting tool has retracted from the stock so that it can be positioned for the next cut. The controllers that perform the work must be guided by sensors that measure the critical parameters of the job being performed.

Sensors are classified as either discrete or continuous based on the type of signal that they produce. Discrete sensors produce a bistate input corresponding to some bistate observation — for example, closed or open, full or not full, on or off, running or stopped. Because most discrete devices normally close or open a switch on the basis of the condition they observe, they are commonly referred to as contact sense inputs. Limit switches are a very common example of a contact sense input. For example, a numerically controlled machine can be inhibited from moving the stock until a limit switch closes, signifying that a drill bit has fully retracted. House thermostats are another type of discrete sensor. When the house temperature falls below a preset limit, a contact closure is made that turns on the heating system. Once the temperature rises, the system is again turned off by the thermostat switch opening. These home heating systems then are either fully on or fully off, and there is no means of heating the house at some intermediate level. Therefore, raising the thermostat's temperature set point heats a house to a higher temperature, but not any faster. Even though temperature is a continuous variable, the sensor's output is bistate, making the device a discrete sensor.

By contrast, a continuous sensor can signal any intermediate value between 0% and 100% and is not limited to simple contact closure. The output of the continuous sensor is normally linear with respect to some observed variable. For example, a pressure sensor can be used to measure the level of a tank in which the pressure on the sensor is proportional to the height of the column of fluid above the sensor. The sensor output may be measured electrically as volts or milliamps and then scaled to the appropriate observed variable. At times the variable that exhibits linearity with the sensor output is not the variable of interest. This is typically the case in measuring flows, where the sensing device's output is linear with respect to the pressure difference between two points. The square root of this differential pressure must be obtained to have a signal proportional to the flow. The fuel tank measurement in a car is a common example of a continuous sensor. The sensor output is a measurement of the height of the gasoline in the tank, which, owing to the tank's geometry, is roughly linear with respect to the volume of gas remaining.

Sensor technology continues to advance as new types of measurements augment the old classics such as position, pressure, temperature, and flow. Many of these new sensors involve the measurement of infrared or visual light, radiation, or ultrasonic sound. These more exotic measurement techniques are used when the classical techniques fail or are inadequate. The measurement of coal in a metal hopper is an example of such an application. Because coal dust obscures any attempt to determine a hopper's fullness by light, a means is needed to make this dust transparent. To accomplish this, a radiation source is aimed through the hopper to illuminate an ionization chamber (receiver). The signal level is then calibrated for the full-scale reading with no coal present so that the effect of the metal sides and air gap are removed from the reading. Any additional attenuation will therefore be caused by the coal and will be proportional to the mass of the coal between the source and the receiver. Hence, dusting will cause a minimal drop in the signal level, while a solid mass will provide a more substantial change. This radiation is used to detect when the level in a coal hopper has fallen below the preset limit to restart the coal-feed mechanism for a power plant. Radiation is thereby used to produce a discrete contact sense output used to trigger a sequence of events. It should be further noted that this more sophisticated measurement is more costly, more difficult to maintain, and requires special licensing in the handling and storage of the radioactive source. Simpler techniques should therefore be used whenever they are practical.

## 3.2 BAR CODES

One of the most common examples of automated data collection and transmission is the use of bar codes. We are all familiar with the coded information stored in an array of wide and narrow bars and spaces that is used at checkout counters to identify and cost the purchased items. With a unique combination of wide and narrow bars and spaces, each bar code represents information (a code number) particular to that item, component, person, or place. With the advances in vision and computer technology, the narrow and wide bars and spaces can be read and decoded directly by appropriate equipment. Thus, a bar code forms a printable machine language that can be read and processed by a computer without the need for manually entering data through a keyboard.

### 3.2.1 Principles of Bar Code Technology

The American Standard Code for Information Interchange (ASCII) suggests binary codes for alphanumeric characters in computer use. For example, the number 7 is represented in ASCII binary code by 0010 0111, and the letter A is represented by 100 0001. In similar fashion, unique patterns of bars and spaces can be assigned to alphanumeric characters to display them in a compact symbol. For example, the number 7 is designated by 00011 in Interleaved 2 of 5 symbology, one of the many bar code schemes in use today.

The binary values 0 and 1 are represented in bar codes in different forms. In some systems, the 0s are designated as white spaces and 1s, as black bars. In other systems, 0s appear as narrow bars or spaces and 1s are shown as wide bars or spaces.

A bar code is translated into machine language by using a bar code reader. In some readers, a beam of red light projected by the reader is manually swept over the bar code. Other bar code readers use oscillating mirrors to reflect a beam of light onto the bar code. The frequency of oscillation is so high that the beam appears as a band of light to the human operator. The light is absorbed by the dark bars and reflected by the light background of the spaces. The intensity and duration of the reflections is detected by a phototransducer within the bar code reader that produces voltage pulses. These pulses are then converted by an analog-to-digital converter so that the computer receives digital 0/1 signals. The 0/1 digital signals provide values in the format necessary for a computer to interpret, or process, the signal. Figure 3.1 illustrates a voltage pulse generated by a bar code, which is then converted to digital 0/1 by an analog-to-digital converter. Figure 3.2 illustrates the schematic diagram of the hardware and software involved in this process.

### 3.2.2 Bar Code Symbologies

Many different bar code systems are in use in industry today. Table 3.1 gives a brief summary of available codes and their characteristics. For example, Universal Product Code (UPC) and its counterpart in Europe, the European Article Number (EAN) system, are mainly used in retail business. The bar codes on grocery items are in UPC code. A very popular code, 3 of 9, or code 39, is used in industry, in the U.S. Department of Defense (DOD), and in many other governmental applications.

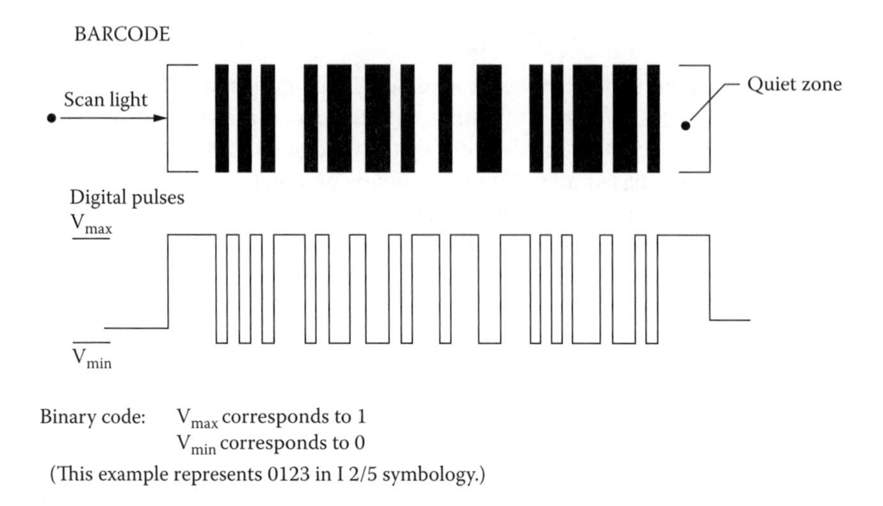


**FIGURE 3.1** Bar code

Interleaved 2 of 5, or I 2/5, is used for shipping containers and heavy industrial applications, warehousing, and in airline ticketing and freight bills. CODABAR is another bar code system; it is used in blood banks, libraries, package tracing, and film processing.

Differences among these systems are attributed to the schemes used to store the data and the way the bars and spaces are arranged to represent the code. A specific predefined combination of 0s and 1s represents a character or a number in each system, and each scheme has its own method of representing 0s and 1s. All these symbologies are governed by a Uniform Symbology Specification (USS) to standardize equipment usage. Each bar code scheme requires a quiet zone before and after the reading. In a quiet zone no markings of any type are allowed. Each bar code also

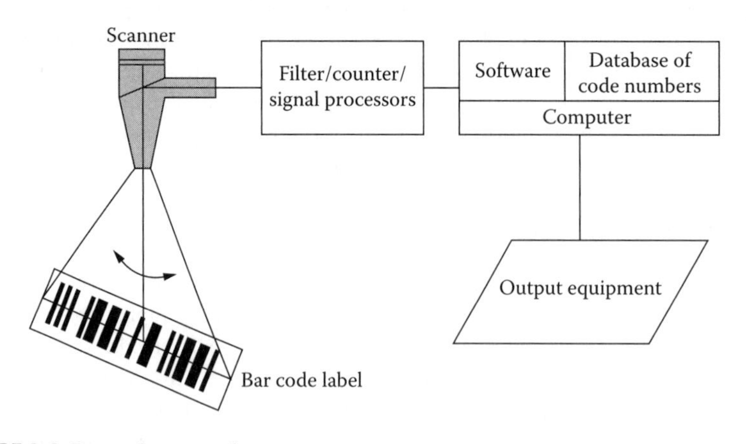


**FIGURE 3.2** Bar code system layout

**TABLE 3.1**
**Bar Code Symbologies**

| Item | Symbol Characteristics | Code 39 | Code 128 | I 2/5 | CODABAR | UPC/EAN |
|---|---|---|---|---|---|---|
| 1 | Encodable character set | Full alphanumeric; 7 special characters (- . space $ / + %) and 1 start/stop character (*) | 128 ASCII characters and 4 nondata function characters; 3 start characters and 1 stop character | Numeric; 10 digits (0 through 9) | 10 digits (0 through 9), 6 special characters (- $ : / . +), and 4 start/stop characters (A B C D) | Numeric; 10 digits |
| 2 | Code type | Discrete; narrow and wide elements used | Continuous; 1, 2, 3, or 4 modules used | Continuous; narrow and wide elements used | Discrete; narrow and wide elements in USS-Codabar | Continuous; 1, 2, 3, or 4 modules used |
| 3 | Symbol length | Variable | Variable | Variable; although it requires an even number of digits | Variable | Fixed |
| 4 | Bidirectional decoding? | Yes | Yes | Yes | Yes | Yes |
| 5 | Character self-checking | Yes | Yes | Yes | Yes | Yes |
| 6 | Use of check character? | Optional; modulo 43 | 1 required — modulo 103 | Optional; modulo 10 | Optional; no standard formula, modulo 16 recommended | 1 required — modulo 10 |
| 7 | Standard smallest "*x*" dimension | 0.0075 inch | 0.0075 inch | 0.0075 inch | 0.0075 inch — USS Codabar 0.0065 inch — "Traditional" | 0.0104 inch |
| 8 | Max. data character density (CPI) | 9.8 | 24.2 numeric 12.1 other ASCII | 18.0 | 12.8 — USS-Codabar 10.0 — "Traditional Codabar" | 13.7 |
| 9 | Options and features | Full ASCII encodation, Concantenation | FNC char 1 and 4 may be used for appl. spec, Concantenation, double-density mode 3 - character subsets | Should use a leading zero if an odd number of digits is encoded, unique start/stop patterns | Concantenation | |
| 10 | Major market where used | Industrial/US DOD/ AIAG/HIBC/ TALC/ FASLINC | Industrial | Industrial/retail/ airlines/UPC SCS | Blood banks/film processing/ package tracking/libraries | Retail |
| 11 | Auto-discriminates with | | | Each with any other | | |

*(continued)*

**TABLE 3.1 (CONTINUED)**
**Bar Code Symbologies**

| Item | Symbol Characteristics | Code 39 | Code 11 | I 2/5 | CODABAR | UPC/EAN |
|---|---|---|---|---|---|---|
| 1 | Encodable character set | 10 digits (0 through 9), 6 special characters additional | Numeric; 10 digits (0 through 9), and dash (-), and start/stop character | Full alphanumeric, 7 special characters, 4 control characters, and unique start/stop pattern | Full ASCII character set | |
| 2 | Code type | Continuous; narrow and wide elements used | Discrete; narrow, medium, and wide elements used | Continuous; 1, 2, 3, or 4 modules used | Continuous | |
| 3 | Symbol length | Variable | Variable | Variable | Variable- | |
| 4 | Bidirectional decoding? | Yes | Yes | Yes | Yes | |
| 5 | Character self-checking? | No | No | No | ? | |
| 6 | Use of check character? | 1 — variety of methods used | 1 or 2 are suggested | 2 are recommended, modulo 47 | 7 | |
| 7 | Standard smallest "*x*" dimension | Varies by version | 0.0075 inch | 0.0075 inch | | |
| 8 | Max. data character density (CPI) | Varies by version | 15 | 14.8 | Said to produce "ultrahigh density" code | |
| 9 | Options and features | Variants of this code use different module lengths to represent characters | | 4 control characters used for ASCII set, concantenation | Two-dimensional structure, using two to eight rows | |
| 10 | Major market where used | Retail/libraries | Telecommunications | Industrial | None yet | |
| 11 | Auto-discriminates with | | | Each with any other | | |

*Courtesy of D. Randall Hicks, Bar Code Systems, Inc., Atlanta, GA.*

requires a specific start symbol and a specific stop symbol to signal the computer to start or stop reading, as well as the direction of reading. It is also common to display the code in readable form below the bar code. A further description of a few commonly used symbologies is given next.

#### 3.2.2.1 2 of 5 Code

The 2 of 5 code, developed in the 1960s, uses a very straightforward approach. Each character in the code is made of five dark bars, and the spaces introduced in between are only to separate the bars. Of the five dark bars representing a character, two are wide and the remaining three are narrow. The wide bar represents 1 and the narrow bar represents 0. Thus the number 4 is denoted by ||▌|▌, representing 0 0 1 0 1.

The start code is denoted by a set of bars that represent 110; the stop code is denoted by 101. The binary code for each of the numbers used for representation in the 2 of 5 code is given in Table 3.2.

When the length of interbar and intercharacter gaps is the same as a narrow bar space, the total length of space required for bar code in 2 of 5 is calculated by

$$L = 2QZ + Z(3 + 8C + 2CR)$$

where

$QZ$ = length of quiet zone (maximum of (0.25 inch or $10X$))
$C$ = number of characters in the code
$R$ = ratio of wide to narrow width (varies from 2 to 3)
$X$ = width of narrow elements (varies from 0.008 to 0.021 inch).

**TABLE 3.2**
**Character Structure for 2 of 5 and I 2/5**

| Character | Binary Code |
|---|---|
| 0 | 00110 |
| 1 | 10001 |
| 2 | 01001 |
| 3 | 11000 |
| 4 | 00101 |
| 5 | 10100 |
| 6 | 01100 |
| 7 | 00011 |
| 8 | 10010 |
| 9 | 01010 |
| Start (2 of 5) | 110 |
| Stop (2 of 5) | 101 |
| Start (I 2/5) | 00 |
| Stop (I 2/5) | 10 |

The recommended height for bar codes is 0.25 inch or 15% of the overall length of code area, whichever is higher. This is a common requirement for all symbology to make the code readable for the bar code readers. Further allowances for the width of the bar to accommodate growth or/and shrinkage in postprinting environmental changes should also be considered in determining the width of bar codes.

Because the white bars in a standard 2 of 5 do not carry any information, the final code results in a long string. This is rectified in the interleaved 2 of 5 code.

#### 3.2.2.2 Interleaved 2 of 5 Code

In the interleaved 2 of 5 code, a single number consists of a total of five 0s and 1s (dark and white bars and spaces). There are no blank spaces between two numbers in the code, which makes it a continuous code (while the standard 2 of 5 is a discrete code). The characters to be represented are grouped by two. Hence, the data to be coded in I 2/5 must contain an even number of characters. The adjacent letters are grouped together and their binary codes (shown in Table 3.2) are inserted (interleaved) into each other.

For example, the number 23 is coded as:

```
Binary code for 2: 0 1 0 0 1

Binary code for 3: 1 1 0 0 0
```

The final representation, 0111000010, is the code for 2 interleaved with the code for 3, which is the code for 23. This binary code is then translated into a combination of white and black bars, with narrow bars for 0s and wide bars for 1s. The final bar code for 23 in I 2/5 is given as

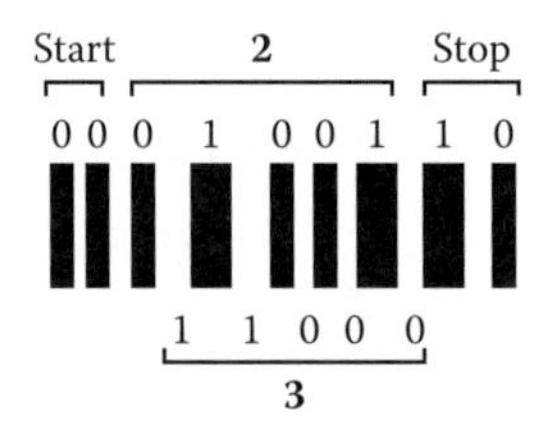


if start and stop codes are included. There is a narrow space between the start symbol and the data characters. A complete symbol consists of four bars more than an integer multiple of 5, allowing for start and stop symbols.

Figure 3.3 shows how the number 123 is coded in the standard 2 of 5 system and in the I 2/5 system. An even number of characters is required for I 2/5, so a zero is added to the beginning of the 123. The standard 2 of 5 and interleaved 2 of 5 coding systems can represent only numeric characters.

The total length of space required for a bar code in I 2/5 is calculated by

$$L = 2QZ + (C(2R+3)+6+R)X$$

The ranges of $QZ$, $R$, $X$, and height are the same as given in the standard 2 of 5.

(a) Encoding the number 123 in standard 2 of 5

The syntax is 110 100010100111000 101.

(b) Encoding 0123 in interleaved 2 of 5

(c) Encoding 123 in code 3 of 9.

**FIGURE 3.3** (a) Encoding the number 123 in standard 2 of 5. (b) Encoding 123 in interleaved 2 of 5. (c) Encoding 123 in code 3 of 9.

### 3.2.2.3 Code 3 of 9

Code 3 of 9 (or 39) allows encoding of numbers, alphabetical characters, and eight special characters, as shown in Figure 3.4. Each character is made of nine elements, three wide and six narrow, of which five are dark and four are white. Between any two characters a space (the same width as that of the narrow bar) is provided, which makes code 3 of 9 a discrete bar code. A narrow bar or narrow space represents 0, while a wide bar or wide space represents 1. As shown in Figure 3.5, the letter X is defined by the binary element 010010001. An asterisk, * is used as a start and stop symbol. Code 3 of 9 is of variable length; that is, it can have any length. However, a practical limitation on length may exist due to the type of scanner available.

The total space required for a bar code in the 3 of 9 code is

$$L = 2QZ + X(3(C+2)R + 7C + 13)$$

The ranges of $QZ$, $R$, $X$, and height are the same as given in the standard 2 of 5 code.

| Char | Pattern | Bars | Spaces | Char | Pattern | Bars | Spaces |
|---|---|---|---|---|---|---|---|
| 1 | | 10001 | 0100 | M | | 11000 | 0001 |
| 2 | | 01001 | 0100 | N | | 00101 | 0001 |
| 3 | | 11000 | 0100 | O | | 10100 | 0001 |
| 4 | | 00101 | 0100 | P | | 01100 | 0001 |
| 5 | | 10100 | 0100 | Q | | 00011 | 0001 |
| 6 | | 01100 | 0100 | R | | 10010 | 0001 |
| 7 | | 00011 | 0100 | S | | 01010 | 0001 |
| 8 | | 10010 | 0100 | T | | 00110 | 0001 |
| 9 | | 01010 | 0100 | U | | 10001 | 1000 |
| 0 | | 00110 | 0100 | V | | 01001 | 1000 |
| A | | 10001 | 0010 | W | | 11000 | 1000 |
| B | | 01001 | 0010 | X | | 00101 | 1000 |
| C | | 11000 | 0010 | Y | | 10100 | 1000 |
| D | | 00101 | 0010 | Z | | 01100 | 1000 |
| E | | 10100 | 0010 | – | | 00011 | 1000 |
| F | | 01100 | 0010 | . | | 10010 | 1000 |
| G | | 00011 | 0010 | Space | | 01010 | 1000 |
| H | | 10010 | 0010 | * | | 00110 | 1000 |
| I | | 01010 | 0010 | $ | | 00000 | 1110 |
| J | | 00110 | 0010 | / | | 00000 | 1101 |
| K | | 10001 | 0001 | + | | 00000 | 1011 |
| L | | 01001 | 0001 | % | | 00000 | 0111 |

**FIGURE 3.4** Character structure in code 3 of 9 (Courtesy of D. Randall Hicks, Bar Code Systems, Inc.)

### 3.2.3 Universal Product Code

The UPC code has many versions. Version A is commonly used in retail applications and can encode numbers with up to 12 digits. Version D is used to encode numbers with more than 12 digits. Version E is commonly found on packages that are too small to hold a normal-size bar code.

Figure 3.6 shows the complete representation of version A of the UPC code. Each code consists of (1) a number system character, (2) a five-digit manufacturer identification code number, (3) a five-digit product code number, and (4) a check

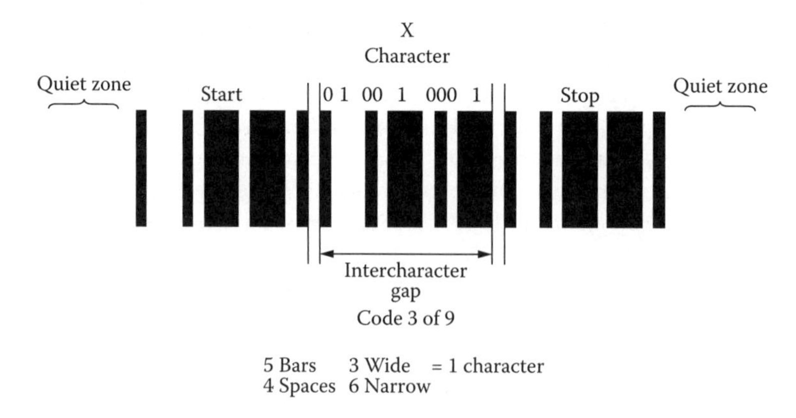


**FIGURE 3.5** Code 3 of 9 (Courtesy of D. Randall Hicks, Bar Code Systems, Inc.)

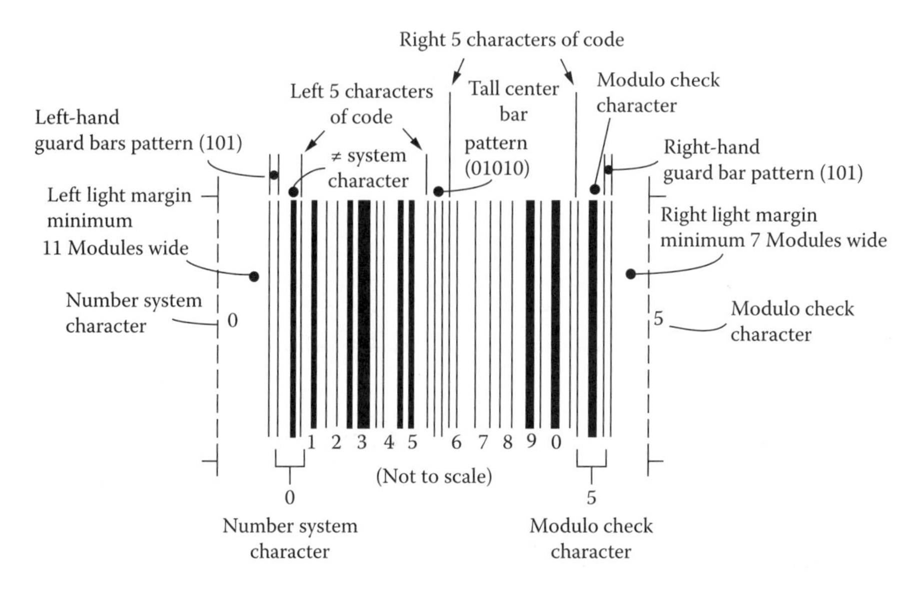


**FIGURE 3.6** UPC-A symbol encoding 1234567890 (Courtesy of D. Randall Hicks, Bar Code Systems, Inc.)

character. The two longer bars in between the 2 five-digit codes divide the symbol into right and left halves. The two longer bars at the beginning and end signal the start and end of reading.

Each digit in the 10-digit number (1234567890) is represented by a unique pattern of two bars and two spaces in a module of seven equal-sized spaces (Figure 3.7). This allows each bar to be of varying width — one, two, three, or four basic widths — and hence represent different characters. Each dark module space is assigned a binary 1, and each light module space a binary 0. The left-hand type characters are mirror images of the right-hand type characters. Left- or right-hand characters having three or five dark module spaces are said to have *odd parity*, while those having two or four dark module spaces have *even parity*. Thus, for each digit, the UPC code provides for two types of left-hand characters and two types of right-hand characters. UPC version A uses odd parity for left-hand characters and even parity for right-hand characters. These 20 characters are shown in Figure 3.7b.

The first digit of the UPC code shown in Figure 3.6 is the number system character that designates the type of application, such as 0 for regular retail items, 2 for variable-weight items, 6 and 7 for industrial products, and 9 for version D coding. The next five-digit manufacturer identification code is assigned by the Uniform Code Council. The second five-digit code is uniquely assigned by the manufacturer for each product. The last digit, the check digit, is calculated using modulo 10 arithmetic. This digit is the remainder that is obtained by dividing by 10 the sum of all the even-position numbers in the 12 digits plus three times the sum of all odd-position numbers. For the code shown in Figure 3.6, position 1 has 0, position 2 has 1, position 3 has 2, and so on. The sum of even-position numbers is $1 + 3 + 5 + 7 + 9 = 25$ and the sum of odd-position numbers is $0 + 2 + 4 + 6 + 8 + 0 = 20$. Thus the sum

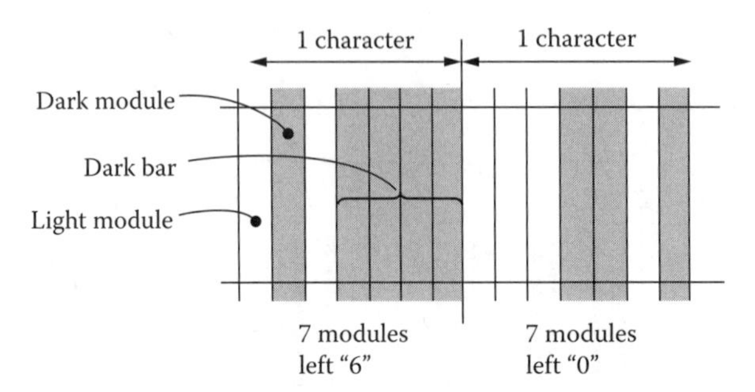


A part of left-hand code showing only two characters (6 and 0)

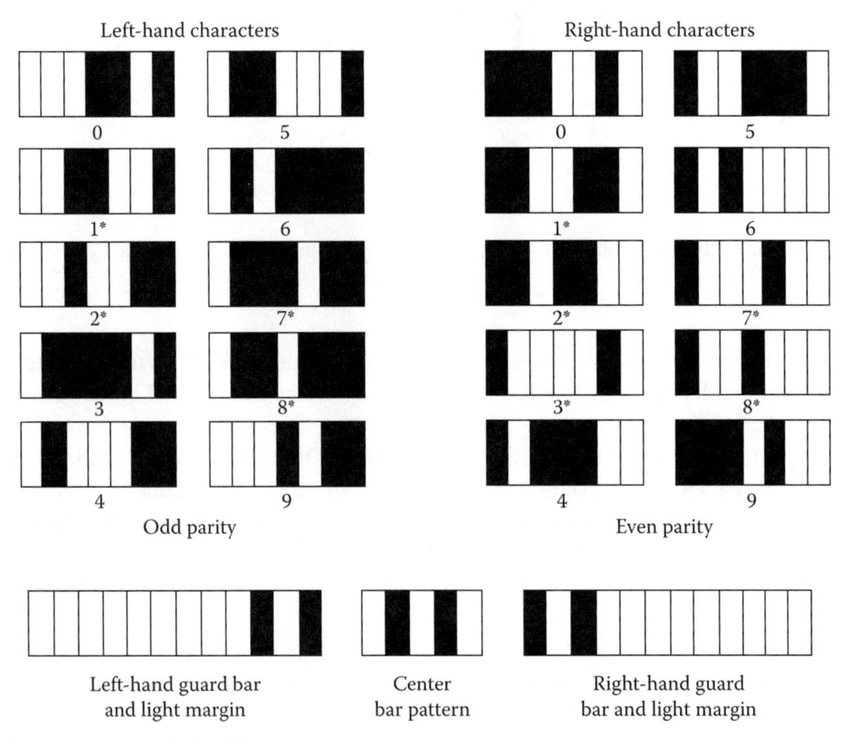


*Note: The numbers 1, 2, 7 and 8 deviate from this modular makeup by 1/13 of a module.

**FIGURE 3.7** Character structure in UPC code (Courtesy of D. Randall Hicks, Bar Code Systems, Inc.)

of even-position numbers plus three times the sum of all odd-position numbers is $25 + 3(20) = 85$. The remainder, obtained by dividing the total sum (85) by 10, is 5, as shown at the end of the bar code.

EAN is a slight variation of the UPC standard. BOOKLAND EAN, which uses EAN-13 (13 digits) and an add-on for price details, is commonly seen on books and periodicals.

One of the characteristics that determine the effective design of a bar code system is the use of a self-checking attribute — that is, the ability to ensure that damage

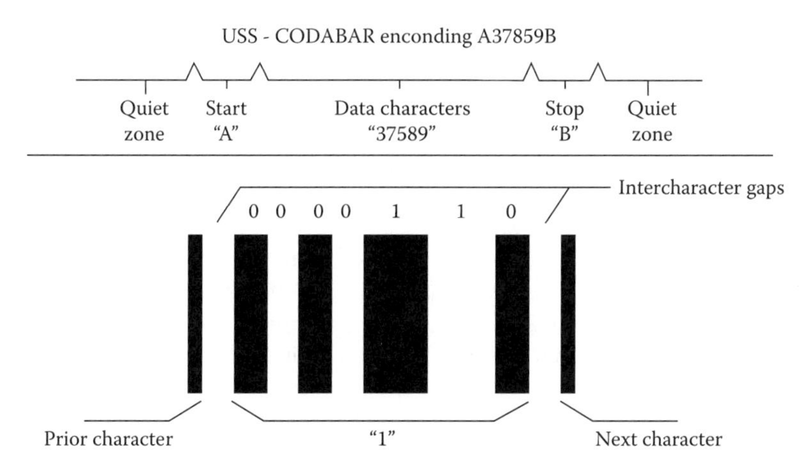


**FIGURE 3.8** USS-CODABAR encoding (Courtesy of D. Randall Hicks, Bar Code Systems, Inc.)

or dust in bars does not result in a faulty read. In this context, standard 2 of 5 is self-checking, because if one of the bars is damaged, it violates the pattern of two wide out of five bars, and the reader will not accept the code. The same principle applies to Interleaved 2 of 5 and Code 3 of 9. The check digit ensures that the UPC code is also self-checking.

There are many other codes as well. Code 128 and code 93 are used in some industries, code 11 is prominent in telecommunications, and code 49 is fairly new and has yet to find a large following. CODABAR is frequently used in blood banks and libraries. Figure 3.8 shows how the character 1 is represented in CODABAR symbology.

Most codes are bidirectional — that is, they can be read in both directions. Code 3 of 9 (Figure 3.5), for example, has asterisks as start and end characters to indicate the starting of the bar code and completion of the bar code. An asterisk is represented by 010010100. When scanning the code from left to right, the asterisk indicates that the message is being read in the forward direction. Reading from right to left results in an inverted asterisk; the decoder (software chip) inside a bar code reader recognizes this as being scanned in reverse direction and converts the message back to proper order.

### 3.2.4 Applications of the Bar Code

The information in the bar code coupled with appropriate computer software can be used in many different ways. We have already mentioned its application in a grocery or retail store. When the bar code on an item is scanned, it is entered into the computer memory, where it is compared with a resident listing of all bar codes. When the two numbers match, the item description and price may be sent to the cash register, the inventory of the item may be reduced to indicate the sale and to record the remaining units in the store, and — if the inventory is below the reorder point — an order for the item may be automatically placed with the supplier.

A bar code may have multiple fields, each separated by a large gap. This gap should be larger than the wide space used to define a bar code. This feature has been used in manufacturing facilities to collect data and update progress. A bar code label may be attached to the document that moves with an order. Such a document is popularly called a shop ticket, work order, or traveler. This label provides a unique identification, such as the order number. Each operator may be given documents, at times referred to as menu cards, with bar codes identifying the operator, the operation number, units of measurements (e.g., standard time), and numbers to describe the quantity processed. When a job arrives at a station, the operator logs in its arrival by scanning the start symbol * in code 3 of 9, his or her identification bar code, his or her station bar code, and the end code *. This information is then readily available on the computer for the production planner and supervisor to plan their assignments. When the job is completed, the operator logs in by scanning the start bar code symbol *, job number, workstation identification numbers, and information on the time spent on the order and units processed, and the end code *. Such timely information keeps the record permanent and up-to-date for accounting. It can also be used to measure the production rate, loss of units in production, and the efficiency of the station. This record is achieved on a real-time basis without spending considerable time on clerical efforts and paperwork.

Another example of the efficient use of bar codes is in storage and warehousing. Accurate data regarding the amount and items stored in each location are very important for a storage operation. A bar code label is attached on the side of each storage location uniquely describing the location. As a tote box is placed in the space, the forklift operator may record the location information by scanning the attached bar code. The bar code found on the traveler associated with the order or a bar code label attached to the tote box is used to identify the unit type and quantity of goods contained in the box. Thus, a computer record can be automatically maintained for each order and its storage location. When the order is retrieved, a similar procedure is performed to indicate removal of the units from storage and availability of the storage location for other orders.

### 3.2.5 Equipment Used in Bar Code Applications

Reading a bar code without error requires that the code is printed properly and read with appropriate reading equipment. Two measurements of label printing quality are used to indicate how accurately a bar code-reading system might work:

1. The print contrast ratio (PCR) is a measure of reflectivity between bars and spaces. It is defined as

$$\text{PCR} = \frac{\text{reflectivity of background} - \text{reflectivity of bars}}{\text{reflectivity of background}}$$

   For a consistent reading on the first attempt, this ratio should be greater than 85%. For bar codes printed by carbon ink on plain paper and read by

**FIGURE 3.9** Bar code densities (Courtesy of D. Randall Hicks, Bar Code Systems, Inc.)

an infrared light source, PCR ranges from 70% to 90%, while for the same codes read by visible red light or incandescent light, the ratio is from 70% to 85%. For noncarbon ink, the ratio may vary between 65% and 85%.

2. Code density is a measure of the width of the bar code. The width of the narrow bar is designated as $X$. The width of the wider bar is then some multiple of $X$. In general, the widths of spaces are the same as the corresponding bars, and all widths remain constant throughout the code. For example, in code 3 of 9, the wider bar ranges from 2.2 to 3.0 times the value of $X$. If the $X$ dimension is less than 0.01 inch, the associated code is called a high-density code. If $X$ is between 0.01 and 0.03, the code is considered a medium-density code, and if $X$ is 0.03 or greater, it is considered a low-density code. The size of the bar code changes, depending on the value of $X$. Figure 3.9 shows examples of the three densities. In selecting the code density, we must consider the space available for the bar code and the capabilities of the bar code-printing and code-reading equipment.

### 3.2.5.1 Bar Code Readers

Bar code readers vary in size and, therefore, in mobility and use. They also differ in the code print they can read, not only in terms of code types, but also in regard to the contrasts and code densities. They also differ in the required proximity of the code for reading.

Contact bar code readers must be close to the bar code in order to read it. Handheld wands or light pens (Figure 3.10) fall into this category. They are small, not very expensive, and can easily be carried from one place to the next. One use of such a bar code reader, along with a small, portable tape recorder, is for taking store inventory. An inventory clerk in a store goes from shelf to shelf recording the bar codes of items and entering the physical inventories of items through a keypad. The keypad can also be used to enter a bar code manually if it cannot be read by contact. The data from the tape are then entered into a computer for further analysis.

The handheld wand has either an incandescent bulb or a light-emitting diode (LED) as the light source. Incandescent light requires high energy input and generates heat but can read over a wide range of PCRs. An LED, on the other hand, operates with very low energy and does not generate heat but cannot read a bar code clearly unless the code has a high PCR. To read a bar code, the wand is held in contact with the code and moved across it in a steady motion. Most wands can read with a wide range of motion velocities, but a change in velocity while reading an individual code can cause errors in reading. The desirable angle for most wands is

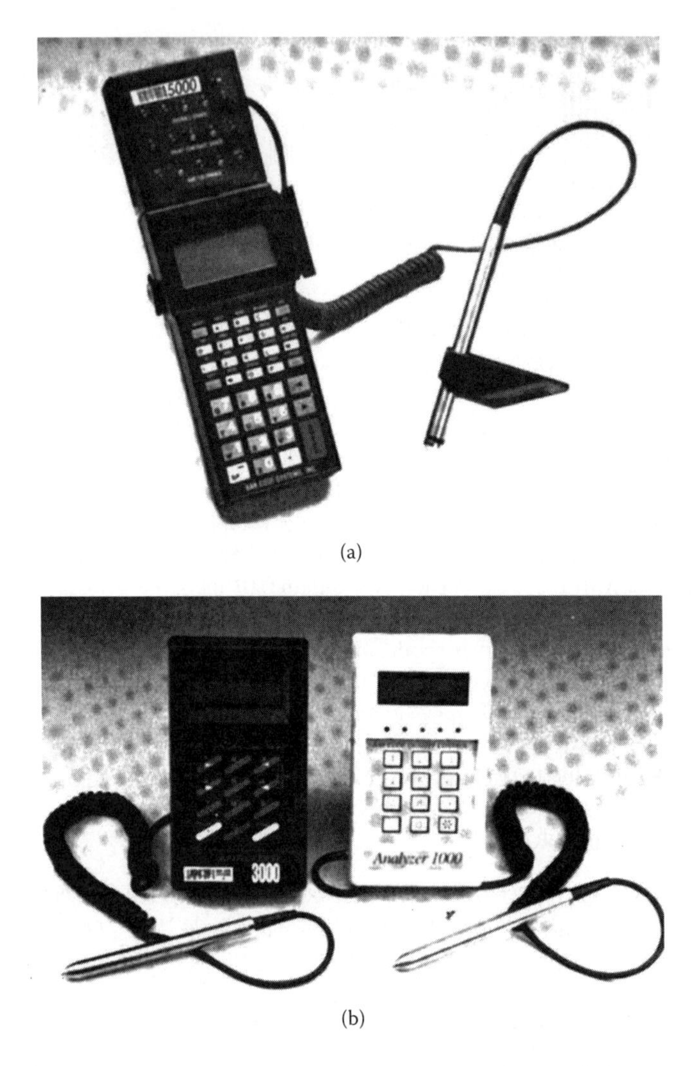


**FIGURE 3.10** Bar code verifiers (Courtesy of D. Randall Hicks, Bar Code Systems, Inc.)

perpendicular to the code; however, there are some wands that work better with a 20° to 30° tilt from vertical. It is a common practice on the shop floor to enclose papers with bar codes in clear plastic sheets for protection. Here the tilt wand performs better, because it is not affected by the reflection from the plastic cover.

Noncontact bar code readers include either handheld or fixed-position laser scanners. They operate by focusing a beam of light from a distance on the entire bar code. A mechanism in the reader directs a small-diameter laser-beam spot to traverse through the code and record the information. The speed of operation allows the scanner to take multiple readings, thus increasing the accuracy of readings and reducing the incidence of no-reads. Laser scanners have the ability to project a non-diffusing, small-diameter ray as far away as 4 feet.

**FIGURE 3.11** Moving-beam scanner in a grocery store

Box-type laser scanners project a continuous beam, and the recording is made by sliding the code in front. They also have beeping devices to indicate a good read. The high speed of operation makes them a popular choice for recording bar codes from packages as they travel on conveyors or on checkout counters in stores where a high volume of items is processed. Figure 3.11 shows the use of a box-type laser scanner in a grocery store.

Laser guns, although faster and more accurate than wands, are slower than box-type scanners. They are portable and somewhat more expensive than wands. They are also used on checkout counters, although at slower speeds, and in most cases where a wand can be used.

The selection of bar code readers further depends on the type of environment in which they are to be used, frequency of usage, the kind of bar code symbologies to be handled, and compatibility of the reading equipment with the computing equipment.

### 3.2.5.2 Bar Code Printers

Bar codes can be printed via a dot matrix, inkjet, electrostatic or laser printer. If care is not taken, printing with a dot matrix printer may result in a gap between two successive dots on the vertical bar. It is then possible for a scanner with a very small beam diameter to interpret the gap as a blank and give a false reading if the symbology does not use a bit-checking structure; where such a check is used, it may result in a no-read. One way to solve this problem may be to select reading equipment with a larger-diameter beam. But this could create another problem. A larger beam may cover two consecutive fields (e.g., a bar and a space together) and again give an incorrect reading or none at all. Thus, it is important that bar code density, contrast, printer, and reader must complement each other. Most often the gap problem is resolved by double striking the dot with slight vertical offset. Electrostatic, laser, and, in particular, inkjet printers give better contrast ratios than dot matrix printers and are, therefore, preferred. Again, care must be taken to see that no stray marks or spots appear in blank spaces, because they may result in the mark being difficult to read.

Printed bar code labels can be obtained from commercial sources. However, in many businesses, labels printed in-house are generated by providing the information through a keyboard and using an on-site printer. For example, a bakery store may produce its own labels for the product made in that day with current pricing policies. A computer program to print code 3 of 9 bar codes was written by James H. Todd and is available in the September 1986 issue of the *Industrial Engineering Journal.*

Bar code printers must consistently maintain the tolerance limits of the bar code widths. The choice of the printer must depend on the quantity of bar codes needed, because the consistency of dimensions varies between different printers over the operating period.

#### 3.2.5.3 Bar Code Verifiers/Analyzers

A bar code verifier or analyzer, put simply, is an advanced version of a bar code reader. These verifiers check the dimensional variations against the specified tolerance limits, the readability of the information, the distance of quiet zones, and other optical characteristics of the symbol to determine whether it meets the specification or standard.

## 3.3 RADIO FREQUENCY IDENTIFICATION SYSTEM

A radio frequency identification (RFID) system is an identification technology in which a base station, known as an RFID reader, communicates with a remote unit attached to an object, known as an RFID tag, through radio waves, and obtains the information stored in the RFID tag's memory. Each RFID tag stores a unique identification code, which is associated with more information concerning the object to which the tag is attached. Thus, the ID code of the RFID tag can be compared to a barcode. However, unlike the barcode system, the reading process in an RFID system does not require direct line of sight, thus enabling automatic product tracking and identification without human involvement. In addition, the RFID system provides more information about the individual product because it can store more fields of information compared to a barcode.

One example of such a system may be the detection and control of a transporter unit as it passes through detection points. The transponders or tags attached to each transporter emit radio signals that are captured by an antenna and relayed to a receiver/transmitter unit. These signals are communicated to the host computer, which may direct a specific response. This response may consist of directing certain programmable logic controllers (PLCs) to function according to the information already on the tag, and/or write additional information to the tag to indicate the action to be taken, such as directing the transporter to a specific station.

The RFID system is preferred over electric eye or other sensor systems when the operational conditions are severe. Radio transmission is possible in most rugged conditions. Nonconducting obstacles, such as cement or wooden structures, or coating the tag with paint, grease, or mud has no effect on the performance of radio signals. Radio transmission also travels over much longer distances compared with most other sensor systems.

Let us look at the use of RFID in a paint shop for automobile bodies. The plant, in general, has multiple booths, each spraying a different color of paint. The cure time and temperature for a paint vary with the type of paint used. The car bodies are delivered to the paint shop on a single monorail conveyor. The tag attached to the body relays the information about the paint type it is supposed to receive. The host computer commands actions from PLCs to direct the body to the appropriate paint shop. When the car body comes out of the paint shop, the tag transmits the information regarding oven temperature and time for cure, and the computer directs the body to an appropriate oven. RFID is used here because it can withstand coatings of auto paint and the extreme oven temperatures, which can reach as high as 400°F.

### 3.3.1 How Does The RFID System Work?

As mentioned earlier, the RFID system consists of tags, readers, and a computer called an application host. An RFID tag consists of a memory in which to store an identification code that is associated with an object the tag is attached to. A reader communicates with RFID tags via radio frequency waves and obtains the information stored in the memory of the tags. The RFID reader sends radio waves from its antenna to the tag and receives the radio waves sent back from the tags. The reader sends command signals by modulating the information into the transmitted radio wave, and receives and demodulates the radio signal from the RFID tags to obtain the ID code. The reader, with the help of application hosts, fetches the detailed information associated with the ID code from a database system. The database system, which is also shared with other organizations, is updated or modified as necessary.

RFID tags can be broadly divided into two categories: active tags or passive tags. Active tags contain an onboard battery, while passive tags operate by using RF power received from the reader. The absence of an onboard battery makes the passive tags inexpensive and long-lasting, and they are therefore more commonly adopted. There are other types of special purpose tags, such as sensor tags, which collect information about the product and the surrounding environment, and are used especially for perishable products.

Use of RFD is encouraged in storage and warehousing. The following stated problems have necessitated computerization and automatic product location and tracking using advanced identification and tracking technologies.

1. Difficulty or lack of adequately managed information associated with location, registration, and operation of products in storage areas.
2. Conventional manual management techniques increase the cost and time of information management and also the probability of errors.
3. Errors occur in product identification.
4. Probability of picking a wrong product is high.
5. The likelihood of older products being left in storage past expiration is high.

RFID-based asset management improves efficiency in the areas of logistics, distribution, and manufacturing. Unlike the simple snapshot provided by conventional techniques such as bar codes, RFID supplies accurate real-time object visibility that enables the continuous identification of items and location of equipment, tools, and other resources in manufacturing and construction areas. RFID can manage complete inventories as well as identification and verification of inbound and outbound shipments without human involvement. Accounting and tracking of goods, along with conditions such as expiration dates, inventory levels, and locations, helps maintain accurate records and deter counterfeiting of products. Work in progress can be obtained by monitoring the system at any time and ensures that corrective steps and quality control measures are being taken. Use of RFID can improve material handling efficiency and eliminate manual inventory counts. It can also be used to automatically detect empty shelves and expired products in a retail store.

At present, a number of large distributors are using RFID. For example, Wal-Mart and the U.S. DOD require all the pallets they receive from manufacturers and suppliers to be RFID-tagged. The DOD has successfully implemented passive RFIDs in the Iraq conflict to constantly track and manage meals and vital supplies.

However, the wide-scale implementation of RFID technology in supply chain and retailer products has been hindered by privacy issues such as the ability of manufacturers to track the buying habits of individual consumers and the cost of the system itself. Currently, an RFID tag costs between and $0.25 and $2.00, and the cost of a reader ranges from $500 to $5000.

Indoor RFID applications require deployment of larger number of readers and good network connections. This necessitates a high initial installation cost. However, pallet and case tagging is economically viable in the present cost environment and RFID has been used by a number of manufacturers, suppliers, and retailers. For RFID tags to be viable at an item level and be a replacement in the future for bar codes, the price of the tag needs to come down to about 5 cents. With the current development trends in the semiconductor industry, there is no reason why the cost of a passive tag cannot come down to a few cents and that of readers to less than $300 in a few years.

### 3.3.2 Radio Data Communication

Radio data communication is another system that communicates with radio signals. It is prominently used in controlling forklift operations in a warehouse and on the plant floor. A unit attached to the forklift consists of a display and special-purpose keyboard. These communicate with the host computer through antennas located in strategic locations. The host computer can direct the forklift to a specific location and request a particular action. Much tighter inventory control and scheduling of forklifts is possible through this technology.

## 3.4 MACHINE VISION

Machine vision is an emerging field in manufacturing. The field attempts to understand the surrounding environment through the acquisition and processing of video images. Just as manual operations are very dependent on the human visual system for information needed in the manufacturing process, machine vision systems provide the sensory input needed in the manufacturing automation environment. Therefore, machine vision is commonly thought of as "automatic acquisition and analysis of images to obtain desired data for interpreting a scene or controlling an activity." It enables the flexible automation that is so essential for the factory of the future.

### 3.4.1 Applications of Machine Vision

Some of the present applications of machine vision in manufacturing include:

- *Quality assurance*: Inspecting the conformance of a part or assembly with respect to dimensions, presence of features, shape, and surface flaws (most of the machine vision applications focus on inspection of parts).
- *Material handling*: Sorting of parts using shape and identification marks such as code number, providing information such as identity to facilitate movement, processing, or assembly of the parts.
- *Guidance*: Providing a robot or other manufacturing tool with the information for adaptive control of its path or function.
- *Monitoring*: Checking the state of a machine or process to verify that the machine tool or the process is functioning in the intended pattern. Also used as a security system in critical-area safety to provide information about any undesirable change in the environment (e.g., intrusion of a person near a robot work area).

Figure 3.12 illustrates the components of a typical machine vision system used in inspection. The camera grabs the image and stores it in a frame grabber, which converts the image from analog to digital form. The frame grabber also sends the

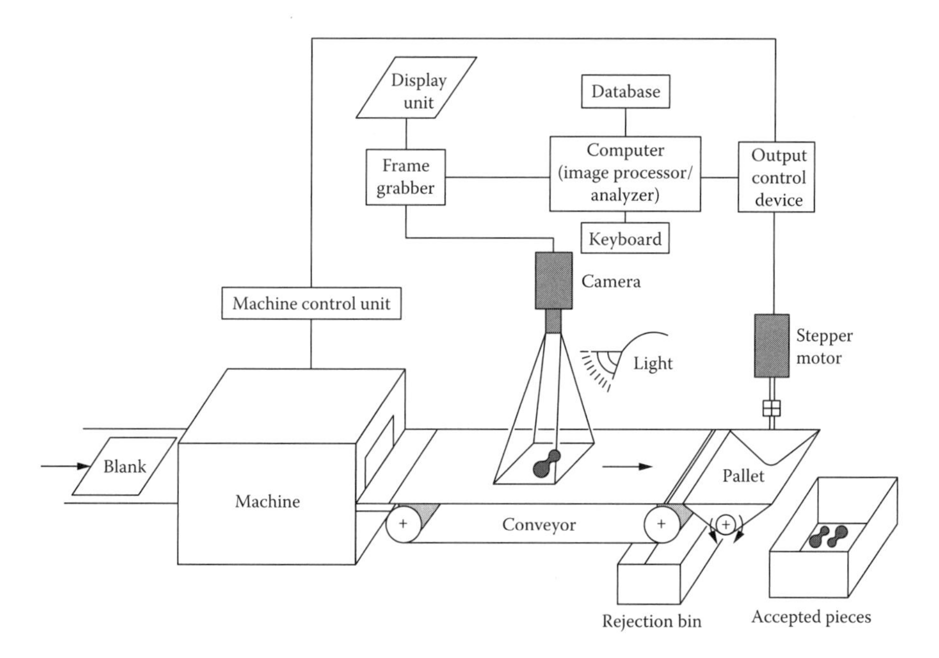


**FIGURE 3.12** A typical machine vision setup in inspection. Machine vision system with light table, camera, computer, and monitor grabbing an image. This is an experimental setup for a machine vision system used in part sorting. In processing, parts such as a plate and two screws may be randomly oriented on a base such as a work table. The vision system grabs the image, identifies each part by comparing it to the database, and commands the robot to pick up and move the appropriate part for the next operation.

Machine vision system with light table, camera, computer, and monitor grabbing an image.

image to a display unit by recoding the digital image back to analog form. The computer accesses the digital image from the frame grabber and compares it to the digital image in the database. Based on the outcome, the computer sends instructions to the material-handling controller (stepper motor), which opens the hopper to either accept or reject the part.

### 3.4.2 Typical Machine Vision Setup

A vision station includes a camera or cameras, typically a solid-state camera such as a charge-coupled device, connected to an acquisition board (also called a frame grabber) residing in a computer. Programs running on the computer can then command the acquisition card to grab images from the camera(s) and store them in the memory residing on the acquisition card. The image is stored in the format of 1 byte (8 bits) of information per picture element (pixel) for monochrome cameras, or 3 bytes per pixel for color cameras. One pixel represents a copying area of the object surface, depending on the focal length of the camera and how close the camera is to the object. The information stored for every pixel is the intensity of the pixel. In monochrome cameras, the intensity is stored as a gray level, while in color cameras three intensity levels are stored. These three levels correspond to the red, green, and blue elements of each color pixel. The control programs can then directly access the acquisition board's memory to process the images in any desirable fashion.

The cameras and acquisition boards can cost as little as several thousand dollars, depending on the desired speed of operation and camera resolution.

Off-the-shelf systems that are capable of plugging into any standard computer are becoming commonplace. Software libraries are also available for the acquisition boards in most popular languages.

### 3.4.3 Uses of Machine Vision in Manufacturing

In facilities planning, the use of machine vision systems should be considered for inspection areas, process monitoring, and assembly areas. Machine vision systems can relieve human inspectors of the drudgery involved in making many accurate measurements throughout a shift. A machine vision system can provide more consistent inspection and take up less room, because cameras can be mounted above the line, requiring no additional movements of the parts off-line to a human inspection station. If the inspection process can be done quickly enough, it is possible to do 100% parts inspection. Using machine vision systems in automated assembly areas where robots perform all assembly operations is very desirable. No human inspectors are then needed to be around the robotics arms during operations. This not only improves the facility's safety but also allows for more efficient use of space because room is not required for human inspectors to move about.

As mentioned before, a major use of a machine vision system is in the area of parts inspection. Here the objects are found — along with their locations and orientation — and important characteristics are automatically measured and compared to design specifications. If the dimensions are not within the allowable tolerances, the part is rejected. The process also includes looking for obvious flaws in the parts, such as missing pieces or cracks in the assemblies. Depending on the parts, the inspection process can be very complicated and involve many steps. Therefore, an especially important factor in the inspection is the precise calibration of the machine vision system.

Machine vision is also being increasingly used in automated manufacturing and assembly of parts. In this application, the system must recognize parts and accurately calculate their positions and orientations. The part, its location, and its orientation are then used by other automation devices, such as robotics arms, to move the part through the manufacturing or assembly process.

Both applications use the basic strength of machine vision — that is, being able to recognize parts or certain patterns that might appear in the sensors' (cameras) field of view. The most common technique used in finding the patterns is through *template matching*. In template matching, a template is created with the desired pattern to be recognized by the system. When an image is acquired, the template is placed over the various areas of the image, and a correlation between the template and the image patch over which the template is placed is calculated. A low correlation indicates no match between the template and image patch, while a high correlation indicates a possible match and thus a possible find of the pattern.

### 3.4.4 Two- and Three-Dimensional Machine Vision

Machine vision systems have traditionally been of only two dimensions, because cameras acquire information only along a two-dimensional (2-D) pixel array.

Different part types are typically distinguishable in a 2-D view (i.e., photograph). Inspection of certain characteristics or the locating of parts and their orientations is usually best done from one particular view. Good placement of the camera or cameras can often provide the necessary information for the tasks without having to resort to more complicated and expensive 3-D vision techniques.

Sometimes, however, it is necessary to obtain a true 3-D model of what the camera is looking at. Although the intensity reading of each pixel location provides some clues to the depth of the object at that point, the true depth cannot be calculated from the intensity information alone. The problem exists because we project a 3-D view onto a 2-D surface (image plane of the camera). Enough information is lost during the projection process that one cannot uniquely recover the 3-D information from the camera view alone. Fortunately, there are existing techniques that allow one to obtain 3-D information in machine vision.

One such technique involves structured light. A sheet of light is projected down upon an object and the camera is placed alongside the light projector, pointed in the direction where the plane of light intersects the object. At the intersection of the sheet of light and the object, there is a line of light. This line of light distorts to match the surface contour of the object. To obtain the depth of the object along this light contour, it is necessary only to measure the length of distortion at every pixel along the profile. Structured light techniques are very straightforward and accurate but can be slow if information along a large surface area is to be obtained. It is possible to project multiple sheets of light on the object to decrease the time it takes to obtain a complete 3-D model.

Another popular technique is called *stereoscopic vision*. With stereoscopic vision, images are acquired of the same scene taken at slightly different camera views. The depth can be recovered by finding the same object point in both camera views and then calculating the vertical offset of both points. The depth of the point is then proportional to the difference in the vertical offsets, which is called the *disparity*. The use of stereoscopic vision is popular because it is one way the human visual system acquires depth information. The most difficult part in this process is attempting to locate the same point in both stereoscopic images.

### 3.4.5 Machine Vision Processing

No matter what the machine vision application is, there are some steps in the processing stage that are almost universal. One of the first steps may be to do noise suppression in the image. There is always the possibility of some disturbance in electrically noisy environments in which the machine vision systems must operate. In the noise-suppression stage, an effort is made to eliminate false readings (pixel intensities) while retaining the good information. The major problem in this process is determining which pixels contain the noisy data and which ones do not. It is often impossible to distinguish between noisy and good signals, so noise-suppression techniques operate on the entire image. The most popular of these techniques involve replacing the pixel value (intensity) based on information obtained from the neighborhood of pixels (pixels touching the pixel being processed). This information is typically statistical measures of the neighborhood, such as the median

**FIGURE 3.13** View of three medicine bottles of different heights

or mean value of the surrounding pixels. In these cases, the pixel intensity would be replaced by the median or mean value. This technique works well in suppressing noise but can also suppress the detail in the image. In factory environments with proper control of lighting conditions, camera angle, and distance, the noise can be reduced, so a noise-suppression step may not be necessary.

Thresholding is a commonly performed function that identifies an object. In this step, a threshold intensity value is selected. Any pixel with an intensity level lower than the threshold value is reset to black. Any pixel with an intensity greater than the threshold level is set to white (the brightest intensity level). The creation of this binary image is useful in separating the objects of interest from the background. As an example of thresholding, consider the three medicine bottles shown in Figure 3.13. Figure 3.14 shows an image obtained from one of the two stereoscopic cameras. The idea of thresholding is to separate the objects of interest from the rest of the image. The selection of the threshold level is extremely important, because it is the threshold level that determines which pixels are black (corresponding to the background),

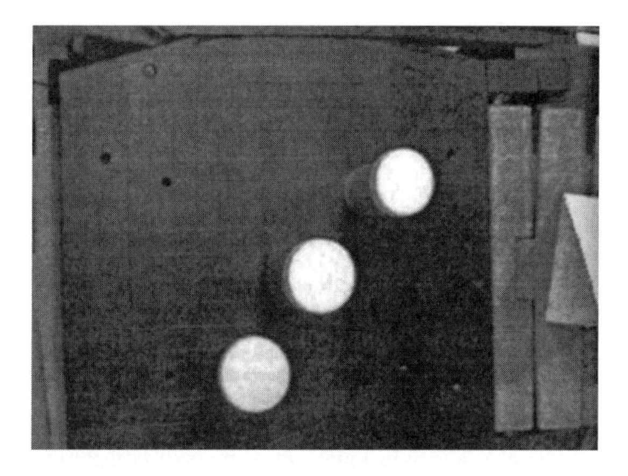

**FIGURE 3.14** Image taken of three medicine bottles before thresholding

**FIGURE 3.15** Three medicine bottles after the thresholding operation.

and which pixels are light (corresponding to part of an object). Figure 3.15 shows the image in Figure 3.14 after the thresholding operation is performed.

Other operations commonly performed on the camera picture depend on the application. These operations include edge detection, edge enhancement, and frequency analysis.

## EXAMPLE—MACHINE VISION PROCESSING

An example of a template-matching operation is illustrated next. Assume that we have taken a snapshot of an irregularly shaped object and converted the picture into a binary image through the thresholding operation. The shape of the object is displayed by the 1s in the image. We want to locate the internal hole of the object. Because the hole pixel would have the intensity of 0 (background) and would be surrounded by the pixels of the object with the intensity of 1, a template is selected that has a 0 surrounded by 1s. The hole should be found at the point in the image where the correlation between the template and the image are greatest.

$$\text{Template: } \begin{matrix} 1 & 1 & 1 \\ 1 & 0 & 1 \\ 1 & 1 & 1 \end{matrix}$$

$$\text{Image: } \begin{matrix} 1 & 1 & 0 & 0 & 0 \\ 1 & 1 & 1 & 0 & 0 \\ 1 & 0 & 1 & 1 & 1 \\ 1 & 0 & 1 & 0 & 1 \\ 0 & 0 & 1 & 1 & 1 \\ 0 & 0 & 0 & 0 & 1 \end{matrix}$$

There are many ways to perform the correlation between the template and the image. In this example, we shall merely determine the number of pixels matching between the template and the pixels of the image on which the template is laid. The process is started by laying the template over the top of the first three rows and first three columns of the image and then doing a pixel-by-pixel comparison, adding up the matching pixels. This number, 6, is stored in the upper left corner of the correlation matrix. Next, the template is shifted over one column so that it lies atop the first three rows and columns 2 through 4. A point-by-point comparison is again done and the resulting sum, 4, is stored in the first row and second column of the correlation matrix. The process continues; the associated complete correlation matrix is shown next.

$$\text{Correlation matrix: } \begin{matrix} 6 & 4 & 5 \\ 8 & 4 & 5 \\ 4 & 4 & 9 \\ 3 & 2 & 5 \end{matrix}$$

From the correlation matrix, we can see that the maximum correlation, 9, is obtained when the template is placed over the third through fifth rows and the third through sixth columns. Thus, the template matches in this location. The higher the correlation values, the closer the image portion matches the template.

As another typical use of a machine vision system, an inspection operation is presented. In this operation an image is acquired of the top of a bottle looking directly down on it. The vision system determines whether the bottle has a cap on it or not. In performing the inspection operation, the image is acquired and then a binary image is created. The binary image is then processed to see if a cap appears in the image. The setup of the system would vary depending on the color of the bottle, cap, and liquid inside the bottle. In this example, we assume a clear bottle with a light-colored cap and a darker liquid. If the cap is on the bottle, the resulting image will show a round bright spot in the image representing the cap, as shown here.

$$\text{Image with cap: } \begin{matrix} 0 & 0 & 0 & 0 & 0 & 0 & 0 & 0 & 0 \\ 0 & 0 & 0 & 0 & 1 & 0 & 0 & 0 & 0 \\ 0 & 0 & 0 & 1 & 1 & 1 & 0 & 0 & 0 \\ 0 & 0 & 1 & 1 & 1 & 1 & 1 & 0 & 0 \\ 0 & 0 & 0 & 1 & 1 & 1 & 0 & 0 & 0 \\ 0 & 0 & 0 & 0 & 1 & 0 & 0 & 0 & 0 \\ 0 & 0 & 0 & 0 & 0 & 0 & 0 & 0 & 0 \end{matrix}$$

If the bottle is missing a cap, the binary image would contain a ring instead of the circle when the cap is present. The following image shows an example of an image obtained of a bottle with the cap missing.

$$\text{Image with cap missing: } \begin{matrix} 0 & 0 & 0 & 0 & 0 & 0 & 0 & 0 & 0 \\ 0 & 0 & 0 & 0 & 1 & 0 & 0 & 0 & 0 \\ 0 & 0 & 0 & 1 & 0 & 1 & 0 & 0 & 0 \\ 0 & 0 & 1 & 0 & 0 & 0 & 1 & 0 & 0 \\ 0 & 0 & 0 & 1 & 0 & 1 & 0 & 0 & 0 \\ 0 & 0 & 0 & 0 & 1 & 0 & 0 & 0 & 0 \\ 0 & 0 & 0 & 0 & 0 & 0 & 0 & 0 & 0 \end{matrix}$$

There are several ways to perform the inspection operation. One way is to calculate the area of the image (number of 1s) and see if the area corresponds to the cap area or the area when there is no cap. This approach, although easy to implement, might not give the best results. A noisy image could give inaccurate readings for the area.

Another method is to again use template matching. Here, though, the templates are created for the entire views. The first template would be for the cap being on the bottle, and the second template would represent the case of a missing cap. Both templates could be placed over the image acquired of the end of the bottle and the correlation is calculated. If the correlation between the template corresponding to the cap being present is higher than the correlation of the template with the missing cap, then it is assumed the cap is present. If the correlation with the template corresponding to the missing cap is higher than the correlation with the cap template, then we assume that the cap is missing. Remember there is only one correlation value per template, because each template covers the entire image.

$$\text{Template for cap inspection:}\quad \begin{matrix} 0 & 0 & 0 & 1 & 0 & 0 & 0 \\ 0 & 0 & 1 & 1 & 1 & 0 & 0 \\ 0 & 1 & 1 & 1 & 1 & 1 & 0 \\ 0 & 0 & 1 & 1 & 1 & 0 & 0 \\ 0 & 0 & 0 & 1 & 0 & 0 & 0 \end{matrix}$$

$$\text{Template for missing cap:}\quad \begin{matrix} 0 & 0 & 0 & 1 & 0 & 0 & 0 \\ 0 & 0 & 1 & 0 & 1 & 0 & 0 \\ 0 & 1 & 0 & 0 & 0 & 1 & 0 \\ 0 & 0 & 1 & 0 & 1 & 0 & 0 \\ 0 & 0 & 0 & 1 & 0 & 0 & 0 \end{matrix}$$

Suppose that, in another example, the clear bottle has clear liquid and is capped with a white cap. Because the images with and without the cap would tend to be light, the thresholding of the associated images is very difficult and the margin of error increases. To continue the inspection using a vision system, one alternative is to change the cap color to a darker color to get better contrast. Other alternatives might include purchasing a better-quality vision system. Yet another alternative is to match the edge of the bottle top. Without the cap the edge would be smooth, while with the cap, as the cap generally has ridges on the edge, the edge in the image should also have ridges. Thus, an edge-detection technique may be an appropriate solution to the problem.

Automatic inspection of glass bottles using machine vision.

## 3.5 VOICE INPUT

Voice recognition is a technology similar to machine vision. Here, human sounds are converted into electric signals, which are then converted into a digital, machine-readable zero-one format. In speaker-dependent systems, templates, associated with a few select words (40–250 words), as pronounced by the speaker, are initially stored in computers. As the operator speaks the key words into a telephone-like handset, they are recognized by a template-matching process similar to the one described for machine vision. If a match is found, associated actions are taken by use of PLCs or other devices. Speaker-independent systems are independent of the speaker but also recognize one word at a time. They cost about four times as much as speaker-dependent systems. Continuous-speech systems are trained for individual speakers but can recognize several words without pauses between the words.

Voice recognition allows an operator with both hands and eyes occupied, such as a forklift operator, to get real-time analysis of the data with desired speed and accuracy. This technology is used mainly in places where the operators are too occupied to enter the observations through keyboards, such as in laboratories or in inventory control, sorting, and quality control functions. The advantages of voice recognition are obvious: real-time data input and verification in any language, requiring no hand or eye involvement. There are also some disadvantages with the voice system. These include the requirement of training the system for each operator, limitations on vocabulary, and the distortions that might be caused because of the noisy background associated with working environments, such as those in manufacturing plants.

## 3.6 PROGRAMMABLE LOGIC CONTROLLERS

PLCs are widely used in industry to achieve automation. They allow input signals based on some logical events to be translated into output signals that can activate associated devices. Furthermore, such response is possible without hard-core wiring connecting input and output gadgets directly. Thus, a great deal of flexibility is obtained in automation, because logical or physical changes in sequencing do not require new wiring. The only change needed is the modification of the logical program that directs actions of the PLC. A typical PLC has a set of "cards" that are inserted into racks connected through a motherboard. The motherboard provides the power and establishes the internal links for sharing data between the cards. The cards provide digital and analog interfaces that allow the flexibility of purchasing only the types and quantities of what are necessary for the specific application. A processor card, which is the brain of the PLC, is always required, but the number of input and output cards can vary, based on the number of input and output devices required. Each input and output device is connected to the rack — and thus, internally, to an input or output card — giving it a unique location address.

On-off–type input signals for a PLC are derived from devices such as push buttons, photoelectric eyes, circuit breakers, motor starters, relay contacts, and limit switches. It is also possible to obtain continuous input signals measuring temperature, pressure, level, and flow. The output devices that may be energized or activated by a PLC include alarms, lights, solenoids, control relays, horns, starters, valve actuators, and chart recorders.

A common programming language for a PLC is a symbolic instruction set called a ladder diagram. The instructions are composed of five categories: relay-type (input/output), time/counter, arithmetic operations, data manipulation, and program control. Table 3.3

**TABLE 3.3**
**Summary of PLC Instructions**

| | |
|---|---|
| | **Relay (Input/Output) Operators** |
| \|\| | Normally open contact; for current to flow, the contact must close. |
| \|/\| | Normally closed contact; for current to flow, the contact must be in nearly closed position. 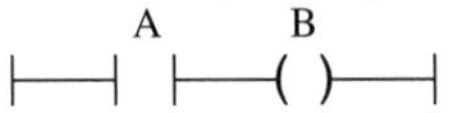 A B ├─┤ ├─( )─┤ |
| ( ) | Energize coil; [e.g., If contact A closes, coil B (output) will energize.] |
| (↑) | Deenergize coil. |
| (L) | Latch coil; coil will remain energized even though the conditions change so that the path that led to energizing the coil no longer needs to be a closed path. |
| (U) | Unlatch coil to reset latched coil. |
| (↑) | Off-on transmission contact: allows one pulse when the contact closes from off to on. The contact will close for one scan and then open, even although the input stays closed. 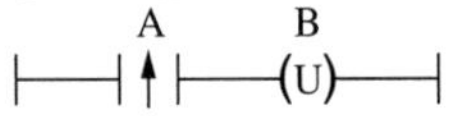 A B ├─┤↑├─(U)─┤ |
| (↓) | [When A closes (goes from off to on), B is unlatched in one scan.] On-off; same as before except pulse flows when contact goes from on to off. |
| | **Timer and Counter** |
| (TON) | Timer on: energizes the output after time delay. Logical continuity of path starts counting time, and when the accumulated time equals a specified value, the output is energized. TB = time base, PR = present value For instance, coil 10 will energize 10 seconds after input A is closed. 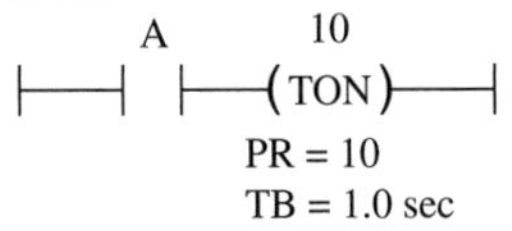 A 10 ├─┤ ├─(TON)─┤ PR = 10 TB = 1.0 sec |
| (TOF) | Timer deenergizes the output after time delay. |
| (RTO) | Retain timer on; the time count is retained even if there is a power failure. When power comes on, the timer continues the count. |
| (RTR) | Reset timer to zero. |

**TABLE 3.3 (CONTINUED)**
**Summary of PLC Instructions**

| | |
|---|---|
| (CTU) | Up counter; counter counts every time the path is closed. When the count reaches a specified value, the output energizes; every time A closes, the counter counts one. When the counter reaches six, coil 10 is energized. |
| (CTD) | Counter is counted down; it will stop at zero or a prespecified negative value. |
| (CTR) | Counter reset; the counter is reset to zero. |

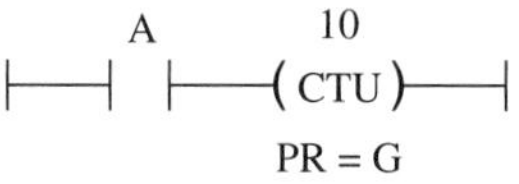

```
        A          10
|---| |---( CTU )---|
              PR = G
```

**Arithmetic Operations**

| | |
|---|---|
| (+) | Add two values stored in the referenced locations. |
| (−) | Subtract two values stored in the referenced locations. |
| (×) | Multiply two values stored in the referenced locations. |
| (÷) | Divide two values stored in the referenced locations. |

**Data Manipulation**

cmp = Compare two referred registers for equal conditions.

< Compare two referred registers for less than condition.

> Compare two referred registers for greater than condition.

**Data Transfer**

GET Accesses the constraints of the referenced address.

(put) Stores result of operation in the location specified.

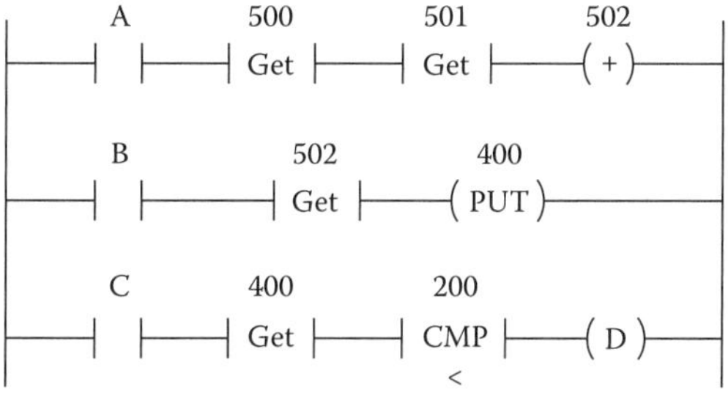

```
      A        500        501        502
|---| |---| Get |---| Get |---( + )---|

      B          502        400
|---| |------| Get |---( PUT )--------|

      C        400        200
|---| |---| Get |---| CMP |---( D )---|
                        <
```

If A is closed, the contents of 500 and 501 are added and the resultant is placed in 502. If B is closed, the contents of 502 are placed in location 400. If C is closed, the contents of 400 are compared with the contents of 200, and if less, D is energized.

*(continued)*

## TABLE 3.3 (CONTINUED)
## Summary of PLC Instructions

**Process (Program) Control**

(MCR) Beginning a micro; used to activate a segment of ladder diagram.

(end MCR) Signifies the end of micro.

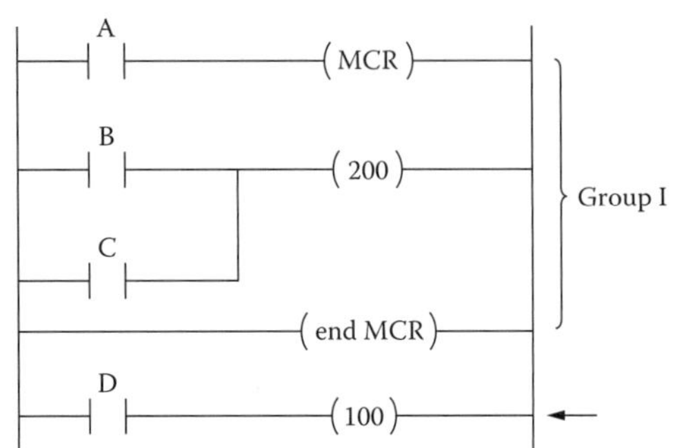


If A is closed, MCR is energized and instructions within group I are executed. If A is open, instructions in group I are ignored and the execution continues on the rung outside the range of the micro — that is, at the rung shown by the arrow above.

(JMP) Jump to the level indicated.

|LBL| Used to label a ladder rung.

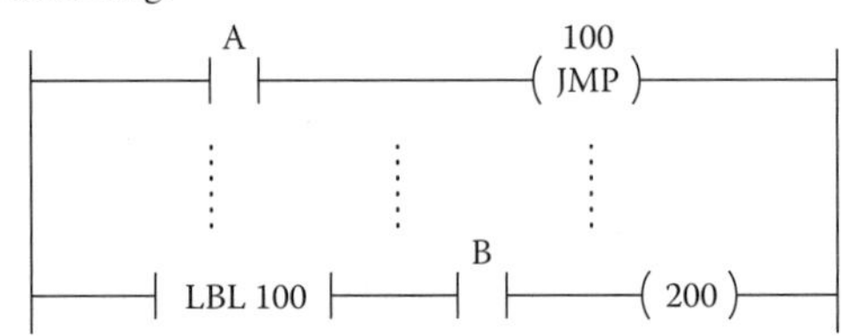


If A is closed, jump to rung labeled 100; where if B is closed, coil 200 is energized. If A is not closed, continue sequential execution.

(JSB) Jump to subroutine

(RET) Return from the subroutine to the rung immediately following the rung where the subroutine is called.

Additional features available in PLC include timers and counters. A timer allows a specified time lag between the closing of the input path and energizing the output device. A counter counts the number of times the input path is closed. When this number reaches a prespecified value, the counter allows the output to energize.

A 10 JSB
B C
10 LBL F 100
RET

As a closes, the subroutine in rung 10 is called for. After the subroutine is evaluated, control returns to rung following (JSB), marked with →.

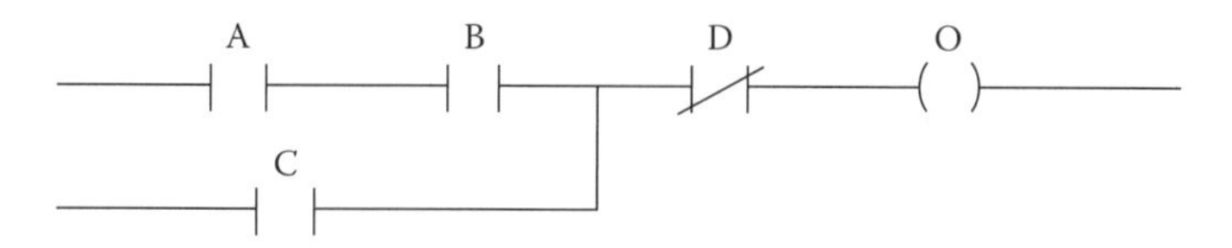


**FIGURE 3.16** A logical path

shows symbols and associated conditions. It is beneficial to understand the basic ladder logic instructions. A rung of the ladder shows how an output, designated by a coil symbol, ( ), is controlled based on inputs, designated by | |. A continuous path from left to right is required for an output to energize. Each logic path may consist of series, parallel, or a combination of input configurations. For example, in Figure 3.16, inputs A, B, and C are normally open and D is normally closed. If both A and B are closed, or alternatively, if C is closed, then because D is closed, there is at least one continuous path from the left to the output, O — therefore, O energizes.

PLCs with cards. The photo shows PLC systems with cards from two different manufacturers. Motherboards with different numbers of plugs to insert cards and with different shapes and sizes are available from different suppliers. A card is an electronic circuit board that can be plugged into a motherboard.

If A, B, C, D, and O are connected to a PLC, then each has a unique symbolic address. While loading the ladder diagram program into a PLC, these devices are identified by their location addresses and not by the letters A, B, C, D, or O. Like a computer program, the ladder diagram is executed sequentially unless otherwise instructed; unlike a computer program, the rungs of the ladder diagram are continuously evaluated.

The evaluation and completion of one rung is known as a scan. The time taken to complete a scan is known as scan time. Most PLCs have scan times measured in microseconds.

Arithmetic operations such as addition, subtraction, multiplication, division, and square root are possible in a PLC. Similarly, data manipulation such as comparing the values, setting constants, and transferring the data within the PLC from one location to another are also feasible. Such operations are necessary mainly for continuous (analog) input and/or output data signals.

### 3.6.1 PLC in Material Handling

PLCs are quite often used in process control and material handling to initiate a response to a change in working conditions. The following is one such example, where versatility in the material-handling setup is obtained by the use of a PLC.

A component is assembled in two stages. After completion of the first stage, the unit is transported to the second-stage operation on a conveyor. Depending on the rate of production in the first stage, the second stage may require two or three operators. The maximum production rate in the first stage is one unit every 30 seconds. The objective is to automatically feed the second-stage operators with the units from the first stage. The schematic diagram of the setup is shown in Figure 3.17.

The push-button inputs at 1, 2, and 3 verify which operators are present. These push buttons are in a normally open state. As an operator comes to the station, he or she pushes the push button down to close the contact. Sensor inputs 4, 5, and 6 are photocells, which are in a normally open state. As an object blocks a photocell, the corresponding photocell input transfers to the closed position. Connections at 11, 12, and 13 represent gates. It takes an object exactly 6 seconds to go from gate

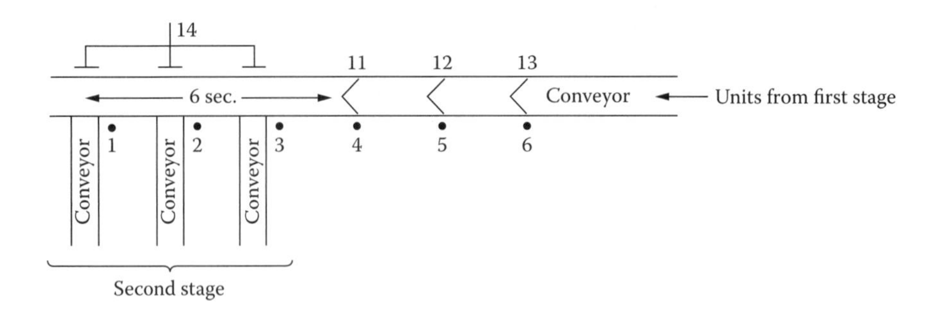


**FIGURE 3.17** Schematic diagram of the workplace.

11 to operator 1, from gate 12 to operator 2, and from gate 13 to operator 3. Point 14 represents the plunger mechanism. When the connection at 14 is closed, the plunger activates, pushing units in front to appropriate operators.

The operational ladder diagram is shown in Figure 3.18. Depending on the operators who are working, the units are held in the appropriate gates and are released simultaneously. After a 6-second delay, the units would be in appropriate positions in front of those second-stage operators who are working. The plunger at position 14 then activates and pushes these units on to the operator feed tables. Because the production rate cannot exceed 1 every 30 seconds, there is at least a 30-second delay between two consecutive units arriving at the gate mechanism.

The details of the ladder diagram are straightforward. An internal input 701 is used as a signaling device; its function is explained later. It is set as a normally closed contact. The ladder diagram can be divided into five distinct segments, or zones, marked as I, II, III, IV, and V. Zone I is associated with conditions when all three operators are present. Rung 1 tests this condition. If all three operators are present, input switches 1, 2, and 3 are closed. Because 701 is a normally closed contact, it is initially closed. This enables the corresponding MCR to energize rungs 2 through 5 for execution.

As the first unit goes through on the conveyor, it closes contact 6, then 5, and then 4. When contact 4 is closed, gate 11 is energized, closing the gate (rung 2). Thus, the first unit is held at gate 11, and contact 4 is in the closed position. When the second unit is delivered on the conveyor, it closes contact 6 and then 5. At that time, the condition that closes both 4 and 5 is satisfied, and based on rung 3, gate 12 is energized. Closing the gate at 12 results in the second unit being held in gate 12 and contact 5 being closed. When the third unit is delivered and it closes the contact at 6, it also satisfies the input requirements for rung 4. This energizes 13, closing the gate and holding the third unit at 13. Rung 6 jumps the control to label 100 associated with rung 24, which is the start of zone V.

Zone V has instructions that are applicable for any operator combination and is therefore executed every time. For convenience in rung 25, any operational condition energizes cell 702, which, in turn, starts the micro logic in rung 26. In rung 27, cell 701 is latched to open the position, thus disabling the first instruction of each micro on zones I, II, III, and IV. There is a reason for this situation. When the units are released through the gates, they change the status of photocells 4, 5, and 6 as they travel toward the push plunger in position 14. They may cause the gates to change their status, closing and opening, if the zones are allowed to function.

Rungs 28 through 30 are used to deenergize and open the gates. This allows the units to proceed in a synchronous fashion, and in 6 seconds they are in front of the appropriate operators. In rung 31, timer 901 is started for 6 seconds. After a 6-second delay, the 701 is unlatched back to the closed position, enabling any of zones I through IV to function (rung 32). At the same time, in rung 33 the plunger mechanism is activated, pushing the units out. In rung 34 the timer is set back to zero, and rung 35 signifies the end of the micro. The control goes to the top of the program and the iteration is repeated.

Zones II, III, and IV are associated with operations when two operators are present. Zone II represents activities when the operators are at stations 1 and 3. Zone III displays operations for operators on stations 1 and 2, and zone IV is associated with the condition when the operators are at stations 2 and 3.

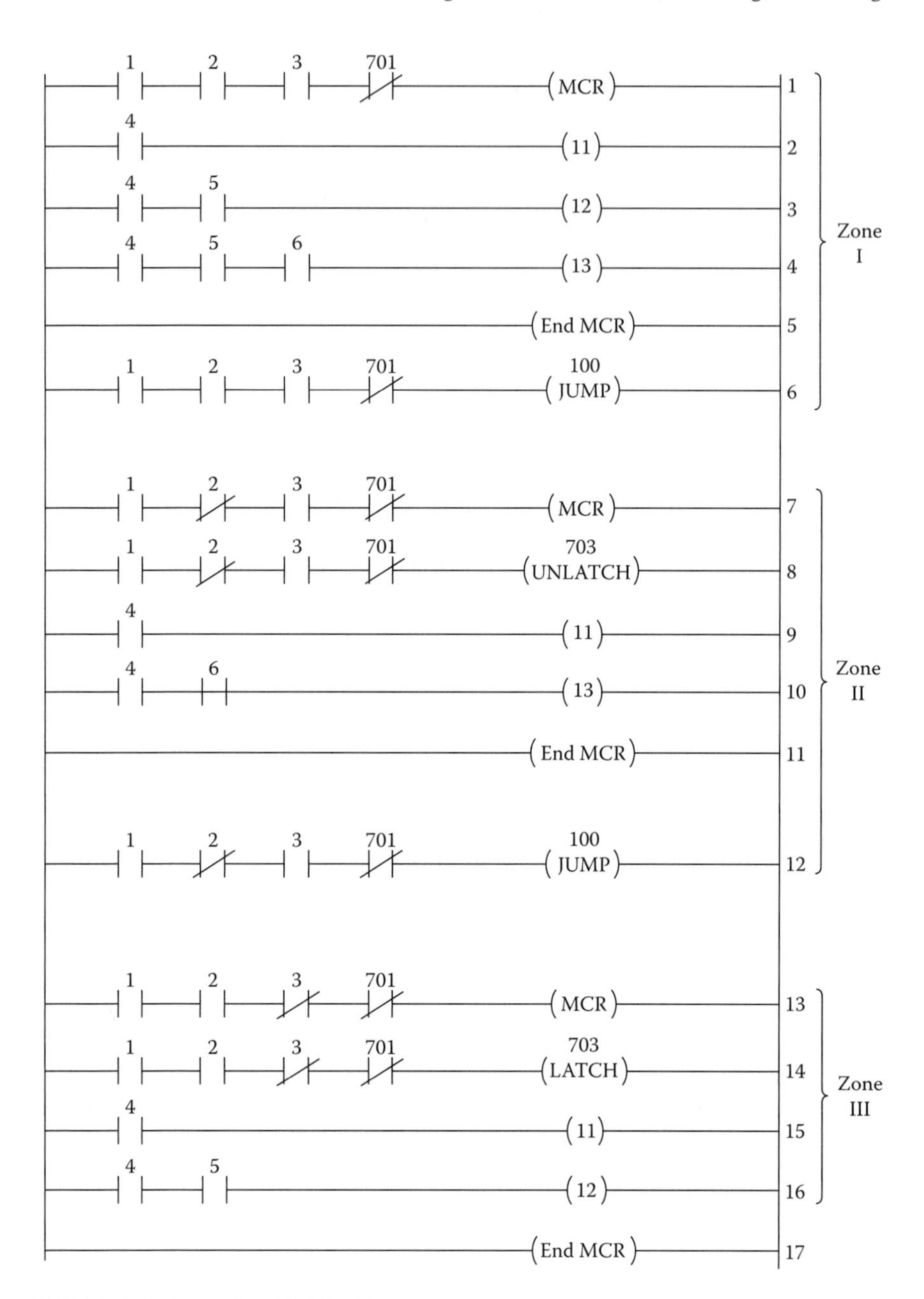


**FIGURE 3.18** Operational ladder diagram

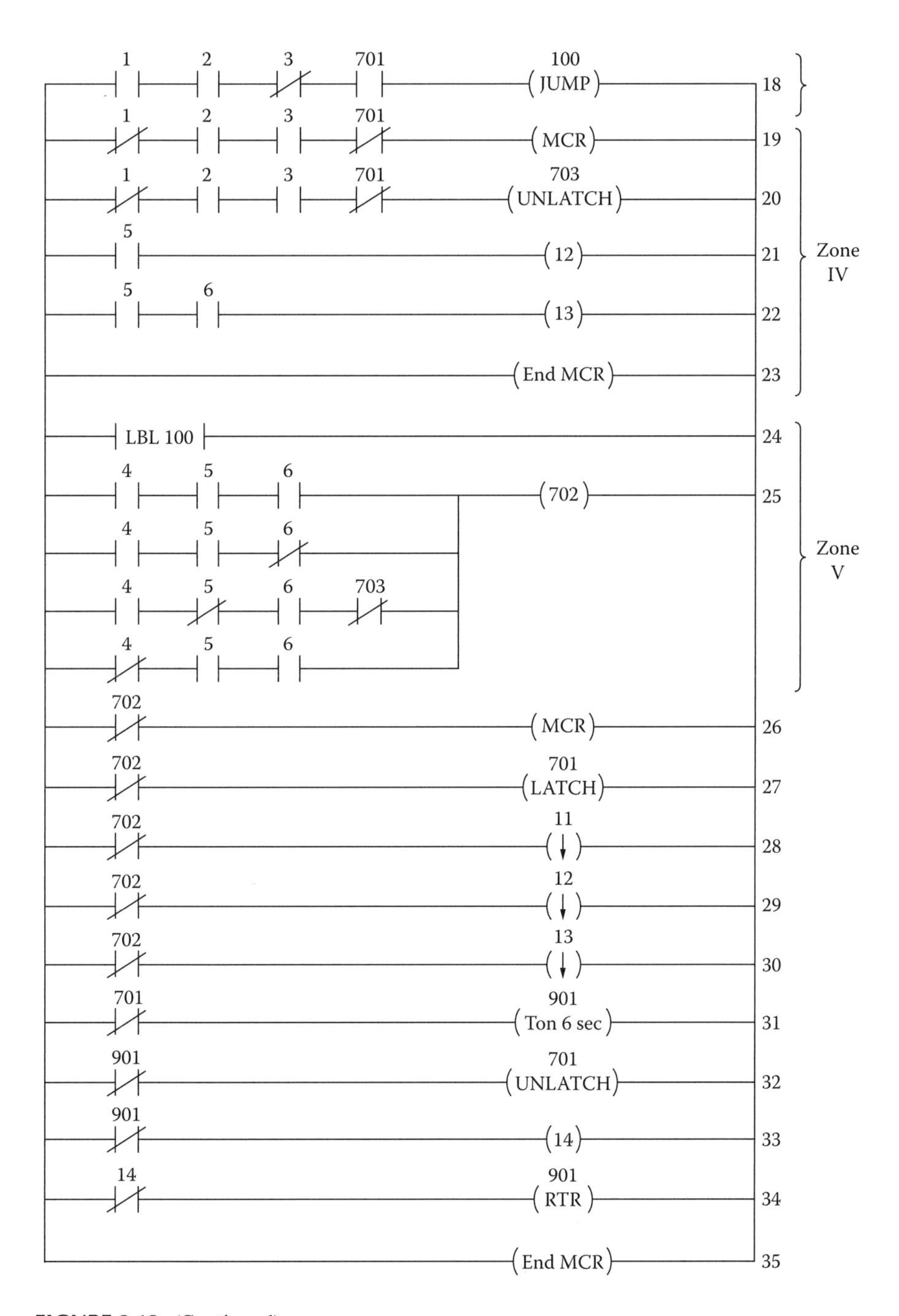


FIGURE 3.18 (Continued)

It is interesting to note that the program readily allows operational flexibility. The number of operators and the stations on which they work can be changed to suit operational convenience, and the PLC will adjust the material-handling mechanism correspondingly. Such flexibility is the major benefit of a PLC system.

## 3.7 NUMERICALLY CONTROLLED MACHINES

Use of computers in manufacturing can be marked by the introduction of numerically controlled (NC) machines to shop floors. These are machines that operate under a given set of instructions, as opposed to the manual manipulations that are used with traditional machines. NC machines are currently used in metal cutting for milling, boring, drilling, turning, grinding, and flame cutting, and in metal joining for riveting and spot and continuous welding. In fact, the total number of NC machines installed for metal working almost doubled from 1978 to 1983 to about 103,000 units. Total sales of NC machines in 1990 and 1991 were 31,104 and 38,480 units, respectively. Because of the growing use of NC machines, we will describe these in some detail.

The earlier development of NC machines is credited to the aircraft industry and the Massachusetts Institute of Technology (MIT). Many identical and complicated parts were needed in manufacturing; and to help avoid manual errors in the duplication of the parts, MIT was asked by the U.S. Air Force in the late 1940s to develop a technology that would cut the required shapes from metal sheets automatically. A machine was developed that would read a series of numerical codes indicating the tool operations and act on these to produce a part. The same operations could be called upon for the same design the next time the same piece needed to be processed. Thus, began the introduction of NC machines in manufacturing.

### 3.7.1 Components of NC Machines

The three basic components in the operation of an NC machine are a set of instructions or software, a machine controller unit (MCU) or controller unit (CU), and machine tools.

#### 3.7.1.1 Instructions

The software for NC machines consists of a set of commands instructing the controller units regarding operation of machine tools. These are very detailed step-by-step directions giving machine tools and their tables the next moves. The instruction code is a binary coded decimal (BCD) system. In a conventional NC machine this can be easily observed by the presence or absence of holes in the correct positions on the punched paper tape that is used to transfer instructions to the controller. In modern systems, instructions are read on floppy disks or entered through control keyboards of the machines.

Like using any other machine language for writing a program for a computer, composing instructions in BCD is a difficult job. Simpler languages have been developed that allow the format of the instructions to be more like commands in English. Automatically Programmed Tools (APT) is one such language developed by MIT.

NC machines. The three basic components in the operation of an NC machine used in manufacturing are a set of instructions (software), an MCU or a CU, and machine tools. Top left: A computer reads step-by-step instructions from software or an operator. Top right: The CU converts the software commands into mechanical actions by tools, such as this NC machine used for milling small metal castings. Bottom: In industrial applications, like the one shown here, NC machines use commands from the MCU to control electronic board assemblies.

This is one of the most widely used languages today, and a variation (ADAPT) developed by IBM for the U.S. Air Force allows application of APT to small computers.

In addition, some manufacturers developed languages for use with the machines they produced. Sundstrand Processing Language Internal Translator (SPLIT), developed for Sundstrand machines, is one example of this type of software. If it was intended to have more than one NC machine in a plant, it was desirable to make them compatible by assuring that all of them could use the same language.

#### 3.7.1.2 Control Unit

It is the hardware, electronic controls, and servomotors that read the software instructions and convert them into mechanical actions by tools and work tables.

Some earlier NC machines also had feedback controls that formed a closed loop system, ensuring the position of the table and the workpieces with respect to the tool. However, experience has proven the reliability of NC machines, and such feedback controls are not used in most modern machines. The components of an individual NC machine may depend on the system of which it is a part (these systems are described next) and may include a tape reader, computers, servomotors, input/output signal controllers, a control panel, and a cathode-ray tube (CRT) monitor.

#### 3.7.1.3 Machine Tool

The machine tool is the third component of an NC machine; it performs the actual work. The tool consists of the work table, fixtures on the table, a turret, spindle(s), and tool(s) on the spindle(s). The tools are designed from materials that enable them to be wear-resistant, as well as having toughness qualities (impact and heat resistance) combined with high strength.

### 3.7.2 Machine System

Four NC machine systems are currently in general use in industries. They range from a single, stand-alone NC machine to a combination of a large mainframe computer and a number of NC machines, each with its own small computer. The mode of instruction transfer and the associated flexibility, which vary from system to system, are described in the following paragraphs.

#### 3.7.2.1 Conventional NC Machine

Instructions for conventional NC machines are read by a tape reader (a part of the controller) from a 1-inch-wide tape. The tape may be made of paper, a low-cost but not very durable medium, or Mylar-reinforced paper, aluminum strip, or even plastic. Manual instructions on the tape are produced by using a typewriter-like machine called the Flexo writer, while a computer-assisted tape (a tape produced by using the computer) is punched by a device called a tape punch.

A tape is fed through the tape reader every time a piece is to be produced by the conventional NC machine. Thus, even for production of identical pieces the same tape of instructions must be read for each piece. The reader deciphers one instruction at a time, saves it in memory, and, while the controller is acting on this instruction, reads the next instruction. Actions of the controller are to control the movements of the tool(s) and the work table according to the instruction. This mode of operation has some drawbacks. Storage and upkeep of the number of tapes needed to produce different parts could become a problem. Also, any change in part design, however small, can mean producing a new tape. The straightforward loading of the machining instructions does not allow adjustments in machining parameters in real time; hence changing the machine tool operation while the work is in progress is difficult at best.

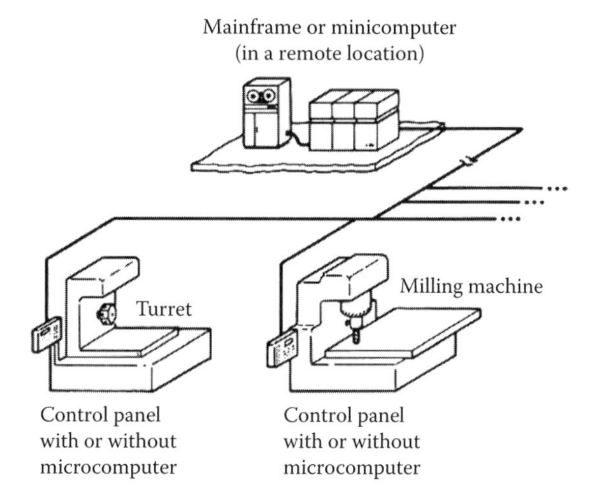


**FIGURE 3.19** Direct numerical control

#### 3.7.2.2 Direct Numerical Control

In a direct numerical control (DNC) arrangement, as illustrated in Figure 3.19, a group of NC machines is connected to a mainframe computer or even a fairly large minicomputer, and each receives its instructions from this computer on a time-sharing basis. The paper tape instructions are no longer required, and the high speed of the computer allows each machine to interrupt and get its instructions on a continuing basis. A number of different programs can be stored on magnetic tape, disk, or main memory, ready to be recalled immediately when needed by any machine.

Part changes can be incorporated in the program with very little effort. However, the system does have its disadvantages. Any failure in the main computer affects all NC machines, perhaps resulting in a serious loss of production. Low-voltage transmission lines from the computer to the machines also occasionally make perfect transmission of instructions difficult, causing errors in manufacturing.

#### 3.7.2.3 Computer Numerical Control

The next step in the development of NC machines is a minicomputer or a microcomputer attached to each NC machine rather than having them share one large mainframe computer (Figure 3.20). The instructions are entered through a floppy disk or a keyboard associated with the computer on the machine. There is greater flexibility here. Program modification can be completed almost immediately through the keyboard; in addition, instructions for the operator and machine data can be displayed on a CRT if needed.

#### 3.7.2.4 Modern DNC

Recent DNC systems take advantage of a computer numerical control (CNC) configuration as it connects a group of CNC units to a central computer. New programs

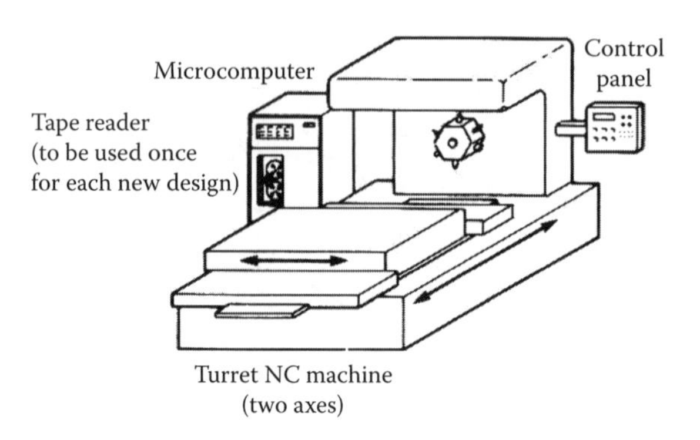


**FIGURE 3.20** Computer numerical control

can be developed with the main computer and then transferred to the individual computer on the particular NC machine.

This also allows storage of large numbers of programs in the main computer, including very long programs or large numbers of short programs that are used infrequently, and then transfers them on an as-needed basis to individual machines. The main computer can also be used for other activities such as controlling inventory, scheduling production and labor resources, and producing work orders for maintenance; all of these lead to overall computerization of the plant, which is called computer-aided manufacturing.

### 3.7.3 Advantages of NC Machines

One of the greatest benefits of the NC machine is the accuracy with which the parts can be produced. Complex geometrical parts can be repeatedly produced to the same very close tolerances. The NC machine can easily accommodate design changes by merely changing the computer program instead of requiring expensive changes in jigs and fixtures. NC machines increase productivity by reducing setup and handling times. These machines have proven to be ideal where greater flexibility is required in terms of production quantity (e.g., one unit or thousands of units), when the processing is expensive, and when the cost of a defective unit is considerable.

### 3.7.4 Classification of NC Machines

An NC machine is categorized on the basis of its control features, and the cost of a machine increases as its performance capability increases. The following list gives the categories in ascending order of complexity.

1. *Point-to-point machine*: Capable of moving from one designated point to another. These points must be defined beforehand. (Figure 3.21.)
2. *Straight-cut machine*: Capable of moving continuously at a controlled speed parallel to one of the major axes ($x$, $y$, or $z$). (Figure 3.22.)

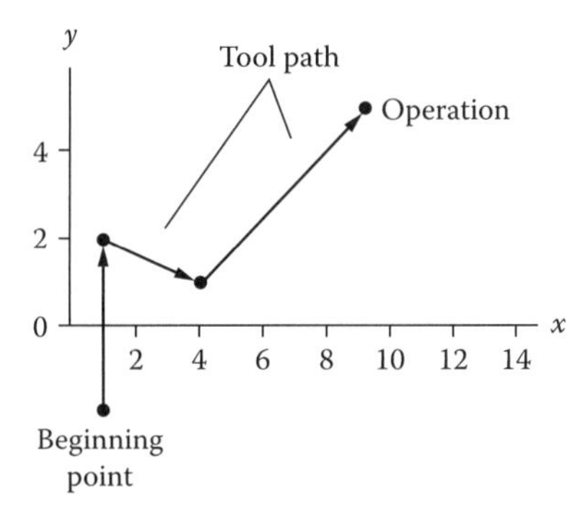


**FIGURE 3.21** Point-to-point operation

3. *Contouring machine*: Capable of following any contours or geometric shapes such as a circle or a curve in two or three dimensions. (Figure 3.23.)
4. *Adaptive control machine*: The next advancement, developed for the U.S. Air Force in 1962, is an adaptive control machine capable of varying its speed and feed based on the conditions of the workpiece or tool wear. For example, if the hardness of a workpiece is reduced at a point, the cutting speed might increase. Even a small air pocket in the cutting path can mean easier cutting, again increasing speed and/or feed. These machines are used when processing time is a large part of the total time of production. (Figure 3.24.)

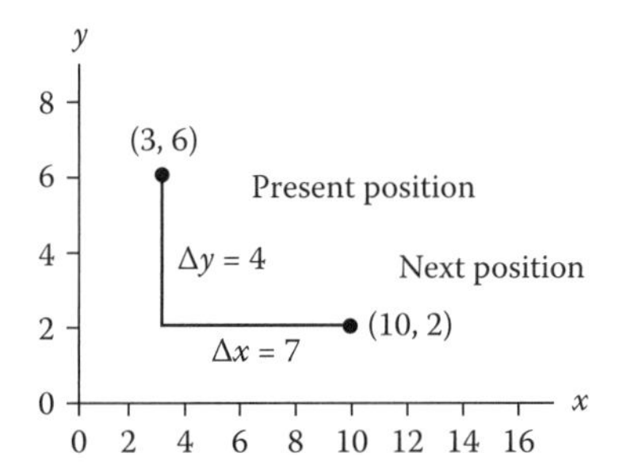


**FIGURE 3.22** Absolute and incremental positioning

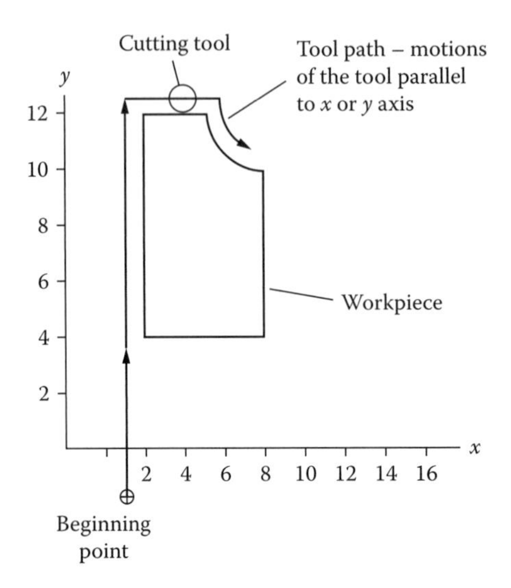


**FIGURE 3.23** Straight-line cutting

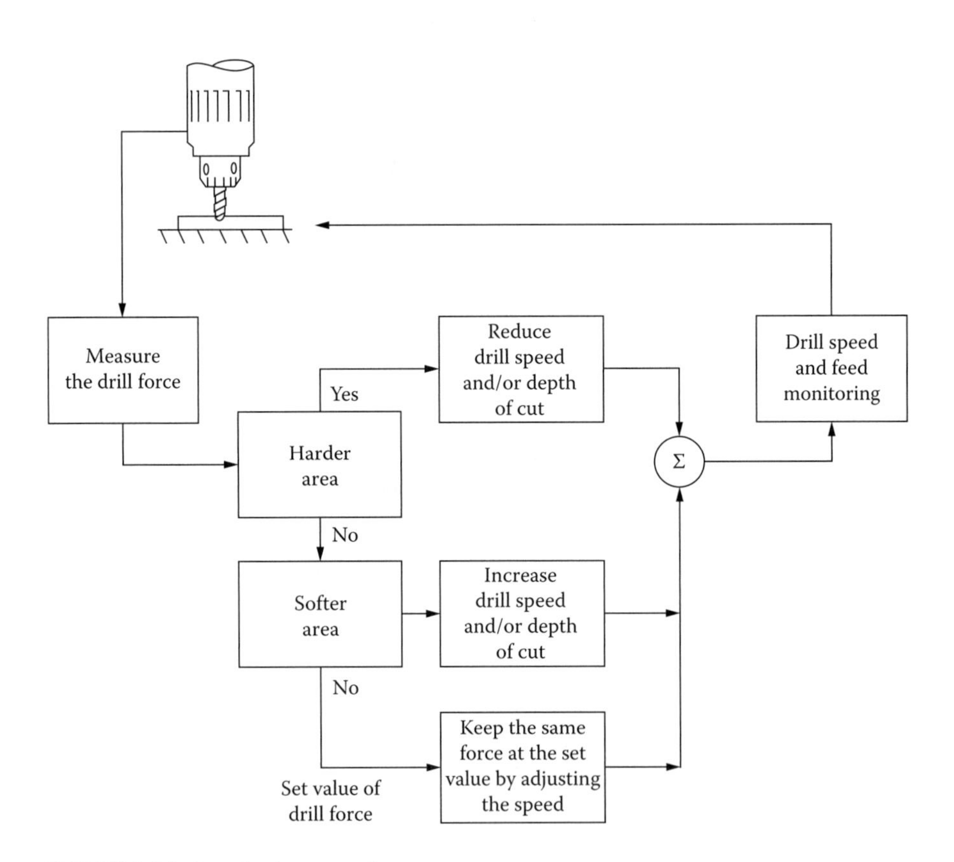


**FIGURE 3.24** An adaptive control system

## 3.8 INDUSTRIAL ROBOTS

The Robot Institute of America defines *robot* as "a programmable, multifunctional, manipulator designed to move material, parts, tools, or specialized devices through variable programmed motions for performance of a variety of tasks." A robot can be programmed and reprogrammed to perform various repetitive tasks. Robots are currently being used in loading and unloading production machines such as stamping, forging, metal cutting, and injection molding. Robots can also perform spray painting, spot and continuous welding, and even adaptive welding, in which the robot scans the joint by laser and then follows it by welding the seam. A robot with vision or an "eye" has been effective in inspection processes.

Operating costs for robots can be economically attractive, being less than \$5.00 per hour for some models. Purchase prices range from \$25,000 to \$200,000 but have been consistently decreasing owing to standardization and increased volume. Robots are reliable, following the same preplanned motions every time, and productive, working about 98% of the time. They can be strong, lifting heavy objects, and will work in hazardous conditions in which high temperatures, dangerous chemicals, or radioactive substances are involved.

### 3.8.1 Degrees of Freedom

To emulate human arm motion requires six degrees of freedom. As shown in Figure 3.25, the arm can move at the shoulder, elbow, and wrist. At the shoulder it can rotate about two axes, giving two degrees of freedom. At the elbow it has one degree of freedom, because it can move only up and down. At the wrist, up-and-down motion, side-to-side motion, and rotational motion give three additional degrees of freedom. Robots are designed to use one or more of these motions on the job. Drives such as a rack and pinion, screws, and hydraulic cylinders are used to achieve transverse motion; angular or rotational motions are obtained by gear drives and timing belts.

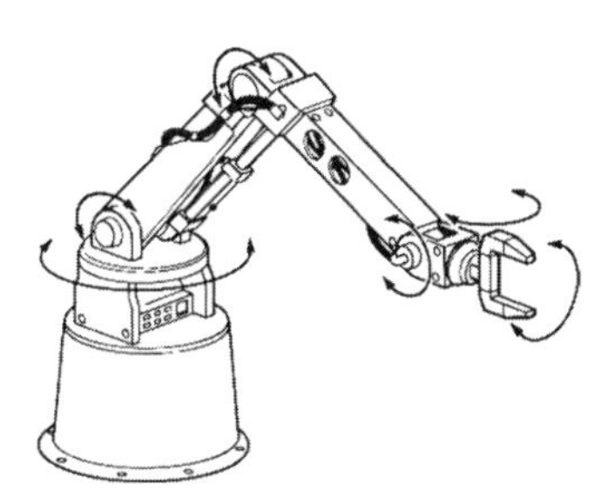

**FIGURE 3.25** A robot with six degrees of freedom

The grasping of an object by a robot is equivalent to human finger motions. Vacuum cups, hooks, electromagnets, clamps, scoops, or even fingerlike mechanisms are used to achieve grasping.

### 3.8.2 Classification of Robots

Like NC machines, robots can be classified with respect to their abilities. A point-to-point robot can go only from one predefined point to another. A continuous-path robot can go to a number of points in a path, but not necessarily continuously following the path. A controlled-path robot can follow a predefined geometric path continuously, and a servo-controlled robot has the ability to sense the points on the path and feed back the information so that corrective actions can be taken by the robot controller.

### 3.8.3 Robot Programming

At present, more than 90% of robots are taught by leading them through a sequence of operations. A robot in the learning mode remembers these operations and repeats the sequence when in the operating mode. The method is popular because an operator on the shop floor can show the robot what to do.

Robots can also be programmed like NC machines by giving them a set of instructions. Robot programming languages are as varied as the number of manufacturers of robots, because each tends to develop a language of its own. These languages are developed as modifications to existing programming languages by adding syntax or subroutines, or by creating a new language. Some of the programming languages are WAVE, RAIL, VAL, PAL, AL, MHI, and HELP.

### 3.8.4 Selecting a Robot

A robot works as an integral part of a production system, and its selection is based on where and how it should function. Environmental conditions such as heat, humidity, and dust determine the surroundings in which the robot must work. The range of motions (<1, 1 to 4, >4 to 10, or 10 feet and greater) and the speed at which it must operate (low speeds of less than 1 foot/minute, medium speeds of from 1 to 5 feet/minute, or high speeds of greater than 5 feet/minute) determine the capabilities of the robot. The type or classification of controls and the needed sensitivity — such as detection of an object by proximity or touch, detection of holes and edges by simple vision, or recognition of objects by complex vision — determine the degree of refinement of the robot. Naturally, as the complexity of the robot increases, the purchase and installation costs increase correspondingly.

Some examples of robot use were mentioned earlier. The important features in selecting the robot for such applications can now be defined. For example, for material handling, a robot must be able to travel from one place to another, perhaps on wheels, rails, or guided paths. For loading and unloading of machines, a robot should be able to manipulate a unit and should also be able to travel, but this travel is mostly in terms of arm motions. Welding and spraying operations require robots that are good at manipulating and following contouring and continuous motions. Robots for machinery and assembly operations should be able to manipulate as well as sense the

Industrial robots. Robots can be programmed to accurately perform repetitive tasks such as welding in automobile manufacturing. A robot with an "eye," such as the one shown here, can detect leaks and take corrective action.

positions of workplaces. The need for sensing becomes more acute when the robots are used in inspection processes.

### 3.8.5 Summary

In manufacturing, automation plays an important part in improving productivity. This chapter discusses two of its components: sensors and numerically controlled machines and robots. Sensors are the eyes and ears of automated systems. They observe what is happening and relay the information to the control unit, which takes appropriate corrective action. Both discrete and continuous sensors have found a number of applications in production facilities.

Bar codes are abundantly used in business and industry. The main application involves keeping up with the information on a part or a product, be it for storage operations, plant operations, or accounting purposes. This chapter presents the major bar code systems, how and where they are used, and the equipment necessary to get a bar code system together. The important characteristics of generating and reading a bar code are also presented as part of making a judicial choice of a bar code system.

RFID systems are expensive but have the advantage of being able to withstand harsh environmental conditions. In places where a bar code operation is not feasible,

RFID can be used to keep up with the product information or even to activate processes based on given information.

Use of machine vision further adds to the automation process, especially in inspection and automatic identification of a part and its orientation. Some basic principles of machine vision are described in this chapter. The examples illustrating template matching should give the reader some understanding of the necessary use of computers for speed and accuracy in most machine vision applications.

A voice activation system follows the same basic principles as that of a machine vision system, except that it processes a voice image rather than a photo image. Although expensive, it is used to record information or create an action based on the information supplied by a human voice. This is especially useful for a person whose eyes or hands are busy and cannot be involved in activating a process.

PLCs are not new. They provide flexible automation. Change in ladder diagram logic is all that is needed to change the sequence, counters, and time delays in operations. The use of PLCs is so dominant in industry today that some knowledge of their programming is almost necessary for any engineer or supervisor who is in charge of a manufacturing plant. The chapter provides the essentials of PLCs, along with an application.

NC machines provide more flexibility and greater accuracy than their conventional counterparts in operations such as milling, turning, drilling, and grinding. These machines follow instructions provided in a binary decimal system or in a language that includes APT, SPLIT, and ADAPT, especially developed for NC machines. Instructions are entered into memory using punch tapes, floppy disk drives, or keyboards, and are then converted by control units into mechanical actions by tools and worktables. When a plant has more than one NC machine, they are controlled by either an individual computer for each machine or a large and small computer combination. Many arrangements are possible and are designated as DNC and CNC. NC machines are classified according to their performance capacities, with point-to-point motion as the lowest classification and adaptive control as the most sophisticated.

Because of their reliability and ability to work in environments that may be hostile to humans, industrial robots are used in operations such as loading/unloading, spray painting, and welding. Freedom of motion is an important trait and is measured by a robot's ability to emulate human motions. Robots are programmed and classified in a manner very similar to that for NC machines.

## PROBLEMS

1. What is automation and what are the components? Give two examples of the types of facilities that can be automated and the degree of automation that can be achieved for each.
2. What are bar codes and bar code symbology? Explain their importance in automation. Give a few commonly used symbologies.
3. Give various constituents of a bar code system in a commonly used application.
4. Explain the following.
   a. Self-checking in bar code scanning. Is UPC sufficiently self-checking?
   b. A continuous and discrete code. Give examples for each.

5. List a few major bar codes in use.
6. What symbology is used or would you recommend to encode the following:
   a. Letters and packages in U.S. Post Office
   b. Automobile part
   c. A code 26SB9
   d. A code 26589
   e. Films
   f. Airline tickets
   g. A can of apple juice
   h. Your textbook

   Is it possible to recommend more than one symbology for these situations?

   Following is a bar code given in code 39. Assume that the scanner for the bar code is properly chosen.
   a. Will the scanner read the data given here? Give reasons for your answer.
   b. Find the possible number of simplest errors that could have occurred in this, and hence the possible information that could be decoded if the error is rectified. (Hint: refer to Table 3.1, code 3 of 9).

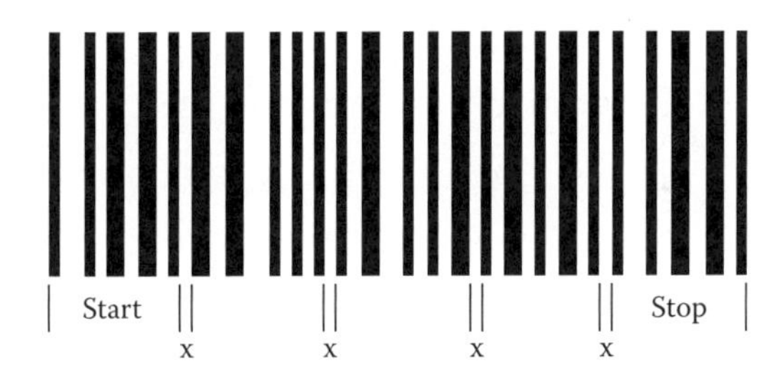


7. The following bar code is given in I 2 of 5. Decipher the code number.

8. Write a complete bar code in I 2 of 5 for 47.
9. Determine the total length required for specifying the following information in bar codes through 2 of 5,1 2/5, code 3 of 9, and UPC-A symbologies. Code number: 4310005222. Assume the value of $X$ (the width of the

narrow dark bar) for all the symbologies to be 0.013 inch. Use 3 as a value of $R$ wherever required. Also calculate the number of characters per inch that could be represented in each symbology for this value of $X$, and the value of the check digit using the modulo 10-digit method.

10. In choosing the bar code readers, does the light-beam diameter depend on the density of the bar code? List a few other characteristics required for the light beam.
11. Are there any common features in representation of numbers by binary numbers in standard 2 of 5, I 2/5, and code 3 of 9, as far as the binary pattern is concerned?
12. Write an algorithm for a bar code decodification procedure by a computer for each of the symbologies.
13. Explain why I 2/5 cannot read alphabets, but code 3 of 9 can. If the interleaved coding is to be used for all alphabets, what is the minimum number of bars required for representing one character?
14. Of the techniques mentioned for obtaining 3-D information from image information, which one would provide the highest accuracy? Why?
15. Using the template given here, produce a correlation matrix to locate the template in the noisy image

$$\text{Template: } \begin{matrix} 1 & 0 & 1 \\ 1 & 0 & 1 \\ 1 & 1 & 0 \end{matrix}$$

$$\text{Image: } \begin{matrix} 0 & 0 & 0 & 0 & 0 \\ 1 & 0 & 1 & 1 & 1 \\ 1 & 0 & 0 & 1 & 0 \\ 1 & 1 & 1 & 0 & 0 \\ 0 & 1 & 0 & 1 & 1 \end{matrix}$$

16. Design templates (four) for locating the corners of a rectangular piece of sheet metal. Use templates of size 6 × 6. You can assume that the sheet metal will always be aligned with the pixels.
17. One commonly used technique in handling noise in an image is to replace every pixel in the image with the median value of the surrounding pixels. Perform this operation on the image given here. The image contains intensity values between 0 and 255 at each pixel location.

$$\text{Image: } \begin{matrix} 0 & 3 & 11 & 1 & 13 & 0 \\ 7 & 128 & 130 & 200 & 115 & 0 \\ 0 & 157 & 175 & 210 & 148 & 12 \\ 6 & 35 & 215 & 240 & 222 & 10 \\ 5 & 144 & 189 & 214 & 231 & 34 \\ 0 & 7 & 16 & 34 & 21 & 18 \end{matrix}$$

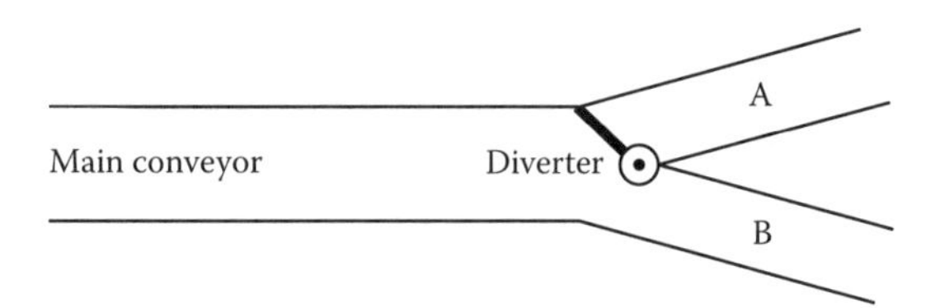


**FIGURE 3.26** The two-position diverter

18. Two types of items are supplied to the packaging area by the main conveyor, traveling at 10 feet/minute. Item 1 is 1 foot long, and item 2 is 1 foot 6 inches long. There is a minimum space of 1 foot between any two successive items. The items are diverted to conveyors A or B by a diverter. The diverter has a two-position plunger. When coil 1 is energized, the item is diverted to conveyor A; when coil 2 is energized, the item is diverted to conveyor B. A sketch of the layout is shown in Figure 3.26.
    a. Determine the positions of electronic eye sensors on the main conveyor that would be used to distinguish the two parts.
    b. Write a PLC program that will operate the diverter based on the signals from the sensors.
19. A robot is used to feed an NC machine. The parts come from two input stations, A and C. The first type of part, from input station A, is processed in the machine for 5 seconds and then released to the output section B. It takes 2 seconds to load a part from A on to the machine and unload the part from the machine to station B.

    The second part from input station C requires 3 seconds of machining time; then the robot unloads the part to station D. The loading and unloading operations require 2 seconds each.

    The parts arrive at random; however part type 1 has priority over part type 2 in processing.

    The work area is shown in Figure 3.27.

    Develop a PLC ladder diagram for the robot operations.
20. PLCs can be used to control an agriculture irrigation system. Here, the water arm swivels around a central pivot point with the help of a diesel engine. The ground sensors signal the PLC to start the engine if the

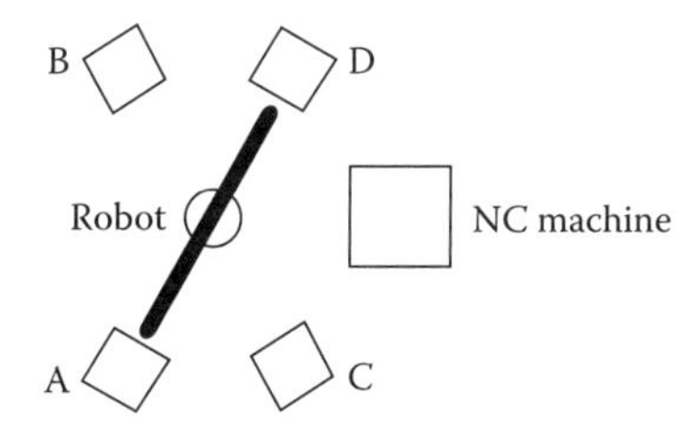


**FIGURE 3.27** The work area

ground is dry (below a certain moisture level). The engine is allowed to warm up for 5 minutes to prevent moisture buildup in the crank case, after which the PLC activates a controlling relay to start the rotation of the arm. It takes 16 h for the arm to make two revolutions, at which point the engine is shut off. The arm rotation can also be stopped either by the emergency switch or by a rain sensor. If it starts raining, there is no need to water, and the rain sensor switches the engine off. The layout is shown in Figure 3.28.

Write a ladder diagram for these operations, defining the coils as need be.

21. What is an NC machine? When and where is it preferable to a traditional machine?
22. Describe the basic components in the operation of an NC machine. How do they vary within different NC machine systems?
23. Distinguish between CNC and DNC systems.
24. Define *robot*. Where can one be used?
25. It is said that a robot provides flexible automation as opposed to hard automation. Explain this statement.
26. What is a degree of freedom for a robot? How many degrees of freedom are needed to emulate a human arm? Describe each.
27. How are robots classified? Identify at least two applications for each type of robot.
28. What are the attributes of a robot that should be considered before acquiring one?
29. A robot costing $50,000 is considered for use in a welding operation that is currently being performed by two welders. Each welder is paid $15 per hour and can produce 15 parts per hour. The robot can weld at a rate of 30 parts per hour and has an operating cost of $5 per hour (including the average cost for setups). If the required production rate is 200 units/day, how long will it take before the robot can "pay for itself"? Assume an interest rate of 12%.

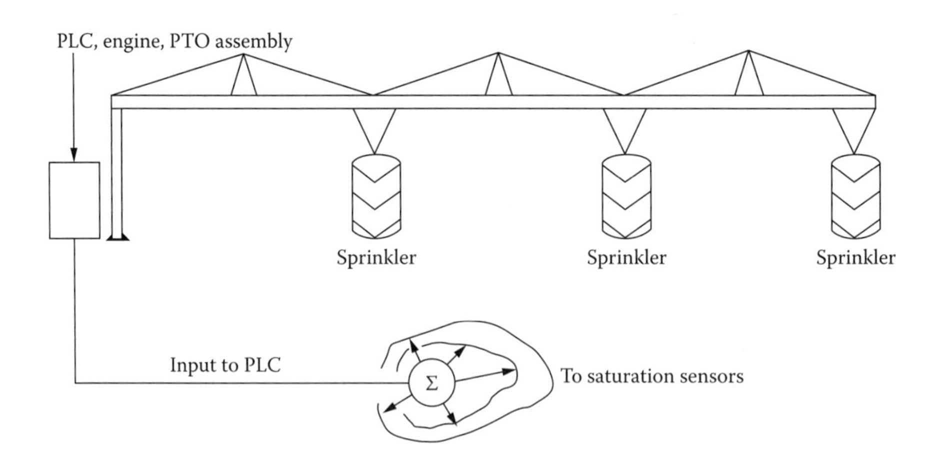


**FIGURE 3.28** PLC engine, power take-off assembly

## SUGGESTED READINGS

### Automation

Asfahl, C.R., *Robots and Manufacturing Automation*, $2^{nd}$ edition, John Wiley, New York, 1992.

Groover, M.P., *Automation, Production Systems, and Computer-Aided Manufacturing*, Prentice Hall, Englewood Cliffs, NJ, 1980.

Knebusch, M., "New software simplifies CNC," *Machine Design*, August 1984, pp. 63–66.

Krouse, J.K., *What Every Engineer Should Know About Computer-Aided Design and Computer-Aided Manufacturing, The CAD/CAM Revolution*, Marcel Dekker, New York, 1982.

"A look at NC today," *Manufacturing Engineering*, October 1975, pp. 24–28.

Winship, J., "CNC lasers cut out saw blades," *American Machinist*, May 1984, pp. 92–94.

### Bar Codes

Fales, J.F., "Bar coding makes document tracking easy," *Modern Bar Coding*, September 1990.

Harmon, C., and Adams, R., *Reading Between the Lines*, Helmers Publishing, Peterborough, NH, 1989.

### Machine Vision

Ballard, D.H., and Brown, CM., *Computer Vision*, Prentice Hall, Englewood Cliffs, NJ, 1982.

Davies, E.R., *Machine Vision: Theories, Algorithms, Practicalities*, $3^{rd}$ edition, Morgan Kaufmann, San Francisco, 2005.

Horn, B.K.P., *Robot Vision*, MIT Press, Cambridge, MA, 1986.

### Programmable Logic Controllers

Bryan, L.A., and Bryan, E.A., *Programmable Controllers, Selected Applications*, Vol. I, An Industrial Text Co. Publication, Atlanta, 1987.

Jones, C.T., and Bryan, L.A., *Programmable Controllers*, An IPC/ASTEC Publication, Atlanta, 1983.

# 4 Production Charts and Systems

Manufacturing a product requires transformation of design ideas into reality, which ultimately results in a useful item. This is accomplished through the efficient completion of operations by a group of well-trained workers in a properly developed plant. Managers and engineers must carefully analyze different ideas regarding the production of an item and then select the alternative that is most cost-effective and feasible given the limitations under which the plant must operate. Consideration of several factors is warranted in many cases — for example, the production of one or several items, volume of production of each item, preferred skills of workers, and flexibility to be incorporated in the plant design for product and/or product mix changes. All these factors contribute heavily in overall plant development.

The discussion that follows is intended to illustrate the available means for converting production ideas into tangible results. It is divided into three phases. First, production charts are illustrated. These charts are the backbone of planning and improving activities. Next, the discussion on production systems leads to our appreciation of various types of production arrangements and their associated advantages and disadvantages. The topics include a broad spectrum, starting with the traditional job shop arrangement and moving toward more modern flexible and cellular manufacturing. The discussion also includes techniques for forming an efficient machine and job cells using group technology. Lastly, we illustrate some labor planning models that are used in improving efficiency in the workplace.

## 4.1 PRODUCTION CHARTS

To illustrate the activities associated with production, engineers often use charts to represent the process graphically. Such an approach increases understanding of the course of action that is needed in manufacturing and helps to resolve many problems related to the design of the production layout. Assembly charts, operation process charts, and process charts are important representations that significantly contribute to this goal. All three charts are drawn by using symbols standardized by the American Society of Mechanical Engineers (ASME) in 1947.

### 4.1.1 Symbols and Descriptions

The symbols representing the five basic activities in manufacturing are shown next, along with definitions that are commonly used throughout industry.

- o: Operation. The item is intentionally changed in one of its characteristics. Examples are filling a bottle with a soft drink, bending a metal sheet, and writing a letter.
- →: Transportation. The item is moved from one place to another (except when the movement occurs as an integral part of an operation or inspection). Examples are moving an item on a conveyor belt between operations and moving an item to storage.
- □ : Inspection. The unit at an inspection point is compared with the quality standard established for that point.
- D: Delay. The next planned action does not take place. Examples are delay in moving a unit and delay in performance of the next operation.
- ∇: Storage. An item is stored in a place such that its withdrawal requires authorization.

### 4.1.2 Assembly and Operation Process Charts

In the planning stages of developing a production system, the layouts of machines and the plant are not yet known. One can only visualize, from parts and product designs, the operations that are necessary and the sequence in which they must be performed. Assembly charts and operation process charts are the graphical representations of such methods. An assembly chart gives a broad overview of how several parts manufactured separately are to be assembled to make the final product, such as an internal combustion engine. It may also show a flow in reverse — how a product that is in unit form is disassembled and distributed in different processes, such as raw milk in a dairy. Figure 4.1 shows the assembly chart for the teakettle mentioned in Chapter 2.

Another important use for an assembly chart is in scheduling production, especially in a job shop. As shown in Figure 4.2, such charts must be drawn to scale by using some convenient unit of time along the horizontal. The graph is a set of bars, each representing the starting and completion times for the manufacture or installation of components and/or subassemblies. Knowing the promised delivery date, a scheduler can work backward and determine the time for processing component work orders.

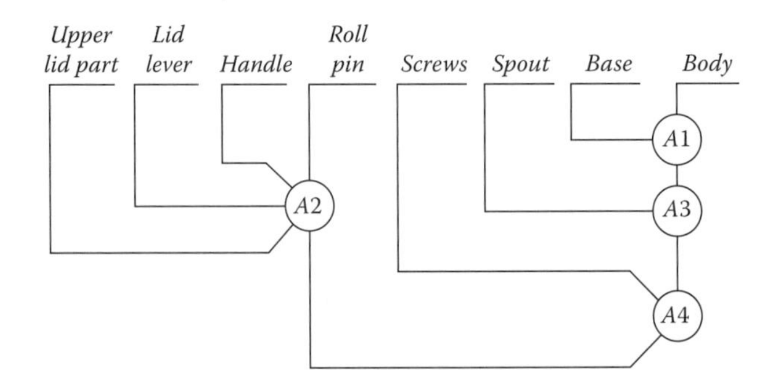


**FIGURE 4.1** Assembly chart for teakettle

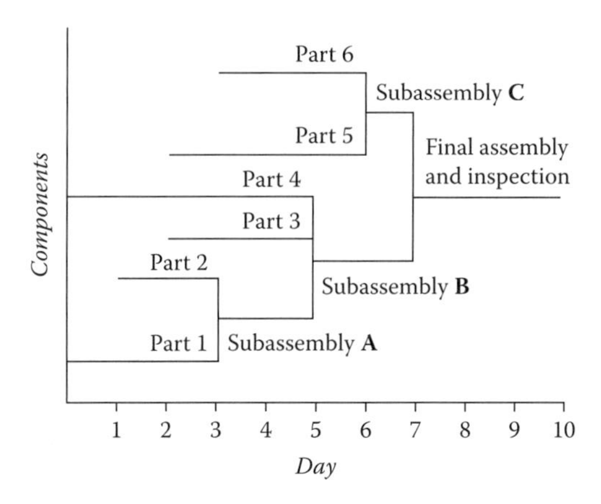


**FIGURE 4.2** Assembly chart for scheduling production

Returning to the teakettle example, Figure 4.3 presents the operating process chart for that item. Preparing such a chart is essentially the task of detailing the assembly chart (Figure 4.1), showing every operation and inspection that each part needs as it progresses from raw material to the final assembly.

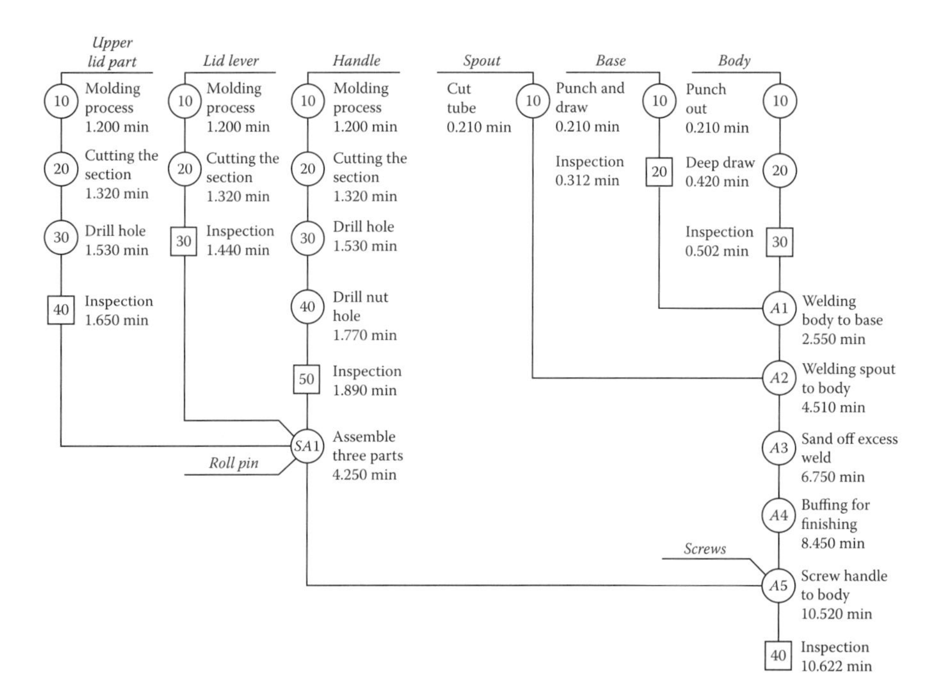


**FIGURE 4.3** Operation process chart for teakettle

### 4.1.3 Time Estimates

For each activity (operation or inspection) in the operation process chart, an estimate should be made of the time it will take to complete the activity. Developing such estimates is a complex task. The time to perform an operation depends on the machine and the jigs and fixtures used. Automatic machines, for example, would probably not take as long to finish a given job as would manually operated machines. Furthermore, the speed and feed of the machine also affect the time to perform the activity, and the degree of automation in the material-handling equipment should be reflected in the setup time.

In the case of an established manufacturer introducing a new product, time estimates can be based on experience and records for the processes, machines, and procedures. For a new enterprise, however, such company data are seldom available. Standard predetermined timetables, such as work factors, brief work factors, methods time measurement, and element time data, may be used to the extent possible in developing time estimates. However, these tables give time values mainly for manual operations. One such effort, the use of brief work factors, is illustrated for a welding job in Figure 4.4.

The machine operation times could be established by using standard machine formulas obtained from any handbook on manufacturing. For example, cutting time on a milling machine is given by the following equation:

$$\text{Cutting time/piece in minutes} = \frac{\text{table or tool travel per cut in inches/cut}}{\text{table or tool feed in inches/minute} \times \text{number of cuts required per piece}}$$

**Part Name:** Body and base assembly
**Operation Name:** Welding
**Tools:** Welder
**Fixture:** Body and base
**Analyst:** Chong

**Part Numbers:** 1 and 2
**Personal Allowance:** 20%

| Element | Description | | Time (0.001 min) |
|---|---|---|---|
| 1 | P 10 | Pick up the body. | 10 |
| 2 | MR 10 | Move to the right. | 10 |
| 3 | MR 10 | Move hand to the welder. | 10 |
| 4 | G 15 | Grasp the welder. | 15 |
| 5 | P 20 | Pick up the welder. | 20 |
| 6 | ML 20 | Move the welder to the left. | 20 |
| 7 | 15A40 | Weld the body to the base. | 600 |
| 8 | MR 20 | Move the welder to the right. | 20 |
| 9 | R 10 | Release the welder in position. | 10 |
| 10 | LR 30 | Lift the welded parts with care. | 30 |
| 11 | MR 30 | Move the welded parts aside. | 30 |

Total: 775
1 unit = 0.001 min
Total calculated time: 0.775 min (120%)
Total operation time: 0.930 min (100%)
Standard time: 0.016 hr
Units per hour: 64.516

**FIGURE 4.4** Brief work factor table

The total distance traveled by the table or tool is equal to the length of the milling surface plus overtravel on each side. The table feed might change somewhat from machine to machine; therefore, the time estimate might have to be modified based on the machine used.

Using the methods just described, one could develop the time values for most of the manual and machine operations. For some other elements, such as inspection, the times might still have to be estimated. The accuracy of these time values will determine the initial operational setup. Naturally, the industrial engineer will be asked to update these time values and perhaps redesign the operations once production has begun.

### 4.1.4 Routing Sheet or Production Work Order

The next step in planning is to draw a routing sheet (sometimes called route sheet or production routing). It shows how a part is to be produced, which machines are needed, the tools to use, estimated setup times for the machines, and production in terms of the number of units expected per hour from each machine. Figure 4.5 shows a routing sheet. One routing sheet is required for each part in the assembly.

The information collected from all the routing sheets (remember that there might be more than one product to manufacture) is extremely important and used in many phases of future planning. As we will observe in the following chapters, this information is needed for determining the number and types of machines to be purchased to produce certain output rates, the number and skill of the employees needed, the production system to use, and indeed how the entire plant should be laid out. Routing sheets and the bill of materials form the major database for future planning.

**Product:** Teakettle **Part:** U. lid part **Part No.:** 6
**Prepared by:** Chong **Date:** Jan 28 **Sheet:** 1 of 1

| Operation | | Machine | Aux. Equip. | Setup Time (hr) | Hr/Pc | Pc/Hr |
|---|---|---|---|---|---|---|
| **No.** | **Description** | | | | | |
| 10 | Molding process | Injection molding machine | | 0.05 | 0.02 | 50 sections of 6 parts |
| 20 | Cutting out of sections (combined with lower part) | Hand cutter | | | 0.002 | 500 |
| 30 | Drill roll pin hole | Drill press | | 0.02 | 0.0035 | 285 |
| 40 | Inspection | | | | 0.0017 | 580 |

**FIGURE 4.5** Production routing sheet

### 4.1.5 Other Charts

A description of some of the other charts used in manufacturing and control follows.

#### 4.1.5.1 Left-Hand, Right-Hand Chart

The left-hand, right-hand chart lists the work performed simultaneously by both the left hand and the right hand of an operator at a specific workstation. There are two columns, one for each hand, which list the sequential elements of the task. There is an additional column for each hand that, by means of two symbols, indicates whether the hand is involved in transporting a part or is performing such actions as grasping, positioning, using, or releasing the item. A small circle designates transporting a part or moving the hand, while a large circle indicates an operation.

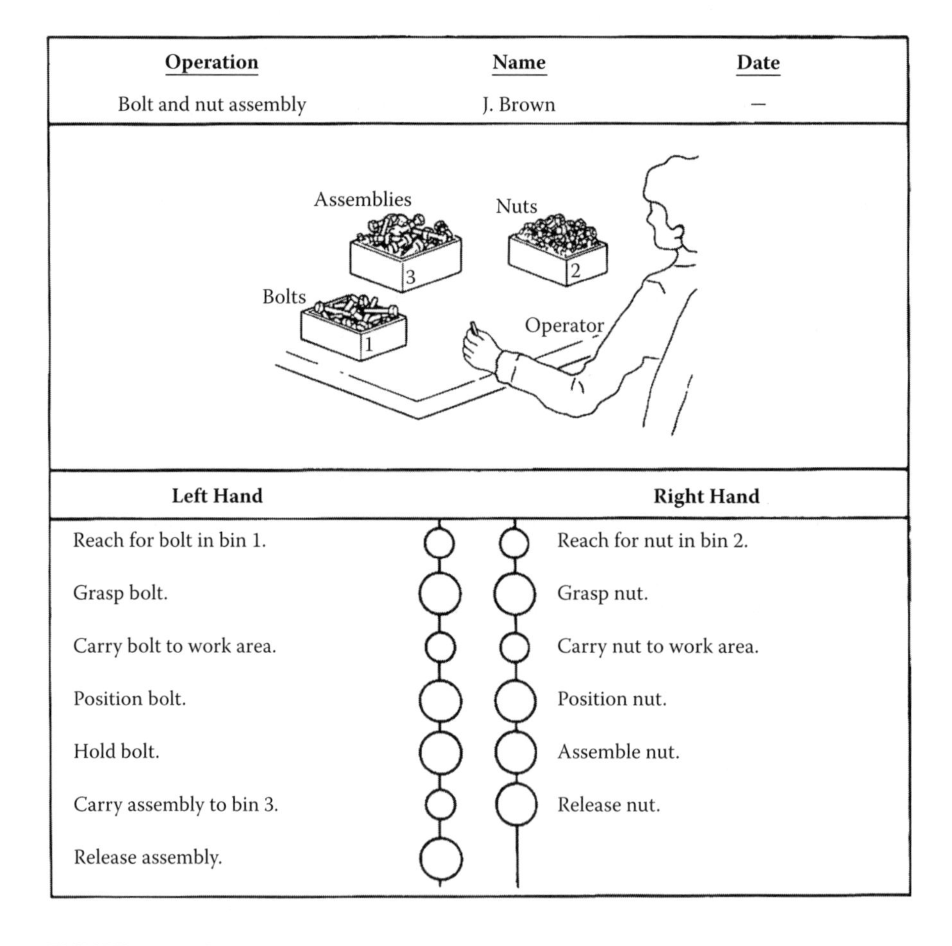


**FIGURE 4.6** Left-hand, right-hand chart

Figure 4.6 is a small portion of a left-hand, right-hand chart for the assembly of a bolt and a nut.

The objective is to design the workstation and the work sequence so that both hands are fully utilized working rather than doing less productive tasks such as holding, excessive delays, or transporting.

#### 4.1.5.2 Gang Chart

The gang chart, also called the multiple-activity chart or the worker-machine chart, depicts the simultaneous activities of all members in a gang or team, the activities performed by a combination of one or more persons and one or more objects, such as people and machines; or one person and one machine. The purpose is to visualize all details of the work being performed by a team to eliminate or minimize any nonproductive element for an individual and to obtain a good balance of work among all team members. Figure 4.7 is a sample gang chart depicting the steps performed by a three-person team in completing a simple task.

**Operation:** Loading finished stock
**Subject:** Warehouse operation
**Operation No:** 632
**Part No:** 54
**Date:** 10-14-08
**Department** shipping **Location** K **Present** **Proposed** X
**Plant** 32 **Chartered by** R. Lewis **Sheet** 1 **of** 1

**No. of Groups** 3

| *Person 1* | *Person 2* | *Person 3* | Steps | Description |
|---|---|---|---|---|
| 1 | 2 | 2 | 1 | Turn the unit. |
| 3 | 3 | 3 | 2 | Wait for work. |
| 4 | 4 | 4 | 3 | Load unit on truck. |
| 5 | 5 | 5 | 4 | Move 50 ft to loading dock. |
| 6 | 6 | 6 | 5 | Unload unit. |
| | | | 6 | Return 50 ft to stockroom. |

**Remarks**

**Summary**

| | Present | Proposed | Reduction |
|---|---|---|---|
| **Total Units** | | 1 | |
| **Steps per Unit** | | 6 | |

**FIGURE 4.7** Gang chart

### 4.1.5.3 Gantt Chart

In a Gantt chart, horizontal bars indicate the proposed (and perhaps actual) times spent in performing each activity or task. These times could be reduced if resource allocations, such as manpower, were increased.

The objective is to assign the limited resources to different activities so as to minimize the total time required for completion of all activities. Figure 4.8a presents a basic Gantt chart for four activities involved in a particular contract. This could be for a major subdivision, or it could be the entire project. The dashed line is used to represent a sliding bar, which is moved horizontally to any date to help compare the progress of the project with the schedule of that date. The actual progress is marked by an individual slide for each task, shown in the figure as a filled dot on the line.

For example, as of the date shown in Figure 4.8a, tasks B and D are shown to be behind schedule.

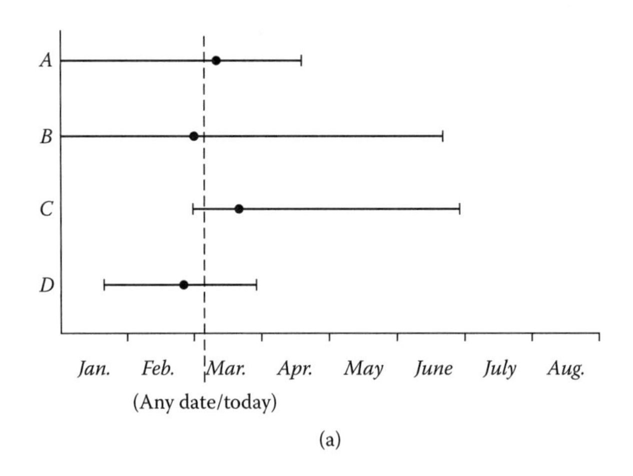


(a)

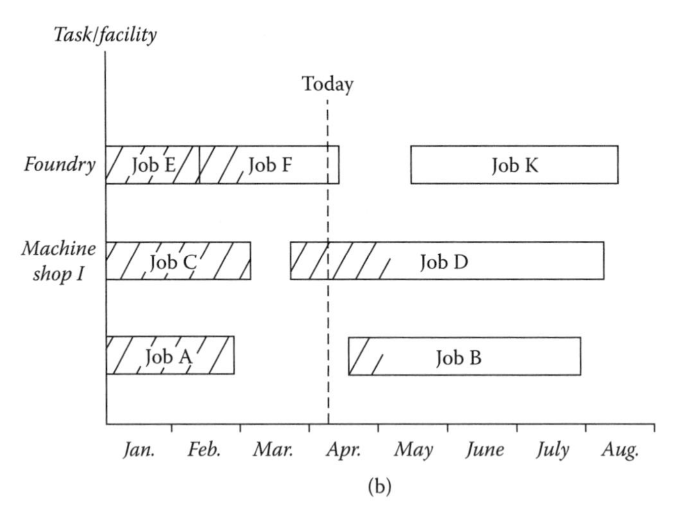


(b)

**FIGURE 4.8** Gantt charts

Figure 4.8b represents another version of a Gantt chart in which the facilities are shown with the operations in the sequence in which they are scheduled. The bar length represents the planned time, while the shaded area represents the actual progress on each activity.

The chart may be extended to that of the Critical Path Method and Program Evaluation Review Technique networks, details for which may be found in the work of Muther and Hale (1979).

#### 4.1.5.4 Flow Process Chart

The flow process chart is another device that is very helpful in analyzing an existing flow layout. It graphically displays every step a unit follows in the plant, starting with raw material and continuing until the product is completed. All the operations, transportations, delays, inspections, and storages are noted symbolically on the chart. The objective is to determine the method of producing the unit that uses the least number of such symbols. For example, it might be possible to combine or eliminate some operations, reduce transportation by relocating machines or finding alternative routes, and eliminate delays by more efficient scheduling. The analyst should ask the *what*, *why*, *when*, *who*, *where*, and *how* questions for each step to see whether any beneficial changes can be made. Each inquiry should lead to improved plant operations and/or layout. A flow process chart is prepared for keeping track of either a product flow or a person but not both simultaneously. Figure 4.9 illustrates a flow process chart.

## 4.2 PRODUCTION SYSTEMS

The next logical step in the development of a manufacturing facility is to determine the type of production system that would be best used. We will therefore discuss here typical production systems and labor planning methods. Production systems are classified according to the arrangement of machines and departments within manufacturing plants. The spectrum of production systems ranges from job shops with mainly manual operators to completely automated assembly lines. The number of different products a company makes, the types of orders (made to stock or made to order), the volume of its sales, and the frequency of reorders strongly influence what the most efficient production system would be for a given firm. How certain the demand is and how long the production will run must also play an important part in this decision. As a general rule, high volume favors automation and low volume favors manual operations. Numerically controlled (NC) machines, however, can produce both high and low volumes with almost equal efficiency. They do need a high initial investment that could be difficult to justify if the machines are not fully utilized. It is not unusual to have different modes of production in the same plant for different products because of the products' individual sales volumes. The four major categories of production systems are job shop production, batch production, mass and continuous production, and cellular and flexible manufacturing.

### 4.2.1 Job Shop Production

A job shop is a suitable mode for a company that produces many different products with a relatively small volume for each. The various items being produced to customer

| Summary | Present | | Proposed | | Difference | | Job | | Question Each Detail | Analysis | |
|---|---|---|---|---|---|---|---|---|---|---|---|
| | No. | Time | No. | Time | No. | Time | Manufacture of a Tissue Box | | | What? | When? |
| | | | | | | | | | | Why? | Who? |
| ○ Operations | 5 | | | | | | | | | Where? | How? |
| ⇨ Transportations | 9 | | | | | | □ Operator | | Date | | |
| □ Inspections | 1 | | | | | | ⊠ Material | | Number | | |
| D Delays | 2 | | | | | | Chart Begins | Receiving (raw materials) | Page 1 of 1 | | |
| ▽ Storages | 3 | | | | | | Chart Ends | Shipping (finished product) | | | |
| Distance Traveled | 1485 FT | | FT | | FT | | Charted by | T.P.C. | | | |

| Details of (Present/Proposed) Method | Operation / Transport / Inspection / Delay / Storage | Distance in feet | Quantity | Time (min.) | Possibilities: Eliminate | Changes: Combine | Changes: Sequence | Changes: Place | Changes: Person | Improve | Safer? | $ Saved? | Notes: |
|---|---|---|---|---|---|---|---|---|---|---|---|---|---|
| 1. Receive raw materials | ○ ⇨ □ D ▽ | 50 | 5 | 7.1 | | | | | | | | | 5 cartons of 1000 sheets of cardboard |
| 2. Inspect | ○ ⇨ □ D ▽ | | | 1.3 | | | | | | | | | |
| 3. Move by forklift | ○ ⇨ □ D ▽ | 40 | | 3.1 | | | | | | | | | |
| 4. Store | ○ ⇨ □ D ▽ | | | 12.3 | | | | | | | | | |
| 5. Move by forklift | ○ ⇨ □ D ▽ | 45 | | 3.7 | | | | | | | | | |
| 6. Set up and print | ○ ⇨ □ D ▽ | | | 120.1 | | | | | | x | | x | |
| 7. Moved by printer | ○ ⇨ □ D ▽ | 120 | | 5.2 | | | | | | | | | |
| 8. Stack at end of printer | ○ ⇨ □ D ▽ | | | 15.3 | | | | | | | | | |
| 9. Move to stripping | ○ ⇨ □ D ▽ | 165 | | 3.9 | | | | | | | | | |
| 10. Delay | ○ ⇨ □ D ▽ | | | 2.2 | x | | | | | | | | |
| 11. Being stripped | ○ ⇨ □ D ▽ | | | 18.9 | | | | | | | | | |
| 12. Move to temp. storage | ○ ⇨ □ D ▽ | 150 | | 10.8 | | | | | | | | | |
| 13. Storage | ○ ⇨ □ D ▽ | | | 20.1 | | | | x | | x | x | x | |
| 14. Move to folders | ○ ⇨ □ D ▽ | 200 | | 15.4 | | | | | | | | | |
| 15. Delay | ○ ⇨ □ D ▽ | | | 6.3 | x | | | | | | | | |
| 16. Set up, fold, glue | ○ ⇨ □ D ▽ | | | 210.5 | | | | | | | | | |
| 17. Mechanically moved | ○ ⇨ □ D ▽ | 90 | | 3.8 | | | | | | | | | |
| 18. Stack, count crate | ○ ⇨ □ D ▽ | | | 45.0 | | | | | | | | | |
| 19. Move by forklift | ○ ⇨ □ D ▽ | 525 | | 22.8 | | | | | | | | | |
| 20. Storage | ○ ⇨ □ D ▽ | | | | | | | | | | | | |
| | ○ ⇨ □ D ▽ | | | | | | | | | | | | |

**FIGURE 4.9** Flow process chart (courtesy of the Material Handling Institute). This chart tabulates steps and moves in a process, in this case for the manufacture of a small box with a printed label. Standard ASME process elements and symbols are used. The first column from the left is the step number, followed by a brief description of the activity. In the next column, the symbol that best describes the activity is used in tracking the step-by-step flow from top to bottom. Improvement opportunities can be noted in the columns headed "Possibilities," and additional information can be entered under "Notes." A single chart can be used to track the flow of either an object or a person, but not both simultaneously.

specifications require the collection of a variety of general-purpose equipment, highly skilled labor to operate this equipment, and general-purpose tooling and fixtures. Similar machines can be grouped together within the plant to form a department (Figure 4.10). The product moves from one department to another based on its operational sequence.

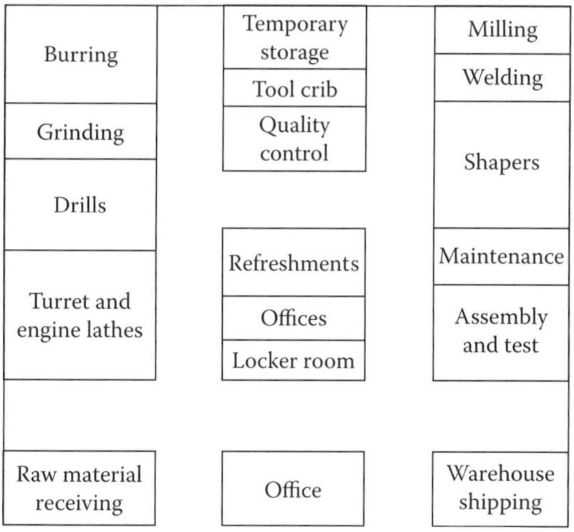

Burring
Grinding
Drills
Turret and engine lathes
Temporary storage
Tool crib
Quality control
Refreshments
Offices
Locker room
Milling
Welding
Shapers
Maintenance
Assembly and test
Raw material receiving
Office
Warehouse shipping

**FIGURE 4.10** Job shop or batch production floor plan

Many problems are encountered in this type of operation. The expense associated with the variety of tools and fixtures can be large, and achieving machine-load balancing through proper job sequencing is very difficult. There is excessive material handling as parts are moved from one department to another; some might pass through the same department several times. Items might have to be held in temporary storage awaiting further processing, resulting in the need for a large area devoted to inventory. Each item or batch of parts requires its own production shop orders, which can result in excessive bookkeeping.

The productivity of a job shop — that is, the percentage of time spent in actual production — is low, mainly because of excessive setup time and material movement. The plant has a minimum of automation; however, it does attain a high degree of flexibility to produce a variety of products. The system is very responsive to changes in the market. It is estimated that about 30% to 50% of the manufacturing systems in the United States are of the job shop type.

### 4.2.2 Batch Production

Batch production is a production system that falls between a job shop and an assembly line in its complexity. The system is suitable for a firm that has numerous items to produce but not so large a variety as to require a job shop type of production. Unlike the situation in the job shop, the products to be made throughout the year are known, each having a stable and continuous demand. The production capacity of the plant is larger than the demand for an item, and hence the products are manufactured in batches. An item is produced once and stored at a preplanned inventory level to meet both present and future demands, and the facility is then switched to produce the next item in the production sequence. A product is rescheduled for production when its stock level has diminished to a predetermined level.

The production machinery and equipment used in the batch shop are somewhat specialized and of high speed, but the skills of the labor force might not be as highly developed as in a job shop. There is some automation within the plant, and the machines and equipment are grouped together to perform certain tasks, which may mean that different types of machines are placed together. The productivity in a batch shop is higher than that in a job shop.

Typical users of batch shop production systems are manufacturers of furniture, appliances, and mobile homes.

### 4.2.3 Mass Production

The mass production system is used for high-volume production. Often the entire plant and its equipment are dedicated to manufacturing a single product. The equipment is very specialized and fast, and the investment in special tools, jigs, and fixtures is large. The work content is broken down into many small groups; a high degree of efficiency is obtained by perfecting the tooling and method of work in each group. The overall labor skills required are minimized by such an approach, but the assigned tasks for the individual can become repetitious and, in some cases, uninteresting. The productivity in mass production is very high and achieved to a large extent by automation.

In mass production, controls of the line output and labor scheduling can readily be achieved. One can control the output rate by the design of the line (stations) and/or the manpower assigned to these stations. (These concepts are explained further by an example later in the chapter.) Periodic checks of the inventory level and product demand might suggest a change in the output rate.

There are two ways of further classifying mass production systems: assembly line and flow line. An assembly line is used to produce a discrete product. Typically, partial assemblies are moved from one workstation to the next in sequence by a fast-moving material-handling system (e.g., conveyor belts), going past each station, advancing the product toward its final assembly. Travel of the assembly within the stations is continuous or intermittent, depending on the nature of the job to be performed. The total work for a job is distributed among the workstations that form the assembly line such that all stations complete their assigned tasks in approximately the same time, called the cycle time. Automobile manufacturing is a prime example of an assembly line operation.

The term *flow line* is typically used to describe a continuous production process such as that of chemicals, liquids, gaseous products, and paper, as well as die operations such as wire manufacturing.

### 4.2.4 Cellular and Flexible Manufacturing

Cellular manufacturing is a system in which a large number of common parts are grouped together and produced in a cell consisting of all the machines that are needed to produce that group. When large quantities of each part are required, cells can be made almost completely automated, leading to the designation of that cell as flexible manufacturing.

**Production systems in manufacturing facilities.** *Top left:* A job shop, like the one shown here, processes several different items in small numbers and has the flexibility needed in a tool room, which must make or obtain the various tools a plant needs as well as store and maintain them. *Top right:* In the assembly line operation here, partial assemblies of electronic boards are moved by conveyor from one work station to the next so workers can use specialized equipment to assemeble the product in small steps. *Bottom left:* In cellular manufacturing, similar parts are made together in areas called cells, which can often be completely automated. Automated guided vehicles can perform material-handling tasks accurately and efficiently. Pictured here is the recharging area for such vehicles. *Bottom right:* Automated guided vehicles like this one can transport parts and/or material from robot-controlled storage systems to automated production machines without direct human control.

**Strategic cell design.** A highly automated cell is displayed in the photo. Six machines are arranged in a circle to form the cell. Jobs processed in the cell may use some or all of the machines. Setup is automated and so is the material handling between machines. (Courtesy of Manufacturing Engineering.)

The major advantage of the mass production system is that it lowers the unit cost of production, which is composed of two components: a fixed cost (for setup) and a variable cost (primarily labor and material). The portion of fixed cost assigned to a unit depends on the quantity produced with the same setup (same fixed cost). The larger the volume of production, the lower the share that each unit must carry. In mass production, the quantity produced with a particular setup is large, and therefore the unit cost of production is much lower than it would be in a job shop with a small production lot. But what can be done if the required quantity for each item does not warrant the use of a mass production system? Cellular and flexible manufacturing and group technology (GT) might provide the answer.

The advantages of cellular manufacturing are well documented. By collecting and processing similar parts together in a cell, machine setup times are reduced and throughput rates are increased. By eliminating many duplicate tools, investment in tools is reduced and — because workers can focus their skills on particular types of parts — the quality and productivity in the workplace is improved.

GT refers to the concept of identifying similar parts and grouping them to take advantage of their common characteristics in design and production methods. For example, from 1000 parts that are produced in a plant, perhaps 20 groups could be formed, each needing similar machines, jigs, and fixtures so that the items within a single group could be produced with very little change in the setup. Thus, the major setup cost is distributed among a large number of items, however small the quantity produced for each individual item may be.

One could attain all these benefits of GT by conceptually forming the machine and part type grouping, and then sequencing the various parts through appropriate machines by following good production planning. In other words, the machines within a group need not be physically close together to achieve the advantages stated so far. Thus, it is even possible to gain some of the benefits of GT in an existing job shop setup with proper planning, but without physically modifying the facility if some part types are regularly produced in the plant. In most practical cases, however, the machines within a group are also physically placed together in a group. This arrangement further reduces material handling by reducing distances between machines that are most often used by the parts.

## 4.3 CELL FORMATION IN GROUP TECHNOLOGY

There are numerous methods available for machine grouping in GT. The chapter references contain a short list of articles on this subject. Here, we discuss a simple method called the tabular method because it involves successive calculations that can be tabulated easily.

The method starts with a 0–1 table called the machine-component matrix table or incident matrix. This table shows the machine that each component (e.g., part or job) needs in its production, with "1" indicating the use of the machine and "0" (or a blank) indicating the nonuse of the machine. For example, in Table 4.1, job 1 needs machines C, D, I, and J in its production sequence, as indicated by corresponding 1s in the table. The objective is to develop machine cells such that each component can (as far as possible) be fully processed in a single cell.

**TABLE 4.1**

| | Machines | | | | | | | | | | |
|---|---|---|---|---|---|---|---|---|---|---|---|
| | A | B | C | D | E | F | G | H | I | J | K |
| Job | 1 | 2 | 3 | 4 | 5 | 6 | 7 | 8 | 9 | 10 | 11 |
| 1 | | | 1 | 1 | | | | | 1 | 1 | |
| 2 | | | 1 | 1 | | | | | 1 | 1 | |
| 3 | | | 1 | 1 | | | | | 1 | 1 | |
| 4 | | | | | | 1 | | 1 | | | 1 |
| 5 | 1 | 1 | | | 1 | | 1 | | | 1 | |
| 6 | | 1 | | | | 1 | 1 | | | 1 | |
| 7 | | | | 1 | | | | | 1 | 1 | |
| 8 | 1 | | | | 1 | | | | | 1 | |
| 9 | | | | | | 1 | 1 | 1 | | | 1 |
| 10 | | | | | | | 1 | | | | 1 |
| 11 | 1 | | 1 | | | | | | | 1 | |
| 12 | 1 | 1 | | | 1 | | 1 | | 1 | 1 | |
| 13 | | 1 | | | | | 1 | | | | |
| 14 | | | | | | 1 | 1 | 1 | | | |
| 15 | | | 1 | 1 | | | | | | 1 | |
| 16 | | | 1 | 1 | | | | | | 1 | |
| 17 | | | | | | 1 | 1 | 1 | | | 1 |
| 18 | | | | | | 1 | 1 | 1 | | | 1 |
| 19 | 1 | 1 | | | 1 | | 1 | | | | |
| 20 | 1 | | 1 | 1 | | | | | 1 | | |
| 21 | 1 | | | | | | | | 1 | | |

### 4.3.1 Solution Procedure

The tabular method given here explains how machine grouping is done. In the first phase of the method, a machine is assigned to a group based on its affinity to all the machines that are presently in the group. It automatically identifies the bottleneck machines (i.e., machines that are required in more than one cell) and distributes them to appropriate cells. The second phase distributes the jobs in the cells generated in the first phase. We illustrate the procedure by applying it to the 11-machine and 21-component example represented in Table 4.1. The two phases of the solution procedure are described next.

#### 4.3.1.1 Phase I

Here are the steps of the first phase of the tabular method:

Step 1: Develop a machine-to-machine relationship table. A machine-to-machine relationship table indicates the number of jobs that are processed on both machines (these jobs may not necessarily be processed in sequence). The table is developed by comparing columns of the machine-component matrix and counting the number of times 1 is common to two columns. This total indicates the number of parts (jobs) the two machines are producing in common, which is also a measure of the relationship between the two machines. Note that the machine-to-machine relationship table is symmetrical about the diagonal; therefore, only the elements below the diagonals are needed.

Step 2: Select the initial value of the relationship counter (RC). The RC defines the value of relationship being used in the present calculation. These values are selected from the table constructed in step 1. The values are ranked in descending order by magnitude. Pick the largest element in the machine-to-machine table and designate it as the present value of RC. The value of RC will change as we go through the iterative process (in step 8).

Step 3: Define a value of minimum percentage. A measure of the effectiveness of joining a machine to a group, such as 50% ($P = 0.5$), is defined by the analyzer at the beginning of the problem. It states the closeness that an entering machine (ENT) must have with all the existing machines within a group in order for the ENT to join that group. Based on the percentage value an analyzer chooses, the solution may have a different number of cells and a different number of total machines. Note that this percentage, designated as $P$, is defined only once during the process of finding a solution to the problem.

Step 4: Starting with the first row in the machine-to-machine table, examine each row for an elemental value that equals RC. Note the machines in the corresponding row and column.

Step 5: If the associated machines in the row and column are not already in a group, then form a group consisting of these two machines and go to step 7. If both machines are already assigned to the same group, ignore the observation and go to step 7. If one of the machines in the pair is in a group and the other one has not been assigned yet (this machine is called the ENT), go to step 6a. If both machines are assigned, but to different groups, go to step 6b.

Step 6a: Calculate the closeness ratio (CR) of the ENT with each group that has already been formed. A CR is defined as the ratio of the total of all relationships the ENT has with the machines that are currently in a group to the total number of machines that are presently assigned to that group.

The entering machine is placed in a group that has the maximum CR (MCR), as long as this maximum is greater than or equal to the minimum threshold value (MTV), which is $P$ multiplied by the present value of RC. If the value of MCR is less than the MTV, then a new group is formed consisting of the two machines noted in step 4, which have the relationship value that equals the present value of RC. Go to step 7.

Step 6b: Duplication of one or more machines is suggested. There are two possible alternatives, and they are checked sequentially because of cost considerations. The first alternative is to add one additional machine of either type (for illustrative purposes, designate the machines in the pair as machine A and machine B) and place it in the appropriate cell; the second alternative is to add two additional machines, one of each type, and form a new group or place the appropriate one in each of the existing groups. The following rules are suggested:

- Duplication of a single machine
  Calculate the effect of duplicating one machine. Check machine A as the ENT for the groups where machine B exists and machine B as the ENT for the groups where machine A exists. Determine the MCR from all the groups that are checked and note the associated group and ENT.
  If MCR > RC × *P*, the noted machine is selected and assigned to the associated group. Go to step 7.
- Addition of two new machines
  If the MCR in the previous calculation (in step 1) was less than RC × P, then a check must be made to see if both machines are to be selected. From the previous calculations, determine the MCR for the groups where A is the ENT (MCRA) and the MCR for the groups where B is the ENT (MCRB). Calculate the index value as maximum RC × P/2. If both MCRA and MCRB are greater than the index value and |MCRA − MCRB| < P × RC/2, then select both machines and place each in an appropriate group. If either MCRA or MCRB is greater than the index value, regardless of the value of |MCRA – MCRB|, then form a new group consisting of machines A and B. Go to step 7.
- No Additional Machine
  If none of the preceding conditions exists, we could ignore this observation and go to step 7, because the contribution of new machines is very limited. (However, if phase III, which is the analysis with additional objectives, is to be performed, then to get a better understanding of the problem structure, we could select one machine of each type and place it in each of the appropriate groups. In phase III, an evaluation would be made on the effectiveness of this step.)

Step 7: Continue the checking of the machine-to-machine table with the present value of RC proceeding sequentially in rows. If an element that is equal to the present value of RC is found, go to step 5. If no such element is found, go to step 8.

Step 8: Check if all positive values of RC greater than 0 are checked. If they are, go to step 9; otherwise, reduce the value of RC to the next value in its descending order of magnitude and return to step 5. Another termination rule is to stop when all the machines are assigned to at least one group. This stopping rule keeps machine duplications to a minimum.

Step 9: This step involves group consolidation. Compare the groups formulated so far. If one group — say, G1 — contains machines such that it is a subset of another group, then eliminate G1 from any further considerations.

#### 4.3.1.2 Phase II

Assign each job to an appropriate group. This is accomplished by examining each job and assigning it to a group that has the maximum number of machines needed to perform the job. Construct a job grouping table similar to Table 4.7.

We shall now apply the two-phase tabular method to the data presented in Table 4.1. It can be easily observed how the successive tables are constructed and appropriate decisions are made.

## AN ILLUSTRATIVE EXAMPLE

Machine component data from Table 4.1 are used to obtain the machine-to-machine table shown in Table 4.2. For example, the entry "AB" is obtained by comparing columns A and B of Table 4.1 and counting the number of components requiring both machines, namely, 5, 12, and 19.

To begin the grouping procedure, let the value of *P* be set to 0.5 (step 3). The maximum value among the elements in Table 4.2 is 6 (i.e., RC = 6), and associated machines C and D are joined together to form the first group, G1 (step 5).

For the second iteration, searching through Table 4.2 for next value equal to RC, we find that machines J and C also have a 6 relationship. Because C already belongs to G1, J becomes the ENT, and we must determine whether it should be included in G1 or a new group consisting of J and a duplicate C should be formed. Step 6a is applied; the associated calculations are shown in Table 4.3.

The explanation for the entries in Table 4.3 is: There is only one group formed so far, G1. It consists of machines C and D. The ENT J has a 6 relationship with machine C and a 6 relationship with machine D. In total, we have two machines in the group, and the sum of the relationships is 12. The CR for the group is therefore 12/2 = 6. Because we have only one group at this stage, the ratio is also the MCR.

MTV in this calculation is obtained by multiplying the present value of RC by *P*; that is, $6 \times 0.5 = 3$. Because MCR > MTV, machine J is joined with G1, which now consists of C, D, and J.

Continuing the search through Table 4.2 for the third iteration, we find that J and D also have a 6 relationship, but because both machines are already in the same group, the observation is ignored (step 5). (As there is no other 6 relationship, RC is set to the next value in the order, i.e., RC = 5.)

The summary of calculations for phase I steps is shown in Table 4.6 (details of four typical iterations).

**TABLE 4.2**
**Machine-to-Machine Relationship Table**

| | A | B | C | D | E | F | G | H | I | J | K |
|---|---|---|---|---|---|---|---|---|---|---|---|
| A | — | | | | | | | | | | |
| B | 3 | — | | | | | | | | | |
| C | 2 | 0 | — | | | | | | | | |
| D | 1 | 0 | 6 | — | | | | | | | |
| E | 4 | 3 | 0 | 0 | — | | | | | | |
| F | 0 | 1 | 0 | 0 | 0 | — | | | | | |
| G | 3 | 5 | 0 | 0 | 3 | 5 | — | | | | |
| H | 0 | 0 | 0 | 0 | 0 | 5 | 4 | — | | | |
| I | 3 | 1 | 4 | 5 | 1 | 0 | 1 | 0 | — | | |
| J | 4 | 2 | 6 | 6 | 3 | 0 | 2 | 0 | 5 | — | |
| K | 0 | 0 | 0 | 0 | 0 | 4 | 4 | 4 | 0 | 0 | — |

**TABLE 4.3**
**Check for J as Entering Machine: RC = 6**

| Entering Machine (ENT) | Existing Groups | | MTV |
|---|---|---|---|
| | Group 1 | Relationship | |
| J | C | 6 | 6 × 0.5 = 3 |
| | D | 6 | |
| Total | 2 | 12 | |
| Closeness ratio, CR | 12/2 = 6 | | |
| Maximum closeness ratio, MCR | 6 | | |

**TABLE 4.4**
**Check for F as the ENT: RC = 5**

| ENT | Existing Groups | | | | MTV |
|---|---|---|---|---|---|
| | G1 | Relationship | G2 | Relationship | |
| F | C | 0 | G | 5 | 5 × 0.5 = 2.5 |
| | D | 0 | B | 1 | |
| | J | 0 | | | |
| Total | 3 | 0 | 2 | 6 | |
| CR | 0/3 = 0 | | 6/2 = 3 | | |
| MCR | | 3 | | | |

**TABLE 4.5**
**Machine J or A Duplication Check: RC = 4**

| ENT | G3 | Relationship | ENT | G1 | Relationship | MTV |
|---|---|---|---|---|---|---|
| J | E | 3 | A | C | 2 | 4 × 0.5 = 2 |
| | A | 4 | | D | 1 | |
| | | | | J | 4 | |
| | | | | I | 3 | |
| Total | 2 | 7 | | 4 | 10 | |
| CR | 7/2 = 3.5 | | 10/4 = 2.5 | | | |
| MCR | | 3.5 | | | | |

**TABLE 4.6**
**Summary of Calculations**

| Iteration | RC | Machine(s) Under Consideration | Group Where the Machine is Assigned or New Group Formed | Present Groups and Associated Machines | Machine Duplicated |
|---|---|---|---|---|---|
| 1 | 6 | C, D | G1 | G1: C, D | |
| 2 | 6 | J | G1 | G1: C, D, J | |
| 3 | 6 | J, D | Same group | | |
| 4 | 5 | G, B | G2 | G1: C, D, J<br>G2: G, B | |
| 5 | 5 | F | G2 | G1: C, D, J<br>G2: G, B, F | |
| 6 | 5 | H | G2 | G1: C, D, J<br>G2: G, B, F, H | |
| 7 | 5 | I | G1 | G1: C, D, J, I<br>G2: G, B, F, H | |
| 8 | 5 | J, I | Same group | | |
| 9 | 4 | E, A | G3 | G1: C, D, J, I<br>G2: G, B, F, H<br>G3: E, A | |
| 10 | 4 | H, G | Same group | | |
| 11 | 4 | I, C | Same group | | |
| 12 | 4 | J or A | G3 | G1: C, D, J, I<br>G2: G, B, F, H<br>G3: E, A, J | J |
| 13 | 4 | K | G2 | G1: C, D, J, I<br>G2: G, B, F, H, K<br>G3: E, A, J | |
| 14 | 4 | K, G | Same group | | |
| 15 | 4 | K, H | Same group | | |
| 16 | 3 | B or A | G3 | G1: C, D, J, I<br>G2: G, B, F, H, K<br>G3: E, A, J, B | B |
| 17 | 3 | E, B | Same group | | |
| 18 | 3 | A or G | G3 | G1: C, D, J, I<br>G2: G, B, F, H, K<br>G3: E, A, J, B, G | G |
| 19 | 3 | G, E | Same group | | |
| 20 | 3 | I or A | G1 | G1: C, D, J, I, A<br>G2: G, B, F, H, K<br>G3: E, A, J, B, G | A |
| 21 | 3 | J, E | Same group | | |
| 22 | 2 | J, B | Same group | | |
| 23 | 2 | J, G | Same group | | |
| 24 | 1 | C, A | Same group | | |
| 25 | 1 | F, B | Same group | | |
| 26 | 1 | I or B | GI | G1: C, D, J, I, A, B<br>G2: G, B, F, H, K<br>G3: E, A, J, B, G | B |
| 27 | 1 | I or E | G1 | G1: C, D, J, I, A, B, E<br>G2: G, B, F, H, K<br>G3: E, A, J, B, G | E |

*Iteration 4*: RC = 5, G-B (the notation indicates the relationship between machine G and B). Because presently neither G nor B is a member of any group, form a new group, G2, consisting of machines G and B.

*Iteration 5*: RC = 5. F-G. At present we have two groups in the solution (G1 and G2). The check for ENT F must be made with the machines in both groups. The CR for each group is calculated, and the maximum among them defines the value of the MCR. The calculations are shown in Table 4.4. Note that the MTV is calculated as the present value of RC, which is $5 \times P$.

Because MCR > MTV, assign F in G2. The present assignments are G1: C, D, J, and G2: G, B, F.

*Iteration 9*: RC = 4. E-A. Because neither E nor A belongs to any group, form a new group, G3, consisting of these two machines.

*Iteration 12*: RC = 4. J-A. Because machine J belongs to G1 and machine A belongs to G3, a duplication check must be made. Check J as the ENT for the group where A exists, G3, and A as the ENT where J exists, G1. Calculations are shown in Table 4.5. Because MCR > MTV, add another unit of J in G3. The present groups are G1 (C, D, J, I), G2 (G, B, F, H, K), and G3 (E, A, J).

Calculations are illustrated for 21 iterations in Table 4.6. The final arrangement consists of three groups with machine distribution:

Group 1: C, D, J, I, A, B, E
Group 2: G, B, F, H, K
Group 3: E, A, J, B, G

We could have continued the procedure until RC = 0, but we will terminate it here to illustrate the concepts of separating machines. Note that using the alternate termination rule we could have stopped after iteration 13 (because all the machines are assigned). Table 4.7 presents the distribution of jobs to each appropriate cell. Each job's machine requirements are checked against the available machines in each cell, and the cell that has the maximum numbers of machines to process the job is the cell to which the job is assigned. For example, job 1 requires machines C, D, I, and J, which are all in group 1; therefore, job 1 is assigned to G1. One could also place another unit of G, B, and E group 1 to process job 12 completely in that cell. These units would then be noted as the cell-separating units.

#### 4.3.1.3 Phase III: Further Group Evaluation with Additional Objectives

Machine grouping can be improved if additional secondary goals are also used in group formation. The analysis suggested next provides the analyzer with information for making adjustments based on the secondary goals. In most instances, there are only finite alternatives to evaluate and therefore the procedure is quite practical. We designate this as phase III of the calculations.

#### 4.3.1.4 Phase III Procedure

Backtrack through the solution tables in phase I and construct a table for duplicated and cell-separating machines. A machine that has the maximum number of jobs

**TABLE 4.7**
**Job Grouping**

| | Jobs | | | | | | | | | | | | | | | | | | | |
|---|---|---|---|---|---|---|---|---|---|---|---|---|---|---|---|---|---|---|---|---|
| **Machines** | **1** | **2** | **3** | **7** | **11** | **15** | **16** | **20** | **12** | **4** | **6** | **9** | **10** | **13** | **14** | **17** | **18** | **5** | **8** | **19** |
| C | 1 | 1 | 1 | | 1 | 1 | 1 | 1 | | | | | | | | | | | | |
| D | 1 | 1 | 1 | 1 | | 1 | 1 | 1 | | | | | | | | | | | | |
| J | 1 | 1 | 1 | 1 | 1 | 1 | 1 | | 1 | | | | | | | | | | | |
| I | 1 | 1 | 1 | 1 | | | | 1 | 1 | | | | | | | | | | | |
| A | | | | | 1 | | | 1 | 1 | | | | | | | | | | | |
| B | | | | | | | | | 1 | | | 1 | | | | | | | | |
| E | | | | | | | | | 1 | | | 1 | | | | | | | | |
| G | | | | | | | | | 1 | | 1 | 1 | 1 | 1 | 1 | 1 | 1 | | | |
| B | | | | | | | | | | | 1 | | | 1 | | | | | | |
| F | | | | | | | | | | 1 | 1 | 1 | | | 1 | 1 | 1 | | | |
| H | | | | | | | | | | 1 | | 1 | | | 1 | 1 | 1 | | | |
| K | | | | | | | | | | 1 | | 1 | 1 | | | 1 | 1 | | | |
| E | | | | | | | | | | | | | | | | | | 1 | 1 | 1 |
| A | | | | | | | | | | | | | | | | | | 1 | 1 | 1 |
| J | | | | | | | | | | | | | | | | | | 1 | 1 | |
| B | | | | | | | | | | | | | | | | | | 1 | | 1 |
| G | | | | | | | | | | | | | | | | | | 1 | | 1 |

assigned to it is considered the primary machine, and any machines of the same type in other cell(s) are considered to be duplicated machines. A cell-separating machine is the one that can be added to the solution from phase I to make a job completely self-contained within a specific cell. The table should show each machine that is duplicated or is cell separating, the group to which it is presently assigned, the job(s) it presently processes, and alternative group(s) in which the same type of machine exists. The information indicates the alternatives available for assigning the job(s) if a decision is made to eliminate the duplicating or separating machine from the present group.

Backtracking through Table 4.7 on duplicated machines leads to the information presented in Table 4.8. The table indicates the marginal importance of each of the duplicated and cell-separating machines. Using this information, we can answer questions such as these:

1. If there is limitation on funding or space, which machines, if any, should be duplicated?
2. Which machines are necessary to reduce intercellular trips between groups or between any two specific groups?
3. If an item or group of items is to be produced in a single cell, what would be the utilization of machines within that cell?
4. If the total number of any machine type is to be restricted to some value, what is the effect of such policy on the overall performance?

**TABLE 4.8**
**Backtracking on Duplicated Machines**

| Duplicated/ Separating Machine | Group in Which It Is (Can Be) Assigned | Job It Serves | Alternative Group(s) to Which Jobs Can Be Assigned |
|---|---|---|---|
| E (separating) | 1 | 12 | 3 |
| B (separating) | 1 | 12 | 2 or 3 |
| A | 1 | 11, 20, 12 | 3 |
| G | 3 | 5, 19 | 2 |
| B | 3 | 5, 19 | 2 or 1 |
| J | 3 | 5, 8 | 1 |
| G (separating) | 1 | 12 | 2, 3 |

For example, elimination of machine E from group 1 will cause transfer of job 12 to group 3 to get the services of machine E. There are presently 2 units of machine A in the solution (one basic and one duplicated). If the total number is to be limited to 1, we might eliminate A from group 1, causing transportation of items 11, 20, and 12, or eliminate A from group 3 (basic unit), causing transportation of items 5, 8, and 19. If it is intended to process job 12 completely in cell 1, then it would require inclusion of machines E, B, A, and G in cell 1, although these machines might also be available in some other cells. Based on the number of units to be processed, volume and weight of the items, and associated difficulty of transportation, we can make an informed decision.

If machine $i$ costs $C_i$ dollars and there are only $f$ additional dollars available after buying one basic unit for each machine type, we might solve the following 0–1 integer programming problem (where the variables take either of the values 0 or 1) to maximize job processing in a single cell or reduce intercellular trips.

$$\text{Maximize} \quad \sum_{i=A}^{z}\sum_{k=1}^{g} j_{ik}X_{ik}$$

$$\text{Subject to} \quad \sum_{i=A}^{z}\sum_{k=1}^{g} C_iX_{ik} \le f$$

where $J_{ik}$ represents the number of jobs that the $i$th duplicating or separating machine serves in the assigned cell $k$

$$X_{ik} = \begin{cases} 1 & \text{if machine } i \text{ is selected in cell } k \\ 0 & \text{otherwise} \end{cases}$$

Note that index $i$ is associated with machines; because the machines are designated as A, B, C, and Z, $i$ will take values A, B, C, …, Z.

This problem is known as a *knapsack problem* in operations research [see Taha (1992) or Winston (1991)]. There are some well-known solution techniques, such as dynamic programming, that can be used to solve such problems.

For example, suppose the costs of the duplicated machines (on the same time base) are: A = \$3000, B = \$2000, E = \$2000, G = \$1000, and J = \$500, and \$7500 is available to buy duplicated machines. Because in group 1, machine E serves one job (job 12), machine B serves one job, etc., the following formulation can be made if the objective is to maximize the number of jobs served with the available funds.

$$\text{Maximize}\quad 1X_{E,1} + 1X_{B,1} + 3X_{A,1} + 2X_{G,3} + 2X_{B,3}2X_{J,3} + 1X_{G,1}$$

$$\text{Subject to } 2000X_{E,1} + 2000X_{B,1} + 3000X_{A,1} + 1000X_{G,1} + 1000X_{B,3}$$

$$+ 500X_{B,3} + 1000X_{G,3} \leq 7500$$

The cost per job served for each machine in each group is: E, 1 = 2000; B, 1 = 2000; A, 1 = 1000; G, 3 = 500; B, 3 = 500; J, 3 = 250; and G, 1 = 1000. If we start choosing the alternatives by the magnitude of their costs, with the least-cost alternative being selected first, and continue the process until all the available money is spent, the following selection results: J, 3; B, 3; G, 3; G, 1; and A, 1. The solution serves 10 jobs. Because the solution spends exactly \$7500, it is the best solution. However, if such a perfect match of the available funds and spending is not possible while applying this procedure, then the solution may not be optimum and a procedure such as dynamic programming or 0/1 integer programming may have to be applied.

Finally, consider a capacity problem. Suppose the present job assignment requires capacities of 2.25 units of B in G1, 0.25 units of B in G2, and 0.25 units of B in G3. It is clear that at least three units of B are required and as many as five units of B could be used, three in G1 and one each in G2 and G3. At least two units of B must be placed in G1. Here, again, the alternatives needing evaluation are rather limited. Calculate the cost associated with five units. If four units are contemplated, we can place three in G1 and one in either G2 or G3, or two in G1 and one each in G2 and G3. Similarly, for three units, we might evaluate all three units in G1 or two units in G1 and one either in G2 or G3. Knowing the number of units to be transported and their characteristics, as well as other cost features, such evaluations are quite simple.

As an illustration, suppose that each unit of machine B costs \$1000/year and the minimum cost of intercellular transportation for 0.25 unit of capacity B is given in Table 4.9. The corresponding analysis is shown in Table 4.10.

With five machines distributed as three to group 1 and one each to groups 2 and 3, no intercellular transportation is required; hence, the total cost of the policy is the machine cost, \$5000/year. With four machines, three options are available, each requiring some intercellular transport. For example, three machines assigned to G1, one to G2, and none to G3 would require transportation of 0.25 unit of capacity of B from G3 to either G2 or G1. The minimum cost is associated with G3 to G2 transport, and hence that policy is chosen. The total cost of the policy is \$4000 for machines and \$300 for transportation, for a sum of \$4300.

**TABLE 4.9**
**Intracellular Transportation Cost/Year**

| For 0.25 Unit of B | | | | | |
|---|---|---|---|---|---|
| G1 to G2 | G1 to G3 | G1 to G3 | G2 to G1 | G3 to G1 | G3 to G2 |
| 500 | 800 | 300 | 800 | 500 | 300 |

With only three machines (minimum required), intercellular transportation for two groups is required. For example, with a policy of G1 = 2 and G2 = 1, 0.25 unit of B capacity from group 1 must go to group 2 and 0.25 unit of B from group 3 must also go to group 2, giving a total transportation cost of $800 and an overall cost of $3800.

The best alternative is the G1 = 2, G2 = 1, and G3 = 0 policy, with a total cost of $3800. It should be noted that it is not advantageous to have five units of B, two (duplications) more than what is minimally (and in this case also optimally) needed, to avoid all intercellular transportations.

Suppose after a review it is determined that due to space limitation, the building in which group 2 is to be installed can accommodate no unit of B. In this case, the best alternative is G1 = 2 and G3 = 1, with a total cost of $4100. The increased cost of $300 can be attributed to the space constraint.

To further analyze the machine utilization by this policy, two units of B in cell 1 are 100% utilized, while one unit in cell 3 would be serving at 0.25 unit capacity for jobs assigned in cell 1, 0.25 unit for jobs assigned in cell 2, and 0.25 unit for jobs in cell 3, for a utilization of 75%. If we wish to normalize the utilization in all cells for machine B to about the same level, 2.75/3 = 0.916, additional jobs requiring (1 − 0.916)2 = 0.168 units of capacity would have to be transported from cell 1 to cell 3. Again, cost will dictate which jobs must be transported between which cells.

**TABLE 4.10**
**Capacity Analysis 161 Imaging**

| | Assignments | | | | | |
|---|---|---|---|---|---|---|
| Policy | G1 | G2 | G3 | Demands to Move | Minimum Cost of Transportation ($) | Total Cost ($) |
| 5 | 3 | 1 | 1 | None | 0 | 5000 |
| 4 | 3 | 1 | 0 | G3 to G2 or G3 to G1 | 300 | 4300 |
| 4 | 3 | 0 | 1 | G2 to G1 or G2 to G3 | 300 | 4300 |
| 4 | 2 | 1 | 1 | G1 to G2 or G1 to G3 | 500 | 4500 |
| 3 | 3 | 0 | 0 | G2 to G1 or G3 to G1 | 1300 | 4300 |
| 3 | 2 | 1 | 0 | G1 to G3 or G3 to G2 | 800 | 3800 |
| 3 | 2 | 0 | 1 | G1 to G3 or G2 to G3 | 1100 | 4100 |

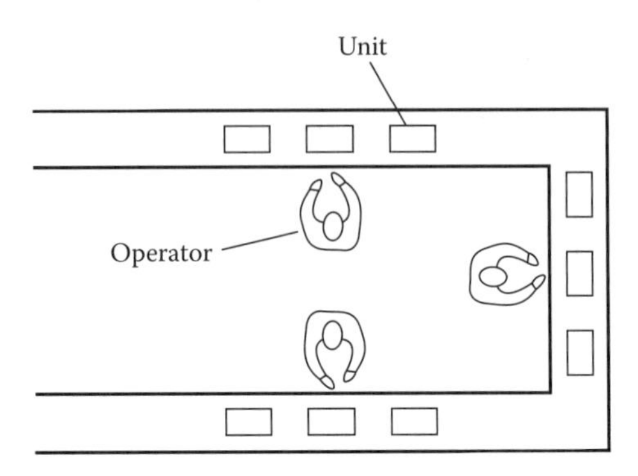


**FIGURE 4.11** U-shaped arrangement

### 4.3.2 Layout Considerations with Cellular Manufacturing

Small- or medium-sized lots (varying from 1 to 200 units) of each item can be made with a minimum of changeover in setups. Anywhere from 1 to 15 machines are grouped together, typically in a U-shaped layout, as shown in Figure 4.11. These machines may be general-purpose or highly automated. The U-shaped layout has many advantages. It provides flexibility in that the number of workers who are trained to perform a range of tasks can easily be increased or decreased on the basis of the amount of work assigned to that cell. Workers are made responsible for more than one machine, a situation called *machine coupling*, so that the utility of a worker is not limited by the utility of a machine. Close contact and cooperation of workers in a U-shaped layout can contribute to an increase in productivity by decreasing idle time, poor quality, and in-process inventory.

Flexible manufacturing is cellular manufacturing in highly automated form. The cells are made very efficient to produce large quantities rapidly by including NC machines, robots, and computer-controlled conveyor systems. A cell in this case might be arranged in a U shape, as before, or even in a straight line, which can increase the efficiency of material movement. This system of manufacturing is used mainly to produce a family of parts in lots of 200–1000 units, because the NC machines require almost no changeover in setups to produce the items in a family. A high degree of automation means high efficiency and very little need for manpower. It is not unusual to find only one or two workers controlling the entire production.

## 4.4 LABOR ASSIGNMENTS

Let us consider two prominent production layout types in which the labor needed can change, depending on how the total production is distributed among different stations. The production line, which is mainly a batch-processing arrangement, and the assembly line, where the units are to be produced at a constant rate, are examined next.

### 4.4.1 Labor Assignment in Production Lines

In some job shops and batch shops, production facilities are referred to as *production lines*. Although similar to assembly lines, production lines differ from them primarily in the rate of flow. Units are processed through many different stations (work areas) before finally being transformed into complete assemblies. Within a given production line, each station might consist of a single machine and/or a worker, an entire department, a group of machines and their operators, or even a number of manual laborers assigned to perform a specific task. The rate of production in each station depends on the type of work assigned and quite often varies from station to station. The production rate can be controlled by the labor allotted to that station; the more labor provided, the greater the production, up to a certain limit. There might also be partially finished assemblies left over from a previous production run of the same product, called *inventory*, in front of a station. For a specific station, such inventory consists of units that have received all the operations by preceding stations and can be made into the final assembly by starting with the operations of the specific station. The problem is to determine the number of workers needed in each station to achieve the required output from a given production line. The following example illustrates this procedure.

Consider the production line shown in Figure 4.12, requiring 20 stations to complete the assembly. For convenience, the stations are numbered from left to right in the direction of material flow. Subassemblies 1 and 2 are combined in station 5, which is also the common input station for the three parallel branches 6–9, 7–10, and 8–11. These branches are required in this example because of limitations on space, equipment, and labor available to work at the stations.

The output from the three branches — that is, from stations 9, 10, and 11 — goes into common storage, from which parts for stations 12 and 13 are drawn. In station 17, two subassemblies are joined at the ratio of 2:5. Finally, station 18 supplies both stations 19 and 20, where the final processing is accomplished. Figure 4.12 also shows the input/output requirements per worker at each station; for example, in station 18, a worker will require 16 units of input and will produce 12 good units.

Table 4.11 presents the tabular approach. The information contained in columns 1, 2, 3, 4, 5, and 10 is known for the given product. By way of illustration, consider station 17. Production per worker at this location is 9 units, but the next workstation, station 18, requires 16 units per worker as its input. The maximum number of workers that can be assigned to station 17 is eight.

Continuing with the analysis, we now examine a few specific stations of interest in the development of Table 4.11. Station 18 provides the input for two stations, 19 and 20, and column 5 for station 18 lists the input requirement for each of these stations (19 and 20).

Stations 9, 10, and 11 provide the input to a common storage area. They are grouped together and are marked as such by a brace. Stations 12 and 13 also form a similar group.

Station 17 is the succeeding station for both stations 16 and 14; however, the units demanded are different, because two different subassemblies are involved. The appropriate input is shown for each station in column 5.

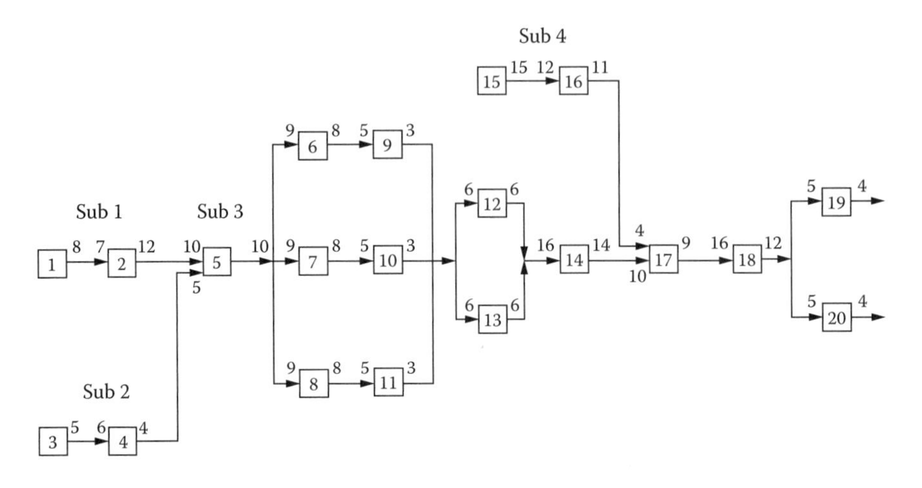


**FIGURE 4.12** Production line. Production systems in manufacturing facilities. Top left: A job shop, like the one shown here, processes several different items in small numbers and has the flexibility needed in a tool room, which must make or obtain the various tools a plant needs as well as store and maintain them. Top right: In the assembly line operation here, partial assemblies of electronic boards are moved by conveyor belt from one workstation to another so workers can use specialized equipment to assemble the product in small steps. Bottom left: In cellular manufacturing, similar parts are made together in areas called cells, which can often be completely automated. Automated guided vehicles can perform material-handling tasks accurately and efficiently. Pictured here is the recharging area for such vehicles. Bottom right: Automated guided vehicles like this one can transport parts and/or material from robot-controlled storage systems to automated production machines without direct human control. Strategic cell design (courtesy of Manufacturing Engineering). A highly automated cell is displayed in the photo. Six machines are arranged in a circle to form the cell. Jobs processed in the cell may use some or all of the machines. Setup is automated and so is material handling between machines.

In this example, the required output is 16 units from station 20 and 18 units from station 19. After accounting for present inventories by completing column 2, the procedure is to work backward from the last work area (station 20), filling in all the columns for each station (or each group of stations). For example, for station 17, the number of workers assigned to the succeeding station — that is, station 18 — is four. The output requirement is therefore 64 (16 × 4) units. (Column 7 is column 6 multiplied by column 5.) Because there is an initial inventory of 7 units, the net production should be 57 (64 – 7) units. (Column 8 is obtained by subtracting column 2 from column 7.) The necessary worker power is calculated by dividing the net production required by the rate of production per worker in this station. (Column 9 results when column 8 is divided by column 3.) Because this must be a whole number and cannot be negative, the minimum value is zero, and fractions are rounded off to the next higher integer. For station 17, we need a minimum of seven workers (column 9) to operate the machines; the maximum number of workers that can be assigned is limited to eight (column 10). Suppose the production manager decides to assign seven workers (column 11). The total output from this station then becomes (7 × 9) = 63 units

**TABLE 4.11**
**Production Line Planning**

| Station (1) | Initial After-Station Inventory (2) | Production Rate of the Station (3) | Succeeding Station (4) | Input Required on Succeeding Station per Worker in that Station (5) | Workers Assigned in Succeeding Stations (6) | Output Required (6) × (5) (7) | Net Production Required (7) × (2) (8) | Minimum Manpower Required (9) | Limit on Manpower Assignment (10) | Workers Assigned (11) | Output (3) × (11) (12) | Final After-Station Inventory (12) – (8) (13) |
|---|---|---|---|---|---|---|---|---|---|---|---|---|
| 20 | 0 | 4 | — | — | — | 16 | 16 | 4 | 6 | 4 | 16 | 0 |
| 19 | 0 | 4 | — | — | — | 18 | 18 | 5 | 6 | 5 | 20 | 2 |
| 18 | 5 | 12 | 19, 20 | 5, 5 | 5, 4 | 25 + 20 = 45 | 40 | 4 | 8 | 4 | 48 | 8 |
| 17 | 7 | 9 | 18 | 16 | 4 | 64 | 57 | 7 | 8 | 7 | 63 | 6 |
| 16 | 10 | 11 | 17 | 4 | 7 | 28 | 18 | 2 | 4 | 2 | 22 | 4 |
| 15 | 15 | 15 | 16 | 12 | 2 | 24 | 9 | 1 | 4 | 1 | 15 | 6 |
| 14 | 34 | 14 | 17 | 10 | 7 | 70 | 36 | 3 | 4 | 3 | 42 | 6 |
| 13 } | 5 | 6 | 14 | 16 | 3 | 48 | 43 | 8 | 4 | 4 | 24 | 5 |
| 12 } | | 6 | 14 | | | | | | 5 | 4 | 24 | |
| 11 } | 1 | 3 | 12, 13 | 6, 6 | 4, 4 | 24 + 24 = 48 | 47 | 16 | 8 | 5 | 15 | 1 |
| 10 } | | 3 | 12, 13 | | | | | | 8 | 5 | 15 | |
| 9 } | | 3 | 12, 13 | | | | | | 10 | 6 | 18 | |
| 8 | 6 | 8 | 11 | 5 | 5 | 25 | 19 | 3 | 4 | 3 | 24 | 5 |
| 7 | 4 | 8 | 10 | 5 | 5 | 25 | 21 | 3 | 4 | 3 | 24 | 3 |
| 6 | 7 | 8 | 9 | 5 | 6 | 30 | 23 | 3 | 4 | 3 | 24 | 1 |
| 5 | 10 | 10 | 6, 7, 8 | 9, 9, 9 | 3, 3, 3 | 27, 27, 27 | 71 | 8 | 10 | 8 | 80 | 9 |
| 4 | 15 | 4 | 5 | 5 | 8 | 40 | 25 | 7 | 9 | 7 | 28 | 3 |
| 3 | 5 | 5 | 4 | 6 | 7 | 42 | 37 | 8 | 8 | 8 | 40 | 3 |
| 2 | 10 | 12 | 5 | 10 | 8 | 80 | 70 | 6 | 8 | 6 | 72 | 2 |
| 1 | 5 | 8 | 2 | 7 | 6 | 42 | 37 | 5 | 7 | 5 | 40 | 3 |

(column 12 = column 3 × column 11). The final remaining inventory is what is produced minus what is needed (column 13 = column 12 – column 8), or 6 in this case.

Now consider some other stations of interest. For station 18, column 6 contains the number of workers assigned to stations 19 and 20. Thus, the output required (column 7) is the total output necessary to supply both stations.

For the group consisting of stations 9, 10, and 11, the succeeding work areas are stations 12 and 13. For each of the latter, the input requirement is 6 units per worker assigned. Therefore, the total output required from stations 9, 10, and 11 is 48 units; however, with the initial inventory of 1 unit, the net production required is 47 units. Each worker in stations 9, 10, and 11 can produce 3 units; therefore, we will need a minimum of 16 workers (column 9). These workers are distributed 5, 5, and 6, respectively, among stations 9, 10, and 11, keeping within the maximum labor assignment limit. In the common storage area, units arrive from all three stations (a total of 48 units), leaving one unit on hand at the end of the production run.

The analysis is complete when the entries associated with all the stations have been made. In this example, station 1 is the last station for which the data are entered. When the analysis is finished, the method tells what the labor assignments should be and what the final inventory will be for each station.

#### 4.4.1.1 Labor Planning

Quite often, each station represents a stage in a production process; however, with a limited crew, it becomes impossible to work on all stages simultaneously. Thus, there is a need to develop a labor scheduling system that will indicate which station is to receive how many workers in what period, perhaps daily.

One approach would be to complete all work for an entire order at a particular station before moving to the next station in the sequence. This method would almost certainly create excessive in-process inventory. The other extreme is to process a single unit in one station and then move the worker to the next station. This requires costly setup changes and worker movements that might be difficult to plan practically. It is quite difficult, for example, to keep a worker 4 minutes in one station, 6 minutes in the next, and so on. (In fact, the worker might spend more time changing stations than working.)

One method of making assignments would be to follow the rules stated below:

1. Move the workers from one station to another only after the end of a period (one day in our example).
2. Assign a worker to the succeeding stations in the production sequence if the input to these stations is sufficient to meet the target production for the period.

We leave it as an exercise to the reader to develop the daily labor schedule for the example in the preceding section.

### 4.4.2 Assembly Line Balancing

An *assembly line* is a method of production in which the parts are assembled and made into the final product as the unit progresses from station to station. Developing a balanced assembly line requires logical planning and involves the distribution of the total job among the workstations so that all stations can complete their designated tasks at approximately the same time. If the line were perfectly balanced, the time required in each station would be identical. However, such balance is rarely possible, and the longest station time dictates the actual cycle time for the entire assembly.

Various techniques are available for developing an assembly line, but the basic requirements for all are the same: break the total job content into small tasks or work elements and determine the precedence relationships — that is, decide which tasks must be performed before the next one begins. These relationships also indicate the tasks that can be performed simultaneously. For example, in paintbrush manufacturing, glue must be poured into a clip before the bristles can be inserted; however, at the same time, someone (or a machine) could polish and stamp the handle. Next, determine the time that will be required for each work element. Any other restrictions, such as the requirement that none of a group of work elements may be placed together in the same workstation, called negative zoning, or the requirement that some tasks must be placed in the same workstation, called positive zoning, should also be considered and clearly stated.

Mathematical methods are available to balance assembly lines in terms that are mainly the application of optimization techniques such as integer programming and branch-and-bound methods. However, these exact methods are very tedious and cumbersome to apply to a large assembly line containing a multitude of tasks. We will show two heuristic methods that are popular because of their simplicity.

#### 4.4.2.1 Balancing Procedures

The methods are illustrated by applying them to the following sample problem. The elements are represented in Figure 4.13 by nodes, and the precedence of each is indicated by an arrow connecting the nodes.

In the diagram, the element (task) times are listed above their corresponding nodes. The same data are also presented in Table 4.12.

The total time needed to complete all tasks is 5.31 minutes.

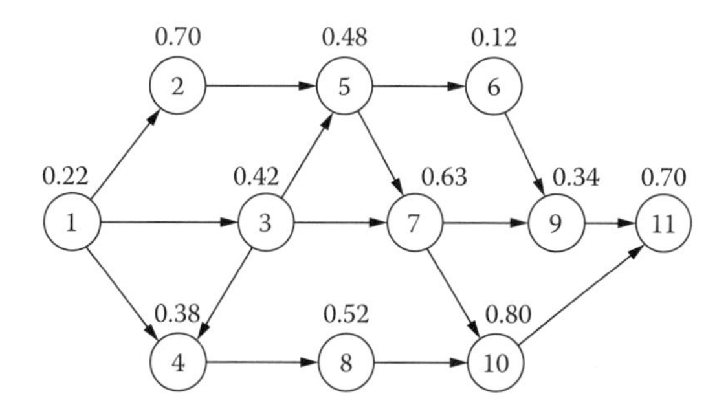


**FIGURE 4.13** Precedence diagram

**TABLE 4.12**
**Predecessor List**

| Element | Time | Immediate Predecessor(s) |
|---|---|---|
| 1 | 0.22 | — |
| 2 | 0.70 | 1 |
| 3 | 0.42 | 1 |
| 4 | 0.38 | 1, 3 |
| 5 | 0.48 | 2, 3 |
| 6 | 0.12 | 5 |
| 7 | 0.63 | 3, 5 |
| 8 | 0.52 | 4 |
| 9 | 0.34 | 6, 7 |
| 10 | 0.80 | 7, 8 |
| 11 | 0.70 | 9, 10 |

##### *4.4.2.1.1 Procedure I — Largest Candidate Rule*

The procedure determines the minimum number of stations necessary to obtain the desired *cycle time*, which is deemed as the total available time per period divided by the required production per period. The objective is to distribute the work among the stations. The steps involved in the procedure are:

1. List the tasks in decreasing order of magnitude of task times, the task requiring the largest time being first. Also list the corresponding immediate predecessor task(s) for each.
2. Designate the first station in step 1 as station 1 and number the remaining stations consecutively.
3. Beginning at the top of the task list, assign a feasible task to the station under consideration. Once the task is assigned, all reference to it is removed from the predecessor task list. A task is feasible only if it does not have any predecessors or if all predecessors have been deleted. It may be assigned only if it does not exceed the cycle time for the station, and this condition can be checked by comparing the cumulative time of all the jobs so far assigned to that station, including the task under consideration, with the cycle time. If the cumulative time is greater than the cycle time, the task under consideration cannot be assigned to the station. If no task is feasible, proceed to step 5.
4. Delete the task that is assigned from the first column of the task list. If the list is now empty, go on to step 6; otherwise, return to step 3.
5. Create a new station by increasing the station count by one. Return to step 3.
6. All jobs are assigned, and the present station number reflects the number of stations required.

**TABLE 4.13**
**Task Time**

| Task | Task Time | Immediate Predecessor(s) |
|---|---|---|
| 10 | 0.80 | 7, 8 |
| 11 | 0.70 | 9, 10 |
| 2 | 0.70 | 1 |
| 7 | 0.63 | 3, 5 |
| 8 | 0.52 | 4 |
| 5 | 0.48 | 2, 3 |
| 3 | 0.42 | 1 |
| 4 | 0.38 | 1, 3 |
| 9 | 0.34 | 6, 7 |
| 1 | 0.22 | — |
| 6 | 0.12 | 5 |

The procedure also shows the job assignments for each station. The largest of the cumulative times for the individual stations is the true cycle time.

We will demonstrate the application of the above-mentioned procedure to the sample problem presented in Figure 4.13 and Table 4.12 with a planned cycle time of 1 minute or with a production rate of 400/day, assuming that there are 400 working minutes in a day shift.

In the first step we list the tasks (elements) in the order of decreasing task time (Table 4.13). Also listed are the immediate predecessor(s) of each task. This information is needed in step 2 and can be obtained either from the data table or from the precedence diagram.

The procedure begins with the element listed first in Table 4.13, task 10. The accumulated task time would be equal to 0.80, which is less than the cycle time; but because neither task 7 nor 8 is completed, task 10 cannot yet be considered. The same procedure is repeated for the next element in the list, task 11, with the same result. As we go down the list, task 1 is the first feasible task to be encountered, and it is assigned to the station list. We delete all reference to task 1 in the predecessor list. In reality, these operations are performed on the same table; however, for clarity, we show the results in Table 4.14.

Next, we return to the top of the list, that is, tasks 10 and 11. Task 2 is the next feasible task found with a cumulative time of 1.0 or less, namely, 0.22 + 0.70 = 0.92.

It is therefore included with task 1 in the station. Inasmuch as no other task can be added to that station without exceeding the cycle time, we proceed to step 5 and create a new station. Task 3 has had its preceding task scratched, and its time is less than the cycle time; therefore, it is assigned to station 2. The procedure continues until all the tasks have been assigned. The final result is shown in Table 4.15.

**TABLE 4.14**
**Task Time After Assignment of Task 1**

| Task | Task Time | Immediate Predecessor(s) |
|---|---|---|
| 10 | 0.80 | 7, 8 |
| 11 | 0.70 | 9, 10 |
| 2 | 0.70 | $\chi$ |
| 7 | 0.63 | 3, 5 |
| 8 | 0.52 | 4 |
| 5 | 0.48 | 2, 3 |
| 3 | 0.42 | $\chi$ |
| 4 | 0.38 | $\chi$, 3 |
| 9 | 0.34 | 6, 7 |
| 1 | 0.22 | — |
| 6 | 0.12 | 5 |

We need seven stations with this arrangement. The actual cycle time is 0.92, the longest time among the stations, in this case the time required by station 1. One measure of efficiency ($e$) is $(1 - p) \times 100$, where $p$ is defined as

$$\frac{\text{(Total worker time required/unit)} - \text{job time/unit}}{\text{Total worker time required/unit}}$$

is known as balanced delay.

**TABLE 4.15**
**Final Task Assignments**

| Station | Element | Task Time | Cumulative Task Time |
|---|---|---|---|
| 1 | 1 | 0.22 | 0.22 |
| | 2 | 0.70 | 0.92 |
| 2 | 3 | 0.42 | 0.42 |
| | 5 | 0.48 | 0.90 |
| 3 | 7 | 0.63 | 0.63 |
| | 6 | 0.12 | 0.75 |
| 4 | 4 | 0.38 | 0.38 |
| | 8 | 0.52 | 0.90 |
| 5 | 10 | 0.80 | 0.80 |
| 6 | 9 | 0.34 | 0.34 |
| 7 | 11 | 0.70 | 0.70 |

In this case, seven stations are used, each worked by one person, with a cycle time of 0.92 minute. Therefore, efficiency is

$$e = 1 - \left[\frac{(7 \times 0.92) - 5.31}{7 \times 0.92}\right] \times 100$$

$$= 82.5\%$$

$$\text{or simply } 5.31(7 \times .92) = 82.5\%$$

#### *4.4.2.1.2 Procedure II — Ranked Positional Weight Method*

In the previous method, the tasks were listed in decreasing magnitude of time requirements. In this method, they are ranked according to their importance to the completion of all the tasks that depend on them. The importance is measured by the ranked positional weight (RPW) of each element, which is the sum of the times for all elements that directly follow it in the precedence diagram plus the time for the particular task itself. For example, the RPW for element 5 is the sum of the times for elements 6, 7, 9, 10, and 11 plus the time for element 5 — that is, 0.12 + 0.63 + 0.34 + 0.80 + 0.70 + 0.48 = 3.07.

Once the elements have been listed in descending order of RPW along with their predecessor elements (as shown in Table 4.16), the procedure for developing an assembly line is exactly the same as for procedure I (steps 2–6).

Carrying out steps 2–6 of procedure I (largest candidate rule) results in the final assignments listed in Table 4.17.

Seven stations are still required, but the cycle time is increased to 0.98 minute, the cumulative time associated with station 4. The efficiency of this arrangement is

$$e = 1 - \left[\frac{(0.98 \times 7) - 5.31}{7 \times 0.98}\right] \times 100$$

$$= 77.4\%$$

The present production rate is 400/0.98 ≈ 408 units/day.

**TABLE 4.16**
**Ranked Positional Weight Listing**

| Task | Ranked Positional Weight | Task Time | Immediate Predecessor(s) |
|---|---|---|---|
| 1 | 5.31 | 0.22 | — |
| 3 | 4.39 | 0.42 | 1 |
| 2 | 3.77 | 0.70 | 1 |
| 5 | 3.07 | 0.48 | 2, 3 |
| 7 | 2.47 | 0.63 | 3, 5 |
| 4 | 2.40 | 0.38 | 1, 3 |
| 8 | 2.02 | 0.52 | 4 |
| 10 | 1.50 | 0.80 | 7, 8 |
| 6 | 1.16 | 0.12 | 5 |
| 9 | 1.04 | 0.34 | 6, 7 |
| 11 | 0.70 | 0.70 | 9, 10 |

**TABLE 4.17**
**RPW Assignments**

| Station | Element | Task Time | Cumulative Task Time |
|---|---|---|---|
| 1 | 1 | 0.22 | 0.22 |
| | 3 | 0.42 | 0.64 |
| 2 | 2 | 0.70 | 0.70 |
| 3 | 5 | 0.48 | 0.48 |
| | 4 | 0.38 | 0.86 |
| | 6 | 0.12 | **0.98** |
| 4 | 7 | 0.63 | 0.63 |
| | 9 | 0.34 | 0.97 |
| 5 | 8 | 0.52 | 0.52 |
| 6 | 10 | 0.80 | 0.80 |
| 7 | 11 | 0.70 | 0.70 |

It should be noted that these heuristic procedures develop task assignments that might not be very satisfactory, as is the case here with an efficiency of 77.4%. One way to increase the efficiency is to try the procedure again with a different cycle time. For example, the cycle time of 0.9 minute results in the task distributions shown in Table 4.18.

Assembly line operations in tractor production

With an efficiency of 88%, the production rate is now increased to 400/0.86 ≈ 465 units/day.

It is also possible to increase the cycle time, thus reducing the production rate but increasing the efficiency by decreasing the number of required stations.

Assembly line operations in car production

### 4.4.3 Parallel Grouping of Stations

In some instances, the required production rate can necessitate having a cycle time that is even less than the time required for the completion of one of the tasks. For instance, in the previous case, if the output needed was 800 units per day, the cycle time would have to be 400/800 = 0.50 minute. Tasks 10, 11, 2, 7, and 8 require more time than that for their completion.

In such an instance, the previous methods, in which the stations were operated in series only, would have to be modified to allow for a parallel setup.

When two or more stations are in parallel and all are assigned to perform the same tasks, the permitted time for completion of the tasks in a station is longer than

**TABLE 4.18**
**Final Task Assignment**

| Station | Elements Assigned to the Station | Cumulative Station Time |
|---|---|---|
| 1 | 1, 3 | 0.64 |
| 2 | 2 | 0.74 |
| 3 | 5, 4 | 0.86 |
| 4 | 7, 6 | 0.85 |
| 5 | 8, 9 | 0.86 |
| 6 | 10 | 0.80 |
| 7 | 11 | 0.70 |

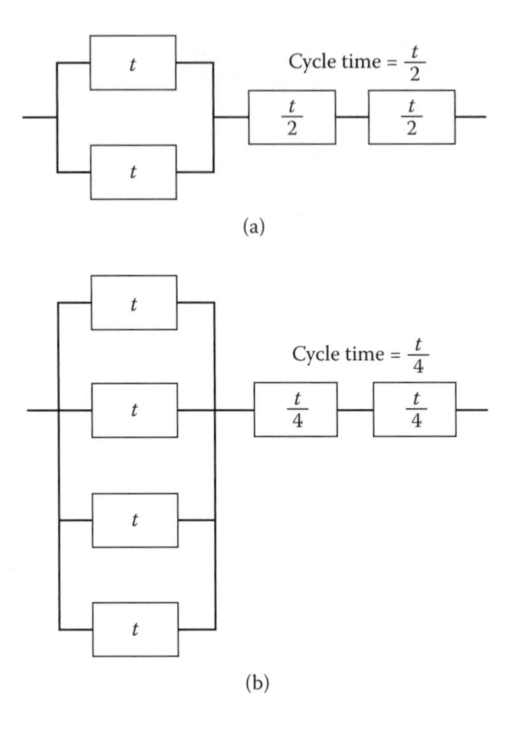


**FIGURE 4.14** Parallel stations

the cycle time. As an example, suppose there are two stations in parallel, as shown in Figure 4.14a, each requiring time $t$ to perform the tasks assigned to it; then two units are produced per $t$ time interval. Other stations in series may therefore operate with $t/2$ as their cycle time. Similarly, if there are $n$ stations parallel to each other, the cycle time for the remaining stations could be $t/n$, where the cycle time for the stations in parallel is $t$.

A question might be raised as to why we do not have six parallel stations in our example each performing all operations. This arrangement should obtain a production rate that is better than the one per minute needed and at the same time would reduce the labor requirement from seven to six. There are several reasons why such a grouping may not be possible. An additional task assigned to a workstation might mean additional investment at that station for the machines and tools that are necessary to perform the task. In multiple stations, the same group of equipment would have to be available at each station, an investment that could be prohibitive. The task time could increase because, as the work area is enlarged, the worker must move farther to perform more tasks. Workers with more experience and training would be needed to perform all tasks. A larger floor space would generally be required if the same tasks were to be repeated in each station.

It should be recognized that in a practical layout of an assembly line, two parallel stations need not be physically parallel. They can, for example, be two successive stations fed by a common conveyor. It is possible to dedicate alternate units on the same conveyor to each of the stations in various ways, from using a sophisticated mechanical, push/pull arrangement to simply marking alternate positions on the conveyor by using differently colored carriages such as trays.

### 4.4.4 Computerized Assembly Line Programs

Numerous computer programs have been designed for assembly line balancing. In most cases, each was developed for a particular situation, but all are based on one of the following techniques.

- Enumerating all possible sequences
- Selecting the best from randomly chosen sequences
- Selecting a sequence that is similar to one already used in the company
- Applying a mathematical programming technique such as linear programming, integer programming, or the branch-and-bound method
- Using heuristic rules, such as the two we have just seen
- Applying rules that are specifically developed for the problem, possibly including among other items, restrictions on tasks and assignments among the tasks

#### 4.4.4.1 Applicable Programs by Others

It is beyond the scope of this chapter to go into the details of all known computer assembly line balancing programs. The following are some of the more popular programs.

- NULISP: Nottingham University Line Sequencing Program, developed by Nottingham University in 1978
- ALPACA: Assembly Line Planning And Control Activity, developed by General Motors in 1967
- GALS: Generalized Assembly Line Simulator, developed by ITT Research Institute in 1968
- COMSOAL: Computer Method of Sequencing Operations for Assembly Line, developed by Chrysler Corporation in 1966

## 4.5 COMPUTER PROGRAM DESCRIPTION

Four programs for this chapter are included on the program disk. For assembly line balancing, the programs include one on the largest candidate rule and one on the RPW method. The third and fourth programs are associated with forming the groups for flexible manufacturing systems. All programs are in interactive mode, and data input is displayed on the screen.

### 4.5.1 Assembly Line Balancing (in GW-BASIC)

#### 4.5.1.1 Largest Candidate Rule

The input is in interactive mode. The questions asked are:

1. *What is the required cycle time*? (Input the desired cycle time.)
2. *How many tasks are to be assigned*? (Input the total number of tasks.)
3. *What is the time for task I/number of predecessors for task I*? (Enter individually the task time and the number of predecessors for each task.)
4. *Task J must be preceded by the following task; give the # K predecessor for task J*. (Enter individually each predecessor task number for task *J*.)

The output consists of: (1) each station's task assignment, duration of each assigned task, and total station time; (2) the cycle time for the balanced assembly line; and (3) the efficiency of the line.

#### 4.5.1.2 RPW Method

The data can be entered into the program interactively or by data statements. An answer to the first question below determines the mode.

1. *If you want interactive mode, enter number 1.*
2. *If you want data mode, enter number 0.*

If the data are to be entered interactively, use the same information and procedure as in the case of the largest candidate rule. If the information is entered in data mode, it should be typed as data statements in the same sequence as one would in interactive mode.

### 4.5.2 GT Computer Programs

Two programs (GROUP and LARGROUP) that form the machine grouping and assign components to the groups are included on the program diskette. In both programs, input can be from a keyboard or through a file, When using the keyboard, press RETURN or ENTER to indicate entry completion for each data field. Input can be stored and reused or modified for the next pass. The number of options is given on the screen. The GROUP program is meant for a smaller-size problem (the actual size depends on the screen width of the monitor), the output of which is displayed on the screen. The hard copy can be obtained by using the PRINT SCREEN command.

The LARGROUP program can handle a large-size problem, where the output is stored on a file specified by the user. The results can be documented by using a PRINT file statement.

#### 4.5.2.1 GROUP

The program GROUP will group machines and assign jobs using the algorithm described in Section 4.3. There is a modification included in the program on the steps described in the algorithm. The program stops the grouping procedure when *all* machines are assigned to at least one group. This happens, in most cases, before RC assumes its lowest possible value. This stopping rule keeps machine duplications to a minimum. From the resulting solution we can easily see, if need be, what additional duplicating or cell-separating machines are necessary to keep the cells completely separated.

Data can be recorded in the program from a file or through a keyboard. The data file may be a file that was previously entered through a keyboard and then stored in the memory or a file created by a spreadsheet, such as Excel 123. The machine-component data table must be keyed in so that jobs or components are in rows and machines are in columns. In each job, for each machine 1 indicates use of the machine and 0 or a blank indicates nonuse of the machine. When entering data via the keyboard, the cursor will prompt for an answer for each job under each machine. Enter 1 or leave blank as appropriate and then press ENTER. Other data prompts are self-explanatory.

A data correction feature is provided, but it is essential that the information for job and machine be given correctly. The cursor will go to the job/machine position in the machine-component matrix for the user to enter either a 1 or a blank. If the information on the job or machine was incorrectly entered in the correction step, it is still advisable to go ahead and enter the appropriate blank or 1 at that cursor position rather than to try to correct the job/machine information at this stage. This will prevent the screen from rolling. Once the screen rolls, the cursor loses its ability to find the right slot, so information will not be recorded correctly. The correct information on job and machine can be entered next time when the opportunity to make the correction is offered again.

The output is displayed in Figure 4.15. First, note that all machines were assigned in iteration 13 in Table 4.6; that is where the program has stopped. The computer solution therefore does not include the duplication of machines from iteration 14 on, namely, machines A, B, and E in group 1 and machines B and G in group 3. As a result, job 12 is assigned to group 3 rather than group 1. Job 19 requires two machines from group 2 and two from group 3, and could therefore be assigned either to group 2 or 3. In the computer output, it is assigned to group 2. The computer output also has an asterisk below a job that requires between-cell transportation. The intergroup machine table displays the machines that are needed from other cells for a job. For example, job 11, which is assigned to group 1, requires machine 1 (machine A), but it is not available in group 1. Machine 1 is available in group 3.

INPUT THE MACHINE-COMPONENT MATRIX

| | MACHINE NUMBER | | | | | | | | | | |
|---|---|---|---|---|---|---|---|---|---|---|---|
| JOB NUMBER | 1 | 2 | 3 | 4 | 5 | 6 | 7 | 8 | 9 | 10 | 11 |
| 1 | 0 | 0 | 1 | 1 | 0 | 0 | 0 | 0 | 1 | 1 | 0 |
| 2 | 0 | 0 | 1 | 1 | 0 | 0 | 0 | 0 | 1 | 1 | 0 |
| 3 | 0 | 0 | 1 | 1 | 0 | 0 | 0 | 0 | 1 | 1 | 0 |
| 4 | 0 | 0 | 0 | 0 | 0 | 1 | 0 | 1 | 0 | 0 | 1 |
| 5 | 1 | 1 | 0 | 0 | 1 | 0 | 1 | 0 | 0 | 1 | 0 |
| 6 | 0 | 1 | 0 | 0 | 0 | 1 | 1 | 0 | 0 | 0 | 0 |
| 7 | 0 | 0 | 0 | 1 | 0 | 0 | 0 | 0 | 1 | 1 | 0 |
| 8 | 1 | 0 | 0 | 0 | 1 | 0 | 0 | 0 | 0 | 1 | 0 |
| 9 | 0 | 0 | 0 | 0 | 0 | 1 | 1 | 1 | 0 | 0 | 1 |
| 10 | 0 | 0 | 0 | 0 | 0 | 0 | 1 | 0 | 0 | 0 | 1 |
| 11 | 1 | 0 | 1 | 0 | 0 | 0 | 0 | 0 | 0 | 1 | 0 |
| 12 | 1 | 1 | 0 | 0 | 1 | 0 | 1 | 0 | 1 | 1 | 0 |
| 13 | 0 | 1 | 0 | 0 | 0 | 0 | 1 | 0 | 0 | 0 | 0 |
| 14 | 0 | 0 | 0 | 0 | 0 | 1 | 1 | 1 | 0 | 0 | 0 |
| 15 | 0 | 0 | 1 | 1 | 0 | 0 | 0 | 0 | 0 | 1 | 0 |
| 16 | 0 | 0 | 1 | 1 | 0 | 0 | 0 | 0 | 0 | 1 | 0 |
| 17 | 0 | 0 | 0 | 0 | 0 | 1 | 1 | 1 | 0 | 0 | 1 |

DO YOU WANT TO MAKE CHANGES (Y/N)? N Percentage = 0.5

**FIGURE 4.15** Computer output

To get a hard copy of the screen display, press SHIFT and PrtSc simultaneously. The input data for the problem illustrated in Section 4.3 are:

#### 4.5.2.2 LARGROUP

The program LARGROUP should be used if the job-machine matrix is large and the output cannot be displayed on the screen because of screen-width limitation. The output of the program must be stored in a file (e.g., a: output), and a hard copy is obtained by a print statement (e.g., print a: output). The output for the same example is displayed in Figure 4.16.

## 4.6 SUMMARY

Working with production charts is a manager's way of visualizing a manufacturing process. The charts allow one to design a sequence of operations that will produce a unit efficiently and economically. These charts evaluate the process in terms of movement and length of time required for a task and permit examination of each operation

Machine Number

| JOB NUMBER | 1 | 2 | 3 | 4 | 5 | 6 | 7 | 8 | 9 | 10 | 11 |
|---|---|---|---|---|---|---|---|---|---|---|---|
| 1 | | | | | | | | | | | |
| 2 | | | | | | | | | | | |
| 3 | | | | | | | | | | | |
| 4 | | | | | | | | | | | |
| 5 | | 1 | | | | | 1 | | | | |
| 6 | | | | | | | | | | | |
| 7 | | | | | | | | | | | |
| 8 | | | | | | | | | | | |
| 9 | | | | | | | | | | | |
| 10 | | | | | | | | | | | |
| 11 | 1 | | | | | | | | | | |
| 12 | | 1 | | | | | 1 | | 1 | | |
| 13 | | | | | | | | | | | |
| 14 | | | | | | | | | | | |
| 15 | | | | | | | | | | | |
| 16 | | | | | | | | | | | |
| 17 | | | | | | | | | | | |

Press Any Key To Continue Percentage = 0.5

Machine Number Inter Group Machining Table

| JOB NUMBER | 1 | 2 | 3 | 4 | 5 | 6 | 7 | 8 | 9 | 10 | 11 |
|---|---|---|---|---|---|---|---|---|---|---|---|
| 18 | | | | | | | | | | | |
| 19 | 1 | | | | 1 | | | | | | |
| 20 | 1 | | | | | | | | | | |
| 21 | 1 | | | | | | | | | | |

**FIGURE 4.15** (Continued)

to see whether it is functioning efficiently. With a questioning mind, an analyst may make the necessary improvements should the operation be deemed unsatisfactory. Various charts are commonly used in industry, but among them the most common are assembly and operation process charts; routing sheets; left-hand, right-hand charts; gang charts; Gantt charts, and flow process charts.

A production arrangement within a plant may be designed to give different degrees of flexibility and production rates. A job shop arrangement is highly flexible and is suitable when a plant produces many different products, each in relatively small quantity. The efficiency in a job shop, however, is relatively low. The other extreme is an assembly line arrangement where each operation is carefully designed and balanced to provide maximum production rate. The sequential stations in an assembly line perform the next required operation on the job as the

MACHINES ASSIGNED TO GROUP 1

3 4 10 9

COMPONENTS ASSIGNED TO GROUP 1
1
2
3
7
11
15
16
20

MACHINES ASSIGNED TO GROUP 2

2 7 6 8 11

COMPONENTS ASSIGNED TO GROUP 2
4
6
9
10
13
14
17
18
19

MACHINES ASSIGNED TO GROUP 3

1 5 10

COMPONENTS ASSIGNED TO GROUP 3
5
8
12

**FIGURE 4.16** Output for example

item moves from one station to the next on a fast-moving material-handling system such as a conveyor belt. An assembly line is generally designed for a single product and often has very high efficiency, but very little flexibility. A batch production system falls between a job shop and an assembly line in its complexity and efficiency.

With cellular and flexible manufacturing, a production system can be developed that has the production rate of an assembly line, yet the flexibility of a job shop. Application of GT is a key ingredient in forming appropriate machine cells for cellular manufacturing, and the tabular method of analysis is very helpful in this regard. Arranging the cell in a U-shaped form has many advantages; it provides flexibility for worker movement and provides the opportunity for machine coupling. Flexible manufacturing is cellular manufacturing that is highly automated, utilizing NC machines, robots, and a computerized material-handling system, and using minimum manpower. Flexible manufacturing requires a fairly large initial investment.

The chapter concludes the analysis of systems by describing two labor assignment models: one for a production line and another for an assembly line. An efficient tabular method is presented which makes it easy to determine the required manpower at each station when the scheduled production rate is known.

Two methods are described for balancing an assembly line: (1) the largest candidate rule and (2) the RPW method. Any problem should be solved by both methods, and the solution giving the greatest efficiency should be selected for planning. Occasionally, cycle times and production rates may be adjusted to increase the efficiency. Parallel stations are needed when a cycle time is less than the sum of the task times; however, there may be a practical limit to the number of parallel stations.

## PROBLEMS

1. Identify the following symbols:
   a. ○
   b. ▽
   c. □
2. Develop an assembly chart for making a pot of coffee.
3. How are time estimates used?
4. Use standard data tables to estimate a standard time for
   a. Replacing a headlight in your car
   b. Sharpening a pencil
5. Describe the left-hand, right-hand chart. What is the objective in using the chart to design a workstation?
6. Develop a left-hand, right-hand chart for putting on and tying a tennis shoe.

7. How are production systems classified? What factors influence a production system?
8. Define the following:
   a. Job shop production
   b. Batch production
   c. Mass and continuous production
   d. Cellular and flexible manufacturing
9. Using the production schedule in the accompanying table, develop a final production matrix and cell arrangement. Use $P = 0.7$

| Item | Required Machines |
|---|---|
| 1 | A, C, E, G |
| 2 | B, D |
| 3 | D, F, G |
| 5 | F, G |
| 6 | C, F |
| 7 | A, B, C, E, G |
| 8 | D, F |
| 9 | A, B, C |
| 10 | C, D, E |
| 11 | D, E, F |
| 12 | A, B, C, D |
| 13 | C, D, E, F |
| 14 | B, C, D, E |
| 15 | A, B |
| 16 | B, G |

10. a. Obtain machine groupings and job assignments for a 9-machine 10-job problem with the following data. Use $P = 0.6$.

| Jobs | Machines Used |
|---|---|
| 1 | 1, 3, 4 |
| 2 | 1, 2, 3 |
| 3 | 3, 4 |
| 4 | 8, 9 |
| 5 | 5, 8, 9 |
| 6 | 5, 7 |
| 7 | 5, 6, 7 |
| 8 | 2, 3, 4 |
| 9 | 1, 2 |
| 10 | 5, 7 |

b. How does the grouping change if $P = 0.9$?

c. Transferring jobs between cells costs \$1000/job/month. The cost of within-cell material handling and administration per month depends on the number of machines and jobs assigned to the cell and is approximated by the formula:

$$\text{Cost/cell} = 500 + 200 \times \text{No. of jobs assigned to the cell} + 100 \times \text{no. of machines assigned to the cell}$$

11. The following data represent the production sequence for 16 jobs. As can be noted, 13 machines are used in the production.

| Job | Machines |
|---|---|
| 1 | 1, 3, 5, 7, 10, 12 |
| 2 | 2, 4, 5 |
| 3 | 1, 3, 6, 11 |
| 4 | 2, 4, 8, 9, 10, 13 |
| 5 | 1, 2, 3, 5, 7, 9, 11, 12 |
| 6 | 1, 6, 13 |
| 7 | 1, 5, 7, 11, 12 |
| 8 | 5, 9, 13 |
| 9 | 1, 11 |
| 10 | 1, 6, 7, 10, 12 |
| 11 | 3, 4, 5, 9, 13 |
| 12 | 2, 7, 10, 12 |
| 13 | 5, 6, 7 |
| 14 | 3, 9, 11 |
| 15 | 2, 5, 7, 8 |
| 16 | 1, 5, 10, 12 |

Transportation cost for a job between any two cells is \$500. Machine costs (in thousands) are shown below.

| Machine | 1 | 2 | 3 | 4 | 5 | 6 | 7 |
|---|---|---|---|---|---|---|---|
| Cost | 8 | 5 | 50 | 15 | 6 | 5 | 5 |
| Machine | 8 | 9 | 10 | 11 | 12 | 13 | |
| Cost | 6 | 8 | 10 | 8 | 5 | 7 | |

a. Apply the tabular approach with $P = 0.5$ and $P = 0.7$ (use the computer program GROUP) and determine the machine cells and job assignments. (In the computer solution, some machines may be supplanted by other machines of the same type in different groups, because no job is assigned to such machines. These machines should be eliminated from any further consideration. Sometimes, as in this case for $P = 0.7$, the entire group may be eliminated with this step. The step is called *machine/group consolidation*.)

b. Determine the cost of each solution.
c. Identify the primary, duplicated, and cell-separating machines in the solution for $P = 0.5$.
d. Suppose only one machine of each type is required to process all jobs. After purchasing the basic machines, $50,000 is available to purchase additional machines. Which additional machines should be acquired to minimize the intercellular transportation of the jobs in the solution associated with $P = 0.5$?
e. Machine 3 is very expensive and hence only one unit of the machine is to be purchased. Assuming that the capacity of one unit is adequate to support all jobs, in which cell should the machine be located? (Use $P = 0.7$.)
f. In the solution with $P = 0.7$, machine 1 is serving jobs with the following capacity.

| | | | | |
|---|---|---|---|---|
| Cell | 1 | 1 | 1 | 1 |
| Job | 1 | 5 | 7 | 10 |
| Capacity | 0.05 | 0.05 | 0.1 | 0.05 |
| Cell | 1 | 2 | 2 | 4 |
| Job | 16 | 3 | 9 | 6 |
| Capacity | 0.1 | 0.2 | 0.1 | 0.3 |

The cost of intercellular transport of any job is $2000 per 0.05 unit of capacity transferred. Determine the optimum number of machines and their placement.

12. Develop an assembly line to produce one chair every 20 minutes for the job described by the schematic diagram in Figure 4.17. The time in minutes required for each element is indicated under the element.
13. On what factors does the rate of production depend?
14. An electronics plant is to produce 500 calculators at 85% efficiency by one line. On the basis of the accompanying table, develop the balanced workstation configurations required to meet this production goal.

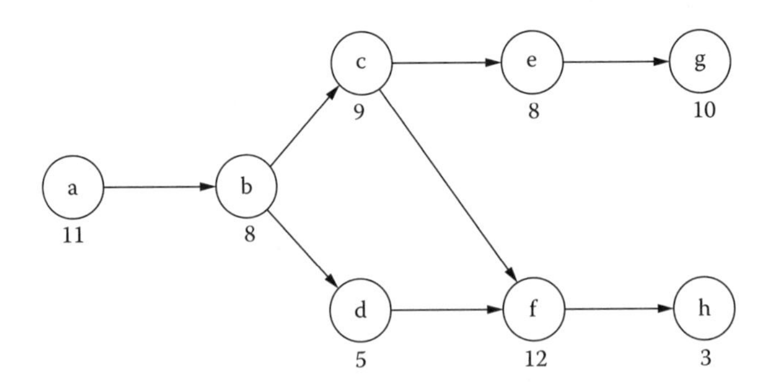


**FIGURE 4.17**

| Task | Time (in minutes) | Immediate Predecessor task(s) |
|---|---|---|
| 1 | 1.10 | — |
| 2 | 0.45 | — |
| 3 | 0.38 | 1 |
| 4 | 0.25 | 2, 3 |
| 5 | 2.50 | 4 |
| 6 | 0.50 | 5 |
| 7 | 0.10 | 5 |
| 8 | 0.05 | 6, 7 |

15. Plan a balanced assembly line for a spring scale production department on the basis of the accompanying table. The plant averages 92% efficiency and would like to produce 750 units/day.

| Element | Time | Immediate Predecessor(s) |
|---|---|---|
| 1 | 0.15 | — |
| 2 | 0.27 | — |
| 3 | 0.32 | — |
| 4 | 0.45 | 1, 2 |
| 5 | 0.30 | 2, 3 |
| 6 | 0.19 | 3 |
| 7 | 0.24 | 4 |
| 8 | 0.38 | 4, 5 |
| 9 | 0.17 | 5 |
| 10 | 0.42 | 5, 6 |
| 11 | 0.21 | 6 |
| 12 | 0.35 | 7, 8 |
| 13 | 0.25 | 9 |
| 14 | 0.44 | 10, 11 |
| 15 | 0.33 | 11 |
| 16 | 0.20 | 12, 13 |
| 17 | 0.40 | 14, 15 |
| 18 | 0.22 | 16, 17 |

16. Use Figure 4.18.
    a. Develop the labor requirement to produce 12 units each from stations 7 and 8. Assume the initial inventory after each station from the accompanying table.

| Station | 1 | 2 and 3 | 4 | 5 | 6 |
|---|---|---|---|---|---|
| Inventory in units | 5 | 8 | 4 | 3 | 6 |

    b. Develop a daily work schedule if the number of workers is limited to five.
    c. Develop a daily work schedule if the number of workers is limited to two.

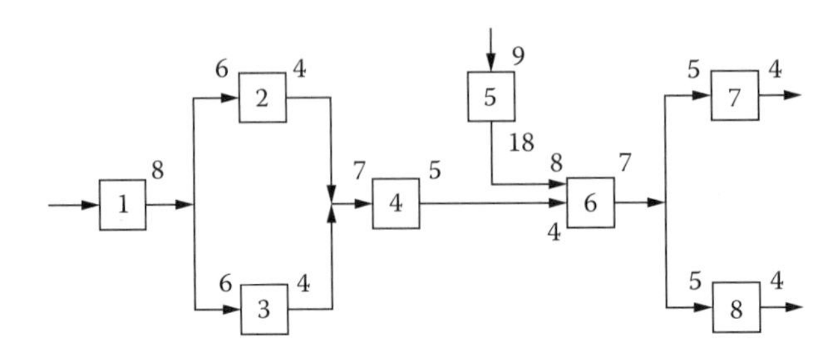


**FIGURE 4.18**

## SUGGESTED READINGS


### Charts

Galgut, P.E., *Production Planning and Control*, Elsevier, New York, 1987.

Muther, R., and Hale, L., *Systematic Planning of Industrial Facilities*, Vols. 1 and 2, Management and Industrial Research Publications, Kansas City, 1979.

### Production Systems

Burbidge, J.L., "Production flow analysis," *The Production Engineer,* No. 50, 1971, p. 139, *Industrial Engineering,* 15(11), 1983.

Kusiak, A., *Modeling and Design of Flexible Manufacturing Systems*, Elsevier, New York, 1986.

Seifoddini, H.K., and Wolfe, M.P., "Application of the similarity coefficient method in group technology," *IIE Transaction,* 18(3), 221–277, 1986.

Seifoddini, H.K., "Comparison between single linkage and average linkage clustering techniques in forming machine cells," *Proceedings of the 10th Annual Conference on Computers and Industrial Engineering*, 210–216, 1988.

Sule, D.R., "A systematic approach for machine grouping in cellular manufacturing," Toronto *International Industrial Engineering Conference and Societies Manufacturing and Productivity Symposium Proceedings*, 619–624, 1989.

Tabucannon, M., and Ojha, R., "ICRMA — A heuristic approach for intercell flow reduction in cellular manufacturing systems," *Material Flow,* No. 4, 187–197, 1987.

Vannelli, A., and Kumar, K.R., "A method for finding minimal bottleneck cells for grouping part-machine families" *International Journal of Production Research,* 24(2), 387–400, 1986.

### Operations Research

Taha, H.A., *Operations Research: An Introduction*, 8th edition, Prentice Hall, Upper Saddle River, NJ, 2006.

Winston, W.L., *Operations Research: Application and Algorithms,* 4th edition, Duxbury Press, Boston, 2003.

### Labor Assignments

Barnes, R.M., *Motion and Time Study Design and Measurement of Work*, 9th edition, John Wiley, New York, 1980.

Bowman, E.N., "Assembly-line balancing by linear programming," Operations Research, 385–389, 1960.

Held, M., Karp, R.M., and Shareshian, R., "Assembly line balancing dynamic programming with precedence constraints," Operations Research, 442–459, 1963.

Sadowski, R.P., "Manpower scheduling method simplifies production line assignment through graphics," Industrial Engineering, 13(8), 34–41, 1981.

Sule, D.R., "Manpower assignment on a production line using tabular approach," Industrial Management, 25(1), 11–13, 1983.

Talbot, F.B., and Patterson, J.H., "An integer programming algorithm with network cuts for solving the assembly line balancing problem," Management Science, 30(1), 85–99, 1984.

Tonge, F.M., *A Heuristic Program of Assembly Line Balancing*, Prentice Hall, Englewood Cliffs, NJ, 1961.

# 5 Requirements and Selection of Machines and Labor

Machines and labor are the backbone of an industry, and their selection requires careful evaluation of what is needed and what is available. This chapter describes the systematic steps involved in specifying machine and labor requirements. The description is divided into three main topics: machine selection, labor selection, and machine coupling.

## 5.1 MACHINE SELECTION

Machines are an integral part of manufacturing, yet seldom do we find an engineer developing production arrangements who also possesses expertise in machines. Some basic data such as types of machines, the names of suppliers, the range of costs, and the associated capacities are needed to begin the machine selection process. Where can we obtain this information? How can we use it? These are the types of questions that we answer in the following sections.

### 5.1.1 MAKE-OR-BUY ANALYSIS

Before one proceeds to select the equipment that is essential for manufacturing and assembling the designed product, it is necessary to perform a preliminary make-or-buy analysis to determine what can and should be produced in the plant. The analysis is preliminary because the data required for the decision are based for the most part on experience and not necessarily on complete production knowledge. At some later date, further analysis might be made that will reflect more reliable data along with a better understanding of the manufacture and sale of the product itself.

The bill of materials (Figure 2.12) provides information on the parts that are necessary to make a finished product. The questions the engineer should ask are: which of these parts should be made in the plant and which ones should be bought from outside vendors. Two determining factors might suggest the answers: expertise and economics.

Parts that require expertise in manufacturing technology other than what the firm possesses should be bought from vendors who have technical know-how in such fields. It is a common practice in the automobile industry, for example, to buy tires from tire manufacturers. Bottling plants that dispense soft drinks buy their bottles from glass manufacturers. Table lamp manufacturers hardly ever attempt to produce light bulbs, without which their products will not function.

The economics of manufacturing also plays an important part in this decision. Production in a small quantity generally results in a larger unit cost than if the same parts are produced in large volumes. It might be cheaper to purchase 2000 units of 1/2-inch springs than to buy the equipment and use the necessary labor to produce them in the plant. An economic analysis such as the one shown in the following example illustrates this point further.

## EXAMPLE—COMPONENT PART SOURCE SELECTION

A prospective kitchen blender manufacturer has a design that requires hard plastic connecting gears between the electric motor and the cutting blade assembly. As shown in Table 5.1, there are three alternatives for obtaining such parts.

### *Alternative A*

A molding specialty house can supply the parts for a price of $500 per thousand units. The price includes the cost of designing and building the tools necessary to manufacture the gears; however, the minimum-order quantity is 20,000 units. The company must also spend $2000 on an engineering effort to review the design before allowing the supplier to begin production.

### *Alternative B*

Plant engineers can design, build, and perform initial testing of a single-cavity tool for $50,000. The gears can then be manufactured in the plant on a small automatic mold press at a cost of $200 per thousand. The unit costs include all of the variable costs: labor and material, as well as all normal overhead operating costs prorated per unit of output.

### *Alternative C*

It is also possible to design and build a multiple-cavity tool for $100,000. This tool would be designed to run on a larger automatic mold press at a cost of $150 per thousand units.

Determine the preferred alternative, given a specific requirement level. Assume that the period over which production will be required is short enough to eliminate the need to consider the time value of money.

**TABLE 5.1**
**Data for the Three Alternatives**

| Alternative | Initial Cost | Cost per 1000 |
|---|---|---|
| A: Purchase mold tool, minimum order of 20,000 | $2000 | $500 |
| B: Manufacture with a single-cavity tool | $50,000 | $200 |
| C: Manufacture with a multiple-cavity tool | $100,000 | $150 |

Neglecting the time value of the money, the break-even point ($Y_1$) for purchasing the parts versus molding them with a single-cavity tool can be computed from the above data:

$$\$2{,}000 + \$500 \times Y_1 = \$50{,}000 + 200 \times Y_1$$

$$\$300 \times Y_1 = \$48{,}000$$

$$Y_1 = 160 \text{ thousand parts}$$

The break-even point ($Y_2$) for molding the parts in a single-cavity tool versus a multiple-cavity tool is

$$\$50{,}000 + \$200 \times Y_2 = \$100{,}000 + \$150 \times Y_2$$

$$\$50 \times Y_2 = \$50{,}000 Y_2$$

$$Y_2 = 1000 \text{ thousand (1 million) parts}$$

It is cheaper to purchase quantities of parts up to 160,000 units from a supplier. From 160,000 to 1,000,000 parts, it is better to build a single-cavity tool and mold the parts in the factory; to produce more than 1,000,000 parts, a multiple-cavity tool should be used (Figure 5.1).

Another important aspect of the analysis is the availability of initial capital. If it is not possible to raise $100,000 immediately, alternative C is not feasible. In this case, even though production of more than 1,000,000. Parts would be cheaper with alternative C. Only the single-cavity mold, alternative B, could be used.

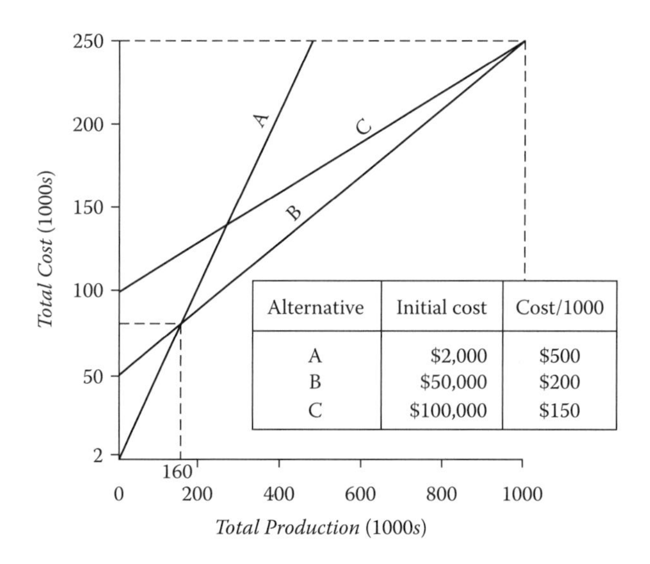


| Alternative | Initial cost | Cost/1000 |
|---|---|---|
| A | $2,000 | $500 |
| B | $50,000 | $200 |
| C | $100,000 | $150 |

**FIGURE 5.1** Total cost versus production level (molded plastic parts)

### 5.1.2 Sources of Information

Once the manufacturer has decided which parts are to be produced in the plant, the next question concerns which machines should be used in the production. This complex and important chore of machine selection will determine how efficiently and economically the unit can be produced. Machine cost contributes a large portion of the fixed cost that cannot be easily changed. Where and how can we obtain the necessary information about the machines to make a selection that meets our plant and product needs? Fortunately, there are many sources of such information.

#### 5.1.2.1 Books and Periodicals

Experienced engineers have a broad background in dealing with suppliers, processes, and materials and generally know where to go for additional information to expand, modify, or build new facilities. Buyers, production clerks, maintenance personnel, and fellow engineers can frequently furnish the names of current and recent suppliers of different types of equipment. In most cases, this knowledge is acquired from on-the-job training, associating with professional societies, and visiting trade shows. Trade journals and professional magazines that feature advertisements by suppliers provide considerable information about processes, materials, and equipment. Managers generally have access to information presented by current and new suppliers who are trying to demonstrate that their latest products are better than those currently being used.

The following discussion is intended to provide a means of locating a few additional records in many fields: information on professional societies, technical data on material and processes, and extensive listings of suppliers.

One of the best-known listings of suppliers is the *Thomas Register of American Manufacturers and Thomas Register Catalog File*, www.thomasnet.com, which is published annually by the Thomas Publishing Company (New York, NY). It contains several listings. The register has an alphabetical listing of products and services of almost 21,000 pages in 12 volumes. (Some advertising is also included in this section.) A list of company names, addresses, telephone numbers, brand names, and trademarks is included in the next two volumes. The last seven volumes contain catalogs from some of the companies listed in the first section. In addition, the publisher has recently added a section in the last volume that provides considerable information on how to import products using the most cost-effective methods. In total, the *Thomas Register* has a listing of 14,000 U.S. companies, 5000 Canadian companies, and 500 trademarks.

Several other directories and indexes are available on the Internet. These are of value when the concern is for sources of information about a product that is relatively new. These allow one to find what information is available about the product, who is trying to sell the equipment and materials, and what technical societies are involved in the field.

The IMS *Ayres Directory of Publications* contains a comprehensive listing of U.S. publications and a small number of foreign publications. The directory has several cross-indexes, including one of more than 100 pages that lists trade and technical publications. This directory has been published annually for many years and is available from IMS Press (Fort Washington, PA).

*Ulrich's International Periodicals Directory* (available in hard copy, or online at www.ulrichsweb.com) a lists domestic and foreign periodicals with more than 1000 pages of titles and subjects along with the current price of each. The search feature cross-index allows one to guess at the name of a publication or topic and, if successful, find a cross-reference to all other publications on the same topic. *Ulrich's International Periodicals Directory* also lists two other sources that may be of value in researching information: Abstracting and Indexing Services and Micropublishers (suppliers of microfilm databases). The directory is revised on a frequent basis by R.R. Bowker Company (New York, NY).

Another index of periodicals is the *Applied Science and Technology Index*, which is available at www.hwwilson.com/Databases/applieds.htm, and is updated daily online & monthly on CD by H.W. Wilson Company (Bronx, NY).

All these indexes and directories make it possible to take advantage of the research work that had already been done by others and to direct a manufacturer to information on the product of interest. Through the articles and advertisements in technical journals, one can gain access to the most current information on major aspects of the business, because these technical journals are one of the first places through which suppliers announce their new offerings. The magazines also list all the regional and national trade shows where additional information can be obtained.

One of the features of Ulrich's International Periodicals Directory is its listing of professional society journals. In addition to the informative articles and listings of suppliers, it provides opportunities to make contact with other individuals who might have similar interests.

#### 5.1.2.2 Computer Databases

Computer databases are a major source that can be accessed by search engines such as Google. These databases provide diverse technical data on almost any topic, including those of interest here: selecting locations, suppliers, processes, and equipment.

Dialog Information Services (Palo Alto, CA) is one company that currently provides computerized database systems. It has access to more than 200 databases containing more than 100 million items. Some of these provide a "Yellow Pages" service for trade and industry.

Some other databases include:

- *Economic Literature Index* (www.econlit.org) and *Trade and Industry Index* are useful in market research and the management areas of business, industry, and economics.

- *Donnelly Demographics* can be used in the area of statistical and demographic data.
- *Encyclopedia of Associations* and *Ulrich's International Periodicals Directory* provide listings of technical associations and periodicals.
- *Compendex* (engineering and technological literature abstract) and *SciSearch* supply scientific and technological information.

### 5.1.3 Production Arrangement

The next concern in machine selection is the production arrangement. Although the final layout of the plant might be unknown at this point, the anticipated production arrangement should be decided upon, because it has a major bearing on machine selection. If a single product is to be produced in mass quantity with an assembly line arrangement (sometimes referred to as product arrangement), highly specialized machines will be needed, each capable of performing a specific task with great speed and accuracy. The other extreme is the job shop arrangement (sometimes referred to as process arrangement), in which machines capable of working on multiple jobs are used to achieve flexibility.

The number of products assigned for production and their required production rates might also dictate the choice of plant layout and ultimately determine the machine selection.

### 5.1.4 Cost Consideration

The decision as to whether to undertake labor-intensive or capital-intensive production is another problem that must be addressed. At present, individual robots and groups of machines with fully automated material-handling systems are being used as attractive alternatives for reducing labor costs. Before purchasing expensive equipment to replace manual operations, however, a manufacturer should perform a careful analysis of real costs associated with automatic equipment: costs of design, purchase-or-build, prove-in, and maintenance. Most automatic processes take longer than expected to produce quality products and are expensive to modify if design changes are necessary.

In most cases, machines are available in different capacities with different operating speeds and feed rates. The available options also vary from one model to the next, and the initial investment and operating cost can differ considerably among the available alternatives. The initial investment is especially important, because it is money committed that cannot be easily recovered. Again, an economic analysis can suggest the best alternative.

### 5.1.5 Available Capacity

Machines can seldom be used at full capacity throughout the production period. There are several reasons why this is so.

The need for setup, preventive maintenance, tool sharpening, and unpredicted failures and repairs reduce the time available for production. Older machines are more susceptible to breakdown, resulting in lower productivity, than newer machines.

Operator-related factors, such as absence, personal time, and time spent in bookkeeping, machine adjustments, and material preparations, reduce machine availability.

As a result of quality requirements in some processes, a certain loss of production is inherent even though the equipment is properly set up and is being operated correctly. Some production loss is also incurred as units are scrapped when a machine malfunctions and produces parts of undesirable quality.

At times an operator is assigned the responsibility for many machines, a situation called machine coupling. This is made possible by dividing the total cycle time into two parts, machine time and worker time. Machine time is the time when the machine is operating without any assistance from the operator, while worker time is the time the operator spends with the machine when the machine is not working — for example, the time for load. If machine time is larger in comparison to worker time, it might be possible for the operator to load another machine while the first machine is in its machine time. One operator therefore can frequently be assigned responsibilities for several machines.

In such a system, a problem with one machine in the group can at times affect the production of all the other machines. An operator who is engaged in identifying and correcting the problem might have to neglect the other machines needing his/her time in their natural cycle of production.

Available time for production does not increase in the same proportion as the number of shifts; the addition of a second shift does not add another 8 hours of operation. The loss of production time increases because the free time available for repairs decreases. For example, in a single-shift operation, a machine that breaks down almost at the end of the shift would normally be repaired during the second (maintenance) shift if the required repair time were expected to be long, say 4 hours or more. Because the machine is now scheduled for operation, however, such repairs must be accomplished during time scheduled for production in the second shift. Often, factors such as shortage of material, absence of technical help, or absenteeism of production workers also affect the overall performance in the second shift. In general, more workers are absent from the second shift than the first, and more are absent from the third shift than the second.

Official holidays, such as Labor Day, New Year's Day, and Christmas, also reduce the available working hours during the year.

The loss of useful time is accounted for by many different allowances; setup time is measured or estimated for each type of operation and is incorporated into the time necessary to produce the units; the personal allowance, ranging from 5% to 15%, depending on the physical strength and dexterity required to perform the job, is given to the operator to accommodate his/her personal needs. Generally, this is accomplished by including the allowance in calculating the standard time for the job. Shrinkage allowance is established to account for the loss of production as a result of rejection of parts due to poor quality even though the machine is set and operating correctly.

All other factors contributing to lost time are grouped together and measured against the efficiency of operation. For example, if one-shift operation is 95% efficient, then the time available for production, excluding the allowances we have considered before, is 456 minutes (0.95 × 480). In multiple-shift operation, it is a normal practice to indicate the efficiency of the additional shift. The terminology commonly used is frequently misleading. It might be said that in our example the addition of a second shift would add 75% to the efficiency of the plant. This does not mean that the first shift would continue at 95% and that the second shift would add 75% for a total of 170% efficiency. It indicates that the two shifts together will operate at a rate of 1.7 times that which one shift would if it were at 100% efficiency. In our example, the first shift might drop to 90%, and the second would be at 80%, for the total of 170%.

### 5.1.6 Required Capacity

The capacity needed depends on the operations the machine is assigned to perform and the required production rate for each item. The following example illustrates this point.

On the basis of the production routing sheets of the items manufactured in the plant, it is decided to perform operations 10, 12, and 15 of item A, operation 101 of item B, and operation 157 of item C on the new machine. The estimated time for completion of an operation on the machine is noted along with the estimated time for setup, and a personal allowance is assigned to compensate for personal needs. The quality requirement for the parts and the tolerance at which the machine and process can operate give an estimate of shrinkage allowance per operation. The required production per day is obtained from the scheduled deliveries. The data are recorded in Table 5.2.

The entries in column 4 are the numbers of units that must be processed to obtain the required number of good parts per day when shrinkage is considered. An entry is obtained by dividing the parts required, column 2, by the yield or 1 minus shrinkage allowance. The time allowed per operation, column 7, is the estimated time of column 5 multiplied by the quantity 1 plus the personal allowance. The total time, column 9, needed to produce the items is column 4 multiplied by time allowed per operation, column 7, plus the setup time, column 8, for that operation. The grand total is the time the new machine must operate each day, in this case, 1308.7 minutes.

### 5.1.7 Number of Machines Needed

Once the required capacity and the available productive time per machine have been established, the calculation for the required number of machines is quite straightforward. In the preceding example, if the plant is to operate on one shift with 95% efficiency, the available time per machine is 456 minutes (480 × 0.95). The total capacity needed is 1308.72 minutes, requiring 2.87 (1308.72/456) or 3 machines.

**TABLE 5.2**
**Production Date**

| (1) Item and Operation | (2) Units Required/Day | (3) Shrinkage Allowance | (4) Items to Produce/ Day = (2)/(1 – (3)) | (5) Time per Operation (minutes) | (6) Personal Allowance (%) | (7) Time Allowed = (5)(1 + (6)) | (8) Setup Time/ Day (Minutes) | (9) Total Time (Minutes/day) |
|---|---|---|---|---|---|---|---|---|
| A–10 | 1000 | 0.03 | 1031 | 0.13 | 5 | 0.1365 | 15 | 155.73 |
| A–12 | 2000 | 0.00 | 2000 | 0.15 | 5 | 0.1575 | 18 | 333.00 |
| A–15 | 1000 | 0.01 | 1011 | 0.08 | 5 | 0.0840 | 12 | 96.92 |
| B–101 | 100 | 0.10 | 112 | 2.2 | 5 | 2.3100 | 10 | 268.72 |
| C–151 | 300 | 0.08 | 327 | 1.3 | 5 | 1.3650 | 8 | 454.35 |
| | | | | | | | Total | 1308.72 |

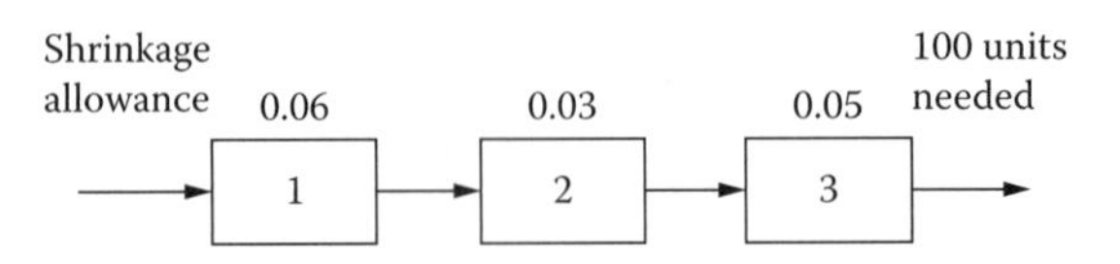


**FIGURE 5.2** Yield and cost models

It should be noted that the data in the table were applicable to the machine under consideration. If a different type of machine were to change the data (e.g., setup, processing time, or shrinkage allowance), then a new required capacity would have to be calculated, perhaps giving a different answer to the problem.

Let us now go one step further and determine the quantity that should be processed at each stage as a product proceeds through a series of stations in its manufacturing sequence. We illustrate the calculations by considering an example. Suppose the production of a part requires a three-stage operation with a shrinkage rate for each stage as indicated in Figure 5.2. Assume further that 100 good units are required from this setup.

The input to any stage must allow for rejects in that stage. For example, only 95% of the products in stage 3 are good; therefore, $100/0.95 = 105.26 \approx 106$ units must be started in that stage. If we proceed backward to stage 2, to obtain 106 good units the input to stage 2 must be $106/0.97 = 109.3 \approx 110$ units. Similarly, in stage 1, $110/0.94 \approx 117$ units must be started. By using appropriate input values, the time requirement for the product in each stage can be calculated by using the procedure described earlier. As the reader has perhaps observed, the numbers are rounded up at each stage; this is mainly to make certain that at least the required number of good units is produced in the three stages.

### 5.1.7.1 Extensions of Variations on Stage Analysis

To make the formulation general, let $P_i$ denote the percent defectives in state $i$. Then, if $N$ is the number of units at the beginning of the process, the number of good units after the first state is $[N(1 - P_1)]$, where the brackets stand for the integer portion of the result only. The number of good units after the second stage is $[N(1 - P_1)(1 - P_2)]$.

One can extend the analysis to all the stages. It is also possible to determine the value of $N$ if the required number of final good products is known. A fairly good approximation of the input/output relationship is given by calculating the yield of the line. Such calculations ignore stage-by-stage integerization to give

$$\text{Line yield} = (1 - P_1)(1 - P_2)(1 - P_3)(1 - P_n)$$

and input quality $N$ = required output/yield of the line. In the previous example, the line yield is $(1 - 0.06)(1 - 0.03)(1 - 0.05) = 0.866$; therefore, producing 100 good units would require starting production with $100/.866 = 115.4 \approx 116$ units at stage 1.

Another interesting aspect of this analysis can be investigated by calculating the cost per good unit produced. Let $C_i$ be the cost of processing a unit in state $i$. Then

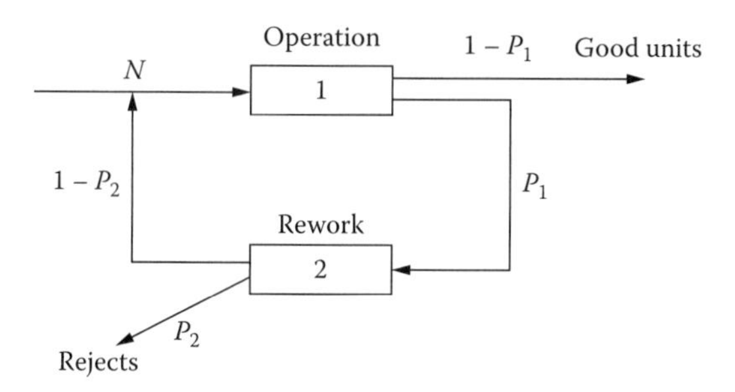


**FIGURE 5.3** Recycling of rejected units to backstream

the number of units processed at each stage multiplied by the cost of processing each unit in the stage gives the cost of each stage.

Accumulating the individual stage cost and dividing it by the total number of good units produced gives cost per good unit. In our example, if $C_1$ = \$5, $C_2$ = \$8, and $C_3$ = \$10, then the cost of a good unit is given by

$$\frac{5+117+8\times 110+10\times 106}{100}=\$25.25$$

The cost can be approximated by adding the cost of each serial stage and dividing it by the yield. In our case, it would be (5 + 8 + 10)/0.866 = \$26.56.

The analysis may be further expanded to include nonserial systems formed by rework and/or repair operations. For example, consider a system such as that shown in Figure 5.3, in which a portion of the units rejected in operation 1 can be reclaimed in operation 2 and again become an input to operation 1.

If $P_1$ and $P_2$ denote the percent defectives at each stage, respectively, then the number of good units produced for $N$ units of input would be

$$N(1-P_1)+\underbrace{NP_1(1-P_2)(1-P_1)}_{\text{Units reworked once}}+\underbrace{NP_1(1-P_2)P_1(1-P_2)(1-P_1)}_{\text{Units reworked twice}}+\cdots$$

$$=N(1-P_1)\left[1+P_1(1-P_2)+P_1^2(1-P_2)^2+\cdots\right]$$

If we denote $a = P_1\,(1 - P_2)$, and since both $P_1$ and $P_2 \leq 1$, we have

$$N(1-P_1)[1+a+a^2+\cdots]$$

Because the terms in the brackets follow the form of a geometric series, the expression transforms to $N(1 - P_1)/(1 - a)$. Hence, the yield of the system is $(1 - P_1)/(1 - a)$.

The value of $N$ can be determined for a required number of good units. For example, if $P_1 = 0.2$, $P_2 = 0.1$, and 100 good units are required, then

$$a = 0.2(1-0.1) = 0.18$$

and the yield is $(1 - P_1)/(1 - a) = 0.975$. Therefore, $N = 100/0.975 = 102.56 \approx 103$ units.

The total cost of processing $N$ units is the sum of the costs at each stage multiplied by the number of units processed at that stage. Because we have a recycling process, the terms in the sum are infinite. However, they are decreasing in value successively:

$$\begin{aligned}\text{Total cost} &= C_1N + C_2NP_1 + C_1NP_1(1-P_2) + C_2NP_1(1-P_2)P_1\\ &\quad + C_1NP_1(1-P_2)P_1(1-P_2) + C_2NP_1(1-P_2)P_1(1-P_2)P_1\\ &\quad + \cdots\\ &= C_1N[1 + P_1(1-P_2) + P(1-P_2)^2 + \cdots]\\ &\quad + C_2NP_1[1 + P_1(1-P_2) + P_1^2(1-P_2)^2 + \cdots]\end{aligned}$$

since, $a = P_1(1 - P_2)$

$$\begin{aligned}\text{Total cost} &= \frac{C_1N}{1-a} + \frac{C_2NP_1}{1-a}\\ &= \frac{N(C_1 = C_2P_1)}{1-a}\end{aligned}$$

Suppose that in the previous illustration $C_1 = \$5$ and $C_2 = \$3$. The cost to produce 100 good units would be

$$\frac{103[5 + (3 \times 0.2)]}{1 - 0.18} = 703.41$$

or \$7.03 per unit.

The recycling could also be of the form in which the unit is as good after repair as a unit coming from the operation itself (Figure 5.4).

The total good units produced would be the sum of good units produced in operation 1 and in operation 2, that is,

$$\text{Number of good units} = N(1 - P_1) + NP_1\,(1 - P_2)$$

The system yield is $(1 - P_1) + P_1(1 - P_2)$, and the total cost of the operation is $NC_1 + P_1NC_2$, from which the cost per good unit of output can be determined.

#### 5.1.7.2 System Decomposition

A system consisting of a complex structure can be broken down step by step into a familiar pattern, which can then be analyzed with the result illustrated in the previous section. The following example demonstrates this approach.

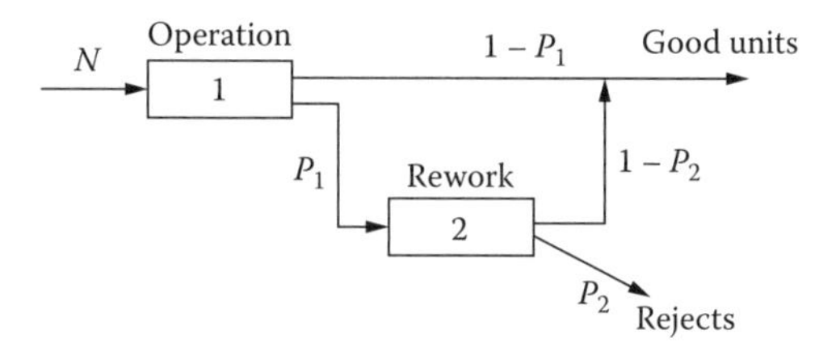


**FIGURE 5.4** Recycling to rejected units upstream

## EXAMPLE—A COMPOSITE PRODUCTION LINE

Consider the following production pattern with the data as given in the accompanying table.

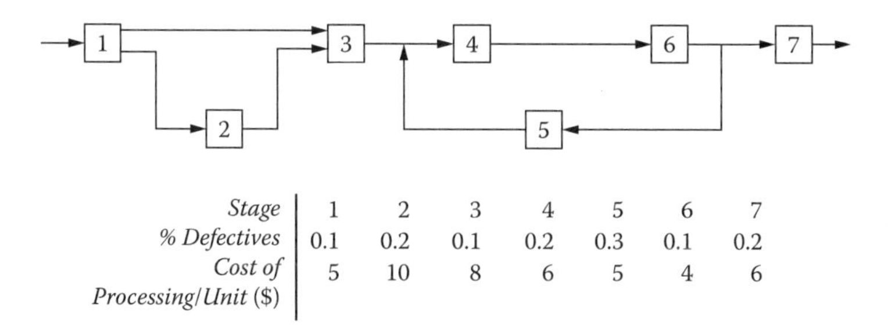


| *Stage* | 1 | 2 | 3 | 4 | 5 | 6 | 7 |
|---|---|---|---|---|---|---|---|
| *% Defectives* | 0.1 | 0.2 | 0.1 | 0.2 | 0.3 | 0.1 | 0.2 |
| *Cost of Processing/Unit* ($) | 5 | 10 | 8 | 6 | 5 | 4 | 6 |

The objective at this point is to reduce the system into a serial system. We shall combine different stages into a single stage and determine its equivalent percent defectives and cost.

Combine operations 4 and 6 into a single stage, and denote it as stage 8.

Because operations 4 and 6 form a serial system, the yield is $(1 - P_4)(1 - P_6) = 0.8 \times 0.9 = 0.72$ or percent defectives $P_8 = 1 - 0.72 = 0.28$. The unit cost consists of the costs for the operation in stage 4 on all units plus the operation in stage 6 on the good units coming from stage 4. That is,

$$\text{Unit cost} = C_8 = \$6 + (1 - 0.2)4 = \$9.20$$

The resulting system is as follows:

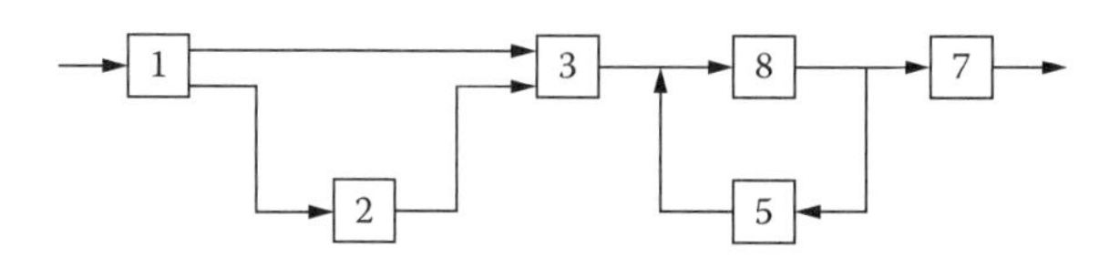


Now let us combine stage 8 and stage 5 into one operation, designated operation 9. We have

$$a = P_8(1 - P_5) = 0.28(1 - 0.3) = 0.196$$

and hence

$$\text{yield} = \frac{1-P_8}{1-a} = \frac{1-0.28}{1-0.196}$$

$$= 0.895$$

$$\%\ \text{defectives} = P_9 = 1 - \text{yield}$$

$$= 1 - 0.895 = 0.105$$

and

$$\text{Cost/Unit} = \frac{C_8 + C_5 P_8}{1-a}$$

$$= \frac{9.2 + 5(0.28)}{1-0.196} = \$13.18$$

The resulting system is as follows:

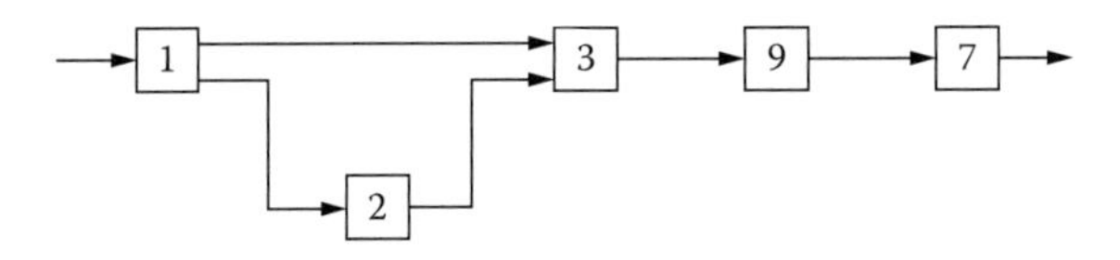


Now let us combine stages 1 and 2 into stage 10.

$$\text{Yield of stage } 10 = (1-P_1) + P_1(1-P_2)$$

$$= 0.9 + (0.1 \times 0.8)$$

$$= 0.98$$

Therefore,

$$\%\ \text{defectives} = P_{10} = 1 - 0.98 = 0.02$$

and

$$C_{10} = (C_1 + P_1 C_2)/(1-P_{10})$$

$$= \frac{5 + (0.2 \times 10)}{0.98} = \$7.14$$

The resulting system is as follows:

Because it is in serial arrangement, we have

$$\text{Yield of the line} = (1-P_{10})(1-P_3)(1-P_9)(1-P_7)$$
$$= 0.98 \times 0.9 \times 0.895 \times 0.8$$
$$= 0.6315$$

and

$$\text{Estimated cost/good unit} = \frac{C_{10}+C_3+C_9+C_7}{yield}$$
$$= \frac{7.14+8+13.8+6}{0.6315}$$
$$= \$54.35$$

### 5.1.8 Machine Specifications

A decision to buy a machine must now be documented for further action by the shop supervisor and the maintenance department. For example, appropriate locations for placement of the machines must be determined, the required utilities must be provided, and foundations must be planned so that when the machines arrive, they can be installed without further delay. The models developed in Chapters 17–19 can be utilized profitably to make the placement efficient. Tables 5.3 and 5.4 display one of the more common forms of specifications.

**TABLE 5.3**
**Forming Press Specifications**

| Item | Specification |
|---|---|
| Machine | 150-ton forming press |
| Manufacturer | Hess |
| Model | SC2 |
| Condition | New |
| Cost | $240,000 |
| Crew | 1 |
| Foundation | 8-inch steel-reinforced concrete |
| Motor horsepower | 20 hp at 1200 rpm |
| Voltage | 440 V at 60 Hz, 3-phase |
| Dimensions | 50 × 123 inches |
| Work rate | 35 strokes/minute |
| Die change time | 3 hours |
| Number required | 1 |
| Usage | Model C — operation 330: plate forming |

**TABLE 5.4**
**Drying Oven Specifications**

| Item | Specification |
|---|---|
| Machine | Paint-drying oven |
| Manufacturer | Monroe Industrial Machinery |
| Model | Custom built |
| Crew | 1 for 1 hour/day |
| Cost | $98,000 |
| Foundation | 6-inch steel-reinforced concrete |
| Motor horsepower | 20 hp at 1800 rpm |
| Fuel | Natural gas |
| Requirement | 1,850,000 cubic feet/hour[a] |
| Voltage | 440 V at 60 Hz, 3-phase |
| Dimensions | 40 × 60 feet |
| Work rate | 8 feet/minute |
| Number required | 1 |
| Usage | Bake enamel paint onto parts |

[a] Overall for paint system.

**TABLE 5.5**
**Air Compressor Specifications**

| Item | Specification |
|---|---|
| Machine | Air compressor |
| Manufacturer | Redman |
| Model | 500-H type SSR |
| Condition | New |
| Cost | $27,500[a] |
| Foundation | Gravel with treated wood runners |
| Motor horsepower | 133 hp; 152 A at full load |
| Voltage | 460 V at 60 Hz, 3-phase |
| Dimensions | 110 × 69 × 58 inches |
| Capacity | 125 psi each 500 cubic feet/minute |
| Number required | 1 |
| Usage | (1) Air for assemble work<br>(2) Reserve for paint system |

[a]This includes the compressor, outdoor protection package, air dryer, tank, cooler, etc.

In most cases, the necessary information can be obtained from the supplier of the machine. It should be remembered that although the supplier will provide such information as the utility and foundation requirements, the ultimate responsibility for installing and running the machine normally rests with the user.

### 5.1.9 Auxiliary Equipment

Air compressors, tool sharpeners, and water treatment facilities are other machines, equipment, and facilities that are not directly related to production but are necessary to achieve the production objectives. These are referred to as auxiliary equipment. An analysis similar to the ones performed for the main equipment should be conducted to determine their required capacities and the number of units needed. Table 5.5 shows a specification for an air compressor.

Material-handling equipment can also be categorized as auxiliary equipment. More detailed descriptions of this kind of equipment are given in the chapter on material handling (Chapter 8).

## 5.2 LABOR REQUIREMENT AND SELECTION

Few plants can operate very long without a qualified workforce that can do its job well. A careful analysis is required to obtain a good match between jobs and workers. Inappropriate assignments generally lead to problems of inefficiency in production, poor-quality parts, disgruntled employees, and low morale.

The employees in a manufacturing facility either are engaged in production or provide the necessary services to facilitate production. The people involved in immediate production activities, such as operating the machines, working at assigned stations, and operating a forklift truck or other material-handling equipment, are referred to as direct laborers. The people who provide the supporting services such as maintenance, janitorial services, and cafeteria operation are termed indirect laborers. We shall study indirect labor further in Chapter 9; however, whether the labor is direct or indirect, an appropriate job evaluation should be conducted to determine the duties and responsibilities of the job and who can best perform them. In the following sections, we will concentrate on job evaluation, which defines the factors that should be considered in hiring new employees, and on-the-job description, which indicates what tasks are assigned to an employee.

### 5.2.1 Job Evaluation

One of the tasks in developing a production design is to determine the number and qualifications of workers needed in each step of the production. The production arrangement described in Chapter 4 plays an important part in this decision. For example, in an assembly line arrangement, in which a worker is mainly responsible for work in one station, job skills and responsibilities are not as demanding as they would be in a job shop environment. Each position should be evaluated in terms of the level of skill, effort, responsibility, working conditions, and safety involved in the job content. This will enable a manager to hire a person with the right qualifications and pay him/her the salary that is appropriate for the job.

The process of job evaluation is begun by comparing each attribute of the job in question with corresponding standards established within the company, and then assigning a point value to each attribute to indicate the degree of its variation from the standard. Because not all the attributes are equally significant, the relative importance of each is reflected by its allowed point spread. The total point value, obtained by adding the points assigned to each attribute, reflects the job level and perhaps even the step in that level for the job.

A company should develop a work factor point scale similar to that shown in Table 5.6. In each category, level 1 corresponds to simple and routine duties, level 2 indicates well-defined duties that may require some decision making, level 3 factors require working independently and devising or modifying methods based on policy, and level 4 is assigned to highly technical or complex work involving independent decisions.

Job evaluation allows management to group all jobs requiring the same types of skills together; thus, an employee with a given set of skills might be qualified to work

**TABLE 5.6**
**Job Evaluation Point Scale**

| | Level | | | |
|---|---|---|---|---|
| **Work Factor** | **1** | **2** | **3** | **4** |
| **I. Skill** | | | | |
| Job knowledge | 40 | 60 | 80 | 100 |
| Experience | 30 | 35 | 40 | 45 |
| Education | 15 | 18 | 21 | 24 |
| Dexterity | 8 | 10 | 12 | 14 |
| **II. Effort** | | | | |
| Mental effort | 25 | 30 | 35 | 40 |
| Physical effort | 20 | 24 | 28 | 32 |
| Fatigue | 5 | 7 | 9 | 11 |
| Monotony | 2 | 3 | 4 | 5 |
| **III. Responsibility** | | | | |
| Equipment | 17 | 20 | 23 | 26 |
| Others | 20 | 25 | 30 | 35 |
| Errors | 12 | 15 | 18 | 21 |
| Coordinating activities | 19 | 20 | 21 | 22 |
| **IV. Working Conditions** | | | | |
| Hazards | 3 | 4 | 5 | 5 |
| Environment | 1 | 2 | 3 | 4 |
| Total | 217 | 273 | 329 | 384 |

in a variety of jobs. This provides information that permits greater flexibility in labor assignments. Management can also achieve a uniform salary structure throughout the plant by comparing point values of each job with the pay scale. Besides assisting in matching a worker's qualifications to the job, the evaluation provides information for training, promotions, transfers, and performance-rating evaluations.

#### 5.2.1.1 Sample Job Evaluation: Machine Shop Supervisor

We illustrate the technique of job evaluation by applying it to a machine shop supervisor for the company mentioned earlier, with a work factor chart as in Table 5.7.

For each factor, the level of difficulty for the job is judged, and corresponding points from the table are assigned to that category.

The total points assigned to a job indicate its relative importance. For example, a job worth 280 points would be twice as important to a company as one worth 140 points.

#### 5.2.1.2 Sample Job Evaluation: Salary Determination

Uniform salary determination is another benefit of job evaluation. For example, the possible points assigned by this company range from 217 to 384. These numbers can be used to assign salaries for the firm. The graph in Figure 5.5 shows the relationship between the point value of a job and the corresponding salary. One might note that as the number of points assigned to a job increases, the salary also increases. In the job evaluation for the machine shop supervisor, the various work factors were found to have a total value of 283 points. The graph equates this to a salary of $25,000, which

**TABLE 5.7**
**Job Evaluation (Machine Shop Supervisor)**

| Work Factor | Level | Points |
|---|---|---|
| 1 | 2 | 60 |
| 2 | 3 | 40 |
| 3 | 1 | 15 |
| 4 | 2 | 10 |
| 5 | 2 | 30 |
| 6 | 3 | 28 |
| 7 | 2 | 7 |
| 8 | 1 | 2 |
| 9 | 2 | 20 |
| 10 | 3 | 30 |
| 11 | 2 | 15 |
| 12 | 2 | 20 |
| 13 | 2 | 4 |
| 14 | 2 | 2 |
| Total | | 283 |

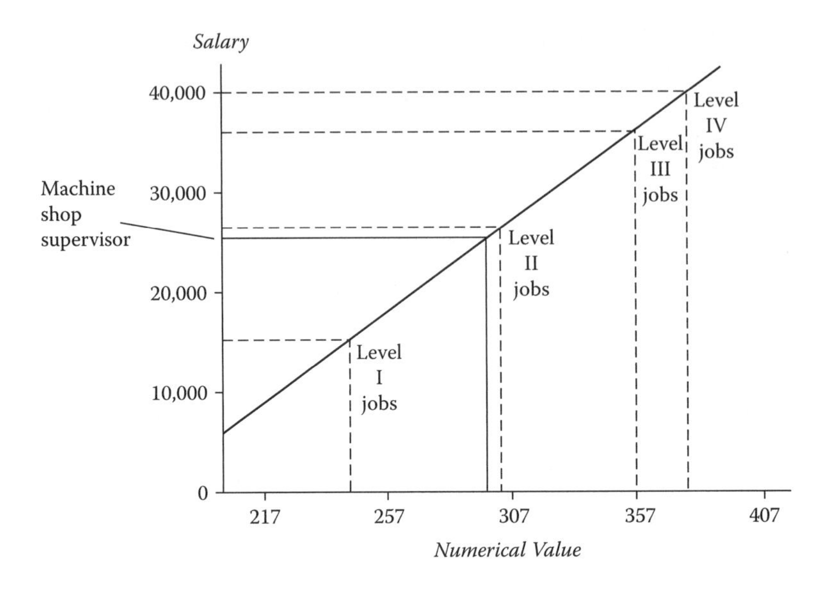


**FIGURE 5.5** Salary determination.

in practice could be the midpoint of the salary range for the supervisor, allowing for periodic raises based on performance.

### 5.2.2 Standard Job Analysis

The *Dictionary of Occupational Titles* (DOT) is a 1400-page listing of 20,000 occupations that is published by the U.S. Department of Labor. The fourth edition was issued in 1991. The information is currently made available through O*NetOnline (http://online.onetcenter.org). The occupations are grouped into classes based on the interrelationship of tasks and requirements. The dictionary is designed to aid in matching job requirements with worker skills. It begins by listing the occupational categories as a quick reference for identifying the numerical classifications. The major part of the book lists the many occupations in detail according to the numerical classifications, categorizing the jobs and giving brief descriptions. The last section is an alphabetical listing of the jobs that refers back to the numerical listing.

The DOT is an example of job analysis on a large scale. All aspects of jobs are broken down into three broad categories: data, people, and things. Each of these categories is subdivided further, and each division is assigned a point value according to intensity. The lower the number, the more time and know-how are required by that job trait. For example, instructing is assigned a 2 while supervising is assigned a 3, because teaching requires more intense interacting with people. Likewise, setting up has a lower "things" value than tending a machine, and computing has a lower "data" value than copying information.

The DOT begins by breaking down jobs into nine primary occupational categories. Numbered from 1 to 9, they are: Professional and Technical; Clerical and Sales;

Service; Agricultural, Fishery, Forestry, and Related; Processing; Machine Trades; Bench Work; Structural Work; and Miscellaneous Occupations. The numbers identifying these categories are the first numerical in the code for each job, and the two that follow indicate one of the many divisions within each category.

The "Data," "People," and "Things" categories are each divided and numbered accordingly. For Data, the subdivisions range from 0 to 6 as follows: Synthesizing, Coordinating, Analyzing, Compiling, Computing, Copying, and Comparing. For People, they range from 0 to 8 as follows: Mentoring, Negotiating, Instructing, Supervising, Diverting, Persuading, Speaking-Signaling, Serving, and Taking Instructions-Helping. The categories 0 to 7 for Things are Setting up, Precision Working, Operating-Controlling, Driving-Operating, Manipulating, Tending, Feeding on/off Machines, and Handling.

The last three digits of the code are assigned to distinguish between jobs for which the first six digits are identical. When there is no other job with the same first six numerals, the last three are 000. If the first six digits are the same, then the last three digits are assigned to categorize in alphabetical order the different jobs in steps of four, beginning with 010. For example, the Fruit Farmworker, Fig Caprifier, Fruit Harvest Worker, and Vine Pruner all have 403.687 as the first six digits. The last three digits assigned to distinguish the four jobs are 010, 014, 018, and 022, respectively.

In using the number code, consider the job of Export Manager, which is listed as 163.117-014. The *1* in the first part identifies it as a professional/technical job, and the *63* classifies it as a sales and distribution management occupation. The second set of numbers, *117*, corresponds to Data, People, and Things; the first *1* shows that the export manager must be able to analyze and coordinate data, the second *1* indicates that he/she must instruct and negotiate with people, and the *7* is indicative of the fact that the export manager has very little to do with things.

Another example is the occupation listed as 559.382-054. The 5's are processing occupations, and all of the 559's are occupations in the processing of chemicals, plastics, synthetics, rubber, paint, and related products. The number 054 identifies this occupation as a soap maker. By looking at the 382, we find that the soap maker must be able to compile data (3), take instructions (8), and operate and control machinery (2). Following the number classification is a brief description of the job, emphasizing some of the major qualifications and responsibilities.

There are several advantages to using this system. It can be considered a standard for job analysis, allowing different companies with similar jobs to compare these jobs, such that identical jobs would result in the same analysis and description. The DOT can be used as a basis for job evaluation by any company, saving the considerable amount of time and money required to start from the beginning in gathering information that is already available and contained in the DOT. Finally, the numerical codes can be of help in developing pay scales. The lower the numbers in a particular category, the higher the value of the job. A pay scale may be based on the emphasis a particular company places on the three traits: Data, People, and Things.

### 5.2.3 Job Definition and Description

Job definition and evaluation go hand in hand. A job cannot be evaluated unless it is well defined. Different analysts might very well develop tasks as they divide the total work required to manufacture a product into jobs. An analyst may assign one worker per station, assign one operator per machine, or make a group of workers collectively responsible for a large task or for operating a group of machines. In a teamwork concept, a job can be defined in a manner that allows for crossover and cooperation between the workers to complete the total task. The initial job description might have to be modified as one gains work experience, leading to the concepts of job reevaluation and job enlargement.

In all cases, the job description is important. It gives a complete listing of an employee's duties and responsibilities and his/her line of command. It helps form an understanding between an employee and an employer as to what is expected from the employee and what the employer will pay in return. Table 5.8 illustrates an example from a job description chart.

**TABLE 5.8**
**Job Description**

| (1) Title (DOT) | (2) Salary | (3) Educational and Experience Requirements | (4) Job | (5) Reports to |
|---|---|---|---|---|
| Supervisor, Plastics 556.130-010 | $49,500 | 2 years of college, 8 years experience in injection molding | Supervises activities of workers in molding and plastics operations. Trains workers in job duties and production techniques. | Plant Engineer |
| Supervisor, Metal Parts Line 619.130-030 | $49,500 | 2 years of college, 8 years experience in press line management | Supervises activities of workers engaged in punch and form press operations. Trains workers on metal parts assembly line. | Plant Engineer |
| Injection Molding Machine Tender 556.685-038 | $35,000 | High school diploma, 3 years experience on injection molding machine | Tends injection, molding machine. Fills hopper of molding machine. Starts up machine to liquefy material to be shot into mold. Observes gage sand controls temperature to ensure good molding. | Supervisor, Plastics |
| Automatic Punch Press Operator 615.482-026 | $35,000/ | High school diploma, 2 years experience on automatic punch and forming presses | Sets up and operates, line power presses that will automatically feed rolls of sheet aluminum into the punch. Positions and clamps feed guides. | Line Supervisor |
| Welder 811.684-014 | $20 /hour | Trade school,1 year welding experience | (1) Welds base to body. (2) Welds spout to body. | Supervisor, Metal Parts Line |

## 5.3 MACHINE COUPLING

The concept of machine coupling was explained in Section 5.1.5. Recall that in machine coupling a single operator is made responsible for operating a group of machines. In production planning, it is necessary to decide whether machine coupling is feasible so that an appropriate job description for the operator can be written.

The use of worker-machine charts plays an important part in this development. The chart graphically displays on the same scale the time when a person is working or idle and when a machine is working or idle. The person could be made responsible for more than one machine if he/she has sufficient idle time during which it would be possible to attend to another machine. The following example illustrates this concept. It should be noted that machine coupling is possible only if the machine, when it is working, is capable of performing its task without any assistance from the operator.

### 5.3.1 Example of Machine Coupling

Suppose we have three machines for which machine coupling is to be investigated, one *A* and two *B*'s. The loading, processing, and cycle times (the sum of loading and processing) in minutes for the parts scheduled for production on these machines are given in Table 5.9.

The company estimates that the cost of the operator is $6.00/hour, while the costs for machines A and B are $12.00/hour and $18.00/hour, respectively. It will take about 30 seconds for the operator to move from one machine to another.

#### 5.3.1.1 Alternative A

Let us begin by assigning one operator per machine. As shown in Figure 5.6, in each chart the working time and the idle time for both worker and machine are plotted simultaneously. The time is represented by the vertical distance, with the bar representing the cycle time for the corresponding machine. The load time, which is the integral part of an operation, represents the time to load and unload a unit from the machine and as such is not considered part of the idle time for the machine. The symbols used are M/C for machine, L for load, W for work, I for idle, and M for move.

**TABLE 5.9**
**Data for an Illustrative Example**

| | Machine | |
|---|---|---|
| | A | B |
| Loading | 2 | 5 |
| Processing | 8 | 7 |
| Cycle time | 10 | 12 |

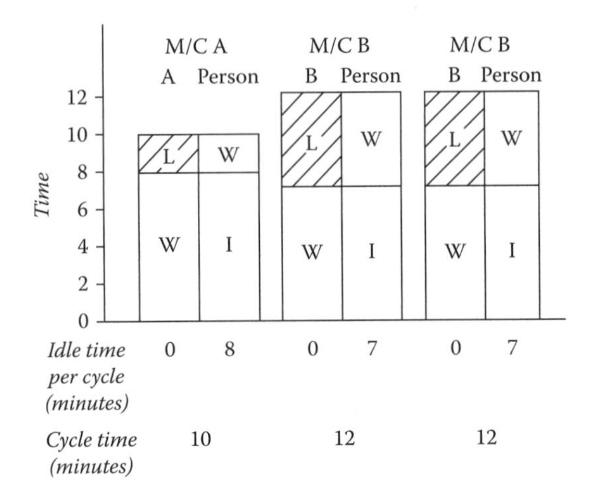


**FIGURE 5.6** Worker-machine chart showing three operators

The total cost consists of two components: machine cost and operator cost. In making comparisons between the alternatives, we will include the cost of idle time for the machine as one component, along with the operator cost. This is because when machines are in use, they are performing desirable, productive activities; however, in the idle state they are nonproductive and an expense. The idle condition can be reduced by increasing the number of operators, thus reducing the machine component of the cost while increasing the operator cost. We are seeking a balance that will minimize the total cost. The total cost of the present alternative is

$$0 + 3(\$6) = \$18/\text{hour}$$

### 5.3.1.2 Alternative B

Now let us examine the time distribution when one operator is assigned to machines A and B and another to the remaining machine B. The time bars are illustrated in Figure 5.7. The minimum cycle time when machines A and B are tended by the same operator is the maximum of cycle times for either machine A or machine B.

$$\begin{aligned} \text{Total cost/hour} &= \text{idle cost of M/C A} + \text{cost of operators} \\ &= (2/12)\$12 + (2 \times \$6) = \$14 \\ &= \$14 \end{aligned}$$

### 5.3.1.3 Alternative C

If one operator is assigned to both B machines and another to machine A, the time bars are as shown in Figure 5.8. The cycle time on the two B machines is 12 minutes;

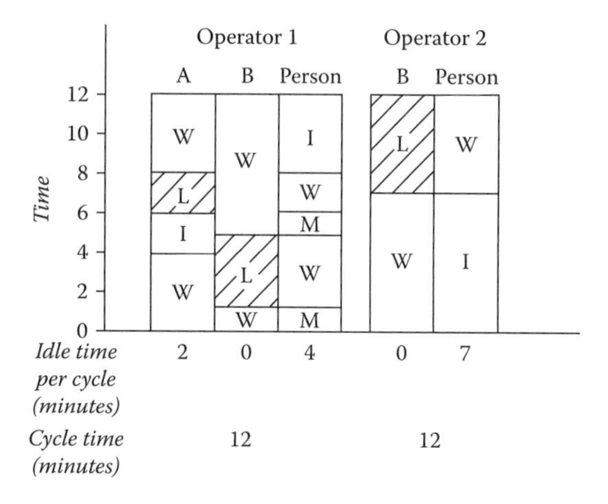


**FIGURE 5.7**

on machine A, it is 10 minutes. There is no idle machine time, and so the total cost per hour consists only of the cost of the operators, that is,

$$\text{Total cost/hour} = 2 \times \$6$$

$$= \$12$$

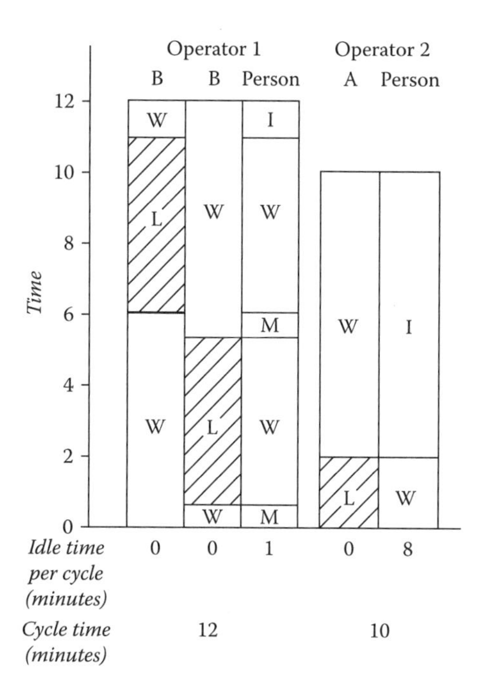


**FIGURE 5.8** Worker–machine chart showing two operators (one worker for both B machines and another worker for machine A)

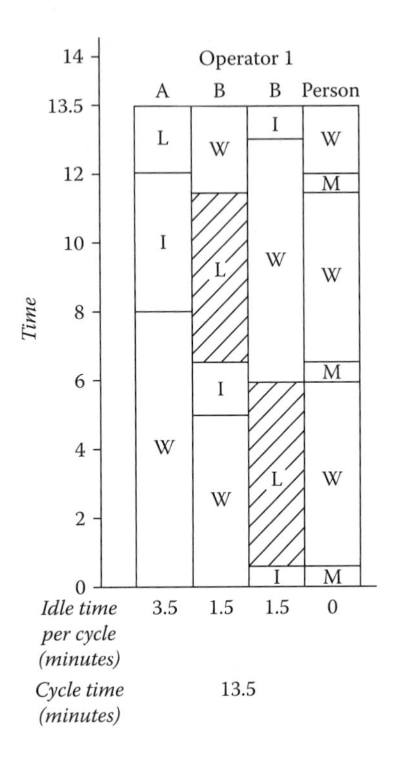


**FIGURE 5.9** Worker-machine chart showing one operator

### 5.3.1.4 Alternative D

If one operator is assigned to all three machines, the time bars are as shown in Figure 5.9. The minimum cycle time must allow the setup of each machine and the time for an operator to go from one machine to the next in the job sequence. There are three "walks," one setup for A and two setups for B, giving a minimum cycle time of 3(1/2) + 2 + 2(5) = 13 1/2 = minutes.

$$\text{Total cost/hour} = \text{Cost of idle time on each M/C} + \text{cost of operation}$$

$$= \frac{3.5}{13.5} \times 12 + \frac{(1.5+1.5)}{(13.5)} \times 18 + 6$$

$$= \$13.11$$

Alternative C has the lowest cost of all the alternatives; therefore, it is the preferred alternative. One operator should be made responsible for two B machines and one operator for machine A.

Machine coupling is increasingly used with a robot working with a number of automated machines. The cycle times can be controlled to very precise values, and the repetitious tasks of loading and unloading each machine are assigned to the robot.

## 5.4 TOTAL PERSONNEL REQUIREMENT

The next step is to list all the personnel requirements in the plant. The number of assembly stations and the amount of manual effort required, along with the machines and their associated operators, determine the number of direct factory workers needed. It should be understood that more than one machine might be assigned to an operator through machine coupling, and it might be necessary to assign more than one operator to a machine, as is the case in a paper-manufacturing facility. Organization and factory layout determine the number of direct supervisory personnel required. For example, in a job shop environment consisting of a large number of machines and operators in each department, a supervisor might be appointed for each department, such as gear cutting, heat treatment, and milling. One supervisor might be sufficient for the entire machine shop if it consists of a much smaller-scale operation, such as only 8–10 machines and 15 total operators.

Although management theorists claim that assigning 10–12 people per supervisor is most efficient, in an assembly line environment, the number of workers reporting to one supervisor may be considerably higher. It is not uncommon to see a ratio of 25:1 in an assembly line in which there are similar work patterns, more manual work than machine operations, and a layout that permits the supervisor to easily see all workers.

If the plant is spread over a large area, the ratio of 12:1 might have to be reduced accordingly. Similarly, if in addition to supervision the person has other responsibilities, such as long-range planning, new product development, and sales analysis, then the time available for direct supervision is limited and the number of personnel assigned per supervisor should be modified correspondingly.

A proposed or current organization chart also plays a role in defining responsibilities and developing the number of required personnel. (More detail is given about organization charts is given in Chapter 7.) For example, the maintenance and quality control departments might be directly supervised by the plant manager in a small operation; or a technician could be hired having broad responsibilities for process control, quality control, and incoming and outgoing inspections in a medium-sized plant. A large plant might require a number of technicians and supervisors to perform similar tasks. Matrix organization, regardless of size, requires an independent manager for each product as well as for each department.

The number of staff appointments in such areas as production and material planning, accounting, computer operations, purchasing, trucking and transportation operations, labor relations, and material handling will depend on the size of the company. For example, each administrator might hire his/her own administrative assitant if the clerical work involved justifies it, or if the confidential nature of the work demands it. Whenever possible, an administrative assistant is shared between departments if the floor plan permits. With today's technology many companies allow technical people to develop their own letters and memos. The secretaries are assigned broad responsibilities in communications, computer operation and data entry, and computer report generation. An administrative assistant might handle the clerical needs of three different departments with a total of 50–60 engineering personnel.

**TABLE 5.10**
**Direct Labor Table**

| Position | Salary | Number Needed | Total Salary |
|---|---|---|---|
| Rip saw operator | $25,000/year | 2 | $50,000/year |
| Radial arm saw operator | $25,000/year | 2 | $50,000/year |
| Variety saw operator | $22,000/year | 2 | $44,000/year |
| Belt sander operator | $18,000/year | 1 | $18,000/year |
| Machine sander operator | $18,000/year | 4 | $72,000/year |
| Drill press operator | $18,000/year | 3 | $54,000/year |
| Cutter | $18,000/year | 2 | $36,000/year |
| Upholstery sewer | $18,000/year | 2 | $36,000/year |
| Chair padder | $18,000/year | 2 | $36,000/year |
| Spray varnisher | $20,000/year | 5 | $100,000/year |
| Inspector | $15,000/year | 4 | $60,000/year |

The number of work shifts has an obvious impact on the requirements for both direct and indirect labor. However, it is a common practice to limit indirect personnel on the second and third shifts. Only the indirect personnel who are responsible for production-related operations are generally working during second and third shifts — for example, a shop supervisor and a skeleton maintenance crew.

The direct and indirect personnel requirements for a small firm working a one-shift operation are listed in Tables 5.10 and 5.11. Note that the plant manager has been given major responsibilities in different areas such as personnel management, sales, and planning along with overall supervision of the plant.

**TABLE 5.11**
**Support Personnel Table**

| Position | Salary | Number Needed | Total Salary |
|---|---|---|---|
| Plant manager | $60,000/year | 1 | $60,000/year |
| Bookkeeper | $30,000/year | 1 | $30,000/year |
| Administrative assistant | $25,000/year | 1 | $25,000/year |
| Maintenance person | $25,000/year | 2 | $50,000/year |
| Cleaner | $18,000/year | 1 | $18,000/year |
| Plant foreman | $30,000/year | 1 | $30,000/year |
| Forklift operator | $25,000/year | 3 | $75,000/year |
| Utility worker | $18,000/year | 2 | $36,000/year |
| Packager | $22,000/year | 3 | $66,000/year |

## 5.5 SUMMARY

To begin manufacturing, we require, among others, two important resources: appropriate machines and a good labor force. The judicious selection for both can be a challenging task. Make-or-buy evaluations and capability or expertise studies help identify the components that could be produced within a plant, but then proper machines must be selected to make this possible. This selection is influenced by needed capabilities, capacities, and prices of various machines, as well as the mode of operation within the plant. To make this determination, we must first gather the information on all available machines in the market that would satisfy our needs. Within the plant, good sources for such information are the people with experience — fellow workers and engineers, buyers, maintenance personnel, and supervisors. We may also acquire this information from outside sources, such as technical periodicals, the *Thomas Register*, *Ayers Directory of Publications*, and *Ulrich's International Periodicals Directory*. Computer databases are also becoming more economical and popular. Information services, such as Dialog, for example, provide access to more than 200 databases containing more than 100 million items.

Once the machines are identified, the next step is to determine the capacities and the number of machines that would be needed. For that, we must first perform a detailed analysis of all production routing sheets and augment the calculations by taking into account the shrinkage factor for each stage in the entire production facility. Such analysis establishes the estimate of the total required capacity for each type of machine operation; after knowing the capability of each machine, their required numbers can be easily established. Economics plays an important part in determining the final selection.

All machines, including the auxiliary equipment, should be specified in detail so that both buyers and suppliers are aware of what the requirements are. These specifications are also helpful to plant engineers as they develop the floor plan and install utilities for the new machines.

Proper labor selection is another important element that should be critically reviewed. Each job should be evaluated for factors such as required skills, efforts, responsibilities, and working conditions. Such an evaluation goes a long way in hiring a person who is suitable and qualified for the job. Job evaluation also helps in determining the salary range for the position, which is compatible with the salary structure within the plant. The DOT provides standardized job analyses that can save a considerable amount of time and effort by eliminating individual job evaluation within the plant.

Coupling machines, which allows one worker to tend more than one machine, is a technique that can be profitably used to reduce the total workforce. Finally, constructing a total labor chart, listing all jobs and indicating for each its title, salary range, and duties, leads us to realize the relative importance and value of each job. It also displays the total labor force and total labor cost for the plant.

## PROBLEMS

1. What basic data are needed to begin the machine selection process?
2. How do expertise and economics affect the make-or-buy decision?

3. In performing central nervous system research, a laboratory is using microelectrodes, which it believes will last 1 year. Microelectrodes can be made in the laboratory or purchased. The equipment to produce the electrodes would cost $800.00 and could be used for 6 months. The research technician could manufacture the electrodes and is paid $12.00 an hour. A research assistant could be trained to produce the electrodes and is paid only $6.80 an hour. However, the assistant would require 2 weeks of training, during which time the electrodes would have to be bought from an outside source. The cost of purchasing the electrodes is $25.00 apiece. The technician could produce 10 electrodes a day, as could the assistant after proper training. The material cost for each electrode is $4.00. If the laboratory will need 500 electrodes in total, determine how they should be acquired.
4. A toy factory is planning to make an additional item, which would be processed in four departments. The company is proposing to produce 600,000 units per year, working 250 days per year. The estimates of times made by an industrial engineer are shown in the accompanying table. The parts are produced within the departments in a sequential manner (from department 1 to 2 to 3 to 4).
   a. Determine the number of units that must be produced each day in each department if all defective units are scrapped.
   b. Determine the number of machines needed in each department to meet the production goals.
   c. Determine the efficiency of each departmental operation if only one shift is operated from 7:30 a.m. to 4:30 p.m., with a 40-minute break for lunch and two 10-minute coffee breaks.
   d. Adding a second shift would reduce the efficiency of the first shift by 10%, and the efficiency of the second shift would be 5% less than that of the first shift. How many machines of each type would be required?

| | Departments | | | |
|---|---|---|---|---|
| **Times** | **1** | **2** | **3** | **4** |
| Average processing time (seconds/part) | 8 | 13 | 10 | 50 |
| Average maintenance, time/day (minutes) | 20 | 30 | 20 | 15 |
| Average daily setup, time (minutes) | 20 | 15 | 30 | 20 |
| Average bookkeeping, time/day (minutes) | 10 | 8 | 12 | 5 |
| Machine adjustment, time/day (minutes) | 15 | 10 | 15 | 15 |
| Defective items, (% of production) | 5 | 6 | 4 | 4 |

| | Numerically Controlled (NC) Machine | Traditional Milling Machine |
|---|---|---|
| Cost | $15,000 | $2100 |
| Useful life manufacturing | 5 years | 3 years |
| Operating cost/year | $500 | $300 |
| Labor cost | $10/hour | $10/hour |

| | | |
|---|---|---|
| **Product I** | | |
| Output rate/hour | 25 units | 10 units |
| % Defective | 0.1% | 6.0% |
| Cost of a defective unit | $10 | $10 |
| Setup time needed | 5 minutes | 25 minutes |
| Output/day | 80 | 80 |
| **Product 2** | | |
| Output rate/hour | 30 units | 12 units |
| % Defective | 0.1% | 5.0% |
| Cost of a defective unit | $8 | $8 |
| Setup time needed | 10 minutes | 30 minutes |
| Output/day | 100 | 100 |

5. In a small machine shop, the manager must decide how many and what types of machines to buy. She has collected data, as shown in the accompanying table, on an NC machine and a traditional milling machine on which she plans to produce two products per day. All of item 1 will be produced first, and then all of item 2 will be manufactured.
   a. Write the job specifications for each.
   b. Construct a list of 10 factors to be considered in the evaluation.
   c. Determine the value and assign points to each factor.
   d. Evaluate the three jobs.
   e. Develop the salary scales.
6. a. Machine operators cost $9.50 an hour. The cost to run the machines in the accompanying table is $10, $12, and $15, respectively. If it takes 30 seconds to get from one machine to the next, investigate the possibility of machine coupling.

| | Machine | | |
|---|---|---|---|
| **Process** | **A** | **B** | **C** |
| Loading | 8 | 5 | 10 |
| Processing | 14 | 10 | 12 |

   b. If two more machines are added with operating costs of $18 per hour, investigate coupling with all five machines. The setup and processing costs are as given in the accompanying table.

| | Machine | |
|---|---|---|
| **Process** | **D** | **E** |
| Loading | | 6 |
| Processing | | 15 |

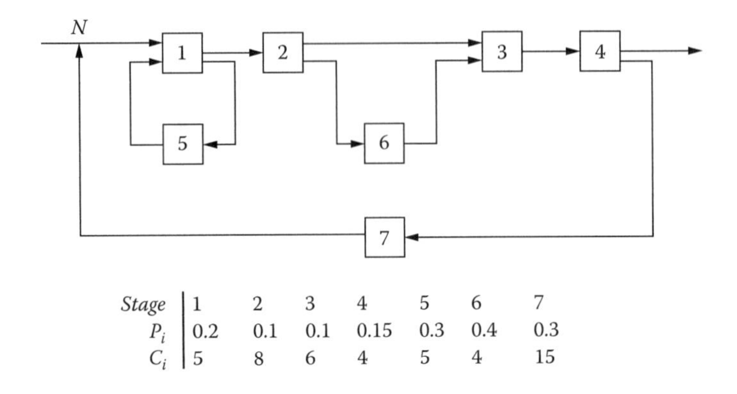


| *Stage* | 1 | 2 | 3 | 4 | 5 | 6 | 7 |
|---|---|---|---|---|---|---|---|
| $P_i$ | 0.2 | 0.1 | 0.1 | 0.15 | 0.3 | 0.4 | 0.3 |
| $C_i$ | 5 | 8 | 6 | 4 | 5 | 4 | 15 |

**FIGURE 5.10** Production of a unit follows the schematic diagram.

7. Production of a unit follows the schematic diagram shown in Figure 5.10. $P_i$, the probability of generating a defective unit in stage $i$, and $C_i$, the unit cost of processing in stage $i$, are given in the accompanying table. Determine the cost of producing 100 good units and the number of units that must be started (i.e., the value of $N$) in stage 1 to obtain 100 good units.
8. Product A and by-products B and C can be produced with the production line shown in Figure 5.11. The average percentage of the input product that will be channeled into a branch is shown by the numerical value above that branch. The cost data for each unit of input in each stage are given in the accompanying table. Determine the average cost of each unit of products A, B, and C.

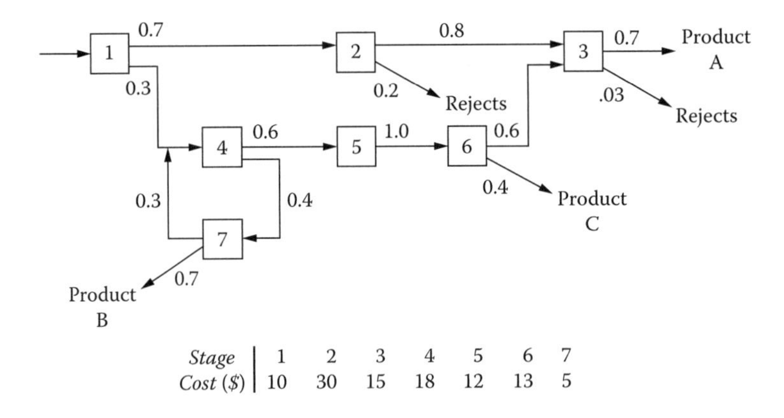


| *Stage* | 1 | 2 | 3 | 4 | 5 | 6 | 7 |
|---|---|---|---|---|---|---|---|
| *Cost ($)* | 10 | 30 | 15 | 18 | 12 | 13 | 5 |

**FIGURE 5.11** Product A and by-products B and C can be produced with the production line.

## SUGGESTED READINGS

### Machine Selection

*Applied Science and Technology Index*, H.W. Wilson Company, New York, available at: www.hwwilson.com/Databases/applieds.htm.

*Thomas Register of American Manufacturers and Thomas Register of Catalog File*, Thomas Publishing Company, New York, 1980.

*Ulrich's International Periodicals Directory*, R.R. Bowker Company, New York, 1980.

### Labor Requirement and Selection

*Directory of Occupational Titles*, U.S. Department of Labor, Washington, DC, 1977.

# 6 Lean Manufacturing and Supply Chain

Lean manufacturing and supply chain are currently the two dominant concepts in the manufacturing industry. Both lead to substantial improvement in productivity and, ultimately, efficient utilization of resources. In this chapter, we briefly discuss the basic ideas under these concepts and what make them so useful.

## 6.1 LEAN MANUFACTURING

Lean means no waste, no misuse, no unwanted use of resources, no overloads, no shortages, no spares, no leftovers, and no surpluses or extras, no delays, no wasted motions, no poorly designed system, and no poorly designed or managed supply chain operation. Lean manufacturing is a means to produce the required product in the most efficient manner using the least amount of resources in the most productive fashion. Lean means constantly improving each segment of the organizational operation, and involving everyone within the organization, that is, workers, management, and other partners.

The basis for this concept is not necessarily new. In the 1800s, Eli Whitney, although more popularly known for inventing the cotton gin, also advanced the idea of interchangeable parts to reduce production cost. When he designed assembly lines, Henry Ford was developing a lean system. Components of auto assembly were broken down into elements and were distributed among workstations in a continuous production process, so that people, machine, and tools could be utilized in a most efficient manner. Industrial engineers have been studying and applying the scientific management concepts of Frederick Taylor, and Frank and Lillian Gilbreth since they were first introduced in the early twentieth century. Popularly known as *work measurement* and *time study*, these methods standardized the way work should be done by eliminating all (as far as possible) unnecessary motions and designing the workplace so that these motions can be routinely performed with ease and without exhausting the worker. Optimum time to perform a particular task was established and designated as a time standard. These times were then used to develop wage plans. Frank Gilbreth also advanced the idea of visual representation of a process via a process chart. These charts clearly identify the nonproductive or bottleneck work elements, which are then studied and marked for improvement or elimination.

However, there was a problem with the concept of scientific management. It was used by organizations to utilize workers as a resource mainly to perform assigned tasks in the way they were assigned to them. Workers were rarely asked for suggestions as to how they could improve their work environment even though they were probably the most familiar with it. Workers were valued for their physical contributions but were rarely asked for mental input.

Furthermore, fixed assembly lines are not able to respond to frequent changes in the marketplace such as new product features, changes in demand pattern, or changes in consumer tastes. Toyota, in contrast, can produce automobiles quickly and economically by using a system known as the Toyota Production System (TPS), from which the term *lean manufacturing* originated.

Lean manufacturing encompasses much more than mere production. It includes all aspects of processes between receiving a customer's order to shipping that order in a most efficient manner. It therefore includes multiple topics on how a customer's order should be managed. This process must continuously respond to a changing environment. In lean manufacturing, one must constantly evaluate and improve. Lean is a way of thinking, developing new ways to improve efficiency, and reduce waste, and responding to changes in production and customer environments. There are a number of recommended techniques in identifying problems and making improvements. These techniques vary in their applicability, and from merely improving methods that improve the workplace to procedures that may result in large-scale changes in thinking, thereby leading to major procedural modifications. We describe some of these procedures in the following sections.

### 6.1.1 5S System

The system was first introduced in TPS. In Japanese and the corresponding rough English translations in parenthesis, the 5 Ss stand for seiri (sort), seiton (straighten), seiso (shine), seiketsu (standardize), and shitsuke (sustain). Each step adds to improvements in a workstation.

#### 6.1.1.1 Sort

Sort means removing all unneeded tools, materials, outdated equipment, and even workers from the workplace. This results in a clean and productive work environment. Items that are not used are discarded, creating a clean and neat workplace. This also reduces inventory loads and deadweight. The guiding principle is "when in doubt, move it out."

To implement *sort*, we can use a Red Tag Technique (RTT). Here, potentially unneeded items are marked with a red tag. These items are then removed to a temporary holding area. If the items are not missed, they are not necessary and can be disposed of. If they are required only on an infrequent basis, a separate storage for all such items can be created away from the workstation.

A checklist for RTT items can be constructed indicating search spaces, storage spaces, and unneeded items to look for. One example of RTT inspection sheet is:

**Red Tag Inspection Sheet**

| Search Spaces | Search Storage Spaces |
|---|---|
| Floor | Shelves |
| Aisles | Racks |
| Workstations | Closets |

| | |
|---|---|
| Stairs | Sheds |
| Behind/beside equipment | Other |
| Small rooms | |
| Loading docks | Look for materials in |
| Office areas | Sign boards/bulletin board |
| | Walls |
| | Other |

| Red Tag Inspection Sheet: Look For Unneeded | | | |
|---|---|---|---|
| Equipment | Materials | Furniture | Items |
| Tools | Raw materials | Chairs | Boxes |
| Jigs | Supplies | Tables | Skids |
| Fixture | Parts | | Oil cans |
| Dies | Finished product | | Work shoes |
| Wires/electrical equipment | | | Trash cans |
| Pipes/plumbing | | | |

| | | | Reason For Disposition | | | | | | | |
|---|---|---|---|---|---|---|---|---|---|---|
| Date | Item | # | Obsolete | Defective | Scrap | Unneeded | Seldom Used (Number of Times per Month) 0 1 2 3 4 | Use Unknown | Other | Signature |

| Item Disposition | | | | | | | | | | | |
|---|---|---|---|---|---|---|---|---|---|---|---|
| | | | Action Taken/Proposed | | | | | | | | |
| Date | Item | No. | Sell | Hold for Depreciation | Give to Charity | Junk | Store in Area | Return To Sender | Keep In Place | Other | Signature |

### 6.1.1.2 Straighten

After sorting, organize and straighten the remaining items for easy access, making sure that key items are within reach. These guidelines may indicate where to place things:

1. An item that is often used should be placed closer to its use.
2. Every item should have a designated place and name.
3. Make items easy to get and replace.
4. Keep the floor clean and safe by not placing things on the floor as much as possible.

Start straightening by placing large items such as machines, large equipment, and benches in their proper places. Develop the storage areas around them for necessary storing of articles such as racks and shelves, pallet areas, and inventory. Then place smaller items such as stools, chairs, and carts.

There are three techniques that can be used so people can easily understand the placement of things. We can choose one or all, depending on the audience and ease of perception.

##### *6.1.1.2.1 Lines*

There are various types of lines that can be effectively used to straighten out the workplace. Dividing lines denote aisles and work areas. Within the work area, positions of equipment and other items can be marked off with marker lines. Range lines are drawn to indicate the range of motion of each equipment (this is required for certain activities such as opening a machine door). Limit lines define maximum height, maximum and minimum inventories, and features that have limitations. Tiger lines draw attention to safety hazards. Arrow lines display directions and shadow lines display placement of hand tools.

##### *6.1.1.2.2 Labels*

Good labels are also useful. They can be color-coded to distinguish different items, different areas for storing these items, and so on.

##### *6.1.1.2.3 Signboards*

Signboards are another means of identifying things. They indicate the location of equipment, work areas, and inventories. They can also display directions to different areas (Figure 6.1).

**FIGURE 6.1** Five S equipment part signs (inventory)

#### 6.1.1.3 Shine

As soon as the required items are maintained with minimum possible inventory and are set in the most efficient places, then clean again. Clean everything so that there is no dirt or grime or any other form of contamination. Start from the top and proceed to the floor. Cleaning should include ceilings, light fixtures, pipes, machines, tables, cabinets, drawers, walls, machines, mats, and floors. In this way, cleaning may reveal underlying issues such as reasons for low worker productivity, existence of old and outdated equipment, or material-handling bottlenecks. For example, it may reveal worn-out wires, hoses, and tubing. This step should result in fewer breakdowns in the future and safer working conditions (Figure 6.2).

#### 6.1.1.4 Standardize

It very seldom happens that setting procedures and processes in motion will have immediate success. Setting unrealistic goals can only create dissatisfaction. Multiple trials may be needed to develop a satisfactory method and workplace. Once an efficient process is established, standardize it. Maintain the workplace as you want it: items in specific locations, items easy to locate and replace, a clean workplace, and so on. Doing so will maintain the gains obtained so far. Monthly cleaning audits or preventive maintenance schedules are some examples for standardization.

#### 6.1.1.5 Sustain

The success achieved by applying the previous steps develops a new awareness and set of skills within the organization. Continue applying it in other areas of the organization. Videotape the success. Develop training sessions for new teams. Get managers, workers, and other members of the organization involved for constant and continuous improvements.

**FIGURE 6.2** Five S system applied to office

### 6.1.2 Six Sigma

This is a data-driven procedure for enhancements of business performance. It measures and implements process improvements and is a systematic approach to problem solving. It consists of six phases:

1. Define
2. Measure
3. Analyze
4. Improve
5. Control
6. Replicate

#### 6.1.2.1 Define

In the define phase, the problem is identified and boundaries are established. Problem definition should be specific, observable, measurable, and manageable. It should be relevant and should result in a major benefit. Definition should not entail who is at fault or imply blame or a cause for the problem, or what the solution should be. If we already know the solution, then there is no need to perform the analysis. Most often such solutions will not be optimal.

Here is an example of a good problem definition: There is a 12-day interval between the time the order is received and the date of actual delivery, while competitor A has a lead time of 6 days only. Objective: to reduce our lead time to less than 6 days.

In the above statement, the problem is identified, is measurable, assigns no blame, and requires an outcome that is also identified in measurable terms.

#### 6.1.2.2 Measure

Collect and analyze the present operational data to set the initial baseline. Determine what is happening now. Collect only the data and information on the problem that we are interested in and determine the causes of the problem.

Various tools are used in this phase. They include:

##### *6.1.2.2.1 Process or Flow Diagrams*

Process or flow diagrams are charts that display (in symbolic format) the total operation. It is therefore necessary to know the boundaries of the problem and collect a team of people who can trace the entire process. Any bottleneck and unnecessary steps in the process are easily discovered in the act of drawing these diagrams.

##### *6.1.2.2.2 Pareto Diagram*

Next, the operation is discussed in detail with the help of a process diagram, and causes that may have resulted in the problem are identified. These causes are ranked and plotted based on their relative frequencies, which results in a Pareto diagram. A few vital sources that contribute the most can be identified for improvements by using a Pareto diagram. A sample Pareto diagram is shown in Figure 6.3.

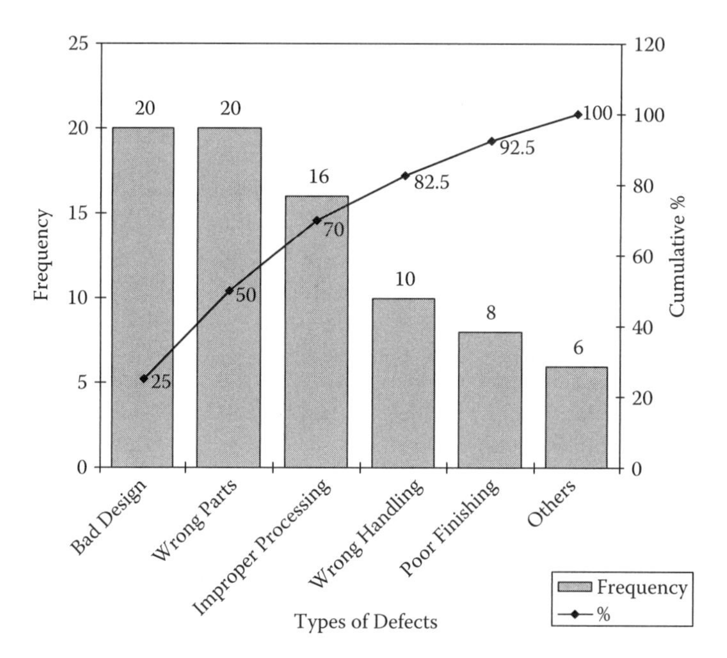


**FIGURE 6.3** Pareto diagram for types of defects

Data stratification based on some criteria and then drawing a Pareto diagram for each, may also add to the clarity of the data. The classification may be based on process, costs, services, steps, manpower, time, place, or type of problems.

#### *6.1.2.2.3 Function Deployment Matrix*

A function deployment matrix is similar to Quality Function Deployment (QFD), where causes that may have the greatest influence on different problems are identified. A matrix is created where a column may display problems, and rows the perceived reasons for these problems, which presently exist in the observed process. The severity of each problem is estimated by assigning a weight for it. A weight is assigned based on how we think each cause may influence a specific problem. The product of these two numbers gives the relative weight of a present procedural activity and the problem. Add all these values for each cause (each row value). The one with the largest total is the most critical cause and so on. Identify the few critical cause corrections that may substantially improve the process. An example of QFD is illustrated in Chapter 2.

### 6.1.2.3 Analyze

In this phase, we try to identify the root cause of the problem. There could be many possible reasons that resulted in the problem. Brainstorming and cause-and-effect (CE) diagrams are useful tools in this phase. Brainstorming is where a group of interested people get together as a team to discuss the range of possible causes of the problem. Every team member is encouraged to contribute to the discussion. This should generate a broad range of alternatives. There are a few guiding principles for a good brainstorming session. They include:

#### *6.1.2.3.1 No Premature Judgment*

Innovative ideas are welcomed — more ideas, however wild, are generated, and more alternatives are available for analysis. Innovative thinking is required because, it is most likely, that all the easy and well-known causes have been previously examined. And even if they have not, all the ideas will be part of the group input that is discussed in detail at this stage. The objective of the group should be to generate a quantity of ideas and not to worry about the quality. Team members should build upon each other's suggestion as the discussion is progressing to expand on the trend of the team members' thoughts.

#### *6.1.2.3.2 Cause-and-Effect Diagram*

The CE diagram, also called the fishbone diagram, is another means of trying to understand the causes of the problem (Figure 6.4). It displays different reasons, including causes for those reasons, that may ultimately create the problem. Diagramming of CE provides a means to focus the discussion on the specific problem, as well as to encourage team members to be creative and contribute to the discussion, resulting in new ways of thinking about the problem.

The details of a CE diagram can be analyzed by asking questions (4Ws): what, when, where, who. Analysis can be further enhanced by asking questions relating

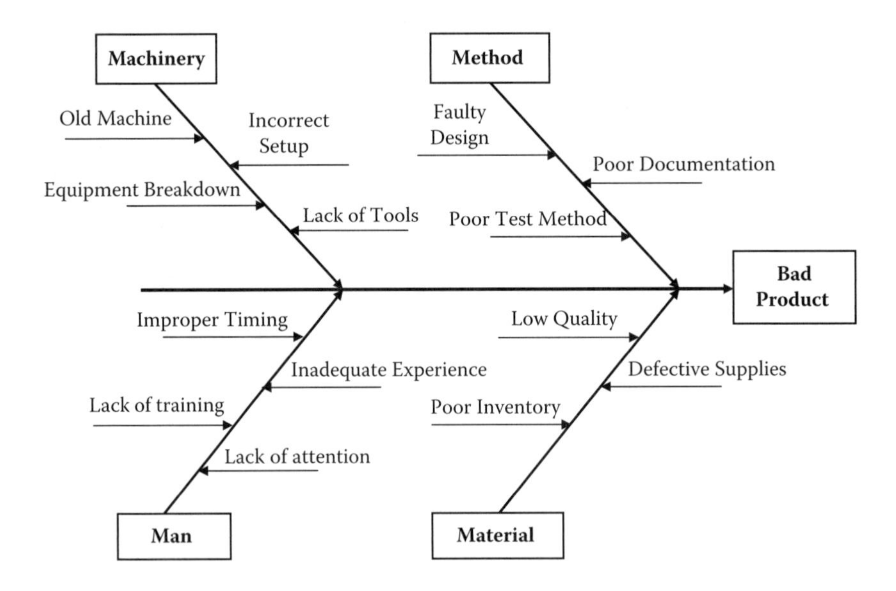


**FIGURE 6.4** Fishbone diagram (also known as cause-and-effect diagram) for a bad product

to 5Ms: manpower, material, methods, machines, and measurement; and/or 5 Ps: people (workers), place, procedure, patrons (customers), and provisions.

There are some cautionary notes on CE evaluation. CE diagrams are not a substitute for collection of data. They are theories that must be evaluated further by compiled data. Too many theories, should they all need evaluation, could require substantial data collection and analysis.

##### *6.1.2.3.3 Data Analysis*

Data analysis can be performed by applying basic statistical methods such as scatter diagrams, histograms, and/or curve fitting. Somewhat advanced methods of analysis of variance may also be necessary in some analyses. The purpose of this analysis is to identify the root causes that we can rectify. These are causes that are within our control and over which corrective actions can be taken.

#### 6.1.2.4 Improve

Among the different solutions to the problem, select the best option and try to implement it. It may involve identifying and acquiring additional resources, changing existing procedures, altering the tasks performed by people, and even shifting the number of workers utilized. This may generate resistance to change from those who are comfortable with the present operation, but with skill, such resistance can be overcome. Try some minitests, dry runs, and even computer simulations to show the effectiveness of the new methods to convince others to make changes.

#### 6.1.2.5 Control

Implement the best solution and make sure it performs according to expectations. If there are deviations between what is expected and what is observed, determine the cause and take corrective actions (Figure 6.5). Develop controls using automatic controls or control charts and continue to observe the process to make sure that it remains productive.

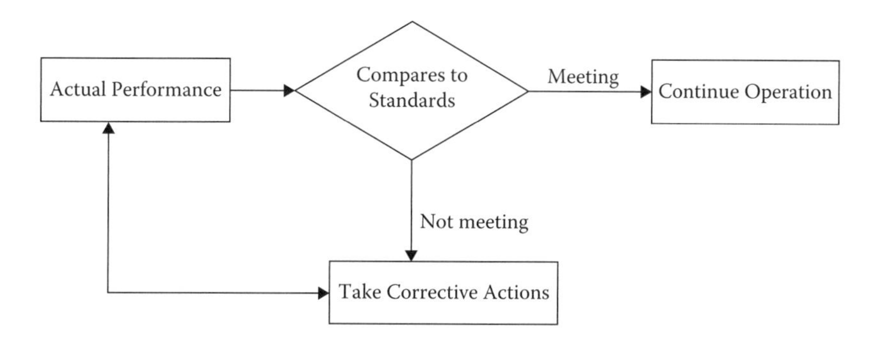


**FIGURE 6.5** Control process

A control spreadsheet can be developed with all the relevant information (Figure 6.6). One such spreadsheet may be:

| Variables or Attribute to Control | Standard Value of the Variable/ Attribute (V/A) | Measured Value of V/A | Who Measured | How and When Measured | What Action to Follow | Who Takes the Action | What is Actually Done |
|---|---|---|---|---|---|---|---|
| | | | | | | | |

#### 6.1.2.6 Replicate

Replicate the success to other processes in the organization. Once a success is achieved, it is easier to get involved in other projects, even the projects that were initially considered difficult. Both the analyzer and management will have more confidence, and further cooperation between them is now possible. It is also easier to sell the new analysis to both management and coworkers once a successful track record is established.

### 6.1.3 Kaizen Philosophy

Kaizen is Japanese for gradual and continuous improvement. "Kai" means change and "zen" means good. Together, it means "change for good." All employees are empowered so that each day they perform their task a little better — better than what and how they had done earlier. Small but continuous improvement results in increased productivity and cost reduction. There are two types of kaizen: (1) flow kaizen, which encourages continuing improvement; (2) point kaizen, which eliminates waste in the work environment.

The basic features of kaizen are:

1. Employee empowerment: Each person is encouraged, indeed helped, to make suggestions for improvement and carry those ideas through.
2. Discipline: When at work, as part of a team or at an independent station, a worker must have discipline or follow the rules that require trust in coworkers and management. It also means the working environment is conducive to new suggestions without fear of retribution.
3. Recognize: Each person contributing to improvement should be recognized monetarily or in some other way. There are various nonmonetary means available that encourage creativity. They include techniques such as verbal recognition within the group or, even better, by giving certificates and awards, giving publicity within and outside the organization, placing a story in a newspaper, and so on. Recognition encourages people to contribute more and more to the improvement of the organization.

Kaizen advocates several basic principles that are similar to other methods that we have seen so far. For example, here is one suggested list.

1. Say no to status quo.
2. Do not believe in old assumptions.

3. Do not make excuses as to why it cannot be done. Think how it can be done.
4. Do not worry about being perfect. Even if you get only half the improvement, it is better than no improvement.
5. Money is not necessarily a basic requirement for every possible improvement. Solve the problem with wisdom and not money.
6. If you see something that needs to be fixed, fix it. Correct mistakes immediately.
7. Good ideas only come after all the easy ideas are dispensed.
8. Ask questions constantly to get to the root cause.
9. Listen to more knowledgeable people for suggestions. It will result in a better ultimate plan.
10. Never stop kaizen.

Kaizen involves creating a culture that will continuously encourage improvement and reduce waste.

A few examples of Kaizen improvement are

1. In a small manufacturing concern, a bar code system that is tasked to track employee attendance is used to develop an automated inventory system.
2. A transfer between two workstations of a very heavy workpiece was initially done by a cart pushed by two men. Next, a forklift was used, but there was a problem of getting the forklift on time because it was shared by the entire plant. A new solution was found, install a rail track between stations on which a cart can be easily pushed by one person.
3. A template is created to make it easy to drill holes in a workpiece with a one shaft drill bit.
4. Use energy-saving bulbs instead of a regular lighting fixture on each machine.

In each case, small improvements are a result of continuing observation of the system.

### 6.1.4 ANDON

*Andon* is an ancient Japanese vertically collapsible paper lantern. It is open on top and illuminated by placing a candle in the center. The lighted lantern was used as a signaling device. In modern days, people also use light as a method of communication. For example, stores display whether they are open or not by using outside lights. In department and grocery stores, a consumer is made aware that a checkout line is open by a lane light. It is a method that informs the presence or absence of a service.

In a manufacturing environment, the same technique is used to indicate if a machine is operating or not. In fact, three colored lights — green, yellow, and red — are used to indicate normal operation, changeover or planned maintenance, and machine breakdown or abnormal condition, respectively. These signals could be generated by the machine operator or without human intervention (by automation). In some instances, orange and red are also tied to distinct sounds to attract the attention of a maintenance operator or a supervisor. A station with an andon system may be allowed to stop production of the entire line until the problem in the station is taken care of. This may avoid creation of excess inventory in other stations. Other reasons for using andon signals are part shortage, machine malfunction, other defects found, tool malfunction, or similar reasons that may require immediate assistance. In computerized control systems, andon signals can also be directly displayed on computers or on report cards. In more advanced computerized systems, it is possible to display where the signal is coming from, the reason for the signal, the nature of the stoppage, and who should attend to it: maintenance, material handling, or supervisory personnel.

### 6.1.5 Takt Time

Takt is a German word for "baton" and refers to beating at regular intervals. In lean manufacturing, takt time is defined as the maximum time allowed to produce a unit. The station time for each work cell, which is a part of the assembly line, is not allowed to exceed the takt time. For example, if there are 8 hours of operation time in a day and there is a daily demand of 16 units, then takt time is 8/16 = 0.5 hour. Each workstation in the assembly line must complete its operation in 0.5 hour for the assembly to work continuously without excess inventory or backlog at any point. It should be noted that takt time is not the cycle time. Cycle time is the actual time it takes to produce one unit. When customers demand changes, corresponding production rates change and takt should also be modified accordingly. This is not necessarily an easy task, but can be accomplished by speeding up the process, adding machines, and using overtime and/or temporary workers.

When all stations do not follow takt time, imbalance develops. If the station requires more time than the takt time, the required production rate cannot be met. If station time is much shorter than takt time, inventory develops at the station.

The advantages of establishing takt time are

1. It builds a balanced assembly line.
2. Each workstation or cell has a target time to build internal procedures and manpower to meet the takt time.
3. Psychologically, it gives awareness of the rate required of each worker/cell, and if a problem arises, it can be immediately addressed.
4. It minimizes work-in-process inventory.

It must be recognized that takt time is a suitable measure in a continuous production environment. It can not be used in a job shop situation because the production may or may not be continuous.

In Figure 6.6, the present takt time is 15 minutes. The process is obviously not balanced. Balance can be achieved in many ways. Reduce the working capacity at each station that has a time requirement of less than 15 minutes so all stations can produce with a cycle time of 15 minutes. Or make takt time of 10 minutes by combining operations of processes 1 and 3 and providing additional resources in process 4 so that the time there can be reduced to 10 minutes. At any rate, the diagram provides a clear picture and suggests alternatives to achieve a balanced line.

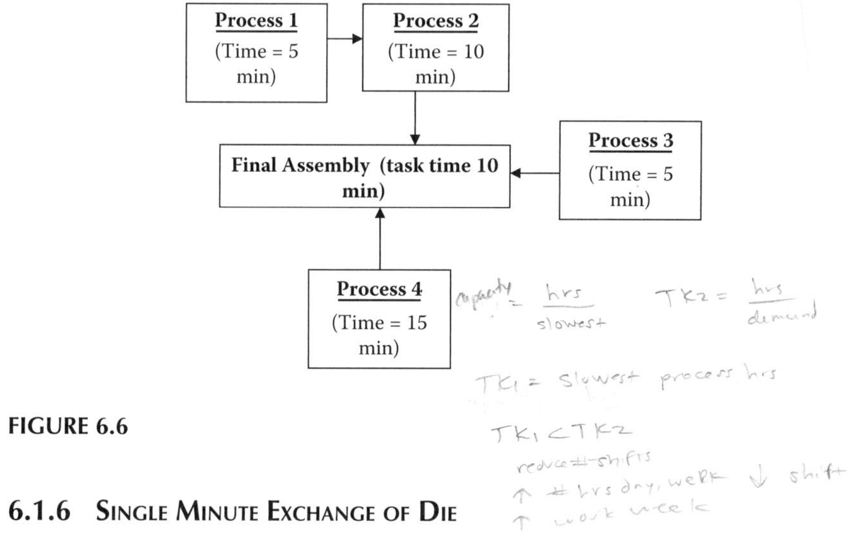


**FIGURE 6.6**

### 6.1.6 Single Minute Exchange of Die

The objective of implementing single minute exchange of die (SMED) is to reduce downtime that occurs when a machine is stopped to set up for a new product. In other words, it is a process to reduce the changeover time. In most cases, the changeover time can be reduced in terms of minutes and hence the name single minute exchange (of dies).

The setup times on a machine are due to a number of activities. Some are independent of the machine and are therefore called external setup, and others require the use of the machine and are called internal setup. Activities that are independent of the machine may include such events as collection of proper "paperwork" to know what and how much to produce; collection of auxiliary equipment and jigs and fixtures needed for the new task; and other setup activities that may not need the machine. For example, an item going into the machine may need some preoperation stages such as preheating and cleaning. The setup activities that require both machine and items together are activities such as tightening of the item in a fixture and adjusting the machine for operation are part of internal setups.

The SMED procedure proposes steps to reduce setup time:

1. Reduce the number of internal activities by either eliminating some requirements or converting them into external activities, if possible.
2. Reduce the time required to perform internal activities by simplifying activities, clamping rather than tightening, autoadjusting tools rather than using manual adjustments, and using standardized tools rather than special-purpose tools. The times can also be reduced by redesigning fixtures such as a quick-action wrench and/or pneumatic hold. Install one turn attachment. Use the product's dedicated fixture.
3. Group workers together as a team to do all setup activities. A good example of achieving efficiency is a crew working in a pit stop in a car racing sport.
4. Perform setup activities in parallel.
5. Use computers for documentation.

### 6.1.7 Value Stream Mapping

Value stream mapping (VSM) is a process of examining the entire chain of events that must happen from the time an order for a product is received until it is delivered. VSM maps the process as it presently exists, and suggests, via examination, improvements that are needed. It displays material flow, product flow, and information flow for the process. The mapping starts from when the order for the product is received and ends when the product is delivered to the customer. Ideally, all in-between stages that a product or a partial product must pass through must add value to the product and any nonvalue activities should be eliminated. VSM is drawn at a level higher than the process chart, generally consisting of about 10 to 12 blocks. The objective is to identify waste and try to find ways to eliminate it.

But what is waste? Sometimes it is difficult to distinguish waste from a necessary operation. Is inspection necessary or wasteful? As far as customers are concerned, they want a product with no defects and they do not care how it is made or inspected. Inspection assures the quality, but manual inspection adds nonproductive time to the process. A solution may be, first, to analyze if inspection can be avoided by improving the process, and if not, to determine if inspection can be automated. Similarly, is moving a unit manually from one station to another a waste? As far as customers are concerned, such an activity does not add value to the product. Perhaps, a conveyor will automate the transfer and will not add cost to the product.

A value stream map starts by creating a current state diagram. For each product, we can draw an independent value stream diagram or combine all the information in one diagram. VSM is appropriate where production is high volume and predictable with limited variety and limited components. Most products are produced on dedicated machines.

Suppose we are developing an independent diagram (Figure 6.8). Start with a production control block. Then, add a customer block to that. The demand from customers will determine the requirements within the plant and ultimately the raw material supply requirements. Start with customer requirements per month (or some other time unit) in terms of shipments made; for example, 300 units per shipment two times a month, and 500 units per shipment two times a month for total of 1600 units. Also, mark the trips needed and frequencies to deliver the product to the customer.

Next, draw a box for each operation in the process. Display order frequencies from production planning to each operation box. Attach as much information or data as needed to each operational box. This information may include such details as cycle time to produce a unit, time the operator spends in making a unit, machining or equipment time, changeover time, and lot size.

Takt time is used to see if production balance can be achieved. If there is imbalance in production between work centers, what can be done to smooth the production rate? Can resources be shuffled? In particular, check if operators can be shared with multiple machines by using machine coupling. Can batch size be reduced? Is cellular manufacturing possible?

Display the storage points and the minimum and maximum inventory maintained there. This can be in terms of days of production until the next operation. These points may need further analysis. Is inventory necessary? Is the amount appropriate? Can *kanban* be used?

Calculate the amount of raw materials needed from the suppler and note it in the diagram; again, display the frequency and quantity of containers that will be shipped from each supplier to the manufacturing facility. All relevant information is transmitted by the production department to each department within the plant and to both the supplier and the customer. It is easy to draw a value stream map using graphical software such as Microsoft Visio. The drag-and-drop function makes it convenient and easy to build the map. Typical symbols used in VSM are shown in Figure 6.7.

| Icon | Description |
|---|---|
| | This icon is used to represent the supplier, in the upper left-hand side, and to represent the customer in the upper right-hand side. |
| | This icon represents the storage or inventory of raw materials, finished products, or semi-processed goods between processes. |
| Process | This icon represents process, the operation through which the material is processed. |
| | This icon represents external shipment method of materials from suppliers, and to customers. |
| | Represents electronic means of information flow. |
| | This straight, thin arrow represents general flow of information. |
| | This "pull" arrow is used to represent removal of material from inventory. |
| | This "push" arrow is used to represent production or processing of materials, with or without need of subsequent processes. |
| | Represents shipment of raw materials from supplier to the inventory, or to the customers from the shipping. |
| | Represents operator/s for the process or operation. |
| | This represents temporary storage against the unexpected fluctuation in the demand, downtimes. |

**FIGURE 6.7**

After drawing the initial diagram, see if any improvements are possible in the existing setup before the final plan is made. Reduction of waste and ensuring that the raw material (semifinished product) is available when needed and in the quantity required are some of the considerations. The analysis may include:

1. *Improve the present process.* The present diagram may be examined for improvements in multiple areas. A flag should be raised if there is large production quantity, excessive inventories, lengthy setups, too many defective items, limited machine (process) availability due to frequent breakdowns, and information delays.
2. *Determine optimum production quantity.* Determine optimum production quantity by knowing the setup cost, inventory carrying cost and setup time, and required production rate. Apply inventory planning methods to determine the optimum production quantity.
3. *Identify bottleneck process (machine) and develop a satisfactory task time.* Find the cycle time for each machine; the one with the longest cycle time is the bottleneck machine. With that cycle time as the takt time, see if the required production rate can be met. If not, reduce the takt time by redistributing tasks from the bottleneck machine to machines where time requirement are less than the takt time or equal the ideal takt time. This should reduce cycle time for the bottleneck machine. If this action does not reduce the task time, reduce the process cycle time for the machine(s) that exceeds the ideal task time by either increasing process speed or purchasing a faster machine, using additional manpower, or sanctioning overtime.
4. *Make sure that plant and machine layout is optimum.* Plant layout should minimize material handling and wasted motions. Material-handling equipment should be properly chosen. In process inventory, storage areas should be in optimum places. A proper production environment such as assembly line, batch processing, and cell manufacturing should be investigated. Details of principles for plant layout and material handling will be presented in the rest of the book, but they should be applied in developing an efficient production facility.
5. *Determine production schedule.* Based on the production system used, planned production quantities for different products, their due dates, and penalties for deviating from due dates, an optimum production schedule can be established. Raw material deliveries can be planned. In-process inventories and an associated number of kanbans can then be established.

## EXAMPLE

Suppose we have a product that is produced by using processes IA, IB, II, and III. The value map diagram is shown below. A customer has ordered 1000 units per week and we need to establish a production and shipping schedule along with inventory points and in-process inventory space (Figure 6.8).

Based on demand, the minimum production per day is 1000/5 = 200 units. To produce this final output, we must make calculations for each process, input and takt time. Proceeding from the last process to the first, we can develop the following calculations.

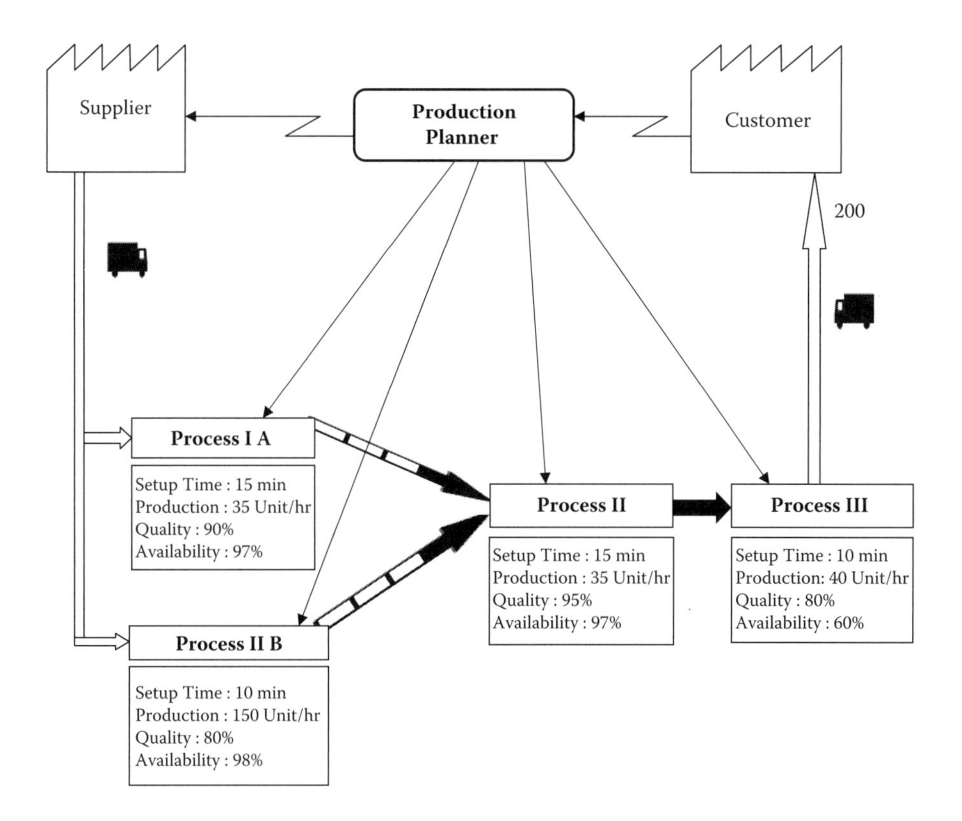


**FIGURE 6.8** Initial value stream mapping

For process III,

$$\text{Daily input required} = \text{output/quality} = 200/0.8 = 250 \text{ units}$$

The time necessary to process 250 units, in minutes, is setup time [in minutes] + units needed/(production rate [in minutes] × process availability).

$$= 10 + 250/(40/60) = 388.78$$

$$\text{Takt time} = \text{total time/input} = 388.78/250 = 1.55$$

Similarly, for process II,

$$\text{Input needed} = 250/0.95 = 263.15$$

$$\text{Time needed} = 15 + 263.1/(35 \times 0.97) = 479.9$$

$$\text{Takt time} = 479.9/263.15 = 1.824$$

For process IA,

$$\text{Input required} = 263.1/0.9 = 292.3$$

$$\text{Time required} = 15 + 292.3/(100 \times 0.99) = 192.1$$

$$\text{Takt time} = 192.1/292.3 = 0.657$$

And for process IB,

$$\text{Input required} = 263.1/0.80 = 328.87$$

$$\text{Time required} = 10 + 328.87/(150 \times 0.98) = 144.27$$

$$\text{Takt time} = 0.4385$$

The takt times are summarized in the following chart.

| | Process | | | |
|---|---|---|---|---|
| | **IA** | **IB** | **II** | **III** |
| Takt Time | 0.657 | 0.4385 | 1.824 | 1.555 |

Because these takt times are not similar, no smooth assembly line can be developed. The maximum takt time is 1.824 and is associated with process II, which makes process II the bottleneck process. Processes IA and IB can produce at a faster rate and thus build inventory upstream for process II, but process III is starved by process II and will have to pull the units from it.

In summary, to supply 200 units per day to a customer, each process will work as follows:

| | **IA** | | **IB** | | **II** | | **III** | |
|---|---|---|---|---|---|---|---|---|
| | **Input 328.8** | **Output 292.3** | **Input 292.3** | **Output 263.15** | **Input 263.15** | **Output 250** | **Input 250** | **Output 200** |
| Time needed | 144.27 | | 192.1 | | 479.9 | | 388.78 | |
| Time available | 335.73 | | 287.9 | | 0 | | 91.22 | |

Process II is the bottleneck. We should make every effort to see if we can relieve this bottleneck. The best that we can do is to reduce the time needed in process II to 388.78 or make a takt time of 1.555 before process III becomes the bottleneck. Suppose that, after a careful evaluation, it was decided there was no easy way to increase the production rate in process II and therefore it was concluded not to spend additional funds in increasing the capacity at process II. Let us assume that excess capacities at IA and IB can be used to produce a different product.

Input rate per station is obtained by dividing the total input needed by a stage by total time needed at that stage. For example, the input rate for stage IA is (292.3/144.27) × 60 = 121.56 per hour. Similarly, for IB, the input rate is (328.8/192.1) × 60 = 102.69 per hour. Similarly, input rates are calculated for all other processes.

The input rates of all stages are not equal (or nearly equal). This is another indication that a smooth assembly line cannot be developed and in-between temporary storages would have to be developed. The difference between the input rate of the succeeding stage and the output rate of the previous stage is the rate at which this inventory is built. Calculations for maximum inventory buildup are shown in the table below:

| Process | Minutes of Operation | Input Rate per Hour | Output Rate per Hour | Input Rate of the Next Process | Inventory Buildup per Hour | Total Inventory Buildup |
|---|---|---|---|---|---|---|
| IA | 144.27 | 100 | 90 (89.1) | 35 | 90 − 35 = 55 | 55/60 × 144.27 = 132.2 |
| | | | | Or | 89.1 − 35 = 54.1; | 54.1/60 × 144.27 = 130 |
| IB | 192.1 | 150 | 120 (114) | 35 | 120 − 35 = 85 | 85/60 × 192.1 = 272.1 |
| | | | | Or | 114 − 35 = 79; | 79/60 × 192.1 = 252.93 |
| II | 480 | 35 | 33.25 (32.25) | 33.25 (32.25) | — | — |
| III | 480 | 33.25 (325) | 26.6 (25.54) | | — | 32.25 × 8 = 258 |

The values for output rate are obtained by multiplying input rate by quality factor. The values within parentheses are obtained by multiplying the previous number by process availability, which is a true output of the process. However, for development of storage space, we shall consider the higher production number. The actual number of units available for shipment at the end of one 8-hour shift is 8 × 25.54 = 204.32 (Figure 6.9).

After eliminating or reducing all wastes from the current map state, the future state map is created. This map should closely match with what is desired.

### 6.1.8 Kanban

*Kanban* is Japanese for card, and is used as a visual record. It is used in a production system to pass a signal when a part or a box goes from one workstation to next. A kanban system is especially useful when a number of different jobs are produced on a machine or in a facility and demand for each product varies from time to time. When setup time is smaller compared to production time, setups can be easily switched between different products, because downtime of setup becomes negligible. It becomes unnecessary to produce a large quantity in one setup because inventory carrying cost becomes more dominating than the setup cost.

This is similar to an inventory system in a supermarket. A small stock of every item is kept on the self. The number of items is large; however, the stock for each item is small. Here, inventory cost is very high compared to resupply cost. When

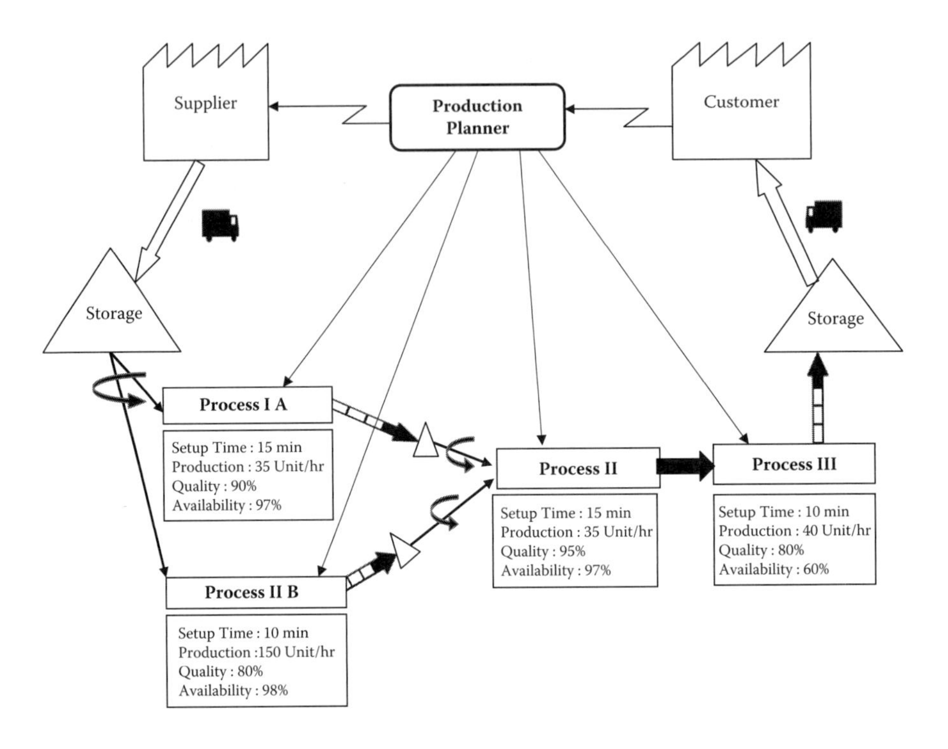


**FIGURE 6.9** Value stream with storage added

an item is purchased, it is scanned at the checkout counter. At the end of a specified period (e.g., 1 day or 1 week), orders are placed for items that have been sold in high quantities. Thus, inventory within the store is maintained at a specific level.

A kanban or circulating card is attached to a part (or box of parts). As the part moves upstream to the next station, the card is sent back to the previous station. When operation in the current station is completed, another card is attached, which will then be returned when the part goes onto the next station. And so on. When a production station collects a predetermined number of kanbans, it is equivalent to getting a signal to produce the next batch (Figure 6.10).

A word of caution: Make sure that sufficient input is available at each station to produce the quantity necessary. For example, let us assume that there are two stations in line and product goes through station 1 and then to station 2. Suppose that the optimum quantity to produce in station 2 is 2. Station 2 will pull 2 units from station 1 at a time as an input. To accommodate this, station 1 must have at least 2 units in the output side storage; this means that it must produce at least 2 units when it is in production mode. If we decide to produce 3 units as an optimum quantity in station 1, then the station will require three kanbans to return from station 2. This becomes a problem because station 2 can send back, at most, only two kanbans. Because these rates do not match, we will have an imbalance in the production line. The production quantity at station 1 should be 2 or a multiple of 2 such as 4, 6, etc.

**FIGURE 6.10** Rack with location identified for different job kanbans.

The advantages of the Kanban system are

1. It gives a visual signal to each station as to when to produce.
2. It reduces inventory.
3. It makes scheduling simple and automatic.

### 6.1.9 Total Production Maintenance

Total Production Maintenance (TPM) is a process of keeping machines productive and operating efficiently so that the throughput is improved. Throughput, or its equivalent, overall equipment effectiveness (OEE), is defined as

$$\text{OEE} = \text{available time on the machine} \times \text{machine speed} \times \text{yield of the machine}$$

$$\text{(i.e., good units of product produced/total units input to the machine).}$$

There are three terms in the right-hand side of the equation. If we try to increase each, overall effectiveness should increase. We shall examine each term in the ensuing discussion.

#### 6.1.9.1 Available Time on Machine

Available time on the machine is when the machine is working. Assuming there is sufficient work available to keep machines engaged all the time, the machine is not working when it is down or has failed. Three types of failures are possible. They are:

##### *6.1.9.1.1 Preventive Maintenance*

Maintenance activities performed to prevent failures and keep the machine in good working condition are called preventive maintenance (PM) activities. These activities might include regular oiling, cleaning the machine and the surrounding area, tightening nuts and bolts, etc. PM can be further classified as time-based and diagnostics-based. Cleaning equipment every so often, such as daily oiling of the machine, are time-based maintenance activities. Examples of PM activity that are condition based are: if a bearing starts making noises, it is time to replace it; if a battery-powered light is dimmer than usual, it is time to replace the battery. These conditions can also be measured conditions; for example, if the pressure does not reach the required level in a pressure hose, then we check the compressor. Vibrations may need corrective action in balancing. This is also called "conditioned monitoring" maintenance.

##### *6.1.9.1.2 Corrective Maintenance*

Corrective maintenance involves improving the design of machines and/or equipment so that they will have a longer working life, for example, replacing a motor with a higher horsepower motor if it will improve the performance (retiring weak components whenever possible). Corrective maintenance requires analyzing the system to determine how spending of a few dollars may lead to substantial improvements in maintainability and performance of the equipment.

##### *6.1.9.1.3 Breakdown Maintenance*

This is the worst type of maintenance: repairing the machine whenever it fails. If a machine has failed, it means some major problem has developed that was not foreseen or could have been prevented by PM. These problems tend to be expensive both in terms of repairs and in terms of loss in production.

To increase the available time of the machine, maintenance time should be reduced. It can be done by distributing maintenance activities among production workers and the maintenance crew. For example, PM activities can be assigned to production workers. These are workers who are close to the machines and are intimately involved in keeping the machine in good working order. They should be allowed to perform maintenance activities whenever required.

Breakdown maintenance should be assigned to maintenance crews who have expertise and have authority to obtain necessary parts from outside vendors. Corrective maintenance may be shared by both production and maintenance workers.

An additional means of reducing machine downtime is designing of tools and fixtures that minimize setup time. Doing 5S analysis in advance also improves productivity.

#### 6.1.9.2 Machine Speed

The second term in the OEE equation is machine speed. Machine speed can be increased by acquiring machines with as much high speed and automation as funds will allow.

#### 6.1.9.3 Increase Yield

There are various concepts that, if followed, will increase the yield. They involve paying attention to quality, which in turn can be improved by using some of the quality control techniques such as zero defects, poka-yoke, preventing defects at source, and quality assurance.

#### 6.1.9.3.1 Zero Defects

This Six Sigma concept involves producing no defective products. This is possible if we have a product assigned to a machine such that

1. Machines have the production accuracy that can produce the product on a continual basis within specification limits. One way to achieve this is to select a machine that has a specification limit of the product > 6 which is the standard deviation of the unit produced by the machine; and
2. Having well-trained operators working on the machine so that machine settings and quality of the product can be continuously monitored and controlled.

#### 6.1.9.3.2 Poka–yoke

The Japanese quality control guru Dr. Shingo developed the concept of *poka-yoke*, which means "mistake-proofing." The concept stresses developing technical features such that first mistakes are prevented and then, if necessary, mistakes are corrected in real time. We see many instances of poka-yoke in the products that we buy. Packaged goods requiring customer assembly have poka-yoke built into them so that the product can only be assembled in the right way. For example, computer cables for connecting speakers, printer, keyboard, and mouse are color-coded and cannot be inserted into the wrong socket.

Jigs and fixtures can be developed in a particular production environment so that no mistakes are possible. Performing 100% inspection and making immediate corrections for any deviation is also a part of poka-yoke. For example, all glass bottles produced in a glass plant go through 100% inspection. The inspection is done by machine vision. All defectives bottles (bottles with crack, air bubble, nick, etc.) are automatically taken off the line and recycled in the glass furnace. Only good units go to the customers.

#### 6.1.9.3.3 Source Inspection

This concept means 100% inspection made at the source. Purchased items are inspected 100%, either by the supplier or by incoming inspection at the plant. Other inspections are performed immediately after an operation is completed. Source inspection involves identifying defects in each stage of operation while the product is passing through various machines. This inspection can be made at the station where the unit had the operation performed either by the operator or by some automated inspection process. Another means of achieving 100% inspection is to have the operator of the next operation in the sequence perform the inspection (Figure 6.11).

#### 6.1.9.3.4 Point of Use

The idea of this technique is to keep everything that you need for the operation as close as possible to where it is used. This includes tools, instructions, fixtures, measuring equipment, and visual charts. If there are multiple stations using the same equipment, buy additional units or bring the stations as close together as possible with the equipment.

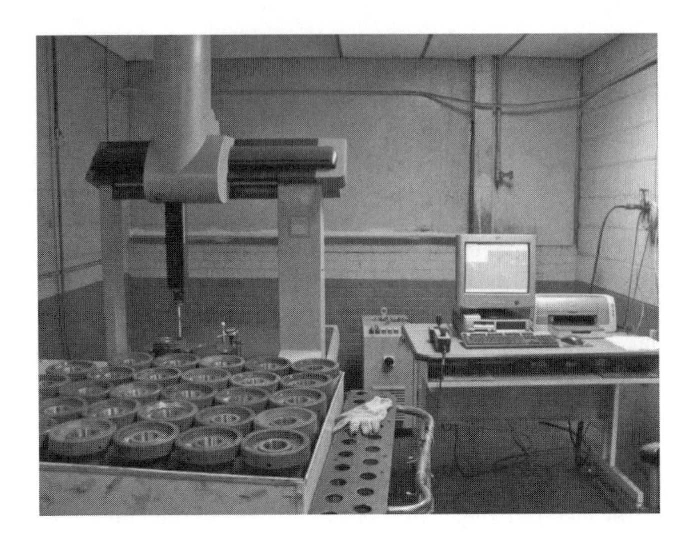

**FIGURE 6.11** Automatic inspection: computerized mold measurement machine measures blank mold volume

### 6.1.10 Visual Management

Visual aids are easily understood by everyone. Charts, graphs, hyetographs, or pie charts can display data, present performance, and expected performance. The data may represent production, safety record, attendance record, defective/scrap rate, or any other information. The data should be displayed in a prominent location so that they can be quickly communicated with everyone. It should be modified frequently to match the current data.

Visual aids are used in training, safety awareness, analysis of performance failures, and recognition of achievements to boost employee morale (Figures 6.12–6.16).

**FIGURE 6.12** Forklift and man aisle boundaries are clearly marked

**FIGURE 6.13** Visual management: defining hearing protection zone

### 6.1.11 Multiskill Workers

Lean systems encourage workers to have multiple skills so they can be productive in any area of expertise that is currently needed. Total productivity, including maintenance activities, encourages multiskill operators who are independent thinkers with good technical knowledge and skills. It places the responsibility of improving production on all employees.

**FIGURE 6.14** Visual management: job sequence on a machine for the next operation is clearly identified by a magnetic number

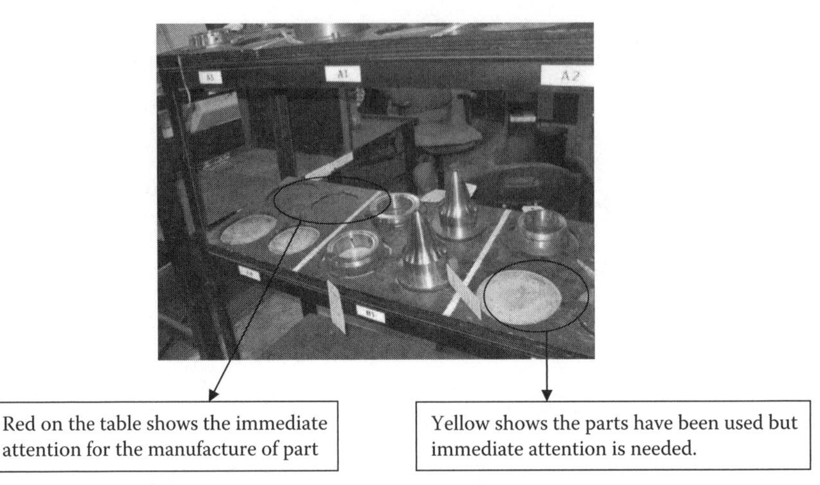
A1
A2
Red on the table shows the immediate attention for the manufacture of part
Yellow shows the parts have been used but immediate attention is needed.
Visual Job Sequencing on a CNC Machine

**FIGURE 6.15** Visual job sequencing on a CNC machine

### 6.1.12 A Case Study in Lean

The following is a case study of the application of Lean Principles in a machine shop where pistons and piston rings are made. The presentation shows the sequential steps involved in applying the Lean Process. The study itself is self-explanatory and hence no further detailed comments are provided except to note a few additional techniques.

The PICK chart lists all projects in terms of the efforts involved and the corresponding payoff. P stands for projects that are possible, I are the projects that should be implemented, C displays projects that are challenging, and K displays projects that are difficult with no significant payoff and should therefore be discontinued.

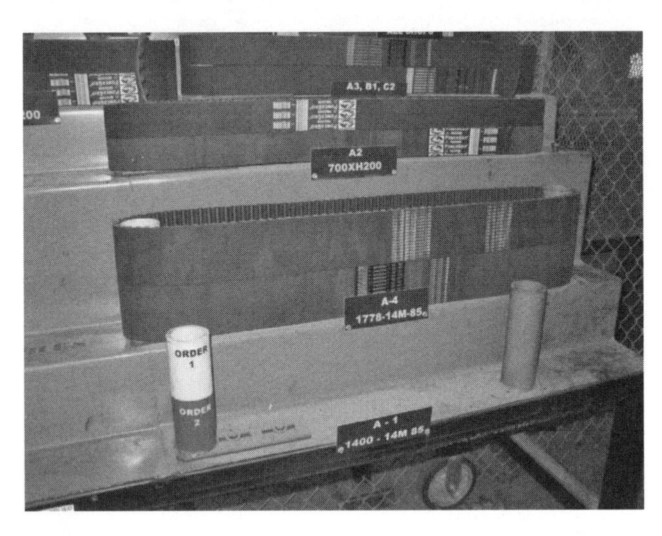
A3, B1, C2
A2
700XH200
A-4
ORDER
1
ORDER
2
A-1

**FIGURE 6.16** Simple visual management system displaying the number of belts to order based on the empty space on the belt rack

The spaghetti diagram is used to display or sketch the existing material handling process and observe any possible improvements in material flow.

The study also shows another common use of VSM in making process improvements. First, VSM is drawn for the present status to see where improvements may be possible and a new VSM is drawn that displays the effects of the new plans (Figures 6.17–6.68).

Machine Shop
Lean Processing

**FIGURE 6.17** Title page for the case study

*Machine Shop - Plungers and Rings*

RIE Mission:

1. Review current state VSM and identify opportunity for improvement
2. Improve overall quality to the customer
3. Eliminate waste and improve flow through the shop
4. Implement sustainable systems to ensure quality

Objectives:

1. Reduce lead time by 50% (from 30.7 to 15.4 hrs. for plungers, from 5.3 to 2.5 hrs. rings)
2. Consistent quality
3. Reduce cost

Prework:

1. Establish product families and consumption rates
2. Define inventory levels and remove excess
3. Relocate machinery
4. Train team
5. Create layout
6. Perform task analysis

Monuments: Modifications must be in alignment with other LEAN projects

Event Dates:

Champion:

Sponsor:

Facilitator:

Coach:

Team Members:

**FIGURE 6.18**

**The 3 p's**

- ***Purpose*** – "Why are we here?"
  - To Create a Plunger and Ring Cell
- ***Process*** – "How are we going to get there?"
  - Establish Customer Expectations
  - Review Current State VSM
  - PICK Chart Opportunities
  - LEAP Assessment
  - Establish Standard Work
  - Implement Visual Management
- ***Payoff*** – "What is in it for me?"
  - Reduced Cost
  - Increased Profitability
  - Ensuring Our Future

**FIGURE 6.19**

**Deliverables**

- Production
  - Consistent Quality
  - Readily Available Supply
  - Zero Defects
- Machine Shop
  - Reduced Waste (Inventory)
  - Five S
  - Standard Work
  - Visual Management
  - Kanbans

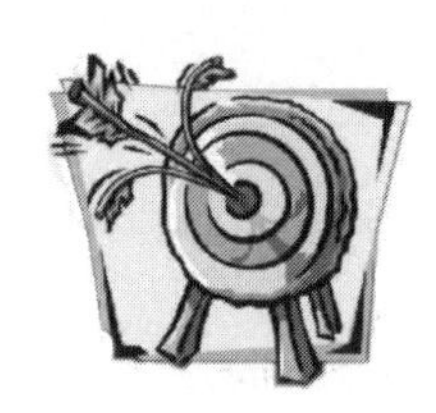

**FIGURE 6.20**

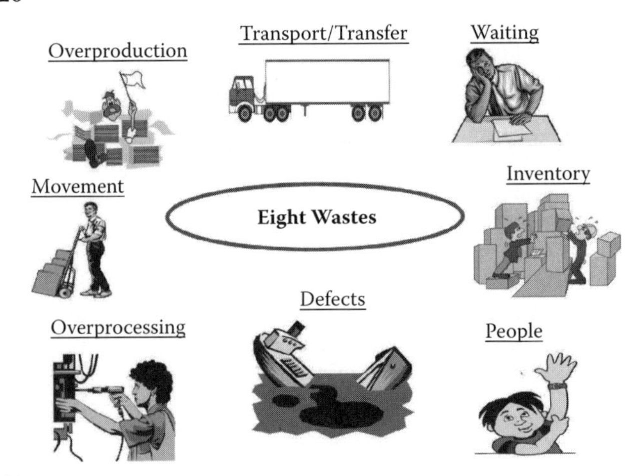


**FIGURE 6.21**

**First Day Agenda**

- Review the Charter/The 3 P's
- Deliverables
- Part Families and Categories
- Plunger and Rings – Shop Layout Spaghetti Diagram
- Value Stream Maps (VSM) – Current State
- Waste Walk
  - Identify Opportunities
  - Discuss/Brainstorm Opportunities

**FIGURE 6.22**

**Plunger and Ring Cell Part Families**

- Plungers
  - Category 1 (< 4" Profile Height)
  - Category 2 (> 4" and < 9" Profile Height)
  - Category 3 (> 9" Profile Height)
  - Category 4 (Optic Plunger)
- Rings
  - Welded
  - Bushed
- HE Plunger (Non-optic)

**FIGURE 6.23**

**Number of Pieces in Inventory by Category (Press Plungers)**

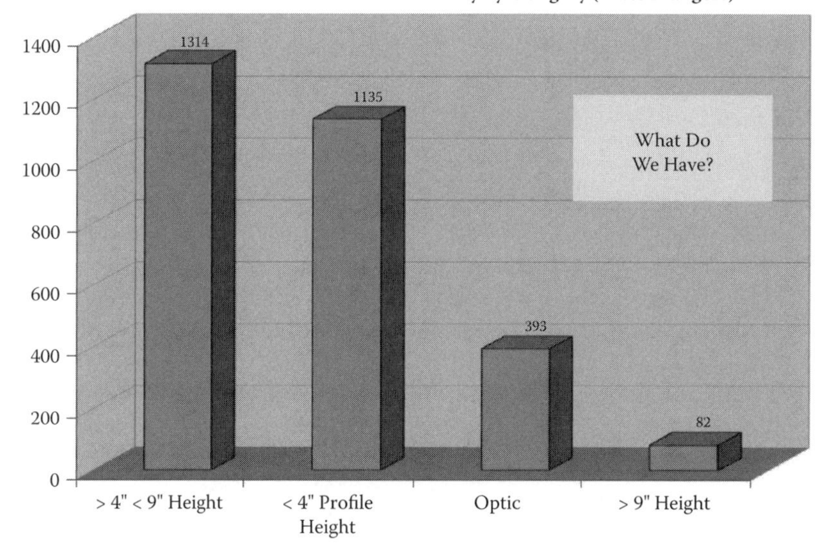


**FIGURE 6.24**

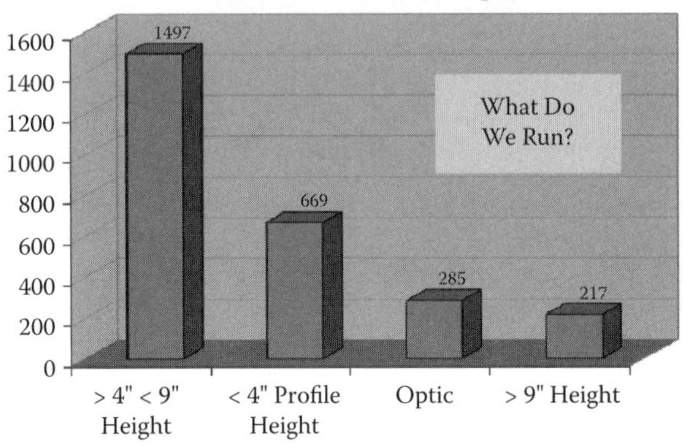


**FIGURE 6.25**

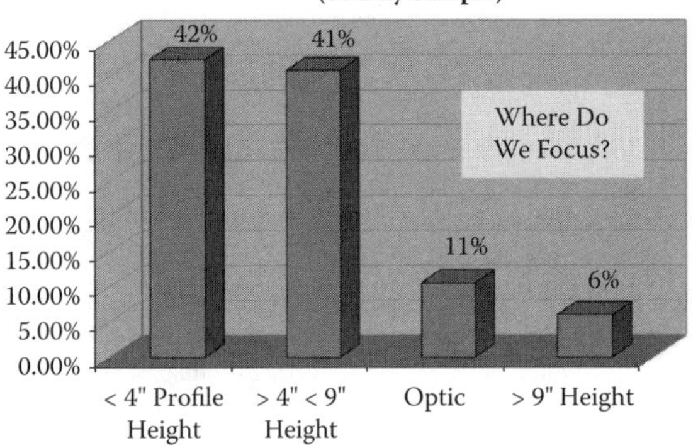


**FIGURE 6.26**

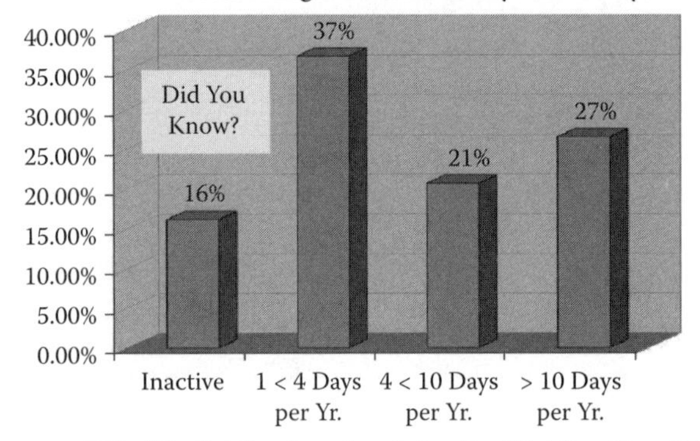


**FIGURE 6.27**

**Current Plunger Inventory Level**

- Total Inventory = 2924
- Inactive Items = 481
- Double Gob Block Mold (143 Items)
  - Average Number in Set = 10.9
  - Min = 3, Max = 32, STD = 4.1
- Single Gob Split Molds (100 Items)
  - Average Number in Set = 8.7
  - Min = 4, Max = 17, STD = 2.7

FIGURE 6.28

**Current Plunger Inventory Level**

- Total Inventory = 3890
- Inactive Items = 472
- Double Gob Block Mold (143 Items)
  - Average Number in Set = 15.5
  - Min = 4, Max = 67, STD = 9.8
- Single Gob Split Molds (100 Items)
  - Average Number in Set = 12.2
  - Min = 4, Max = 38, STD = 5.7

FIGURE 6.29

**Press Rings**

- Welded Rings
  - 82% are Repaired in this Manner
- Cast Iron Bushed Rings
  - Typically Required for Tall Straight Items
  - 28 Items Currently Defined

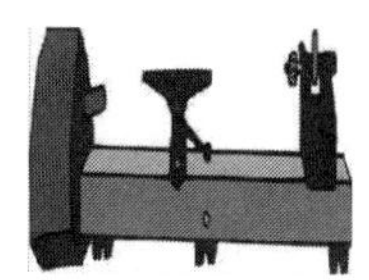

FIGURE 6.30

**Pareto Analysis (80/20 Rule)**

- Exclusions (Not Included in the 80%)
  - Plungers > 9" in Profile Height
  - Optic Plungers
  - Inactive Plungers
  - HE Plungers
- Other Exclusioins
  - Welding Operations

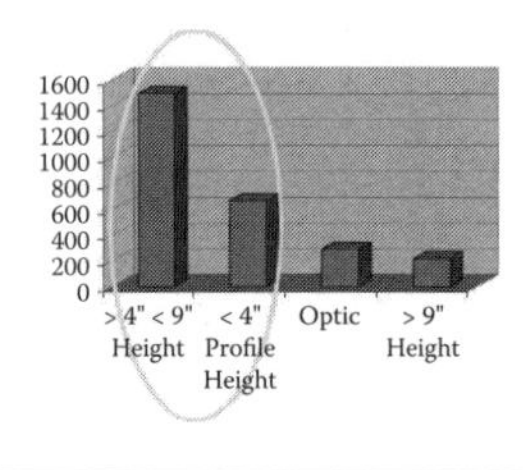


**FIGURE 6.31**

**The Focus of this RIE**

- Plungers
  - Category 1 (< 4" Profile Height)
  - Category 2 (> 4" and < 9" Profile Height)
- Rings
  - Welded and Machined
  - Cast Iron Bushings

**FIGURE 6.32**

**Customer Feedback**

- Plunger Threads Not Cleaned
- Clearances Out of Spec
- Tags Not Readable (Need Seq. #)
- Better Quality
- Equipment Placed in Correct Shop

**FIGURE 6.33**

## Layout – Current State

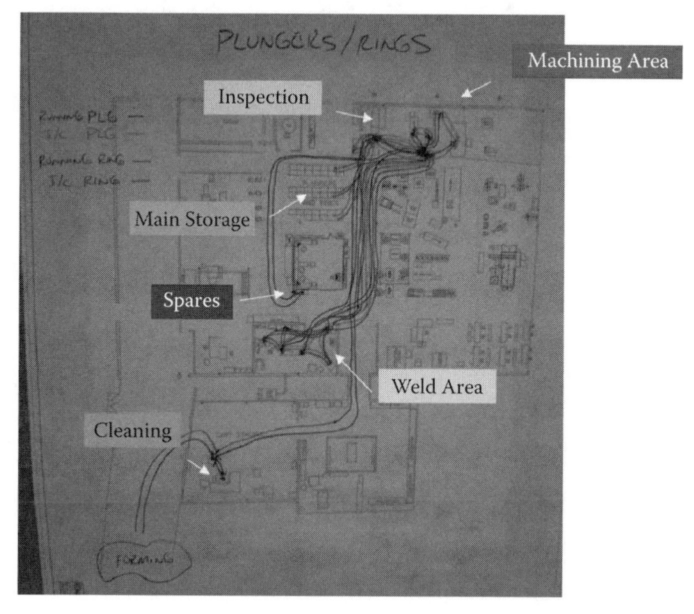


**FIGURE 6.34**

## Processing Data (Current State)

| Daily Consumption Rates | | | | | | |
|---|---|---|---|---|---|---|
| | Plungers Used | | Welded Rings Used | | Bush Rings Used | |
| Item Categories | Total | Per Day | Total | Per Day | Total | Per Day |
| < 4" Profile Height | 182 | 6.5 | 163 | 5.8 | 16 | 0.6 |
| > 4" < 9" Height | 161 | 5.8 | 121 | 4.3 | 35 | 1.3 |
| > 9" Height | 29 | 1.0 | 4 | 0.1 | 20 | 0.7 |
| Optic | 40 | 1.4 | 32 | 1.1 | 0 | 0.0 |
| **Total** | **412** | **14.7** | **320** | **11.4** | **71** | **2.5** |

18 Percent of Rings Consumed Are Bushed

Note: This information is based on a 28-day sample of what the shop actually worked, not what was necessarily consumed

**FIGURE 6.35**

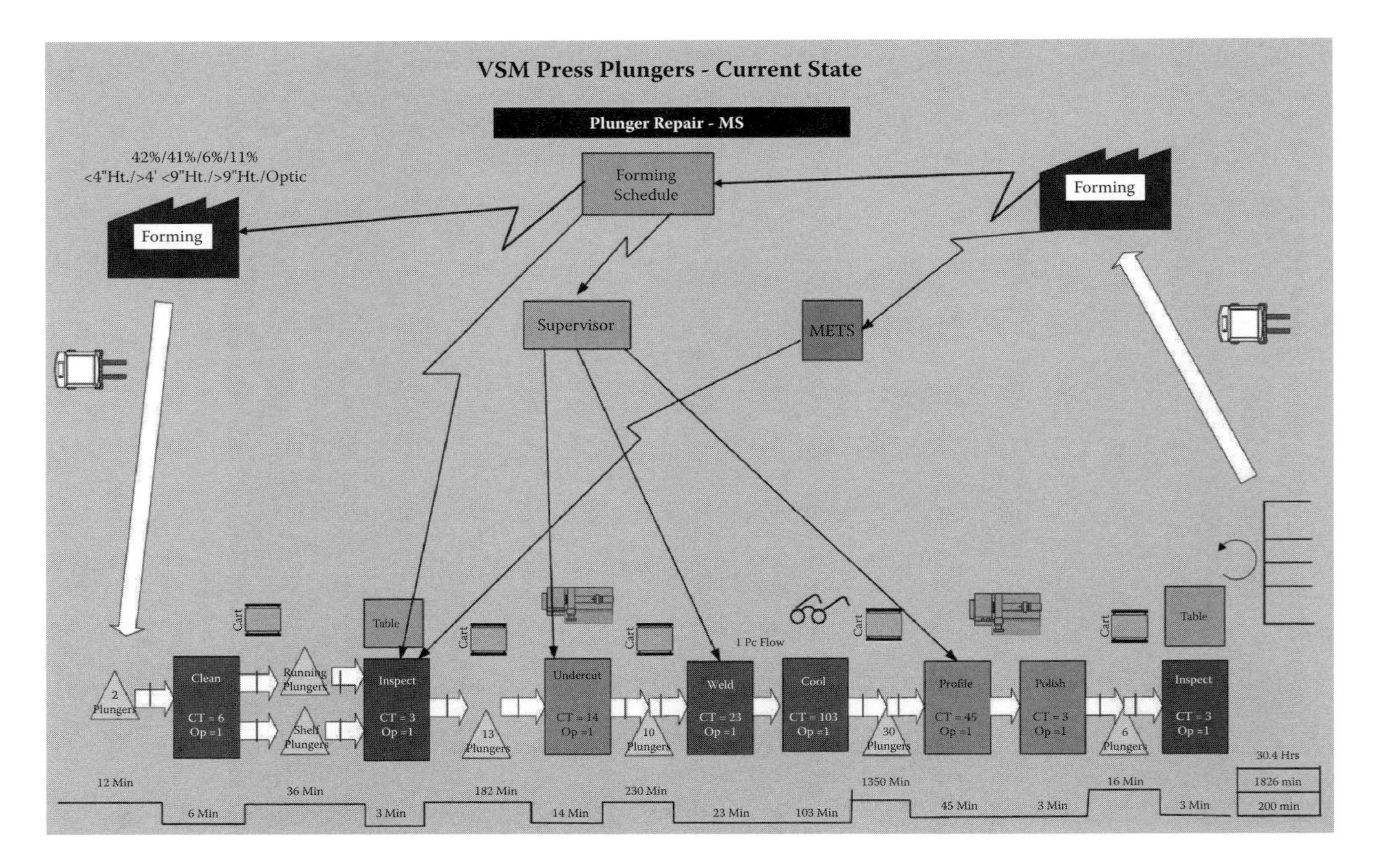
VSM Press Plungers - Current State
Plunger Repair - MS
42%/41%/6%/11%
<4"Ht./>4' <9"Ht./>9"Ht./Optic
Forming
Forming Schedule
Forming
Supervisor
METS
Cart
Table
Cart
Cart
1 Pc Flow
Cart
Cart
Table
2 Plungers
Clean
CT = 6
Op =1
Running Plungers
Shelf Plungers
Inspect
CT = 3
Op =1
13 Plungers
Undercut
CT = 14
Op =1
10 Plungers
Weld
CT = 23
Op =1
Cool
CT = 103
Op =1
30 Plungers
Profile
CT = 45
Op =1
Polish
CT = 3
Op =1
6 Plungers
Inspect
CT = 3
Op =1
30.4 Hrs
12 Min
6 Min
36 Min
3 Min
182 Min
14 Min
230 Min
23 Min
103 Min
1350 Min
45 Min
3 Min
16 Min
3 Min
1826 min
200 min

**FIGURE 6.36**

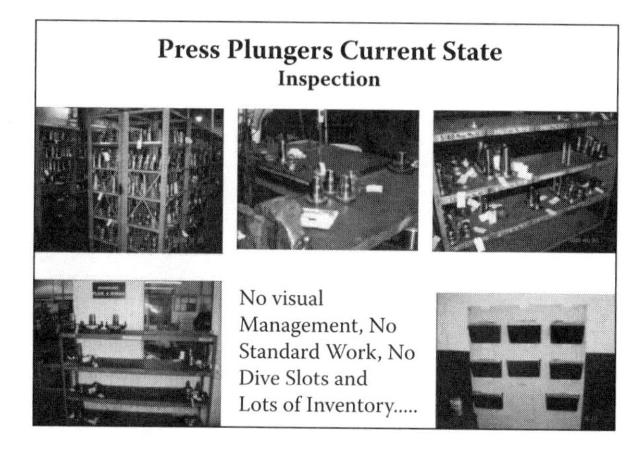


**FIGURE 6.37**

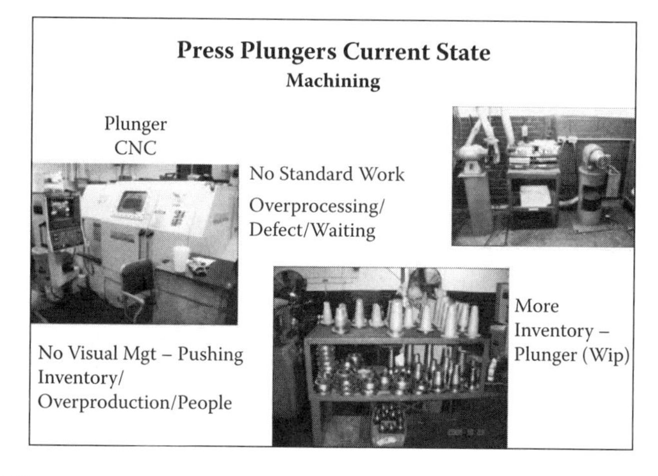


**FIGURE 6.38**

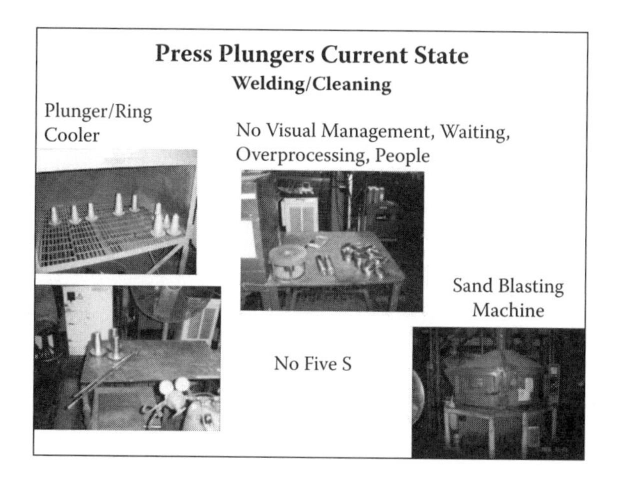


**FIGURE 6.39**

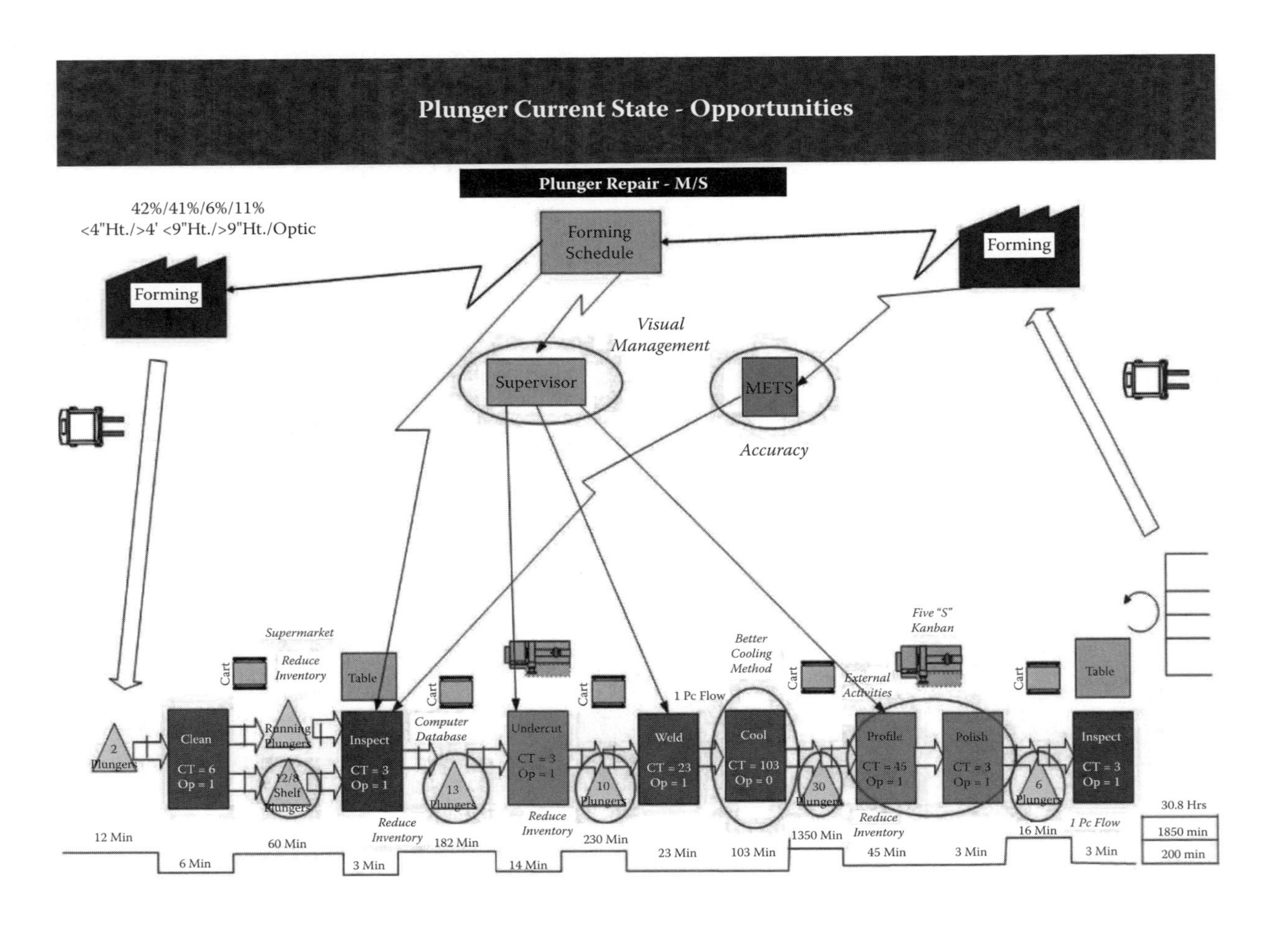
Plunger Current State - Opportunities
Plunger Repair - M/S
42%/41%/6%/11%
<4"Ht./>4' <9"Ht./>9"Ht./Optic
Forming
Forming Schedule
Forming
Visual Management
Supervisor
METS
Accuracy
Supermarket
Reduce Inventory
Cart
Table
Cart
Computer Database
Cart
1 Pc Flow
Better Cooling Method
Cart
External Activities
Five "S" Kanban
Cart
Table
2 Plungers
Clean CT = 6 Op = 1
Running Plungers
12/8 Shelf Plungers
Inspect CT = 3 Op = 1
13 Plungers
Undercut CT = 3 Op = 1
10 Plungers
Weld CT = 23 Op = 1
Cool CT = 103 Op = 0
30 Plungers
Profile CT = 45 Op = 1
Polish CT = 3 Op = 1
6 Plungers
Inspect CT = 3 Op = 1
Reduce Inventory
Reduce Inventory
Reduce Inventory
1 Pc Flow
12 Min
6 Min
60 Min
3 Min
182 Min
14 Min
230 Min
23 Min
103 Min
1350 Min
45 Min
3 Min
16 Min
3 Min
30.8 Hrs
1850 min
200 min

FIGURE 6.40

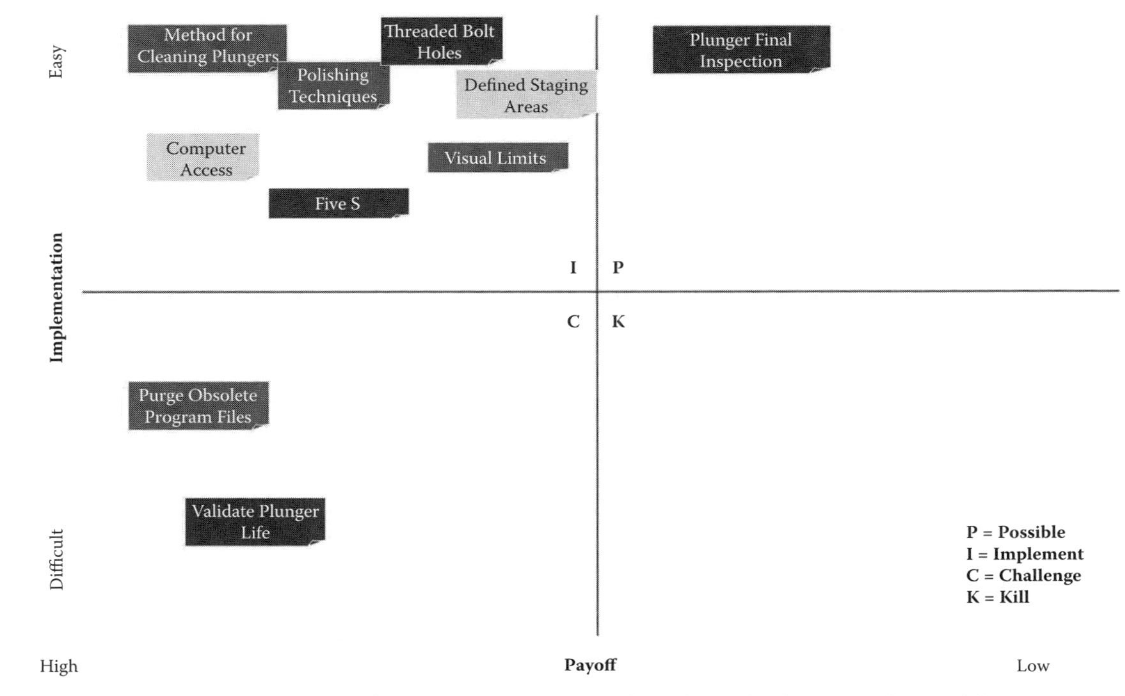


Initially select the project which has the highest payoff and is the easiest to implement (in the first quadrant). Success is shown in different quadrants the more difficulty is the project

**FIGURE 6.41**

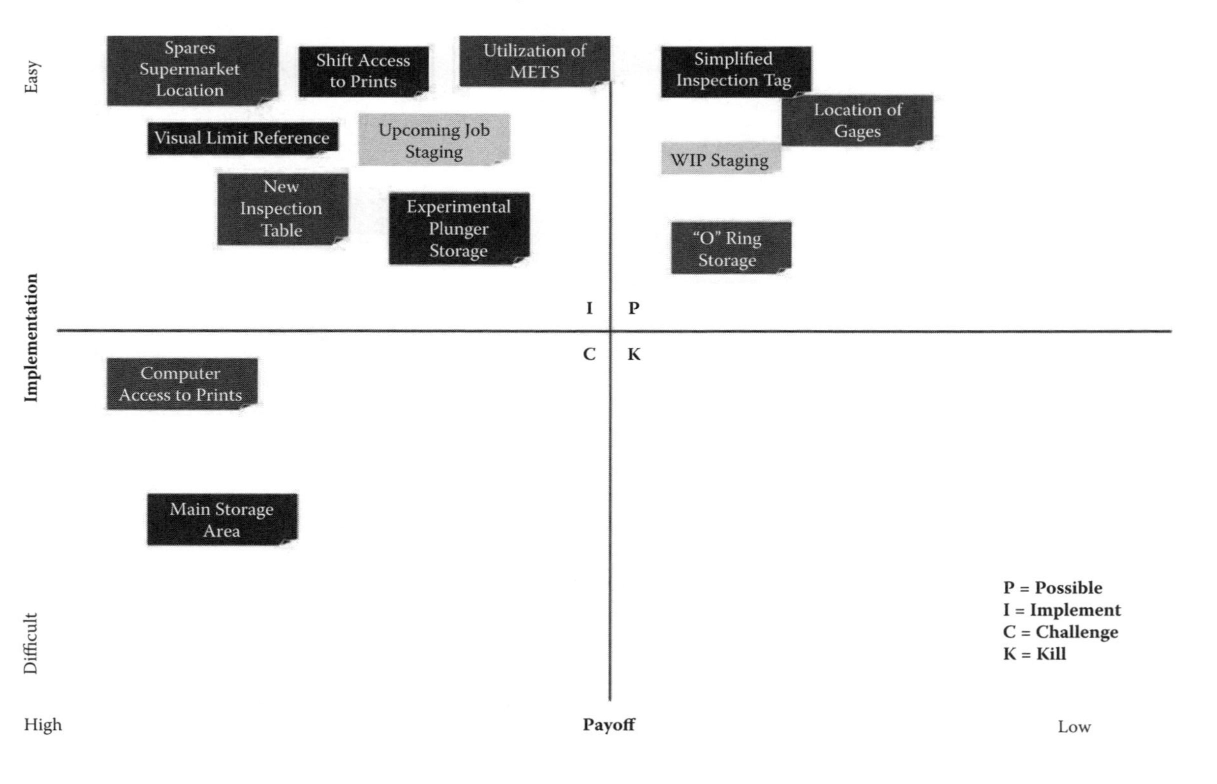


**FIGURE 6.42** No divide slots and lots and lots of inventory.

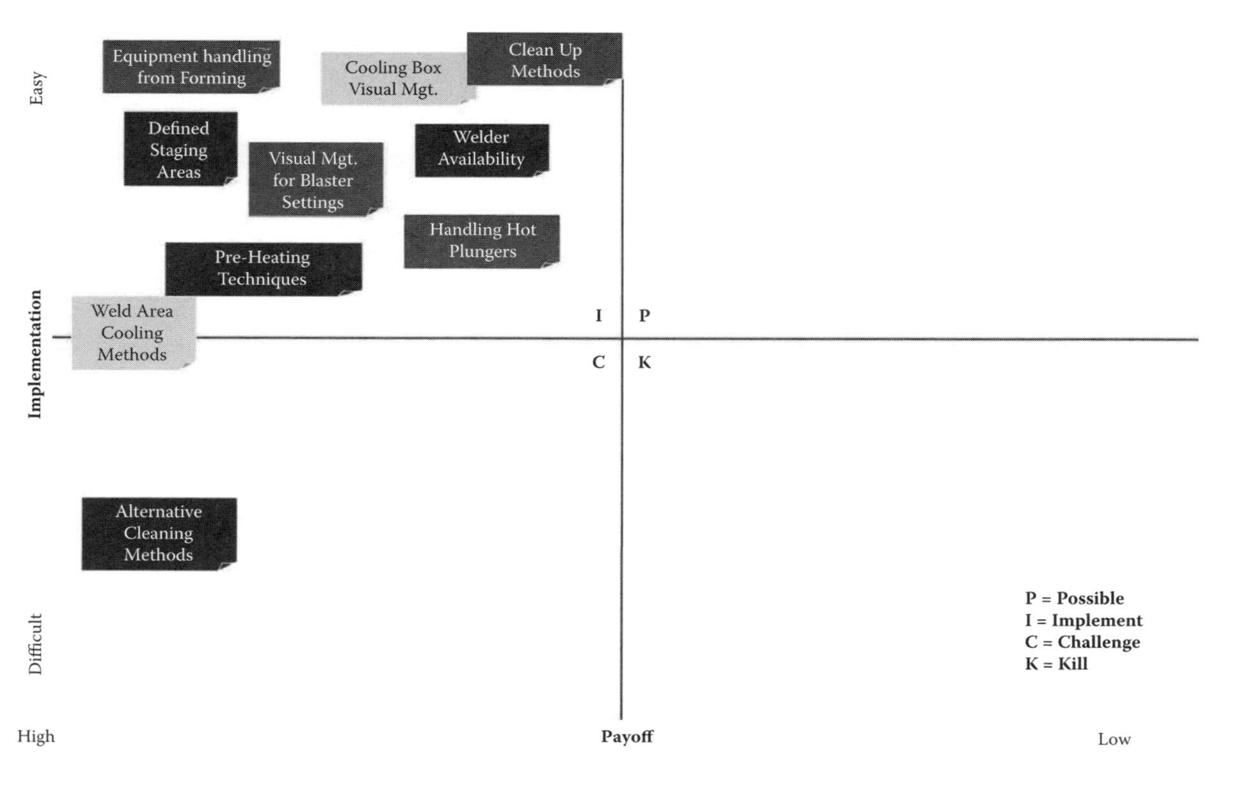


**FIGURE 6.43** More inventory—plunger work in progress (WIP).

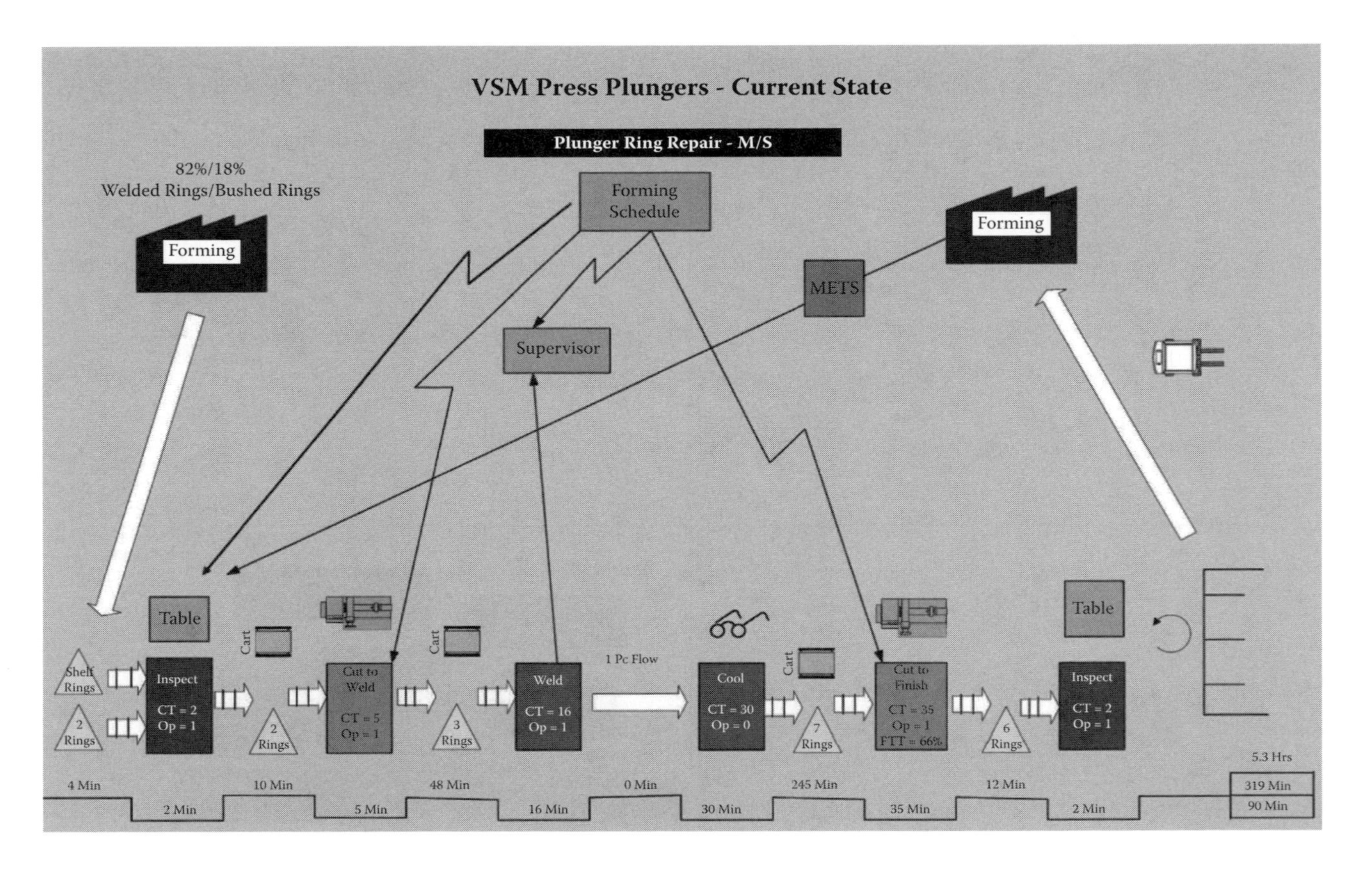
VSM Press Plungers - Current State
Plunger Ring Repair - M/S
82%/18%
Welded Rings/Bushed Rings
Forming
Forming Schedule
Supervisor
METS
Forming
Table
Shelf Rings
2 Rings
Inspect
CT = 2
Op = 1
Cart
2 Rings
Cut to Weld
CT = 5
Op = 1
Cart
3 Rings
Weld
CT = 16
Op = 1
1 Pc Flow
Cool
CT = 30
Op = 0
Cart
7 Rings
Cut to Finish
CT = 35
Op = 1
FTT = 66%
6 Rings
Table
Inspect
CT = 2
Op = 1
4 Min
2 Min
10 Min
5 Min
48 Min
16 Min
0 Min
30 Min
245 Min
35 Min
12 Min
2 Min
5.3 Hrs
319 Min
90 Min

**FIGURE 6.44**

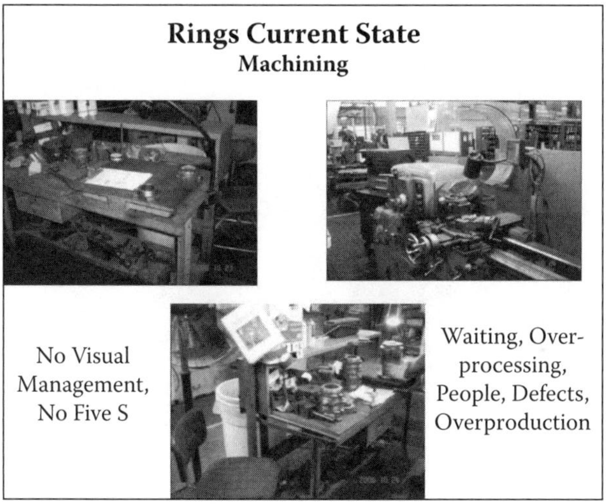

**Rings Current State**
**Machining**

No Visual Management, No Five S

Waiting, Over-processing, People, Defects, Overproduction

**FIGURE 6.45**

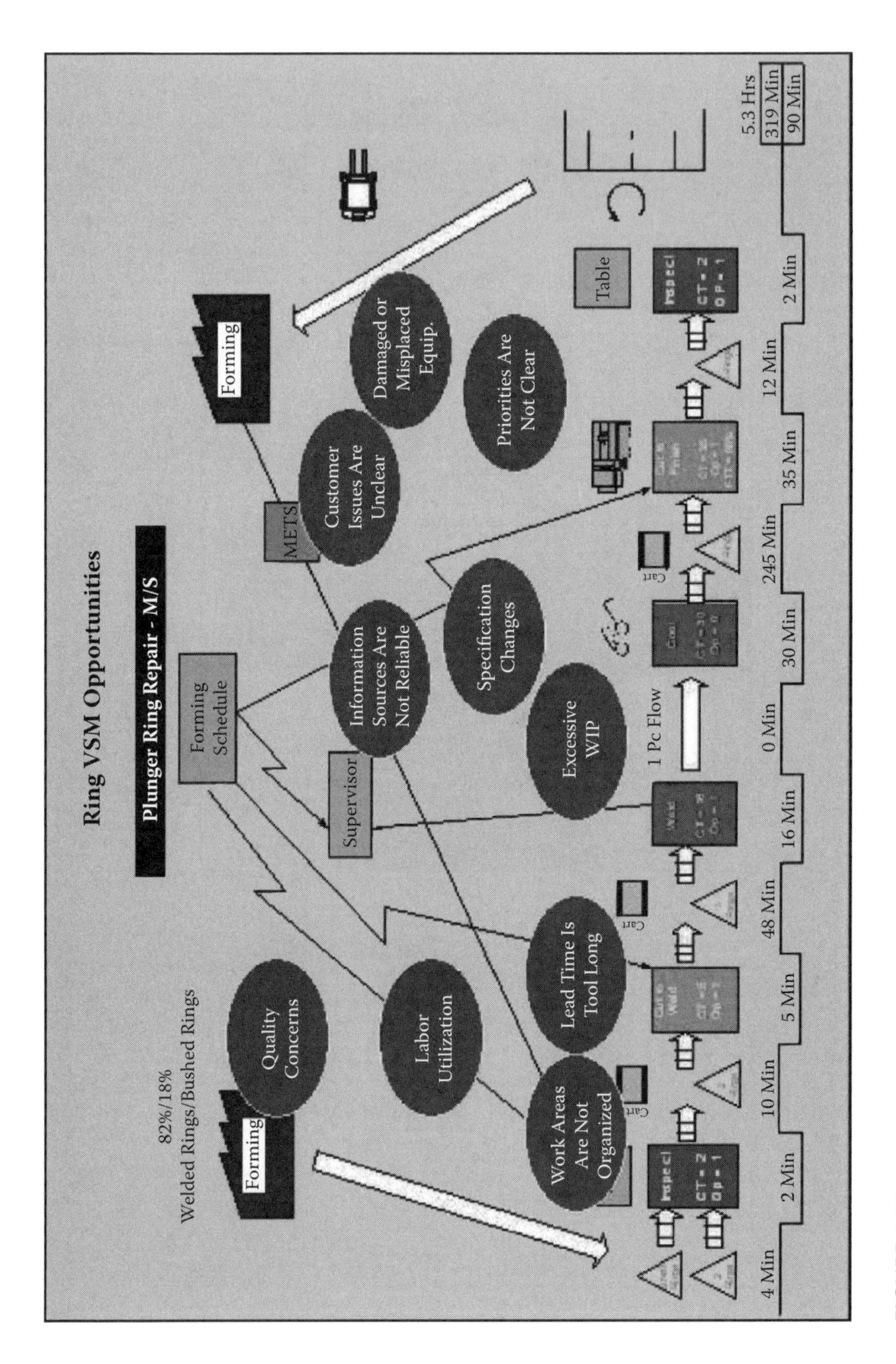
Ring VSM Opportunities
Plunger Ring Repair - M/S
82%/18%
Welded Rings/Bushed Rings
Forming
Forming Schedule
Supervisor
METS
Forming
Quality Concerns
Labor Utilization
Work Areas Are Not Organized
Lead Time Is Tool Long
Information Sources Are Not Reliable
Specification Changes
Excessive WIP
Customer Issues Are Unclear
Damaged or Misplaced Equip.
Priorities Are Not Clear
Table
1 Pc Flow
Cart
Cart
Cart
4 Min
2 Min
10 Min
5 Min
48 Min
16 Min
0 Min
30 Min
245 Min
35 Min
12 Min
2 Min
5.3 Hrs
319 Min
90 Min

**FIGURE 6.46**

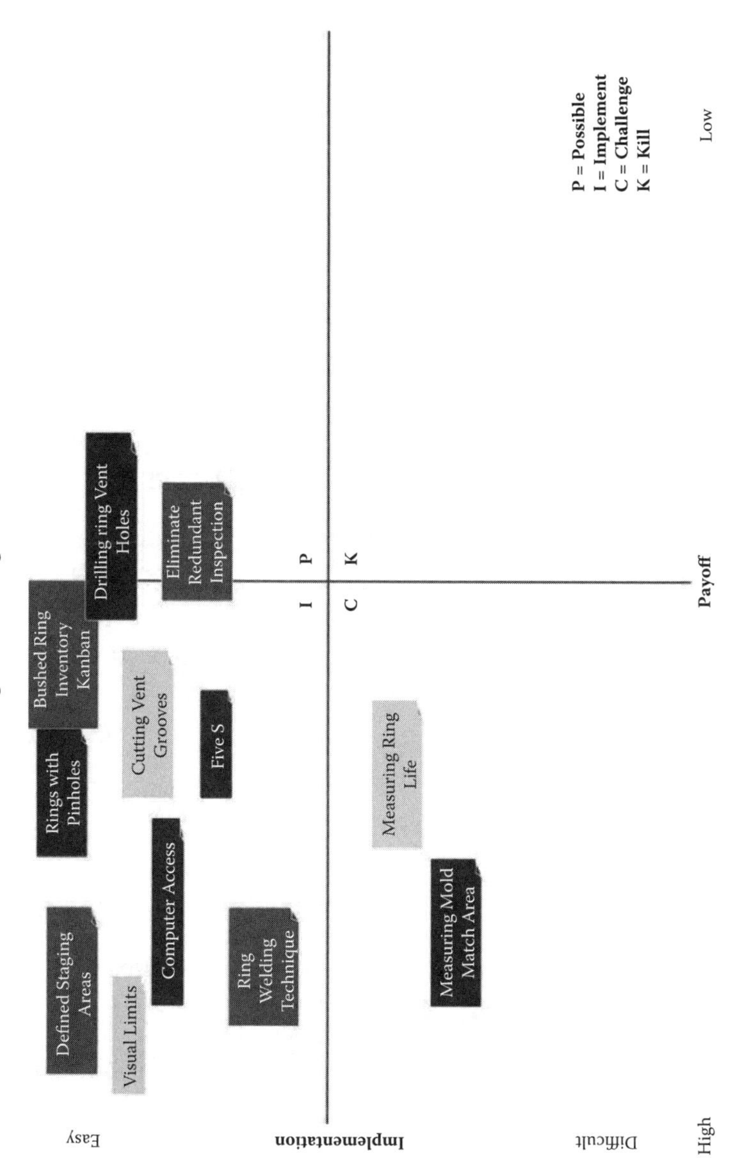
Ring Machining PICK Chart
Defined Staging Areas
Visual Limits
Computer Access
Ring Welding Technique
Rings with Pinholes
Cutting Vent Grooves
Five S
Bushed Ring Inventory Kanban
Drilling ring Vent Holes
Eliminate Redundant Inspection
Measuring Ring Life
Measuring Mold Match Area
I
P
C
K
Easy
Implementation
Difficult
High
Payoff
Low
P = Possible
I = Implement
C = Challenge
K = Kill

**FIGURE 6.47**

**Teams**

*Inspection*
*Plunger Machining*
*Ring Machining/Cleaning/Weld*
*Customer Representative*

FIGURE 6.48

**Improvements - Inspection**

- Improved Inventory Storage/Reduced Set
- Relocation of Ready Table (Supermarket)
- Better Access to Specifications/Prints
- Eliminated Redundant Inspection
- Improved Inspection Table
- Improved Upcoming Job Preparation
- Five S
- Standard Work

FIGURE 6.49

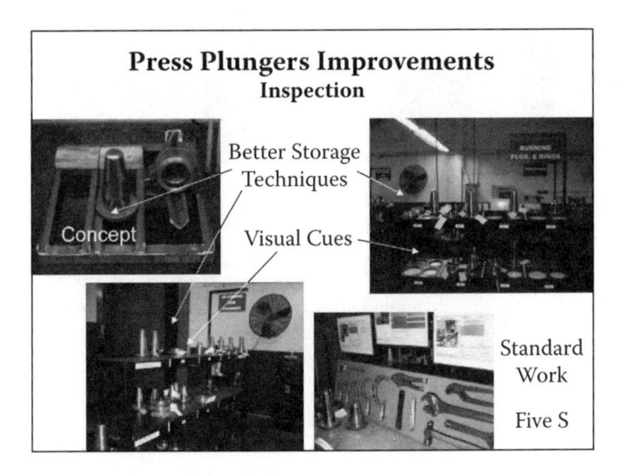


FIGURE 6.50

**Improvements - Plunger Machining**

- Reduced Cycle Time (polishing methods)
- Defined Staging Areas
- Better Access to Specifications
- Standard Work
- Five S
- Visual Limits/Customer Expectations
- Visual Management

**FIGURE 6.51**

**Press Plungers Improvements**

**Plunger Machining**

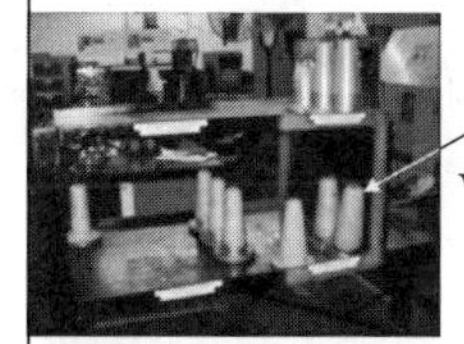

Better Storage Techniques

Visual Cues

Five S

Needed Information is Readily Available

Quality

Specifications

**FIGURE 6.52**

**Improvements - Clean/Weld Area**

- Designated Storage Areas
- Defined Flow of Running Equipment
- Cooling Box Visual Management
- Cooling Box Re-design
- Blaster Visual Management

**FIGURE 6.53**

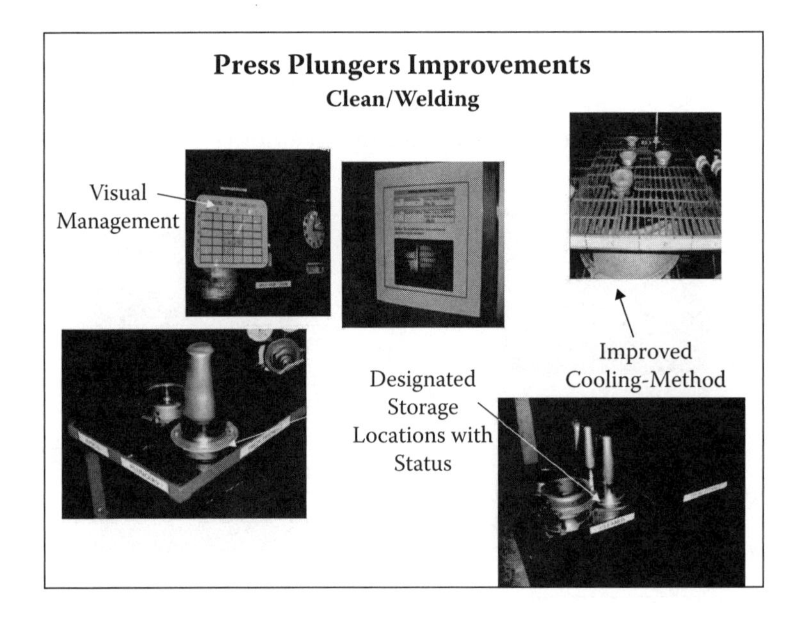
Press Plungers Improvements
Clean/Welding
Visual Management
Improved Cooling-Method
Designated Storage Locations with Status

**FIGURE 6.54**

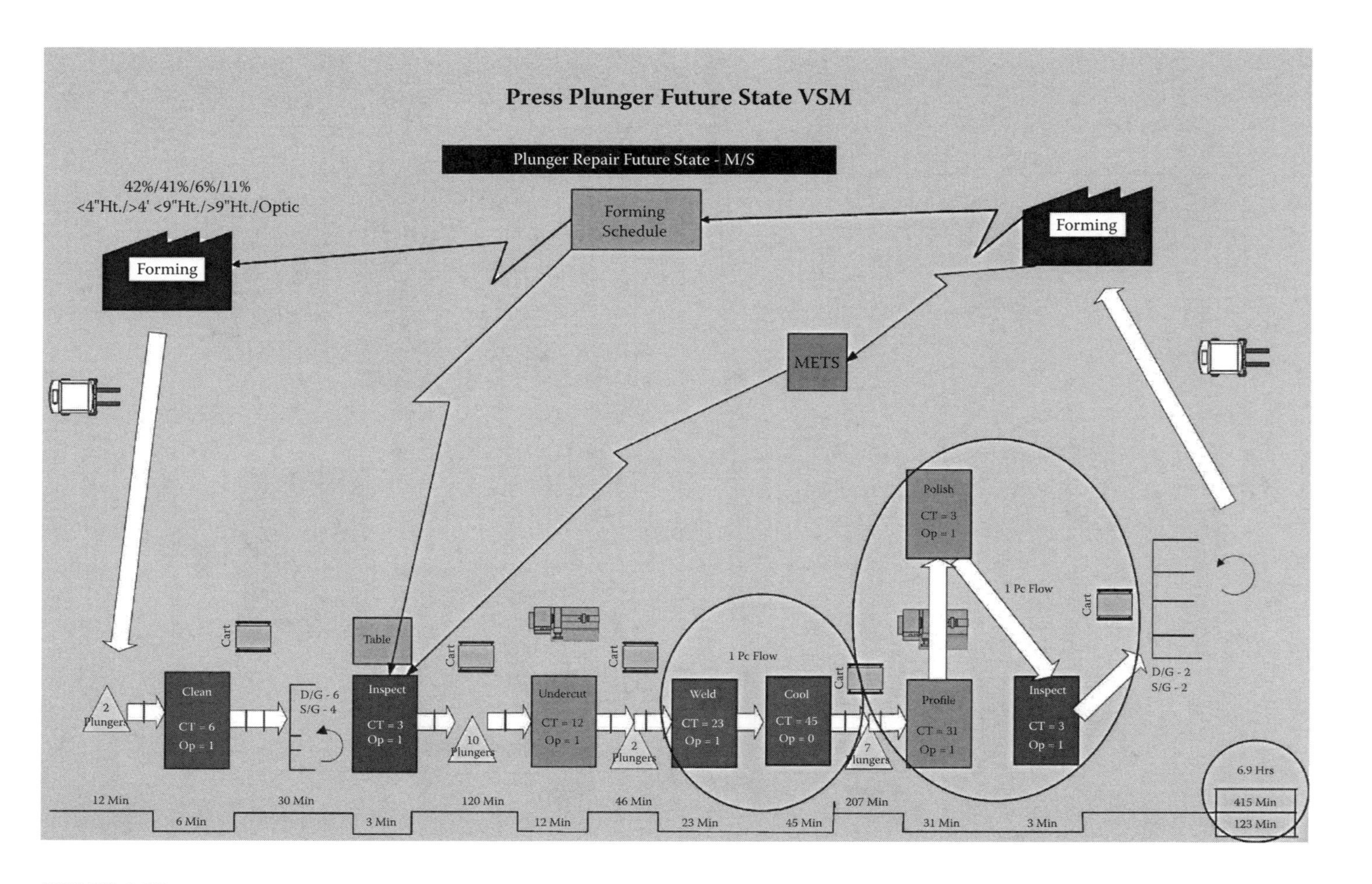
Press Plunger Future State VSM
Plunger Repair Future State - M/S
42%/41%/6%/11%
<4"Ht./>4' <9"Ht./>9"Ht./Optic
Forming
Forming Schedule
Forming
METS
2 Plungers
Clean
CT = 6
Op = 1
Cart
D/G - 6
S/G - 4
Table
Inspect
CT = 3
Op = 1
Cart
10 Plungers
Undercut
CT = 12
Op = 1
Cart
2 Plungers
1 Pc Flow
Weld
CT = 23
Op = 1
Cool
CT = 45
Op = 0
Cart
7 Plungers
Polish
CT = 3
Op = 1
1 Pc Flow
Profile
CT = 31
Op = 1
Inspect
CT = 3
Op = 1
Cart
D/G - 2
S/G - 2
12 Min
6 Min
30 Min
3 Min
120 Min
12 Min
46 Min
23 Min
45 Min
207 Min
31 Min
3 Min
6.9 Hrs
415 Min
123 Min

FIGURE 6.55

**Improvement - Ring Machining**

- Defined Staging Areas
- Drilling Vent Holes
- Visual Prioritization
- Better Access to Specifications
- Standard Work
- Five S
- Visual Limits/Customer Expectations
- Visual Management

**FIGURE 6.56**

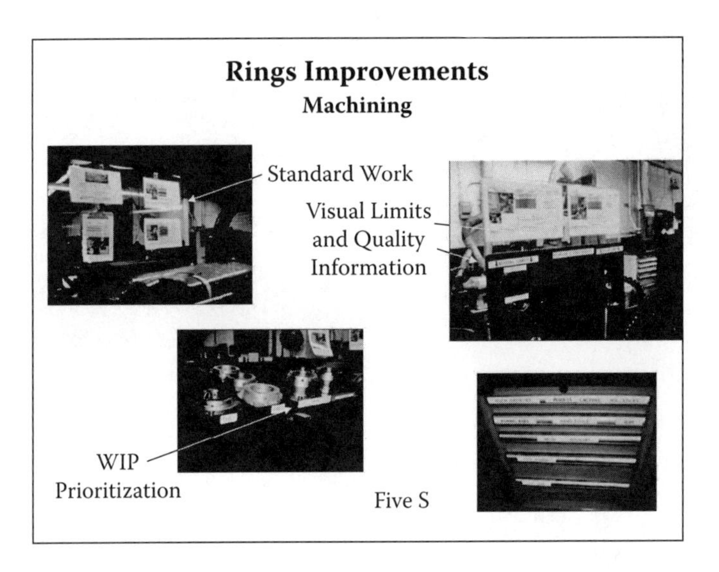


**FIGURE 6.57**

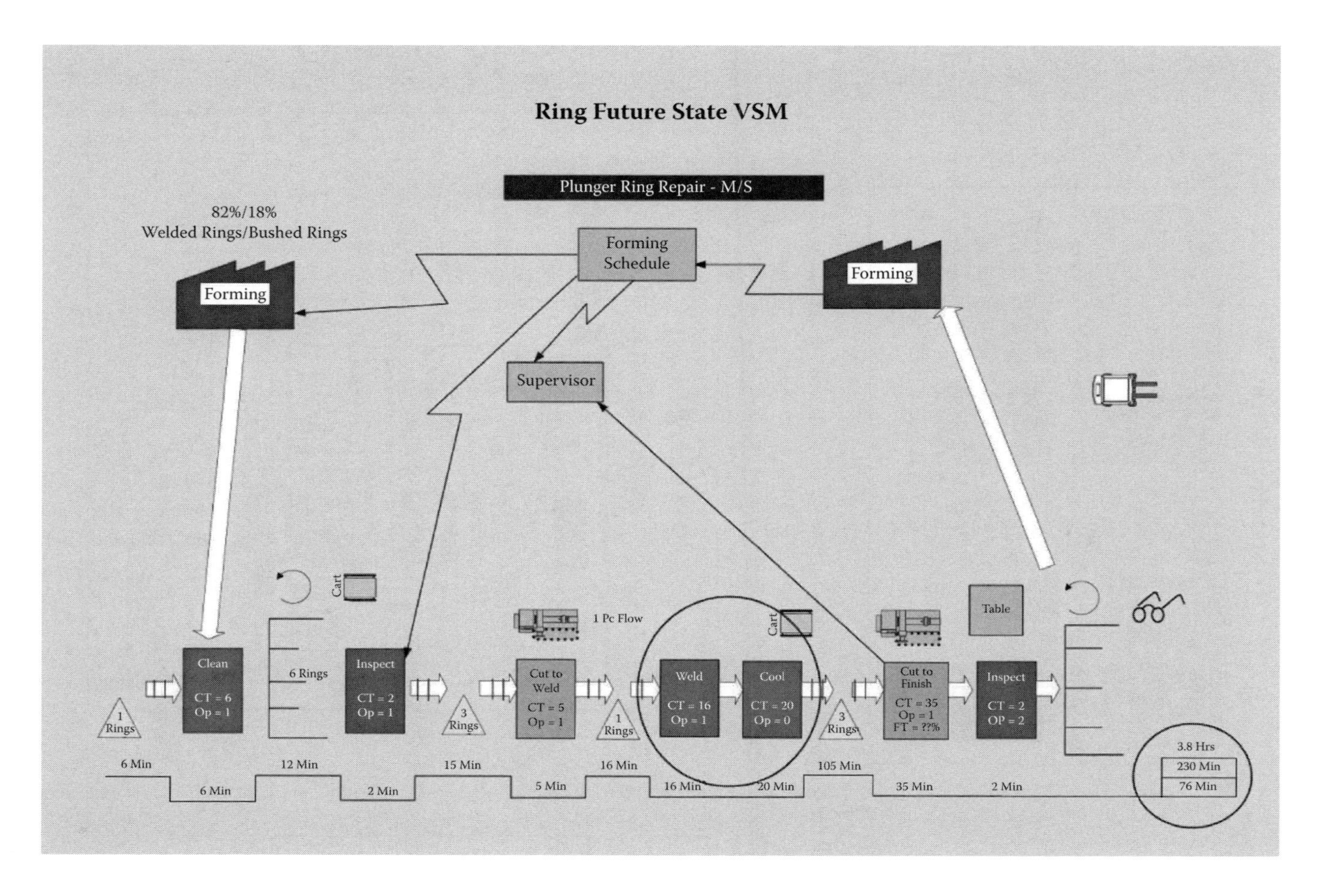
Ring Future State VSM
Plunger Ring Repair - M/S
82%/18%
Welded Rings/Bushed Rings
Forming
Forming Schedule
Forming
Supervisor
Cart
1 Pc Flow
Cart
Table
1 Rings
Clean
CT = 6
Op = 1
6 Rings
Inspect
CT = 2
Op = 1
3 Rings
Cut to Weld
CT = 5
Op = 1
1 Rings
Weld
CT = 16
Op = 1
Cool
CT = 20
Op = 0
3 Rings
Cut to Finish
CT = 35
Op = 1
FT = ??%
Inspect
CT = 2
OP = 2
6 Min
6 Min
12 Min
2 Min
15 Min
5 Min
16 Min
16 Min
20 Min
105 Min
35 Min
2 Min
3.8 Hrs
230 Min
76 Min

FIGURE 6.58

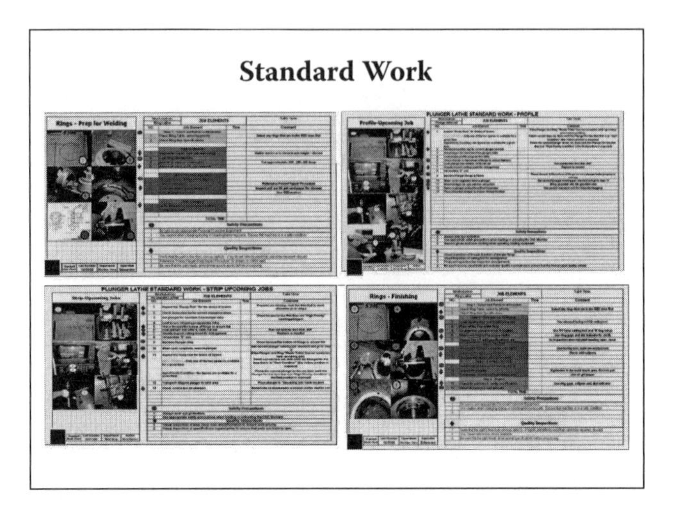


**FIGURE 6.59**

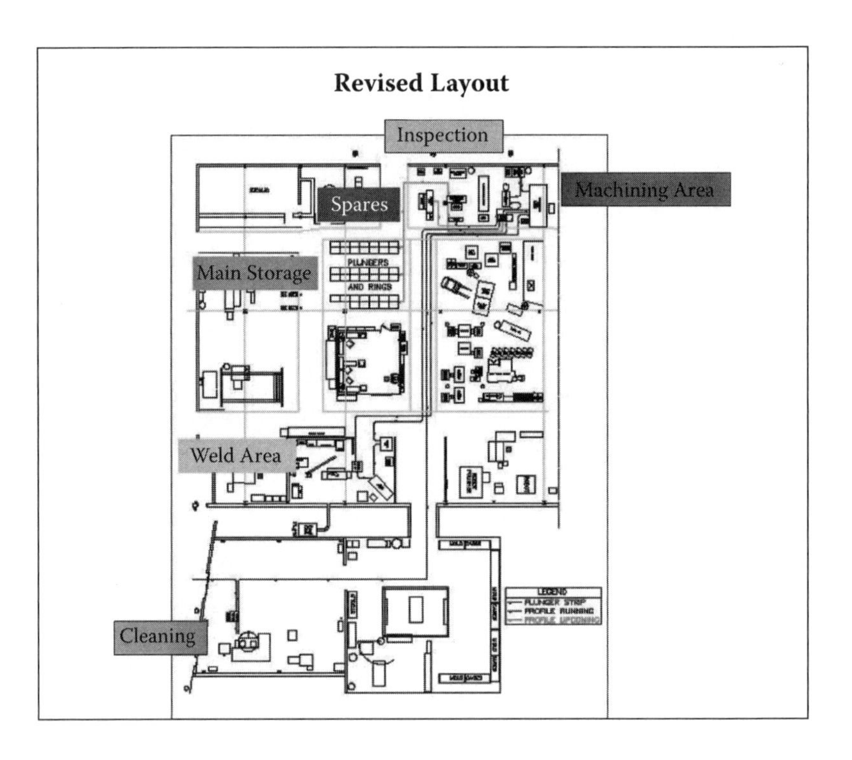


**FIGURE 6.60**

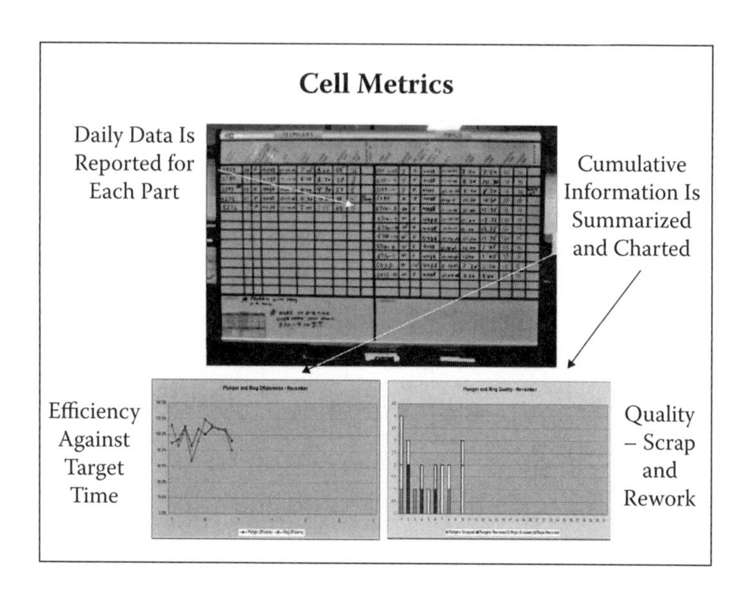


**FIGURE 6.61**

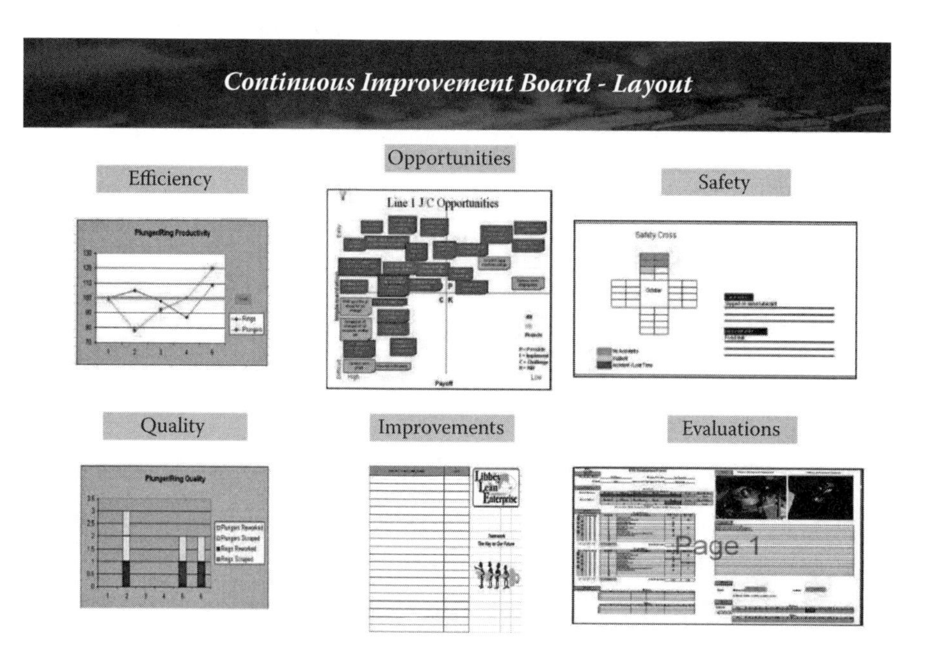


**FIGURE 6.62**

**Continuous Improvement Board**

**FIGURE 6.63**

**Training Game Plan**

- Review
  - Standard Work
  - Cell Flow
  - Five “S” Checklist
  - Visual Mgt. Boards/Cell Metrics
  - Continuous Improvement Game Plan

**FIGURE 6.64**

**Communication Game Plan**

- Debrief with Management
- Debrief with All Affected Operators
  - Forming
  - Machine Shop
- Post RIE Summary on Libbey Website

**FIGURE 6.65**

**Future State Improvements**

| | Current State | Future State | Change | Percent Reduction |
|---|---|---|---|---|
| Plunger Inventory | 2924 | 1258 | 1666 | 57% |
| Plunger Cycle Time | 200 | 123 | 77 | 39% |
| Plunger Lead Time | 1826 | 415 | 1411 | 77% |
| | | | | |
| Ring Inventory | 3890 | 1214 | 2676 | 69% |
| Ring Cycle Time | 90 | 76 | 14 | 16% |
| Ring Lead Time | 319 | 230 | 69 | 28% |
| | | | | |
| | | | | |

Inventory Reduction Cost Savings - $950,000

**FIGURE 6.66**

**Lessons Learned**

- Having Customer on Team Very Valuable
- Team/Lean Training Prior to Event Helpful (2 Days of Training)
- Creating VSM of Cell Helped Focus Effort
- Need to Complete Data Mining Prior to Event
- LEAP expert on Team Very Helpful
- Lack of Five "S" Slowed Down Team

**FIGURE 6.67**

**Next Steps**

- Continue Coaching Work Cell Personnel
- Follow-up on Open Assignment
- Enhance Cell Metrics
- Evaluate Bushed Rings
- Replace Cooling Box—Permanent
- Work Continuous Improvement Opportunities

**FIGURE 6.68**

### 6.1.13 Toyota Production System

At this point, it may be interesting to briefly note one of the prominently discussed production systems, the TPS, which is used to reduce stresses and failures and eliminate waste. The three main categories targeted by TPS are

- Overburdens (*muri*)
- Inconsistency (*mura*)
- Waste elimination (*muda*)

The leaders of Toyota describe these as the easiest for American companies to adapt because these principles can all be quantitatively measured.

The muri principles are concerned with people working within the organization. The concept of muri is to develop leaders who thoroughly understand the work, live with the organization's philosophy, and teach it to others. Workers and managers are required to work as a team so that no member can be overburdened. Meeting partners such as part suppliers, service technicians, and sales franchises in an open atmosphere encourages discussion of and creation of solutions for any problems that may arise to mutual satisfaction.

The mura principles refer to consistency. This concept requires workers and managers to continuously solve the root problems within the organization. The managers, including the upper management, are encouraged to see the problems for themselves and are expected to make decisions based on all available options and based on their past experiences. There is no problem relating to a manager using *genchi genbutsu*, or managing by wandering around the plant. Managers thus grow within the organization and provide a consistent performance. This system believes in continuous training for every member so that he/she can gain with motivation and support from every member of the organization. Leaders (*sensei*) are always coaching the subordinates and leading by example.

The muda principles are concerned with policies and procedures put in place for efficient operations. These principles address issues such as overproduction, inventory, motion and speed of operators, and machine and service processes. All the principles of lean manufacturing can be used here.

The TPS works because employees maintain a sense of inclusion; with good communication horizontally and vertically, all parts of the organization are involved in problem solving and attaining efficiency. In this process, decisions are made through consensus and continuing proactive improvements are expected to eliminate waste.

## 6.2 SUPPLY CHAIN

The procedures illustrated so far lead to an efficient and productive work environment. All "fat" or waste is taken out of the system and the system becomes lean. People who are knowledgeable about the workings of the system, along with other experts, trace the present activities and suggest means to improve. Most of these improvements, however, are in terms of what can be observed and improved relatively quickly and without much expense. This often leads to excellent improvements in local areas. However, at times much larger savings can be obtained via analysis of the entire system and improving such areas as estimating future demands by forecasting, planning a minimum cost production schedule, creating good inventory control and storage policies, planning preventive and emergency maintenances, scheduling machines, evaluating transportation networks, and planning an efficient supplier/customer relationship network. These topics are typically discussed under an umbrella called supply chain management.

There are number of good books on supply chain management, and it is difficult to illustrate all aspects of the supply chain in a few pages. We illustrate here several techniques within the supply chain to show how it can reduce the costs of production and distribution.

### 6.2.1 Distribution Problem

We have three plants that can produce items required in four markets. The cost of producing and shipping a unit quantity from each plant $i$ to each market $j$ is $C_{ij}$, and is given in the table below. The table also shows the maximum level that each plant can produce, as well as demand from each market. If the plant is set up for production, a fixed cost is incurred (also displayed in the table). If a unit sells for \$20, determine the optimum shipping policy.

| Customer | 1 | 2 | 3 | 4 | Plant Capacity | Fixed Cost of Setup |
|---|---|---|---|---|---|---|
| Plant 1 | 3 | 5 | 6 | 8 | 30 | 35 |
| Plant 2 | 4 | 5 | 4 | 6 | 40 | 20 |
| Plant 3 | 3 | 7 | 7 | 8 | 60 | 50 |
| Demand | 30 | 50 | 40 | 20 | | |
| Sale Price | 6 | 7 | 8 | 10 | | |

Those who are familiar with linear programming (LP) will immediately recognize this as an LP problem. If we define $X_{ij}$ as the amount shipped from plant $i$ to customer $j$ and use binary variables $K_i$ to indicate if plant $i$ is producing the product ($K_i = 1$) or not ($K_i = 0$), then we can write the following equations:

Quantity shipped from each plant cannot exceed plant capacity if it is open:

$$X_{11} + X_{12} + X_{13} + X_{14} - 30K_1 \leq 0$$
$$X_{21} + X_{22} + X_{23} + X_{24} - 40K_2 \leq 0$$
$$X_{31} + X_{32} + X_{33} + X_{34} - 60K_3 \leq 0$$

Quantity received at each market should not be more than what can be sold

$$X_{11} + X_{21} + X_{31} \leq 30$$

$$X_{12} + X_{22} + X_{32} \leq 50$$

$$X_{13} + X_{23} + X_{33} \leq 40$$

$$X_{14} + X_{24} + X_{34} \leq 20$$

Total profit is equal to sold units minus the cost of production and transportation minus the fixed cost of setting up a production line

$$6(X_{11}+X_{21}+X_{31})+7(X_{12}+X_{22}+X_{32})+8(X_{13}+X_{23}+X_{33})+10(X_{14}+X_{24}+X_{34})$$
$$-3X_{11}-5X_{12}-6X_{13}-8X_{14}-4X_{21}-5X_{22}-4X_{23}-6X_{24}-3X_{31}$$
$$-7X_{32}-7X_{33}-8X_{34}-35K_1-20K_2-50K_3$$

The solution to the problem is

$$X_{12}=30 \quad X_{22}=20 \quad X_{24}=20 \quad K_1=1 \quad \text{and} \quad K_2=1$$

with a maximum profit of $305.

We note that the optimum solution does not call for production in all four plants to supply the overall demand. Had this move been followed, which would have seemed like the logical option, the maximum profit would have been reduced.

### 6.2.2 Aggregate Planning Model

Consider another typical example: that of making decisions regarding workload and inventory policies from day to day. This process is commonly referred to as aggregate planning. Suppose we know the product requirements for the next 6 days or for the next 6 periods:

| | Period | | | | | |
|---|---|---|---|---|---|---|
| | 1 | 2 | 3 | 4 | 5 | 6 |
| Requirements | 300 | 400 | 530 | 360 | 400 | 200 |

We can use a full-time worker at a cost of $15/hour or a part-time employee at $10/hour. A full-time worker can produce 5 units per hour or $5 \times 8 = 40$ units/day and a part-timer can work 4 hours/day to produce 4 units/hour for a total of 16 units/day. There are at most four full-time workers that can be assigned to this product. At any time, no more than 25 workers can work in the plant on this line. In addition, to have proper supervision, for every full-time employee no more than two part-time employees can be hired. The product can be purchased from an outside supplier for $20/unit. The unit sells for $25 a piece. A unit stored to carry from one day to the next costs $2/unit, and at the most 50 units can be stored in the warehouse. Each day's demand must be met so that there is no backorder or shortage. The problem is to determine the optimum production policy from day to day.

Again, the problem can be formulated and solved by using LP. Let us define the following variables.

$F_t$ = number of full-time employees working on day $t$
$P_t$ = number of part-time employees working on day $t$

$D_t$ = demand on day $t$
$I_t$ = initial inventory for day $t$
$O_t$ = the amount purchased on day $t$ from outside suppliers

Because all demands must be met, the maximum income is 25 × total demand, or 25 × 2190 = $54,750. The profit is maximized if we can minimize the cost. Therefore, the objective is to minimize cost, that is,

$$\sum (15 \times 8)F_t + \sum (10 \times 4)P_t + \sum 20O_t + \sum 2I_t$$

Subject to material balance equation that states that for each day, $t$

Initial inventory + Total production + Units Purchased – Units sold
= Initial inventory for next day

$$I_t + 40F_t + 16P_t + O_t - D_t = I_{t+1} \quad \text{for } t = 1, 2, \quad , 6.$$

Because, initially, there is no inventory and there is no inventory required at the end of the planning period, we have

$$I_1 = 0 \quad \text{and} \quad I_7 = 0$$

No more than 25 workers can work at a time leads to

$$F_t + P_t \leq 25 \quad \text{for} \quad t = 1, 2, \quad , 6.$$

At the most, 50 units can be stored in warehouse leads to

$$I_t \leq 50 \quad \text{for } t = 1, 2, \quad , 6.$$

For every full-time employee, no more than two part-time employees can be hired

$$F_i - 0.5P_i \leq 0 \quad \text{for } t = 1, 2, \quad , 6.$$

#### 6.2.2.1 Formulation and Solution

1. min $120F_1 + 40P_1 + 20O_1 + 2I_1 + 120F_2 + 40P_2 + 20O_2 + 2I_2$
$+ 120F_3 + 40P_3 + 20O_3 + 2I_3 + 120F_4 + 40P_4 + 20O_4 + 2I_4$
$+ 120F_5 + 40P_5 + 20O_5 + 2I_5 + 120F_6 + 40P_6 + 20O_6 + 2I_6$

Subject to

2. $40F_1 + 16P_1 + O_1 + I_1 - I_2 - D_1 = 0$
3. $40F_2 + 16P_2 + O_2 + I_2 - I_3 - D_2 = 0$
4. $40F_3 + 16P_3 + O_3 + I_3 - I_4 - D_3 = 0$
5. $40F_4 + 16P_4 + O_4 + I_4 - I_5 - D_4 = 0$
6. $40F_5 + I_6P_5 + O_5 + I_5 - I_6 - D_5 = 0$
7. $40F_6 + I_6P_6 + O_6 + I_6 - D_6 - I_7 = 0$
8. $I_1 = 0$
9. $I_7 = 0$
10. $F_1 + P_1 \leq 25$
11. $F_2 + P_2 \leq 25$
12. $F_3 + P_3 \leq 25$
13. $F_4 + P_4 \leq 25$
14. $F_5 + P_5 \leq 25$
15. $F_6 + P_6 \leq 25$
16. $I_1 \leq 50$
17. $I_2 \leq 50$
18. $I_3 \leq 50$
19. $I_4 \leq 50$
20. $I_5 \leq 50$
21. $I_6 \leq 50$
22. $D_1 = 300$
23. $D_2 = 400$
24. $D_3 = 530$
25. $D_4 = 360$
26. $D_5 = 400$
27. $D_6 = 200$

END

The solution to the problem is

Objective Function Value 5583.333

With the following values for the variables:

| | | | | | | | | | |
|---|---|---|---|---|---|---|---|---|---|
| $F_1$ | 4.0 | $P_1$ | 8.0 | $O_1$ | 12.0 | $I_1$ | 0.0 | $D_1$ | 300.0 |
| $F_2$ | 6.0 | $P_2$ | 12.0 | $O_2$ | 0.0 | $I_2$ | 0.0 | $D_2$ | 400.0 |
| $F_3$ | 7.0 | $P_3$ | 14.0 | $O_3$ | 0.0 | $I_3$ | 32.0 | $D_3$ | 530.0 |
| $F_4$ | 5.0 | $P_4$ | 10.0 | $O_4$ | 0.0 | $I_4$ | 6.0 | $D_4$ | 360.0 |
| $F_5$ | 6.0 | $P_5$ | 12.0 | $O_5$ | 0.0 | $I_5$ | 6.0 | $D_5$ | 400.0 |
| $F_6$ | 2.0 | $P_6$ | 4.0 | $O_6$ | 18.0 | $I_6$ | 38.0 | $D_6$ | 200.0 |
| | | | | | | $I_7$ | 0.0 | | |

The problem was a good illustration of effective planning. If operations are planned and devised in an optimum manner, there is efficiency built into the system. If, on other hand, lean processes are introduced to improve poor or neglected planning, substantial improvement may not be possible. Examples are numerous.

Consider another problem. Suppose we want to schedule four jobs in the facility. Each job has a processing time, that is, the time it will use our production facility for a due date when a customer expects a delivery of the product, as well as penalties in terms of $/period if the job is completed early and has to be stored before the due date, called early penalty, or if the job is late and shipped after the due date, called late penalty. The data are displayed in the table below.

| Job Number | Processing Time (days) | Due Date | Early Penalty ($/day early) | Late Penalty ($/day late) |
|---|---|---|---|---|
| 1 | 3 | 5 | 1 | 2 |
| 2 | 5 | 7 | 1 | 3 |
| 3 | 2 | 4 | 2 | 3 |
| 4 | 4 | 10 | 3 | 5 |

Because all jobs are available right now, the first-come, first-serve rule does not apply. If we schedule the jobs based on earliest due date, the sequence is 3-1-2-4 with a penalty of $33. If we apply least processing time rule, the sequence is 3-1-4-2 with a penalty of $30. The best sequence is (reference: Industrial Scheduling by Sule) 3-2-4-1 with a penalty of $27. Not an obvious sequence, but this can be derived by using some techniques that are not obvious in lean manufacturing methods.

Similarly, no amount of improvement in warehouse operations can provide as much savings as eliminating the need for storage altogether by planning a just-in-time (JIT) process. The number of in-process storage can be eliminated or combined by simulating the entire facility through software such as Arena and asking what-if questions. A number of U.S. manufactures have shifted their operations to countries with lower labor costs to reduce the unit price of a product. With improved infrastructure, even in low-cost locations such as China or India, for example, plants are shifted from cities to the countryside, where labor cost is 50% lower. These are strategic decisions that produce substantial savings beyond just trying to improve localized operations within manufacturing facilities.

### 6.2.3 Just-in-Time Production/Pickups

This is an example of how the development and application of a heuristic procedure can lead to substantial savings in some problems. In the JIT production system, it is important that raw materials are available when needed and in the quantity required. As a result, only the immediate required quantity is produced. This also gives one of the major advantages of the JIT system: it can respond to changing demands in a most efficient manner without creating an inventory of raw or finished products. Inventory of raw materials or finished goods may result in a substantial inventory cost or obsolesce cost if there is a danger of the product becoming obsolete in the near future.

One simple example of supply chain management technique in the JIT environment is illustrated here. Suppose we are making a product that incorporates three subproducts. These subproducts are supplied by three nearby suppliers but we are required to collect these units from these suppliers. We have a truck (capacity: 20 units) that can pick up these units. If there is a need for more than 20 units from any supplier, then we make a direct trip to pick up these units. But when the total requirement from all suppliers is less than 20, we combine pickups by making one round trip from our plant to each supplier where we must pick up the units.

Being a JIT production facility, the demand for our product may change on a daily basis (period) and hence the required supply may change everyday. Whether or not pickups from each supplier are made on a daily basis, every day we must have on hand the necessary raw materials (parts from suppliers). The problem is to develop a pickup policy that minimizes cost of pickup for a specific period (e.g., 1 week).

There are two types of costs involved: cost of pickup and cost of inventory. One policy would be to visit every supplier everyday and pick up only the daily needs. In this case, the truck must visit every supplier everyday. This is feasible policy but may be expensive. This pickup cost may be reduced if we can pick up several days' worth of supply from a supplier in one trip, and eliminate the necessity of a daily visit to each supplier. However, this requires considerable truck capacity and incurs an inventory cost.

Let us consider an example using the data in Table 6.1. Suppose we have a truck with a 20-unit capacity; when we pick up for the first day, it leaves $20 - (9 + 6 + 3) = 2$ of excess capacity. Because the truck is visiting each supplier, it is possible to pick up additional supply from any one or more of the suppliers. We should only pick up additional units if it will avoid a service call some other day and create savings even with increased inventory cost. How should we develop our pickup policies to minimize the cost?

#### 6.2.3.1 Linear Heuristic Programming Formulation

A heuristic procedure is stated next, and illustrated by applying it to the example. Although, initially, these steps may seem somewhat confusing, they will become clearer when we follow their application.

1. Determine the available slack for each day. This is equal to truck capacity minus pickup required for that day.
2. Start from the day before the last day of the planning period, and proceed backwards toward day 1 by applying the following analysis one day at a time.
3. Calculate savings by moving pickup from the following day(s) to the present day, and knowing the available slack capacity. Available slack is the

**TABLE 6.1**
**Data for an Example**

| | | | Daily Pickup Requirement | | | | |
|---|---|---|---|---|---|---|---|
| Product Number | Procurement Cost | Carry Cost/day | Day 1 | Day 2 | Day 3 | Day 4 | Day 5 |
| 1 | 10 | 1 | 9 | 2 | 5 | 4 | 2 |
| 2 | 12 | 1 | 6 | 6 | 3 | 7 | 3 |
| 3 | 14 | 1 | 3 | 4 | | 6 | 5 |

slack available for the current day plus all slack that can be utilized from the following days. Savings will be realized only if a trip to that location can be avoided. It requires transfer of all the demands from that location to previous day(s). Savings is equal to the cost of a visit that is eliminated minus the additional cost of inventory.
4. Determine the "moving combination" that gives the maximum savings and make associated new assignments by moving these demands.
5. Recalculate the slacks. Check if the total slack available from days analyzed so far can be used to pick up one or more demand from the days that were examined so far. This may be achievable by breaking a pickup into more than one load on different days. If there is such savings, make the necessary transfers.

Assume that the inventory cost per unit is $1/unit/day. A savings table is calculated next. The savings table gives the savings if we pick up additional units that can be picked up at the same time when we are visiting a supplier. Because on day 1 we are visiting all suppliers, we can pick up additional units from any or all. However, there would be savings only if by picking up additional units, we can avoid visiting the supplier on that day or by picking up partial supplies in every trip so that we can avoid making another visit on some other day.

### Iteration 1: Day 4

First iteration starts with day 4. There are 3 units of excess capacity in the truck or slack available in the truck. Using this slack, we can move in entire requirements for item 1 or 2. Corresponding savings are: for item 1, $10 - 1 \times 2 = 8$; for item, 2: $12 - 1 \times 3 = 9$. These quantities are calculated via a simple formula: Cost of the trip saved − days inventory picked up earlier × quantity picked up.

Maximum savings is associated with item 2, so pick up three additional units of item 2 on day 4. There is no more slack, hence we can terminate day 4 analysis. The results are displayed in the day 3 table, because analysis for day 3 starts with the end results of day 4. Note that savings for the day analyzed, day 4 in this case, are not shown to avoid any confusion with the next day's (day 3) analysis. Savings in iteration 1 is $9.

### Day 4 Analysis

| | | 1 | | 2 | | 3 | | 4 | | 5 | |
|---|---|---|---|---|---|---|---|---|---|---|---|
| Item | Pickup Cost | Pickup Quantity | Savings | Pickup Quantity | Savings | Pickup Quantity | Savings | Pickup Quantity | Savings | Pickup Quantity | Savings |
| 1 | 10 | 9 | | 8 | | 5 | | 4 | 8 | 2 | |
| 2 | 12 | 6 | | 6 | | 3 | | 7 | 9 | 3 | |
| 3 | 14 | 3 | | 4 | | 1 | | 6 | | 5 | |
| Slack | | 2 | | 2 | | 11 | | 3 | | 10 | |

### Iteration 2: Day 3

Recalculate slack rates. There is no slack in day 4. In day 3, there are 11 units of slack. We can pick up different items from both day 4 and day 5 with this slack.

For example, for item 1, if 4 units are picked up from day 4, savings amounts to $10 - 1 \times 4 = 6$. If 2 units are picked up from day 5, the savings is $10 - 2 \times 2 = 6$. Similarly, savings are shown for other items. Select a combination that gives the maximum savings and still fits within the total slack available. In our case, the combination is: for item 1, 4 units from day 4 and 2 units from day 5; and for item 3, 5 units from day 5, giving a savings of $6 + 6 + 4 = 16$. The results are shown in the table for day 2 analysis.

## Day 3 Analysis

| | | 1 | | 2 | | 3 | | 4 | | 5 | |
|---|---|---|---|---|---|---|---|---|---|---|---|
| Item | Pickup Cost | Pickup Quantity | Savings | Pickup Quantity | Savings | Pickup Quantity | Savings | Pickup Quantity | Savings | Pickup Quantity | Savings |
| 1 | 10 | 9 | | 8 | | 5 | 6, 6 | 4 | | 2 | |
| 2 | 12 | 6 | | 6 | | 3 | 2 | 10 | | — | |
| 3 | 14 | 3 | | 4 | | 1 | 8, 4 | 6 | | 5 | |
| Slack | | 2 | | 2 | | 11 | | 0 | | | |

## Iteration 3: Day 2

Recalculate the slack rates. There are only 2 units of slack in day 2, and 4 units in day 4 cannot be used to make any pickups. Go to day 1 with present day 2 table.

## Day 2 Analysis

| | | 1 | | 2 | | 3 | | 4 | | 5 | |
|---|---|---|---|---|---|---|---|---|---|---|---|
| Item | Pickup Cost | Pickup Quantity | Savings | Pickup Quantity | Savings | Pickup Quantity | Savings | Pickup Quantity | Savings | Pickup Quantity | Savings |
| 1 | 10 | 9 | | 8 | | 11 | | — | | — | |
| 2 | 12 | 6 | | 6 | | 3 | | 10 | | — | |
| 3 | 14 | 3 | | 4 | | 6 | | 6 | | — | |
| Slack | | 2 | | 2 | | 0 | | 4 | | | |

## Iteration 4: Day 1

There are only 2 units of slack for day 1, and this cannot accommodate any additional pickups. However, when it is combined with slack for day 2, we get 4 units of slack. Pickup quantity for day 3 for item 2 is 3 units, so check if it is possible to make this pickup in two loads. To minimize inventory cost, we should pick up additional units as close to their use as possible. So check if picking up 1 unit in day 1 and 2 units in day 2 will result in any savings. One trip is eliminated and therefore the savings are $12 - (2 \times 1 + 1 \times 2) = 8$. Because it is a joint pickup, it is by pickup quantity in ( ) and total savings by an *. Make the associated assignment.

| | | 1 | | 2 | | 3 | | 4 | | 5 | |
|---|---|---|---|---|---|---|---|---|---|---|---|
| Item | Pickup Cost | Pickup Quantity | Savings | Pickup Quantity | Savings | Pickup Quantity | Savings | Pickup Quantity | Savings | Pickup Quantity | Savings |
| 1 | 10 | 9 | | 8 | | 11 | | — | 8 | — | |
| 2 | 12 | 6 | (1) | 6 | (2) 8* | 3 | | 10 | 9 | — | |
| 3 | 14 | 3 | | 4 | | 6 | | 6 | | — | |
| Slack | | 2 | | 2 | | 0 | | 4 | | | |

Savings in iteration 4 is $8.

The final pickup policy is shown below. The total cost of this policy is: (total pickup cost if we had to visit all sites on every day) – (total savings obtained by combining pickups). Or it is equal to 5 × (10 + 12 + 14) – (9 + 12 + 0 + 8) = 180 – 29 = $151.

**Final Pickup Table**

| | | 1 | | 2 | | 3 | | 4 | | 5 | |
|---|---|---|---|---|---|---|---|---|---|---|---|
| Item | Pickup Cost | Pickup Quantity | Savings | Pickup Quantity | Savings | Pickup Quantity | Savings | Pickup Quantity | Savings | Pickup Quantity | Savings |
| 1 | 10 | 9 | | 8 | | 11 | | — | | — | |
| 2 | 12 | 7 | | 8 | | — | | 10 | | — | |
| 3 | 14 | 3 | | 4 | | 6 | | 6 | | — | |
| Slack | | 1 | | 0 | | 3 | | 4 | | | |

## 6.3 SUMMARY

The chapter briefly presents two important topics: lean manufacturing and supply chain management. The principles from both approaches are used in improving existing facilities and can be easily incorporated when developing new layouts and facilities. Thinking ahead of the problems that may arise and then applying the principles described in this chapter can save considerable headaches and expenses later. Naturally, these concepts are the foundations for making current systems efficient.

## PROBLEMS

1. One of the principles of the 5S system is to "reduce, if possible eliminate, clutter." List the reasons why a workplace becomes cluttered.
2. How can computers and/or computerized systems be used in implementing 5S?
3. Why a procedure is named Six Sigma when there are six steps involved in the procedure?
4. Kaizen calls for continuous improvement. Do you think such improvements may follow a decreasing rate of return in terms of time spent and dollars saved? Justify your answer.

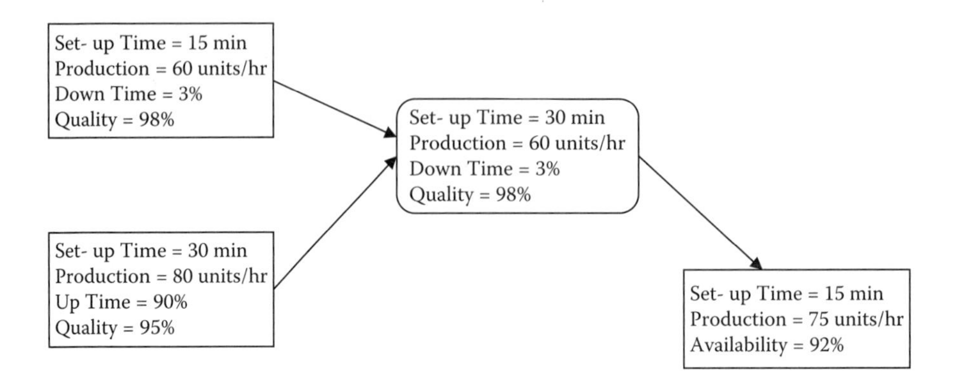


**FIGURE 6.69**

5. Andon is visual signaling. List other forms of signaling you think can be implemented and under what conditions?
6. Is there any difference between takt time and cycle time?
7. How is takt time established? How does the customers' buying frequency affect takt time?
8. Calculate takt time in following process (Figure 6.69).
9. Suppose a product follows the process as shown in problem 8. Develop the value map diagram if a customer has ordered 1500 units per week by establishing production and shipping schedules along with inventory points and in-process inventory space.
10. In TPM, various concepts are applied to increase production. Is it possible to list these factors in order of importance so that we can concentrate our efforts on the few important ones, rather than all of them simultaneously?
11. Does source inspection add cost or provide savings to the product? Why?
12. Perform a Poka-yoke for a four-way stop on a highway intersection.
13. What are the PM procedures that should be followed with automobile tires?
14. Distribution cost from plants to customers is illustrated in the following table, along with other pertinent information. Develop the plans for production and distribution.
15. Suppose we know that the product requirements for the next 10 days are as follows:

| Period | 1 | 2 | 3 | 4 | 5 | 6 | 7 | 8 | 9 | 10 |
|---|---|---|---|---|---|---|---|---|---|---|
| Requirements | 200 | 250 | 230 | 160 | 200 | 200 | 100 | 120 | 150 | 130 |

Full-time workers can be used at a cost of \$20/hour and part-time employees at \$10/hour. A full-time worker can produce 3 units/hour and a part-timer can produce 2.5 units/hour. A part-time employee can work

20 hours/week (5 consecutive days) and must have a full-time employee's supervision. A full-time worker can supervise at the most 4 part-timers. At any time, no more than 30 workers can work in the plant on this line. The product can be purchased from an outside supplier for $30/unit. The unit sells for $50 a piece. A unit stored to carry from one day to the next costs $2/unit and at the most, 50 units can be stored in the warehouse. Each day's demand must be met so that there is no back order or shortage. The problem is to determine optimum production policy from day to day.

16. Four products are picked up from three suppliers by a truck that can carry 20 units. Suppose the inventory carrying cost is $1/unit/day. Develop an optimum pickup policy for the demands shown in the table, where the cost of visiting suppliers is stated as procurement costs.

**Data for an Example**

| Product Number | Procurement Cost | Maximum | Daily Pickup Requirement | | | | |
|---|---|---|---|---|---|---|---|
| | | | Day 1 | Day 2 | Day 3 | Day 4 | Day 5 |
| 1 | 12 | | 5 | 5 | 4 | 4 | 2 |
| 2 | 15 | | 3 | 6 | 3 | | 3 |
| 3 | 18 | | 3 | | 6 | 6 | 5 |
| 4 | 10 | | 2 | 2 | 2 | 2 | 2 |

# 7 Building, Organization, Communications, and Selected Support Requirements

This chapter discusses a variety of topics that contribute to the overall design and workings of a production facility. A manufacturing unit must be housed in a building, and Section 7.1 discusses a few traits of the more popular industrial buildings. The facility must be operated efficiently, requiring an organization structure such as that described in Section 7.2. As we have seen, the nature of an organization affects the total labor resources that are needed in the plant. It also influences the office and plant layouts detailed in Chapter 13. Section 7.3 is a short statement on the communications needed to operate a plant coherently. The mode of communication also defines the facilities required in the plant. For example, storing documents in the quality control room might require a stacked cabinet, or the documents could be stored on a USB drive if a desktop computer is available; written memoranda could be transmitted on paper or through desktop computers connected in a network. Section 7.4 discusses the services and facilities that are necessary in any production operation, including shop offices, inspection and maintenance departments, locker rooms, lavatories, a cafeteria, and even a parking area. Essentially, this chapter provides information on integrated plant design.

## 7.1 BUILDING

More and more industrial engineers, along with architects and civil engineers, are getting involved in the basic phases of designing and constructing industrial buildings. An industrial engineer is concerned with the efficient operation of a plant and is therefore responsible for arranging machines, departments, and material-handling facilities within it. Many times the engineer can obtain a superior arrangement if the building is constructed to fit the layout rather than a layout being developed to accommodate the existing building. Layout considerations must also play an important part in enlarging or revising an existing structure. We now study some building features in industrial plants.

### 7.1.1 Conventional Building Characteristics

The following subsections discuss the major components of industrial buildings, their properties, and their requirements.

#### 7.1.1.1 Structure

The structural considerations of a plant are influenced by the activities to be conducted within the building. For heavy manufacturing, which includes iron smelting, steel fabrication, ship building, and automobile manufacturing, the support structure is large and heavy. This is mainly due to the demands placed on the structure by gantries, cranes, and other heavy equipment.

Light industrial buildings are generally single story (unless the land is very expensive) and are constructed of steel frame. The spacing of columns and bays within the building plays an important role in the overall layout. It is common practice to define major aisles and partition walls for various departments along the columns. The sizes of "bays" (between columns) range from 20 × 20 to 35 × 160 feet. The longer the span, the more expensive it is to construct the building; however, a long span also reduces the number of columns that are necessary to support the entire structure. It is estimated that each column effectively uses up between 9 and 16 square feet of floor space, an area that could be used for productive activity. Long spans also require deeper steel frame construction, and this permits the overhead space within the frame to be used for compressed-air piping, air-conditioning ducts, and electrical conduits. A square building is preferred over a rectangular one, with the latter type being more flexible for expansion. The clear height of the building differs depending on the products, generally ranging from 11 to 38 feet, with 18–20 feet being most common.

#### 7.1.1.2 Walls

Outside walls provide protection and security to the plant. In conventional structures, most walls are made of masonry or concrete which provides good fire protection and presents a neat appearance. Such walls are not moveable, however, and expanding or revising the plant is therefore not easy. Temporary walls of corrugated aluminum or sheet metal with fiberglass insulation are normally used when expansion is envisioned.

Interior partition walls are generally non-load-bearing, and they can be prefabricated or built on-site using various materials. In a food plant, for example, walls might be tiled for a smooth and sanitary interior.

#### 7.1.1.3 Floor

The floor of a plant is almost always a concrete slab. Such a floor can be made strong enough to support heavy equipment loads and can be leveled to an accuracy ranging from ±4 mm in 2 m (normal) to ±1 mm in 2 m (super flat). The flatness is important, because it relates to the vehicle dynamics and mast deflection of a floor-riding material-handling device such as a forklift truck.

In general, the floor and its covering should be durable, impact- and vibration-resistant, sound-absorbent, nonskid, nonsparking, and easy to clean. It should not be affected by changes in temperature or humidity, or by accidental spills of chemicals, oils, or other industrial substances.

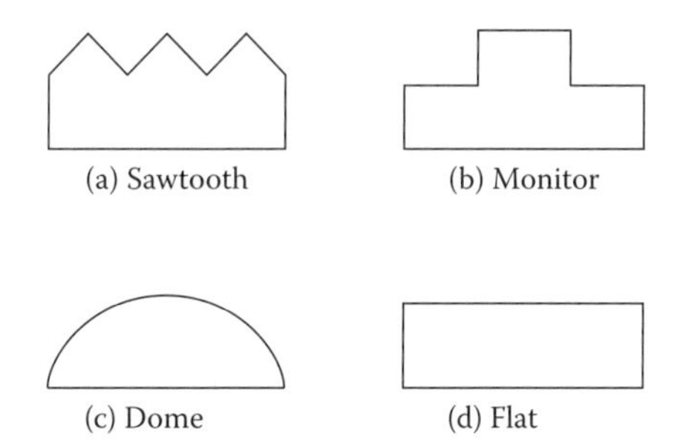


**FIGURE 7.1** Roof types

### 7.1.1.4 Roof

The choice of roof shape from several shapes that are generally used (Figure 7.1) can give distinct character to the building. A sawtooth roof with translucent glass on one side of the teeth allows natural light to penetrate into the plant without causing glare. A sawtooth roof can also allow an opening for fresh air and ventilation. A monitor roof is useful when an overhead crane is needed at the center of the plant. A dome shape encloses a large volume when considerable height in the center is desired or essential. A flat roof is most common in modern buildings because of its low cost of construction, as well as ease of installing insulation and acoustic ceilings.

The materials used in roofing can vary from long spans of aluminum or steel, costing as much as $600 per 100 square feet, to a combination of fire resistance material and cement sheeting, costing about $150 per 100 square feet. A flat, reinforced concrete roof is often covered outside with an insulating material similar to cork to reduce temperature variations in the plant.

### 7.1.1.5 Interior

A pleasant interior environment can add considerably to the worker's morale and productivity. It is the responsibility of engineering and layout personnel to provide such an environment.

Temperature in the plant should be maintained between 68°F and 72°F at a relative humidity of 50%. Air-conditioning and heating units might be necessary to prevent the ambient temperature from varying too much. Flow of air within the facility is desirable to avoid a feeling of stuffiness; in addition, dust, odor, and the vapors as well as fumes generated by machines or workplaces should be controlled. This can be accomplished by providing vacuum ducts and tubing that transport air and particles to a centralized location away from the work spaces. Depending on the nature and amount of the contamination, any of numerous collecting devices might be used; if no hazard is involved, air could be discharged to the atmosphere. Appropriate drainage should also be provided whenever necessary.

Noise within the plant should be minimized by isolating the sources (if possible) or controlling them. Proper maintenance and sound-absorbing tiles on high ceilings and walls contribute considerably toward noise reduction. Hearing protection (earplugs or muffs) should be worn by people who work in an environment where an unsafe, high-noise level cannot be reduced. The Occupational Safety and Health Administration (OSHA) has issued regulations governing the maximum noise level permitted.

Good illumination within the plant is important, and between 50 and 60 foot-candles of light should be maintained throughout the facility. At times a workstation for delicate or close manual work might need a much greater amount of illumination, perhaps as much as 100 foot-candles or more.

Interior colors also contribute toward making the workplace enjoyable. Dark colors on the walls might hide dirt, but they also absorb light, making the place look dull. Cream, peach, and light shades of green and blue are favorite colors for walls, and white is most common for ceilings. Yellow is used to indicate the need for caution (e.g., low overhead pipes) and other problem areas that might need special attention. Black-and-white stripes generally define traffic areas, and aisles are marked with white borders. Red is associated with danger and is used to indicate fire protection apparatus, stop signs, traffic lights, and dangerous areas where admission might be restricted.

### 7.1.2 Prefabricated Buildings

Many sources are now available for prefabricated buildings of almost any desired shape. Prefabrication is a cost-effective and time-saving method of construction, mainly using a corrugated or flat-channel steel body shell. The modular approach to a building also allows expansion or alteration. Interior walls or offices can be moved from one location to another with ease to modify the layout within an existing plant.

## 7.2 ORGANIZATION

Proper organizational planning eases the coordination of all personnel in a manufacturing concern by showing how their efforts contribute to a common purpose. It also helps in facility planning to outline the number and types of personnel who will be working in the organization and their relationships with each other. The method of structuring is determined by the aims of the company, what resources are available, and the amount of time the organization has to accomplish a task.

Employees within a business who share one or more important common factors should be grouped together. Some of these factors are skills, processes, geographical locations, types of customers, and goals. The chain of command should facilitate decision making at the lowest possible level at which the required information is available, but such decisions should be reviewed and modified, if necessary, by people at higher levels in the organization. The structure should be as simple as possible to avoid overlap, redundancy, and stifling of creative efforts. Required coordination time and communication difficulties increase drastically with each added layer of supervision.

### 7.2.1 Organizational Concepts

Classic U.S. and European organizational theories are based on principles derived from military and religious orders. In contrast, the typical Japanese manufacturing corporation does not even have an organization chart, often using project teams whose managers may readily be moved from one area to another. The individual worker not only feels responsible for performing his or her assigned task, but also sees to it that all tasks assigned to his or her project team are carried out. Japanese workers view employment as a lifetime arrangement and coworkers as lifetime colleagues; as a result, they make every effort to get along well together. Japanese factory workers usually tend to be generalists, trained in most aspects of a manufacturing process.

In the United States, even though nowadays teamwork is encouraged, workers are more frequently specialists, and businesses are designed with as little redundancy as possible. Management attempts to make lines of communication easily identifiable tend to push employees toward specialization and compartmentalize functions. A well-defined organization helps decrease operating costs, reduces organizational friction, and consequently minimizes the number of management problems.

### 7.2.2 Organization Charts

How an organization is structured at a particular time can be depicted by a chart that shows the formal relationships among functions and provides the titles (and frequently the names) of the people who are responsible for those activities. Some companies avoid using such charts because they feel that organization charts can cause inflexibility; other companies like them because they show who is responsible for what function in the lines of direct authority. Charts can help a company prevent duplication of effort or having a responsibility not be properly assigned. The charts can also be used in expansion planning and to show employees who reports to whom throughout the organization.

### 7.2.3 Organizational Structures

The structure of an industrial plant can be set up in one of several ways depending on the size of the company, the type of business being carried out, and the complexity of difficulties encountered in day-to-day operations. The structure can range from a straight line to a multidimensional matrix organization.

#### 7.2.3.1 Line Organization

The oldest and simplest format is the line form. A straight-line organization most often exists in small companies in which authority extends from the highest to the lowest level of concern. A simple line organization chart is shown in Figure 7.2. The head of the company oversees all functions, such as personnel, sales, engineering, and purchasing. An organization of this type has a unity of command in which each person reports to and receives orders from only one other person. It has the advantage of having easily understood relationships at all levels. Conversely, it has the disadvantage of requiring personnel at most levels to be capable in several functional areas. This may result in some duplication of effort; for example, each supervisor might find it necessary to keep a set of equipment catalogues. As such a

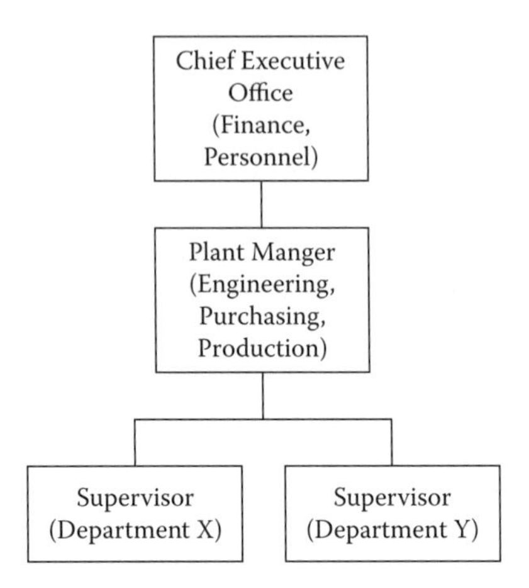


**FIGURE 7.2** Line organization

company increases in size, managers become overwhelmed by the details that must be handled. In this situation, the organization will probably evolve into a staff-line organization.

### 7.2.3.2 Staff-Line Organization

A staff-line organization evolves when the need for specialists in different areas becomes so great that managers cannot handle the burden of details. Figure 7.3 shows an example of a typical staff-line organization. It is similar to a line organization, except that specialists have been added to contribute technical expertise in

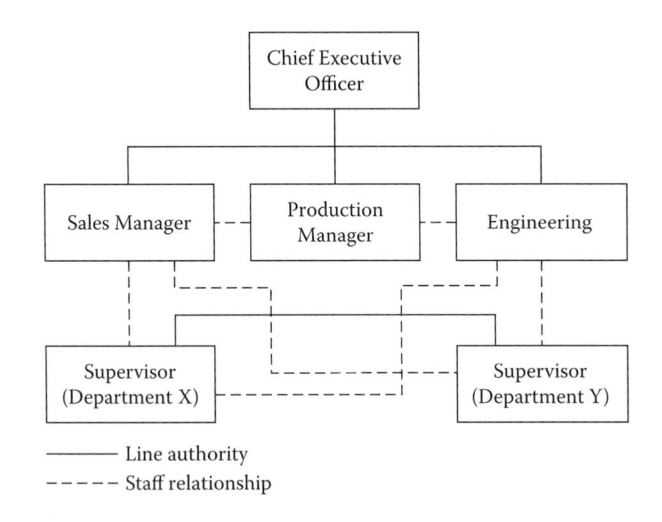


**FIGURE 7.3** Staff-line organization

particular areas. In theory, authority and responsibility still lie within the line, but the staff can exert considerable influence while acting in its advisory capacity. A modification of this type of structure is the line and functional staff organization in which the specialists have full authority over matters relating to their particular areas of expertise. Staff functions are usually one or a combination of service, control, advisory, or coordination types.

### 7.2.3.3 Product Organization

When a business becomes a multiproduct company, organizing it into a product-type structure might prove to be more efficient. This is especially true if the products are easily distinguishable, in either form or market, from each other. Figure 7.4 shows

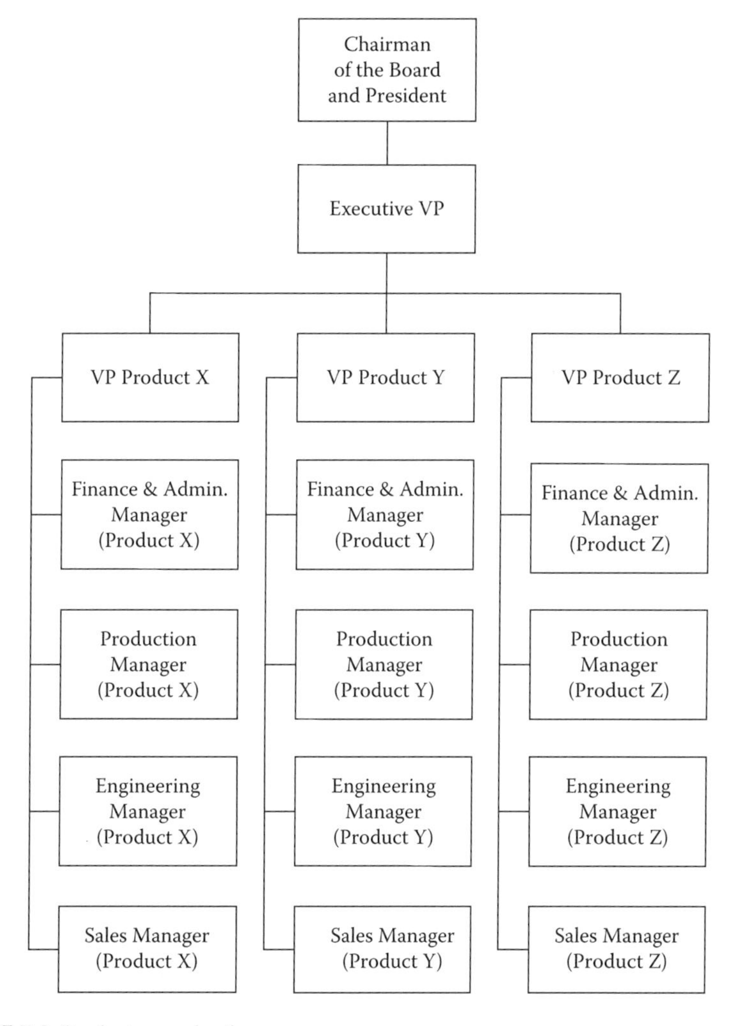


**FIGURE 7.4** Product organization

how the chart of such a firm might look. The disadvantage of this arrangement is that several people with the same general type of skill might be required, but this may be overcome in part by centralizing some functions, such as engineering and administration. As the business grows, this classification becomes quite complex, and a more modern concept, the matrix organization, might be used to alleviate the problems.

#### 7.2.3.4 Matrix Organization

In a matrix chart, such as that shown in Figure 7.5, activities are shown vertically and products are shown horizontally. In this arrangement, financial, engineering, and other methods can be standardized through function managers who can assign personnel and resources as needed to the different products. (In practice, the function managers are actually staff managers.) However, the matrix organization does have some inherent disadvantages. The more significant ones are that it violates the principle of unity of command and that it makes communication and coordination even more critical, both vertically and horizontally.

### 7.2.4 Selection of Organization Structure

Of the organizational forms discussed above, choosing the one that is best for a particular business depends on a thorough understanding of (1) the goals of the organization, (2) the product or service to be offered, and (3) the capabilities and

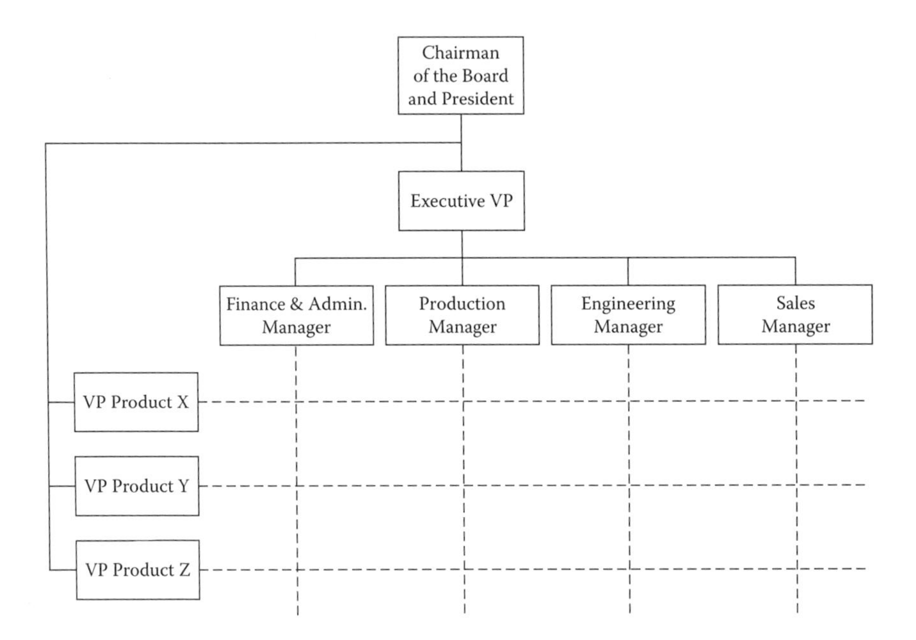


**FIGURE 7.5** Matrix organization

willingness of the individuals who work in the organization. In reaching this understanding, one must determine and analyze the types of manufacturing facilities and departments required for efficient and profitable operation. Once the appropriate structure is selected, implementation will depend in part on effective communication throughout the organization.

## 7.3 COMMUNICATIONS

The methods of communication are changing in modern plants and offices, and so are the necessary facilities. It is therefore beneficial to understand the need and means of communication as we develop the facilities.

Communication is an essential function because it is the foundation for transmitting management objectives and provides the means for understanding those objectives. Communication can have one or more of several purposes:

- to furnish information or transmit data;
- to persuade;
- to give directives or direction; or
- to gather information or receive input.

In order for the people who receive a communication to take the desired action or retain the desired knowledge, they must accept and understand the message. Communications can be written or spoken; each method has inherent advantages and disadvantages.

### 7.3.1 Written Messages

It is often difficult to maintain attention to a written communication that is very long. E-mails are commonly used to communicate with written messages. A written communication has several advantages: it ensures that the same message is received by everyone involved, it is easily disseminated, and it provides a permanent record to which one can refer at a later time. It can also confirm the message is delivered and read.

### 7.3.2 Spoken Messages

Spoken communications offer the chance for immediate feedback to clarify or confirm understanding of a message. Psychologically, however, people tend to remember less of what they hear than of what they see. If a communicator has a poor delivery, even a good message might be improperly received and accepted.

### 7.3.3 Methods of Presentation

A message might be intended for one, two, or many people. When information is intended for general distribution, one of several means can be used.

Group meetings can be held in a conference room with the speaker present, or the message can be delivered live via a teleconference or a broadcast. Delayed presentations can be made via video recordings. One of the more common methods of in-plant mass communication is the use of bulletin boards. In plants with large workforces, employee periodicals, handbooks, and policy manuals are effective means of using mass communication. Although the manuals might not be read by everyone, they do serve as written records of the information made available.

Person-to-person communications can be carried out face to face, by telephone, or by computer.

### 7.3.4 Documentation

We stated earlier that the design and operation of a manufacturing plant is a team effort. Each member of the team is an expert in one field of operation, yet coordination between all the members is essential to accomplish overall integration in the plant. Documentation is a means of achieving a common objective without duplication, confusion, and chaos.

Three principal factors contribute to the need for documentation in achieving integration of work: the number of workers (people) involved, the physical distribution of the workers and their activities, and the time span over which the activities are performed.

For a small organization with one-person supervision and a small area of operation, oral communication might serve well, but for most other organizations, written rules, instructions, and records are required. The red tape created by paperwork is not meant to impede the activities of a creative individual, but to aid in the team effort by maintaining orderliness in the information flow and by reducing mistakes. Care must be taken to avoid overburdening employees with preparing and reading unnecessary reports. That is, such documents should not be required simply for the sake of having reports. Ideally, the system will provide everyone with accurate, timely, and useful information required for efficient operation of the organization.

When workers are located in widely separated areas (e.g., different rooms or buildings), oral communication requires the assembly of all workers in a conference room, or the speaker must telephone or travel from one employee or group to the next. All these alternatives are time-consuming and not cost-effective, especially if only routine matters are involved. A memorandum or a message transmitted via the company's intranet is preferable.

This also resolves problems associated with remembering instructions over an extended period, as well as ensuring that all people concerned will receive the message even if some are not present when the message is transmitted.

Printed forms provide a convenient means of recording standard information. They prevent wordiness and display information in the same manner every time, making it easy to read and recognize. They also help a person who is new on the job to understand information that must be collected and recorded.

Engineers also frequently record their calculations, notes, and communications on a notepad, which becomes a legal document that should be readily available to other authorized personnel. While working on a project, engineers should carefully record all calculations, assumptions, instructions, and dates. They are often asked to produce these records in support of their presentations of project reports.

At times, owing to an emergency or another critical event(s), people might have to deviate from normal procedures by "cutting the red tape." An organization in which such deviation is possible without causing confusion, hard feelings, or disorder has two principal traits: trust between the employees who will help one another, and people who are very knowledgeable about the workings of the system. Any shortcuts taken by one person can add to the time demanded from another; in a well-organized and well-run plant, this procedure should not become a common practice.

An example of good documentation is shown in Appendix B. The instructions are for a quality control operation in a spring manufacturing facility.

## 7.4 SUPPORT FACILITIES AND REQUIREMENTS

The facilities and services that are not directly necessary for production but are still required for the operation of a business are termed support facilities and services. Their availability and timeliness contribute considerably to improving worker satisfaction and productivity. We discuss some of the more important support services in this section; several others, including utilities, will be discussed in Chapter 15. We have excluded managerial activities such as supervision, personnel management, accounting, sales, and advertising. These activities, for the most part, are self-explanatory with respect to their purposes.

### 7.4.1 Shop Offices

In its operation, a manufacturing facility is normally divided into several shops or departments, each managed by a supervisor and supporting staff. Although many support functions, such as payroll, engineering, and personnel, are directed through the central office, an area must be set aside for office space for the people who are directly responsible for supervising the activities of a department.

### 7.4.2 Inspection

There might be one centralized location for all inspection activities if the products are small and can be easily transported, or there might be multiple inspection stations, perhaps more than one for each department, if the product is large or inspection frequency demands it. The necessary space for the inspectors and their equipment must be planned. The personnel should be trained in statistical methods of quality control, should be familiar with the production processes and characteristics of the product, and must be able to communicate effectively, both orally and with written reports.

### 7.4.3 Maintenance

All production plants need good maintenance facilities. There is a large investment in the building, equipment, and material-handling system, and it is important to obtain continuous and well-planned use from them. Any failure in the conveyor system, for example, can halt production throughout the plant. The maintenance department anticipates such potential failures and takes action through preventive maintenance programs. It also makes emergency repairs to avoid excessive loss of production when failures occur. Maintenance personnel are generally skilled in a variety of trades. They are trained in electrical wiring, hydraulic and pneumatic systems, controls, heating and air-conditioning, welding, and working with specific machines. The primary skills required vary from plant to plant depending on the equipment and processes installed. Most maintenance departments include one or more people with specialized training, such as electricians, welders, and mechanics. In small plants, one person might be trained in more than one trade. This is especially true when maintenance of the building is involved: the same individual might be required to paint and clean, as well as do plumbing, electrical, and air-conditioning work.

For highly skilled work such as maintenance of computers, or when the demand for certain skills is infrequent, it might be better to contract the maintenance work to outside specialists. Routine maintenance, such as lubrication and general inspection, can be assigned to the operator of an individual machine.

**Support facilities.** *Top left:* Tools that workers need to do their jobs are kept in a tool crib like the one shown here. This secure enclosure has racks and cabinets that hold anything from simple wrenches to special drilling or grinding tools. *Top right:* In plants that involve messy work, washing facilities like this foot-operated hand-washing facility are located conveniently near the work area and can serve several people at one time. *Bottom left:* In-plant food services are beneficial to both workers and the company. Workers can get good, inexpensive meals without having to leave the plant, and the company avoids problems of drinking and absenteeism. *Bottom right:* Coffee service in a work area contributes to employee morale, and cart delivery saves employee travel time during breaks.

Facilities for minor or emergency repairs might be provided in each shop; a centralized location might be needed for major repairs and overhauls. At times, heavy machines require service within the plant, and the maintenance department must be located near the production area. The shop should contain hand and power tools, workbenches, storage areas and racks, and light machines such as small lathes.

### 7.4.4 Toolroom

The purpose of a toolroom is to procure (or make), store, and repair tools that are commonly used in the plant. Typically, the tools consist of mold dies, punch press tools, milling cutters, drills, and cutting tools for lathes, as well as jigs and fixtures. As part of a normal maintenance program, toolroom workers inspect tools, molds, and jigs and fixtures to see that they comply with the specifications and drawings. When necessary, they restore, clean, grind, sharpen, polish, and repair the items. A toolroom is a job shop operation; depending on the activities in the plant, it might contain high-precision equipment such as a Bridgeport drill press, boring machines, planers, a surface grinder, a lathe, and various hand tools. Some toolrooms are also equipped with specialized steel stock used in making tools, a heat treatment furnace, and high-speed cutoff equipment. A toolroom should have all the utilities that are available in the plant, such as compressed air, 220/110-V electric outlets, and lubrication facilities. Material handling in a toolroom is generally achieved by using handcarts and monorail and jib cranes. The layout of a typical toolroom is shown in Figure 7.6.

### 7.4.5 Tool Crib

Necessary tools are issued to the workers in a plant through a tool crib, which serves as a storage center for all tools and accessories. A tool crib might consist of cabinets

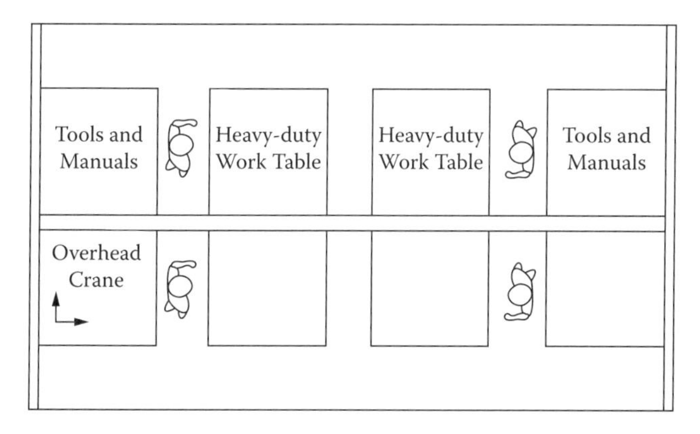


**FIGURE 7.6** A typical room layout

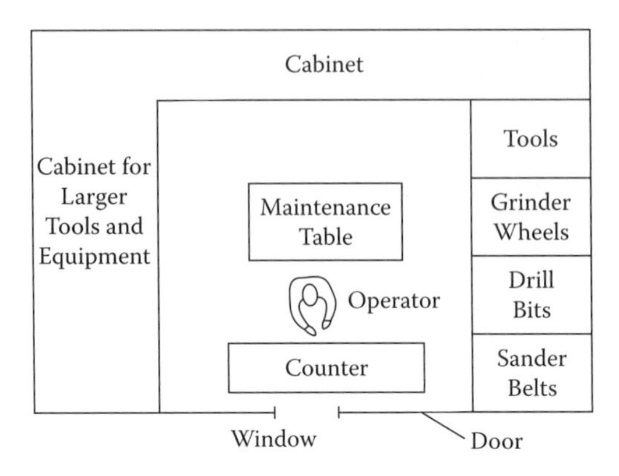


**FIGURE 7.7** Tool crib

and racks for holding such items as emery cloth, steel wool, sander belts, socket sets, drill bits, milling tools, cam followers, grinding wheels, special clamps, and general-purpose dowel pins. Some security must be provided for a tool crib area; it generally consists of an 8- to 12-foot area with a 10-gauge chain-link fence around it, with a window opening for issuing the tools. The layout of a typical tool crib is shown in Figure 7.7.

### 7.4.6 Lockers/Changing Area

Each employee might be provided with a locker to store his or her belongings. If a uniform is required or if the work environment makes protective clothing necessary, a changing area should be provided in the plant. Layouts for locker rooms are shown in Figure 7.8 and Figure 7.9.

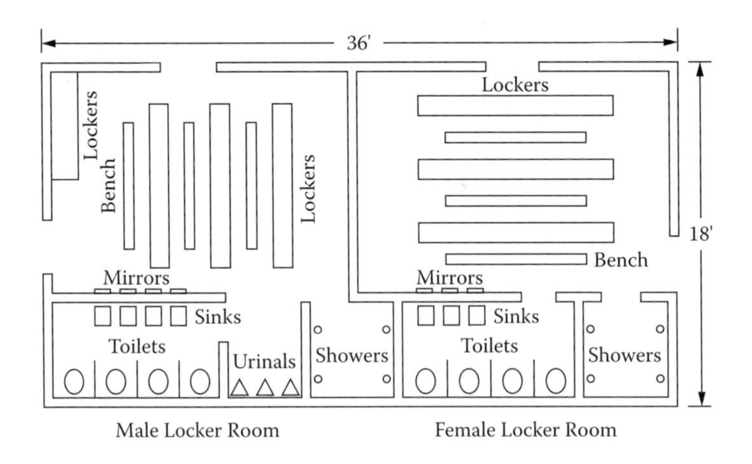


**FIGURE 7.8** Locker room plan 1

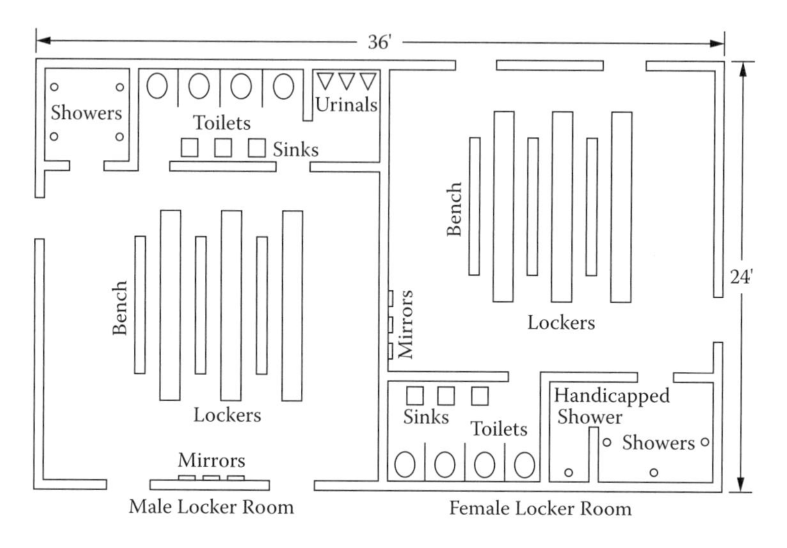


**FIGURE 7.9** Locker room plan 2

### 7.4.7 Lavatories

In most plants, locker rooms and lavatories are combined into one area. However, in a large plant, there might be a need for lavatories in several locations; and to minimize time spent away from the job, they should be close to the work space. Individual states and local authorities might have regulations defining the number of lavatories required for a specific number of workers. OSHA has regulations regarding the requirements for toilets, lavatories, and showers in an industrial setting. Depending on the number of employees, the minimum number of toilets required is specified as given in Table 7.1.

**TABLE 7.1**
**Number of Toilets Required**

| Number of Employees | Minimum Number of Toilets |
|---|---|
| 1–15 | 1 |
| 16–35 | 2 |
| 36–55 | 3 |
| 56–80 | 4 |
| 81–110 | 5 |
| 110–150 | 6 |
| >150 | 1 additional fixture for each additional 40 employees |

For dividing toilets based on the male/female population, OSHA has the following regulation: "Where toilet rooms will be occupied by no more than one person at a time, can be locked from the inside, and contain at least one water closet, separate toilet rooms for each sex need not be provided." Otherwise, "toilet facilities, in toilet rooms separate for each sex shall be provided…."

An average of 4.5 feet per employee for washroom space is recommended; generally, 1.2–1.5% of the working area should be set aside for lavatories.

Lavatories are required in all places of employment; showers are required when certain materials, such as paints, herbicides, and insecticides, are handled or produced. One shower is required for every 10 employees who will be using such a facility. Both hot and cold water should be provided.

In addition we must comply with Americans with Disability Act (ADA). According to the ADA, when either a drinking fountain, urinal or toilet is provided, at least one of each must be accessible by a handicap person. The details are available at site: http://www.access-board.gov/ada-aba/final.htm#pgfId-1010419. In general a handicap toilet or bathroom requires almost twice as large an area as a standard facility.

### 7.4.8 Janitorial and Custodial Services

Upkeep and cleaning of the facility are the responsibility of the janitorial group. Most of the work must be performed at night (after regular working hours) to avoid disrupting plant activities during busy hours. Some companies prefer to subcontract these services to businesses that can obtain labor at less cost than an industrial plant's contract with its labor union might permit.

A storage area must be provided for custodians to maintain brooms, mops, cleaning powders, fluids, and other needs. This area is nothing more than a small closet in many cases.

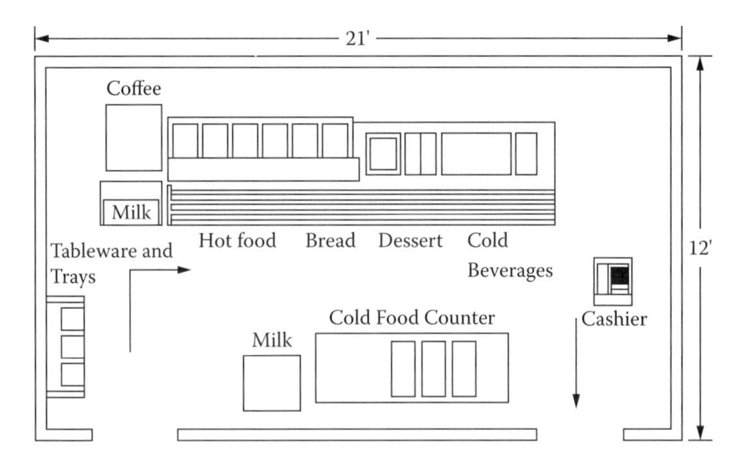


**FIGURE 7.10** Cafeteria plan 1

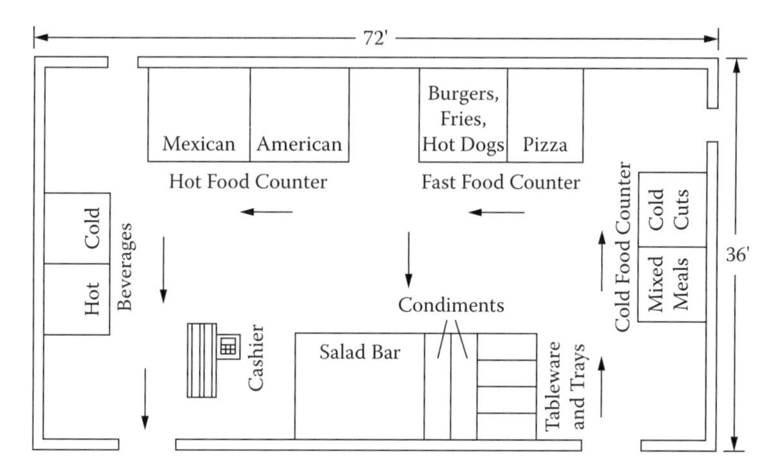


**FIGURE 7.11** Cafeteria plan 2

## 7.4.9 Eating Area

A plant should provide an area where employees can relax and eat during their breaks; the size obviously depends on the number of employees. Such places might provide anything from a few tables and benches, vending machines, and a microwave oven to a full-size cafeteria serving a variety of hot lunches. See Figure 7.10, Figure 7.11, and Figure 7.12 for various cafeteria layouts.

A rough estimate of the area needed for a lunchroom is 17 square feet per employee, for employees who will be eating at one time. For example, if it is expected that at most 200 people will eat at one time, the area assigned to the cafeteria would be $17 \times 200 = 3400$ square feet. This includes space for tables, aisles, counters, and serving tables.

Kitchen area allocation depends on the number of meals to be prepared and served. Table 7.2 provides an estimate of the area required for kitchen activities.

If the company does provide a cafeteria, associated workers such as cooks, helpers, and dishwashers are also needed. Some companies choose to subcontract the entire cafeteria activity. In many instances, the meals provided are subsidized by the firm as a fringe benefit to its workers.

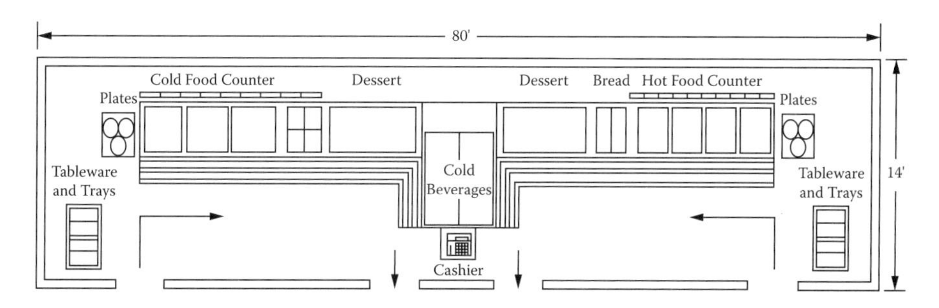


**FIGURE 7.12** Cafeteria plan 3

**TABLE 7.2**
**Kitchen Area Requirements**

| Number of Meals | Kitchen Area (square feet) |
|---|---|
| 200 | 1000 |
| 400 | 1800 |
| 800 | 2800 |
| 1000 | 3000 |
| 2000 | 5000 |
| 3000 | 6000 |

A wide range of food can easily be distributed by vending machines. Soup, sandwiches, hot food entrees, desserts, ice cream, milk, soft drinks, tea, coffee, fruit, and candy are some of the most popular items that are thus sold. The machines range in dimension from 1 foot 4 inches × 1 foot 9 inches × 3 feet 10½ inches for a small hot chocolate dispenser to 3 feet 1 inch × 2 feet 6 inches × 5 feet 2 inches for a candy bar dispenser. One or more microwave ovens in vending areas can quickly warm food if desired. The eating area can be provided adjacent to the machines and might include tables, chairs, and benches. See Figure 7.13 for a typical layout.

A snack bar or sandwich line can be accommodated in a small area to provide a variety of quick hot and cold food such as hamburgers, hot dogs, potatoes, a variety of fresh salads, and drinks.

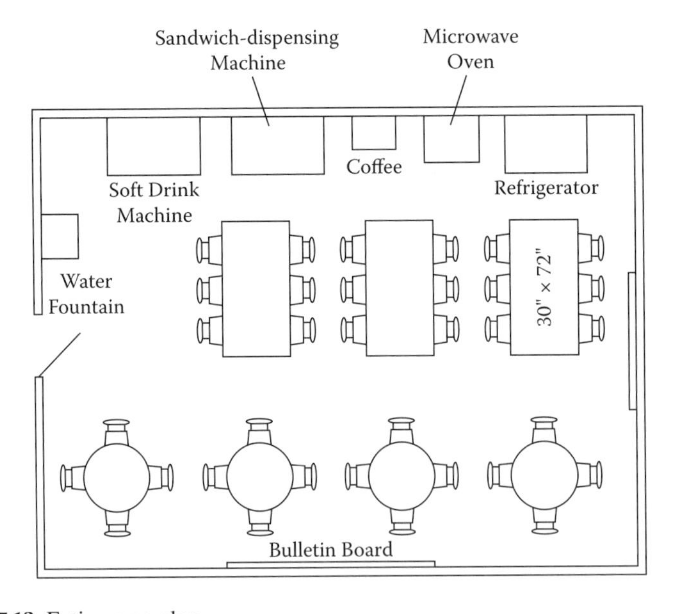


**FIGURE 7.13** Eating area plan

The equipment area might include a refrigerator, coffee maker, stove, drink dispenser, sink, and counter with a cash register. The sitting area might include stools, tables, and chairs. Although a facility with a sandwich line requires more space than vending machines, it is considerably more compact than a cafeteria. It also provides more flexibility in food preparation for an individual than what vending machines offer.

Certain benefits can be reaped from providing food services within the plant. It will no longer be necessary for employees to leave the facility. The company still maintains supervision over the employees, reducing or even eliminating the problems of drinking, returning late, or not returning at all. Of course, it might not always be economically feasible for a plant to have such facilities.

### 7.4.10 Security Force

Plant security requirements might require using a full-time security force. This could be because of danger present within the plant, such as producing chemicals, or because of the nature of the products — for instance, secret defense materials. Most plants, however, require nothing more than a night watchperson or an alarm system.

Many plants issue passes to all their employees. A pass usually contains a person's picture and must be presented to gain entry to the plant. A guard or receptionist might be assigned to check each pass. As an alternative, automatic doors and gates are available that can be opened only by using a coded magnetic strip on the pass.

If a plant uses a security force, worker areas must be provided for the people involved. These might be a guardhouse at each entrance and/or a security office within the plant.

### 7.4.11 Parking Lot

A parking area is needed whenever public parking is not available in the immediate vicinity; many states require it by law. However, a parking area can mean considerable investment, and an analysis should be performed to determine the details of its development. For example, lots with parking spaces at 90° or 60° are quite popular; however, each requires a different average area per car. To provide parking for full-sized cars, the stall dimensions frequently used for a 90° arrangement are 9 × 19 feet (although any width between 8.5 and 10.0 feet is common), with a 24-foot-wide driveway. Thus, an average of 300 square feet of parking area is needed per car (including driveway allocation). If the 60° angular parking is chosen, the stalls become elongated parallelograms with sides of 10.5 and 21 feet. The width of the driveway can be safely reduced to 18 feet, but the total allocation of space per automobile increases to 360 square feet. The advantages of the 60° arrangement over the 90° setup are that parking and backing out are easier as well as safer, and a narrower lot can be used for the same number of rows of stalls. Figure 7.14 illustrates parallel 60° and 90° parking areas.

Small cars require a much smaller area compared with large cars. It is therefore possible to obtain many additional parking spaces in the same area if the lot is divided into two segments: one for small cars and another for large cars. Such

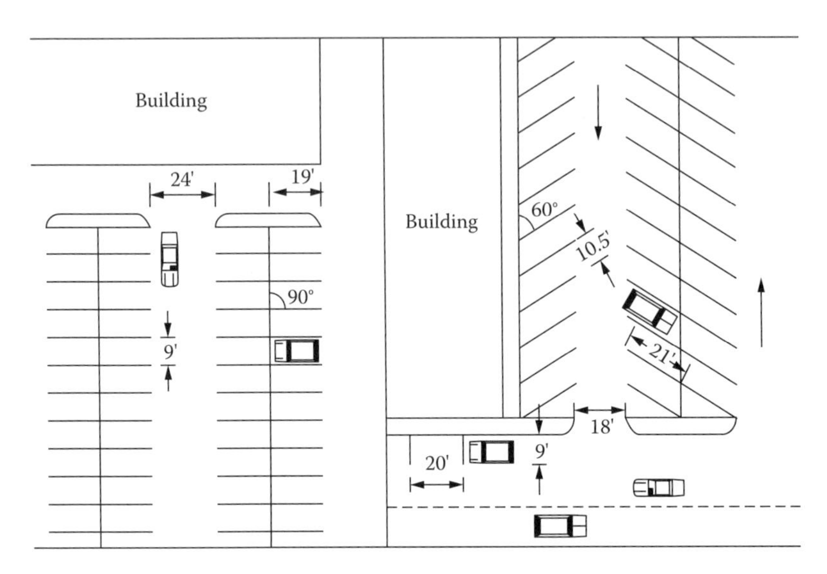


**FIGURE 7.14** Layout for two parking lots

a division should be based on the proportion of the small and large cars expected in the total automobile population.

Many physical characteristics of a parking lot must be taken into consideration. The parking area should be firm and built with a weather-resistant surface such as asphalt or concrete. There should be adequate lighting in the area and security provided if needed. Parking spaces and the direction of travel should be clearly marked on the surface. Entrance, exit, and speed limit signs should be posted.

The number of parking spaces needed might be dictated by state and local regulations. Generally, one parking space for every one to three employees is required. Locations that have adequate public transportation will not need as many spaces as plants in areas where all the employees must drive or ride in private cars.

If shipping and receiving is by truck, an additional area must be available for parking and maneuvering trucks. The exact area required is determined by the type, size, and number of trucks that the plant will handle at one time. The surface used for trucks must be much stronger than that for automobiles. Concrete with heavy reinforcing bars or rods is the most widely used paving material. Loading and receiving docks should be properly located with regard to other traffic in the plant vicinity. The opening to the plant should be of sufficient area to enable mechanical equipment such as forklift trucks to operate. Adjustable lift docks might be necessary when the sizes of trucks vary widely. Loading docks are discussed in more detail in Chapter 11.

### 7.4.12 Medical Facilities

Medical facilities in a plant might range from strategically located first aid kits to an infirmary capable of handling minor medical emergencies. In any case, the company should provide the necessary personnel and equipment for handling sickness and

**TABLE 7.3**
**Permissible Noise Exposures**

| Duration Per Day (hours) | Maximum Sound Level (decibels) |
|---|---|
| 8 | 90 |
| 6 | 92 |
| 4 | 95 |
| 3 | 97 |
| 2 | 100 |
| 1.5 | 102 |
| 1 | 105 |
| 0.5 | 110 |
| ≤0.25 | 115 |

accident. If a major medical facility is located close to the plant, a well-stocked first aid kit and someone trained in administering first aid might be all that is necessary. When the environment is particularly dangerous or hazardous materials are being handled, the availability of medical treatment should be increased accordingly.

### 7.4.13 Noise Exposure

Exposure of employees to high noise levels or noises over a long period can result in permanent hearing loss. Any noise exceeding 90 decibels is considered a noise hazard, and hearing protection is strongly suggested. Table 7.3 lists the permissible noise exposures over a defined period that OSHA has developed for various intensities of noise. All manufacturing facilities are required to adhere to these limits.

## 7.5 SUMMARY

This chapter provides information for integrated plant operation and discusses different topics that contribute to the design and workings of a production facility. The building in which a plant is located can have many traits. The structure of the building may differ depending on whether heavy or light industry is to be accommodated. Building spans can vary from 20 × 20 to 35 × 160 feet. The walls can be made of masonry (for permanent use) or corrugated aluminum (for temporary use). Although plant floors are almost always made of concrete, their smoothness depends on the material-handling equipment that is to be installed. The roof can be of different shapes — dome, sawtooth, monitor or flat — each having its own set of characteristics. The interior of a building should be pleasant and should have temperature and noise control, as well as proper illumination.

Organization within a plant dictates the command structure, which, in turn, may influence plant and office layouts. This chapter briefly discusses various organizational forms, such as line, staff-line, product, and matrix. Each form has both benefits and drawbacks.

Communication is important in operating a plant efficiently, and a brief discussion of the subject is presented. Facilities available for communication can also influence the design and development of a plant.

Support facilities are necessary for the smooth operation of a plant, although they require valuable space within it. For example, offices are needed for departmental supervisors and support staff; inspection and maintenance departments require a place to work in; employees need lockers, lavatories, and eating areas; and room for automobile parking is essential and must be well planned. Basic information and necessary data for the various facets of planning support activities are provided in this chapter. The analysis performed using this information is essential in developing a total plant layout.

## PROBLEMS

1. List the basic structural considerations that would be called for in a building constructed for:
   a. heavy equipment use
   b. light equipment use
2. Discuss the factors of importance in building:
   a. walls
   b. floors
   c. roof
3. What are the advantages of the following types of roofs?
   a. sawtooth
   b. dome
   c. flat
4. A pleasant environment should be maintained for the comfort of workers. Discuss this in terms of:
   a. temperature
   b. ventilation
   c. noise
   d. lighting
   e. colors
5. In the interior of a manufacturing plant, the colors white, yellow, and red are used to indicate certain features. Specifically, what do they point out?
6. What is "good" illumination within a plant? How does the type of work affect this requirement?
7. Contrast U.S. and Japanese organizational theories.
8. Use an organization chart to show the chain of command in the College of Engineering at a state university.
9. What are the advantages and disadvantages of line organization? What benefits does a matrix organization have over a line organization?
10. What is the purpose of communication? Why should a facility planner consider the need for communication?
11. A manager of a canned food company has a list of things to do. He needs to request a financial report from accounting, make inquiries concerning

a new piece of equipment the company might be interested in purchasing, and distribute information to the workers under his supervision concerning a change in policy. This appointment calendar with his secretary. How should each of these matters be communicated?

12. What are support facilities and why are they necessary? Describe each of the following support facilities, indicating why it is necessary and how it should be planned.
    a. eating area
    b. maintenance
    c. parking
    d. custodial service
13. How is the number of spaces required in the parking lot determined? How does transportation by trucks affect the parking lot requirement? (See also Chapter 11.)
14. a. A company employs 350 people, 105 of whom are women. One quarter of the employees work in conditions in which they need to shower and change clothes before leaving work. How many showers and toilets should be provided?
    b. The same company has a cafeteria that serves lunch. There is only one lunch shift, and approximately 90% of the employees eat in the cafeteria. How much space should be provided for the kitchen and the lunchroom?
    c. Design parking lots for each of the following conditions for the same company if 75% of the employees drive to work. Include all dimensions. First, use 90° parking spaces, then use 60° parking spaces. Finally, consider 40% of the cars to be small cars requiring a space 1 foot less in length standard. Use either a 60° or a 90° arrangement.

## SUGGESTED READINGS

Estall, R.C., and Buchanan, R.D., *Industrial Activity and Economic Geography*, Hutchinson Co., London, England, 1980.

Lewis, B.T., and Marron, J.P., *Facilities and Plant Engineering Handbook*, McGraw-Hill, New York, 1973.

Spradlin, W.M., *Walker's Building Estimator's Reference Book*, 28th edition, Frank R. Walker & Company, Lisle, IL, 2008.

### Organization

Miller, R.W., "IE's face challenges in managing human resources to improve declining productivity," *Industrial Engineering*, 17(1), 72–79, 1985.

Shannon, R.E., *Engineering Management,* John Wiley, New York, 1980.

### Support Facilities and Requirements

Occupational Health and Safety Administration, *OSHA Safety and Health Standards,* U.S. Department of Labor, Washington, D.C., 1983.

## Communication

Borman, E.G., Howell, W.S., Nichols, R.G., and Shapiro, G.L., *Interpersonal Communication in the Modern Organization,* Prentice Hall, Englewood Cliffs, NJ, 1982.
Fallon, W.K., *Effective Communication on the Job,* AMACON, New York, 1981.
Verderber, R.F., *Communicate!,* 11th edition, Wadsworth, Belmont, CA, 2004.
Williams, F., *The New Communications,* Wadsworth, Belmont, CA, 1984.

# 8 Material Handling
## *Principles and Equipment Description*

One of the most important aspects in new plant development or in modification of an existing handling plant is thorough analysis of the material system. Material handling can account for 30–75% of the total cost, and efficient material handling can be primarily responsible for reducing a plant's operating cost by 15–30%. How material is handled can determine some of the building requirements, department arrangements, and time needed to produce a unit. When an employee handles an item, he or she adds nothing to the product's value but does add to its cost. Planning the handling, storage, and transportation associated with manufacturing can considerably reduce the cost of material handling. In an assembly line system, for example, properly designed equipment will space the production along the assembly lines, bringing material to an operator at a steady rate and sending the part or subassembly to the next station after the operator has finished the task.

In this chapter, an introduction to material handling is presented. The purpose of the chapter is twofold: to help the reader understand the relationship between material handling and plant layout as well as the complexity of designing such a system, and to describe the more commonly used material-handling equipment. The information provides the basics of material handling as applied to different environments, such as warehouses and manufacturing plants.

## 8.1 DEFINITION OF MATERIAL HANDLING

Several definitions are available for material handling. The most comprehensive definition is the one provided by the Material Handling Institute, Inc. (MHI), which states: "Material handling embraces all of the basic operations involved in the movement of bulk, packaged, and individual products in a semisolid or a solid state by means of machinery, and within the limits of a place of business."

Even a cursory examination of the statement reveals that material handling involves much more than just moving the material by using machinery; several additional functions are implied in the system.

First, material handling involves the movement of material in a horizontal (transfer) and a vertical (lifting) direction, as well as the loading and unloading of items. Second, specifying that the movement of materials is "within a place of business" implies that the movement includes raw materials to workstations, semifinished products between workstations, and removal of the finished products to their storage locations. It also distinguishes material handling from transportation; the latter involves moving materials from suppliers to places of business or from places of business to customers.

Third, the selection of handling equipment is another activity in designing material-handling systems (MHSs). Fourth, the term *bulk* indicates that the materials to be moved may be in large, unpackaged volumes, such as sand, sawdust, and coal. Fifth, and lastly, using machinery for handling material is the preferred method even though the initial cost might be high. Employment of human beings on a continuous basis is inefficient and can be costly; material-handling equipment soon pays for itself, especially in societies in which the cost of labor is high.

## 8.2 OBJECTIVES OF MATERIAL HANDLING

The need for the study and careful planning of an MHS can be attributed to two factors. First, as previously mentioned, material-handling costs represent a large portion of production cost. Second, material handling affects the operations and design of the facilities in which it is implemented. These then lead us to the major objective of MHS design — that of reducing production cost through efficient handling or, more specifically

- to increase the efficiency of material flow by ensuring the availability of materials when and where they are needed
- to reduce material-handling cost
- to improve utilization of facilities
- to improve safety and working conditions
- to facilitate the manufacturing process
- to increase productivity

This chapter will clarify how these objectives can be met.

## 8.3 MATERIAL-HANDLING EQUIPMENT TYPES

The backbone of an MHS is the handling equipment. A wide variety of equipment is available, each having distinct characteristics and costs that distinguish it from the others. All such equipment, however, can be classified into three main types: conveyors, cranes, and trucks. Each type has its own advantages and disadvantages, and some equipment is more suitable for certain tasks than others. This is mainly based on the characteristics of the material, the physical characteristics of the workplace, and the nature of the process using the equipment.

A brief discussion of the main equipment types is given here. A detailed description of each equipment type is given in later sections of the chapter.

### 8.3.1 Conveyors

Conveyors are used for moving material continuously over a fixed path. Examples of different types of conveyors are roller, belt, and chute conveyors.

#### 8.3.1.1 Advantages of Conveyors

- Their high capacity permits moving a large number of items.
- Their speed is adjustable.
- Handling, combined with other activities such as processing and inspection, is possible.
- They are versatile and can be on placed the floor or overhead.
- Temporary storage of loads between workstations is possible (with overhead conveyors in particular).
- Load transfer is automatic and does not require the assistance of many operators.
- Straight line paths or aisles are not required.
- Utilization of the cube (entire volume of the workplace) is feasible through the use of overhead conveyors.

#### 8.3.1.2 Disadvantages of Conveyors

- They follow a fixed path, serving only limited areas.
- Bottlenecks can develop in the system.
- A breakdown in any part of the conveyor stops the entire line.
- Because conveyors are fixed in position, they hinder the movement of mobile equipment on the floor.

### 8.3.2 Cranes and Hoists

Cranes and hoists are items of overhead equipment for moving loads intermittently within a limited area. Bridge cranes, jib cranes, monorail cranes, and hoists are examples of this basic equipment type.

Roller and belt conveyor. This sorting conveyor arrangement sorts the products according to their weights.

Rolling conveyors in sorting operation.

Gantry cranes like this provide economical and efficient material handling when an overhead crane system is not available or practical. (Courtesy: Bushman Equipment, Inc.)

10 ton magnetic hoist (Courtesy: Ace World Companies)

#### 8.3.2.1 Advantages of Cranes and Hoists

- Lifting as well as transferring of material is possible.
- Interference with the work on the floor is minimized.
- Valuable floor space is saved for work rather than being utilized for installation of handling equipment.
- Such equipment is capable of handling heavy loads.
- Such equipment can be used for loading and unloading material.

#### 8.3.2.2 Disadvantages of Cranes and Hoists

- They require heavy investment (especially bridge cranes).
- They serve a limited area.
- Some cranes move only in a straight line and thus cannot make turns.
- Utilization may not be as high as desirable because cranes are used only for a short time during daily work.
- An operator has to be available for operating some types, such as bridge cranes.

We find cranes being used in places such as shipyards and heavy equipment production facilities.

### 8.3.3 Trucks

Hand or powered trucks move loads over varying paths. Examples of such trucks include lift trucks, hand trucks, fork trucks, trailer trains, and automated guided vehicles (AGVs).

#### 8.3.3.1 Advantages of Trucks

- They are not required to follow a fixed path of movement and therefore can be used anywhere on the floor where space permits.

- They are capable of loading, unloading, and lifting, in addition to transferring material.
- Because of their unrestricted mobility, which allows them to serve different areas, trucks can achieve high utilization.

Electric counterbalance trucks, with twin AC motor front wheel drive maintenance-free multiple disc brakes, offer for indoor and combined indoor/outdoor operations. (Courtesy: Jungheinrich AG, Hamburg, Germany)

Some trucks are customized for special uses in construction. Here, the truck carries the wooden frame and other materials to different heights in the construction projects.

#### 8.3.3.2 Disadvantages of Trucks

- They cannot handle heavy loads.
- They have limited capacity per trip.

- Aisles are required; otherwise, the trucks will interfere with the work on the floor.
- Most trucks have to be driven by an operator.
- When trucks are used as an option, handling can not be combined with processing and inspection, as with other types of equipment.

## 8.4 DEGREES OF MECHANIZATION

An MHS can be completely manual or fully automated; different degrees of mechanization also exist between these two extremes. Classification of a handling system according to its level of mechanization is based on the source of power for handling and the degree of involvement of humans and computers in operating the equipment. The levels of mechanization can be classified as:

1. *Manual and dependent on physical effort.* This level also includes manually driven equipment such as hand trucks.
2. *Mechanized.* Power, instead of physical effort, is used for driving the equipment. Some trucks, conveyors, and cranes fall into this level. Here, operators are needed for operating the equipment as opposed to providing the power.
3. *Mechanized complemented with computers* (an extension of the second level). The function of the computers is to generate documents specifying moves and operations.
4. *Automated.* Minimal human intervention is used for driving and operating the equipment, and most of these functions are performed by computers. Examples include conveyors, automated guided vehicles (AGVs), and automated storage/retrieval system (AS/RS). The equipment usually receives instructions from keyboards, push buttons, and tape or card readers.
5. *Fully automated.* This level is similar to the fourth level except that computers perform the additional task of on-line control, thus eliminating the need for human intervention.

The cost and complexity of designing the system increase as the degree of mechanization increases. However, efficiency of operations and labor savings can result.

The advantages of using mechanized and higher-level systems include an increase in speed of handling operations, which in turn may decrease the overall production time; a reduction in fatigue and an improvement in safety; better control of material flow; lower labor cost; and better record keeping regarding the inventory status of the material.

There are also some disadvantages as the degree of mechanization increases. For example, mechanization requires a high investment cost, training of operators and maintenance personnel, and specialized equipment and personnel, which reduces flexibility. It is therefore necessary to carefully weigh the advantages and disadvantages before deciding which system to use. Changing from one mode of operation to another is always expensive and time consuming.

Unit loads used in a plant play an important part in defining the items to be moved. The degree of mechanization affects the unit load; conversely, the defined unit load also influences the degree of achievable mechanization.

## 8.5 THE UNIT LOAD CONCEPT

The unit load concept depends on the fact that it is more economical to move items and material in groups than individually. A unit load is defined as a number of items arranged such that they can be handled as a single object. This can be accomplished by palletization, unitization, and containerization.

Palletization is the assembling and securing of individual items on a platform that can be moved by a truck or a crane. Unitization also involves assembling of goods, but as one compact load. Unlike palletization, unitization uses additional materials for packaging and wrapping items as a complete unit. The unit load can be handled by trucks, conveyors, or cranes depending on its size and weight. Containerization is the assembling of items in a box or a bin. It is most suitable for use with conveyors, especially for small items.

Each unit load type is most suitable for certain situations. For example, a pallet is most suitable for stacking similar items that have regular shapes. Items that have different shapes and sizes can be grouped inside a container. In general, the factors that influence the selection of the unit load type are the weight, size, and shape of the material; compatibility with the material-handling equipment; cost of the unit load; and the additional functions provided by the unit load, such as stacking and protection of material.

Using unit loads has both advantages and disadvantages. Among the advantages are the following: Using unit loads allows for moving large quantities of material, which reduces the frequency of movement and therefore reduces the handling cost. The ease of stacking helps achieve better space and cube utilization and promotes good housekeeping. There is greater speed in loading and unloading and a corresponding reduction of handling time. Protection against material damage is provided.

The disadvantages of using unit loads are as follows: Costs of the unit load can be high if a large number are required, especially if the containers are not reusable. Loading and unloading equipment that is different from what is available might be required. When used in shipping to suppliers, there is the problem of returning the empty pallets and containers if they are reusable.

## 8.6 PRINCIPLES OF MATERIAL HANDLING

Designing and operating an MHS is a complex task because of the numerous issues involved. There are no definite rules that can be followed for achieving a successful MHS. However, there are several guidelines that can result in reducing the system cost and in enhancing its efficiency. These guidelines are known as the principles of material handling. They represent the experience of designers who have been working in the design and operations of handling systems. The 20 principles of material handling are listed in Table 8.1.

These principles can also be compiled slightly differently to suggest how the objectives are to be achieved. For example, to lower the handling cost, one should reduce unnecessary handling by properly planning material movement; by

**TABLE 8.1**
**Principles of Material Handling**

| | Principle | Description |
|---|---|---|
| 1. | Planning | Plan all material-handling and storage activities to obtain maximum overall operating efficiency. |
| 2. | Systems flow | Integrate as many handling activities as is practical into a coordinated system of operations covering vendor, receiving, storage, production, inspection, packaging, warehousing, shipping, transportation, and customer. |
| 3. | Material flow | Provide an operation sequence and equipment layout optimizing material flow. |
| 4. | Simplification | Simplify handling by reducing, eliminating, or combining unnecessary movements and/or equipment. |
| 5. | Gravity | Use gravity to move materials whenever practical. |
| 6. | Space utilization | Make optimum utilization of the building cube. |
| 7. | Unit size | Increase the quantity, size, or weight of unit loads or flow rate. |
| 8. | Mechanization | Mechanize handling operations. |
| 9. | Automation | Provide automation to include production, handling, and storage functions. |
| 10. | Equipment selection | In selecting handling equipment, consider all aspects of the material handled, the movement, and the method to be used. |
| 11. | Standardization | Standardize handling methods as well as types and sizes of handling equipment. |
| 12. | Adaptability | Use methods and equipment that can best perform a variety of tasks and applications when special-purpose equipment is not justified. |
| 13. | Dead weight | Reduce the ratio of dead weight of mobile handling equipment to load carried. |
| 14. | Utilization | Plan for optimum utilization of handling equipment and manpower. |
| 15. | Maintenance | Plan for preventive maintenance and scheduled repairs of all handling equipment. |
| 16. | Obsolescence | Replace obsolete handling methods and equipment when more efficient methods of equipment will improve operations. |
| 17. | Control | Use material-handling activities to improve control of production, inventory, and order handling. |
| 18. | Capacity | Use handling equipment to help achieve the desired production capacity. |
| 19. | Performance | Determine the effectiveness of handling performance in terms of expense per unit handled. |
| 20. | Safety | Provide suitable methods and equipment for safe handling. |

*Courtesy of The Material Handling Institute, Inc.*

delivering units to the required place the first time without stopovers; by using proper material-handling equipment such as forklift trucks, pallets, boxes, and conveyors; by replacing obsolete equipment with new and more efficient systems when the savings justify it; and by reducing the ratio of dead weight (pallets, boxes) to payload. One can also use unit loads and move as many pieces at one time as possible.

One can increase productivity by minimizing machine operators' waiting time by delivering raw materials and subassemblies when needed, and by maintaining a steady movement of work at a rate that will match the machine operator's rate.

One can make all workers more productive by eliminating nonproductive activities associated with proper material handling, by removing damaged parts from assembly lines before they reach the workstation, by not using nonstandard and unique equipment, and by coordinating material movement throughout the plant.

One can reduce floor space use by using material-handling equipment and production schedules that will require a minimum amount of stock on the floor; by storing material in spaces that do not hinder production (e.g., stock material should not be piled so close to a machine that it interferes with the operator's ability to perform); and by arranging the plant layout so that that it will permit smooth flow of materials between stations.

Accidents can be minimized by using material-handling equipment that has appropriate safety features for lifting and moving heavy materials and by using gravity to move the material whenever possible.

How do we apply these principles? Some applications are easily recognizable. For example, to apply the *gravity* principle, use chutes. To apply the *safety* principle, reduce or eliminate manual handling that causes injuries. For the *space utilization* principle, stack items and use overhead equipment. The *unit size* principle recommends use of containers and pallets to move groups of items. To apply the *utilization* principle, select equipment that is capable of performing several handling tasks in different areas, thereby preventing it from being idle.

### 8.6.1 Compatibility of the Principles

The principles of material handling are compatible with each other and with the objectives of material handling. Achieving some of the principles will help achieve others. As examples, consider the following.

When applied properly, equipment selection and adaptability principles will help achieve the utilization principle, because in this case only the necessary equipment will be acquired and thus will rarely be idle.

The mechanization principle will reduce manual handling, thus minimizing injuries and help to achieve the safety principle.

The unit size and equipment selection principles help in achieving the space utilization principle. Using a unit load will permit stacking of items, thus reducing the floor space required. Selecting overhead equipment will free some space on the floor for other purposes.

### 8.6.2 Difficulties in Applications of the Principles

The designers of MHS are usually advised to follow these principles. However, in some cases, they might not be able to apply them to the fullest extent because of factors such as limitation on capital, physical characteristics of the building, and capability of the equipment.

Lack of the desired capital might prevent one from realizing a high degree of mechanization or from applying the maintenance principle. Building characteristics such as ceiling height, location of columns, and number and width of aisles can influence the gravity as well as the space utilization and material flow principles. Capability and type of equipment may affect application of the space utilization, unit sizing, and utilization principles.

## 8.7 MATERIAL-HANDLING COST

The main costs involved in designing and operating an MHS are

- Equipment cost, which comprises the purchasing of the equipment as well as auxiliary components and their installation
- Operating cost, which includes maintenance, fuel, and labor cost, consisting of both wages and injury compensation
- Unit purchase cost, which is associated with purchasing the pallets and containers
- Cost due to packaging and damaged materials

Reducing such costs is one of the primary objectives of a handling system. There are several means of achieving this goal. For example, one can minimize the idle time of the equipment. High utilization of equipment will eliminate the need to acquire extra units. One can minimize rehandling of material and backtracking, thus reducing operating cost. One can arrange closely related departments to be near each other so material can be moved across shorter distances. One can prevent excessive repairs by planning maintenance activities in advance. One should use proper equipment to reduce material damage and use unit loads whenever possible. The gravity principle should be used whenever possible, because it can reduce operating cost. One should eliminate unsafe practices by employees such as lifting heavy items; this will reduce injuries and consequent worker compensation. One can minimize variations in equipment types, thus eliminating the need for an inventory of a variety of spare parts and their associated costs. One can replace obsolete equipment with new and more efficient equipment when the savings justify it. As the reader has probably realized by now, we are again applying the 20 principles of material handling.

## 8.8 RELATIONSHIP BETWEEN MATERIAL HANDLING AND PLANT LAYOUT

In a manufacturing system, no two other activities affect each other to the extent that plant layout and material handling do. The relationship between the two involves the data required for designing each activity, their common objectives, the effect on space, and the flow pattern. Specifically, plant layout problems require knowledge of the equipment operating cost in order to locate the departments in a manner that will minimize the total material-handling cost. At the same time, in designing an MHS, the layout should be known in order to consider the move length, move time, and source and destination of the move. Because of this dependency, many designers stress the need to solve the two problems jointly. However, the only feasible way is to start with one problem, use its solution for solving the other, then go back and modify the first problem on the basis of the new information obtained from the second, and so on until a satisfactory design is obtained.

Plant layout and material handling have the common objective of cost minimization. The material-handling cost can be minimized by arranging closely related departments such that the material moves across short distances only. Determining the flow pattern is a common concern in both problems and is addressed in a later section.

In addition, material handling and plant layout influence each other in terms of space requirements and utilization. Trucks that are compact in size and have the capability of side loading do not require wide aisles. Overhead equipment does not occupy any space on the floor of the layout. Stacking items as high as possible by using the appropriate unit load will help reduce the space required by these activities and best utilize the cube. Trucks that have the ability to lift to a high level will assist in achieving cube utilization. It can also be achieved by using mezzanines, carousels, and high-rise storage for storing material.

Finally, the physical characteristics of the building, such as aisle width, ceiling height, and columns, will affect equipment selection and its routing.

## 8.9 MHS DESIGN

In production systems, developing an MHS is one of the most crucial and difficult tasks a designer faces owing to its impact on the cost and efficiency of operations. There are no standard steps that can be followed, although the general procedures of designing a system can be adopted. The procedure is iterative; the analyzer has to go back and forth between the different steps until a satisfactory design has been obtained and can be implemented. Experience and sound judgment are indispensable during this process.

Developing an MHS involves selection of material-handling equipment, selection of a unit load, and assignment of the equipment to the moves and determining their routes. These three components of the design have been expressed in what is known as the material-handling equation. This equation is a nonquantitative relationship between the three components and is expressed as:

$$\text{Material} + \text{moves} = \text{methods}$$

Questions concerning the materials involve their type, size, shape, quantity, and weight. Those concerning the moves involve their origin and destination, length, frequency of movement, and duration of the move. (This part of the design is closely related to plant layout.) Determining the material-handling methods — that is, equipment and unit load — depends on the information obtained from studying the material and the moves. Because of the heavy investment associated with equipment that render irreversible the decisions that have been made and implemented, selection of the equipment is usually the most crucial and difficult task of the design.

## 8.10 DILEMMA OF AN ANALYZER

The difficulties in designing an MHS, particularly in equipment selection, are numerous and can be attributed to several factors associated with subjective analysis and/or application of analytical models to the design. Most important among these factors are

1. The interrelation between material handling and plant layout, as explained previously, is highly significant.
2. The existence of a large variety of equipment with different capabilities and limitations renders evaluating and comparing all of them an impossible task. Undoubtedly, this will result in ignoring some good alternatives.
3. Some of the characteristics of the equipment are difficult to quantify, either for comparison between different types of equipment or for inclusion in an analytical model. Examples of these characteristics are equipment flexibility and maneuverability and the effort required to move the material.
4. Although the most common objective of the design is to ensure a system with minimum cost, other objectives have to be considered. Notable among these are maximum utilization of the equipment, reducing variations in equipment types, and safety of employees.
5. Application of an analytical model to the design problem might require an inordinate amount of calculation with no guarantee of an optimal system design. For example, equipment selection, when performed analytically, will result in a maximum of $pq$ combinations to evaluate, where $p$ and $q$ are the number of equipment types and moves, respectively. If there are many moves and/or several types of equipment to be considered, $pq$ can become quite large.
6. Some of the data required for designing the system, such as move time and equipment-operating cost, cannot be known exactly unless the system is already operating. Therefore, those data must be estimated, which is not an easy task, and a poor estimate can result in several modifications after the system has been installed and is operating.

## 8.11 SPECIFICATIONS OF THE DESIGN

Generally speaking, the steps to be followed in designing a handling system, assuming the layout has already been prepared, are

1. State the intended function of the handling system — whether it is for a warehouse whose function is storing, packaging, inspection, and shipping to customers, or for a manufacturing system where the function is to move items or partial assemblies from station to station. Knowing the type of manufacturing system (product, process, group technology, or any other type) is very helpful here.
2. Collect the necessary data about the material, such as its characteristics and the quantities involved. Data regarding quantity may be summarized in the form of a chart.
3. Identify the moves, their origin and destination, their path, and their length.
4. Determine the basic handling system to be used and the degree of mechanization desired. Here, the idea is to establish whether a conveyor, a truck, or a crane will be best suited for the situation.

5. Perform an initial screening of suitable equipment and select a set of candidate equipment from among them. Evaluate the candidate equipment on the basis of such measures as cost and utilization. Always match the equipment with the material characteristics.
6. Select a set of suitable unit loads and match them with material and equipment characteristics.

These steps are more easily stated than followed and must be repeated several times until compatibility between the components of the material-handling equation is assured. Several factors influence the designer in making decisions at each of these steps. Important factors are stated briefly here.

- Costs of equipment and unit loads and availability of funds. This factor will affect the degree of mechanization achieved in the design.
- Physical characteristics of the building and the available space. Aisle width and number will be affected by the available space, which in turn will influence decisions regarding mobile equipment. Overhead equipment may or may not be considered, depending on the height of the ceiling.
- Management attitude toward safety and employee welfare, which will affect the degree of involvement of material-handling personnel in manual handling.
- Degree of involvement between handling and processing.

During the design of the system, the objectives and the principles of material handling should be kept in mind. Achieving as many of the objectives and principles as possible will result in a satisfactory and efficient design. Although there is no benchmark against which to measure the quality of the design, a good material-handling design should possess most or all of the following characteristics:

- well planned
- handling combined with processing whenever possible
- mechanical whenever possible
- minimum manual handling
- minimum handling by production personnel
- safe
- protection of material provided
- minimum variation in equipment types
- maximum utilization of equipment
- minimum backtracking, handling, or transferring
- minimum congestion or delay
- economical

## 8.12 ANALYZING AN EXISTING MHS

To analyze an existing MHS means to determine whether it is functioning efficiently without creating any bottlenecks or excessive inventories and is transporting the units when and where needed. The problems in an existing MHS will be evident if one can observe one or more of the following symptoms in the system (courtesy of MHI):

- backtracking in material flow path
- built-in hindrances to flow
- cluttered aisles
- confusion at the dock
- disorganized storage
- excess scrap
- excessive handling of individual pieces
- excessive manual effort
- excessive walking
- failure to use gravity
- fragmented operations
- high indirect labor costs
- idle machines
- inefficient use of skilled labor
- lack of cube storage
- lack of parts and supplies
- long hauls
- material piled up on the floor
- no standardization
- overcrowding
- poor housekeeping
- poor inventory control
- product damage
- repetitive handling
- service areas not conveniently located
- trucks delayed or tied up
- two-person lifting jobs

It is also a good idea to examine the entire MHS in a plant with a checklist similar to the one shown in Table 8.2 and identify the problems. Again, the checklist is developed by MHI; however, one can develop a more specific checklist for the particular plant in question.

Once the problem areas have been identified, they must be reexamined for possible improvements. In performing a study, several basic questions must be asked such as *why*, *what*, *where*, *how*, and *who*. A process chart, described in Chapter 5, is very helpful in making such an analysis.

*Why* is this activity taking place? For example, moving an incoming unit from receiving to inspection and then back to receiving could be avoided if the final quality assurance inspection takes place at the supplier and we can rely on the supplier's data.

The *what* question is associated with understanding the type of material to be handled. The material to move and its frequency could define the type of equipment and accessories that might be used in making such material moves. For example, small discrete units could be handled by a conveyor or by a forklift by unitizing,

**TABLE 8.2**
**Material-Handling Checklist**

- — Is the material-handling equipment more than 10 years old?
- — Do you use a wide variety of makes and models that require a high spare parts inventory?
- — Are equipment breakdowns the result of poor preventive maintenance?
- — Do the lift trucks have to go too far for servicing?
- — Are there excessive employee accidents due to manual handling of materials?
- — Are materials weighing more than 50 pounds handled manually?
- — Are there many handling tasks that require two or more employees?
- — Are skilled employees wasting time handling materials?
- — Does material become congested at any point?
- — Is production work delayed owing to poorly scheduled delivery and removal of materials?
- — Is high storage space being wasted?
- — Are high demurrage charges experienced?
- — Is material being damaged during handling?
- — Do shop trucks operate empty more than 20% of the time?
- — Does the plant have an excessive number of handling points?
- — Is power equipment used on jobs that could be handled by gravity?
- — Are too many pieces of equipment being used because their scope of activity is confined?
- — Are many handling operations unnecessary?
- — Are single pieces being handled where unit loads could be used?
- — Are floors and ramps dirty or in need of repair?
- — Is handling equipment being overloaded?
- — Is there unnecessary transfer of material from one container to another?
- — Are inadequate storage areas hampering efficient scheduling of movement?
- — Is it difficult to analyze the system because there is no detailed flowchart?
- — Are indirect labor costs too high?

while large bulky units such as pressure vessels might need a crane. Parts and material lists and production schedules are good sources of this information.

*Where* describes the data associated with the move. The travel path, distance, and any physical limitations should be noted. Process charts, flow diagrams, and scale models of the plant are some of the sources from which these data can be gathered.

*When* defines the time at which the material is to be moved and within what time span the move must be made. Its answer provides the speed and frequency with which the MHS must operate.

*How* indicates the method for the move. By analyzing the method, inefficient and expensive aspects are revealed. Operations charts and time study data are useful in this analysis.

*Who* refers to the person responsible for material handling as well as the labor required for such a task. Sufficient labor is necessary for efficient working, and although the degree of mechanization determines the labor, the material-handling workers should also realize their responsibility and importance in the smooth operation of the plant.

Asking these questions about the MHS, especially the problem areas, reveals the nature of overall difficulties. In most instances, the answers suggest the mode of improvement and/or provide guidance as to where further in-depth study would be appropriate.

## 8.13 PRODUCTIVITY RATIOS

Every system within a manufacturing plant must work efficiently to reduce the production cost and be competitive. Several typical productivity ratios are used as indicators of the performance of a system. Several such ratios are listed below; however, the list is by no means exhaustive, and an organization often develops its own ratios that more clearly identify the factors with which it is concerned. These ratios are monitored periodically to ensure that their values are within acceptable tolerances. Any wild fluctuation from period to period is a clear signal that the system is not working smoothly and needs further investigation and control. Although some ratios do not specifically address productivity associated with material handling, all the ratios are stated here for completeness. They should be reviewed when we discuss topics such as storage and energy consumption later in the text. Most of the factors listed below are defined by MHI and require little or no explanation. Collection of the required information is likely to present problems, at least initially. These can probably be overcome in a given plant once the guidelines are established to identify and quantify the information necessary to calculate each ratio used. Because several departments might be involved, interdepartmental cooperation is essential in forming the guidelines and collecting the data.

### 8.13.1 Material-Handling/Labor (MHL) Ratio

$$\text{MHL} = \frac{\text{personnel assigned to material handling}}{\text{total operating personnel}}$$

The personnel assigned could be measured in terms of the number of full-time workers or dollar expenditure. If a person is not performing a material-handling task full time, then a percentage of the person's work associated with material handling must be estimated and used in the preceding calculations.

The ratio should be less than 1, and a reasonable value would be less than 0.30 in a plant, while in a warehouse a higher value should be expected.

### 8.13.2 Handling Equipment Utilization (HEU) Ratio

$$\text{HEU} = \frac{\text{items (or load weight) moved per hour}}{\text{theoretical capacity}}$$

Ideally, the HEU ratio should be close to 1.0; however, equipment breakdown, poor scheduling, poor housekeeping, and building geography can reduce the load movement.

### 8.13.3 Storage Space Utilization (SSU) Ratio

$$\text{SSU} = \frac{\text{storage space occupied}}{\text{total available storage space}}$$

If the storage areas, such as bins or racks, are only partially full, then the percentage of utilization should be estimated and included in the calculation. A value close to 1 indicates assignment of appropriate space for the storage activities.

### 8.13.4 Aisle Space Percentage (ASP)

$$ASP = \frac{\text{space occupied by aisles}}{\text{total space}}$$

The ASP should have a value between 0.10 and 0.15.

### 8.13.5 Movement Operation (MO) Ratio

$$MO = \frac{\text{number of moves}}{\text{number of productive operations}}$$

The MO ratio indicates the amount of material handling performed. The moves involved may consist of material moved from receiving, from storage to an operation and back to storage, and so on. A high value indicates room for improvement.

### 8.13.6 Manufacturing Cycle Efficiency (MCE)

$$MCE = \frac{\text{time in actual production operations (machine time)}}{\text{time in production department}}$$

Time not spent in production could be caused by delays in material movement, poor scheduling, machine failure, and storage limitation, among other items. For increasing machine utilization, the delay should be eliminated or at least minimized. The performance index should be observed over a period for consistency.

### 8.13.7 Damaged Loads (DL) Ratio

$$DL = \frac{\text{number of damaged loads}}{\text{total number of loads}}$$

The ratio measures the quality performance of material-handling personnel. Damage to the loads during receiving, in-process movement, and shipping should be minimized.

### 8.13.8 Energy Ratio (ER)

$$ER = \frac{\text{total BTU consumption in the warehouse}}{\text{warehouse space}}$$

The ER measures the efficiency of heating and cooling operations. Some of the ways in which this ratio can be improved are reducing heating or cooling of a portion of the warehouse wherein there are no workers, turning lights off when not needed, and using lights on moving vehicles rather than permanent lighting.

### 8.13.9 Equipment Used for Material Handling

A wide variety of equipment is available for material handling. Which equipment to use under what conditions depends on the responsible person's judgment and knowledge of the machines and costs associated with the task of moving the material. An economic analysis can be performed to justify the selection of a particular machine, and we will illustrate this approach by using examples in Chapter 9.

### 8.13.10 Equipment Types

The equipment can be characterized by the area it is intended to serve:

1. Between fixed points over a fixed path
   a. Belt conveyor
   b. Roller conveyor
   c. Chute conveyor
   d. Slat conveyor
   e. Screw conveyor
   f. Chain conveyor
   g. Overhead monorail conveyor
   h. Trolley conveyor
   i. Wheel conveyor
   j. Tow conveyor
   k. Bucket conveyor
   l. Cart-on-track conveyor
   m. Pneumatic tube conveyor
2. Over limited areas
   a. Hoists
   b. Overhead cranes
   c. Hydraulic scissors lift
3. Over large areas
   a. Handcart/truck
   b. Tier platform truck
   c. Hand lift truck/pallet jack
   d. Power-driven hand truck
   e. Power-driven platform truck
   f. Forklift truck
   g. Narrow-aisle truck
   h. Tractor-trailer train
   i. Material lift
   j. Drum truck
   k. Drum lifter
   l. Dolly
   m. AGV system

### 8.13.10.1 Equipment Description

Each of the equipment types listed is discussed in some detail in this section. A brief description is provided, followed by a listing of the equipment's principal characteristics and/or applications.

#### *8.13.10.1.1 Belt Conveyor*

A belt conveyor is an endless belt driven by power rollers or drums at one or both ends and supported by flatbeds or rollers. These rollers can produce a flat conveyor belt or a trough conveyor. The belt can be made of rubber, woven wires, metal, or fabric, depending on the load to be carried. On special occasions, to carry ferrous metal or to separate ferrous from other types of metals, the belt may have a magnetic bed. The belt can be vibrated to feed assembly parts, position items, and deliver small amounts of bulk material.

##### *8.13.10.1.1.1 Characteristics/Applications*

- The belt can operate at a horizontal orientation or on an incline of up to 30°.
- A flatbed belt can be used to carry light objects in assembly lines.
- Roller belts can be used to carry heavy boxes, bags, or other containers in warehousing and storage operations.
- Trough rollers can be used to carry bulk materials such as coal and raw materials.
- Belt speed can be adjusted from 2 to 300 feet/minute.
- Belt width can be from 12 to 36 inches with a capacity of 300–1500 pounds per linear foot.

Belt and roller conveyors in tile transfers. Individual tiles are moved by the belt conveyor while boxes are transferred by roller conveyors.

Plastic mesh belt conveyor on stainless steel frame with wheels (Courtesy: Machinery & Equipment, Inc. San Francisco, CA)

### *8.13.10.1.2 Roller Conveyor*

A roller conveyor consists of rollers attached to side rails supported by a steel frame. The load is carried on the rollers, each of which rotates about a fixed axis. The type of roll (steel, rubber, or wood), its shape (cylindrical or wheel, the latter sometimes

This flat-belt conveyor carries finished goods to packaging. The surface of the belt has a great friction constant to keep the items from sliding.

called a wheel conveyor), and its spacing depend on the load being carried. The conveyor can be either gravity operated or power-driven. The gravity-operated conveyor has a slight downward slope, allowing the material to move because of gravitational force. On the power-driven conveyor, some of the rollers are driven by chains or belts to provide the motion for the material on the conveyor.

###### *8.13.10.1.2.1 Characteristics/Applications*

- Material can be moved between workstations.
- The height can be adjusted to the level of the work area.
- Loads must have a firm, even base.
- Uneven and fragile objects can be carried in boxes, containers, or pallets placed on the conveyor.
- The width can be between 7 and 51 inches, with a capacity of 460–25,000 pounds per linear foot. Conveyors can be bought in sections of 5 to 10 feet.

Sorting operation using mechanical pushers.

##### *8.13.10.1.3 Chute Conveyor*

A chute conveyor is a slide, generally made of metal, that guides materials as they are lowered from a higher-level to a lower-level workstation. The shape of the chute can be straight or spiral to save space.

###### *8.13.10.1.3.1 Characteristics/Applications*

- Materials (boxes, packages) are moved across a short distance because of gravity.
- The chute can have a door that controls the flow of the work items (usually in batches).

- This type of conveyor is very inexpensive and provides an efficient means of connecting conveyors at different levels.
- The chute can occasionally be jammed if the design is not appropriate for the size and shape of the objects being transported.
- The diameter of a spiral can be between 18 and 48 inches, and the length can be customized.

Chute Conveyor carrying a tile raw material batch for further processing (Courtesy of Texas Tile Manufacturing LLC).

#### *8.13.10.1.4 Slat Conveyor*

A slat conveyor is a moving surface that is made of slats attached to power-driven chains. The slats move over rollers at each end of the conveyor to form a closed loop.

##### *8.13.10.1.4.1 Characteristics/Applications*

- Heavy, uneven loads can be placed directly on the slats.
- Operation can be horizontal or inclined by as much as 40°.
- Slat conveyors can be used as a series of platforms in assembly operations.
- Sanitation is better compared to other conveyors. Slat conveyors are mainly used for bottling and canning owing to ease of cleaning.
- Slat size can be 3¼ to 7½ inches for plastic slats and 3½ to 12 inches for steel slats, with slats arranged ½ to 7/8 inch apart.
- This type of conveyor can be used horizontally or on an incline.
- Because of the small volume, they can be used in cramped spaces.

Slat conveyor carrying heavy uneven load for crushing at 35° inclined.

Screw conveyor (Courtesy: Machine & Equipment, Inc.)

#### *8.13.10.1.5 Chain Conveyor*

A chain conveyor is an endless chain directly carrying loads, sometimes located at the bottom of a trough.

##### 8.13.10.1.5.1 Characteristics/Applications

- It is useful for moving tote boxes and pallets.
- A chain conveyor can pull bulk material along a trough.
- Generally, the length can be between 10 and 100 feet with a capacity of 300–3000 pounds per linear foot.

#### 8.13.10.1.6 Overhead Monorail

An overhead monorail is a track to transport carrying devices such as trolleys and hooks. The track itself can form a closed loop, and each trolley can be either powered or manually operated.

##### 8.13.10.1.6.1 Characteristics/Applications

- Trolleys can be independently powered and controlled by a hand device.
- The monorail can be designed to carry heavy objects.
- Monorails are often used in transporting units to a spray paint booth or a baking oven where a uniform rate of travel is necessary with the entire unit suspended in the air.
- Overhead conveyors are generally 8 to 9 feet from the floor.

#### 8.13.10.1.7 Trolley Conveyor

A trolley conveyor is a closed-loop overhead track with an endless chain carrying uniformly spaced trolleys that support the loads.

##### 8.13.10.1.7.1 Characteristics/Applications

- Loads can move horizontally or at a significant incline.
- This type of conveyor can be used for overhead storage, thereby saving floor space.
- Degreasing, spray painting, assembly, and packaging are the most common applications.
- The conveyor can have a free, unpowered track to which trolleys can be shifted when accumulation is important.
- Conveyors are generally 8 to 9 feet from the floor and can be functional in vertical, horizontal, and inclined positions.

#### 8.13.10.1.8 Wheel Conveyor

The wheel conveyor is similar in function to the roller conveyor; but instead of long rollers, wheels are mounted on a shaft.

##### 8.13.10.1.8.1 Characteristics/Applications

- The spacing of the wheels depends on the load to be transported.
- Wheel conveyors are more economical than roller conveyors.
- This type of conveyor is used for light loads.

#### 8.13.10.1.9 Tow Conveyor

A tow conveyor is a tow line, and it pulls trolleys, trucks, or dollies over a fixed path.

*8.13.10.1.9.1 Characteristics/Applications*

- The tow line can be overhead or in the ground.
- The lines can be installed so as to permit automatic switching from one tow line to another.
- This type of conveyor is generally used in frequent trips of long distances.
- Carts are 3 × 5 feet or larger.

### *8.13.10.1.10 Bucket Conveyor*

In a bucket conveyor, buckets are evenly spaced on a chain moving between levels. The buckets automatically tip at the top or bottom of the conveyor to deliver the material.

*8.13.10.1.10.1 Characteristics/Applications*

- This type of conveyor can be used for moving raw materials.
- The chain can be vertical or inclined.

### *8.13.10.1.11 Cart-on-Track Conveyor*

In a cart-on-track conveyor, carts are spaced along a track with a tube beneath the carts to move them along by rotating. A drive wheel rotates the tube.

A very common conveyor belt, used for transporting heavy goods to higher locations. In this picture, the conveyor is used to transport rocks.

*8.13.10.1.11.1 Characteristics/Applications*

- Carts can be accumulated when necessary.
- Independent control of each cart is possible.

### *8.13.10.1.12 Pneumatic Tube System*

A pneumatic tube system consists of a cylinder in which messages or small items are carried over a predetermined path by compressed air or vacuum.

Automatic loop conveyor used ASRS system. Material being loaded into the conveyor, processed and stored again in the storage area.

#### 8.13.10.1.12.1 Characteristics/Applications

- Such a system is designed to transport lightweight objects (small tools, dies, gages, money, and messages) rapidly to and from stations within toolrooms, banks, offices, lumberyards, and similar locations.
- The requirement for a constant pressure or vacuum results in high operating and maintenance costs for large, complicated systems with multiple branches and stations.

### 8.13.10.1.13 Hoist

A hoist is a lifting device attached to monorails, cranes, or a fixed point. A hoist can be powered manually or by electric or pneumatic motors. Strictly speaking, the hoist is the lifting device itself; however, by general usage, it is frequently named by the type of crane to which it is attached. There are three major types:

1. *Chain hoist* is a hoist that serves a fixed spot directly beneath the hoist. A hand chain hoist has a capacity lift ranging from 250 pounds to 30 tons, with a standard lift of 7 to 10 feet. An electric chain hoist has a capacity of 500 pounds to 50 tons with a lift of 10 to 35 feet.
2. *Monorail hoist* is hoist that is free to move along an overhead rail serving any spot under the track.
3. *Jib hoist* is a hoist that serves any area circumscribed by the jib in a 360° rotation (a boom extending from, and free to pivot around, a fixed vertical post). The height can be adjusted from 4 to 30 feet.

#### 8.13.10.1.13.1 Characteristics/Applications

- Hoists are primarily intended for transferring moderately heavy objects across a short distance.

- Hoists are also used to suspend a workpiece while various operations are being performed.

#### 8.13.10.1.14 Overhead Traveling Crane/Bridge Crane

This is an overhead handling unit resembling a bridge that is mounted on a pair of tracks, traveling lengthwise. Within the bridge is a cable and hoist that can be positioned at any point along the bridge.

##### 8.13.10.1.14.1 Characteristics/Applications

- Such a crane covers the entire area within the rectangle over which it travels. It provides three-dimensional coverage while moving up and down, sideways, and lengthwise.
- With various accessories such as buckets, electromagnetic disks, and chains it can handle almost any material, from light tools to heavy, flat metal plates.
- Such cranes have a capacity range of 5 to 50 tons, and can lift anywhere from 10 to 30 feet.

Variations of bridge cranes are:

1. *Stacker crane* is a crane that, instead of hoisting, uses platforms with forks. It is used mainly in storing and retrieving a unit load (palletized materials, boxes, or other containerized loads).
2. *Tower crane* is used mainly on large construction projects. It consists of a hoist that travels on a horizontal boom attached at one end to a vertical post. The other end of the boom is supported by a guy line to the top of the post. The boom itself can be rotated 360° around its post.
3. *Gantry crane* is a large traveling crane supported by towers or side frames running on parallel tracks. It is mainly used in heavy outdoor activities such as loading and unloading on shipping docks, shipbuilding, etc.
4. *Jib crane* is a crane that can travel on a horizontal boom, which may be mounted on a column, or mast. The mast can be fastened to the floor or roof, or the boom can be fastened directly to the wall brackets or rails on the wall. Jib cranes can rotate 360°, are inexpensive and versatile, and are used in loading and unloading individual workstations or material-handling carriers.

#### 8.13.10.1.15 Hydraulic Scissors Lift Equipment

A hydraulic scissors lift consists of scissor legs and hydraulic cylinders with a platform on top to travel in a vertical direction from the ground up to about 10 feet.

##### 8.13.10.1.15.1 Characteristics/Applications

- This equipment is used for raising, lowering, and supporting heavy loads during loading and unloading.
- It is primarily permanently located but may be moved across short distances.

- It is designed to serve in transfer operations using forklift trucks or other powered equipment, or as an industrial heavy-duty work positioner.
- It can also be used as an adjustable loading dock.

3 Ton Monorail hoist. (Courtesy: Ace World Companies)

Overhead crane: scrap handling magnetic crane. (Courtesy: Ace World Companies)

Jib crane: The "Super Jib" shown here is a full cantilevered jib with a 60 foot span and a capacity of 7 1/2 tons. It is mounted onto a large gantry crane at a nuclear power plant in California. The Super Jib required seismic calculations, special surface preparation with epoxy paint, and powered rotation. (Courtesy: Bushman Equipment, Inc)

The economical jib crane model is designed for mounting on to walls or columns to serve a 180-degree arc. It can be used under an overhead bridge crane if adequate head room is provided. (Courtesy: Bushman Equipment, Inc)

Tower cranes like this can reach great heights and are commonly used in construction projects.

Portal crane, as shown, is being used to carry wood logs with the jaws.

Hydraulic scissors lifts eliminate wasted time spent setting up ladders, scaffolds and other temporary work platforms (Courtesy: Bushman Equipment, Inc)

Hydraulic lift table and scissors lifts use all welded steel construction, and flow control device is rigidly attached to the lift cylinders, and a maintenance locking pin is provided to secure the legs in an open position. (Courtesy. West Bend Equipment)

The lift table is used to lift the items to higher locations. It uses a hydraulic jack system, and works manually.

Jib crane with coil ID lifter for manual stacking, and provides electric powered chain or wire rope hoist.(Courtesy: Avon Engineering, a division of Bushman Equipment, Inc.)

Pallet trucks are particularly suited to truck loading and offloading as well as horizontal transport over short distances, and with its short chassis length, can be maneuvered in the smallest of spaces. (Courtesy: Jungheinrich AG, Hamburg, Germany)

#### *8.13.10.1.16 Handcart/Truck*

A handcart or hand truck is a wheel-mounted platform with handles to manually push or pull the unit.

##### *8.13.10.1.16.1 Characteristics/Applications*

- This is the simplest and most inexpensive method of transporting a load.
- It is used to move material across a short distance with frequent stops for loading and unloading.
- It can be used for storage of material between operations.
- It requires smooth and mainly level floors.
- The platform has a capacity ranging from 200 to 10,000 pounds.

#### *8.13.10.1.17 Tier Platform Truck*

A tier platform truck is a hand truck with one or more additional platforms stacked vertically.

##### *8.13.10.1.17.1 Characteristics/Application*

- It is used to carry light parts or materials in large quantities.
- Additional platforms increase both storage and handling capacities.
- There are from one to five trays with a capacity of 150 to 1800 pounds.

#### 8.13.10.1.18 Hand Lift Truck/Pallet Jack

A hand lift truck or pallet jack is a hand-operated truck that can raise loads hydraulically or mechanically to clear the floor before transporting them to the desired destination(s).

##### 8.13.10.1.18.1 Characteristics/Applications

- Operators can transport heavier objects more easily than with a hand truck.
- Applications are similar to those of a hand truck or tier truck.
- It is used mainly in moving material into and out of storage areas on pallets or skids.
- Its capacity can range from 500 to 1000 pounds, with the load center from 8 to 15 inches from the frame. Vertical movement is from 64 to 136 inches.

#### 8.13.10.1.19 Power-Driven Hand truck

A power-driven hand truck is similar to the hand lift truck, except that it is driven by a battery-operated electric motor.

##### 8.13.10.1.19.1 Characteristics/Applications

- It is used when the distances are greater than can be conveniently negotiated by hand lift trucks, about 150 to 300 feet.
- It can be used with a slightly inclined floor.
- It provides greater productivity compared to hand-driven trucks.
- The operator walks behind the truck.
- Its capacity varies from 500 to 6000 pounds, with a lift ranging from 0 to 54 inches and a speed range of 4 to 6 miles/hour. It can operate on 12- or 24-volt batteries.

#### 8.13.10.1.20 Power-Driven Platform Truck

A power-driven platform truck is a much larger device than the power-driven hand truck. It carries both load and operator. Power is supplied by a diesel or gasoline engine or by an electric motor.

##### 8.13.10.1.20.1 Characteristics/Applications

- It can carry a much heavier load for a longer distance; about 400 to 500 feet is the normal application.
- It is often used in maintenance and heavy storeroom work.
- Its capacity varies from 200 to 2000 pounds.

#### 8.13.10.1.21 Forklift Truck

A forklift truck is an operator-ridden, power-driven truck with forks in front that lift and carry heavy loads on skids or pallets.

*8.13.10.1.21.1 Characteristics/Applications*

- It can carry loads weighing thousands of pounds; a load of 100,000 pounds is not unusual.
- It can lift on skids as high as about 25 feet.
- It can be used to load and unload trucks or railroad boxcars.
- It is used extensively in storage and warehousing.
- It generally requires an aisle 10 to 12 feet wide.
- Its capacity is from 1000 to 100,000 pounds.
- It can be operated on gas or electricity.

Some trucks are customized for special uses in construction. Here, the truck carries the wooden frame and other materials to different heights in the construction projects.

### *8.13.10.1.22 Narrow-Aisle Trucks*

Narrow-aisle trucks are variations of industrial trucks that are specifically designed for aisles in which regular industrial trucks are too wide to operate.

*8.13.10.1.22.1 Characteristics/Applications*

- They require less aisle space (5 or 6 feet) to operate.
- Narrow-aisle trucks are more maneuverable than regular industrial trucks.
- They can operate on electricity or gas.
- They have a capacity of 1000 to 100,000 pounds, with a lift height of up to 30 feet.

Variations of the narrow-aisle truck are:

1. *Side-loader truck* is a fork truck with forks on the side rather than the front.
2. *Straddle truck* is a fork truck with out-riggers to balance loaded trucks.

3. *Reach truck* is a fork truck with telescoping forks to reach loads that are set back.
4. *Order-picker truck* is a truck with a platform that lifts the operator to the desired level to choose the load desired. It is used mainly when only part of a pallet is desired.
5. *Turret truck* is a fork truck with forks that can rotate left or right to place or pick up a load without requiring the truck to turn in the aisle.

#### *8.13.10.1.23 Tractor-Trailer Train*

A tractor-trailer train is a series of carts pulled by a self-propelled tractor.

##### *8.13.10.1.23.1 Characteristics/Applications*

- A tractor-trailer train is used for long-distance moves.
- It allows for a variety of load movement, such as platforms, bins, and racks.
- It allows many trains to be driven by the same tractor. The tractor can be disconnected and used to pull another train while the first is being loaded or unloaded.
- It can carry relatively heavy loads.
- It is mainly used for stop-and-go operations carrying loads from different points and delivering them to various destinations.
- It is usually used where the traveling distance is between 200 and 300 feet.

#### *8.13.10.1.24 Material Lift*

A material lift is a hand truck with a winch or hydraulic lifting mechanism for relatively small loads.

##### *8.13.10.1.24.1 Characteristics/Applications*

- It is manually operated.
- It has a variety of uses in warehouses, storerooms, offices, and shops where materials are lifted and moved from one location to another.
- It is easily maneuverable.
- It can be used to stack small items.
- Its lifting capabilities vary from 2 to 6 feet.
- Its capacity varies from 20 to 1000 pounds, with a lift height of up to 78 inches. The width can vary from 15 to 24 inches.

#### *8.13.10.1.25 Drum Truck*

A drum truck carries bulk like a hand truck, but it is specifically designed for drums. It has a rectangular, upright, and open-type frame with two or four wheels. It has nose tips at the bottom for placing under drums.

##### *8.13.10.1.25.1 Characteristics/Applications*

- It is manually operated
- One person can easily handle a full drum

- It is used on smooth, level surfaces.
- Four wheels allow the truck to stand without support when either loaded or unloaded.
- It can move drums onto or off pallets.
- It may have a winch or hydraulic lift to lift the drum.
- It has a capacity of up to 1000 pounds.

#### *8.13.10.1.26 Drum Lifter*

A drum lifter is a drum-handling unit with gripping arms that slide onto the forks of a lift truck.

##### *8.13.10.1.26.1 Characteristics/Applications*

- It allows the lift truck operator to pick up, transport, and deposit drums without assistance and without leaving the operator's seat.
- It provides no-tilt lifting of closed-head or open-head drums, with or without the tops in place.

#### *8.13.10.1.27 Dolly*

A dolly is a horizontal platform or open-type frame with wheels attached to the underside. It is used for transporting relatively light weights and low volumes across short distances.

##### *8.13.10.1.27.1 Characteristics/Applications*

- It is manually operated.
- It is primarily for use on level surfaces.
- It is highly maneuverable.
- Dollies come in various shapes and sizes for specific uses such as moving pallets, drums, cabinets, and heavy machinery.
- It can be designed to pry heavy objects.
- It is inexpensive.
- It provides easy one-person movement of objects.
- It is stable.
- There are two-wheel and four-wheel models with capacities ranging from 500 to 14,000 pounds. Dollies can weigh from 6 to 127 pounds and may be made of aluminum, plastic, or steel. Sketches of wide variety of industrial trucks are shown in Figure 8.1. The trucks have different functions, capacities, and reach. They can be powered by batteries or gas and can even be converted to automatic guided vehicles (discussed next). Small trucks require the operator to walk or stand in the back when the truck is moving, while the larger trucks provide comfortable seats for riding operators.

Electric Pallet Truck
3,500 lbs. Capacity
(Available as AGVS)

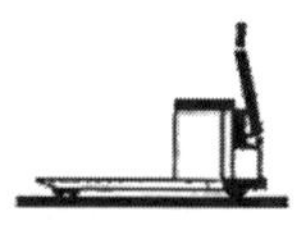

Electric Pallet Truck
Man Ride,
4,400 lbs. Capacity
(Available as AGVS)

Electric Straddle Stacker
Man Ride,
3,300 lbs. Capacity
(Available as AGVS)

Electric Pallet Stacker
Man Ride,
2,750 & 3,500 lbs. Capacity
(Available as AGVS)

Low Level Order Selector
9' Picking Height
4,400 lbs. Capacity
(Available as AGVS)

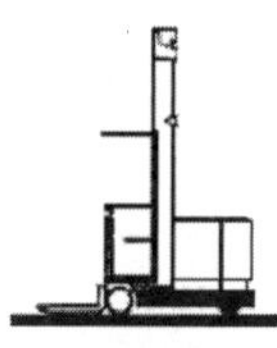

High Productivity Order
Selector. To 30' Lift
3,300 lbs. Capacity
Secondary Lift Mast

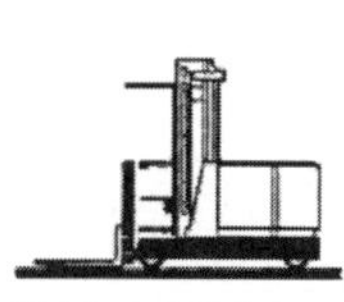

High Productivity Order
Selector with walk through
platform option. To 45' Lift
3,300 lbs. Capacity

Electric Reach Truck
3,000 & 3,500 lbs, Capacity
'Reach Mast' Truck

Three Wheel Electric
2,000 lbs. Capacity
Cushion or Pneumatic Tires
(Available as AGVS)

Three Wheel Electric
24V or 38V
2,500 & 2,200 lbs. Capacity
Cushion or Pneumatic Tires

Four Wheel Electric
3,000 & 4,000 lbs. Capacity
Cushion or Pneumatic Tires

Four Wheel Electric
3,200 & 6,000 lbs. Capacity
Cushion or Pneumatic Tires

Turret Truck
2,200 lbs. Capacity
Rail, Wire or Manually Guided

Turret Truck
2,750 lbs. Capacity
Manual, Rail or Wire Guided

Turret Truck
2,200 - 3,300 lbs. Capacity
Manual, Rail or Wire Guided
(Available as AGVS)

Turret Truck/Order Selector
2,200 - 3,300 lbs. Capacity
Manual, Rail or Wire Guided
(Available as AGVS)

Stacker (Model EMC 110/BMC 110) can be used for any transport distance and lifting height, featuring pedestrian trucks for short distances, combined pedestrian/stand-on equipment for medium-range operations, and stand-on/seat trucks for longer distances. Payload capacities of 1,600 kg to 3,000 kg and lifting heights of up to 5,350 mm provide a wide range of options. (Courtesy: Jungheinrich AG, Hamburg, Germany)

Diesel/gas forklifts offers a wide scope of options in terms of payload capacity and drive systems. Most up-to-date 8-cylinder LPG engine with electronic speed control. (Courtesy: Jungheinrich AG, Hamburg, Germany)

This high performance electronic counterbalance forklift with enclosed, twin AC motor front wheel drive is suitable for indoor and combined indoor/ outdoor operations. (Courtesy: Jungheinrich AG, Hamburg, Germany)

Order Pickers are the most economical way of maximizing picks per hour. Horizontal order pickers, and vertical order pickers have picking heights of up to 10,390 mm (Courtesy: Jungheinrich AG, Hamburg, Germany)

Reach trucks are used for any lifting height. Options include trucks with lifting heights of up to 12,020 mm, as well as–multi directional reach trucks. (Courtesy: Jungheinrich AG, Hamburg, Germany)

#### *8.13.10.1.28 AGV System*

The use of computer controls has been extended to material-handling tasks. A system operator at a station can control the vehicles by causing them to move along a predetermined path and to perform certain duties. Such a setup is known as an AGV system.

Tow tractors offer the best solution for towing any kind of trailer. They are ideally suited for small train supply with trailer payloads of up to 5,000 kg. (Model EZS 130) (Courtesy: Jungheinrich AG, Hamburg, Germany)

**FIGURE 8.1** Equipment used for material handling over large areas. Top left: A hand truck is the simplest, least expensive way to move small loads across short distances. Top right: A forklift truck, which can lift and carry heavy loads on pallets, is especially useful in warehouse storage. Bottom left: A high-rise forklift truck is adapted to hydraulically lift heavy loads for high-level storage. Bottom right: In this AGV system, the computer-controlled vehicle follows a buried guide path according to preprogrammed instructions or coordination by a more sophisticated computer

The guided vehicle can take any of several forms, including a flat load carrier similar to a dolly, pallet trucks, automatic forklifts, side loading vehicles, and robot arms. These machines can load themselves, travel to their destinations, unload, and return to their place of origin or any other desired location. For example, the pallet lift could be programmed to pick up a pallet in storage, bring it to the shipping dock, and return for another load. Human interaction is held to a minimum, thereby reducing labor costs, human error, and the possibility of injury.

AGV systems capable of handling up to 10 tons of weight are commonly available today. In addition, a recently developed system that floats on a microthin film (about 0.003 inches) of air between vehicle and floor can carry heavy loads of up to 35 tons. The frictionless design permits movement of loads on rough floors and can eliminate the use of cranes and bridge structures, which otherwise would be necessary for moving such heavy loads. These vehicles can also be maneuvered manually, requiring only about 1 pound of force for every 2000 pounds of carrying load.

There are two basic types of AGVs: "dumb" and "smart." They are categorized on the basis of how much control is given to the guide paths and other elements outside the vehicle and how much control is in the vehicle itself. A system that places the entire control outside the vehicle is built on a zone arrangement. The routes are divided into zones, and the vehicle is not allowed to enter a region unless other traffic in the zone is cleared. A vehicle can be made smart by using on-board microprocessors and radio transmitters. Different levels of communication between outside control and vehicles can be developed. For example, the destination can be punched into the vehicle's microprocessor, and it can travel along the route, communicating with other vehicles by radio to avoid collision. A more sophisticated system can have a host computer controlling the movement of all vehicles. The computer is instructed as to what is to be moved, and then it selects the available vehicle closest to the job and issues orders to the vehicle to perform the task. The net result is good coordination and utilization of vehicles as well as good record keeping on the movement of the goods.

Kulwiec (1984) listed the following specific advantages of an AGV system:

1. *Material control.* More accurate accounting of the use and transfer of material is possible as it is transferred between designated points, reducing the need for large safety stock.
2. *More efficient use of personnel.* Many forklift truck operators are no longer required; this reduces the overall number of material-handling personnel and the necessary coordination between them.
3. *Efficient work environment.* AGVs allow loading and unloading of each station independently of other stations, thus permitting each operator to work at his or her own pace and not limiting the operator to the speed of the conveyor. Many AGVs have movable platforms that allow the work to be delivered at the desired height.
4. *Flexibility.* The routes can be changed and new ones added with considerably greater ease than can be accomplished with systems of fixed bed conveyors.
5. *Better use of floor space.* Buried guide lines in the floor do not permanently occupy floor space as do conveyors fixed on the floor.
6. *Adaptability to automation.* AGVs can operate efficiently with other automated and computer-controlled systems such as robots, automatic storage and retrieval systems, conveyors, elevators, doors, trappers, wrappers, and automatic production machines.
7. *Integration within plant.* AGVs can provide a link between different cells of automation within the plant, thereby integrating the overall operation.
8. *Adaptability to existing facilities.* A new AGV system can be installed within an existing plant with a minimum of structural changes in the plant building. The guide paths require no more than slots in the floor compared to heavy support structure for cranes and substantial floor or structural modifications for conveyor systems.

#### 8.13.10.2 Accessories

A variety of accessories is currently available to facilitate material handling economically. Investment in these accessories can easily become quite large.

Therefore, it is necessary to carefully analyze each potential *purchase bdm* before a decision is made to obtain a certain type of accessory. Listed next are some of the most widely used accessories.

##### *8.13.10.2.1 Pallet*

A pallet is a platform on which material can be stacked in unit loads and handled by lifting equipment such as the forklift truck. A pallet and its load can then be moved from one place to another by such equipment.

Pallets are made of wood, plastic, metal, or a combination of the three, and are available in various sizes, the more popular of which are:

| | |
|---|---|
| 24 × 32 inches | 42 × 42 inches |
| 32 × 40 inches | 48 × 40 inches |
| 32 × 48 inches | 48 × 48 inches |
| 36 × 36 inches | 48 × 60 inches |
| 36 × 42 inches | 48 × 72, 40 × 48 inches |

The selected size depends on the size and weight of the unit load; the equipment used in moving pallets; the sizes of aisles, doors, and other spaces; and the cost of the pallet and whether it is to be saved or delivered and not returned.

Pallets are constructed so that the forks of the forklift truck can enter from either two or four sides. There are several types of pallets:

1. *Single-faced pallets* are pallets consisting of two or more stringers to which three or more planks comprising the upper deck are attached.
2. *Double-faced pallets* comprise two or more stringers with attached upper and lower decks. These pallets may be reversible and require very little care in storage when not in use.
3. *Box pallets* are single- or double-faced pallets with a framework that forms a box. They are mainly used to contain and stack nonuniform or bulk loads to convenient heights.

#### *8.13.10.2.2 Box*

A box is a portable container in which parts or material can be stored in unit loads. Boxes are made of cardboard, wood, plastic, or metal, and are available in various sizes. Some types may be folded for easy storage. Boxes can store and move parts or materials that are small or vary in size without crushing or dropping the units. Boxes can be moved by either hand-operated or power-driven equipment. The size can vary from 11½ × 2¾ × 2¾ to 71 × 18 × 19 inches (length × width × height).

#### *8.13.10.2.3 Tote Pan*

A tote pan is a portable container that is smaller in size than a box. It is used to carry small parts. Tote pans are made of plastic, metal, or wood and are used to transport small parts from one workstation to another. Their size can vary on the basis of the job requirements. They can be moved either by power-driven or hand-operated devices. Their size can vary from 16¾ × 10¾ × 3 to 46 × 34 × 33 inches.

#### *8.13.10.2.4 Skid*

A skid is similar to a pallet, except that the construction does not permit stacking of loaded skids on top of one another. Skids are made mainly of metal or heavy wood and are used to store and move heavy and/or bulky materials. They can be moved by power-driven or hand-operated devices, and can be made portable by attaching two wheels on one end and a carrying dolly at the other.

### 8.13.10.3 Optical Code, or Bar Code, Reader

An optical or bar code reader is a hand-held device that can read an optical code to identify the product or handling device on which the code is affixed. Optical code readers are small and easily carried between workstations. They can be used to keep track of inventory or products as items are moved from station to station. For example, an operator can acknowledge receipt of material by using an electric pencil to read the optical code affixed to the container (skid, box, etc.) and identify himself or herself by punching in a personal code or reading his or her special wristband with the same electrical pencil. This identifies the material, the operator, and the time when the material was received. A similar procedure is followed to signify completion of work on the material by the operator and its dispatch to the next station.

The type of scanner can be a gun, wand, pistol, pen, or magnetic swipe. The light source can be laser, light-emitting diode (LED), infrared LED, or incandescent. The scan rate is up to 100 scans/second. The depth of the field can range from 10 to 25 inches, while the height of the scan can range from 0 to 1½ inches. The life of the light source is between 3000 and 100,000 hours. The memory capacity can vary from 2K to 512K. Bar code systems are described in detail in Chapter 4.

Adjustable pallet racking is particularly suitable for large quantities of single, mainly palletized articles. Pallets allow direct access to all articles and random position allocation.

## 8.14 SUMMARY

The chapter presents the basic principles for designing a good MHS within a plant. The objectives are to increase efficiency in material flow, reduce material-handling cost, and make the overall system more productive and safe. The equipment available for such purposes can be broken down into three major categories: conveyors, cranes, and trucks. Conveyors are popular when the material is relatively light and is to be moved over a fixed path. Cranes also travel a fixed path but are generally used to handle much heavier loads. Trucks, on the other hand, can travel over varying paths but, as with all other systems, have some advantages and disadvantages.

The degree of mechanization can influence the cost of material handling. The level of sophistication may vary from a manual handling system to a completely automated one. The initial required investment and operating cost in each case are of course considerably different. Moving material in unit loads is one way to reduce the frequency of trips; therefore, development of a unit load is an important facet in overall planning.

There are 20 basic principles that serve as guides in designing and operating an MHS. The chapter lists them all and suggests ways of applying them. Many of these principles are compatible with one another, the application of one being able to achieve the fulfillment of another.

Specifying an MHS for a new plant and developing its layout have an interesting interrelationship. If the layout is known, then the point-of-origin and destination for each material move are known, along with the floor plan; hence, one can design an MHS to suit the layout. But it is also possible to develop a layout around a known MHS. To obtain maximum operating efficiency, both the layout and the MHS should be considered simultaneously. This may require a procedure that goes back and forth between two developments.

The existing MHS should also be audited frequently to ensure that it is functioning efficiently. The chapter lists the symptoms of an inefficient MHS and provides a checklist that may suggest ways for improvement. Productivity ratios can also be used to trace inefficiencies and to incorporate further controls.

A wide variety of equipment is available for material handling. The chapter lists the major types of conveyors, cranes, and trucks. The description of commonly used equipment in an MHS should help the reader understand the advantages, characteristics, and applications of various alternatives. Not all equipment and their features can be described in detail in the limited space available here. The reader should refer to the following monthly publications for recent developments and applications in material handling:

- *Modern Material Handling* (Cahners Publishing, 1350 E. Touhy Ave., Des Plaines, IL 60018).
- *Plant Engineering* (Technical Publishing Corp., P.O. Box 1030, Barrington, IL 60010).
- *Materials Handling Engineering* (614 Superior Ave., Cleveland, OH 44113).

More information on each product can also be obtained by writing directly to the manufacturer, the address for which can be found in the abovementioned publications or in the *Thomas Register of American Manufacturers* referred to in Chapter 5.

## PROBLEMS

1. Define material handling. State at least four principles involved and indicate how you would apply them.
2. Discuss the significance of material handling in the industry. Give some examples of material handling in a light industrial plant that you might have visited.
3. What are the three basic types of MHSs? Give three advantages and three disadvantages of each type.
4. What is a unit load? What factors influence the formulation of a unit load?
5. What are the benefits of automation in material handling?
6. The principles of material handling given in Table 8.1 can be remembered by the following mnemonic:

   The Galactic EMPire Called SPAM was Attacking Uno Over the Seven Stars. Spam's Mother Ship returned Undestroyed. Or by a more down-to-earth phrase:

   GUSS M.D. has A CAMP to teach POSSUM to be C.E.S.

   Identify each principle by the letters capitalized in these mnemonics.
7. Give examples for each principle listed in problem 6.
8. Select an establishment in your community. Discuss its material-handling methods and machinery.
9. For the establishment chosen in problem 8, go through the questioning process for one of its material moves.
10. Identify the productivity ratios that are important to you if you are
    a. the plant manager
    b. the warehouse supervisor
    c. the industrial engineer of the plant
    d. designing a plant layout
11. Discuss the pros and cons associated with replacing a belt conveyor connecting two stations with an automatic guide vehicle. What volume of production justifies such replacement? Will the outlook change if 10 stations distributed within the plant are to be connected by either a conveyor system or an AGV system?
12. Determine the appropriate pallet size required to load packages with a base of $5 \times 12$ inches so as to minimize wasted space.
13. Differentiate among the following material-handling equipment, giving at least two advantages and disadvantages of each:
    a. Roller conveyor and salt conveyor
    b. Belt conveyor and wheel conveyor
    c. Forklift truck and trolley conveyor
    d. Hand truck and forklift truck
    e. Jib hoist and bridge crane
    f. AGV and forklift truck
14. What factors must be considered in choosing the size of pallets to be used in a plant?

## SUGGESTED READING

### Material Handling

Apple, J.M., *Material Handling Systems Design,* John Wiley, New York, 1972.
*Basics of Material Handling,* The Material Handling Institute, Inc., Charlotte, NC, 1973.
"Equipment — the big goal: more work at lower cost," *Modern Material Handling,* 66–69, 1975.
Kulwiec, R., "Trends in automatic guided vehicle systems," *Plant Engineering,* 66–73, 1984.
"Labor — productivity is where the savings are," *Modern Material Handling*, 52–59, 1975.
Rose, W., *Logistic Management: Systems and Components,* William C. Brown Company Publishers, Dubuque, 1979.

# 9 Material Handling *Equipment Selection*

Three important topics are discussed in this chapter. First, the discussion section on equipment selection shows different methods for determining the type and/or number of units required to perform the necessary material moves. Examples involve algebraic methods as well as the more advanced approach of a heuristic solution procedure.

The second section concerns the use of robotics in material handling. A robot could be used for loading and unloading operations with a single machine or with several machines where loading and unloading are done in a sequential manner. Establishing the cycle time for an operation in each case is important in determining the productivity of each setup.

The third section covers the use of automated guided vehicles (AGVs) in material handling. Commonly used rules for the operation of AGVs are discussed. Simulation of the AGV system is the best way of deciding which rules should be used in a particular design. An example of an AGV system illustrates this point.

## 9.1 BASICS OF EQUIPMENT SELECTION

As mentioned in the previous chapter, in general, conveyors are used where flow of material is continuous or very close to being continuous and the path of material movement is fixed. Conveyors have a large capacity for transporting over considerable distances if required. The material or parts can be added to and taken off of the conveyor at intermittent points, but once the conveyor is installed, the position of the conveyor is somewhat permanent since it is very expensive to reposition it. Conveyors can handle parts or material in packages (e.g., boxes, small pallets, or bags), or in loose form (e.g., stamped pieces, forged and/or cast products, individual pieces, plastic parts), or bulk material (e.g., wood chips, ore, or cement).

Trucks, on the other hand, are generally used where delivery in batches is adequate and where flexibility is important. The demand between two stations may change based on the planned product mix, and transportation by truck can respond to capacity change much more readily. The route a truck takes between two departments can be adjusted to fit changing needs. A truck can also be used to load and unload a conveyor, thus taking advantage of the conveyor for transporting material over long distances, and adding the flexibility of a truck for transportation over a short distance between the conveyor and the workstation. When using a truck, the load is generally on a pallet, in a drum, or in a box, and rarely an independent part.

Cranes are mainly used to lift very heavy pieces. The mobility of cranes is rather limited, and they are very expensive. They also require a building with a heavy load bearing foundation, most often with deep piles.

There are many activities in a production plant where material is moved. For example, the most critical and prominent activities in warehousing are receiving, storing, and reissuing material, partial assemblies, and final products. At each stage, considerable material handling is involved; to a large extent, efficient operation of a warehouse depends on the proper selection of equipment for such handling. Some types of equipment have already been described; others will be discussed later.

In this section, we shall concentrate on methods of equipment selection that are dependent on:

- *Material to be moved*: The type, weight, volume, shape, and size of the raw material and other assemblies that are to be moved.
- *Movement*: The frequency, path, aisle space, and loading/unloading mechanism.
- *Storage*: The area, volume, shape, and size of storage facility, columns, and other obstacles; the spacing arrangement of shelves and racks; and company policies regarding storage and issuing.
- *Costs*: The investment and operating expenses of equipment, interest rate, depreciation, and useful life of the equipment.
- *Other factors*: Flexibility to perform multiple tasks and to work on many products, as well as obsolescence of equipment.

Once the data for material handling are established, justification of the material-handling equipment on some routes may be simple, because there is only one physically possible mode of transportation. On some other routes, different alternatives may satisfy the basic needs. For example, either a conveyor or a forklift truck might be used on a particular route. Such alternatives must be evaluated and justified.

The following examples illustrate some of the economic analyses needed in evaluating material-handling equipment requirements.

## EQUIPMENT-OPERATING COST PER UNIT DISTANCE

A forklift truck initially costs \$20,000 and has an expected life of 5 years. Fuel cost is \$10.00 per 8 hours of operation, and maintenance cost is \$1.50 an hour. If the truck travels an average of 10,000 feet per day, determine the cost per foot. Assume that the truck operates 360 days a year and that the operator is paid \$10.00 an hour (including fringe benefits).

### SOLUTION

First, determine depreciation. Using the straight line depreciation method, we have

$$20{,}000/5 \text{ years} \times 1 \text{ year}/360 \text{ days} \times 1 \text{ day}/8 \text{ hours} = \$1.39/\text{hour}$$

The distance traveled per hour is

$$10{,}000 \text{ feet/day} \times 1 \text{ day}/8 \text{ hours} = 1250 \text{ feet/hour}$$

**TABLE 9.1**
**Material Characteristics**

| Item | Volume (L × W × H) (inches) | Distance From Receiving (feet) | Distance to Shipping (feet) | Units Received per Day | Units Shipped per Week |
|---|---|---|---|---|---|
| A | 12 × 6 × 6 | 525 | 125 | 118 | 590 |
| B | 48 × 36 × 24 | 225 | 375 | 165 | 825 |
| C | 24 × 24 × 24 | 400 | 200 | 121 | 605 |

$$\begin{aligned}\text{Total cost per hour} &= \text{maintenance}\\ &\quad + \text{fuel}\\ &\quad + \text{depreciation}\\ &\quad + \text{operator costs}\\ &= 1.50 + 10/8\\ &\quad + 1.39 + 10.00\\ &= \$14.14/\text{hour}\end{aligned}$$

$$\begin{aligned}\text{Operating cost per foot} &= \frac{\text{cost/hour}}{\text{feet/hour}}\\ &= \frac{14.14}{1250}\\ &= \$0.0113/\text{foot}\end{aligned}$$

## EQUIPMENT SELECTION

The warehouse receives daily loads of items A, B, and C from the plant, and shipments are made once a week. Information about these items is given in Table 9.1. The company is considering purchasing a tractor that pulls four trailers, a forklift truck, or a hand truck for use in transporting. Pertinent sizes and costs for these are listed in Table 9.2. The loading and unloading costs include not only the cost for loading and unloading for one trip, but also the cost for making the return trip to pick up another load. Determine the least expensive method for transporting the goods.

### Solution

The number of units that can be moved per trip is given by the capacity of the piece of equipment divided by the volume of the unit to be moved.

**TABLE 9.2**
**Equipment Specifications**

| Equipment | Maximum Volume (cubic inches) | Loading and Unloading Cost ($) | Cost/Foot ($) |
|---|---|---|---|
| Tractor truck | 4 (60 × 27 × 72) | 1.20/trip | 0.01 |
| Forklift truck (pallet) | 48 × 48 × 48 | 0.05/trip | 0.007 |
| Hand truck | 60 × 27 × 72 | 0.40/trip | 0.005 |

For item A, using the tractor truck, the capacity of each truck is the volume of the truck divided by the volume of the item.

$$\text{Capacity} = \frac{60}{12} \times \frac{27}{6} \times \frac{72}{6}$$
$$= 5 \times 4 \times 12$$
$$= 240 \text{ units/truck}$$

Because parts of units cannot be transported, only a whole number answer for each dimension division is used. For example, $27 \div 6 = 4.5$; and since half a unit cannot be transported, 4.5 becomes 4.

Four trailers are pulled by the tractor.

$$\text{Total capacity} = 4(240)$$
$$= 690 \text{ units per trip}$$

Each day, 118 units are moved. The tractor can move 960 units at a time, so only one trip is necessary. At shipping, 590 units are moved, still requiring only one trip.

For the forklift truck,

$$\text{Capacity} = \frac{48}{12} \times \frac{48}{6} \times \frac{48}{6}$$
$$= 4 \times 8 \times 8$$
$$= 256 \text{ units/pallet}$$

With 590 units to be moved at shipping, three trips will be required.

The hand truck is the same size as one of the trailers for the tractor; therefore, it can carry 240 units per trip and will require three trips to move the 590 units at shipping.

Next, using this information and that from Tables 9.1 and 9.2, we can determine the cost for transporting item A using each piece of equipment.

$$\begin{aligned}
\text{Receiving cost} &= \text{Loading and unloading cost per week} \\
&\quad + \text{travel cost per week} \\
&= (\text{Loading and unloading cost per trip})(\text{trips per day})(5 \text{ days}) \\
&\quad + (\text{cost per foot})(\text{number of trips})(\text{feet per trip})(5 \text{ days}) \\
\text{Shipping cost} &= \text{Loading and unloading cost per order} \\
&\quad + \text{Travel cost per order} \\
&= (\text{Loading and unloading cost per trip})(\text{Number of trips}) \\
&\quad + (\text{Cost per foot})(\text{Number of trips})(\text{Feet/trip}) \\
\text{Total cost} &= \text{Receiving cost} + \text{Shipping cost}
\end{aligned}$$

The following abbreviations will be used:

T = tractor
F = forklift truck
H = hand truck

**TABLE 9.3**
**Final Cost Values (in Dollars)**

| | Item | | | |
|---|---|---|---|---|
| **Equipment** | **A** | **B** | **C** | **Total Cost** |
| Tractor truck | 34.70 | 583.05 | 239.20 | 856.95 |
| Forklift tractor | 21.40 | 1779.16 | 338.20 | 2138.77 |
| Hand truck | 19.23 | [a] | 203.60 | |

[a]Not feasible, item is too heavy.

RC = receiving cost
SC = shipping cost
TC = total cost
LT = loading and unloading trip

Tractor:

$$
\begin{aligned}
RC(T) &= \text{(5 days) (Loading and unloading cost per trip) (1 trip per day)} \\
&\quad + \text{(5 days) (Cost per foot)(Feet per trip) (1 trip per day)} \\
&= \text{(5 days) (1.20/trip)} \\
&\quad + \text{(5 days)(0.01/foot) (525 feet)} = \$32.25 \\
SC(T) &= \text{(Loading and unloading cost per trip) (1 trip)} \\
&\quad + \text{(Feet per trip) (1 trip) (Cost per foot)} \\
&= 1.20 \text{ per trip} + \text{(125 feet) (0.01/foot)} = \$2.45 \\
TC(T) &= 32.25 + 2.45 = \$34.70
\end{aligned}
$$

Forklift truck:

$$
\begin{aligned}
RC(F) &= \text{(0.05 per trip) (5 days)} + 5\text{(525 feet) (0.007 per foot)} = \$18.63 \\
SC(F) &= 3 \text{ trips (0.05 per trip)} \\
&\quad + 3 \text{ trips (125 feet) (0.007 per foot)} = \$2.78 \\
TC(F) &= \$21.41
\end{aligned}
$$

Hand truck:

$$
\begin{aligned}
RC(H) &= \text{(0.40 per trip) (5 day)} + 5 \text{ days (525 feet) (0.005 per foot)} \\
&= \$15.13 \\
SC(H) &= 4 \text{ trips (0.4 per trip)} + 4 \text{ trips (125 feet) (0.005 per foot)} \\
&= \$4.10 \\
TC(H) &= \$19.23
\end{aligned}
$$

Similar calculations are made for items B and C, and the results are shown in Table 9.3. The tractor truck is the least expensive and is therefore selected.

### 9.1.1 Work Volume Analysis

We illustrate here a basic tabular method called work volume analysis, which is used to determine the number of handling units needed to perform a material-handling task. This analysis consists of measuring the volume, characteristics, and handling requirements

**TABLE 9.4**
**Work Volume Analysis**

| Dock Identity (1) | Commodity (2) | Transport Type (3) | Loads per Day: Average (4) | Loads per Day: Peak (5) | Loads per Day: Selected Mean (6) | Unload to (7) | Load Quantity (8) | Handling Unit (9) | Handling Units per Load (10) | Estimated Handling Minutes per Handling Unit (11) | Handling Hours per Load (12) | Dock Hours per Load, Including Allowances (13) | Operating Hours per Day (14) | Dock Spot Capacity per Day (15) | Calculated Number of Dock Spots Required (16) | Actual Number of Dock Spots Required (17) | Handling Units per Days (18) | Handling Hours per Day (19) | Number of Trucks Required (20) |
|---|---|---|---|---|---|---|---|---|---|---|---|---|---|---|---|---|---|---|---|
| Dock H | Bulk A | R.R. | 0.25 c/1 | 1.0 | 1.0 | Open bin | 100,000 lb | 800 lb, grab hook | 125 | 3 | 6.3 | 7.0 | 16 | 2.2 | 0.5 | 1 | | | |
| Dock F | Bulk A | R.R. | 0.1 c/1 | 1.0 | 0.5 | CT 626 | 100,000 lb | 400 lb, shovel | 250 | 3 | 12.5 | 13.0 | 16 | 1.2 | 0.5 | | 125 | 6.25 | |
| | Loose A | R.R. | 2.1 c/1 | 5.0 | 3.5 | CT 626 | 100,000 lb | 600 lb, shovel | 167 | 3 | 8.4 | 9.0 | 16 | 1.7 | 2.1 | | 585 | 29.25 | |
| | Package A | R.R. | 0.2 c/1 | 1.0 | 0.75 | CT 540 | 20 pallets | 1 unit load | 20 | 3 | 1.0 | 2.0 | 16 | 8.0 | 0.1 | | 15 | 0.75 | |
| | Package B | R.R. | 2.1 c/1 | 5.0 | 3.5 | CT 610 | 20 loose containers | 1 container full | 20 | 10 | 3.3 | 4.0 | 16 | 4.0 | 0.9 | | 70 | 3.50 | |
| | Total | | 4.5 c/l | 12.0 | 8.25 | | | | | | | | | | 3.6 | 4 | 795 | 39.75 | 2.5 |
| Dock M | Bulk B | Truck | 0.8 t/1 | 5.0 | 4.0 | Underground tank | 3000 gal | | | 1000 gph | 3.0 | 3.5 | 16 | 4.5 | 0.9 | 1 | | | |
| Dock C | Loose B | Truck | 1.25 t/1 | 3.0 | 3.0 | CT 610 | 30,000 lb | 1500 lb per cont. | 20 | 10 | 3.3 | 4.0 | 8 | 2.0 | 1.5 | | 60 | 3.0 | |
| | Package A | Truck | 0.5 t/1 | 2.0 | 1.25 | CT 540 | 10 pallets | 1 unit load | 10 | 3 | 0.5 | 1.0 | 8 | 8.0 | 0.2 | | 13 | 0.7 | |
| | Package B | Truck | 6.5 t/1 | 10.0 | 8.0 | CT 610 | 21,000 lb | 1500 lb per cont. | 14 | 7 | 1.6 | 2.0 | 8 | 4.0 | 2.0 | | 112 | 5.6 | |
| | Total | | 8.25 t/1 | 15.0 | 12.25 | | | | | | | | | | 3.7 | 4 | 185 | 9.3 | 1.2 |

Courtesy of The Material Handling Institute, Inc.

of work that must be moved from one work center to another. The measurements may be divided on the basis of the type of work, such as receiving, shipping, storage, and in-process movement. The data are analyzed to determine the type and number of handling units that would be needed to accomplish the task. Table 9.4 shows an example work volume analysis, taken mostly from *Basics of Material Handling* (1973). It shows an analysis of receiving docks for a plant that is being supplied by both rail cars and trucks.

Column 1 identifies the location where the analysis is performed. Columns 2 and 3 identify the commodity and mode of transportation used. Columns 4, 5, and 6 estimate the loads per day of the transportation unit that will need services, with the value in column 6 being taken as the design parameter. Column 7 indicates where the unloaded material should be stored. Column 8 indicates the load quantity per unit of the transportation (one rail car, one truck, etc.). Column 9 identifies the material-handling unit that will be used to unload and its capacity per operation. Column 10 calculates how many complete operations (strokes) of the handling unit are required to unload the quantity listed in column 8; that is, column 10 = column 8/column 9. Column 11 is an estimate of the time to perform one stroke of the handling unit; the information is used to calculate the time (in hours) needed to unload one transportation unit; that is, column 12 = column 11 × column 10/60. Column 13 modifies the time in column 12 by including allowances. Column 14 states the working hours of the plant (one shift = 8 hours, two shifts = 16 hours). The dock spot capacity, column 15, is calculated by dividing column 14 by column 13. This value indicates how many times during a day's work a single delivery unit (rail car, truck) can be emptied on a single dock. Since the average number of daily loads is given in column 6, the number of dock spots needed, column 16, is obtained by dividing column 6 by column 15. The actual number of dock spots required, column 17, must be an integer equal to or greater than the value calculated in column 16.

To calculate the number of forklift trucks necessary, the unit used in some docks, we must first calculate the handling loads per day, column 18. This is obtained by taking the product of column 6 and column 10. Column 19 indicates the expected number of hours a forklift will work. The numbers are obtained by dividing entries in column 18 by 20, the estimated load a forklift will handle per hour. The number of trucks required is given by column 20, which is obtained by dividing the total hours required, given as the total in column 19, by the number of operating hours per day, column 14.

## APPLICATION OF OPERATIONS RESEARCH TO MATERIAL-HANDLING PROBLEMS

Operations research models have been applied to the design and operations of material-handling systems. These applications involve using mathematical programming, simulation, queuing theory, and network models. Some areas of material handling have benefited more than others from these models. A few examples of the applications are

- Conveyor systems
- Pallet design and loading
- Equipment selection
- Dock design
- Equipment routing
- Packaging
- Storage system design

**TABLE 9.5**
**Candidate Equipment and Move Data**

| | Equipment | | | | | | | |
|---|---|---|---|---|---|---|---|---|
| | 1 | | 2 | | 3 | | ...P | |
| Move | Cost ($W_{1j}$) | Time ($h_{1j}$) | Cost ($W_{2j}$) | Time ($h_{2j}$) | Cost ($W_{3j}$) | Time ($h_{3j}$) | Cost ($W_{pj}$) | Time ($h_{pj}$) |
| 1 | 200 | 0.3[a] | 300 | 0.2 | m[b] | — | | |
| 2 | 500 | 0.5 | 250 | 0.7 | | | | |
| ⋮ | ⋮ | ⋮ | ⋮ | ⋮ | ⋮ | ⋮ | ⋮ | ⋮ |
| $Q$ | $W_{1q}$ | $h_{1p}$ | $W_{2q}$ | $h_{2q}$ | $W_{3q}$ | $h_{3q}$ | $W_{pq}$ | $h_{pq}$ |

[a] Assuming that the total operating time of an equipment unit is 1, the operating time of a move can be expressed as a fraction.

[b] Move 1 cannot be performed by equipment 3, which is shown with a very large operating cost denoted by "m" and an associated operating time denoted by "—."

In this section, a heuristic procedure, first proposed by Hassan et al. (1985), is presented that can assist designers in the selection of material-handling equipment and its assignment to departmental material-handling tasks (moves). Such selection is applicable after an initial screening has been performed by the designer to determine the most promising candidates, from which the final selection is to be made analytically.

In such circumstances, a set of candidate equipment types is generally available. Each move can be performed by most or all of the candidate equipment. Thus, for each move, there are different values for the operating cost and time based on the equipment used (Table 9.5). The problem requires selection of equipment from among the candidate set and assigning moves such that a move is not made by more than one item of equipment unless they are of the same type (i.e., each move is assigned to only one equipment type), and all moves assigned to a piece of equipment can be performed in the available time on the equipment.

The primary objective of the problem is cost (operating and initial) minimization. There are also some secondary objectives such as maximum utilization of equipment and minimum variation in the selected types, but they are most often compatible with our primary objective. The problem can be expressed mathematically as follows:

$$\text{Minimizing} \quad Z = \sum_{i=1}^{p}\sum_{j=1}^{q} W_{ij}X_{ij} + \sum_{i=j}^{p} \lambda_i K_i$$

subject to

$$\sum_{i=1}^{p} a_{ij}X_{ij} = 1 \qquad j = 1,2,3,\ldots,q$$

where

$$\sum_{j=1}^{q} h_{ij}X_{ij} = iHi \qquad \lambda_i = 1,2,3, \quad ,p$$

**TABLE 9.6**
**Analogy Between the Knapsack and Move Assignment to a Piece of Equipment**

| Equipment | Knapsack |
|---|---|
| Moves | Items |
| Operating time of a move | Volume or weight of item |
| Available time on equipment | Capacity of knapsack |
| Operating cost of the moves | Value of the items |

$$X_{ij} = \{0 \text{ or } 1\}\ for\ all\ ij$$

$$\lambda_i = 0 \text{ and integer for all } i$$

$$a_{ij} = \begin{cases} 1 & \text{if equipment type } i \text{ can perform move } j \\ 0 & \text{otherwise} \end{cases}$$

$h_{ij}$ = total operating time required by equipment type $i$ to perform move $j$
$H_i$ = available operating time of one unit of equipment type $i$
$K_i$ = capital cost of one unit of equipment type in the same time unit as the operating cost (e.g., 1 year)
$q_i$ = number of moves that can be assigned on equipment type $i$
$q$ = total number of moves to be assigned
$p$ = number of candidate equipment types
$W_{ij}$ = total operating cost of performing move $j$ by equipment type $i$ in the same time unit as $K_i$ (e.g., 1 year)

$$X_{ij} = \begin{cases} 1 & \text{if equipment type } i \text{ can perform move } j \\ 0 & \text{otherwise} \end{cases}$$

$\lambda_i$ = number of units of equipment of a selected equipment type $i$ that is required

The algorithm presented here for an equipment selection problem is based on an analogy to both the knapsack* and loading problems† (see Hassan et al., 1985). Assuming that an equipment type has been selected, then an assignment to one unit of that equipment is analogous to allocating items to a knapsack (the analogy is illustrated in Table 9.6). When a second unit of the same equipment type is required for a move assignment, the situation becomes analogous to the loading problem.

The algorithm considers the equipment types one at a time. Moves are assigned to a unit of the selected equipment until it is fully utilized or no other move can be assigned. A selection of the second unit or another type is then made, and the moves are assigned until that second unit or type is also fully utilized or no further assignment is possible. The algorithm terminates when all moves are assigned. Both equipment selection and

* The one-dimensional knapsack problem involves selecting some items from a set of candidates. Each item has a specific weight (or volume) and a measure of merit or value associated with it. The objective of the problem is to maximize the total value of the selected items such that the capacity of the knapsack is not exceeded.

† In the loading problem, some items, each with a certain volume, require allocation to boxes with equal capacities such that the total number of boxes or their value is minimized.

move assignment are performed in a manner that helps in cost minimization. The steps of the algorithm are:

1. For each equipment type, calculate the number of units that would be needed if the equipment performs all the moves

$$Y_i = \sum_j \frac{h_{ij}}{H_i}$$

If the division is an exact integer, then

$$\lambda_i = Y_i$$

If the division is not an exact integer, then

$$\lambda_i = [Y_i] + 1$$

where the quantity in brackets is the integer portion of $Y_i$. Usually, $H_i$ is a set equal to 1 and $h_{ij}$ is expressed as a fraction of $H_i$.

2. Calculate the total cost of material handling for each equipment type as

$$Z_i = \lambda_i K_i + \sum_{jEEi} W_{ij}$$

where $\boldsymbol{E}_i$ is the vector of the moves that can be performed by equipment type $i$, and the number of these moves (used in the next step) is $q_i$.

3. Calculate the average cost for each equipment type per move as

$$\bar{Z}_i = \frac{Z_i}{q_i}$$

4. First, select the equipment with the smallest $\bar{Z}_i$. Resolve ties by selecting the equipment with the smallest $Z_i$. If ties persist, resolve them by selecting in order of ascending $\lambda_i K_i$.
5. For the selected equipment type, arrange the moves that can be performed by it in increasing order of operating cost.
6. Assign the moves to the selected equipment starting with the move having the smallest operating cost. After each assignment, check whether the sum of $h_{ij}$ is equal to $H_i$ or within a tolerance $\boldsymbol{E}_i$ of it. If the sum of $h_{ij}$ is equal to $H_i$, go to the next step; otherwise, check either of the following two cases:
   a. If the moves are the only remaining moves or cannot be assigned to another piece of equipment, leave the assignment as it is.
   b. If the sum of $h_{ij}$ is greater than $H_i$ (or a multiple of $H_i$ depending on the number of units required of the equipment so far), check the difference between the least integer multiple of $H$ (making it greater than the sum of $h_i$) and the sum of $h_{ij}$. If the difference, which represents idle time, is less than or equal to $E_2$ (a specified acceptable idle time), leave the assignment as it is. If the difference is larger than $E_2$, remove moves from the equipment starting with the last assigned move, until the acceptable utilization level is achieved.
7. Delete the moves assigned from consideration for the remaining moves, calculate a new value for $Z_i$, as before, and repeat the steps until all the moves are assigned.

To demonstrate the procedure, we use the following example. Four candidate types of equipment have the data shown in Table 9.7. $H_i$ is assumed to be equal to 1 for all $i$, and $\boldsymbol{E}_i$ and $E_2$ are 0.1 and 0.2, respectively. Table 9.8 shows the initial calculation.

**TABLE 9.7**
**Cost per Year and Move Time Data, Cycle I**

| | Equipment Type | | | | | | | |
|---|---|---|---|---|---|---|---|---|
| | 1 | | 2 | | 3 | | 4 | |
| Move | $W_{ij}$ | $h_{ij}$ | $W_{2j}$ | $h_{2j}$ | $W_{3j}$ | $h_{3j}$ | $W_{4j}$ | $h_{4j}$ |
| 1 | 400 | 0.4 | M | — | 200 | 0.5 | M | — |
| 2 | 600 | 0.6 | 400 | 1 | 900 | 0.8 | M | — |
| 3 | 400 | 0.5 | 500 | 1 | 900 | 0.9 | M | — |
| 4 | 500 | 0.4 | 400 | 1 | 800 | 0.9 | M | — |
| 5 | 100 | 0.3 | 300 | 1 | M | — | M | — |
| 6 | 200 | 0.7 | 900 | 1 | M | — | M | — |
| 7 | M | — | 400 | 1 | 400 | 0.7 | 200 | 0.4 |
| 8 | M | — | 100 | 1 | 300 | 0.6 | 200 | 0.4 |
| 9 | M | — | 200 | 1 | 800 | 0.7 | 900 | 0.3 |
| 10 | M | — | 100 | 1 | M | — | 500 | 0.2 |
| Total | 2200 | 2.9 | 3300 | 9 | 4300 | 5.1 | 1800 | 1.3 |

Data for the problem are from the paper by Webster and Reed (1971).

The capital costs for one unit of equipment types 1, 2, 3, and 4 are 5000, 2777.77, 3000, and 4000, respectively.

The smallest $\bar{Z}_i$ is that of equipment type 4; hence, type 4 is selected first, and the moves are arranged according to their operating costs as in Table 9.9. Move 7 is assigned first, and the sum of $h_{4j} = 0.4$. Move 8 is assigned next, and the sum of $h_{4j}$ is now $0.4 + 0.4 = 0.8$. Move 10 renders the sum of $h_{4j} = 1$; therefore, this iteration is terminated, and one unit of equipment 4 is fully utilized.

With the assignment of moves 7, 8, and 10 to equipment 4, the capacity of that unit is utilized to the extent possible (in this case, completely). Moves 7, 8, and 10 are deleted from the set of moves.

For the next iteration, the cost and move time data are shown in Table 9.10. The moves already assigned are not included in this table.

**TABLE 9.8**
**Calculations for Cycle I**

| Equipment Type (1) | Total Number of Possible Moves, $q_t$ (2) | Total Operating Time, $h_{ij}$ (3) | Number of Equipment, $\lambda_i$ (4) | Total Capital Cost, $K_t$ (5) | Total Operating Cost, $\Sigma W_{ij}$ (6) | Total Cost, $Z_i$ (5) + (6) (7) | $\bar{Z}_i$ (7)/(2) (8) |
|---|---|---|---|---|---|---|---|
| 1 | 6 | 2.9 | 3 | 15,000 | 2200 | 17,200 | 2866 |
| 2 | 9 | 9.0 | 9 | 25,000 | 3300 | 28,300 | 3144 |
| 3 | 7 | 5.1 | 6 | 18,000 | 4300 | 22,300 | 3185 |
| 4 | 4 | 1.3 | 2 | 8000 | 1800 | 9800 | 2450 |

**TABLE 9.9**
**Ranked Moves and Their Parameters for a Selected Unit of Equipment 4, Iteration I**

| Move | $W_{4j}$ | $h_{4j}$ | $\Sigma h_{4j}$ |
|---|---|---|---|
| 7 | 200 | 0.4 | 0.4 |
| 8 | 200 | 0.4 | 0.8 |
| 10 | 500 | 0.2 | 1.0 |
| 9 | 900 | 0.3 | |

**TABLE 9.10**
**Cost and Move Time Data, Iteration II**

| | Equipment Type | | | | | | | |
|---|---|---|---|---|---|---|---|---|
| | 1 | | 2 | | 3 | | 4 | |
| Move | $W_{ij}$ | $h_{ij}$ | $W_{2j}$ | $h_{2j}$ | $W_{3j}$ | $h_{3j}$ | $W_{4j}$ | $h_{4j}$ |
| 1 | 400 | 0.4 | M | — | 200 | 0.5 | M | — |
| 2 | 600 | 0.6 | 400 | 1 | 900 | 0.8 | M | — |
| 3 | 400 | 0.5 | 500 | 1 | 900 | 0.9 | M | — |
| 4 | 500 | 0.4 | 400 | 1 | 800 | 0.9 | M | — |
| 5 | 100 | 0.3 | 300 | 1 | M | — | M | — |
| 6 | 200 | 0.7 | 900 | 1 | M | — | M | — |
| 9 | M | — | 200 | 1 | 800 | 0.7 | 900 | 0.3 |
| Total | 2200 | 2.9 | 2700 | 6 | 3600 | 3.8 | 900 | 0.3 |

**TABLE 9.11**
**Calculations for Iteration II**

| Equipment Type (1) | $q_i$ (2) | $h_{ij}$ (3) | $\lambda_i$ (4) | $ZW_i$ (5) | $K_i$ (6) | $Z_i$, (5) + (6) (7) | $Z_i$ (7)/(2) (8) |
|---|---|---|---|---|---|---|---|
| 1 | 6 | 2.9 | 3 | 15,000 | 2200 | 17,200 | 2866.66 |
| 2 | 6 | 6 | 6 | 16,666.6 | 2700 | 19,366 | 3227.66 |
| 3 | 4 | 3.8 | 4 | 12,000 | 3600 | 15,000 | 3900 |
| 4 | 1 | 0.3 | 1 | 4000 | 900 | 4.900 | 4900 |

**TABLE 9.12**
**Ranked Moves and Their Parameters for a Selected Unit of Equipment 1, Iteration II**

| Move | $W_{1j}$ | $h_{1j}$ | $\Sigma h_{1j}$ |
|---|---|---|---|
| 5 | 100 | 0.3 | 0.3 |
| 6 | 200 | 0.7 | 1.0 |
| 1 | 400 | 0.4 | |
| 3 | 400 | 0.5 | |
| 4 | 500 | 0.4 | |
| 2 | 600 | 0.6 | |

Equipment 1 has the smallest $\bar{Z}_i$. Table 9.12 lists the moves that can be made by this equipment in ascending order of their move cost ($W_{ij}$). Moves 5 and 6 are assigned to one unit of equipment 1.

If we continue in the same manner, the final assignment of the moves to the candidate equipment is:

| Move | Equipment |
|---|---|
| 1 | 1 |
| 2 | 1 |
| 3 | 1 |
| 4 | 1 |
| 5 | 1 |
| 6 | 1 |
| 7 | 4 |
| 8 | 4 |
| 9 | 2 |
| 10 | 4 |

The required number of units of each equipment type is 3 of type 1, 1 of type 2, and 1 of type 4. The resulting cost is $24,300. Table 9.13 is a summary of the interations performed in this example.

**TABLE 9.13**
**Summary of the Iterations for the Example Problem**

| Iteration | Smallest $\bar{Z}_i$ | Equipment Selected | Moves Assigned |
|---|---|---|---|
| 1 | 2450 | 4 | 7, 8, 10 |
| 2 | 2866 | 1 | 5, 6 |
| 3 | 2975 | 1 | 1, 3 |
| 4 | 3050 | 1 | 4,2 |
| 5 | 2977 | 2 | 9 |

The algorithm presented is of the construction type. An improvement algorithm has also been developed by Webster and Reed (1971). (For a discussion on construction and improvement types, see Section 13.4.) This algorithm first assigns moves to equipment on the basis of cost alone. The initial solution obtained is then improved by changing the assignment of moves to other equipment in an attempt to improve equipment utilization and thus reduce cost.

## 9.2 ROBOTS IN MATERIAL HANDLING

Robots are increasingly being used in material handling when it is profitable to do so. A discussion on robotics and its use in material handling is included in Chapter 3. In addition to economics, a few technical factors must be considered when selecting a robot. These include weight and the quantity of the items to be moved, the rate at which the items must be moved, the orientation of the object, the ease of handling the object by the robot, and working conditions.

Robots can lift loads ranging from a few grams to thousands of pounds. Naturally, the size and cost of the robot increase as its weight-lifting capacity increases. Robots can work in environments that might be hostile to humans, such as in a poisonous atmosphere and extreme temperatures. They are also used in manufacturing where there is an inert atmosphere, such as the production of medical drugs and high-purity silicon-wafer manufacturing. Robots can perform repeated operations without fatigue and can reach long distances with dexterity to manipulate or rotate an object in three dimensions. Robots can be programmed to perform certain steps logically; therefore, they can work easily with humans in a workstation or with other material-handling equipment.

Prominent examples of the use of robots in material handling include loading and unloading of machines, pallet loading, sorting and accumulating parts, and insertion of components in electronic assemblies.

### 9.2.1 Robot Grippers for Loading and Unloading Operations

Two types of hand grippers can be used in loading and unloading operations: one-handed grippers and two-handed grippers. In one-handed grippers, all operations are performed by one hand. As a result, the machine must wait for all the motions of the robot associated with unloading the finished part and then loading an unfinished part before it can start processing the part. In a two-handed gripper, however, one hand holds the unfinished part as the other hand unloads the finished part from the machine. The unfinished part is then loaded into the machine before the finished part is transported to the final destination. Thus, with two-handed grippers, the operations of moving the finished part from the machine to an unloading platform (e.g., conveyor), releasing the unit, moving to the loading platform, grasping the unfinished unit, and transporting the unfinished unit to the machine can all be performed while the machine is operating.

### 9.2.2 Calculation of Cycle Times

Production rate is an important criterion in any production facility, as is the number of times a robot is used to increase production. To determine what the production rate would

**TABLE 9.14**
**Elemental Operation Times for a Robot**

| Activities | Time (seconds) |
|---|---|
| Machine operation time | 25.0 |
| Times associated with the robot | |
| Unload a unit | 1.2 |
| Move to the finished part conveyor | 1.7 |
| Release the unit | 0.2 |
| Move the arm to the input conveyor | 2.6 |
| Pick up a unit | 0.3 |
| Move to the machine | 1.9 |
| Load the machine | 2.1 |

be using a robot, we must first decide the cycle time, or how long it takes to process an item through the setup consisting of machines and the robot. Operational time available per day divided by the cycle time in the same units gives us the production rate.

The following examples illustrate how cycle time is determined when a robot is used for loading and unloading operations. Table 9.14 displays the operation time of the sequential activities associated with using a robot to load and unload workpieces from a machine. The cycle time with a one-handed gripper is the sum of all activities:

$$25 + 1.2 + 1.7 + 0.2 + 2.6 + 0.3 + 1.9 + 2.1 = 35 \text{ seconds}$$

Cycle time with a two-handed gripper consists of only the processing time and times for unloading and loading a unit:

$$25 + 1.2 + 2.1 = 28.3 \text{ seconds}$$

The number of units produced in an 8-hour-per-day operation with 80% efficiency is:

$$\text{One-hand gripper} = \frac{60}{35 \text{ second/unit}} \times (60 \times 8) \text{ min /day} \times 0.8 = 658 \text{ units}$$

$$\text{Two-hand gripper} = \frac{60}{28.3} \times (60 \times 0.8) \times 0.8 = 814 \text{ units}$$

### 9.2.3 Sequential Loading

More often than not, a robot exceeds the requirements of serving a single machine and is idle for a large percentage of the time. To utilize a robot more fully, it can be programmed to perform loading and unloading operations for several machines. This is also called *sequential machine loading*. In this arrangement, several machines are placed in a circular arc so that the robot's arm can reach and perform loading and unloading operations on each machine. As in an assembly line, the operations to be performed on an item are divided into smaller elements, and each successive element is performed on each sequential machine. Thus, the output of machine $i$ is the input to machine $i + 1$. No in-process inventory is provided in between the

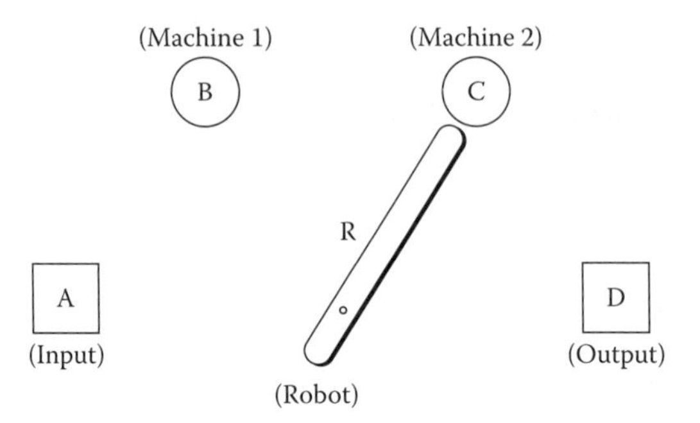


**FIGURE 9.1** Sequential robot operation serving two machines

machines, because if each machine had its own cycle time, an inventory would accumulate between the machines that would have no chance to be depleted. Thus, all machines in sequence have the same cycle time. Figure 9.1 shows an example of a two-machine operation.

A and D are, respectively, the input and output points for the units. Conveyors are universally used for supplying units to point A and carrying them away from point D. Machines 1 and 2 are labeled points B and C, respectively. For example, the machines may be numerically controlled milling, drilling, and/or boring machines. R is a robot that serves these machines.

In a continually operating mode, a fixed sequence of operations must be followed. When the operation in machine C is over, the robot must unload the unit and transfer it to D. Then it must travel back to machine B and — if the operation of machine B is over — unload the unit from B and load it in C. Next, it must travel back to A, pick up a unit, and load it in B. From B, it must go back to machine C and start the next cycle of actions. Thus, in a sequential arrangement, the next sequential station must be available before a processed unit can be moved forward.

In general, the stations are numbered from 1, 2, …, $N$, as in Figure 9.2, where 1 is the input station and $N$ is the output station. Starting with station $N - 1$ and proceeding backward, for any station $i$ from $N - 1$ to 2, when viewed as an independent station, the robot actions are:

**A actions**

1. Unload machine $i$.
2. Move arm to station $i + 1$.
3. Load station $i + 1$ (or for station $N$, place the unit on a conveyor).

**B actions**

1. Move arm to station $i - 1$.
2. Unload station $i - 1$ (or for station 1, pick up the unit from conveyor).
3. Move arm to station $i$.
4. Load station $i$.

However, if all the robot actions are considered together, both A action and B action steps are required for the last machine only. From there on (proceeding backward), the succeeding machine is always in an unloaded or empty state, and so only B actions

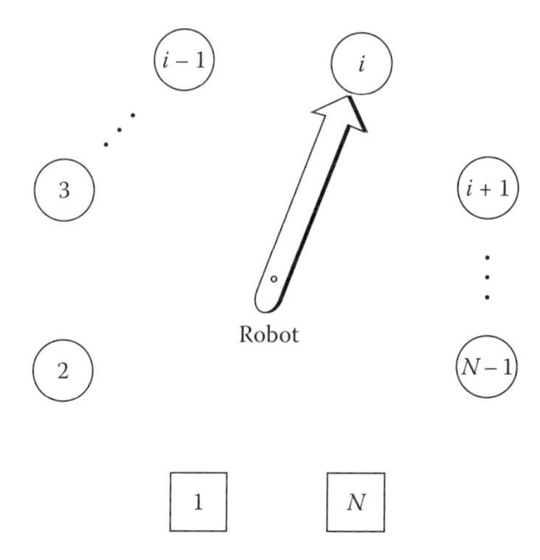


**FIGURE 9.2** Robot in sequential loading/unloading operations

are necessary for all preceding machines. The process continues by decreasing the value of index $i$ by 1 and reapplying the B action steps. When the first machine — that is, station 2 — is reloaded, the robot must go back to the last machine or to station $N-1$ and repeat the cycle.

The cycle time is the total time the unit spends in the system —that is, from the time it is picked up at input station 1 to the time it is released at output station $N$. This time could either be machine-dominant or robot-dominant time. In the case of machine dominance, the robot waits on one or more machines to complete their processing, while in robot dominance, the robot is so busy that one or more machines wait for its services. Time for each dominance is calculated, and the maximum time value among them defines the cycle time.

In determining the machine-dominant time, note that each machine must be served by a robot during its loading and unloading operations. Hence, the minimum time a unit spends with any one machine (assuming there is no delay because of the robot) is the time for processing plus the time for loading and unloading of the unit. When considered as an independent station, the unloading-loading operation involves both A actions and B actions. Once the sum of the process times and the seven-step loading and unloading times for each machine is calculated, the maximum value among them is the machine-dominant time.

The robot-dominant time is obtained by adding times associated with all robot actions in a cycle. Note that for an $N$-station line, there are $N - 2$ machines in a sequence. The robot is busy for all seven unloading-loading actions (A and B actions) with the last machine, and for the remaining $N - 2 - 1$ machines, it performs all B actions. In addition, at the end of the cycle, the robot must go from machine 2 to machine $N - 1$ or move across $N - 1 - 2$ machines. When all these times are added together, the total gives the robot-dominant time.

For instance, suppose we have the following data for an example problem (case I). The loading and unloading times are 0.1 minute each; the move time between two successive stations is 0.3 minute, and processing times on machines B and C are 5 and 3 minutes, respectively.

Here, the total number of stations $N$ is 4. There are $N - 2$, or 2, machines in sequence. The times for A action and B action motions are:

| Motions | Times (minutes) |
|---|---|
| **A actions** | |
| 1. Unload machine $i$ | 0.1 |
| 2. Move to station $i + 1$ | 0.3 |
| 3. Load station $i + 1$ | 0.1 |
| **B actions** | |
| 4. Move to station $i - 1$ (from station $i + 1$) | 0.6 (two-station move) |
| 5. Unload station $i - 1$ | 0.1 |
| 6. Move to station $i$ | 0.3 |
| 7. Load station $i$ | 0.1 |

Robot in production line machining coupling (courtesy of Manufacturing Engineering). A robot is used in two machines' sequential loading and unloading operations. A heavy workpiece (drum) is machined in two sequential machines. Note the U shape of the work cell, which allows the robot's arm to reach all the workstations. The output is placed on a table, from which it is moved by a hand truck.

This gives an A action total of 0.5 minutes, a B action total of 1.1 minutes, and a combined total of 1.6 minutes.

The machine-dominant times are:

For Machine B: $5 + 1.6 = 6.6$ minutes<br>
For Machine C: $3 + 1.6 = 4.6$ minutes

The robot-dominant time is given by the equation

$$(\text{A action times} + \text{B action times}) + (N - 2 - 1) \times \text{B action times} + (N - 2 - 1) \times \text{one-station move time}$$

Using the data, the robot-dominant time is

$$1.6 + 1 \times 1.1 + 1 \times 0.3 = 3.0 \text{ minutes}$$

The maximum of these is 6.6; hence, the cycle time is 6.6 minutes.

The entire operation can be simulated. Table 9.15 displays the simulation of the example problem. The notation for the columns is:

For robot

U = unit number
Ini = initial position
Fin = final position.
St = starting time
Act = activity time
Fin = completion time

**TABLE 9.15**
**Simulation of Robot Operation: Case I**

| Robot | | | | | | Stations | | | | | | | |
|---|---|---|---|---|---|---|---|---|---|---|---|---|---|
| | Position | | Times | | | B | | | C | | | D | |
| U | Ini. | Fin. | St. | Act. | Fin. | U | St. | Fin. | U | St. | Fin. | U | Fin. |
| 1 | A | B | 0 | 0.5 | 0.5 | 1 | 0.5 | 5.5 | | | | | |
| | B | C | 5.5 | 0.5 | 6.0 | | | | 1 | 6.0 | 9.0 | | |
| | C | A | 6.0 | 0.6 | 6.6 | | | | | | | | |
| 2 | A | B | 6.6 | 0.5 | 7.1 | 2 | 7.1 | 12.1 | | | | | |
| | B | C* | 7.1 | 0.3 | 7.4 | | | | | | | | |
| | C | D | 9.0 | 0.5 | 9.5 | | | | | | | 1 | 9.5 |
| | D | B | 9.5 | 0.6 | 10.1 | | | | | | | | |
| | B | C | 12.1 | 0.5 | 12.6 | | | | 2 | 12.6 | 15.6 | | |
| | C | A | 12.6 | 0.6 | 13.2 | | | | | | | | |
| 3 | A | B | 13.2 | 0.5 | 13.7 | 3 | 13.7 | 18.7 | | | | | |
| | B | C* | 13.7 | 0.3 | 14.0 | | | | | | | | |
| | C | D | 15.6 | 0.5 | 16.1 | | | | | | | 2 | 16.1 |
| | D | B | 16.1 | 0.6 | 16.7 | | | | | | | | |
| | B | C | 18.7 | 0.5 | 19.2 | | | | 3 | 19.2 | 22.2 | | |
| | C | A | 19.2 | 0.6 | 19.8 | | | | | | | | |
| 4 | A | B | 19.8 | 0.5 | 20.3 | 4 | 20.3 | 25.3 | | | | | |
| | B | C* | 20.3 | 0.3 | 20.6 | | | | | | | | |
| | C | D | 22.2 | 0.5 | 22.7 | | | | | | | 3 | 22.7 |
| | D | B | 22.7 | 0.6 | 23.3 | | | | | | | | |
| | B | C | 25.3 | 0.5 | 25.8 | | | | 4 | 25.8 | 28.8 | | |
| | C | A | 25.8 | 0.6 | 26.4 | | | | | | | | |
| 5 | A | B | 26.4 | 0.5 | 26.9 | 5 | 26.9 | 31.9 | | | | | |
| | B | C* | 26.9 | 0.3 | 27.2 | | | | | | | | |
| | C | D | 28.8 | 0.5 | 29.3 | | | | | | | 4 | 29.3 |

For stations:

U = unit number
St = time machine starts processing
Fin = time when processing is complete

The cycle time is obtained by subtracting the completion times of two successive units (station D) after the steady state is obtained, namely, 29.3 − 22.7 = 6.6 minutes. The influence of initial empty machines have vanished rather rapidly — that is immediately after the entire system is loaded with one unit in each machine.

**TABLE 9.16**
**Simulation of Robot Operation: Case II**

| Robot | | | | | | Stations | | | | | | | |
|---|---|---|---|---|---|---|---|---|---|---|---|---|---|
| | Position | | Times | | | B | | | C | | | D | |
| U | Ini. | Fin. | St. | Act. | Fin. | U | St. | Fin. | U | St. | Fin. | U | Fin. |
| 1 | A | B | 0 | 0.5 | 0.5 | 1 | 0.5 | 1.0 | | | | | |
| | B | C | 1.0 | 0.5 | 1.5 | | | | 1 | 1.5 | 2.5 | | |
| | C | A | 1.5 | 0.6 | 2.1 | | | | | | | | |
| 2 | A | B | 2.1 | 0.5 | 2.6 | 2 | 2.6 | 3.1 | | | | | |
| | B | C* | 2.6 | 0.3 | 2.9 | | | | | | | | |
| | C | D | 2.9 | 0.5 | 3.4 | | | | | | | 1 | 3.4 |
| | D | B | 3.4 | 0.6 | 4.0 | | | | | | | | |
| | B | C | 4.0 | 0.5 | 4.5 | | | | 2 | 4.5 | 5.5 | | |
| | C | A | 4.5 | 0.6 | 5.1 | | | | | | | | |
| 3 | A | B | 5.1 | 0.5 | 5.6 | 3 | 5.6 | 6.1 | | | | | |
| | B | C* | 5.6 | 0.3 | 5.9 | | | | | | | | |
| | C | D | 5.9 | 0.5 | 6.4 | | | | | | | 2 | 6.4 |
| | D | B | 6.4 | 0.6 | 7.0 | | | | | | | | |
| | B | C | 7.0 | 0.5 | 7.5 | | | | 3 | 7.5 | 8.5 | | |
| | C | A | 7.5 | 0.6 | 8.1 | | | | | | | | |
| 4 | A | B | 8.1 | 0.5 | 8.6 | 4 | 8.6 | 9.1 | | | | | |
| | B | C* | 8.6 | 0.3 | 8.9 | | | | | | | | |
| | C | D | 8.9 | 0.5 | 9.4 | | | | | | | 3 | 9.4 |
| | D | B | 9.4 | 0.6 | 10.0 | | | | | | | | |
| | B | C | 10.0 | 0.5 | 10.5 | | | | 4 | 10.5 | 11.5 | | |
| | C | A | 10.5 | 0.6 | 11.1 | | | | | | | | |
| 5 | A | B | 11.1 | 0.5 | 11.6 | 5 | 11.6 | 12.1 | | | | | |
| | B | C* | 11.6 | 0.3 | 11.9 | | | | | | | | |
| | C | D | 11.9 | 0.5 | 12.4 | | | | | | | 4 | 12.4 |

The element times are:

| | | |
|---|---|---|
| Load/unload transport motions: | A to B, B to C, C to D | $0.1 + 0.3 + 0.1 = 0.5$ |
| Move two stations: | C to A, D to B | $2 \times 0.3 = 0.6$ |
| One station: | B to C* | 0.3 |
| Processing times: | Station B | 5.0 |
| | Station C | 3.0 |

Cycle time = 29.3 – 22.7 = 6.6 minutes

In the preceding problem, if the processing times on machines B and C are 0.5 and 1 minute, respectively (case II), then the cycle time is the robot-dominated time of 3.0. Again, the reader is referred to the simulation in Table 9.16. The cycle time from the simulation is 12.4 – 9.4 = 3 minutes.

The element times are:

| | | |
|---|---|---|
| Load/unload transport motions: | A to B, B to C, C to D | $0.1 + 0.3 + 0.1 = 0.5$ |
| Empty two-station moves: | C to A, D to B | $2 \times 0.3 = 0.6$ |
| Empty one-station move: | B to C* | 0.3 |
| Processing times: | Station B | 0.5 |
| | Station C | 0.5 |

Cycle time = 12.4 – 9.4 = 3.0 minutes

## 9.3 AGVs IN MATERIAL HANDLING

The use of AGVs in some material-handling situations is becoming quite economical. AGVs allow random and independent movement of parts or material between stations. Such flexibility is important for responding to changing work loads and product mix, which are common in a flexible manufacturing system. A system using an AGV can also react efficiently to changes such as when the regular machine is busy or under repair and another machine has to perform the task, and/or the material has to be delivered to a different location following an alternative path.

Types of AGVs that are typically found in manufacturing plants are as follows:

1. *AGV towing vehicles:* These are the earliest types of AGVs, with capacities ranging from 1000 to 50,000 pounds. They are used primarily in the movement of bulk material in and out of warehouse areas.
2. *AGV unit load vehicles:* These AGVs are equipped with decks that carry a unit load. They are generally integrated with conveyor and automatic

storage-and-retrieval systems, making material movement in and out of a warehouse or a distribution system totally automatic.

3. *AGV fork trucks:* These AGVs function as forklift trucks with automatic pickup and drop-off of loads at various heights, ranging from the floor to standing level. These vehicles are especially useful when load-transfer heights vary from station to station.
4. *Light-load AGVs:* These AGVs can handle light loads of up to 500 pounds. The vehicles are light and compact and can maneuver in tight areas. They are used mainly in light manufacturing.
5. *AGV assembly line vehicles:* These vehicles are modifications of light-load AGVs in which major subassemblies, such as motors or transmissions, are carried from station to station and parts are added on at each station to make the complete assembly.

In a plant where the paths traveled by AGVs are not too complex and where multiple decisions are not required, the number of AGVs necessary to perform the material-handling tasks can be readily determined. Following is an illustration of this approach.

## DETERMINING THE REQUIRED NUMBER OF AGVs

Suppose an AGV system is chosen for delivering small parts from storage to four shops in a plant, as shown in Figure 9.3. (Most guided vehicles can carry a load of 1000 to 4000 pounds.) Determine the number of vehicles needed, given the information in Table 9.17. (Assume that each vehicle can carry parts to and from only one shop at a time.) The AGVs under consideration can travel at a rate of 200 feet/minute. (Between 200 and 260 feet/minute is the normal speed for an AGV.)

### Solution

The steps involved in the procedure are as follows. First, determine the total time needed to travel from storage to each destination (round trip). To this we must add the time for loading and unloading. Using the frequency of trips, determine the total vehicle use time. Divide this number by the traffic congestion factor, generally 0.85, to determine the total required time in minutes. Divide the total load by 60 minutes per hour to obtain the number of vehicles necessary.

As shown here, the time to travel to and from a shop to storage is calculated by dividing the total distance between them by the average speed of the vehicles, to which

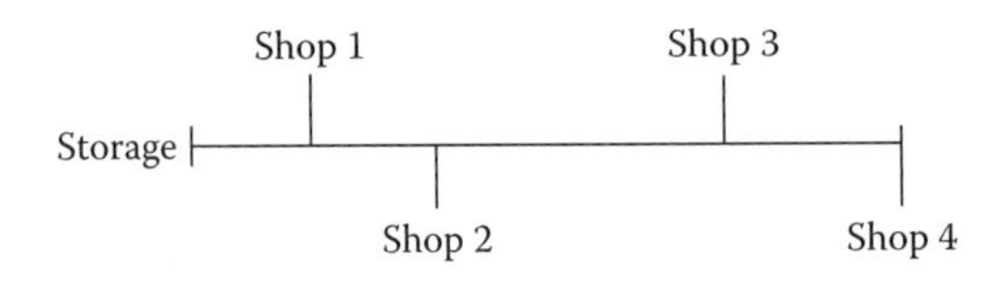


**FIGURE 9.3** Departmental layouts

**TABLE 9.17**
**AGV Distribution System Data**

| Shop Number | Distance from Storage (feet) | Number of trips (per hour) | Loading/Unloading (minutes) |
|---|---|---|---|
| 1 | 300 | 14 | 2 |
| 2 | 500 | 16 | 3 |
| 3 | 600 | 5 | 3 |
| 4 | 800 | 8 | 2 |

we must add loading and unloading times to obtain the necessary time per trip.

| Shop | Travel Time Calculations | Minutes of AVG Use per Hour |
|---|---|---|
| 1 | $\frac{600\times 14}{200}+14\times 2\times 2$ | 98 |
| 2 | $\frac{1000\times 16}{200}+16\times 3\times 2$ | 176 |
| 3 | $\frac{1200\times 5}{200}+5\times 3\times 2$ | 60 |
| 4 | $\frac{1600\times 8}{200}+8\times 2\times 2$ | 96 |
| | **Total** | 430 |

Adding all these times results in the total vehicle use time of 430 minutes. With 0.85 as the traffic congestion factor, the total load is 430/0.85 = 505 minutes. The number of AGVs required would be 505/60 = 8.43 ≈ 9.

### 9.3.1 Introducing an AGV in the Manufacturing Plant

Installation of an AGV system in a large manufacturing plant to serve multiple routes requires many decisions that are not immediately obvious. We must be able to answer questions such as:

- What should be the configuration of the AGV's path?
- How many AGVs should be in the system and what should be their capacities?
- What control system (rules) should be used in dispatching an AGV to a station?
- If more than one station is demanding the services of an AGV, what priority rule should be followed in selecting the first station to visit?

- How large should the in-process storage banks be for incoming and outgoing units at each station?
- Which route should an AGV follow to go from a point of origin to the destination?
- What rules should an AGV follow at the intersection of the routes to avoid collision or blocking of the paths?

Alternatives are numerous. For example, although there is no best method for vehicle dispatching, some suggested rules are:

1. Random vehicle rule: Assign any vehicle randomly.
2. Nearest vehicle (NV) rule: Assign the vehicle that is closest to the demand point.
3. Longest idle vehicle rule: The vehicle that has remained idle for the longest period is assigned first.
4. Least utilized vehicle rule: The vehicle with least utilization at the demand time is assigned next. This rule tries to equalize the work load on all vehicles.

Similarly, there are some well-known rules for prioritizing stations if more than one station requests the services of an AGV at the same time. Some of these rules are:

1. Random work center rule: Select a station at random.
2. Shortest travel time/distance rule: Assign a station that is closest to an available vehicle based on either distance or travel time.
3. Maximum outgoing queue size rule: Dispatch a vehicle to a station with the maximum number of unit loads in its output inventory.
4. Minimum remaining outgoing queue space rule: Dispatch a vehicle to a station that has minimum space remaining in the space for outgoing inventory.

To avoid collision, no two vehicles are allowed to travel on the same segment of the path at the same time. How large this segment should be is another factor that must be considered in detailing the layout of the guided path. Rules can also be developed for path-segment selection or path-intersection selection by the vehicles. For example, if a path segment is required by two or more vehicles simultaneously, it may be assigned to a vehicle based on one of the following:

1. Random selection: A vehicle is randomly selected.
2. Full vehicle (FV): A vehicle carrying a load is given priority over an empty vehicle. Among the loaded vehicles, assignment is random.
3. Critical station response: A vehicle responding to a critical station is given priority.
4. Critical part (CP): A vehicle carrying a critical part is given priority.

## 9.4 SIMULATION OF AN AGV MATERIAL-HANDLING SYSTEM

The selection of the values or attributes of the factors just mentioned must be made simultaneously, because these factors may influence each other. Simulation is the most commonly used technique to duplicate the workings of a system operating under some specific rules. It allows us to experiment with the system without physically building it. If the outcome is not satisfactory, the operating rules can be changed and the system can be simulated again. Because no physical system has been put in existence yet, no substantial loss is incurred. The process may be continued until we obtain a set of rules that satisfies our requirements.

Many commercially developed software languages can be used to model the physical layout and evaluate different operating rules. For example, simulation languages such as SLAM II and ARENA have special blocks or commands for AGV simulation. For those not proficient in simulation techniques, a black-box approach called *automatic simulation* is suggested by some developers. Such simulators create a generic simulation code for AGV operations.

Although a computer is commonly used for this purpose, we present here an example of a paper-and-pencil simulation to illustrate the basic principles. If required, it is not too difficult to write a computer program once the procedure is understood.

Suppose an AGV material-handling system is installed in a flexible manufacturing system (FMS) environment. The layout consists of three machines, M1, M2, and M3, connected by an AGV track, as shown in Figure 9.4. For stations M1, M2, and M3, respectively, stations a, c, and e serve as the input points, and stations b, d, and f serve as the output points. Input to the system arrives on a conveyor at station A, and output from the system is carried away at station B.

There are two AGVs, V1 and V2, used in material handling. There are 12 path segments; some are bidirectional, but others are unidirectional, as shown by the

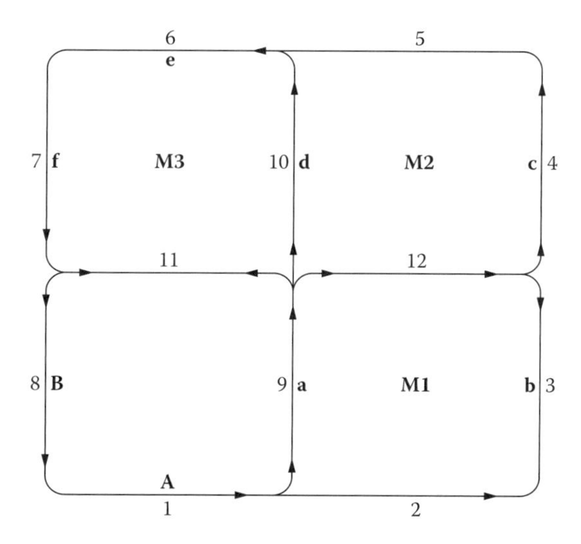


**FIGURE 9.4** Layout of paths for the AGV system

**TABLE 9.18**
**Machine and Time Requirements for Jobs**

| Job | Machine Sequence | Processing Times |
|---|---|---|
| J1 | M1, M3 | 5, 7 |
| J2 | M1, M2 | 6, 3 |
| J3 | M3 | 8 |

arrows in Figure 9.4. It takes AGVs 2 minutes to travel each path segment. All input and output stations are in the center of the corresponding path segment; for instance, station a is in the center of segment 9.

There are three jobs, J1 J2, and J3, being processed in the system. The machine requirements and corresponding processing times (including setup) for each job are as shown in Table 9.18. Jobs arrive at station A with a predetermined time between arrivals, shown in Table 9.19. Loading and unloading takes 2 minutes on any station.

These rules are used in operating the system:

1. The nearest vehicle (NV) responds to a demand. If there is no demand for the vehicle, the vehicle returns to the home base at station A.
2. If a machine output station and the system input station A are waiting for an AGV at the same time, preference is given to transport the partially or fully completed job at the machine output station. If more than one output station is waiting for an AGV, the shortest travel distance (STD) station preference rule is applied.
3. On a path segment, a full vehicle (FV) gets preference.

A segment of the simulation is shown in three separate tables. We could have displayed the entire simulation in one table with proper choice of columns. However, for convenience of space, the information is divided into three main categories and displayed in separate tables. For connectivity between tables, however, line numbers and clock times are repeated in each table. The table associated with vehicle loading

**TABLE 9.19**
**Time Between Arrivals**

| Job | Time Between Arrivals (minutes) |
|---|---|
| J1 | 10 |
| J2 | 15 |
| J3 | 20 |

shows the travels of the two vehicles. The travel paths must be checked to ensure that no two vehicles travel the same path segment simultaneously. This is checked in the path-loading table (PLT). The machine-loading table tabulates the activities associated with the three machines.

The nomenclature used in the columns of the tables is:

L = line number
CC = current clock

The L and CC columns are common for all tables: the vehicle-loading table, Table 9.20; the PLT, Table 9.21; and the machine-loading table, Table 9.22.

Two columns are associated with new job arrivals in the vehicle-loading table, Table 9.20:

JA = jobs arriving at station A
JQA = jobs in queue for pickup by an AGV at station A

Vehicle-related columns in the vehicle-loading table (Table 9.20) are

Job = job loaded on the vehicle
ST = start station for the vehicle
EN = end station for the vehicle
STT = operational start time for the vehicle (If loading is involved, this is the start time of the loading; if there is no loading, this is the start time of the trip.)
L/UL = load/unload time, if loading and unloading is required
PT = path segments traveled by the AGV
TrT = travel time on the path
Comp = completion time for the use of the AGV (This time is equal to EndT from the PLT plus the unloading time if the trip involves transportation of a job.)

The columns in the PLT (Table 9.21) are:

V = vehicle used on the path
ST = starting station of the travel path
End = ending station of the travel path
Path = the path segments on which the vehicle will travel
SST = scheduled starting time for the vehicle (For a loaded trip, this time is STT + loading time [1 minute in our case].)
EndT = scheduled completion time of the trip (The time should include any delay due to path blockage.)

Also recorded for each path are the starting time of the vehicle on the path (S) and the finishing time of the travel of the vehicle on that path (F). Table 9.22 includes

**TABLE 9.20**
**Vehicle-Loading Table (VLT)**

| Common | | Job | | Vehicle 1 Related | | | | | | | | Vehicle 2 Related | | | | | | | |
|---|---|---|---|---|---|---|---|---|---|---|---|---|---|---|---|---|---|---|---|
| L | CC | JA | JQA | Job | ST | EN | STT | L/UL | PT | TrT | Comp | Job | ST | EN | STT | L/UL | PT | Trt | Comp |
| 1 | 10 | 1 | 1 | 1 | A | a | 10 | 2 | 1,9 | 2 | 14 | | | | | | | | |
| 2 | 14 | | | | a | A | 14 | | 9, 11, 8, 1 | 6 | 20 | | | | | | | | |
| 3 | 15 | 2 | 2 | | | | | | | | | 2 | A | a | 15 | 2 | 1, 9 | 2 | 19 |
| 4 | 19 | | | | | | | | | | | | a | b | 19 | | 9,12<br>3 | 4 | 23 |
| 5 | 19 | | | | | | | | | | | | | | | | | | |
| 6 | 20 | 1,3 | 1,3 | | | | | | | | | | | | | | | | |
| 7 | 20 | | 3 | 1 | A | a | 20 | 2 | 1, 9 | 2 | 24 | | | | | | | | |
| 8 | 23 | | | | | | | | | | | 1 | b | e | 23 | 2 | 3, 4, 5, 6 | 6 | 31 |
| 9 | 24 | | | | a | A | 24 | | 9, 11, 8, 1 | 6 | 30 | | | | | | | | |
| 10 | 25 | | | | | | | | | | | | | | | | | | |
| 11 | 30 | 1, 2 | 3, 1<br>2 | | | | | | | | | | | | | | | | |
| 12 | 30 | | 1, 2 | 3 | A | e | 30 | 2 | 1, 9, 10, 6 | 6 | 38 | | | | | | | | 39 |
| 13 | 30 | | | | | | | | | | | | | | | | | | |
| 14 | 31 | | | | | | | | | | | | e | b | 31 | | 6, 7, 11, 12, 3 | 8 | 39 |
| 15 | 38 | | | | | | | | | | | | | | | | | | |
| 16 | 38 | | | | e | f | 38 | | 6, 7 | 2 | 40 | | | | | | | | |
| 17 | 39 | | | | | | | | | | | 2 | b | c | 39 | 2 | 3, 4 | 2 | 43 |
| 18 | 40 | 1, 3 | 1, 2<br>1, 3 | | | | | | | | | | | | | | | | |

**TABLE 9.21**
**Path-Loading Table (PLT)**

| Common | Vehicle | | | | | | | Starting and Finishing Time of Travel on Path Segments | | | | | | | | | | | | | | | | | | | | | | | |
|---|---|---|---|---|---|---|---|---|---|---|---|---|---|---|---|---|---|---|---|---|---|---|---|---|---|---|---|---|---|---|---|
| | Travel Stations | | | | | | | 1 | | 2 | | 3 | | 4 | | 5 | | 6 | | 7 | | 8 | | 9 | | 10 | | 11 | | 12 | |
| L | CC | V | ST (From) | END (To) | PATH | SST | ENDT | S | F | S | F | S | F | S | F | S | F | S | F | S | F | S | F | S | F | S | F | S | F | S | F |
| 1 | 10 | 1 | A | a | 1, 9 | 11 | 13 | 11 | 12 | | | | | | | | | | | | | | | 12 | 13 | | | | | | |
| 2 | 14 | 1 | a | A | 9, 11, 8, 1 | 14 | 20 | 19 | 20 | | | | | | | | | | | | | 17 | 19 | 14 | 15 | | | 15 | 17 | | |
| 3 | 15 | 2 | A | a | 1, 9 | 16 | 18 | 16 | 17 | | | | | | | | | | | | | | | 17 | 18 | | | | | | |
| 4 | 19 | 2 | a | b | 9, 12, 3 | 19 | 23 | | | | | 22 | 23 | | | | | | | | | | | 19 | 20 | | | 20 | 22 | | |
| 7 | 20 | 1 | A | a | 1, 9 | 21 | 23 | 21 | 22 | | | | | | | | | | | | | | | 22 | 23 | | | | | | |
| 8 | 23 | 2 | b | e | 3, 4, 5, 6 | 24 | 30 | | | | | 24 | 25 | 25 | 27 | 27 | 29 | 29 | 30 | | | | | | | | | | | | |
| 9 | 24 | 1 | a | A | 9, 11, 8, 1 | 24 | 30 | 29 | 30 | | | | | | | | | | | | | 27 | 29 | 24 | 25 | | | 25 | 27 | | |
| 12 | 30 | 1 | A | e | 1, 9, 10, 6 | 31 | 37 | 31 | 32 | | | | | | | | | 36 | 37 | | | | | 32 | 34 | 34 | 36 | | | | |
| 14 | 31 | 2 | e | b | 6, 7, 11, 12, 3 | 31 | 39 | | | | | 38 | 39 | | | | | 31 | 32 | 32 | 34 | | | | | | | 34 | 36 | 36 | 38 |
| 16 | 38 | 1 | e | f | 6, 7 | 38 | 40 | | | | | | | | | | | 38 | 39 | 39 | 40 | | | | | | | | | | |
| 17 | 39 | 2 | b | c | 3, 4 | 40 | 42 | | | | | 40 | 41 | 41 | 42 | | | | | | | | | | | | | | | | |

**TABLE 9.22**
**Machine-Loading Table (MLT)**

| Common | | Machines | | | | | | | | | | | | | | | | | |
|---|---|---|---|---|---|---|---|---|---|---|---|---|---|---|---|---|---|---|---|
| | | M1 | | | | | | M2 | | | | | | M3 | | | | | |
| LN | CC | JIQ a | JM | ST | PR | CMP | JOQ b | JIQ c | JM | ST | PR | CMP | JOQ d | JIQ e | JM | ST | PR | CMP | JOQ f |
| 1 | 10 | | | | | | | | | | | | | | | | | | |
| 2 | 14 | | 1 | 14 | 5 | 19 | | | | | | | | | | | | | |
| 3 | 15 | | | | | | | | | | | | | | | | | | |
| 4 | 19 | | | | | | 1 | | | | | | | | | | | | |
| 5 | 19 | | 2 | 19 | 6 | 25 | | | | | | | | | | | | | |
| 6 | 20 | | | | | | | | | | | | | | | | | | |
| 7 | 20 | | | | | | | | | | | | | | | | | | |
| 8 | 23 | | | | | | | | | | | | | | | | | | |
| 9 | 24 | 1 | 2 | | | | | | | | | | | | | | | | |
| 10 | 25 | | 1 | 25 | 5 | 30 | 2 | | | | | | | | | | | | |
| 11 | 30 | | | | | | | | | | | | | | | | | | |
| 12 | 30 | | | | | | | | | | | | | | | | | | |
| 13 | 30 | | | | | | 2 | | | | | | | | | | | | |
| 14 | 31 | | | | | | | | | | | | | | 1 | 31 | 7 | 38 | |
| 15 | 38 | | | | | | | | | | | | | | | | | | 1 |
| 16 | 38 | | | | | | | | | | | | | | 3 | 38 | 8 | 46 | |
| 17 | 39 | | | | | | | | | | | | | | | | | | |
| 18 | 40 | | | | | | | | | | | | | | | | | | |
| 19 | 40 | | | | | | | | | | | | | | | | | | |
| 20 | 43 | | | | | | | | 2 | 43 | 3 | 46 | | | | | | | |

the following columns for each machine:

JIQ = job(s) in associated input station queue waiting for the machine
JM = job being processed on the machine
ST = starting time for processing
PR = processing time for the job in the machine
CMP = job completion time on the machine
JOQ = job(s) waiting in the associated output station queue for an AGV

It is also convenient to work with a time line graph (Figure 9.5) at the same time. The graph shows the clock in increments of 1 and, as the simulation progresses, lists the arrivals of the jobs at appropriate clock times. It also lists the resources — that is, the vehicles and machines. When the resources are engaged, they are as such marked by a bar. The end arrow for the bar also indicates when the resource will be free and when it can start the operation on the next possible activity. The least clock value among the arrival times for the jobs that are not scheduled so far, and ending arrow points signifying the times of release of one or more resources, is the next time value for the clock in our simulation. If one or more activities are possible, they are performed at this clock value, and the appropriate time lines are drawn on the graph to signify the associated engagement of the resources. If no action is possible, the clock is advanced to the next immediate value when either arrival occurs or a resource is freed, and action is taken if possible. This process is repeated throughout the simulation.

### 9.4.1 Explanation of Simulation Logic

#### 9.4.1.1 Line I

At clock 10, job type 1, or J1, arrives at A (JA = 1). It immediately goes in the job queue at A (JQA = 1). Because both vehicles, V1 and V2, are available, V1 is chosen at random to carry the job to the input buffer of the first machine in sequence, station a. V1 starts at clock 10, takes 1 minute for loading and 1 minute for unloading, and takes 2 minutes to travel on paths 1 and 9 from A to a. Plotting the AGV travel on the PLT shows no conflict.

#### 9.4.1.2 Line 2

At clock 14, V1 starts its travel from station a back to station A. There is no loading or unloading involved in this activity. In addition, there is no conflict in PLT, and it takes 6 minutes to return on paths 9, 11, 8, and 1. M1 is loaded with job 1. It takes 5 minutes to process the job, so the machine finishes its task at clock 19.

#### 9.4.1.3 Line 3

At time 15, J2 arrives in the system. It joins the job queue at A; because V2 is available at A, V2 is assigned to transport the job to the input for the first machine in the sequence, station a. Starting time is 15, and ending time of the travel is 19. There is no path conflict based on the path table.

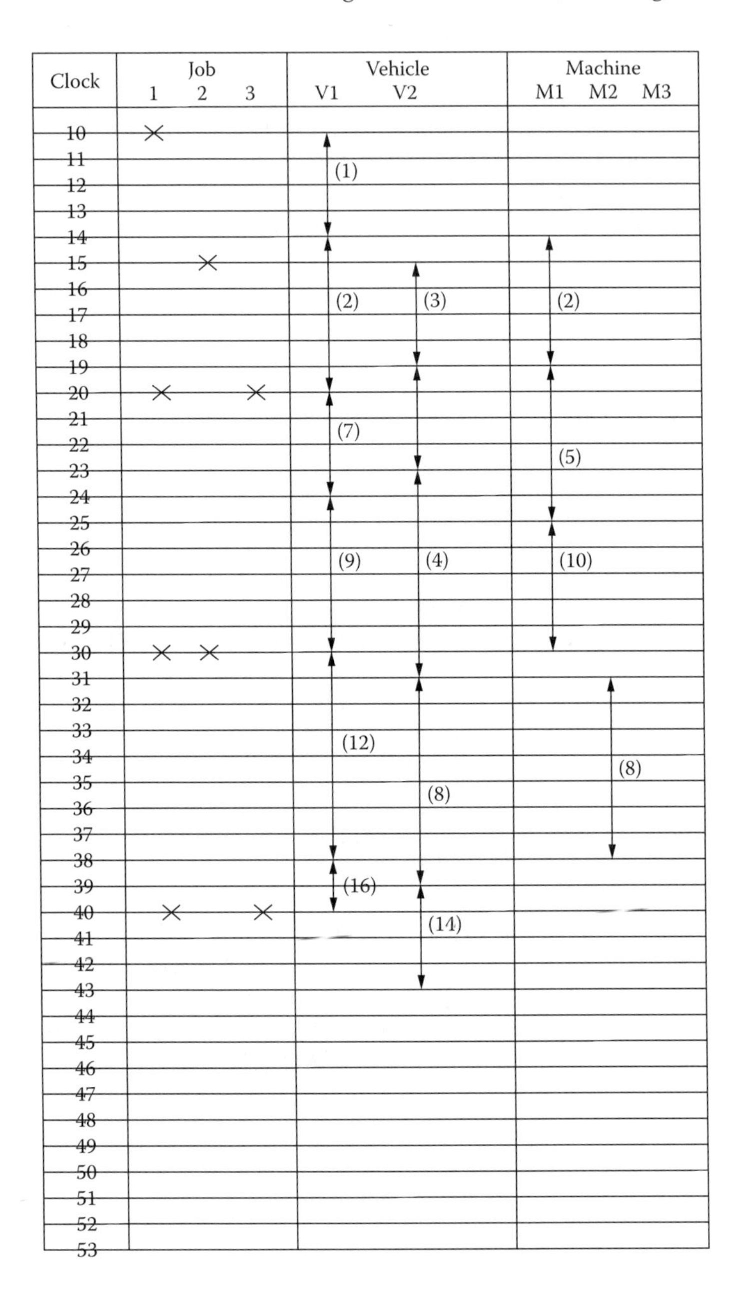


**FIGURE 9.5** Time line graph

### 9.4.1.4 Lines 4 and 5

Both lines describe the events that occur at clock 19. These two events have been separated in to two lines only for clarity. J1 on M1 completes its task at 19, at the same time V2 reaches station a with J2. J2 enters immediately into M1, since it becomes available. V2 is freed at station a, and being nearest to the output station b, where J1 is available, V2 is dispatched to b. There are no path conflicts.

#### 9.4.1.5 Lines 6 and 7

There are arrivals of J1 and J3 to the system. Because V1 is available at A, it picks up J1 and proceeds to a. There are no path conflicts.

It should be clear that if there is any path conflict, a resolution based on the path-selection rule should be made. Sometimes it may require the altering of completion times for a previously scheduled vehicle that has not yet completed its travel. Such changes are then entered in the associated tables and timeline diagram. From now on, although the path is checked for every vehicle move, we will note it only if conflict develops.

#### 9.4.1.6 Line 8

At time 23, V2 begins transfer of J1 from station b to station e.

#### 9.4.1.7 Line 9

After delivering J1 to station a, since there is no job waiting on any output queue, V1 is scheduled to return to its home station A.

#### 9.4.1.8 Line 10

J2 completes processing in M1; because there is no AGV scheduled to arrive at station b at clock 25, J2 is moved to the queue in output station b. J1 is waiting in the input station queue a, and hence it is loaded onto M1.

#### 9.4.1.8 Line 11

At time 30, two new jobs enter the system: J1 and J2. The job queue in station A now consists of J3, J1, and J2.

#### 9.4.1.9 Line 12

V1 is available at station A, and it is loaded with J3 and dispatched to station e.

#### 9.4.1.10 Line 13

At clock time 30, J1 completes processing in M1. Because there is no AGV scheduled to arrive at station b at clock 30, J1 is moved to the queue in output station b.

#### 9.4.1.11 Line 14

At clock 31, two jobs — J1 and J2 — are in the queue of output station b. V2 has completed its delivery and becomes free in station e. Because it is the closest free vehicle to the demand station b, it is scheduled to go to station b. Note that station b is chosen over station A, which also has J1 and J2 waiting in it, because station b has a partially completed job (rule 2). In PLT, path segment 6 is occupied by V1 from time 29 to time 30 and by V2 from time 31 to time 32. Thus, there is no conflict.

The processing in M3 starts on J1 just delivered by V2. It takes 7 minutes to process; hence, M3 is engaged till clock 38.

#### 9.4.1.12 Line 15

At clock time 38, M3 completes J1, and it is brought into output queue at station f.

#### 9.4.1.13 Line 16

At time 38, V1 unloads J3 in M3; being closest to station f, where a job is waiting, it starts toward station f.

#### 9.4.1.14 Line 17

At time 39, V2 starts the transportation of J2 from station b to station c.

#### 9.4.1.15 Line 18

At time 40, jobs 1 and 3 enter the system, enlarging the job queue at A to J1, J2, J1, and J3.

By now you have probably visualized how to advance the clock and perform the next required action. Look at the end times of all events — that is, events associated with the AGVs V1 and V2 and the processing on M1, M2, and M3. In addition, look for the arrival times of new jobs in the system. The least among these values is the time value for the next clock. The time line graph is the visual presentation of this situation. For example, to go from line 7 to line 8, look at the completion time of V1 (which is 24), V2 (which is 23), and M1 (which is 25); M2 and M3 are idle. The least of these time values is 23; therefore, the clock is incremented at 23. Look for all actions that are pending and/or possible. At time 23, for example, jobs 1 and 3 are in queue at station A, and job 1 is waiting at station b. V1 is at station b, and hence job 1 from station b is loaded onto V1; it is then scheduled for transporting the job to the input station of the next required machine, which is station e.

We stopped our simulation at time 40 (you may wish to continue further). It is obvious that the job queue at the input station A is growing at a faster rate than what the two AGVs can handle working under the present rules. We could add more AGVs, but that would be expensive. Perhaps, we should check the working policies. For example, having unidirectional paths on path segments 1 and 2 is adding to the travel times. Not looking ahead in scheduling AGVs is introducing inefficiencies. For example, J1 completes its service on M1 at time 25 (line 5), so rather than bringing the empty AGV back to A at time 24 (line 9), we should dispatch it to station b. Thus, our rule of bringing all empty AGVs back to the home station A does seem to be contributing to the delays in the system. Also, the fact that machines are two stations away from each other for input and output encourages empty AGV travel. This situation could be avoided by redesigning the layout of the system.

Even this small paper-and-pencil simulation has contributed to our understanding of the operation of the AGV system. It is not overly difficult to write a computer simulation program to evaluate a larger system for a longer period. We might also use commercially available programs, such as SLAM, that are especially user-friendly in AGV simulations.

## 9.5 SUMMARY

Selecting the right equipment for moving material is a challenging task. One must know what is to be moved, how frequently it is to be moved, what the physical limitations are, and what costs are involved. This chapter illustrates how this can be done with numerous examples. It begins with a simple model showing basic cost calculations and proceeds toward the more complex problem of equipment selection.

Work volume analysis is one technique that is used. It decides the number of handling units required to perform all necessary material moves if the type of equipment needed for each move is known. The heuristic procedure goes one step further and determines both equipment types and their number such that it minimizes the total cost of operation. Selection of AGVs involves a slightly different analysis and is also shown by an example.

More and more robots are being used in material handling, mainly for loading and unloading operations. A robot may work in a station consisting of a single machine or in a work cell with multiple machines arranged in a sequential manner. They are untiring workers and can be used in hostile and unpleasant environments. Their operating capacities change considerably, depending on whether they have a one-handed or a two-handed gripper. Evaluation of the production cycle time is very important in a robotic workstation. If a robot is a bottleneck, it can be replaced by a faster robot. If, however, the robot is not the cause of the bottleneck, then funds should not be wasted on a newer and better robot.

In flexible manufacturing, considerable versatility may be obtained with the use of AGVs in material handling. This chapter presents common rules that are used in designing an AGV system. The efficiency of an operation can be tested by using simulation. The chapter presents an example of manual simulation. A considerable amount of bookkeeping is required for such a simulation, but it is not an impossible task. There are commercially available computer programs that may be preferable because of ease of use. However, the example in the chapter demonstrates the fundamentals of such a simulation.

## PROBLEMS

1. What factors affect the selection of material-handling equipment?
2. A $10,000 tractor truck has an expected life of 8 years. The truck operates 280 days a year and travels an average of 15,000 feet per day. If the fuel costs about $9.00 a day and maintenance costs are $0.75 an hour, determine the cost per foot for operating this vehicle. Operator cost is $8.00/hour.
3. A plant produces 10,000 telephones per day. The dimensions of each telephone are $8 \times 3 \times 3$ inches. Management wishes to have a 1-week (5 days) supply in stock. One hundred telephones are packed in a box that forms a

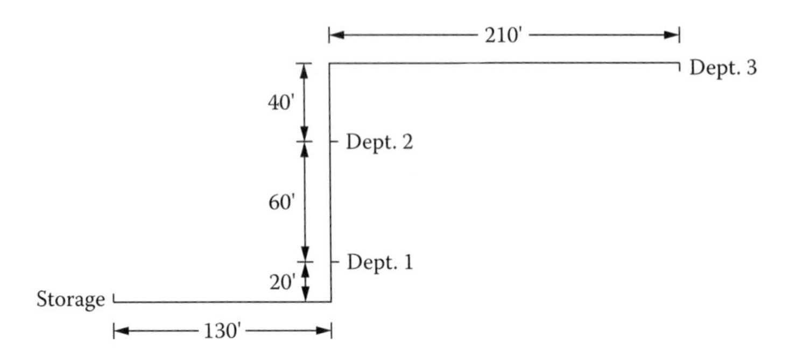


**FIGURE 9.6**

unit that may be stacked five high. Determine the floor space required to store the desired stock. Allow aisles of 10%.

4. Three different material-handling units are available for moving items of material. The costs of these lifts are \$3000, \$2500, and \$3300, respectively. These units must make six moves. Data on the operating cost, $W_{ij}$, and the operating time, $h_{ij}$, are given in the accompanying table. Determine which type of equipment should be used.

| Move | $W_{1j}$ | $h_{1j}$ | $W_{2j}$ | $h_{2j}$ | $W_{3j}$ | $h_{3j}$ |
|---|---|---|---|---|---|---|
| 1 | 50 | 0.8 | 40 | 0.5 | 80 | 0.7 |
| 2 | 60 | 0.7 | 70 | 0.3 | 50 | 0.8 |
| 3 | 70 | 0.4 | M | — | 30 | 0.9 |
| 4 | 80 | 0.5 | 100 | 0.3 | 60 | 0.2 |
| 5 | 100 | 0.1 | 50 | 0.2 | 40 | 0.1 |
| 6 | 40 | 0.2 | M | — | M | — |

5. Parts needed for production in departments 1, 2, and 3 are moved from storage by AGVs. Using the distance from the storage to each department as shown in Figure 9.6, and given the number of trips that will be made per hour to each department and the time needed for loading and unloading given in the accompanying table, determine how many vehicles are needed. Assume an AGV speed of 200 feet/minute and a congestion factor of 0.85.

| Department | Number of Trips per Hour | Loading/ Unloading (minutes) |
|---|---|---|
| 1 | 10 | 2 |
| 2 | 15 | 4 |
| 3 | 20 | 3 |

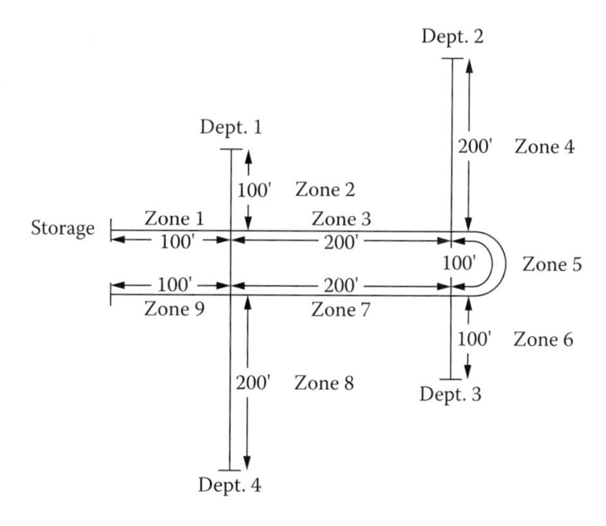


**FIGURE 9.7**

6. AGVs are being considered for use in moving supplies from storage to each of four departments. Using Figure 9.7 and given the number of trips made from storage each day and the loading time in the accompanying table, determine how many machines will be needed.

   Assume that we are working with a zone system allowing only one AGV to visit a zone at one time and that the demands are uniformly distributed; for example, for department 2, demand occurs at 8:00 a.m., 8:20 a.m., 8:40 a.m., and so on. The first demand from each department can be staggered according to number of trips; for example, the first demand from department 2 may be at 8:00 a.m. or 8:20 a.m., and so on. However, the earliest possible delivery is preferred. (The problem might involve scheduling.) Zone right-of-way rule: first come, first served (FCFS). Storage loading rule: 1 AGV at a time FCFS.

| Department | Number of Trips per Hour | Loading/Unloading (minutes) |
|---|---|---|
| 1 | 4 | 2 |
| 2 | 3 | 3 |
| 3 | 3 | 1 |
| 4 | 3 | 3 |

7. A robot is used for loading and unloading operations in a single machine-processing application. The linear speed of the robot is 0.1 meter/second. The perpendicular distances of the machine from input and output conveyor belts are 0.5 and 0.8, respectively. Also, the loading and unloading points subtend a right angle at the machine. The time required by the

robot to unload and load the unit is 2.5 seconds for each. The machine operation time is 30 seconds with 90% efficiency. Assuming the machine is not operating continuously, calculate the cycle time for the system. Suggest a method to reduce this cycle time and calculate the relative efficiency of this method.

8. A robot is used for material handling in a system of three machines, B, C, and D. A and E are input and output stations. The loading and unloading times are 0.2 minutes each. Movement between successive stations takes 0.3 minutes. Processing times on machines B, C, and D are 2, 3, and 4 minutes, respectively. Calculate the cycle time for the system.
9. Verify the cycle time calculated for the system in problem 8 by simulating the entire operation of the system.
10. An AGV system is installed to handle material handling in an FMS environment. The system consists of two machines connected by an AGV tract. Stations a and c serve as input for machines 1 and 2 and stations b and d serve as output for machines 1 and 2, respectively. Input to the system arrives on a conveyor belt at station A, and output is carried away from station B. There are three AGVs (V1, V2, and V3) in the system, Of the 12 path segments, only 9, 10, 11, and 12 are assumed to be bidirectional. V1, V2, and V3 take 2 minutes each on path segments. All input and output stations are assumed to be in the center of corresponding path segments (e.g., station a is in the center of segment 9). See Figure 9.8.

    There are two jobs being processed in the system:

| Job | M/C Sequence | Processing Time (minutes) |
|---|---|---|
| J1 | M1, M2 | 5, 10 |
| J2 | M2 | 6 |

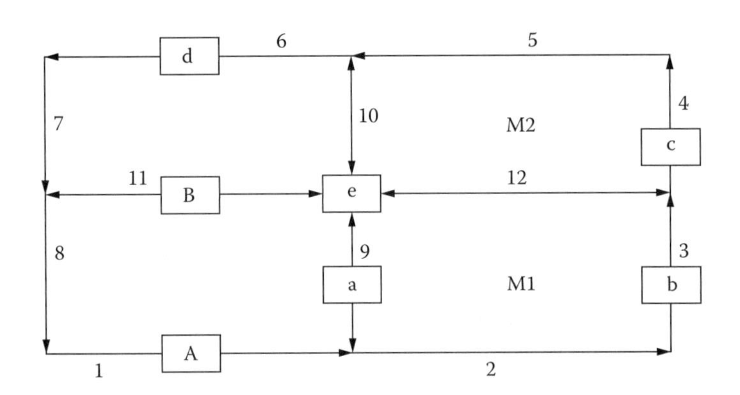


FIGURE 9.8

Jobs arrive at station A with the given times between arrivals:

| Job | Time (minutes) |
|---|---|
| J1 | 10 |
| J2 | 5 |

Loading and unloading takes 2 minutes on any station.
The following rules are used in the operating system:

1. The NV responds to demand.
2. If machine output station and system input station A are waiting for an AGV, then preference is given to the former. If more than one output station is waiting for an AGV, the STD station-preference rule is applied.
3. On a path segment, an FV gets preference.
   Simulate the system. Suggest measures to make the system practical.

13. Solve problem 10 with the following additional assumptions:
    1. There is a common buffer station e at the intersection of path segments 9, 10, 11, and 12.
    2. There are only two AGVs, V1 and V2, in the system instead of three.
    3. V1 has access to travel only in path segments 1, 2, 3, 12, 11, 8, and 9. V1 is normally housed at input station A and returns to input station A when no job is waiting for it in M1.
    4. V2 has access to travel only in path segments 5, 6, 7, 11, 12, 4, and 10. V2 is normally housed at common buffer station e and returns to station e when no job is waiting for it in M2.
    5. Output station B is shifted to the center of path segment 11 instead of path segment 8.

       Compare the results of problems 10 and 11 and discuss their relative advantages and disadvantages.

## SUGGESTED READINGS


Apple, J.M., *Material Handling Systems Design,* John Wiley, New York, NY, 1972.

*Basics of Material Handling,* The Material Handling Institute, Inc., 1973.

"Can you computerize equipment selection?," *Modern Materials Handling*, 21(1), 46–50, 1966.

Hassan, M.M.D., Hogg, G.L., and Smith, D.R., "A construction algorithm for the selection and assignment of materials handling equipment," *International Journal of Production Research*, 23(2), 381–392, 1985.

Keller, H.C., *Conveyors: Application and Design,* Ronald Press, New York, NY, 1967.

Matson, J.O., and White, J.A., "Operational research and material handling," *European Journal of Operational Research,* 11, 309–318, 1982.

Ravindran, A., Phillips, D.T., and Solberg, J.J., *Operations Research—Principles and Practices,* John Wiley, New York, NY, 1987.

Taha, H., *Operations Research,* 8th edition, Prentice Hall, Upper Saddle River, NJ, 2006.

Webster, D.B., and Reed, R., Jr., "A Material Handling System Selection Model," *AIEE Transactions,* 3(1), 13–21, 1971.

# 10 Material Handling *Flow Lines, Grouping, and Packaging*

This chapter introduces two topics. The initial discussion is concerned with concepts for arranging machines and in-process storage in order to make smooth material flow possible. Later in the chapter, we discuss issues associated with packaging.

Conveyors are extensively used in continuous manufacturing systems. The first section addresses the issues associated with development of flow and assembly lines. If the lines are not perfectly balanced, buffer spaces or banks must be provided. This section also presents helpful analytical techniques for balancing flow and assembly lines.

The second section introduces a method for machine grouping to reduce total material handling (MH) in flexible manufacturing. The procedure is a simple extension of the grouping procedure introduced in Chapter 5.

The third section introduces a heuristic method to arrange machines in a cell so that MH within the cell is minimized.

Finally, the chapter concludes with a discussion on packaging. Packaging can aid in the selling of the product in addition to influencing the MH cost. The need for packaging, the equipment available, and several means of reducing packaging costs are discussed.

## 10.1 FLOW PATTERN IN ASSEMBLY LINES

Assembly lines and process lines are both developed with continuous and smooth material flow in mind. Each workstation is arranged in a sequence based on the production flow to perform the next necessary operation. Conveyors play an important part in delivering material to and carrying it from each station. Additional automation should be sought to make MH as easy as possible within the workstations. In the following sections, we discuss in some detail how job stations might be arranged in different patterns to obtain a continuous and smooth flow of material among them.

### 10.1.1 Serial and Modular Conveyor Systems

As illustrated in Figure 10.1, an assembly line can be developed along a single conveyor or consist of a number of separable segments connected together, physically or functionally. There is a considerable degree of flexibility for changing the floor plan, machine arrangement, and operation sequence when the segments are thus connected.

A set of conveyors can also be arranged to attain a modular manufacturing system (MMS). An MMS uses general-purpose and even portable equipment within it, allowing for the production of many different items in the same physical setup.

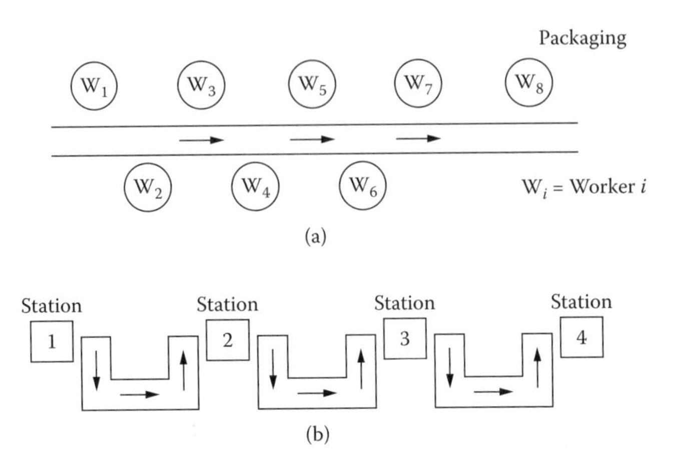


**FIGURE 10.1** Serial systems

For instance, as shown in Figure 10.2, by including a loop and thus providing a return path within the system, jobs requiring the machining sequence of 2, 1, 3, 4 can be as easily produced as those requiring a sequence of 1, 2, 3, 4. A robot-and-conveyor combination, shown in Figure 10.3, can also achieve an MMS. The robot can be programmed to select different loading/unloading sequences on the machines, based on the job requirements.

In an assembly operation requiring highly skilled workers, the unit completion time can vary widely, and roller conveyors are frequently used to connect the stations. The units are moved manually from one station to the next by pushing. Banks, which will be discussed in the next subsection, are a must for efficient operation of such a system.

A power roller, a belt, a chain, or a slat conveyor is often used in a self-paced assembly line. The conveyor serves stations by supplying an incoming unit or material and by carrying the finished item to the next station. In many instances, the final operation on the line is that of packaging.

Conveyor speeds can be adjusted to the desired level; however, there are two types of movement: intermittent and continuous.

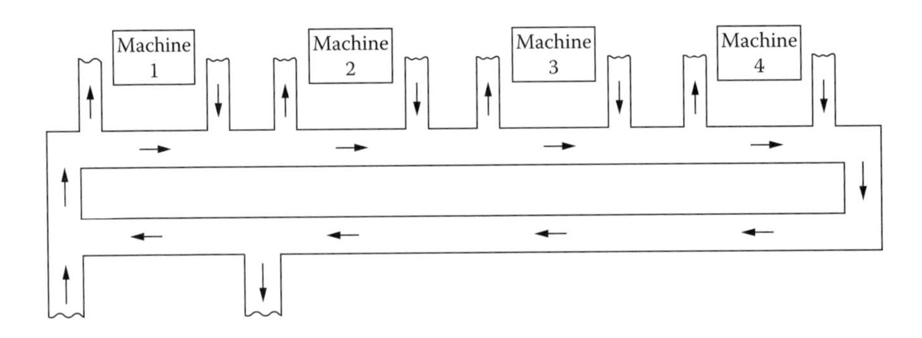


**FIGURE 10.2** A modular manufacturing system

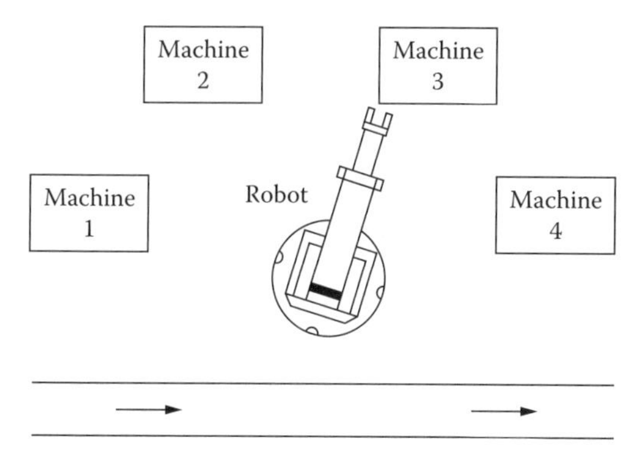


**FIGURE 10.3** Another formation of MMS

In a simple serial system in which the conveyor moves intermittently, the speed of the conveyor and the spacing of the items on the conveyor are adjusted to allow a part or subassembly to remain at a station for a cycle time. For example, with three workstations, each 3 feet long, and with a cycle time of 1 minute, the conveyor belt feeding the input and carrying the units to the next station requires a speed of 3 feet/minute. If the stations cannot be placed immediately adjacent to each other, the conveyor length should be 3 feet, or some integral multiple of that, between adjacent stations to permit smooth and continuous material flow. The units should be placed 3 feet apart on the conveyor. Figure 10.4 shows how the system might be arranged. Figure 10.4a shows one buffer item between workstations, and Figure 10.4b shows three buffer items. For a continuous-flow conveyor, such strict spacing might not be necessary as long as stations can work out of phase with each other; that is, stations can start and complete the work at times that might not be the same for each station.

### 10.1.2 Banking

Quite commonly, buffer spaces or banks are provided in conveyor systems to separate or decouple stations, allowing for the absorption of fluctuations in production rates within the workplaces. Such an arrangement also permits the cycle time to be the average production time of the slowest station rather than the slowest cycle time of the slowest station. In addition, a bank provides a space where in-process inspection between stations can be performed.

Banking can be achieved by either of two methods: by changing the product flow or by providing for overlapping operations. Each method has certain advantages and disadvantages, as will be indicated in the following discussion.

Figure 10.5 shows two methods of providing on-line space. In the first case, a rotating table serves as a temporary storage space on the conveyor. The operator selects a unit at random from the table, performs operations, and releases the unit on the other side of the conveyor (downstream). In the second case, a dam is created by placing across the conveyor a barrier such as a wooden plank ($2 \times 4$) or a steel bar, which stops the incoming units. The operator picks up a unit, completes his or her task, and places the item on the downstream side, allowing it to proceed to the next station.

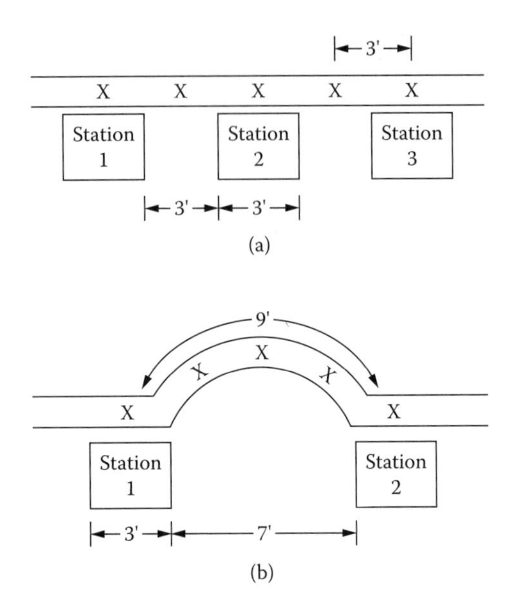


**FIGURE 10.4** Intermittently operated serial conveyor

Other means for providing banking are by off-line product flow. Three alternatives are illustrated in Figure 10.6. A U-shaped arrangement provides storage in the curve of the U, the accumulation line furnishes auxiliary conveyor storage, and a completely detached line separates the input and output areas and generates storage at each point.

Supplying such banks increases the time a unit is available for the required operation as well as the number of units available at any one time. On-line banks reduce minor shock due to employee absenteeism, machine breakdown, or other problems

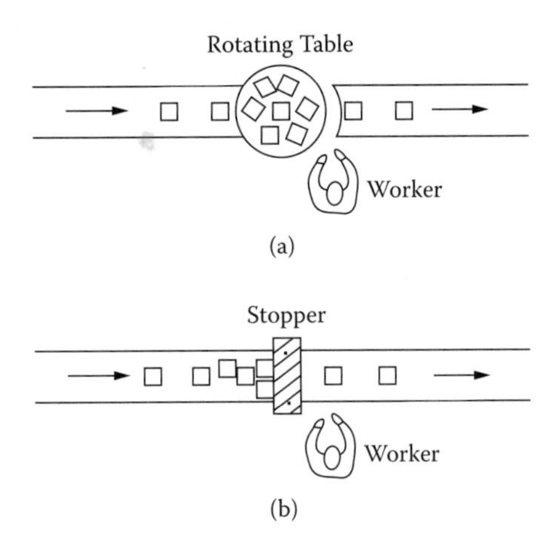


**FIGURE 10.5** Changing the product flow

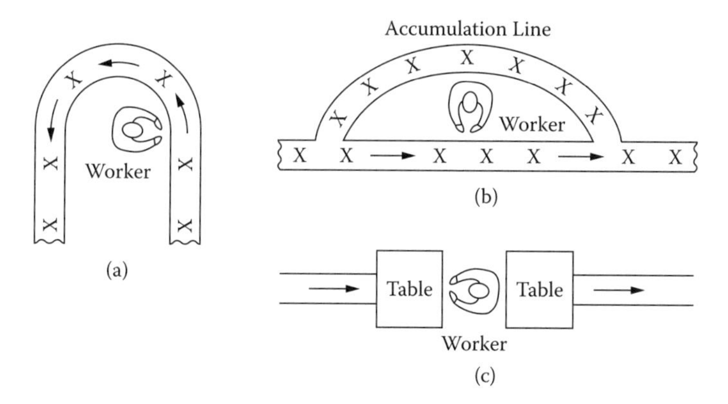


**FIGURE 10.6** Off-line banking

in the preceding station. Off-line banks require more floor space, increase MH, and perhaps require more complex scheduling.

Another method of providing banking is using overlapping operators by either one of the following procedures.

1. *Moving operator.* One operator, generally the supervisor, helps out in a group wherever and whenever assistance is required. In the auto industry, for example, it is common practice to provide seven operators for a six-station assembly line (Figure 10.7). The seventh person is used for relieving an operator, helping as needed, and training operators on the six stations in addition to making minor repairs and doing paperwork.
2. *Overlapping operation.* An example of overlapping operations, shown in Figure 10.8, occurs when the duties of a station are increased substantially and two additional workers are assigned there. The middle operator is responsible for the overall operation of the station; however, the first operator assists on the upstream portion of the task, while the last operator helps on the downstream portion. The overlap of duties permits cooperation between operators and irons out differences in their working speeds. The increase in operator cost is offset by an increase in efficiency and a reduction in the downtime of the assembly line. In addition, by

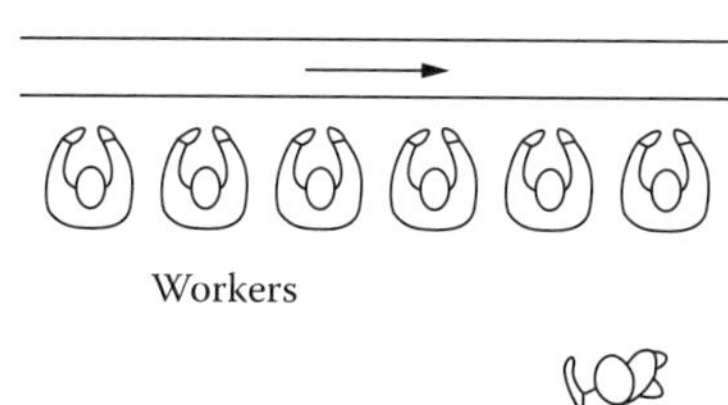


**FIGURE 10.7** A moving operator

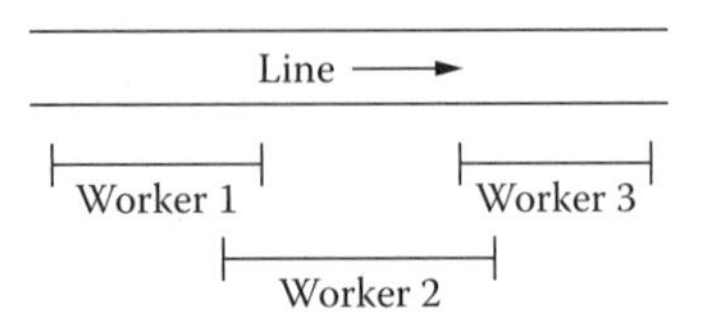


**FIGURE 10.8** Banking by overlapping operator

performing the tasks on the conveyors themselves, the operators require no increase in working space. The costs are normally monitored periodically to determine whether the arrangement continues to be efficient.

Many quantitative methods can be used to analyze layout and banking problems. Prominent among them are linear programming queuing theory and simulation. For example, banking is nothing more than providing a waiting space for the units as they arrive for the service at the facility. The question of how many waiting spaces are appropriate can be answered analytically either by applying queuing models or simulation, based on the nature of the available data. Queuing models, being analytical in nature, would give results that do not contain the sampling errors that the results from simulation may have. But the interactions between factors affecting the model are often too complex, so simulation may be the only way for analysis. The following examples illustrate a few situations.

#### 10.1.2.1 Machine Layouts

Location problem results in determining the optimum place for each machine to minimize the total material movement cost. The quadratic assignment problem (QAP) is a method by which a job shop layout is developed. The machine layout problem is a type of layout problem in which the optimum locations of machines are determined if they must be placed in one of the following arrangements (Figures 10.9–10.12):

1. Linear or single row
2. Loop or circular row
3. Double or multiple row serpentine flow
4. Backtracking

The mathematical formulation of the machine layout problem can be formed as:

$$\text{Minimize} \quad \sum\sum C_{ij}F_{ij}\,|X_i - X_j|$$

$$\text{Subject to} \quad |X_i - X_j| \geq \frac{1}{2}(l_i - l_j) + S$$

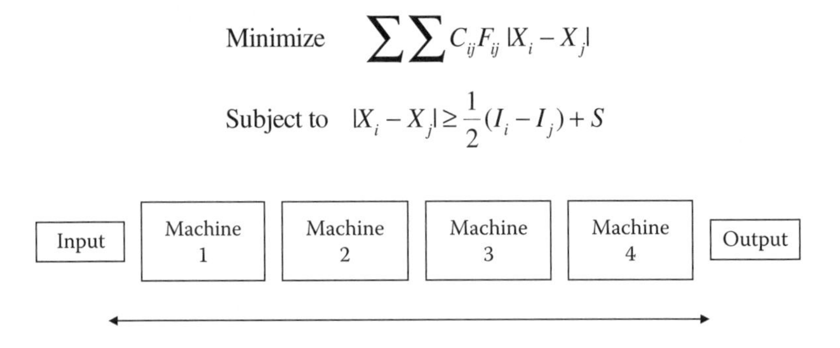


**FIGURE 10.9** Linear bidirectional flow

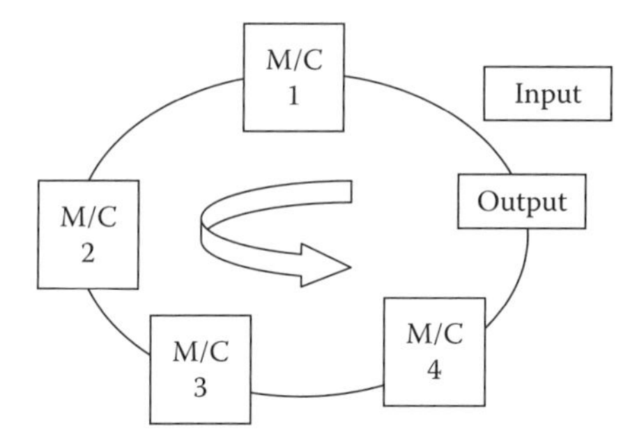


**FIGURE 10.10** Loop flow

where

$X_i$ = $x$-coordinate for the center of machine $i$
$F_{ij}$ = material flow between machines $i$ and $j$
$d_{ij}$ = distance between machines $i$ and $j$
$C_{ij}$ = cost of transporting a unit of material for unit distance between machines $i$ and $j$
$l_i$ = length of machine $i$ [MW]
$S$ = minimum space required between two machines.

The problem can be solved using linear programming. However, a simple heuristic procedure is promoted below.

Procedure:

1. Compute a matrix that shows the cost $C_{ij}F_{ij}$ from each machine $i$ to each machine $j$.
2. Select the largest value element from the matrix and place associated machine $i^*$ and $j^*$ adjacent to each other with $1/2(l_i{}^* + l_j{}^*) + S$ distance apart. The sequence is $i^*$-$j^*$.

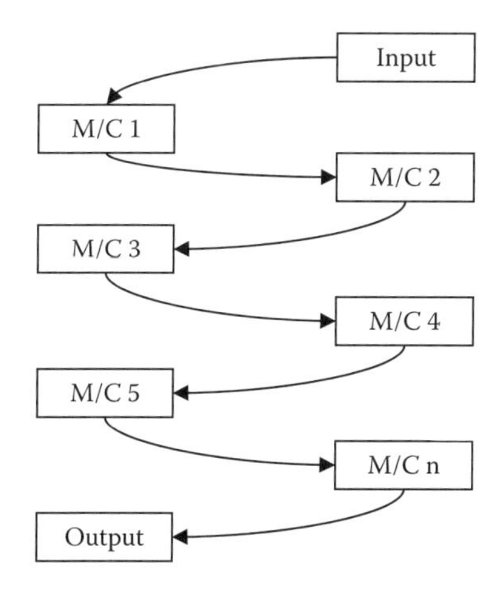


**FIGURE 10.11** Serpentine flow

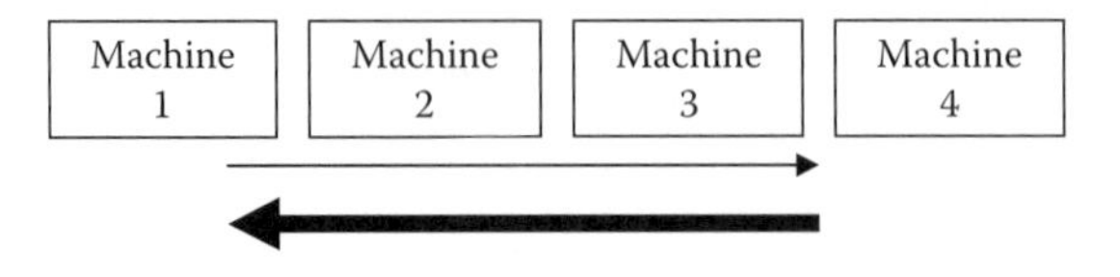


**FIGURE 10.12** Backtracking

3. Follow the remaining steps for each machine that is not yet placed. Select a machine, say *k*, and place it alternately on each side of the sequence developed so far. (Initially, the sequence is $i^*$-$j^*$ and, therefore, we need to try $k$-$i^*$-$j^*$ and $i^*$-$j^*$-$k$.) In each case, determine the flow cost by maintaining the minimum necessary separation between the machines. From among all the sequences thus tested, select the sequence that gives the minimum cost. The associated machine $k^*$ is then placed in the position that gave the minimum cost sequence. With the newly developed machine sequence, repeat step 3 again.
4. Continue the process until all machines are placed.

##### 10.1.2.1.1 Single-Row Layout Example

Let us consider four machines in a single-row problem. The following assumptions are made for the problem:

- The flow of the materials is bidirectional, that is, travel is permitted in either direction.
- The clearance between any two machines is 2 units.
- The cost of transporting a unit material for unit distance between two machines is $C_{ij} = 1$.

The following data gives the dimension of the machines:

| Machine | A | B | C | D |
|---|---|---|---|---|
| Dimensions | 2 × 2 | 4 × 4 | 4 × 4 | 6 × 6 |

The loaded MH trips between the machines are:

$$[F_{ij}] = \text{machine} \begin{array}{c} \\ A \\ B \\ C \\ D \end{array} \begin{array}{c} \begin{array}{cccc} A & B & C & D \end{array} \\ \begin{bmatrix} - & 10 & 15 & 12 \\ 5 & - & 20 & 18 \\ 3 & 15 & - & 21 \\ 6 & 3 & 17 & - \end{bmatrix} \end{array}$$

###### 10.1.2.1.1.1 Solution

Step 1: The first step is to construct a table, as shown below, showing the total flow cost $\{C_{ij} \times F_{ij}\}$ from each machine $i$ to each machine $j$. The matrix is symmetrical about the diagonal and, therefore, only the top or the bottom half needs to be displayed.

$$[F_{ij}] = \text{machine} \begin{array}{c} \\ A \\ B \\ C \\ D \end{array} \begin{array}{c} \begin{array}{cccc} A & B & C & D \end{array} \\ \begin{bmatrix} - & 15 & 18 & 18 \\ & - & 35 & 21 \\ & & - & 38 \\ & & & - \end{bmatrix} \end{array}$$

Step 2: The two machines with maximum flow cost are entered in the layout first. The selection of the machine pair with the maximum value of $C_{ij} \times F_{ij}$ which results in the maximum cost saving when the distance between them is set close. Here, the maximum flow cost is 38, between machines *C* and *D*. The placement of these two machines (C-D or D-C) does not affect the solution. This is because the measuring points are located at the midpoint of the machine edge along its width, resulting in no difference in total cost between the placement orders. The distance between the two machine centers is $[(4 + 6)/2] + 2 = 7$ units.

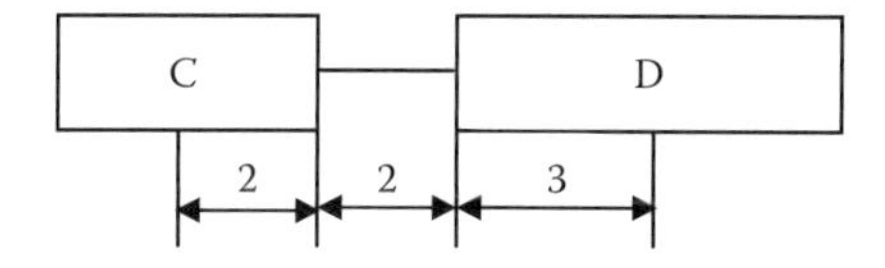


Step 3: The next step will be the placement of the third machine. The effectiveness of each machine that is yet to be placed is evaluated. In each case, the appropriate distances from the center of one machine to the center of another are determined. The sum of the flows between the test machine and the existing layout multiplied by the appropriate distances gives the MH cost. An arrangement that gives the minimum cost is selected for the machine placement. In our case, two machines, A and B, are yet to be placed. Placing them will generate four possible alternatives. They are:

3(a) Place machine A on the left-hand side of machine C.
3(b) Place machine A on the right-hand side of machine D.
3(c) Place machine B on the left-hand side of machine C.
3(d) Place machine B on the right-hand side of machine D.

The associated costs are:

Case 3(a)

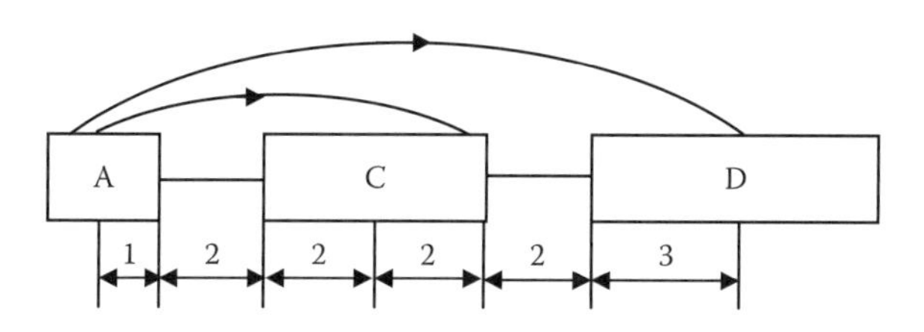

$$\text{Total cost} = F_{AC} \times d_{AC} + F_{AD} \times d_{AD}$$
$$= 18 \times 5 + 18 \times 12$$
$$= 306$$

Case 3(b)

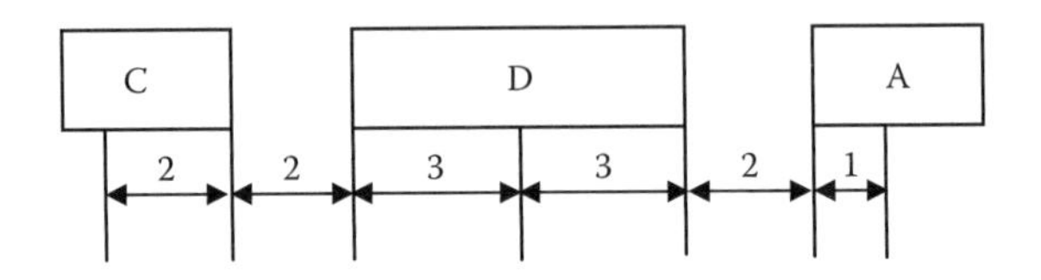


$$\text{Total cost} = F_{CD} \times d_{CD} + F_{CA} \times d_{CA}$$
$$= 38 \times 7 + 18 \times 13$$
$$= 500$$

Case 3(c)

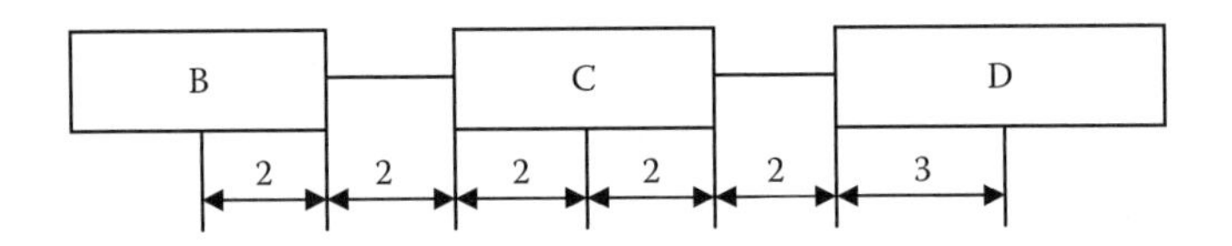


$$\text{Total cost} = F_{BC} \times d_{BC} + F_{CD} \times d_{CD}$$
$$= 35 \times 6 + 38 \times 7$$
$$= 476$$

Case 3(d)

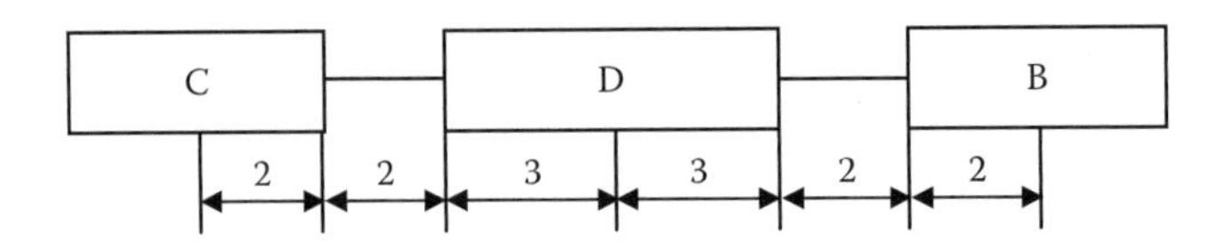


$$\text{Total cost} = F_{CD} \times d_{CD} + F_{CB} \times d_{CB}$$
$$= 38 \times 7 + 35 \times 14$$
$$= 756$$

The minimum placement cost is \$306, and the corresponding layout is A-C-D.

Step 4: The next step is to place the fourth machine with the best layout obtained from step 3. Two alternatives are available for placing machine B are:

4(a) Place machine B on the left-hand side of machine A.

4(b) Place machine B on the right-hand side of machine D.

Case 4(a)

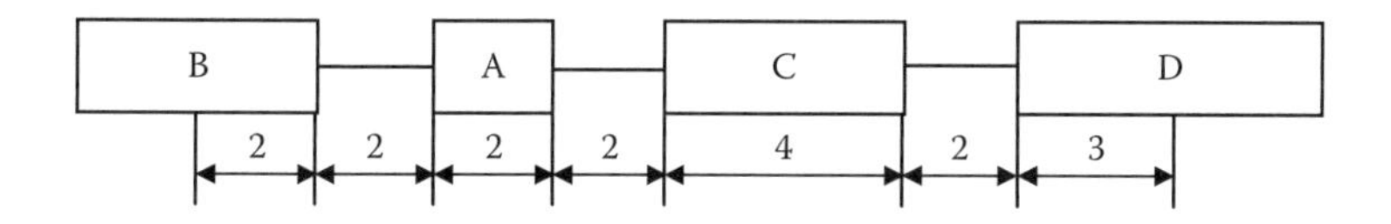


$$\text{Total cost} = F_{BA} \times d_{BA} + F_{BC} \times d_{BC} + F_{BD} \times d_{BD}$$

$$= 15 \times 5 + 35 \times 10 + 21 \times 17$$

$$= 782$$

Case 4(b)

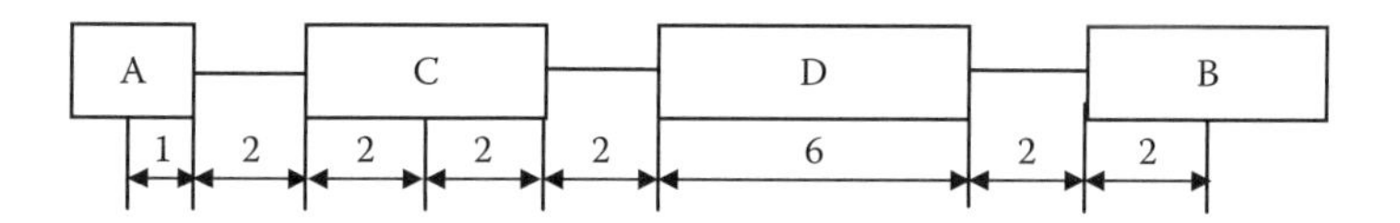


$$\text{Total cost} = F_{AC} \times d_{AC} + F_{AD} \times d_{AD} + F_{AB} \times d_{AB}$$

$$= 18 \times 5 + 18 \times 12 + 15 \times 19$$

$$= 591$$

The lowest cost is associated with alternative 4(b), and thus the final layout will be A-C-D-B.

### *10.1.2.1.2 Loop Layout*

Consider the problem in which machines are placed in a loop so that the material flow is only in one direction. This will result in variation in travel distance between the two machines based on the direction of the flow. Travel from machine $i$ to machine $j$ may not have the same distance as that from $j$ to $i$. If we have the total loop length, $l$, and if travel distance in the $i$-$j$ direction is $d(ij)$, then travel distance in the $j$-$i$ direction $d(ji) = l - d(ij)$.

The following assumptions are made for this problem:

- The flow of the materials is unidirectional, that is, travel is permitted in one direction only.
- The clearance between any two machines is 2 units.

- The cost of transporting a unit material for unit distance between two machines is $C_{ij} = 1$.

The following data gives the dimension of the machines:

| Machine | A | B | C | D |
|---|---|---|---|---|
| Dimensions | $2\times2$ | $4\times4$ | $4\times4$ | $6\times6$ |

The loaded MH trip between the machines is:

$$[F_{ij}] = \text{machine} \begin{array}{c} \\ A \\ B \\ C \\ D \end{array} \begin{array}{c} \begin{array}{cccc} A & B & C & D \end{array} \\ \begin{bmatrix} - & 10 & 15 & 12 \\ 5 & - & 20 & 18 \\ 3 & 15 & - & 21 \\ 6 & 3 & 17 & - \end{bmatrix} \end{array}$$

*10.1.2.1.2.1 Solution*

The loop length is Σ length of all machines + Σ distance between machines.
Here,

$$\text{loop length} = (2+4+4+6)+(2+2+2+2)$$
$$= 24 \text{ units}$$

Step 1: Start with the flow table (Figure 10.13). Machines C and D, with flow being maximum between them, are selected for placement in the layout first. Following the same procedure described earlier, we have following alternatives:

B-C-D

$$\text{Total cost} = F_{BC} \times d_{BC} + F_{BD} \times d_{BD} + F_{CB} \times d_{CB} + F_{DB} \times d_{DB}$$
$$= 20\times6+18\times13+15\times(24-6)+3\times(24-13)$$
$$= 657$$

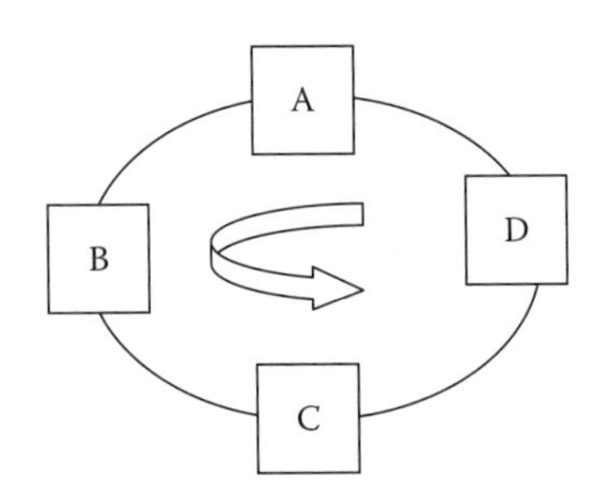


**FIGURE 10.13** Loop flow

C-D-B

$$\begin{aligned}\text{Total cost} &= F_{CB} \times d_{CB} + F_{DB} \times d_{DB} + F_{BC} \times d_{BC} + F_{BD} \times d_{BD} \\ &= 15 \times 14 + 3 \times 7 + 20 + (24 - 14) + 18 \times (24 - 7) \\ &= 737\end{aligned}$$

A-C-D

$$\begin{aligned}\text{Total cost} &= F_{AC} \times d_{AC} + F_{AD} \times d_{AD} + F_{CA} \times d_{CA} + F_{DA} \times d_{DA} \\ &= 15 \times 5 + 12 \times 12 + 3 \times (24 - 5) + 6 \times (24 - 12) \\ &= 354\end{aligned}$$

C-D-A

$$\begin{aligned}\text{Total cost} &= F_{CA} \times d_{CA} + F_{DA} \times d_{DA} + F_{AC} \times d_{AC} + F_{AD} \times d_{AD} \\ &= 3 \times 13 + 6 \times 6 + 15 \times (24 - 13) + 12 \times (24 - 6) \\ &= 456\end{aligned}$$

The minimum cost is associated with layout A-C-D.

Step 2: The next step is to place machine B. The two alternatives associated with it are:

B-A-C-D

$$\begin{aligned}\text{Total cost} &= F_{BA} \times d_{BA} + F_{BC} \times d_{BC} + F_{BD} \times d_{BD} + F_{AB} \times d_{AB} + F_{CB} \times d_{CB} + F_{DB} \times d_{DB} \\ &= 5 \times 5 + 20 \times 10 + 18 \times 17 + 10 \times (24 - 5) + 15 \times (24 - 10) + 3 \times (24 - 17) \\ &= 952\end{aligned}$$

A-C-D-B

$$\begin{aligned}\text{Total cost} &= F_{AB} \times d_{AB} + F_{CB} \times d_{CB} + F_{DB} \times d_{DB} + F_{BA} \times d_{BA} + F_{BC} \times d_{BC} + F_{BD} \times d_{BD} \\ &= 10 \times 19 + 15 \times 14 + 3 \times 7 + 5 \times (24 - 19) + 20 \times (24 - 14) + 18 \times (24 - 7) \\ &= 952\end{aligned}$$

The minimum cost is associated with both sequences, B-A-C-D and A-C-D-B. Hence, machine B can be placed in either way without affecting the optimum cost.

#### *10.1.2.1.3 Backtracking*

In some instances, backtracking cost is considerably larger than forward flow cost. For example, the flow in the forward direction may be achieved via a conveyor, while bringing units back in the reverse direction of flow may require special handling.

In such instance, the objective may be to minimize backtracking rather than to minimize the total flow cost.

For the backtracking process, determine the machines with the maximum or minimum backtracking flow. For example, consider machines A and B. The flow from A-B is 10 and that from B-A is 5. Thus, the minimum backtracking flow between these two machines is the minimum of 10 and 5 = 5. Similarly, calculate the minimum backtrack in all other pairs of machines. Select the pair of machines with the maximum value. If there is a tie, select all tied pairs.

From the above data, the minimum backtrack flow for the pair of machines is given below:

A-B = 5
A-C = 3
A-D = 6
B-C = 15
B-D = 3
C-D = 17

Form these machine pairs, select the one with the maximum backtracking flow — in this case, 17 for the pair C-D.

The machines can be placed either as C-D or D-C. In the case of C-D, the backtracking cost is 119, which is lower than that of D-C. Thus, choice C-D is taken as the layout.

Steps 2 and 3 require a similar calculation process, placing the candidate machine in front of the fixed sequence and then in the back of the fixed sequence to determine the backtracking costs. A sequence is selected at each step as shown in the following table, which gives the minimum cost.

The backtracking cost is calculated as:

| Step | Sequence | Backtrack Cases | Backtrack cost | Optimum Sequence |
|---|---|---|---|---|
| 1 | C-D | D-C | $17 \times 7 = 119$ | C-D |
| | D-C | C-D | $21 \times 7 = 147$ | |
| 2 | A-C-D | D-A, C-A | $6 \times 12 + 3 \times 5 = 87$ | A-C-D |
| | C-D-A | A-C, A-D | $15 \times 13 + 12 \times 6 = 267$ | |
| | B-C-D | D-B, D-C | $3 \times 13 + 17 \times 7 = 158$ | |
| | C-D-B | B-C, B-D | $20 \times 14 + 18 \times 7 = 406$ | |
| 3 | B-A-C-D | D-B, C-B, A-B | $3 \times 17 + 15 \times 10 + 10 \times 5 = 251$ | B-A-C-D |
| | A-C-D-B | B-A, B-C, B-D | $5 \times 19 + 20 \times 14 + 18 \times 7 = 501$ | |

Finally, the sequence B-A-C-D, which is an optimum one, is selected.

The total backtracking cost is = 119 + 87 + 251 = 457.

### 10.1.3 Queuing Models

#### EXAMPLE 1 ROTARY TABLE CAPACITY

Items arrive at a station following a Poisson distribution with a mean of 2 units/minute. The station can service these units at a rate of 3 units/minute. The incoming items are stored on a rotary table when the worker is busy. We wish to determine the required capacity of the table so that the probability of not being able to accommodate a unit on the table is less than 3%.

The situation can be analyzed as a single-channel queuing model. We refer the reader for a detailed discussion on queuing theory to any standard operations research book (Ravindran, 2006; Taha, 2006). Appendix C supplies a few well-known formulas, which are used here in our analysis.

For a Poisson arrival and exponential service with the number of services being 1, we have

$$P_o = 1 - \frac{\lambda}{\mu}$$

$$P_n = \left(\frac{\lambda}{\mu}\right)^n P_0$$

In our example, $\lambda = 2$/minute and $\mu = 3$/minute.

The values of $P_0, P_1, P_2, \ldots, P_n$ are listed in Table 10.1. Each indicates the probability of being in that state. For example, $P_o = 0.333$ implies that the probability of not having a unit on the table is 33.3%.

The cumulative probabilities indicate the percent of time the number of units in the system would be less than or equal to the state value. For example, for state 4 the cumulative probability is 86.9, implying that the number of units in the system would be less than or equal to 86.9% of the time.

**TABLE 10.1**
**Calculations of State and Cumulative Probabilities for Each State**

| State | Probability | Cumulative Probabilities |
|---|---|---|
| 0 | $P_o = (1 - 2/3) = 0.333$ | 0.333 |
| 1 | $P_1 = (2/3)^1 (0.333) = 0.222$ | 0.555 |
| 2 | $P_2 = (2/3)^2 (0.333) = 0.149$ | 0.704 |
| 3 | $P_3 = (2/3)^3 (0.333) = 0.099$ | 0.803 |
| 4 | $\mathrm{P}_4 = (2/3)^4 (0.333) = 0.066$ | 0.869 |
| 5 | $P_5 = (2/3)^5 (0.333) = 0.0439$ | 0.9129 |
| 6 | $P_6 = (2/3)^6 (0.333) = 0.0293$ | 0.943 |
| 7 | $\mathrm{P}_7 = (2/3)^7 (0.333) = 0.0195$ | 0.9625 |
| 8 | $P_8 = (2/3)^8 (0.333) = 0.0130$ | 0.9755 |

To accommodate all units at least 97% of time, the minimum size of the table should be sufficient to place eight units, because the associated cumulative probability is 0.9755.

It might be of interest to find how many units, on the average, would be on the table. This is given by $L = \Sigma_i iP_i$, which is equal to 1.7165 units.

## EXAMPLE 2 ROTARY TABLE CAPACITY WITH MODIFICATION

Now, suppose that in the previous example the machine periodically requires adjustment that stops the work for at most 3 minutes. What is the probability of not accommodating an incoming unit if the table capacity is set to 8?

During the time that the machine is down there is no service, and therefore the model only has the arriving pattern. The probability of having $n$ units in the system in period $t$ is then given by

$$P_n(t) = \frac{(\lambda t)^n}{n!} e^{-\lambda t}$$

Here, $\lambda t = 2 \times 3 = 6$; and with the assumption of no units being on the table, the likelihood that the table would be full is the probability that we would have eight arrivals in 3 minutes. It can be calculated as

$$P_8(3) = \frac{(6)^8}{8!} e^{-6} = 0.1032, \quad \text{or } 10.32\%$$

Because on the average, there are 1.7165 units on the table, the probability that the table would be full is the probability that we have $8 - 1.7165 = 6.28 \approx 6$ new arrivals. The value of $P_6$ is

$$P_6(3) = \frac{(6)^6}{6!} e^{-6} = 0.1606, \quad \text{or } 16\%$$

If this is too high for the plant under consideration, the table capacity should be increased accordingly.

## EXAMPLE 3 ACCUMULATION LINE

Two workers receive units for operation from an accumulator roller conveyor. The main conveyor feeds the accumulator as long as it has room to accommodate a unit. Figure 10.9 shows the arrangement.

If the accumulator is full, the units continue on the main conveyor and are processed later by a relief worker. However, to the extent possible, we prefer not to have any unit going past the accumulator (Figure 10.14). If units arrive at a rate of 3 units/minute and each worker can process 2 units/minute, determine the size of the needed accumulator.

The problem can be analyzed by applying the queuing model with Poisson arrival, exponential service, and two workers (Appendix C). The values of probabilities for having $n$ units in the system — that is, $n - 2$ units on the accumulator — are listed in Table 10.2. Also listed are the values of the corresponding effective arrival rates.

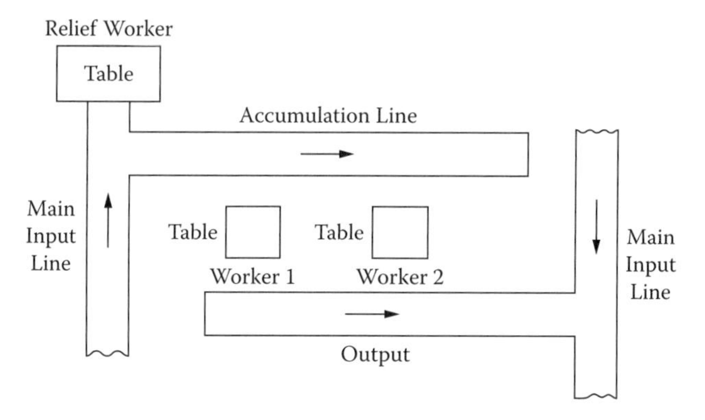


**FIGURE 10.14** Layout of the work area

It is apparent that if the accumulation line is made long enough to accommodate eight units ($n$ = 10), the probability that the system will be full is less than 2% ($P$ = 1.7%). In effect, the two workers can process 2.949 units/minute, leaving about $(3 - 2.949) \times 480 = 24.48$, approximately 24 or 25 units/day, to be served by the relief worker. This number being reasonable, the decision is made to provide for an accumulator with an eight-unit capacity.

## EXAMPLE 4 SIMULATION MODEL

We demonstrate the use of simulation in conveyor bank design by means of an example.

The final phase of production of a microwave oven consists of two operations. The first station finishes the assembly, and the second tests the product. There is a proposal to arrange them in the sequence shown in Figure 10.15.

Station 1 receives units on skids, and a gravity roller conveyor is to be used between stations 1 and 2. The conveyor will also serve as a temporary storage for the units. The

**TABLE 10.2**
**Calculations for Accumulation Line Example**

| Number of Units in System, $n$ | Probability, $P_n$ | Effective Arrival Rate, $\lambda_t$ |
|---|---|---|
| 2 | 0.3103 | 2.069 |
| 3 | 0.1888 | 2.434 |
| 4 | 0.1240 | 2.628 |
| 5 | 0.0851 | 2.745 |
| 6 | 0.0600 | 2.820 |
| 7 | 0.0430 | 2.871 |
| 8 | 0.0313 | 2.906 |
| 9 | 0.0229 | 2.931 |
| 10 | 0.0170 | 2.949 |
| 11 | 0.0130 | 2.961 |
| 12 | 0.0090 | 2.973 |

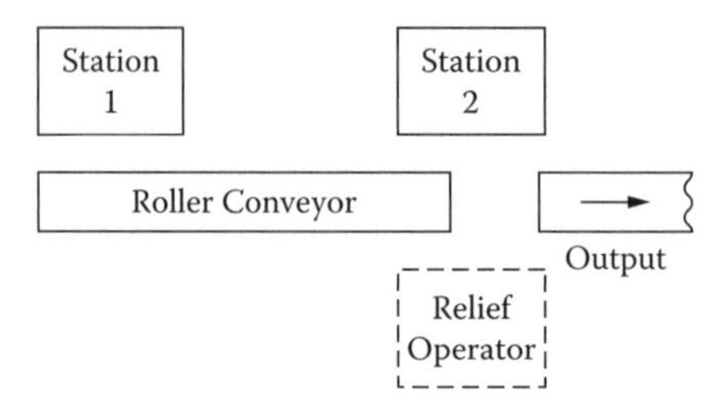


**FIGURE 10.15** End stations arrangement

service times required at stations 1 and 2 follow uniform distributions with the following parameters:

- Station 1: 10 ± 3 minutes
- Station 2: 12 ± 5 minutes

Because, on average, station 2 takes 2 minutes longer than station 1, it has been decided to assign an operator to work on station 1 for 400 minutes and then transfer to work with the operator on station 2 for the remaining 80 minutes of regular time. It is a policy of the company to finish testing all the units that are on the feeder conveyor before allowing workers to go home (even if it means overtime) because work held over would create a backlog that could not easily be eliminated the next morning. Station 1 essentially requires no setup time, while station 2 takes about 5 minutes before testing can begin. Thus, there is a small built-in time delay already in the system.

Simulation results for one day of operation are presented in Tables 10.3, 10.4, and 10.5. Table 10.3 shows the time of completion of a unit in the first station.

Service times are generated by using a uniformly distributed variable generator, namely, $X_1 = a + b(\text{RAN})$, where $a$ is the lower limit, $b$ is the range of the distribution, and RAN is a random number between 0 and 1. (In our case, $a = 7$ and $b = 6$.) The numbers are rounded to the nearest minute because that unit of time is the largest that will allow the desired accuracy.

The actual time of release of a unit on the conveyor is shown in the third column.

For station 2, the arrival time of a unit is the release time of the unit from station 1 (neglecting time of travel on the conveyor). The service times, $X_2$, are again uniform and are generated by using $X_2 = 7 + 10(\text{RAN})$. Column 2 in Table 10.4 indicates the random numbers used, and column 3 shows the corresponding service time for the unit. Column 4 is the departure time obtained by adding the service time to the arrival time, or the previous unit release time, whichever is greater. If the arrival time is greater than the previous unit release time, the operator is waiting on the unit; and if the condition is reversed, the unit is waiting on the operator. For example, the unit that arrived at time 59 must wait until the server completes his or her task on the previous unit that arrived at time 51. This completion is scheduled at 71. Hence, the unit will begin to be serviced at 71 and, needing 9 minutes of service time, will be released at 80. Because the server is busy until 80, all of the units that arrive after 59, through and including 80, must wait in the queue. There are two units in this category: one arriving at 67 and another at 80. Therefore, the queue length for time 71–80 is 2, as indicated in columns 5 and 6 of the table.

At time 400, the operator at station 1 ceases work on that station and transfers to help the operator at the second station. Table 10.5 indicates the disposition of the units

**TABLE 10.3**
**Simulation of Station 1 (First 400 Minutes of Operation)**

| Random Number (RAN) | Time for Service, $X_1 = 7 + 6 \times$ (RAN) | Time of Release |
|---|---|---|
| 0.25 | $8.5 \approx 9$ | 9 |
| 0.59 | $10.54 \approx 11$ | 20 |
| 0.19 | $8.14 \approx 8$ | 28 |
| 0.91 | $12.46 \approx 12$ | 40 |
| 0.61 | $10.66 \approx 11$ | 51 |
| 0.14 | $7.84 \approx 8$ | 59 |
| 0.15 | $7.9 \approx 8$ | 67 |
| 0.92 | $12.52 \approx 13$ | 80 |
| 0.45 | $9.7 \approx 10$ | 90 |
| 0.82 | $11.92 \approx 12$ | 102 |
| 0.50 | $10.0 \approx 10$ | 112 |
| 0.72 | $11.32 \approx 11$ | 123 |
| 0.76 | $11.44 \approx 11$ | 134 |
| 0.53 | $10.18 \approx 10$ | 144 |
| 0.26 | $8.56 \approx 9$ | 153 |
| 0.85 | $12.1 \approx 12$ | 165 |
| 0.39 | $9.34 \approx 9$ | 174 |
| 0.41 | $9.46 \approx 9$ | 183 |
| 0.80 | $11.8 \approx 12$ | 195 |
| 0.33 | $8.98 \approx 9$ | 204 |
| 0.46 | $9.76 \approx 10$ | 214 |
| 0.48 | $9.88 \approx 10$ | 224 |
| 0.60 | $10.6 \approx 11$ | 235 |
| 0.75 | $11.5 \approx 12$ | 247 |
| 0.68 | $11.08 \approx 11$ | 258 |
| 0.03 | $7.18 \approx 7$ | 265 |
| 0.98 | $12.88 \approx 13$ | 278 |
| 0.88 | $12.28 \approx 12$ | 290 |
| 0.01 | $7.06 \approx 7$ | 297 |
| 042 | $9.52 \approx 10$ | 307 |
| 0.36 | $9.16 \approx 9$ | 316 |
| 0.21 | $8.26 \approx 8$ | 324 |
| 0.20 | $8.2 \approx 8$ | 332 |
| 0.87 | $12.22 \approx 12$ | 344 |
| 0.16 | $7.96 \approx 8$ | 352 |
| 0.64 | $10.84 \approx 11$ | 363 |
| 0.54 | $10.24 \approx 10$ | 373 |
| 0.29 | $8.74 \approx 9$ | 382 |
| 0.91 | $12.46 \approx 12$ | 394 |
| 0.94 | $12.64 \approx 13$ | 407[a] |

[a] Data not used because it exceeds 400.

**TABLE 10.4**
**Simulation of Station 2 (First 400 Minutes of Operations with One Operator)**

| Arrival (1) | RAN (2) | Time for Service, $X_2 = 7 + 10 \times$ (RAN) (3) | Departure Time (4) | Time (5) | Number in Queue (6) |
|---|---|---|---|---|---|
| 9 | 0.67 | $13.7 \approx 14$ | 23 | 20–23 | 1 |
| 20 | 0.70 | $14.0 \approx 14$ | 37 | 23–37 | 1 |
| 28 | 0.30 | $10.0 \approx 10$ | 47 | 37–47 | 1 |
| 40 | 0.46 | $11.6 \approx 12$ | 59 | 47–59 | 2 |
| 51 | 0.52 | $12.2 \approx 12$ | 71 | 59–71 | 2 |
| 59 | 0.16 | $8.6 \approx 9$ | 80 | 71–80 | 2 |
| 67 | 0.80 | $15.0 \approx 15$ | 95 | 80–95 | 2 |
| 80 | 0.65 | $13.5 \approx 14$ | 109 | 95–109 | 3 |
| 90 | 0.95 | $16.5 \approx 17$ | 125 | 109–125 | 3 |
| 102 | 0.96 | $16.6 \approx 17$ | 143 | 125–143 | 3 |
| 112 | 0.38 | $10.8 \approx 11$ | 154 | 143–153 | 3 |
| 123 | 0.33 | $10.3 \approx 10$ | 164 | 153–164 | 3 |
| 134 | 0.96 | $16.6 \approx 17$ | 181 | 164–181 | 4 |
| 144 | 0.11 | $8.1 \approx 8$ | 189 | 181–189 | 4 |
| 153 | 0.76 | $14.6 \approx 15$ | 204 | 189–204 | 5 |
| 165 | 0.66 | $13.6 \approx 14$ | 218 | 204–218 | 5 |
| 174 | 0.12 | $8.2 \approx 8$ | 226 | 218–226 | 5 |
| 183 | 0.76 | $14.6 \approx 15$ | 241 | 226–241 | 5 |
| 195 | 0.37 | $10.7 \approx 11$ | 252 | 241–252 | 5 |
| 204 | 0.62 | $13.2 \approx 13$ | 265 | 252–265 | 6 |
| 214 | 0.24 | $9.4 \approx 9$ | 274 | 265–274 | 5 |
| 224 | 0.91 | $16.1 \approx 16$ | 290 | 274–290 | 6 |
| 235 | 0.41 | $11.1 \approx 11$ | 301 | 290–301 | 6 |
| 247 | 0.23 | $9.3 \approx 9$ | 310 | 301–310 | 6 |
| 258 | 0.64 | $13.4 \approx 13$ | 323 | 310–323 | 6 |
| 265 | 0.98 | $16.8 \approx 17$ | 340 | 323–340 | 7 |
| 278 | 0.83 | $15.3 \approx 15$ | 355 | 340–355 | 8 |
| 290 | 0.56 | $12.6 \approx 13$ | 368 | 355–368 | 8 |
| 297 | 0.95 | $16.5 \approx 17$ | 385 | 368–385 | 9 |
| 307 | 0.97 | $16.7 \approx 17$ | 396 | 385–402 | 9 |
| 316 | 0.03 | $7.3 \approx 7$ | | | |
| 324 | 0.12 | $8.2 \approx 8$ | | | |
| 332 | 0.16 | $8.6 \approx 9$ | | | |
| 344 | 0.07 | $7.7 \approx 8$ | | | |
| 352 | 0.92 | $16.2 \approx 16$ | | | |
| 363 | 0.48 | $11.8 \approx 12$ | | | |
| 373 | 0.88 | $15.8 \approx 16$ | | | |
| 382 | 0.52 | $12.2 \approx 12$ | | | |
| 394 | 0.73 | $14.3 \approx 14$ | | | |

that have accumulated on the conveyor at time 400. It shows the service time for each unit, which operator services the unit, and how long that operator is then occupied. For example, the unit that arrives at time 332 needs 9 minutes of service, and it is serviced by operator 2, who then is occupied until time 416. The number of units in storage at

**TABLE 10.5**
**Simulation of Station 2 (After 400 Minutes With Two Operators)**

| Arrival | Service Time | Operator 1 | Operator 2[a] | Number in Queue |
|---|---|---|---|---|
| 307 | | 402 | | 9 |
| 316 | 7 | | 407 | 8 |
| 324 | 8 | 410 | | 7 |
| 332 | 9 | | 416 | 6 |
| 344 | 8 | 418 | | 5 |
| 352 | 16 | | 432 | 4 |
| 363 | 12 | 430 | | 3 |
| 373 | 16 | 446 | | 2 |
| 382 | 12 | | 444 | 1 |
| 394 | 14 | | 458 | 0 |

[a] Starts work at 400 minutes.

that point is 6. The operators continue to work until all of the units arriving on that day are serviced.

The maximum queue length is 9 units; therefore, the conveyor must be long enough to accommodate at least 9 units. Simulation being a sampling technique, we might obtain different queue lengths if other random numbers are used. From a practical standpoint, the trials should be repeated several times to obtain a better estimate of the maximum queue length. In this case, we will add a safety factor of about 20% and increase the storage space to 11.

Here, the analysis of the problem was done with simulation because it was easier to model the changes in service rates in stations 1 and 2 using logic rather than mathematics. No simple queuing model could describe these workings.

### 10.1.4 Closed-Loop Conveyor Systems

We studied an assembly line balancing technique in Chapter 4. The flow of material between stations was independent of the stations, even though common cycle time was maintained. An assembly line can also be developed using a closed-loop conveyor system.

A basic closed-loop conveyor is a nonreversible, continuous operating system with carriers such as hooks, trays, and clamps attached to the conveyor that are not removed during normal operations. As shown in Figure 10.16, the parts with their fixtures are loaded at one station, and the final assemblies are unloaded at the same station or close to its location. Other stations in the assembly process are placed along the conveyor and perform their tasks as the conveyor is moved at a constant speed. The carriers are uniformly spaced, and the distance between them is controlled by the size of the family of parts that is scheduled for production. The spacing must be large enough to accommodate the largest part and its fixture.

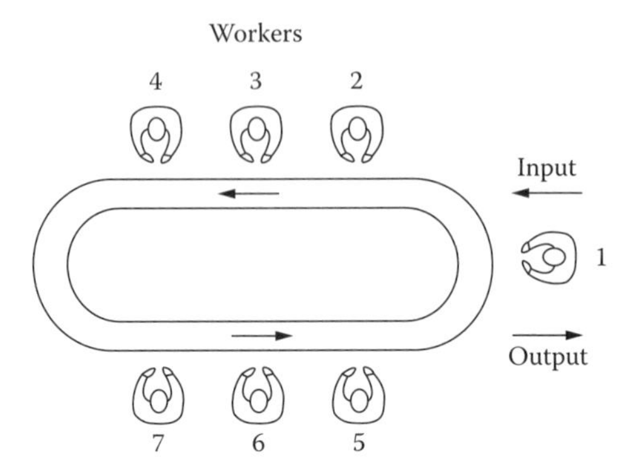


**FIGURE 10.16** Closed-loop conveyor system

The closed-loop conveyor has the following advantages and characteristics:

- MH is considerably reduced by depositing the parts fixtures at the starting point, conveniently allowing the operators to use them in reloading the conveyor.
- Workers are paced by the speed of the conveyor as they work off it.
- MH between the stations by operators is eliminated.
- The speed of the conveyor can be adjusted to suit different production rates and various products.
- The conveyor speed can be adjusted to compensate for the learning curve effect on the line, allowing it to run more slowly initially and at an increased speed as learning takes place.
- Conveyors can be (and often are) used as temporary storage.
- In the case of job changeover, the conveyor speed is the slower of the two job speeds until the old job is completely replaced by the new job on the line.

For a basic closed-loop conveyor system, three principles must be followed (Kwo, 1960).

1. *Uniformity principle* suggests that the conveyor be loaded and unloaded uniformly over the entire length of the loop.
2. *Capacity principle* states that the unit spacing (capacity) on the conveyor should be large enough for the number of units to be stored on the conveyor.
3. *Speed principle* determines the permissible speeds of the conveyor. The lowest speed is defined by the highest of the loading and unloading rates of the stations on the conveyor; the maximum speed is governed by the mechanical/electrical conveyor equipment or the initial and final stations' stock input or clearance rates.

A variation of the closed-loop conveyor is one in which each operator removes the material (unit) from the conveyor, places the work at the station in front of him

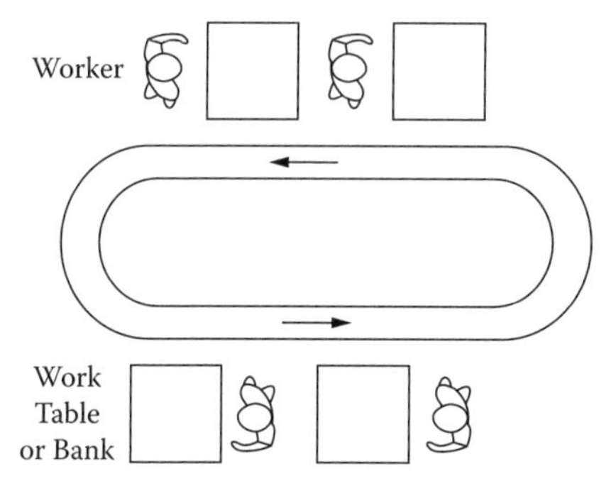


**FIGURE 10.17** Closed loop conveyor with banks

or her, performs operation(s), and loads the unit back on the conveyor (Figure 10.17). Carousel or tray conveyors are most often used in this arrangement, each level being reserved for the unit in a particular state of completion.

Another extension of the closed-loop conveyor allows for multiple input/output stations (Figure 10.18). This is especially useful if an input/output station is a bottleneck because of its low speed of operation or lack of operating space.

To design a closed-loop conveyor system that is able to produce a necessary output requires an analysis that includes determination of the speed of the conveyor, the number of stations, assignment of the task to the stations, physical dimensions (lengths) of the stations, and spacing between successive stations. Such information, in turn, determines the total length of the conveyor and the area required to install it. The following example illustrates such analysis.

### EXAMPLE 5—A CLOSED-LOOP CONVEYOR DESIGN

In a closed-loop conveyor system, the conveyor is 90 feet long, and the carriers are placed 3 feet apart. Thirty-eight feet of the floor space along the conveyor is available for installing assembly stations. Each operator is comfortable working within a maximum size of 6 feet work station, 3 to 4 feet being the preferred space. There is only

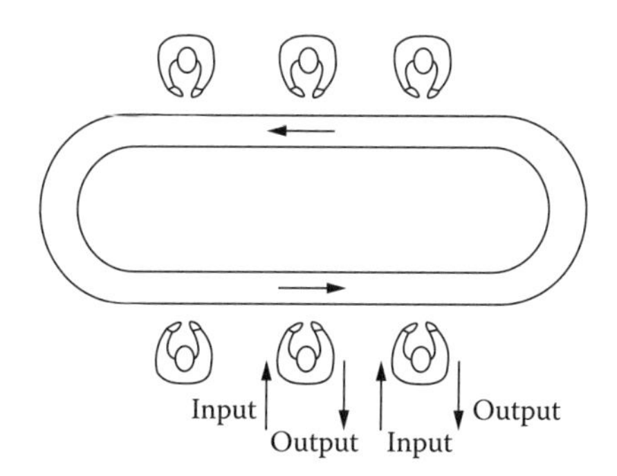


**FIGURE 10.18** Closed loop conveyor with multiple input/output stations

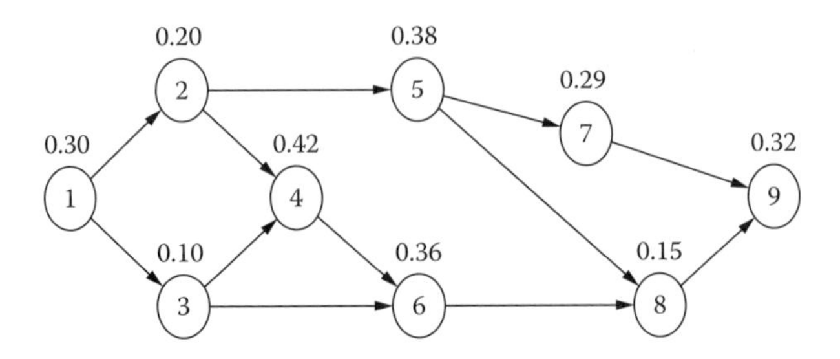


**FIGURE 10.19** Time and precedence relationship

one station for the input/output operation, and it takes a total of 0.4 minute to load and unload the new part that is being considered. The precedence relationships and time data for work elements are given in Figure 10.19.

If a production rate of 100 units/hour is required, determine the conveyor speed, workstation length, number of work stations, and load distribution.

The production rate of 100 units/hour indicates a cycle time of 60/100 = 0.6 minute. If each station is made 4 feet long, and each item must travel from the beginning to the end of a station in 0.6 minute, the speed of the conveyor must be 4/0.6 = 6.67 feet/minute, The maximum possible number of stations, $n$, on the conveyor with at least 2 feet of clearance between each station is

$$4'n + 2'(n - 1) \leq 38', \quad \text{or} \quad n = 6$$

Hence, the part must be produced with six or fewer stations.

By applying the largest candidate rule, we obtain the task distribution shown in Table 10.6.

Because there is room for six stations, the required production can be achieved.

Now, suppose the production rate is to be increased to 133.33 units/hour, requiring a reduction of the cycle time to 60/133.33 = 0.45 minute. This results in the task distribution shown in Table 10.7.

Because there is room for only six stations, the seventh station cannot be accommodated unless the station dimensions are changed.

With a 4-foot working space per station, the velocity of the conveyor can be fixed to 4/0.45 = 8.88 feet/minute. Station 2, however, requires only 0.2 minute to perform its

**TABLE 10.6**
**Task Distribution (Cycle Time: 0.6 minute)**

| Station | Task | Total Time on Station (minutes) |
|---|---|---|
| 1 | 1, 2, 3 | 0.6 |
| 2 | 4 | 0.42 |
| 3 | 5 | 0.38 |
| 4 | 6,8 | 0.51 |
| 5 | 7 | 0.29 |
| 6 | 9 | 0.32 |

**TABLE 10.7**
**Task Distribution (Cycle Time: 0.45 minute)**

| Station | Tasks | Total/ Time on Station (minutes) |
|---|---|---|
| 1 | 1, 3 | 0.4 |
| 2 | 2 | 0.2 |
| 3 | 4 | 0.42 |
| 4 | 5 | 0.38 |
| 5 | 6 | 0.36 |
| 6 | 7, 8 | 0.44 |
| 7 | 9 | 0.32 |

task, and therefore the working space at that station could be reduced to $8.88 \times 0.2 = 1.78$ feet $\approx 2$ feet. Similarly, the required dimensions for other stations could be recalculated, as shown in Table 10.8.

The total is 24 feet. To this, we must add 2 feet for clearance between each two stations; there are six such gaps involved, giving the total required length of $24 + 12 = 36$ feet. Because the available working space is greater than the required space, the assembly line can be formed with the length of each station set to the value calculated before.

### 10.1.5 Automated Controls and Transfers

Productivity in operations can be substantially improved by using automatic control and transfer mechanisms. Automatic controls on conveyor systems allow units to be counted; to be separated on the basis of quality, size, shape, weight, color, and desired destination; and to be collected from multiple conveyors onto one conveyor. The sensors discussed in Chapter 3 play an important part in developing the desired system of operation.

Accessories such as diverters, pushers, turntables, and positioners permit a load to change directions and/or be oriented for a machine tool, as well as change its

**TABLE 10.8**
**Space Reallocations**

| Station | Total Time | Working Space Required |
|---|---|---|
| 1 | 0.4 | $3.55 \approx 4$ |
| 2 | 0.2 | $1.78 \approx 2$ |
| 3 | 0.42 | $3.73 \approx 4$ |
| 4 | 0.38 | $3.37 \approx 3.5$ |
| 5 | 0.36 | $3.19 \approx 3.5$ |
| 6 | 0.44 | $3.90 \approx 4$ |
| 7 | 0.32 | $2.84 \approx 3$ |

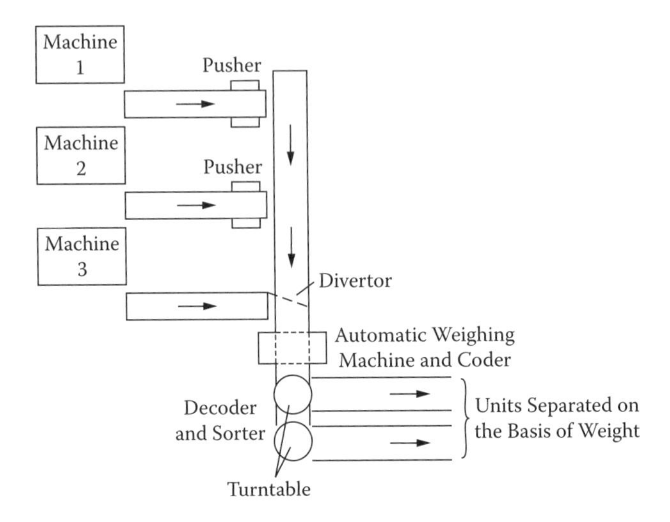


**FIGURE 10.20** Use of automatic controls

position while still on the conveyor (Figure 10.20). Such accessories are quite common in most conveyor systems.

Automatic transfer mechanisms not only transport a unit from one station to the next, but also position it for the next operation with very little, if any, human interface. There are many ways of achieving such transfer, including the use of robots and automated guided vehicles (AGVs). We will describe here very briefly two of the most commonly used methods.

Wheel-conveyor carriage assembly is used to direct a package on the second roller conveyor if desired. Lifting of the carriage induces the wheel conveyor to contact the package, thus changing the direction of travel.

#### 10.1.5.1 Automated Transfer Machines

A transfer machine performs almost all operations to manufacture a specific product to a certain degree of completion. The machine consists of several stations, each performing a distinct operation. This part is automatically transported between the stations, positioned, and clamped, and then the next operation is carried out.

There are two general configurations of automated transfer machines: the in-line type and the rotary type. The in-line arrangement has its work stations in a straight line, while the rotary configuration has a circular flow of work, usually around a rotating table. Basically, the difference is only in the physical grouping and not in the method of operation.

Automated transfer machines offer many advantages, such as improved part quality and accuracy, reduced labor requirements, reduced floor space requirements, and reduced in-process inventory.

A common problem associated with automated transfer machines, however, is the low working efficiency of the line compared to that of the individual machines. This is mainly because of the linkage between the stations; if one station breaks down, then all stations must be shut down.

The problem can be partially overcome by providing buffer spaces between the different stations. If a station is then forced to shut down, the preceding station can continue to work until the buffer space is full. Also, the succeeding station can continue to operate if its buffer contains any materials.

#### 10.1.5.2 Monorails

With the ability to automatically pick up, transfer, and place loads, automated monorails are becoming popular in U.S. industries as another means for automated transfer. Computerized monorails operating on lightweight tracks generally handle from 1000 to 2000 pounds per carrier. These monorails are often powered by current-carrying conductor bars designed into the tracks. Several manufacturers offer systems with a pair of channels that allow loads of up to 3000 pounds per carrier.

The prompt delivery and accurate positioning of material by monorails provide productivity boosts in parts fabrication, finishing, and assembly operations.

Self-powered carriers can automatically transport, lift, lower, pick up, and place loads. These systems are also capable of sorting loads and holding them in buffer storage. Monorails can easily be interfaced with AGV systems, automated storage and retrieval systems, and other automated equipment.

Monorails offer several advantages. They are quiet, clean, and individually powered and controlled. They can receive and/or change commands while en route. Furthermore, monorails can operate at slow or high speeds (up to 500 feet/minute) and are capable of quick acceleration. Finally, monorails provide greater flexibility for expansion and layout changes because of overhead track systems with a single track and no chains.

### 10.1.6 Horizontal and Vertical Flow

A question might arise as to how the assembly line or main production flow should be laid out in a plant. The answer depends mainly on the physical structure and

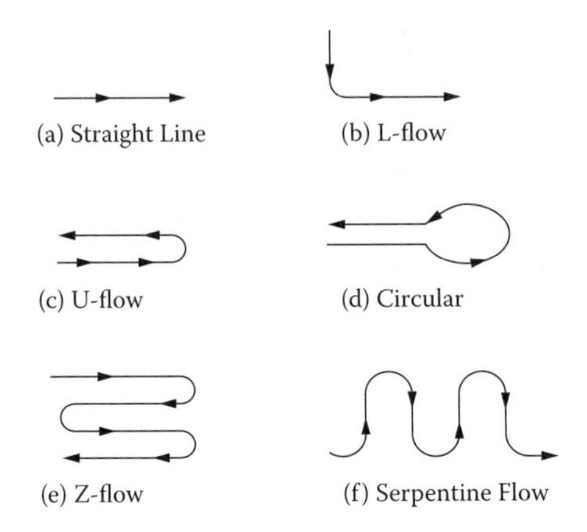


**FIGURE 10.21** Horizontal material flow lines

location of the receiving and shipping departments. For example, in a narrow building with facilities for receiving at one end and shipping at the other, a straight-line flow would be suitable. A U-shape allows the shipping and receiving departments to be on the same side of the building. The circular shape allows the same crew to receive and ship, while a serpentine flow allows for an assembly line of many stations. Other flow shapes could easily be developed to suit particular needs. Some of the more popular are shown in Figure 10.21.

A plant in a multistoried building presents a special problem. Material must flow vertically as well as horizontally on each floor. Again, many flow patterns could be developed, some of which are shown in Figure 10.22.

Vertical conveyors can play an important role in transporting material from one level to the next. One type, known as a vertical reciprocating conveyor, carries a car up and down in its guides. The car is specifically designed to carry a particular type of load (not people) by providing an appropriate bed (floor). For example, a roller conveyor bed can be used for a flat load, a chain conveyor bed might be appropriate to carry a truck or trolley, and a V-shaped bed might be appropriate to carry a round load.

As shown in Figure 10.23, the arm tray or shelf conveyor can also be used to carry a number of relatively small loads continuously between floors. Arms, trays,

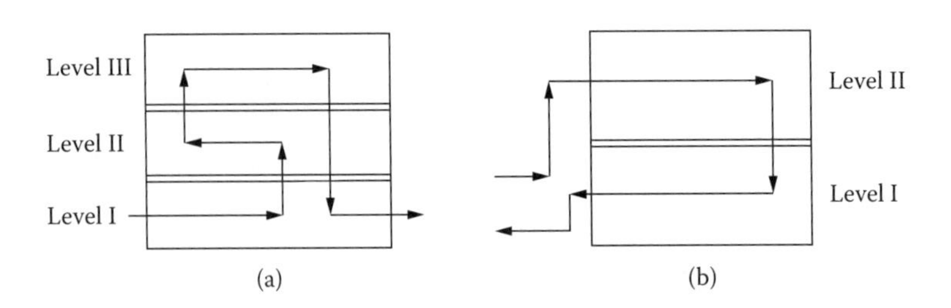


**FIGURE 10.22** Vertical material flow lines

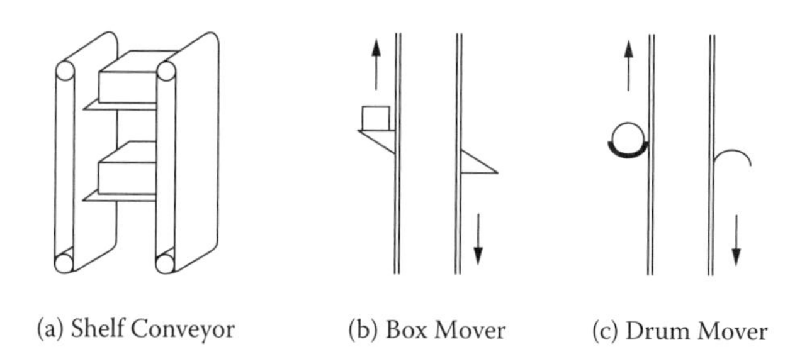


**FIGURE 10.23** Vertical Conveyors

and shelves may fold when traveling in the reverse direction, providing economy in space utilization.

The vertical flow pattern can be especially challenging when old, multistoried buildings are converted to productive use. Previous tax incentive schemes encouraged such conversions, and many companies were prompted to remodel old vacant buildings and develop manufacturing plants within them.

The existing walls, many of which are load-bearing and not easily modified, create the need to develop an efficient production facility within the existing structure. Figure 10.24 shows several possible arrangements of conveyor systems with lifting or turning tables. (These systems were designed by PFLOW Industries, Milwaukee, WI, and are presented here through their courtesy.)

### 10.1.7 Material Flow in Cellular and Job Shops

In cellular and job shop environments there is no single fixed path over which all parts and materials are moved. This is because the work sequences of jobs may differ from one another, requiring different machines in a variety of sequences of operations. It is therefore not possible to define a single material flow pattern that would be ideal in all cases. Trucks, cranes, and conveyors frequently play an important part in material moves that define the flow of patterns. A reduction in MH is frequently possible, however, by planning the floor layout so that the machines and/or departments with the most frequent interchange of parts are placed next to one another. The following two sections, along with the study of plant layout in Chapter 12, demonstrate this idea.

## 10.2 MACHINE GROUPING IN CELLULAR MANUFACTURING WITH REDUCTION OF MATERIAL HANDLING AS THE OBJECTIVE

Here, we investigate how to reduce MH in a cellular manufacturing layout. We concentrate on determining the effective machine cells and analyzing the economic feasibility of duplicating machines.

As discussed in Chapter 4, in flexible manufacturing machines and jobs are grouped in coherent fashion so that most of the operations on a job are performed in

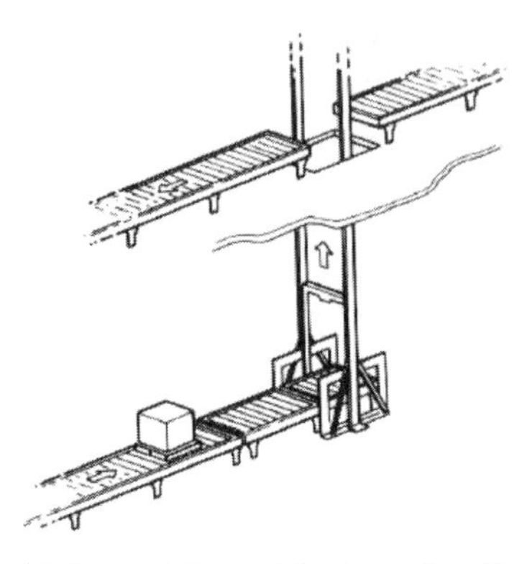

(a) System 1: Power infeed to and on lift, gravity discharge second level.

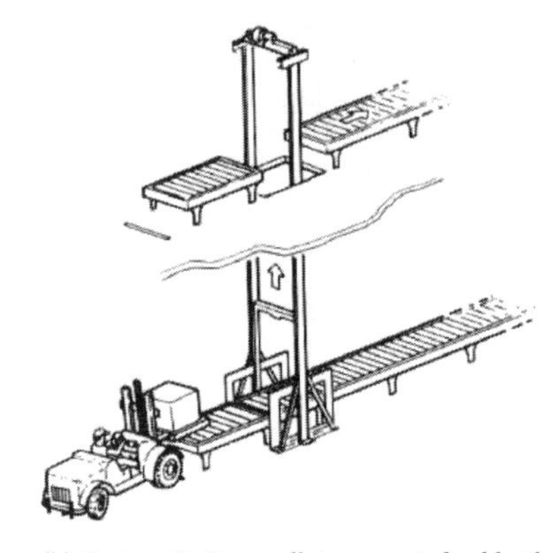

(b) System 2: One pallet power infeed both levels and multiple pallet power discharge both levels.

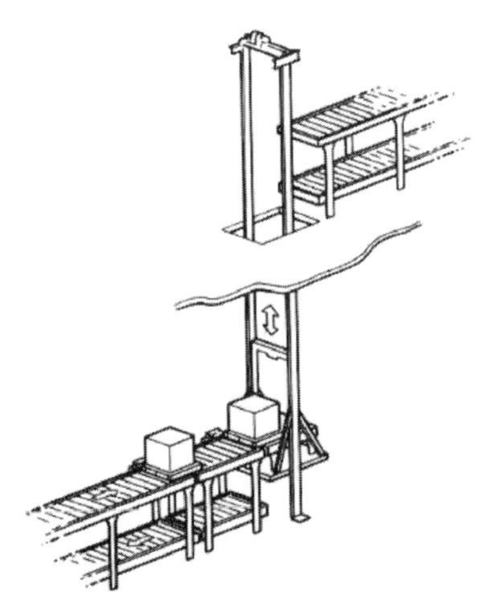

(c) System 3: Floor space saving system. Upper power infeed and lower gravity discharge both levels.

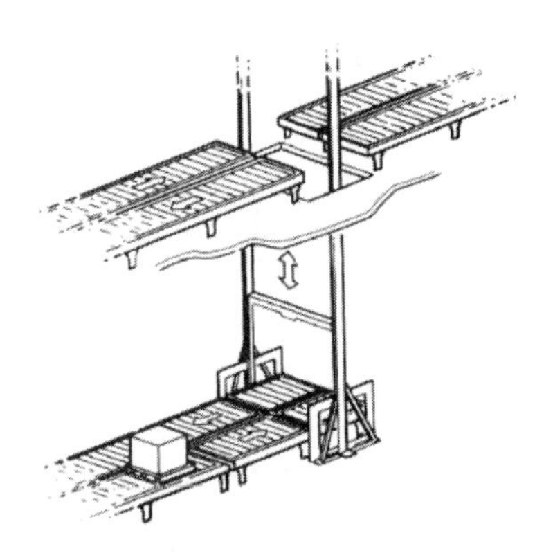

(d) System 4: Double-width conveyor carriage. Each side handles pallets in one direction.

**FIGURE 10.24** Possible conveyor arrangements (redrawn courtesy of PFLOW Industries, Inc., Milwaukee, WI)

a machine cell to which the job is assigned. The idea is to reduce MH for the job by concentrating it only in the cell to which the job is assigned. Because the machines within the cell are also placed close together physically, there is considerable reduction in the amount of MH.

However, the machine-grouping analysis in Chapter 4 does not consider either the production or the associated MH difficulties in machine grouping. These can be important factors. The main emphasis of the algorithm in the previous chapter was on grouping machines so that a cell can process a part completely.

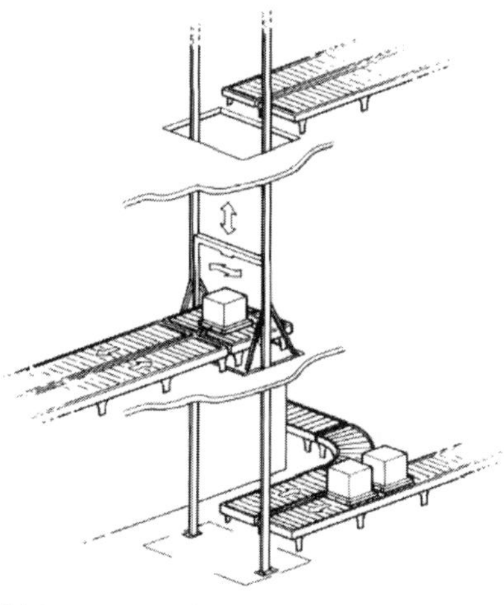

(e) System 5: Double-width conveyor carriage with drag chain transfer. Allows unit to handle two pallets at one time and still feed and discharge from the same side and level. Simultaneous feed and discharge is also possible.

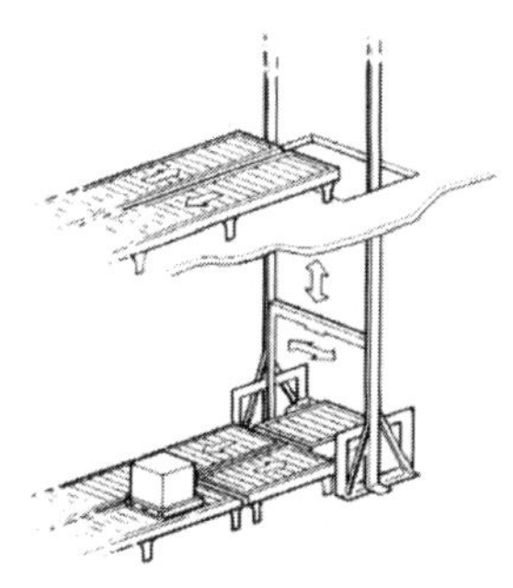

(f) System 6: Double-width carriage with a single-width sliding conveyor. Allows infeed and discharge from the same side and level.

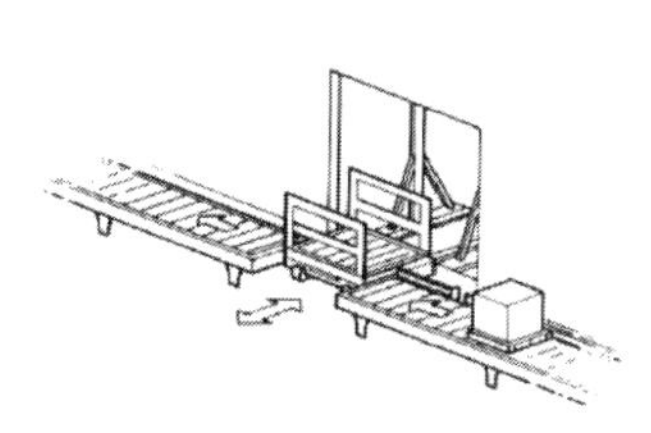

(h) System 8: Multi-travel-carriage (MTC) system. Carriage extends to load or discharge and retracts to move to another level.

(g) System 7: Single pallet conveyor lift with traveling transfer cart on first floor. Allows one lift to service many conveyors and locations on the same floor.

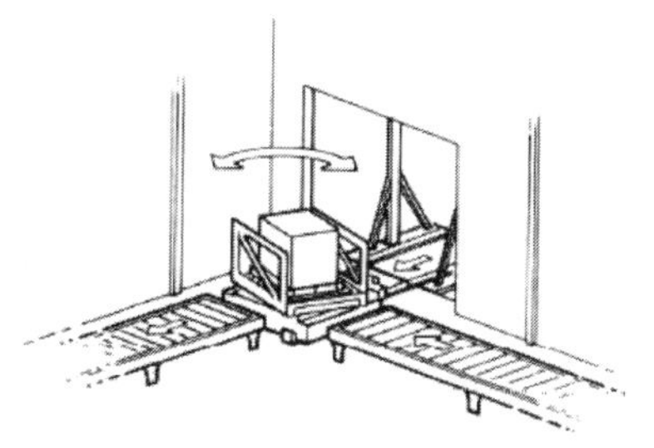

(i) System 9: MTC system incorporating a turntable on carriage for through-wall two-direction corner application.

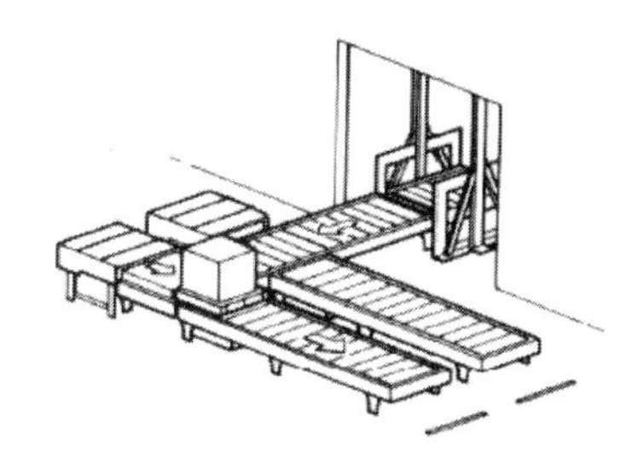

(j) System 10: Single-load conveyor lift discharging to 90° pushoff points. Used in systems handling different sizes or types of materials needing multiple unloading locations.

**FIGURE 10.24** (Continued)

The incidence matrix on which the method is based depends on the use or nonuse of a machine for a part, and each part is given the same importance in cell formation. If the parts vary in their production volume, identical weighting does not seem logical. It seems obvious that the machines used in producing a large-volume part should be physically closer together than the machines associated with a part that has a very small demand.

Along with volume of flow between machines that defines MH, there may also be some difficulties due to varying weights and volumes of the parts, and special care that may be needed in transportation (e.g., glass, explosives). In addition, unitization of a part in handling may produce different proportions of pallet loads than the relative proportions associated with the part demands.

The new solution procedure that incorporates MH is a slight modification of the tabular method illustrated in Chapter 5. The modification involves the addition of steps 0a and 0b before step 1, and modification of step 1. The new procedure is as follows.

### Step 0a: Determine MH for Each Type of Part

For each part type $i$, based on the volume of production, determine the unit loads of MH necessary per unit time ($u_i$). Also note the degree of difficulty in transportation associated with part type $i$. The degree of difficulty is a relative measure of difficulty in MH between two part types. The unit load of a part that is easiest to handle is given a degree of difficulty of 1. All other parts are compared with this part to determine (perhaps subjectively) their relative degree of difficulty in handling, $d_i$.

### Step 0b: Construct an MH Table

To develop the MH table, start with a machine-component matrix in which 1 indicates use of the machine and a blank or zero indicates nonuse of the machine in processing a specific part type. Multiply each element 1 for part $i$ by its $u_i$ and $d_i$ values. This gives the MH table. The elements in the MH table indicate the weighted MH trips between two machines to produce a specific part.

### Step 1: Develop the Machine-to-Machine Material Handling Table (MM)

The table indicates the total of all modified MH trips that would occur between two machines. Each element in the table is obtained by adding, from the MH table, all the MH trips that occur between the associated two machines. Note that the MM table is symmetrical about the diagonal; therefore, only the elements above or below the diagonal are needed.

## AN ILLUSTRATIVE EXAMPLE

Machine-component data from Table 10.9 is used to illustrate the procedure. In addition, Table 10.10 shows the unit loads of material movement per month for each part. These unit loads are calculated by knowing the demands and the number of units that would go into making a 1-unit load for MH. The table also shows the degree of MH difficulty for the part based on the physical characteristics of the unit load and the part itself. Table 10.11 is the MH table obtained by applying step 0b of the

**TABLE 10.9**
**Machine-Component Chart**

| | Machine Number | | | | | | | | | | |
|---|---|---|---|---|---|---|---|---|---|---|---|
| Part Number | 1 | 2 | 3 | 4 | 5 | 6 | 7 | 8 | 9 | 10 | 11 |
| 1 | 1 | 0 | 0 | 1 | 1 | 0 | 0 | 0 | 0 | 1 | 0 |
| 2 | 1 | 0 | 0 | 1 | 1 | 0 | 0 | 0 | 0 | 1 | 0 |
| 3 | 1 | 0 | 0 | 1 | 1 | 0 | 0 | 0 | 0 | 1 | 0 |
| 4 | 0 | 0 | 0 | 0 | 0 | 0 | 1 | 0 | 1 | 0 | 1 |
| 5 | 0 | 1 | 1 | 0 | 0 | 1 | 0 | 1 | 0 | 1 | 0 |
| 6 | 0 | 0 | 1 | 0 | 0 | 0 | 1 | 1 | 0 | 0 | 0 |
| 7 | 0 | 0 | 0 | 0 | 1 | 0 | 0 | 0 | 0 | 1 | 0 |
| 8 | 0 | 1 | 0 | 0 | 0 | 1 | 0 | 0 | 0 | 1 | 0 |
| 9 | 0 | 0 | 0 | 0 | 0 | 0 | 1 | 1 | 1 | 0 | 1 |
| 10 | 0 | 0 | 0 | 0 | 0 | 0 | 0 | 1 | 0 | 0 | 1 |
| 11 | 1 | 1 | 0 | 1 | 0 | 0 | 0 | 0 | 0 | 1 | 0 |
| 12 | 0 | 1 | 1 | 0 | 0 | 1 | 0 | 1 | 0 | 1 | 0 |
| 13 | 0 | 0 | 1 | 0 | 0 | 0 | 0 | 1 | 0 | 0 | 0 |
| 14 | 0 | 0 | 0 | 0 | 0 | 0 | 1 | 1 | 1 | 0 | 0 |
| 15 | 1 | 0 | 0 | 1 | 1 | 0 | 0 | 0 | 0 | 1 | 0 |
| 16 | 1 | 0 | 0 | 1 | 1 | 0 | 0 | 0 | 0 | 1 | 0 |
| 17 | 0 | 0 | 0 | 0 | 0 | 0 | 1 | 1 | 1 | 0 | 1 |
| 18 | 0 | 0 | 0 | 0 | 0 | 0 | 1 | 1 | 1 | 0 | 1 |
| 19 | 0 | 1 | 1 | 0 | 0 | 1 | 0 | 1 | 0 | 0 | 0 |
| 20 | 1 | 0 | 0 | 1 | 1 | 0 | 0 | 0 | 0 | 0 | 0 |
| 21 | 1 | 0 | 0 | 1 | 1 | 0 | 0 | 0 | 0 | 0 | 0 |
| 22 | 1 | 0 | 0 | 1 | 1 | 0 | 0 | 0 | 0 | 0 | 0 |

procedure. Table 10.12 is the machine-to-machine MH table following step 1. Entries in Table 10.12 are obtained by comparing two machines; for example, the entry 1, 5 indicates number of trips between machines 1 and 5, which is the sum of the following (compare columns 1 and 5): 10 units for part 1, 55 units for part 2, 30 units for part 3, 60 units for part 15, 30 units for part 16, 100 units for 20, 50 units for 21, and 110 units for item 22, for the total of 445.

To begin the grouping process (the steps are explained in Chapter 4), consider an illustrative case where the value of $P$ (the ratio) is set to 0.5 (step 5). The maximum value among the elements in Table 10.12 is 1515 [i.e., relationship counter (RC) = 1515], between machines 7 and 8. These two machines form the first group. In the next iteration, RC = 1345 for machines 3 and 8. Because machine 8 is already in group 1, machine 3 could be joined to that group, or a new group consisting of machines 3 and 8 could be formed. A check is made as shown in Table 10.13. Because the closeness ratio (CR) is greater than the minimum threshold value (MTV), machine 3 is added to G1.

The summary of all calculations is shown in Table 10.14. We describe a few iterations here for clarification. In iteration 3 the value of RC is 1315, and the associated machines are 7 and 9. A check similar to the one shown previously is made before adding machine 9 to group 1.

In iteration 5, RC is 1300. The corresponding machines, 9 and 11, are already in group 1; therefore, the observation is ignored.

**TABLE 10.10**
**Material Movement and Difficulty Index**

| Part Number | Unit Load Required, $u_i$ | Difficulty Index for Part, $d_i$ |
|---|---|---|
| 1 | 10 | 1.00 |
| 2 | 50 | 1.10 |
| 3 | 30 | 1.00 |
| 4 | 300 | 1.00 |
| 5 | 800 | 1.00 |
| 6 | 400 | 1.25 |
| 7 | 60 | 1.00 |
| 8 | 100 | 2.20 |
| 9 | 50 | 1.00 |
| 10 | 90 | 1.00 |
| 11 | 80 | 1.00 |
| 12 | 10 | 1.00 |
| 13 | 5 | 1.00 |
| 14 | 15 | 1.00 |
| 15 | 20 | 3.00 |
| 16 | 30 | 1.00 |
| 17 | 50 | 1.00 |
| 18 | 900 | 1.00 |
| 19 | 30 | 1.00 |
| 20 | 40 | 2.50 |
| 21 | 50 | 1.00 |
| 22 | 110 | 1.00 |

Consider iteration 8 with RC = 1060 between machines 2 and 6. Machine 6 is the entering machine. A check is made to see if it would join G1 or G2, as shown in Table 10.15. Because max CR = 1045 > present RC × $P$ = 1060 × 0.5 = 530, we can join 6 to group 2.

In iteration 11 (Table 10.16), RC = 840; the corresponding machines are 2 and 3. At present, machine 3 is in G1, while machine 2 is in G2. Because both machines are already assigned, but to different groups, it suggests either duplicating machine 3 or 2 or duplicating both machines and forming a new group consisting of machines 3 and 2. A check is made in the following manner. Because maximum CR = 830 > MTV = 840 × 0.5, duplicate machine 3 and add it to G2. The present arrangement of the machines is group G1, machines 7, 8, 3, 9, 11, and group G2, machines 2, 10, 6, 3.

In iteration 17, with RC = 525, the corresponding machines 1 and 4 are not part of any group, therefore a new group, G3, is formed consisting of these two machines.

All machines are assigned in iteration 19; therefore, the procedure is terminated following the alternate termination rule. The final group consists of three groups, with the following machine arrangements:

| | |
|---|---|
| G1 | 3, 7, 8, 9, 11 |
| G2 | 2, 3, 6, 8, 10 |
| G3 | 1, 4, 5 |

**TABLE 10.11**
**Material Handling for Parts ($u_i \times d_i$)**

| | Machine Number | | | | | | | | | | |
|---|---|---|---|---|---|---|---|---|---|---|---|
| **Part Number** | **1** | **2** | **3** | **4** | **5** | **6** | **7** | **8** | **9** | **10** | **11** |
| 1 | 10 | 0 | 0 | 10 | 10 | 0 | 0 | 0 | 0 | 10 | 0 |
| 2 | 55 | 0 | 0 | 55 | 55 | 0 | 0 | 0 | 0 | 55 | 0 |
| 3 | 30 | 0 | 0 | 30 | 30 | 0 | 0 | 0 | 0 | 30 | 0 |
| 4 | 0 | 0 | 0 | 0 | 0 | 0 | 300 | 0 | 300 | 0 | 300 |
| 5 | 0 | 800 | 800 | 0 | 0 | 800 | 0 | 800 | 0 | 800 | 0 |
| 6 | 0 | 0 | 500 | 0 | 0 | 0 | 500 | 500 | 0 | 0 | 0 |
| 7 | 0 | 0 | 0 | 0 | 60 | 0 | 0 | 0 | 0 | 60 | 0 |
| 8 | 0 | 220 | 0 | 0 | 0 | 220 | 0 | 0 | 0 | 220 | 0 |
| 9 | 0 | 0 | 0 | 0 | 0 | 0 | 50 | 50 | 50 | 0 | 50 |
| 10 | 0 | 0 | 0 | 0 | 0 | 0 | 0 | 90 | 0 | 0 | 90 |
| 11 | 80 | 80 | 0 | 80 | 0 | 0 | 0 | 0 | 0 | 80 | 0 |
| 12 | 0 | 10 | 10 | 0 | 0 | 10 | 0 | 10 | 0 | 10 | 0 |
| 13 | 0 | 0 | 5 | 0 | 0 | 0 | 0 | 5 | 0 | 0 | 0 |
| 14 | 0 | 0 | 0 | 0 | 0 | 0 | 15 | 15 | 15 | 0 | 0 |
| 15 | 60 | 0 | 0 | 60 | 60 | 0 | 0 | 0 | 0 | 60 | 0 |
| 16 | 30 | 0 | 0 | 30 | 30 | 0 | 0 | 0 | 0 | 30 | 0 |
| 17 | 0 | 0 | 0 | 0 | 0 | 0 | 50 | 50 | 50 | 0 | 50 |
| 18 | 0 | 0 | 0 | 0 | 0 | 0 | 900 | 900 | 900 | 0 | 900 |
| 19 | 0 | 30 | 30 | 0 | 0 | 30 | 0 | 30 | 0 | 0 | 0 |
| 20 | 100 | 0 | 0 | 100 | 100 | 0 | 0 | 0 | 0 | 0 | 0 |
| 21 | 50 | 0 | 0 | 50 | 50 | 0 | 0 | 0 | 0 | 0 | 0 |
| 22 | 110 | 0 | 0 | 110 | 110 | 0 | 0 | 0 | 0 | 0 | 0 |

**TABLE 10.12**
**Machine-to-Machine MH Table**

| | Machine Number | | | | | | | | | | |
|---|---|---|---|---|---|---|---|---|---|---|---|
| **Part Number** | **1** | **2** | **3** | **4** | **5** | **6** | **7** | **8** | **9** | **10** | **11** |
| 1 | 0 | 80 | 0 | 525 | 445 | 0 | 0 | 0 | 0 | 265 | 0 |
| 2 | | 0 | 840 | 80 | 0 | 1060 | 0 | 840 | 0 | 1110 | 0 |
| 3 | | | 0 | 0 | 0 | 840 | 500 | 1345 | 0 | 810 | 0 |
| 4 | | | | 0 | 445 | 0 | 0 | 0 | 0 | 265 | 0 |
| 5 | | | | | 0 | 0 | 0 | 0 | 0 | 245 | 0 |
| 6 | | | | | | 0 | 0 | 840 | 0 | 1030 | 0 |
| 7 | | | | | | | 0 | 1515 | 1315 | 0 | 1300 |
| 8 | | | | | | | | 0 | 1015 | 810 | 1090 |
| 9 | | | | | | | | | 0 | 0 | 1300 |
| 10 | | | | | | | | | | 0 | 0 |
| 11 | | | | | | | | | | | 0 |

**TABLE 10.13**
**Check on Entering Machine 3: RC = 1345**

| Entering Machine | Existing Machines in Group G1 | Relationship | MTV |
|---|---|---|---|
| 3 | 8 | 1345 | $1345 \times 0.5 = 672.5$ |
| | 7 | 500 | |
| Total | 2 | 1845 | |
| **Closeness ratio** | $1845 \div 2 = 922.5$ | | |

The corresponding part assignments are shown in Table 10.17. The solution has two duplicate machines: machines 3 and 8. There are also eight intercellular trips (jobs 1, 2, 3, 7, 11, 15, and 16), for a total of 405 units of material transfer. If only one duplicate machine is allowed, the duplicate machine serving the least number of units should be dropped. In this case, machine 3 in G1, which presently processes 505 units, is dropped. The result is 910 units (405 + 505), of MH.

If no duplication of the machines is allowed, both 3 and 8 must be dropped. Eliminating machine 3 from G1 and machine 8 from G2 results in 1750 units of MH.

It is interesting to compare the solution based on MH with the incident matrix (Chapter 5) for grouping. For example, the tabular method, based on an incident matrix when applied to the present problem, gives seven intercellular trips and 1760 units of MH if machine 10 is duplicated. If duplication is disallowed, it requires 10 intercellular trips and 2085 units of MH when machine 10 is removed from group 1, or 8 intercellular trips and 2790 units of MH if machine 10 is removed from group 3.

Note that the final grouping is sensitive to the value of $P$ chosen for the process. This value can vary between 0 and 1 ($0 < P < 1$). If a different value of $P$ produces a different grouping, then what is the best value to use? This question is answered by the economic analysis that follows.

### 10.2.1 Economic Analysis

The machine grouping obtained in the procedure just discussed may have duplicated machines that are used solely for reducing intercellular material movement. Are these duplications economically justifiable?

The cost expression used in such analysis consists of four components. They are within-cell MH cost, between-cell MH cost, cost of duplication of the machines, and cost of cell management. An explanation of each cost component follows.

1. *Cost of within-cell MH.* The primary objective in machine-cell formation is to group machines and jobs in cells such that, as far as possible, the MH associated with a job is concentrated in the single cell to which the job is assigned. This is in contrast to transporting the part all over the plant when no cells are formed and all jobs are processed in a single cell. How to measure the total MH is a problem that cannot be addressed until the

**TABLE 10.14**
**Summary of Steps in Tabular Method**

| | | | | State | | | | | Action Taken | | | Present Group and Machine Assignment | | |
|---|---|---|---|---|---|---|---|---|---|---|---|---|---|---|
| | | | | Enter MC | | | Both MC in Sol. but Different Group | | New Group Form | MC Assign to Exist Group | MC Duplicated Assign Group | | | |
| Iteration Number | RC Value | Pair | Exist MC | One | Both | Both MC in Same Group | | Observation Ignore | | | | G1 | G2 | G3 |
| 1 | 1515 | 7, 8 | | | × | | | | G1 | 7, 8 | | 7, 8 | | |
| 2 | 1345 | 3, 8 | 8 | 3 | | | | | | G1 | | 7, 8, 3 | | |
| 3 | 1315 | 7, 9 | 7 | 9 | | | | | | G1 | | 7, 8, 3, 9 | | |
| 4 | 1300 | 7, 11 | 7 | 11 | | | | | | G1 | | 7, 8, 3, | | |
| | | | | | | | | | | | | 9, 11 | | |
| 5 | 1300 | 9, 11 | | | | × | | × | | | | 7, 8, 3 | | |
| | | | | | | | | | | | | 9, 11 | | |
| 6 | 1100 | 2, 10 | | | × | | | | G2 | | | 7, 8, 3 | 2, 10 | |
| | | | | | | | | | | | | 9, 11 | | |
| 7 | 1090 | 8, 11 | | | | × | | × | | | | 7, 8, 3 | 2, 10 | |
| | | | | | | | | | | | | 9, 11 | | |
| 8 | 1060 | 2, 6 | 2 | 6 | | | | | | G2 | | 7, 8, 3 | 2, 10, 6 | |
| | | | | | | | | | | | | 9, 11 | | |
| 9 | 1030 | 6, 10 | | | | × | | | | | | 7, 8, 3 | 2, 10, 6 | |
| | | | | | | | | × | | | | 9, 11 | | |
| 10 | 1015 | 8, 9 | | | | × | | | | | | 7, 8, 3 | 2, 10, 6 | |
| | | | | | | | | × | | | | 9, 11 | | |
| 11 | 840 | 2, 3 | | | | | 3/G1 | | | | 3, G2 | 7, 8, 3 | 2, 10, 6, 3 | |
| | | | | | | | 2/G2 | | | | | 9, 11 | | |
| 12 | 840 | 2, 8 | | | | | 8/G1 | | | | 8, G2 | 7, 8, 3 | 2, 10, 6, | |

*(Continued)*

**TABLE 10.14 (CONTINUED)**
**Summary of Steps in Tabular Method**

| | | | State | | | | | | Action Taken | | | Present Group and Machine Assignment | | |
|---|---|---|---|---|---|---|---|---|---|---|---|---|---|---|
| | | | | Enter MC | | | | | | | | | | |
| Iteration Number | RC Value | Pair | Exist MC | One | Both | Both MC in Same Group | Both MC in Sol. but Different Group | Observation Ignore | New Group Form | MC Assign to Exist Group | MC Duplicated Assign Group | G1 | G2 | G3 |
| | | | | | | | 2/G2 | | | | | 9, 11 | 8, 3 | |
| 13 | 840 | 3, 6 | | | | × | | × | | | | 7, 8, 3 | 2, 10, 6, | |
| | | | | | | | | | | | | 9, 11 | 8, 3 | |
| 14 | 840 | 6, 8 | | | | × | | × | | | | 7, 8, 3 | 2, 10, 6, | |
| | | | | | | | | | | | | 9, 11 | 8, 3 | |
| 15 | 810 | 3, 10 | | | | × | | × | | | | 7, 8, 3 | 2, 10, 6 | |
| | | | | | | | | | | | | 9, 11 | 8, 3 | |
| 16 | 810 | 8, 30 | | | | × | | × | | | | 7, 8, 3 | 2, 10, 6 | |
| | | | | | | | | | | | | 9, 11 | 8, 3 | |
| 17 | 525 | 1, 4 | | | × | | | | G3 | | | 7, 8, 3 | 2, 10, 6 | 1, 4 |
| | | | | | | | | | | | | 9, 11 | 8, 3 | |
| 18 | 500 | 3, 7 | | | | × | | × | | | | 7, 8, 3 | 2, 10, 6 | 1, 4 |
| | | | | | | | | | | | | 9, 11 | 8, 3 | |
| 19 | 445 | 1, 5 | 1 | 5 | | | | | | G3 | | 7, 8, 3 | 2, 10, 6 | 1, 4, 5 |
| | | | | | | | | | | | | 9, 11 | 8, 3 | |

**TABLE 10.15**
**Check on Entering Machine 6**

| Entering Machine | Existing Machines | | | |
|---|---|---|---|---|
| | G1 | Relationship | G2 | Relationship |
| 6 | 7 | 0 | 2 | 1060 |
| | 8 | 840 | 10 | 1030 |
| | 3 | 840 | | |
| | 9 | 0 | | |
| | 11 | 0 | | |
| Total | 5 | 1680 | 2 | 2090 |
| **CR** | 1680 ÷ 5 = 336 | | 2090 ÷ 2 = 1045 | |

layouts of the plant and of an individual cell are known. However, it is clear that the *denser* a cell is, the more efficient it is in MH. Density is a measure of how compact a cell is in terms of the number of machines and jobs allotted to the cell.

The measure of MH, called the *material-handling density index* (MHDI), is defined as

$$\text{MHDI} = \frac{\text{total material flow in cell } i}{\text{no. of machines in cell } i \times \text{no. of jobs in cell } i} \Big/ G$$

where $G$ is the total number of cells in the solution. The cost of within-cell MH is inversely proportional to MHDI. The more compact the cell, the more efficient the MH.

2. *Cost of between-cell MH.* This cost includes two components: the cost determined by the number of units transported between cells and the cost determined by the number of jobs that are transferred from one cell to another. It is assumed that the transportation cost per unit includes the cost of handling the unit in the parent cell, between the parent cell and the host cell, and within the host cell. The job-related expense is

**TABLE 10.16**
**Check for Duplication**

| Entering Machine | Existing Group Machines | | Entering Machine | Existing Group Machines | |
|---|---|---|---|---|---|
| | G1 | Relationship | | G2 | Relationship |
| 6 | 7 | 0 | 8 | 2 | 840 |
| | 8 | 840 | | 10 | 810 |
| | 3 | 840 | | 6 | 840 |
| | 9 | 0 | | | |
| | 11 | 0 | | | |
| Total | $n = 5$ | 1680 | | 3 | 2490 |
| **CR** | 1680 ÷ 5 = 336 | | | 2490 ÷ 3 = 830 | |

**TABLE 10.17**
**Group for MH, $P = 0.1–0.6$**

| | Machine Number | | | | | | | | | | | | | |
|---|---|---|---|---|---|---|---|---|---|---|---|---|---|---|
| | Group 1 | | | | | Group 2 | | | | | Group 3 | | | |
| Part Number | 3 | 7 | 8 | 9 | 11 | 2 | 3 | 6 | 8 | 10 | 1 | 4 | 5 |
| 4 | | 300 | | 300 | 300 | | | | | | | | |
| 6 | 500 | 500 | 500 | | | | | | | | | | |
| 9 | | 50 | 50 | 50 | 50 | | | | | | | | |
| 10 | | | 90 | | 90 | | | | | | | | |
| 13 | 5 | | 5 | | | | | | | | | | |
| 14 | | 15 | 15 | 15 | | | | | | | | | |
| 17 | | 50 | 50 | 50 | 50 | | | | | | | | |
| 18 | | 900 | 900 | 900 | 900 | | | | | | | | |
| 5 | | | | | | 800 | 800 | 800 | 800 | 800 | | | |
| 7 | | | | | | | | | | 60 | | | 60 |
| 8 | | | | | | 220 | | 220 | | 220 | | | |
| 11 | | | | | | | | | | 80 | 80 | 80 | |
| 12 | | | | | | 10 | 10 | 10 | 10 | 10 | | | |
| 19 | | | | | | 30 | 30 | 30 | 30 | | | | |
| 1 | | | | | | | | | | 10 | 10 | 10 | 10 |
| 2 | | | | | | | | | | 55 | 55 | 55 | 55 |
| 3 | | | | | | | | | | 30 | 30 | 30 | 30 |
| 15 | | | | | | | | | | 60 | 60 | 60 | 60 |
| 16 | | | | | | | | | | 30 | 30 | 30 | 30 |
| 20 | | | | | | | | | | | 100 | 100 | 100 |
| 21 | | | | | | | | | | | 50 | 50 | 50 |
| 22 | | | | | | | | | | | 110 | 110 | 110 |

attributable to additional clerical accounting when the job is transferred between cells.

3. *Cost of machine duplication.* A minimum number of machines is needed to complete all jobs. Any additional machines acquired to obtain cell separation are charged as duplicating machines.
4. *Cost of cell management.* This cost is associated with planning and scheduling of jobs and labor within each cell, as well as with the clerical and managerial work otherwise involved within each cell. The cost is considered to be directly proportional to the number of cells. Thus the total cost associated with grouping can be stated as

$$\mathrm{MH} \times \frac{\mathrm{MHDI}}{\mathrm{MHDI}_G} + \sum_j \sum_k u_{jk} \times B + K \times J_G + \sum_i C_i \times D_i + CM \times G$$

where

MH = material-handling cost when all jobs are processed in a single cell — that is, when there is no cellular grouping

$MHDI_1$ = MHDI when the production is distributed in a single cell
$MHDI_G$ = MHDI when the production is distributed in $G$ cells
$u_{jk}$ = number of units transported between cells $j$ and $k$; $j < k$ and $j = 1, 2, \ldots, G - 1$
$B$ = cost of transportation of a unit between cells
$J_G$ = number of jobs requiring between cell transportation
$K$ = administration cost per job that is processed between cell
$C_i$ = cost of machine $i$
$D_i$ = number of units of machine $i$ that are duplicated
$CM$ = cost of management per cell
$G$ = total number of cells in the solution.

### 10.2.2 Evaluation Procedure

For each of the solutions obtained in phase I, perform following steps.

Step 1: Calculate the cost of the solution.
Step 2: Determine if any machines are duplicated in the solution. If there is none, step 1 gives the optimum cost. If one or more machines are duplicated, apply the remaining steps.
Step 3: Determine the increase in intercellular transportation if a duplicated machine is taken off a cell. Perform this analysis for each duplicated machine.
Step 4: Determine the cost for each case in step 3 by using the cost expression stated previously.
Step 5: Select the solution with the minimum cost as the best solution.

#### CONTINUATION OF THE ILLUSTRATIVE EXAMPLE

Consider again the problem illustrated previously. For $P = 0.1$–$0.6$, the solution remains the same as shown in Table 10.17. The solution for $P = 0.7$ and 0.8 is illustrated in Table 10.18, and the solution for $P = 0.9$ is shown in Table 10.19. The solutions are obtained by using either of the computer programs GROUP or LARGROUP provided on the diskette.

Now, suppose the yearly costs are as follows: MH = \$40,000, $B$ = \$5/ unit, $K$ = \$100/ job, $C_3$ = \$3000, $C_8$ = \$4000, CM = \$1500. The following shows an example of the cost calculations for one such machine grouping that was obtained for $P = 0.1$–$0.6$.

#### 10.2.2.1 Solution with a Single Cell

That is, the original problem without machine grouping

$$\text{Total cost} = \text{material-handling cost} + \text{cell-management cost}$$
$$= 40{,}000 + 1500 = \$41{,}500$$

The MHDI for the single-cell solution is:

$$\text{MHDI} = \frac{[(\text{sum of material flow})/(\text{no. of jobs}) \times (\text{no. of machines})}{\text{no. of cells}}$$
$$= \frac{13{,}425/(22 \times 11)}{1} = 55.47$$

**TABLE 10.18**
**Grouping for MH: $P = 0.7$ and $P = 0.8$**

| | Machine Number | | | | | | | | | | | |
|---|---|---|---|---|---|---|---|---|---|---|---|---|
| **Part** | **Group 1** | | | | **Group 2** | | | | | **Group 3** | | |
| **Number** | **7** | **8** | **9** | **11** | **3** | **8** | **6** | **2** | **10** | **1** | **4** | **5** |
| 4 | 300 | | 300 | 300 | | | | | | | | |
| 6 | 500 | 500 | | | 500 | | | | | | | |
| 9 | 50 | 590 | 50 | 50 | | | | | | | | |
| 10 | | 90 | | 90 | | | | | | | | |
| 14 | 15 | 15 | 15 | | | | | | | | | |
| 17 | 50 | 50 | 50 | 50 | | | | | | | | |
| 18 | 900 | 900 | 900 | 900 | | | | | | | | |
| 5 | | | | | 800 | 800 | 800 | 800 | 800 | | | |
| 7 | | | | | | | | | 60 | | | 60 |
| 8 | | | | | | | 220 | 220 | 220 | | | |
| 11 | | | | | | | | 80 | 80 | 80 | 80 | |
| 12 | | | | | | 10 | 10 | 10 | 10 | 10 | | |
| 13 | | | | | | 5 | 5 | | | | | |
| 19 | | | | | | 30 | 30 | 30 | 30 | 30 | | 10 |
| 1 | | | | | | | | | 10 | 10 | 10 | |
| 2 | | | | | | | | | 55 | 55 | 55 | 55 |
| 3 | | | | | | | | | 30 | 30 | 30 | 30 |
| 15 | | | | | | | | | 60 | 60 | 60 | 60 |
| 16 | | | | | | | | | 30 | 30 | 30 | 30 |
| 20 | | | | | | | | | | 100 | 100 | 100 |
| 21 | | | | | | | | | | 50 | 50 | 50 |
| 22 | | | | | | | | | | 110 | 110 | 110 |

#### 10.2.2.2 Solution with $P = 0.1$–$0.6$

The MH in cells 1, 2, and 3 are 6635, 5050, and 1335 units, respectively. Similarly, the number of jobs and machines assigned to cells are

For cell 1: eight jobs and five machines
For cell 2: six jobs and five machines
For cell 3: eight jobs and three machines

Therefore,

$$\text{MHDI}_3 = \frac{[6635/(8 \times 5) + 5050/(6 \times 5) + 1355/(8 \times 3)]}{3}$$

$$= 129.94$$

$$\text{Total cost} = \text{within-cell material handling} + \text{between-cell material handling}$$
$$+ \text{duplicated machines} + \text{cell management}$$

## TABLE 10.19
## Grouping for MH: $P = 0.9$

| Part Number | Machine Number | | | | | | | | | | | |
|---|---|---|---|---|---|---|---|---|---|---|---|---|
| | Group 1 | | | | Group 2 | | | | | Group 3 | | |
| | 7 | 9 | 11 | | 3 | 8 | 6 | 2 | 10 | 1 | 4 | 5 |
| 4 | 300 | 300 | 300 | | | | | | | | | |
| 9 | 50 | 50 | 50 | | 50 | | | | | | | |
| 10 | | | 90 | | 90 | | | | | | | |
| 14 | 15 | 15 | | | 15 | | | | | | | |
| 17 | 50 | 50 | 50 | | 50 | | | | | | | |
| 18 | 900 | 900 | 900 | | 900 | | | | | | | |
| 5 | | | | 800 | 800 | 800 | 800 | 800 | | | | |
| 6 | 500 | | | 500 | 500 | | | | | | | |
| 7 | | | | | | | | | 60 | | | 60 |
| 8 | | | | | | | 220 | 220 | 220 | | | |
| 11 | | | | | | | | 80 | 80 | 80 | 80 | |
| 12 | | | | | 10 | 10 | 10 | 10 | 10 | | | |
| 13 | | | | | 5 | 5 | | | | | | |
| 19 | | | | | 30 | 30 | 30 | 30 | | | | |
| 1 | | | | | | | | | 10 | 10 | 10 | 10 |
| 2 | | | | | | | | | 55 | 55 | 55 | 55 |
| 3 | | | | | | | | | 30 | 30 | 30 | 30 |
| 15 | | | | | | | | | 60 | 60 | 60 | 60 |
| 16 | | | | | | | | | 30 | 30 | 30 | 30 |
| 20 | | | | | | | | | | 100 | 100 | 100 |
| 21 | | | | | | | | | | 50 | 50 | 50 |
| 22 | | | | | | | | | | 110 | 110 | 110 |

$$= 40,000 \times \frac{55.47}{129.94} + 405 \times 5 + 8 \times 100 + 300$$

$$+ 4000 + 1500 \times 3$$

$$= \$31,400$$

Here 405 is the total material flow between cells associated with eight jobs that are transferred between cells. Machines 3 and 8 are duplicated, and there are three cells to manage.

As noted before, in this solution machines 3 and 8 are duplicated. Now let us evaluate the cost benefit of eliminating duplicated machines. Eliminating machine 8 from group 1 would involve the transfer of seven jobs, with a total of 500 + 50 + 90 + 5 + 15 + 50 + 900 = 1610 units to transfer between cells. Even though this would cause an increase in intercellular transportation costs, it would reduce the cost of MH within the cell. This reduction occurs because the units assigned for intercell transportation are handled differently and are included in the cost of between-cell handling. If machine 8 is eliminated from cell 2, it would cause transfer of three jobs with 800 + 10 + 30 = 840 units.

For the first alternative, MHDI is

$$\text{MHDI} = \left[\frac{6635-1610}{4\times 8} + \frac{5050}{5\times 6} + \frac{1335}{8\times 3}\right]\Big/3$$

$$= 127$$

Because machine 8 is eliminated from group 1, the number of machines in that group is reduced from 5 to 4. This is reflected in the denominator of term 1. Similarly, 1610 units are transferred from group 1 and are given special handling. Hence, the numerator of the first term is reduced by 1610.

The cost of this alternative is

$$\frac{40{,}000\times 55.47}{127} + (405+1610)\times 5 + (8+7)$$

$$\times 100 + 3000 + 4500\$$$

Removing machine 8 from group 2 results in the transfer of 840 units associated with three jobs. The corresponding MHDI is 132.3, and the cost is $31,596. These costs are higher than the original cost and hence machine 8 remains in both cells.

Similar analysis is performed with machine 3. However, a question may be raised about why we do not just perform the marginal analysis. For example, removal of machine 3 in cell 1 could cause 505 units of intercellular moves, associated with two jobs. It would save $3000, the cost of machine 3. The marginal profitability is 3000 – 505 × 5 – 2 × 100 = $275. Hence, even without savings in cell MH, it is profitable to remove machine 3 from cell 1. Similar analysis with machine 3 in cell 2 shows 840 units of intercellular moves, associated with three jobs and resulting in $3000 – 840 × 5 – 3 × 100 = –$1500. Because machine 3 is needed in at least one of the cells, it would be profitable to remove it from cell 1. The actual cost is

$$\frac{40{,}000\times 55.47 + (405+505)\times 5 + (8+2)}{137.83}$$

$$\times 100 + 3000 + 4500 = \$29{,}148$$

A similar check on the removal of machine 3 from cell 2 would result in a cost of $31,596. This marginal analysis proved effective in this case, However, because we cannot be certain of the reduction in within-cell MH costs, it is preferable to calculate the total cost for each alternative and then make the decision.

We will leave it to the reader to perform similar economic analyses for different values of $P$ and determine the most economical machine arrangement.

## 10.3 MACHINE PLACEMENT IN JOB SHOP OR CELLULAR MANUFACTURING

The problem of placing machines in a job shop or even in a large cellular manufacturing layout is rather complex. In both these types, multiple jobs with different machining requirements and varied processing sequences are produced using many of the standard machines. In addition, the demand for each product may vary, necessitating different MH requirements. Thus, which machine to place in which location to minimize the total material travel is not immediately obvious. This is unlike an assembly line, where the machining sequence is inflexible, so a very efficient flow arrangement can be developed.

Suppose $N$ machines are to be placed in $N$ locations. There are $N$ possible combinations. Evaluating all possible combinations to select the best option is burdensome even for small values of $N$. For $N = 10$, there are 3,628,800 combinations that must be checked. An exhaustive search is computationally prohibitive, so some other more effective approach must be examined. This problem is called the *quadratic assignment problem* in the literature, and there are a few heuristics that may be used to obtain an optimum solution — or even one very close to the optimum solution. Presented here is one simple heuristic that is quite effective in getting a good answer.

The heuristic is divided into two phases. The first phase obtains a good initial solution, which is further improved in the second phase. The data for the problem may be displayed in two tables. The first gives the distances between the locations where the machines could be placed, and the second gives the flow of units or MH trips between the machines. This flow value between two machines is obtained by adding the expected material flow or trips between the two machines for each product. The procedure is as follows.

### 10.3.1 Phase I: Initial Solution

In this phase, two chains are developed: a facility chain and a location chain. The facilities are then assigned to the appropriate locations to obtain the initial solution.

#### 10.3.1.1 Facility or Machine Chain

The purpose here is to connect all the facilities (machines) in a chain such that those facilities with the maximum flow between them are connected together. Each node of the chain, representing a facility, may have at the most two other nodes connected to it. The details are:

1. Construct a total flow table by adding the flow from machine $i$ to $j$ and machine $j$ to $i$. Because the table is symmetrical about the diagonal, only the upper or the lower half needs to be constructed.
2. From the table, select the maximum-flow element (if there is a tie, break it randomly) and connect the nodes (facilities) associated with this element.
3. Cross off the element used in decision making. From the remaining elements, select the maximum element.
4. Connect the nodes associated with this maximum element.

Remember that a node may be connected to a maximum of only two other nodes. If either of the nodes designated for connection already has two other nodes attached to it, this new connection cannot be made. Also, if connecting two nodes makes a loop and not a chain, the connection cannot be made. In such cases, return to step 3. If all the nodes are connected in a chain, go to the next step, developing the location chain. If not, go back to step 3.

#### 10.3.1.2 Location Chain

The steps involved in development of a location chain are identical to the steps in facility chain development except for:

1. Use the distance table, which shows the distances between the locations, for making decisions.
2. Choose the *minimum* element from the table in every calculation instead of the maximum element, as was done in facility chain development.

Finally, sequentially assign each facility from the facility chain to a corresponding location in the location chain. The first facility in the facility chain is assigned to the first location in the location chain, the second facility to the second location, and so on. This is the initial solution.

In development of a facility chain or location chain, the chain obtained may or may not be unique, and hence it may influence the final solution. But in each case, the final solution obtained by making improvements on the initial solution would be very close to the optimum, if not the optimum.

### 10.3.2 Phase II: Improvement Routine

After obtaining the initial solution, the next step is to determine if any improvement is possible. This is accomplished by interchanging the locations of two facilities while keeping all other facilities in their respective locations. If any improvement in the solution is obtained, this becomes the new interim solution, and the process continues. For $N$ facilities, exchanging two machines at a time would require $N!/(N-2)!2! = N(N-1)/2$ combinations to be checked. The procedure is divided into two stages.

#### 10.3.2.1 Initial Solution

1. Construct a facility-to-facility distance table based on the present placement of the facilities. The order (sequence) of the facilities in the table must be the same as the order of the facilities in the flow table.
2. Multiply each element of the facility-to-facility distance table by the corresponding element of the flow table. Designate this as the initial solution table.
3. Obtain the total of the elements in each column and each row. The value of the initial solution is given by the grand sum of either all the column totals or all the row totals.

#### 10.3.2.1.1 Improvement Check

Check if any improvement of the solution is possible by interchanging any two facilities $i$ and $j$. This interchange is called a two-way interchange; the steps are:

1. Begin with the initial solution table.
2. Interchange facilities $i$ and $j$ in the facility-to-facility distance table by interchanging the columns and rows associated with the $i$th and $j$th facilities. The order is immaterial; either the columns or rows may be switched first. The net effect is to change the elements in the rows and columns of the $i$th and $j$th facilities except for the intersecting elements — that is, elements $C_{ij}$, $C_{ii}$, $C_{ji}$, and $C_{jj}$. Because the $i$th and $j$th rows and columns are the only ones of interest, we shall ignore all other elements. The resulting table is referred to as an *exchange table*.
3. Take the product, element by element, of interchanged rows and columns from the exchange table and corresponding elements from the flow table. Denote this as the *product table*.
4. Take the sum of all the elements in columns $i$ and $j$, and the sum of all the elements in rows $i$ and $j$ from the product table. Calculate the grand total by adding all four totals together.
5. Add the four totals associated with the rows and columns $i$ and $j$ in the solution table.
6. If the grand total in step 4 is less than the grand total in step 5, then the exchange of facilities $i$ and $j$ is profitable; proceed to step 8. If not, continue to the next step.
7. See if all possible combinations have been checked. If they have, go directly to step 10; if not, continue with the next combination by assigning appropriate values for $i$ and $j$. Return to step 2.
8. Form the new facility-to-facility distance table by using the exchanged elements and the remaining elements of the previous facility-to-facility distance table.
9. Construct the new solution table by repeating steps 2 and 3 in the initial solution section using the new facility-to-facility distance table. Return to step 7 in the improvement check section.
10. The final solution is obtained.

## ILLUSTRATIVE EXAMPLE

Consider the decision of placing four machines in four locations in a flexible manufacturing shop. The expected flows between machines per unit of time (week) for all products and the distances between possible machine locations are given in the Table 10.20 and Table 10.21.

### *Initial Solution*

To develop the facility chain, first establish the total flow between the machines. For example, the total flow between machine 1 and 2 (either 1 to 2 or 2 to 1) is equal to $10 + 3 = 13$ units.

Table 10.22 below shows all other values.

**TABLE 10.20**
**Material Flow Between Machines**

| From Machine | To Machine 1 | 2 | 3 | 4 |
|---|---|---|---|---|
| 1 | 0 | 10 | 5 | 6 |
| 2 | 3 | 0 | 3 | 6 |
| 3 | 5 | 8 | 0 | 9 |
| 4 | 1 | 4 | 3 | 0 |

The largest element in Table 10.22 is 13, which is associated with machines 1 and 2; therefore, nodes 1 and 2 are connected to form the portion of the chain 1-2. The element 13 is scratched. The next largest element is 12, and associated machines 3 and 4 are connected to form the next segment, 3-4. The element 12 is scratched. The next largest element is 11, between 2 and 3. Because 3 and 2 each have only one connected node, 3 and 2 can be connected. Thus we have 1-2-3-4 as the machine chain. Because all the machines are now connected, the chain is complete.

The next step is to develop the distance chain. Because the distances are symmetrical about the diagonal in the distance matrix, only the elements above or below the diagonal need to be considered. The smallest element above the diagonal is 2, associated with locations 1 and 3 and locations 2 and 4. We may arbitrarily break the tie by selecting locations 1 and 3 and joining them together: 1-3. The next smallest element is 2, between locations 2 and 4. Thus connect 2 and 4: 2-4. The third smallest distance element is 3, between locations 3 and 4. Connecting 3 and 4 results in the chain 1-3-4-2. Note that connections 2-4 and 4-2 are equally efficient. All the locations are connected, and hence we have completed the location chain. Pairing sequentially from the machine and location chains, the initial assignment in Table 10.23 is obtained.

The next step is to develop the machine-to-machine distance table, Table 10.24. In this table, we must make sure that the machines in the distance table are listed in the same order as the machines in the flow table, namely, 1, 2, 3, 4. The format shown in Table 10.24 for determining the distances might help in achieving that goal. In the table, machines are arranged in the same order as in the flow table. The location for each machine is noted and then the distance elements are filled by noting the distances between locations; for instance, the distance between locations 1 and 3 is 2, which is also the distance between machines 1 and 2.

**TABLE 10.21**
**Distances Between Locations**

| From Machine | To Location 1 | 2 | 3 | 4 |
|---|---|---|---|---|
| 1 | 0 | 6 | 2 | 4 |
| 2 | 6 | 0 | 5 | 2 |
| 3 | 2 | 5 | 0 | 3 |
| 4 | 4 | 2 | 3 | 0 |

**TABLE 10.22**
**Total Traffic Flow Between Machines**

| | Machine | | | |
|---|---|---|---|---|
| **Machine** | **1** | **2** | **3** | **4** |
| 1 | 0 | 13 | 10 | 7 |
| 2 | | 0 | 11 | 10 |
| 3 | | | 0 | 12 |
| 4 | | | | 0 |

Multiplying each element of the machine-to-machine distance table by the corresponding element of the flow table, Table 10.20, gives the results as shown in Table 10.25. Adding either row totals or column totals gives the grand total, which is the cost of the solution. In this case the solution is a unit distance of 215.

### *Improvement Check*

There are four machines, and if we interchange two at a time, there will be $4!/2! \times 2! = 6$ combinations to check. These combinations are 1-2 (i.e., machine 1 exchanged with machine 2), 1-3, 1-4, 2-3, 2-4, and 3-4. Because the order of examination is of no significance, we check 1-2 first.

To get a new machine-to-machine distance table, simply interchange columns and rows associated with the exchanged machines, namely, 1 and 2 in Table 10.24. The resulting values are shown in Table 10.26 (you may verify this by working with the distance table). Again, note that through this process, the resulting table has the machines automatically arranged in the same sequence as the sequence in the flow table, namely, 1, 2, 3, 4.

The flow cost is obtained by multiplying the distance elements by the corresponding flow elements from the flow table, Table 10.20, and adding these products together.

Comparing the modified solution table, Table 10.27, with the present solution table, Table 10.25, we can see that only the rows and column associated with machines 1 and 2 are interchanged; all other elements are the same. Hence, if we just compare the effect of the elements that have changed in the solution matrices, we will find the difference between the solutions.

However, it may be more convenient to add the elements at the intersections of rows and columns that are interchanged and make the comparison. By doing so, we will add C11, C12, C21, and C22 twice each time. Inasmuch as these values do not change in the two solution matrices, the net difference is not affected.

**TABLE 10.23**
**Initial Machine Assignments**

| Machine | 1 | 2 | 3 | 4 |
|---|---|---|---|---|
| Location | 1 | 3 | 4 | 2 |

**TABLE 10.24**
**Machine-to-Machine Distance Table**

| Machine | | 1 | 2 | 3 | 4 |
|---|---|---|---|---|---|
| | Location | 1 | 3 | 4 | 2 |
| 1 | 1 | 0 | 2 | 4 | 6 |
| 2 | 3 | 2 | 0 | 3 | 5 |
| 3 | 4 | 4 | 3 | 0 | 2 |
| 4 | 2 | 6 | 5 | 2 | 0 |

**TABLE 10.25**
**Product Matrix**

| | Machine | | | | |
|---|---|---|---|---|---|
| Machine | 1 | 2 | 3 | 4 | Total |
| 1 | 0 | 20 | 20 | 36 | 76 |
| 2 | 6 | 0 | 9 | 30 | 45 |
| 3 | 20 | 24 | 0 | 18 | 62 |
| 4 | 6 | 20 | 6 | 0 | 32 |
| **Total** | 32 | 64 | 35 | 84 | 215 |

**TABLE 10.26**
**Machine-to-Machine Distance Table**

| Location | Machine | 1 | 2 | 3 | 4 |
|---|---|---|---|---|---|
| 3 | 1 | 0 | 2 | 3 | 5 |
| 1 | 2 | 2 | 0 | 4 | 6 |
| 4 | 3 | 3 | 4 | 0 | 2 |
| 2 | 4 | 5 | 6 | 2 | 0 |

**TABLE 10.27**
**Modified Solution Values: 1–2**

| Machines | 1 | 2 | 3 | 4 | Total |
|---|---|---|---|---|---|
| 1 | 0 | 20 | 15 | 30 | 65 |
| 2 | 6 | 0 | 12 | 36 | 54 |
| 3 | 15 | 32 | 0 | 18 | 65 |
| 4 | 5 | 24 | 6 | 0 | 35 |
| **Total** | 26 | 76 | 33 | 84 | 219 |

**TABLE 10.28**
**Machine-to-Machine Distance, 1–3, Cycle III**

| Location | Machine | 1 | 2 | 3 | 4 |
|---|---|---|---|---|---|
| 4 | 1 | 0 | 3 | 4 | 2 |
| 3 | 2 | 3 | | 2 | |
| 1 | 3 | 4 | 2 | 0 | 6 |
| 2 | 4 | 2 | | 6 | |

For example, the sum of rows 1 and 2 and columns 1 and 2 in the modified solution matrix is 65 + 54 + 26 + 76 = 221. The corresponding sum in the product matrix (Table 10.25) is 76 + 45 + 32 + 64 = 217, indicating that the new solution is 221 − 217 = 4 units higher in terms of cost. Recall that the cost for the initial arrangement was 215 and the cost of the 1-2 exchange is 219, an increase of 4 units; therefore, the change would not be made and the present solution remains the solution at this stage.

Similar analyses with other combinations are made. For example, 1-3 results in the distance matrix in Table 10.28 (only the rows and columns that are interchanged are shown). Multiplying by the corresponding flow values gives Table 10.29.

The comparative value for this solution is 62 + 90 + 31 +44 = 227, and the corresponding total from the present solution matrix is 76 + 62 + 32 + 35 = 205. Because 205 < 227, we will not interchange machines 1 and 3. Checking all other combinations results in no interchanges, giving the present (initial) solution as the best solution.

It must be emphasized that the method presented here is a heuristic. On large problems, there may be several ways of getting the initial solutions, because the ties can be broken in different ways.

This may lead to different final solutions, some better than others. However, most would be in the vicinity of the optimum solution. We could perform three-way, four-way, or higher-order exchanges, wherein three, four, or more machines are simultaneously exchanged, to see if the solution can be further improved. But in most examples, such efforts have not contributed much in further improving the solution.

**TABLE 10.29**
**Solution Values, Cycle III**

| Machines | 1 | 2 | 3 | 4 | Total |
|---|---|---|---|---|---|
| 1 | 0 | 30 | 20 | 12 | 62 |
| 2 | 9 | | 6 | | |
| 3 | 20 | 16 | 0 | 54 | 90 |
| 4 | 2 | | 18 | | |
| **Total** | 31 | | 44 | | |

## 10.4 PACKAGING

Next, we discuss packaging. Packaging is an important aspect in overall production and MH, and requires input from engineering, production, graphics, and advertising personnel. The specifications for the product package very much depend on the product design, and any change in design can cause a significant change in package requirements. It is therefore essential to consider packaging in the designing, production, and MH phases of the product.

Packaging also has a big role in a consumer's decision to purchase. If there are several varieties of the same basic product for about the same price, it is most likely that the one purchased will be the one that stands out the most. The size, shape, and colors of the packaging can be very instrumental in product sales. Even for industrial or commercial products, packaging plays an important role in delivering the product intact at minimal additional cost.

### 10.4.1 Functions

Packaging mainly serves to protect a product from damage caused by handling or exposure to environmental conditions involving heat, moisture, light, and even electronic interference and radiation. It allows a manufacturing firm flexibility in locating its facilities in a site that is most suitable in terms of production-oriented factors such as labor, raw materials, and utilities, without having to be concerned with whether the finished product can be delivered safely to its customers. The type of packaging also contributes to formation of the unit load, which is necessary in the selection and use of the type of MH equipment.

Plastic film wrapping for coils (courtesy of Avon Engineering — a division of Bushman Equipment, Inc.)

Box date and time stamp machine

There are three major categories in packaging: consumer, industrial, and military. Consumer packaging, which can be subdivided into retail and institutional, is characterized by small units of products handled in large numbers. When the packaging is for retail purposes, its appearance should be emphasized. For institutional use, protection, cost, and convenience are much more important than appearance. Quite commonly, large-sized units of a product indicate industrial packaging. Military packaging is specified by the government.

The important aspects of a package include its structure, aesthetic appeal, style, ability to communicate information to the user, and adherence to legal specifications. The development of a package follows steps similar to those of product design. First, the designer should determine whether the packaging is for industrial or retail use to get a sense of the appropriate size and weight of a single package. Then the pallet size for shipping and how high the pallets can be stacked without damage will dictate a load.

The packaging personnel must be very familiar with the product to develop a proper package. This includes its physical specifications, how it is to be used, and details of its promotional information. They must also maintain high ethical

Shrink wrap machine wrapping a box

standards by not using deceptive labeling and should pay attention to consumer needs, which can be identified through market research.

The type of material used for packaging is controlled by the protection needed for the product, which in turn depends on factors such as the sensitivity of the product (electronic instruments are very sensitive, while refrigerators and appliances are moderately rugged), the weight of the product, the method of shipping and handling, the desired shelf life of the material, and whether the packaged material is to be

A wrapping machine wrapping a pallet (courtesy of Texas Tile Manufacturing LLC)

stored indoors or outdoors. There are different materials to be used depending on the protection desired, for example, protection against breakage, moisture, or heat.

In designing the individual package, an existing design that fills all the packaging needs could be used, or the package could be designed entirely from scratch if no suitable modification to the existing design can be made. In any case, customer appeal, the packaging budget of the company, proper product labeling information, and the universal product code should be considered.

When the package is ready to be put into use, several production aspects must be kept in mind. The product manufacturing rate must be the minimum rate of packaging. The procedure should therefore be evaluated to determine the number of machines and personnel that would be needed to achieve this balance.

### 10.4.2 Protection

Shock from handling and transportation can be damaging to products, especially fragile objects. Formed plastic trays or Styrofoam molds, which are lightweight and can be shaped to fit the object, can be helpful. Foam-in-place is very versatile, a light and inexpensive method of packaging that can be partly or fully automated if the volume justifies. Packaging materials such as Styrofoam chips, thermofoam, polyethylene and polyurethane foams, paperboard partitions, air cushion mats, and die-cut corrugated inserts are other means of protecting against shock damage. Packaging the product in large units can help hold each individual unit in its place. Human error in handling that results in damage to the product can be reduced through training or the use of automated handling.

Federal and state regulations and company ethics require that packaging methods enhance the safety of the consumer. Potentially hazardous materials should be properly packaged and handled. In 1970, the Poison Prevention Packaging Act allowed for the formation of the Consumer Product Safety Commission. Among the services provided by this commission is the publication of a list of products requiring childproof packaging. Products containing dangerous chemicals or even radioactive substances should be packed to ensure that no leakage will occur during the roughest handling.

### 10.4.3 Design and Material Considerations

Two major aspects in designing the package are careful consideration and evaluation of all available material that could be used. Of particular importance are the static electricity, humidity, temperature, and barrier qualities; the material should keep out water and moisture, greases, oils, gases, and odors while holding in the product. The most common types of packaging materials are glass, steel, aluminum, paper, cardboard, wood, and petrochemical products such as synthetic rubber and plastics. Various chemicals, adhesives, inks, and solvents are also used in developing the final package. We discuss a few of these materials in the following paragraphs.

#### 10.4.3.1 Glass

One of the oldest packaging materials used is glass. It is formed by melting sand with limestone and soda ash. Glass has the advantage of being strong so as to securely hold the product, but it is relatively heavy for handling and is fragile, breaking easily upon dropping or bumping. Packages formed from glass include bottles, jars, tumblers, jugs, carboys, and vials.

The bottle is the most popular form of glass container. It is generally characterized by its narrow neck and mouth. The bottle is used primarily for holding liquids, and the small mouth minimizes the overall size of the closure. Numerous bottles can be spotted in the average home, storing such items as medicines, carbonated beverages, juices, and spices. A bottle with a wide mouth is classified as a jar, which is commonly used for food in a viscous, semisolid, or granular state. Common kitchen items stored in jars include instant coffee, jelly, mayonnaise, peanut butter, and pickles.

The tumbler and jug are two more glass containers that are variations of the bottle. A tumbler is an inexpensive drinking glass that is frequently used for packaging jams, jellies, and fruit preserves. The top is pressed on instead of being screwed on as in the case of a jar. When the original contents are consumed, the tumbler may become a drinking glass. A large bottle with a handle and a screw-on cap is called a jug and is frequently used to store liquid chemicals and foods.

Carboys and vials are so named because of the relative thickness of the glass. A carboy is a bottle made of very heavy glass used to hold liquid industrial chemicals, while a vial is a small, thin, tubular glass container used for expensive and sensitive drugs.

#### 10.4.3.2 Metal

Steel and aluminum are the most commonly used metals because they are readily formed into cans and drums. Cans are constructed in two or three pieces. In two-piece construction, the metal is formed into the shape of a cup, and the top is added to seal the can. For three-piece construction, the metal is formed into a tube; the top and bottom are produced separately and are then secured to the cylinder to make the can. The inside of the can is frequently coated with tin or lined with a plastic sealant. The light weight of these cans and their resistance to chemical reaction with the product make them attractive, but they dent easily; and if made from steel, they are susceptible to rusting.

#### 10.4.3.3 Paper Products

Cartons and bags are two major paper products used in packaging. They are very common because of their availability and ease of manufacturing, storing, and labeling.

#### 10.4.3.4 Cartons

The folding carton is the most popular form of cardboard packaging because it is economical in terms of both the cost of the material and the cost to produce the finished carton. Their collapsibility makes them easy to ship, because they can be folded flat and may be stacked. Cardboard cartons are versatile, allowing for different styles and numerous printing and labeling methods. Lightweight cartons are

A roller conveyor transporting a box from station to station for filling up the material, sealed, labeled and stamped

relatively flimsy, not giving much protection to the contents; heavy corrugated cardboard boxes, however, rival those made of wood.

### 10.4.3.5 Bags

Bags are a form of packaging that can be made resistant to moisture, are easy to fill and empty, and have low shipping costs. Bags are light and can be folded and stacked. The most common bag materials are paper, plastic, and textiles. Bags have some disadvantages as packages; they are not supportive of the product, and their durability is only average. There are four basic types of bags:

- Pasted open mouth
- Sewn open mouth
- Pasted valve
- Sewn valve

The term *valve* indicates that the bag is secured or sealed at both ends once it is filled.

Examples of these are shown in Figure 10.25.

Plastic is obviously not a paper product, but plastic film can be formed in shapes similar to those of paper bags. The ends can then be closed by sealing or by using a wire tie.

The most common textile bags are made of burlap or cotton, and are sewn together and stitched closed. Burlap bags are commonly used for grains such as oats, while cotton sacks may be filled with finer substances such as flour. Textile bags have the advantage of being reusable.

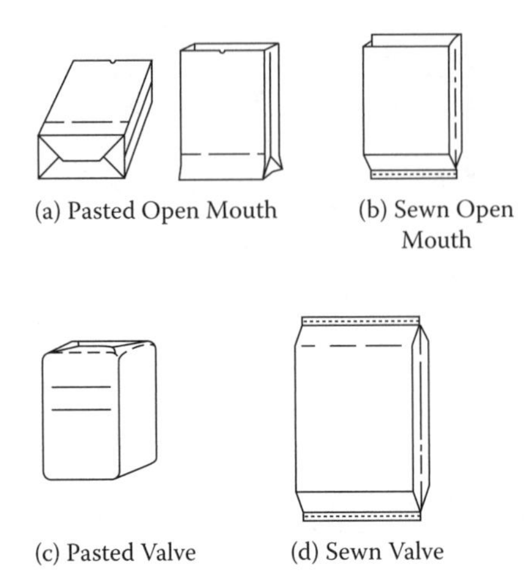


**FIGURE 10.25** Types of bags used in packaging

#### 10.4.3.6 Wood

Boxes and pallets made of plywood and wooden boards are used for packaging and carrying heavy, bulky material. They are comparatively strong and easy to palletize.

### 10.4.4 Consolidation and Palletizing

Protection for individual units can be increased through consolidation or combining several packages into a compact package in which the units support and protect each other. The number of units per container is generally defined by the sales department (e.g., 24 bars of soap in a box), and the packaging department must then develop the size of the pallet. The method of shipping also plays an important part in palletization. Shippers often limit the overall dimensions and weight of shipping containers. Air and ground parcel services quite often do not allow palletization because of size and weight limitations.

Several methods exist for consolidation of units. Adhesives can be used to hold them together, or units can be encased in shrouds, or wraps of stretch or shrink plastics. Boxes, bins, trucks, and rail cars can be used to transport the product in bulk. The large boxes and bins for this purpose are constructed of cardboard, wood, plastic, or metal. Groups of products can also be consolidated by securing them on pallets.

An additional consideration in palletizing is load locking, which is a method used to prevent the various levels stacked on a pallet from sliding apart. One solution is to use a nonslip application on the exterior of the package; however, this can complicate the shipping process. Adhesive can be applied between the levels by spraying, brushing, or rolling. Enclosing the pallet in stretch or shrink wrap can also load-lock the units.

**TABLE 10.30**
**Commonly Used Wrapping Material**

| Material | Advantages | Disadvantages |
|---|---|---|
| Cellophane | Stretchable, gas barrier | Expensive, tears, short shelf life |
| Cellulose acetate | Transparent | Poor barrier for gases and moisture |
| Ethyl acetate | Strong | Expensive |
| Methyl cellulose | Oil barrier | Soluble in water |
| Nylon | Moisture barrier | Expensive, poor gas barrier |
| Polyester | Strong, transparent | Noisy, expensive, average barrier to moisture and gases |
| Polyvinyl chloride | Cheap, transparent, good barrier | Retains odors |
| Rubber hydrochloride | Transparent, strong, oil barrier | Retains odors, average gas barrier |
| Polyvinylidene chloride (trade name: Saran) | Excellent barrier | Expensive |
| Styrene | Transparent | Poor barrier, develops static electrical charge |
| Polypropylene | Transparent, excellent barrier to gases and moisture | Average strength |
| Polyethylene | Low cost, soft, good moisture barrier | Poor gas barrier, slight haze |

#### 10.4.4.1 Shrink/Stretch Film

A common industrial method of holding a pallet of units together is the use of shrink wrap or stretch wrap. The wrap is used with the pallet to prevent shifting of loads and to facilitate the use of automated handling equipment. Shrink wrap requires a heat source to shrink the material around the units. Stretch wrap requires no heat source and therefore has a lower operating cost, but it does not conform as well to irregularly shaped loads as does shrink wrap.

Many different materials are currently in use as stretch and shrink wraps. Table 10.30 lists the most common films in use and their advantages and disadvantages.

Different densities of wraps are commercially available to fill the needs of the company. As density is increased, surface hardness and the heat-seal temperature increase; flexibility is lowered; impact strength decreases; and moisture, gas, and grease resistance increase. Low-density materials are characterized by a lack of stiffness and high-impact strength. The high-density materials are stiffer and provide better barriers.

#### 10.4.4.2 Strapping

Another way to palletize a group of items is to use steel strapping, in which bands of steel, plastic, or other strong materials are wrapped around the units on pallets. The efficiency is greatly improved if such a method can be automated rather than performed manually. However, a large volume of packaging is needed to justify the initial cost of the expensive and sometimes specialized equipment that is needed in such an operation.

#### 10.4.4.3 Labeling

Once the package has been designed and manufactured, its proper labeling must be considered. The function of the label is to relay information to the consumer, information that includes a brief description of the product along with a list of ingredients or contents, instructions for use, warnings, the net weight, and/or the volume, and any other pertinent items. The label should also instruct the user regarding precautions and ways the product should not be used. The manufacturer's name and address should be somewhere on the package, as should the applicable universal product coding.

#### 10.4.4.4 Final Step

A concluding inspection should be made to ensure that all items such as spare parts, mounting screws, and maintenance supplies, as well as all necessary documents are packed. The latter might include the warranty, installation instructions, wiring diagrams as appropriate, shipping papers, and a consumer research questionnaire.

### 10.4.5 Packaging Equipment

Once a product is ready for shipment, it must be securely placed in the appropriate package and sealed for shipping and handling. Box flaps and bag openings have been folded closed and sealed in preparation for palletization. Full pallets and irregularly shaped objects might require strapping to secure them. Bottles must be packed to prevent breakage during shipping. When the production volume is low, packaging can often be done manually, but large companies with high volumes of production depend on semiautomatic equipment for sealing products. The following sections list some of the equipment available.

#### 10.4.5.1 Automatic Adhesive Sealers

An automatic adhesive sealer is a conveyor with an adhesive applicator and a compression unit for folding down and holding the glued flaps of a box. It features top or bottom sealers or simultaneous loader seals. Feeding of boxes can be manual or automatic.

#### 10.4.5.2 Automatic Tape Sealers

An automatic tape sealer is a conveyor with a tape dispenser and a compressor. Tape is measured to the proper length, wetted if necessary, placed on the flaps of the box, and compressed in place to hold the flaps down. The tape sealer can be combined with an adhesive sealer to glue and tape the box at the same time. It features manual or automatic feeding of boxes.

#### 10.4.5.3 Stitchers

In a stitcher, a coil of wire is fed through, cut to proper length, formed into a staple, driven into the box or bag, and tightly clinched in place to secure the flaps. Feeding of boxes can be manual or automatic.

#### 10.4.5.4 Staplers

In a stapler, preformed staples are held in a magazine, pressed into a box or bag, and tightly clinched in place to secure the flaps. The bar-type stapler has a bar on which the box is placed to provide a surface for clinching the staple after it is driven through the box. The anvil-type stapler has two anvils that are driven through the box with the staples to provide a surface for clinching. Staplers can be manually or automatically operated.

#### 10.4.5.5 Strappers

In a strapper, round wire or flat strapping is drawn around the box, bundle, or object, and tightened. The ends are then secured together to close, strengthen, or hold the unit to be shipped. The unit can be passed through the strapper on a conveyor. Semiautomatic operations might require directing of the strapping by the operator. Strapping is secured by twisting or clipping the ends together. Strappers are used primarily to secure pallets.

#### 10.4.5.6 Wrappers

A wrapper is a conveyor-type machine that wraps the unit in paper or film and seals the ends. It is supplied by a roll of paper or film and can combine loading and wrapping.

#### 10.4.5.7 Palletizers

A palletizer is an automatically controlled machine that is capable of stacking a unit load on a pallet. High-speed palletizers (e.g., those used in breweries) are capable of stacking 150 cases/minute, but they are expensive (frequently costing over $100,000) and require a large floor space (about 150 square feet). The cost of low-speed palletizers (12–20 cases/minute) start at about $45,000 and require less than 100 square feet. However, there is a drawback; their stack height is limited to 80 inches or less, while high-speed units can stack to about 150 inches. If the cases being handled are lightweight (less than 40 pounds), robotic palletizers, ranging in cost from $30,000 to $60,000, are available that operate at speeds of 5–15 cases/minute. They can operate in limited spaces and can be programmed for many different patterns.

Advances in the control units used on palletizers have made it possible to integrate palletizers into existing automated systems. In recent years the use of programmable controllers has made it possible to increase the number of available patterns for palletizing from less than 10 with punched tape to about 50. The programmable controller also allows the loading pattern to be changed extremely rapidly.

An automatic palletize machine making a pallet (courtesy of Texas Tile Manufacturing LLC)

## 10.5 REDUCING PACKAGING COSTS

Factors affecting the cost of packaging are similar to those influencing the cost of manufacturing a product. These elements are associated with the design, material, and production phases of packaging. We will discuss several ways in which each factor can be controlled.

### 10.5.1 DESIGN

To the extent possible, select a square shape for a package. This results in minimizing the required surface area, which in turn reduces the material needed for the package. For example, $2 \times 2 \times 2$ gives 8 cubic units of volume with a surface area of 24 square units. Although $1 \times 1 \times 8$ also provides 8 cubic units of volume, it has a surface area of 34 square units. As a case in point, a physical fitness company decided to ship three weights in a box instead of two (when the order permits), which resulted in a square package and reduced the packaging cost by 5%.

Determine how a customer intends to unload the product and, if possible, design the package to accommodate the customer. This might mean developing different packages for different customers, but an increase in sales volume could justify such individualized attention. When an electrical company that supplies products primarily to manufacturers of electronic equipment changed its packaging to suit the automated opening and unloading machines used by such manufacturers, its sales and revenue increased greatly. This offset a small increase in packaging cost, and the net result was increased profit.

Make basic changes in design if they will result in a reduction in cost. Controlling factors could be reduction in the weight of a package, reduction in the number of components in the package, and redesign of the product itself (e.g., by eliminating indented rings around the cans, a soup manufacturer improved its packaging by eliminating wasted space between the cans in packing).

### 10.5.2 Material

The material used in packaging could perhaps be changed to lower the cost. A syrup manufacturer changed its bottles from glass to plastic and realized a saving of 20%, mainly in MH because of the large reduction in total weight requiring transportation. It might be possible to obtain the necessary material for packaging at a lower cost by increasing the volume of purchase, lengthening contracts, or changing from a plant-to-plant contract to a national contract. A supplier that is closer to the plant may be asked to offer better rates, because its freight cost might be considerably lower compared to that of a more distant supplier. Similarly, a supplier with equipment that is better suited for a manufacturer's need could offer a better price. For example, 40-inch paper bags might be provided more cheaply by a supplier with a 40-inch press than by one with a 60-inch press.

### 10.5.3 Production

At times a company can realize a substantial saving by producing packaging material itself. An economic analysis must be performed to evaluate this alternative. Increasing the productivity of the packaging process by increasing utilization of machines, obtaining an efficient and balanced assembly line, and minimizing the idle time should reduce packaging cost. Eliminating or reducing defects and maintenance cost and providing better training for workers are also alternatives that may be considered in the effort to control costs.

## 10.6 DESIGNING A PACKAGING AREA

Packaging is the final task before shipping the product. Like any other job, it must be analyzed to determine an appropriate method of performance. An operation chart and an operation process chart are useful tools in developing the necessary sequence. However, there are a few points that should be noted.

1. Both the means of packaging — for example, strapping, boxing, and stretch wrapping — and the size of the unit load should be established on the basis of the considerations previously discussed.
2. If a machine is to be used, the rate of packaging must be determined. It is possible to buy machines with different speeds — for example, 1 or 10 package(s)/second.
3. The decision as to which machine to buy is also influenced by the variety of packaging the machine will have to perform. Generally, high-speed machines do not adapt as well to changes in specifications as do slower machines.

The layout and floor space required for the packaging area must be given consideration similar to that in designing a work station or a series of work stations for an assembly line. The methods of obtaining a smooth and continuous flow in an assembly line using conveyors were discussed in an earlier section, and the means for calculating area for a work station will be developed in Chapter 12. However, one must keep in mind that packaging most often reflects the product line. It could be an assembly line operation if only one product is involved, a batch-processing operation if there are a number of products manufactured in groups, or in some cases it might even be a job shop arrangement. Because of such possibilities, automation in MH (e.g., a fixed conveyor or skids and forklift truck) should also be established on the basis of the product mix. However, most packaging works well with conveyors; hence, conveyors are normally a dominant feature in packaging areas.

## 10.7 COMPUTER PROGRAM DESCRIPTION

Two interactive programs for queuing analysis are presented for this chapter. One is for a single-server model, and the other is for a multiserver model. The notations used are:

$p(I)$ = probability of units being in the system
$L$ = mean number of units (length) in the system
LQ = mean length of the queue
$W$ = mean waiting time in the system
WQ = mean waiting time in the queue

The programs calculate each of the above system performance measures. Inputs are supplied with answers to the following questions.

1. Single-server model M/M/l: *What is the arrival rate? And the service rate?* (Enter the rate of arrival and rate of service.)
2. Multiserver model M/M/C/K (the model is for $c$ servers and a finite maximum queue length):
   *What is the alrrival rate? And the service rate?* (Enter the total arrival rate and the rate of service per server.)
   *How many servers are there?* (Enter the total number of servers.)
   *What is the queue limit?* (Enter the value of the maximum number of units that are allowed in the queue.)

### 10.7.1 Machine Grouping to Reduce Material Handling

Both the GROUP and LARGROUP programs, introduced in Chapter 4, can be used to perform the grouping up to the economic analysis phase. The only modification that is needed is the input of the data. Rather than elements of an incidence matrix (i.e., 0 or 1) as the input, we should use elements of the machine-to-machine MH table (e.g., 80, 525, …) as the input to the program(s).

### 10.7.2 Machine Placement in Flexible Manufacturing

The program Simultaneous Facility Location (SML) can be found at http://www.crcpress.com/e_products/downloads/download.asp?cat_no=44222. The program is in interactive mode, and the instructions are fairly clear. The program has the ability to evaluate best locations for placement of machines even if there is a fixed cost for a location (e.g., cost of site preparation for a particular machine). Note that this cost may be zero (conforms to the analysis in the book) or could change from machine to machine for each location.

## 10.8 SUMMARY

Selecting the right equipment for moving material is a challenging task. One must know what is to be moved, how frequently it is to be moved, what the physical limitations are, and what costs are involved. The chapter illustrates with numerous examples how this can be done. It begins with a simple model showing basic cost calculations and proceeds toward the more complex problem of equipment selection. Work-volume analysis is one technique that is used. It decides the number of handling units required to perform all necessary material moves if the type of equipment needed for each move is known. The heuristic procedure goes one step further and determines both equipment type and number such that it minimizes the total cost of operation. Selection of AGVs involves a slightly different analysis and is also shown by an example.

The next topic of discussion is flow lines, a study of workplace arrangement to obtain a smooth and continuous flow of material within a plant. Conveyors are most often used for moving material between stations, and they are arranged serially, modularly, or in a closed-loop form.

Conveyors can be made to move intermittently or continuously; both modes are commonly used in industry. However, to provide a safety cushion of unfinished goods in between the stations, decoupling of successive stations is desirable and is obtained by providing banking. Various banking procedures are discussed here, and examples illustrate ways to determine the sizes of the banks. Techniques such as queuing analysis and simulation are helpful in this regard.

A closed-loop conveyor system presents a special challenge, and the basic rules for designing such a system are stated. Two illustrative examples also give some flavor of the design analysis.

Automated control and transfer mechanisms can considerably improve productivity. Controls allow one to perform functions such as count size and weight while units are moving on conveyors. Accessories allow units to change direction and orient the position while still on the conveyors. Automated transfer machines and monorails are two of the most commonly used devices for automated transfer. They transport a unit from one station to the next and position it for subsequent operation with very little, if any, human interaction.

Various configurations are possible in designing an assembly line or a flow line within a plant. Straight line, U-shape, Z-shape, and circular are a few of these arrangements. The contributing factors in such decisions are the physical structure of the plant and the required location for receiving and shipping departments. To develop a plant in a multistory building necessitates planning for vertical flows as well. The chapter presents various arrangements for building such a system.

The next topic discussed is how to group machines in flexible manufacturing to reduce the MH. The procedure is a simple extension of the machine-grouping procedure presented in Chapter 5. Depending on the value of $P$ the analyzer selects, the procedure may provide different groupings. Evaluation of each arrangement with cost minimization as the goal may lead to the selection of the best possible alternative.

Finding the optimal arrangement of machines within a cell to minimize MH is also a challenge. Exhaustive enumeration is impractical in most cases. A simple heuristic, which provides a good solution, is presented in the chapter. The procedure is also computerized, and the program can be found at http://www.crcpress.com/e_products/downloads/download.asp?cat_no=44222.

Packaging is generally the last operation before a product is shipped to a customer. Packaging serves to protect the product from damage in handling and environmental factors. It is important that the final product package has aesthetic appeal if the product is consumer-oriented. For industrial use, protection, cost, and convenience are much more important. Styrofoam molds, trays and chips, air cushion mats, and paperboard partitions are some of the materials used for holding units in place. The material used in packaging the product itself is also important in providing protection. Glass bottles, jars, and tumblers are often used for holding liquids and viscous and semisolid or granular substances. Glass keeps foreign elements such as moisture, gases, or odors from contaminating the product. Steel and aluminum are used in the form of cans and drums to hold liquids. Paper products such as cartons and bags are used to hold granular and small substances, while wooden boxes are used to carry heavy and bulky materials.

Additional protection for the product can be provided by consolidating many units together. For example, a box may contain 12 individual items. Furthermore, the number of such boxes can be packed together on a pallet; this method of shipping is called palletization. Use of a shrink/ stretch film or metal straps is very common in forming such a pallet load or unit load.

Various packaging machines are available to increase the productivity in packaging operations. Automatic adhesive sealers, stitchers, staplers, strappers, wrappers, and palletizers are some of the machines that the chapter describes.

The chapter also presents a discussion on how to reduce packaging costs. Evaluation of design and the material used in packaging, as well as the performance of a make-or-buy analysis are some of the methods suggested for lowering the packaging cost.

Designing the packaging area is a problem similar to designing the rest of the plant. One must define the mode of operation and then develop corresponding work stations and MH facilities. However, a dominant feature in most packaging areas is the use of conveyors, especially if automatic packing machines are used.

## PROBLEMS

1. What is banking in MH? Describe the methods of achieving banking, and the advantages and disadvantages of each method.
2. A rotary table serves as a bank for one workstation. If items arrive at the station following a Poisson distribution with a mean of 3 units/minute and an

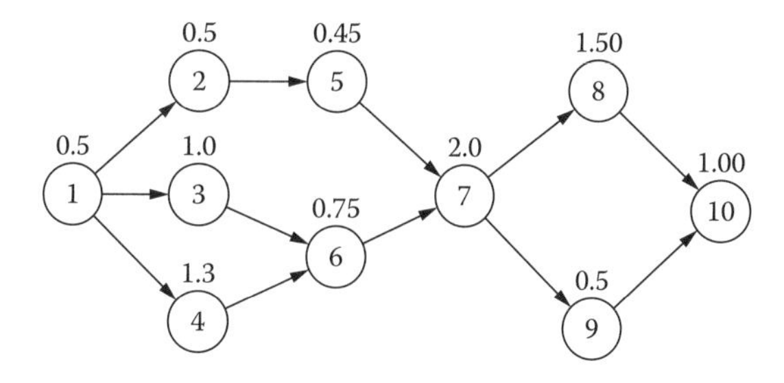


**FIGURE 10.26**

average service time of 0.2 minute/unit, what should be the minimum table size? The probability of not being able to accommodate a unit on the table should be less than 5%. If the operation is to be modified so that the table will now serve as a bank for two operators, each serving the unit with a mean of 0.2 minute, and there are two belts unloading the units on the table, each with a mean of 3 units/minute, what should be the minimum size of the table for keeping the probability of not accommodating a unit to less than 5%?

3. A product has a work distribution as shown by the precedence relationship diagram in Figure 10.26. The number above a node indicates the time in minutes needed to perform that particular operation. A closed-loop hook conveyor system is to be built that will produce 30 items/hour. Four other products are to be built on the same conveyor system with the dimensions (W × L) shown in the accompanying table. If an operator table is to be limited to no more than 6 feet in length, determine the number of stations needed, the length of each station, the spacing of the hooks on the conveyor, and the length of conveyor if the length is to be at least 1½ times the minimum required length for storing the units and the spacing between stations is to be 2 feet. The input/output rates to the system can be 30 and 20 units/hour per person, respectively, and these stations are in addition to the stations necessary for the assembly jobs. The work on the input/output station can be speeded up by increasing the number of persons. For example, two workers can unload 60 units/hour.

| **Product** | **1** | **2** | **3** | **4** | **5** |
|---|---|---|---|---|---|
| Dimensions (inches) | 4 × 2 | 1 × 1 | 3 × 2 | 1 × 2 | 3 × 5 |

4. In the closed-loop conveyor system shown in Figure 10.27, units are loaded automatically from a box onto the conveyor at the input point. The finished products are collected automatically in a box at the discharge point. The input and output boxes can hold 100 and 60 units, respectively. The input source can be replenished every 30 minutes by a forklift operator, while the output can be removed every 20 minutes by another forklift operator. Station lengths can, at the most, be 4 feet.

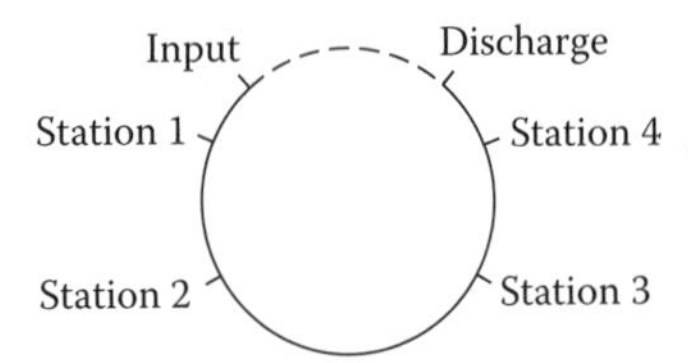


**FIGURE 10.27**

a. Given the information in the accompanying table concerning the time needed at each station to perform each job, determine the best speed for the conveyor.

| Station | Necessary Time (minutes) |
|---|---|
| 1 | 1.5 |
| 2 | 2 |
| 3 | 2 |
| 4 | 1 |

b. What is the minimum capacity of the containers (boxes) needed at the input and output stations for the speed of the conveyor or fixed in part a.
c. What adjustment must be made if the required production rate is to be increased to 60 units/hour?

5. At Pink Dot Battery, an accumulation line is used in conjunction with a roving operator on the acid line. When the flow of empty batteries is too much for one operator to handle, the rover will come and load a batch of batteries (one batch = 32 batteries) on the accumulation line. After the operator catches up, the rover comes and replaces the batch on the line. Because of the flow arrangement shown in Figure 10.28, if the rover did not divert a batch or two, the batteries could back up into assembly. This would cause a slowdown in the assemblers' work and a decrease in their incentive pay (labor problem). Each batch is released by the assembly operator following a uniform distribution with parameters 50 + 8 minutes. The acid line operator takes 3 + 1 minutes to fill each battery. The U-shape within the main conveyor can hold at the most one batch. The rover will take 10 + 3 minutes to load or unload a batch. (The units must be handled in batches for quality control reasons.) The assembly line works 4 hours a day, while the acid line works 8 hours a day. Perform simulation for 2 days of work and decide how frequently the rover should visit the work station. (For simplicity, assume that the rover can visit the station at intervals of 15 minutes only, e.g., 8:00 a.m., 8:15 a.m., 8:30 a.m., and so on.)

6. In problem 5, assume that the time between releasing batches by the assembly line operator follows an exponential distribution (arrival follows a Poisson distribution) with a mean of 50 minutes, and the acid line operator processes a battery following an exponential distribution with a mean of

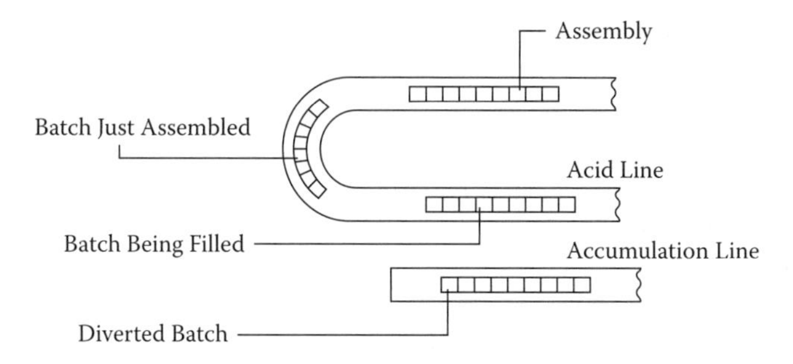


**FIGURE 10.28**

3 minutes. Determine the number of acid line operators needed if no accumulation line is provided. Assume that the assembly line will now work 8 hours/day.

7. Suppose in problem 10, the unit loads of MH and associated material difficult for each job are given in the following table.

   a. Obtain machine groupings and job assignments. Use $P = 0.6$.

| Job | Machines Used | Unit Loads of MH | Degree of Difficulty |
|---|---|---|---|
| 1 | 1, 3, 4 | 100 | 1.2 |
| 2 | 1, 2, 3 | 500 | 1 |
| 3 | 3, 4 | 50 | 1 |
| 4 | 8, 9 | 300 | 1 |
| 5 | 5, 8, 9 | 5000 | 1.2 |
| 6 | 5, 7 | 600 | 1 |
| 7 | 5, 6, 7 | 100 | 1.5 |
| 8 | 2, 3, 4 | 700 | 1 |
| 9 | 1, 2 | 100 | 1 |
| 10 | 5, 7 | 400 | 1.1 |

   b. Use a computer program to solve the problem
   c. How does the grouping compare with the grouping obtained in the solution of problem 10?
   d. Change the percentage in the program to 0.9. How does the new solution compare with the part b solution?
   e. Suppose the cost of operations are as follows:

$$\text{MH} = \$30,000, \quad B = \$1/\text{unit}$$

$$K = 200\,/\,\text{job}, \quad \text{C} = 1000/\text{cell}$$

   The cost of machines (in thousands of dollars) are as follows:

| Machine | 1 | 2 | 3 | 4 | 5 | 6 | 7 | 8 | 9 |
|---|---|---|---|---|---|---|---|---|---|
| Cost | 2 | 3 | 2 | 4 | 3 | 5 | 2 | 1 | 1 |

Determine the cost of operation using the solution in part b and part d. Which is the preferred alternative?"

8. Rework problem 7 assuming you have lost the contract on product 5.
   Three facilities are to be located in three locations. The flow between the facilities and distances between possible locations are given in the following tables:

**Flow Table**

| Facility | Facility 1 | 2 | 3 |
|---|---|---|---|
| 1 | — | 50 | 10 |
| 2 | 5 | — | 15 |
| 3 | 15 | 20 | — |

**Distance Table**

| Location | Location 1 | 2 | 3 |
|---|---|---|---|
| 1 | — | 50 | 10 |
| 2 | 5 | — | 15 |
| 3 | 15 | 20 | — |

   a. Develop the facility chain.
   b. Develop the location chain.
   c. Determine the initial solution.
   d. Perform two-way exchanges to see if any improvement is possible.
   e. Use the computer program SML to determine the best locations.
   f. How does your answer in part d compare to the solution in part e?

9. Five machines are to be placed in five possible locations. The machines are used to produce five products. The sequence of operation and the production volume per day is given in the following table:

| Product | Machining Sequence | Volume (units/day) |
|---|---|---|
| 1 | 1-3-5 | 50 |
| 2 | 1-2-3-5 | 100 |
| 3 | 2-4-5 | 100 |
| 4 | 2-3-4-3-5 | 150 |
| 5 | 1-4-5 | 50 |

   Each machine can be placed in any of the five locations. Distances between the locations are as follows (the distance matrix must be symmetrical about the diagonal):

| | Location | | | | |
|---|---|---|---|---|---|
| Location | 1 | 2 | 3 | 4 | 5 |
| 1 | — | 3 | 5 | 8 | 6 |
| 2 | 3 | — | 4 | 3 | 5 |
| 3 | 5 | 4 | — | 7 | 2 |
| 4 | 8 | 3 | 7 | — | 3 |
| 5 | 6 | 5 | 2 | 3 | — |

a. Determine the best location for each machine.
b. If the distance between machines 1 and 3 is 25 units (and therefore, 5 to 1 is 25 units), how does the answer to part a change?
c. The computer program that can be found at http://www.crcpress.com/e_products/downloads/download.asp?cat_no=44222 also solves problems if there is a fixed cost associated with modifying the location to suit a particular machine. Suppose such modification would be required in some locations and the associated cost data are as follows:

| | Machine | | | | |
|---|---|---|---|---|---|
| Location | 1 | 2 | 3 | 4 | 5 |
| 1 | 10 | 5 | 0 | 3 | 0 |
| 2 | 5 | 8 | 10 | 15 | 4 |
| 3 | 0 | 0 | 0 | 5 | 4 |
| 4 | 10 | 12 | 15 | 5 | 8 |
| 5 | 0 | 0 | 0 | 0 | 0 |

Use the computer program and determine the minimum-cost machine locations using the original flow and distance matrices.
d. If machine 1 cannot be assigned to locations 1 and 3, how would you modify the fixed-cost table to reflect this fact?

10. What are the goals in packaging? Describe how each of the following materials is used in packaging.
    a. Glass
    b. Paper carton
    c. Wood
    d. Styrofoam chips
    e. Air cushion mats
    f. Formed Styrofoam
11. What differences in packaging would be apparent between a household cleaner sold for home use and the same cleaner packaged for industrial use?
12. Why is a label of a product important? What information should it display? What are the commonly used materials for labels? Why are these materials chosen?

13. What are the four types of bag packaging? Describe and give an example of each. Describe the following packaging equipment, specifying their characteristics and how they can be used.
    a. Automatic adhesive sealer
    b. Strapper
    c. Wrapper
    d. Palletizer

## SUGGESTED READINGS

Apple, J.M., *Material Handling Systems Design,* John Wiley, New York, 1972.

*Basics of Material Handling,* The Material Handling Institute, Inc., 1973.

"Can you computerize equipment selection?" *Modern Materials Handling,* 21(1), 46–50, 1966.

Hassan, M.M.D., Hogg, G.L., and Smith, D.R., "A construction algorithm for the selection and assignment of materials handling equipment," *International Journal of Production Research*, 23(2), 381–392, 1985.

Hudson, W.G., *Conveyors and Related Equipment,* John Wiley, New York, 1954.

Keller, H.C., *Conveyors: Application and Design,* Ronald Press, New York, 1967.

Kwo, T.T., "A method for designing irreversible overhead loop conveyors," *Journal of Industrial Engineering*, 11(6), 459–466, 1960.

"Materials handling — the trends to watch," *Modern Materials Handling*, 30(1), 52–59, May 1975.

Matson, J.O., and White, J.A., "Operational research and material handling," *European Journal of Operational Research*, 11, 309–318, 1982.

Ravindran, A., Phillips, D.T., and Solberg, J.J., *Operations Research — Principles and Practices,* John Wiley, New York, 2006 .

Taha, H., *Operations Research,* 8th edn, Prentice Hall, Upper Saddle River, NJ, 2006.

Webster, D.B., and Reed, R., Jr., "A material handling system selection model," *AIEE Transactions*, 3(1), 13–21, 1971.

# 11 Storage and Warehousing

Managers have always sought a method for obtaining a continuous production flow in their plants. Ideally, the raw material coming in should immediately be processed and the final finished products promptly shipped, eliminating any need for storage at either end. This theoretical concept is appropriately called just-in-time (JIT). The idea can be extended within a plant as the product moves from one work station to another. A station should receive the item and the necessary parts just when it is due to process the unit; and upon completing the required task the station should transfer the unit immediately to the next station, which in turn is scheduled to receive the assembly just at that time. In JIT, one makes (produces) only what is needed only when it is required. JIT dramatically reduces the need for storage of raw materials, semifinished assemblies, and finished products. To an appreciable extent, however, the use of JIT is practical only for large, stable manufacturers. The company must be able to predict its raw material requirements in advance so as to coordinate the activities of all its vendors. This requires exact forecasting of the demand, which generally means a firm production schedule and confidence that all the items produced will be sold immediately. The vendors also attempt to implement their own JITs, but whether they achieve JIT in their own plants or not, they still work to deliver the manufacturers' orders on time, the main reason being that the customers will probably take their business elsewhere if the deliveries are not made when promised.

Although one should strive to achieve JIT, perfect implementation of such a system is not possible. Material requirement-planning techniques could be used to reduce inventory and still meet a specific production schedule. Still, manufacturers will always have some storage requirement, however small. The need to store raw materials, partially finished products, and finished goods must then be answered by storage and warehousing. The term storage is generally, but not always, associated with raw material and in-process goods, while warehousing refers to the storing of finished goods. A company might have one or more storage and/or warehouse facilities. In some cases, both storage and warehousing share the same building; in others, the storage might be located near the production facilities and the warehouses might be built separately to serve as distribution centers. Within a plant, however, the terms storage and warehousing are frequently used interchangeably to mean either facility.

## 11.1 WAREHOUSE OWNERSHIP

Once the need for a warehouse has been established, the next step is to decide whether a company-owned warehouse is necessary or a commercial warehouse can better be utilized. Commercial warehouses are built to serve many different customers, and they generally have sufficient personnel, equipment, and storage space to satisfy both the long-term and short-term needs of a customer. Commercial warehouses have two important advantages: flexibility and professional management. The manufacturer is not tied down to a specific warehouse location; as the distribution pattern changes, the company has the option of relocating its distribution centers. Unstable or seasonal demands can make renting the space on an as-needed basis more attractive than building a large warehouse to accommodate the maximum expected demand. If the demand for the company's product(s) is expected to continue for a long time, however, and if its commitment to the present location is strong, a private warehouse might be appropriate.

## 11.2 STORAGE/WAREHOUSE LOCATION

As in the case of plant location, the selection of storage and warehouse sites is very important. Some of the basic conclusions are obvious. If the warehouse is to contain mainly finished products, it should be close to the customers. If the material stored is to be used in manufacturing, the storage facility should be near the production plant. The location must have sufficient land and good transportation available. The site should not be cut off from suppliers and markets by geographical barriers such as rivers, mountains, and lakes. The site itself must be appropriately zoned by the local authorities and should have good fire and police protection. The company must also be able to obtain the necessary utilities and the required labor for operation of the facility. The site must be of sufficient size to accommodate any future expansion; as a general rule, it should be about five times as large as current needs dictate.

As a first step, a table similar to the one that will be shown for the plant site selection (Chapter 15) could be developed for the warehouse site selection as long as it is feasible to place the warehouse away from the plant. Appropriate weights could be assigned to each factor, and the evaluation would be made in a fashion similar to that shown in the plant site selection procedure. Many of the models developed in Chapters 17, 18, and 19 can be applied in such decisions because the necessary information is generally available to make such determination.

## 11.3 BUILDING

Factors to be considered in the construction of a warehouse (storage) building include the following:

- Location
- Size of the site
- Building placement
- Approach roads and railroad sidings
- Layout, dock site, and receiving and dispatching areas
- Column patterns and clear height needed for vertical storage
- Aisle layout and width and number, sizes, and arrangement of stacks
- Equipment to be used in material handling
- Lighting, heating plumbing, and air-conditioning.

### 11.3.1 Building and Layout Considerations

A warehouse building within a plant or on a different site should receive the same basic considerations as a manufacturing plant. The location is important, because it determines the cost of transportation, and the plot should be large enough to handle present truck traffic and future expansions; furthermore, it should offer flexibility to change. The building itself should be large enough to handle present requirements and expected demands in the future for a 5- to 10-year period. A square building shape has proven to be very efficient, giving the shortest average distance to be traveled during pickup and distribution operations when the units having about the same demand are stored over the entire floor evenly. However, the plot size most often dictates a rectangular building. Up to some practical limit, the incremental cost of construction of a building decreases as the height increases. It is therefore more economical to construct a taller building (18–20 feet) than one with a broader base for the same volume requirement. However, a constraint is imposed on usable height by material-handling equipment and its cost, as well as the cost of storage racks. The warehouse can also be multistory, but the cost of operation in such a building is normally greater than that in a single-story structure of equal capacity. The structure of warehouse exterior walls is most often a steel frame with concrete blocks or 4-inch-thick brick walls. The floor of the building is usually made of reinforced concrete to ensure strength, rigidity, and uniformity; and it should be able to support the weight of the stored material and equipment.

Warehouse aisles. The aisles are made just wide enough for the movement of a forklift truck, or other material-handling equipment, and the material is stacked on pallets as high as is practical to maximize the space utilization. (Courtesy of Jungheinrich AG, Hamburg, Germany.)

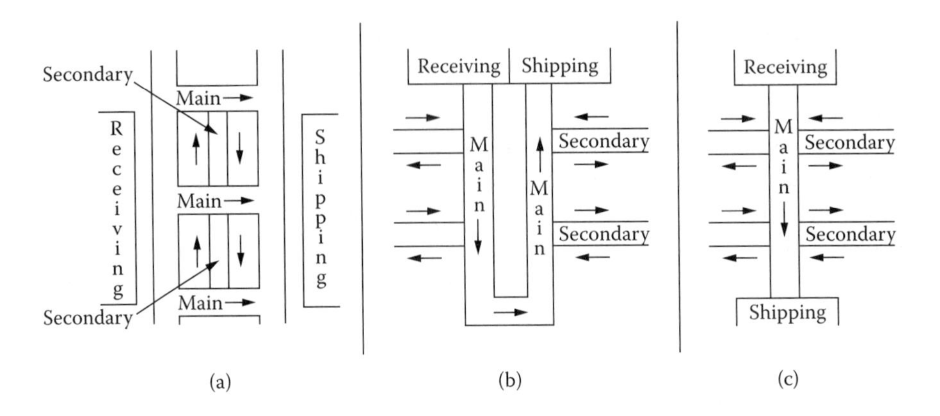


**FIGURE 11.1** Aisle configuration

Aisles are necessary in a warehouse for allowing material-handling equipment to reach different parts of the storage areas. From another viewpoint, however, aisles also constitute wasted space, a space that is not used for the main purpose of the warehouse, which is to store material awaiting use or shipment. To minimize this loss, many warehouses have two types of aisles: main (or working) aisles and utility (or secondary) aisles. The main aisles are wide (generally 10–12 feet) to allow a material-handling unit such as a forklift truck to operate. The utility aisles are used to gain access to racks, offices, elevators, and utility rooms, and they are much narrower — anywhere from 2 to 9 feet — depending on the size and configuration of the units being processed and the material-handling equipment required for this function. The main aisles connect the receiving and shipping areas, and their placement dictates the material flow. Figure 11.1 presents some of the possible aisle configurations.

The aisle space percentage ratio defined in Chapter 8 is a good measure to determine the efficiency of aisle space allocation. This requirement can be considerably reduced by using the automated storage and retrieval (AS/R) systems, which will be described in Section 11.9.

The loading docks for the building are described in detail in Section 11.10. Other service facilities that are generally available in the building include both 110- and 220-volt electricity, water, sprinklers, drains, fire extinguishers and fire exits, telephones, compressed air in maintenance areas, battery chargers, and toilets.

### 11.3.2 Space Determination

It is important that a building has sufficient space to accommodate all the items it is intended to store, and yet is not overly large to keep the cost of construction and maintenance down. The following examples illustrate the necessary procedures to determine the space requirement for the building. These methods should be followed for each item in storage

## EXAMPLE—FLOOR SPACE DETERMINATION I

A plant produces 75 units/hour of an item with dimensions of $0.5 \times 0.5 \times 0.1$ foot. The management wishes to store a 1-week supply in containers measuring $7 \times 7 \times 4$ feet. A minimum of 3 inches of space is required between adjacent units in each direction for packaging and handling. Determine the number of containers needed. If these containers can be stacked two high, determine the floor space required.

### SOLUTION

If there are $n$ items in a row (or column), then $n + 1$ packing spaces must be provided (Figure 11.2).

Thus the number of units that can be stored in each direction ($n_1$, $n_2$, and $n_3$) must satisfy the equation $n_i w_i + (n_i + 1)s_i = d_i$, where $w_i$ is the dimension of the unit, $s_i$ is the packing space required, and $d_i$ is the dimension of the package in the $i$ direction.

The length of the box is 7 feet, and therefore

$$0.5(n_1) + 0.25(n_1 + 1) = 7$$

Hence, the maximum $n_1$ is $9.1 \approx 9$.

The width of the box is also 7 feet; therefore,

$$0.5(n_2) + 0.25(n_2 + 1) = 7$$

Hence, $n_2$ is also equal to 9.

The height of the box is 4 feet; thus

$$0.1 \times n_3 + 0.25(n_3 + 1) = 4$$

Hence, $n_3 = 3$.

Accordingly, the total number of units that can be stored in a box is $n_1 \times n_2 \times n_3$ or 243.

Now, let us calculate the number of units to be stored in a week and the associated container requirements:

$$75 \text{ units/hour} \times 40 \text{ hours/week} = 3000 \text{ units/week}$$

$$\frac{3000 \text{ units}}{243 \text{ units/container}} = 12.34, \text{ or } 13 \text{ containers}$$

$$\frac{13 \text{ containers}}{2 \text{ containers/stack}} = 6.5, \text{ or } 7 \text{ stacks}$$

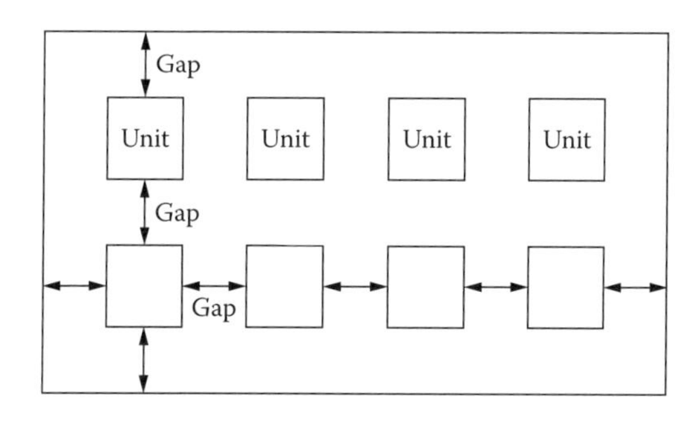


**FIGURE 11.2** Packaging of the units

The floor space needed for each stack is $7 \times 7$, or 49 square feet.

$$49 \text{ square feet} \times 7 \text{ stacks} = 343 \text{ square feet of floor space required}$$

## EXAMPLE—FLOOR SPACE DETERMINATION 2

A plant produces 100 cartons/day to be stored in the warehouse. The company would like to keep a 3-day supply of stock in inventory as a safety cushion to fall back on if production is temporarily interrupted. The maximum time between orders is 30 days. Determine the maximum and average amount of stock to be stored. If each unit is $4 \times 3 \times 2$ feet and if they can be stacked six units high, determine the floor space required for storage. (When small loads are being stacked, aisle space of 5–10% of the total storage area should be allowed; however, the required space would increase with the size of the load to as much as 30%. For this problem, 8% will be used.)

Section 11.4 shows how to optimize order quantity that minimizes the cost of operation. This also give some idea as to maximum quantity that may need storage.

### Solution

$$\begin{aligned}\text{Maximum units in stock} &= \text{inventory for safety} + \text{number of units for a single order}\\ &= (3 \text{ days}) \times (100 \text{ units/day}) + (30 \text{ days}) \times (100 \text{ units/day})\\ &= 300 + 3000\\ &= 3300 \text{ units}\end{aligned}$$

$$\begin{aligned}\text{Average units in stock} &= \text{inventory for safety } 1/2 \text{ of the units for a single order}\\ &= 300 + 1/2\ (3000)\\ &= 1800\end{aligned}$$

The maximum storage required is for 3300 units.

$$\frac{3000 \text{ units}}{6 \text{ units/container}} = 550 \text{ stacks (each 4 feet} \times 3 \text{ feet)}$$

$$550(4 \times 3) = 6600 \text{ square feet}$$

$$6600 \times 0.08 = 528 \text{ square feet for aisles}$$

$$6600 + 528 = 7128 \text{ square feet of floor space required}$$

## 11.4 MATERIAL REQUIREMENT PLANNING

Material requirement planning (MRP) is a technique that determines the timing for ordering and receiving dependent units, such as integral parts or subassemblies of the main products. The production of the main products is based on a specific production plan, called a master production schedule (MPS). The MPS specifies the requirements for all products in each time period, or time bucket. Although the time period for analysis can vary in length, the most common duration for a time bucket is 1 week. Because the number of units received determines the storage space needed, it is appropriate to consider MRP when planning for storage space.

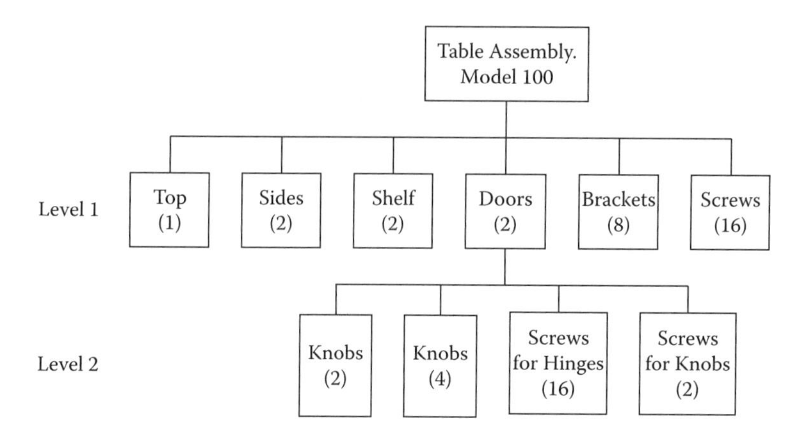


**FIGURE 11.3** Component tree structure for table assembly, model 100

### 11.4.1 Data for MRP

The MRP system is built on the information obtained from the bill of materials (BOM), which was first introduced in Chapter 2. The BOM can be expanded in the form of a structured tree to show the hierarchy of the components. Component dependency in parent-offspring form and the quantities of each component needed to make a complete assembly are illustrated in the tree.

Figure 11.3 shows an example of a typical structure tree and Table 11.1 illustrates the associated BOM for a small desk with closed-door shelves.

It is convenient to display the tree structure in terms of levels, the level 0 component (also called end product) being the parent of level 1 components and level 2 components being the offspring of corresponding level 1 components, etc. Note that in our example table, model 100 is the main product, while all other items are the

**TABLE 11.1**
**Bill of Materials: Table Model 100**

| Part | Description | Number | Level |
|---|---|---|---|
| 1000 | Table assembly | 1 | 0 |
| 1011 | Tabletop | 1 | 1 |
| 1012 | Table side | 2 | 1 |
| 1013 | Shelf | 2 | 1 |
| 1014 | Door | 2 | 1 |
| 1015 | Bracket | 8 | 1 |
| 1016 | Screw | 16 | 1 |
| 1021 | Doorknob | 2 | 2 |
| 1022 | Door hinge | 4 | 2 |
| 1023 | Hinge screw | 16 | 2 |
| 1024 | Knob screw | 2 | 2 |

dependent components. This arrangement makes it easier for an analyst or a computer program to evaluate the quantities of different items needed to meet a known demand for the parent item. For example, suppose 100 units of table model 100 are needed 2 weeks from now. Thus, 100 tabletops, 200 table sides, 200 shelves, 200 doors, and so on, would be needed.

Furthermore, producing 200 doors would require 400 doorknobs, 800 hinges, 3200 hinge screws, and 400 knob screws. The process of doing these calculations is called exploding the tree.

In MRP, the tree for every product that is planned for delivery in each time bucket is exploded, and the common components are combined. For example, suppose the same doorknobs, hinges, and hinge screws, in the same quantities, are part of a tree for another table assembly, model 103, Furthermore, 50 units of model 103 along with 100 units of model 100 are required 2 weeks from now. Then the total requirement for doorknobs 2 weeks from now is: 200 for model 103 plus 400 for model 100, for a total of 600 units. This total is designated as the gross requirement for the product. The net requirement is calculated as

$$\text{Net requirement} = \text{gross requirement} - (\text{inventory on hand} + \text{scheduled receipts})$$

This equation is useful in planning procurement policies.

### 11.4.2 MRP Procedure

The following example illustrates the working of an MRP system. Suppose table models 100 and 103 differ only in the surface polish of the top of the tables. Therefore, the only difference between their tree structures is that the tabletop for model 100 is part 1011, while for model 103 it is part 2011. All other parts in both tables are the same. Next, suppose the demands for the next 8 weeks for both table models are as shown in Table 11.2.

The first step in MRP planning is to explode the tree structure for each table and to find the number of dependent components required at each level in each period. For example, the requirements for doors — that is, component 1014 — are as given in Table 11.3.

Now, suppose these doors are ordered from a supplier for whom the economic order quantity (EOQ) is 60 units per order. The lead time for the order — that is, time between placing an order and receiving it — is 1 week. There is an initial inventory of 50 units in the stock. How should we place the orders to satisfy the production requirements?

**TABLE 11.2**
**Expected Demands**

| | Week | | | | | | | |
|---|---|---|---|---|---|---|---|---|
| | 1 | 2 | 3 | 4 | 5 | 6 | 7 | 8 |
| Demand | | | | | | | | |
| Model 100 | 20 | | 50 | | 30 | | | 10 |
| Model 103 | | 10 | | | 20 | 10 | | |

**TABLE 11.3**
**Gross Requirement Calculation — Component 1014: Doors**

| | Week | | | | | | | |
|---|---|---|---|---|---|---|---|---|
| **For Mode** | **1** | **2** | **3** | **4** | **5** | **6** | **7** | **8** |
| 100 | 40 | | 100 | | 60 | | | 20 |
| 103 | | 20 | | | 40 | 20 | | |
| Gross | 40 | 20 | 100 | 0 | 100 | 20 | 0 | 20 |

The principle applied in MRP is to order only the quantity that is required and only when it is required. At the heart of the MRP system is the development of a table similar to Table 11.4. The first row of the table is the individual time period (bucket) in the planning horizon. In our case, the planning horizon is 8 weeks. The second row indicates the "gross requirements" for the part. This requirement is the actual demand in each period and not the average over all periods. For our example, the gross requirements are calculated in Table 11.3 and are inserted in the second row. The third row displays "scheduled receipts." These are the quantities already ordered and expected to be received in that time period. Once the units are received, they are immediately available to satisfy the requirement of the associated period.

The row entitled "On hand at the end of period" is divided into two parts. The upper values show the inventory on hand at the end of that period if *no* additional order is placed.

If this quantity is negative, there is insufficient inventory on hand to satisfy the demand for that period and an order *must* be placed (perhaps in some previous period, based on the lead time). The lower values indicate the inventory at the end of the period after all the orders that are expected to be received so far (including the order we may have placed to satisfy the demand for the period) have been received and all the demands have been met.

The values in the "Planned Order Release" row are determined directly from the previous row. Whenever inventory level at the end of the period is insufficient to meet the demand of the next time period, an order must be released so that sufficient units are received in that period to satisfy the associated demand. The order must be released in time to allow for the necessary lead time. For example, if the lead time

**TABLE 11.4**
**MRP Planning for Order Release**

| | Period | | | | | | | | |
|---|---|---|---|---|---|---|---|---|---|
| | **0** | **1** | **2** | **3** | **4** | **5** | **6** | **7** | **8** |
| Requirement | | 40 | 20 | 100 | 0 | 100 | 20 | 0 | 20 |
| Receipts | | | 60 | 60 | 60 | 60 | | | 60 |
| On hand at the end of period | 50 | 10 | −10 | −50 | 10 | −90 | 10 | 10 | −10 |
| | | | 50 | 10 | 70 | 30 | | | 50 |
| Planned order release | | 60 | 60 | 60 | 60 | | | 60 | |

is 2 weeks, in order to receive the units in the fifth week, the order must be released 2 weeks in advance, or in the third week. Thus, the MRP system produces accurate and timely information on the quantity and timings for the order releases. The results for our example are displayed in Table 11.4.

For period 1, 40 units are required because 50 units are in inventory; the demand is satisfied with the inventory, leaving 10 units in the stock at the end of period 1. The second period has 20 units of demand and 10 units of stock from the previous period, producing net inventory of –10 at the end of the period 2. In other words, the demand for period 2 cannot be met with the inventory on hand at the end of period 1. To satisfy the demand, we should have an order coming in during this period. For an order to be received in period 2, the order must be released in period 1. Hence, we have entries of 60 units for planned release in period 1 and 60 units for receipt in period 2, which gives the net inventory of 50 units available at the end of period 2.

Calculations continue in a similar manner. Periods 4 and 5 are interesting. With no order release in period 3, there is inventory of 10 units at the end of period 3. Because period 4 requires no units, 10 units are carried through to the end of period 4. However, period 5 needs 100 units, and hence there will be a shortage of 90 units in period 5. Even if one order of 60 units is received in period 5, there still will be a shortage of 30 units. The only way to avoid this shortage, assuming no more than one order can be received in one time period, is to receive one order in period 4, which explains the numbers in the table.

### 11.4.3 Order Quantity

The calculations shown in Table 11.4 determine the order releases needed to satisfy the demands in each time period. However, no cost has been considered so far. Suppose the order cost is 250¢ and the carrying cost is 40¢/period/unit. Then the total inventory cost is

$$\text{Order cost:} \quad 5 \text{ orders}@250/\text{order} = 1250$$

$$\text{Carrying cost:} \quad 10\times 4 + 50\times 4 + 10\times 4 + 70\times 4 + 30\times 4 + 10\times 4 + 10\times 4 = 760$$

Therefore, the total cost is 1250 + 760 = 2010.

If we had a constant and uniform demand in each period, the EOQ would give us the minimum inventory cost. However, the demands in each period are neither constant nor uniform. If we decide to take the average of all the demands as an approximation to the constant demand and use the well-known EOQ formula to determine order quantity, how much will the cost decrease?

The formula for EOQ is $\sqrt{2AD/H}$, where $A$ is the order cost per order, $H$ is the carrying cost, and $D$ is the average demand. Both $D$ and $H$ are in the same time units, generally a week. The value of $D$ is obtained by averaging the actual demands and making the correction for what is on hand. Here, $D$ is equal to

$$(-50 + 40 + 20 + 100 + 100 + 20 + 20)/8 = 31.25$$

giving $EOQ = 62.5$.

**TABLE 11.5**
**EOQ Order Release**

| | Period | | | | | | | | |
|---|---|---|---|---|---|---|---|---|---|
| | 0 | 1 | 2 | 3 | 4 | 5 | 6 | 7 | 8 |
| Requirement | | 40 | 20 | 100 | 0 | 100 | 20 | 0 | 20 |
| Receipts | | | 62.5 | 62.5 | 0 | 125 | 0 | 0 | 0 |
| On hand at the end | 50 | 10 | −10 | −47.5 | 15 | −85 | 20 | 20 | 0 |
| of period | 50 | 10 | −10 | −50 | 10 | −90 | 10 | 10 | −10 |
| | | | 52.5 | 15 | 15 | 40 | 20 | 20 | 0 |
| Planned order release | | 62.5 | 62.5 | 0 | 125[a] | 0 | 0 | 0 | 0 |

[a] Two orders

For demonstration purposes, assuming that a half unit can be ordered, the associated inventory levels and order releases are as shown in Table 11.5.

The associated cost is

$$\text{Order cost:} \quad 4 \times 250 = 1000$$

$$\text{Carrying cost:} \quad 10 \times 4 + 52.5 \times 4 + 15 \times 4 + 15 \times 4 + 40 \times 4 + 20 \times 4 + 20 \times 4 = 690$$

for a total cost of 1690.

This cost is less than for the order-release schedule of Table 11.4, but is it possible to make any further improvements? In particular, observe that the demand per period is not particularly constant! In the ideal condition — that is, when demand is constant and uniform — it is well known that with EOQ, the order cost equals the carrying cost, each being one-half of the total cost. This principle is used in the part-period balancing (PPB) technique described in the next subsection.

### 11.4.4 Part Period Balancing (PPB)

The policy of ordering a fixed quantity, as in EOQ, can produce an excess inventory if the requirements from period to period are not equal. This can lead to an excessive total cost. The PPB technique tries to develop order quantities that result in the order cost and carrying costs being as close to equal as possible.

Let us apply this principle to our example. Because the initial inventory is sufficient to satisfy period 1 demand, there is no need to place an order in period 1. The remaining inventory of 10 units in period 2 is not sufficient to satisfy the demand for 20 units in period 2, so an order must be placed. To determine the size of the order in PPB, two rules should be followed: (1) order a quantity that would make carrying cost as close to order cost as possible, and (2) there is no reason to order a partial requirement for a period, because it would only increase the order cost without decreasing the carrying cost.

**TABLE 11.6**
**Calculations for Iteration 1**

| Iteration | $P$ | Quantity Ordered | Carrying Cost |
|---|---|---|---|
| 1 | 2 | 10 | 0 |
| | 3 | 110 | $100 \times 4 = 400$ |

Using these rules, the order to be released at time 2 to satisfy the requirements for period 2 through $P$ are calculated in Table 11.6. Along with the inventory left over from period 1, an order of 10 units would satisfy the demand in period 2, so the first order quantity checked is for 10 units. Because all the units would be used in the same period, there is no carrying cost. Next, check for the carrying cost if order quantity equals total requirements for periods 2 and 3 — that is, 110 units. From here, 100 units would be carried in period 2 for period 3; hence, the carrying cost is 400.

The limiting period in the calculation is the value that would make the carrying cost exceed the order cost for the first time. Because the order cost is 250, here the limiting value of $P$ is 3.

To determine the actual order quantity, see for which period the carrying cost is closest to the order cost. Comparing 0 and 400 to 250, we conclude that 400 is closer to 250; therefore, the order quantity is 110 units. Because the lead time is one period, the order is released in period 1 so that it can be received in period 2.

Similar calculations for other orders are shown in Table 11.7. The evaluation starts with period 5, because there is no demand in period 4. If we order through period 8, we need 140 units. Twenty units for period 6 must be stored for one period — from period 5 to 6. Also, 20 units must be stored for three periods, 5, 6, and 7, and hence the corresponding carrying cost is 320 units for period 8.

The schedule can be developed and displayed in an MRP-type table. Table 11.8 shows the results for our example.

**TABLE 11.7**
**Calculations for Iteration 2**

| Iteration | $P$ | Quantity Ordered | Carrying Cost |
|---|---|---|---|
| 2 | 5 | 100 | 0 |
| | 6 | 120 | $20 \times 4 = 80$ |
| | 7 | 120 | $20 \times 4 = 80$ |
| | 8 | 140 | $20 \times 4 + 20 \times 3 \times 4 = 320$ |

**TABLE 11.8**
**PPB Order Release**

| | Period | | | | | | | | |
|---|---|---|---|---|---|---|---|---|---|
| | **0** | **1** | **2** | **3** | **4** | **5** | **6** | **7** | **8** |
| Requirement | | 40 | 20 | 100 | 0 | 100 | 20 | 0 | 20 |
| Receipts | | 0 | 110 | 0 | 0 | 140 | 0 | 0 | 0 |
| On hand at the end of period | 50 | 10 | −10 | 0 | 0 | −100 | 20 | 20 | 0 |
| | | | 100 | | | 40 | | | |
| Planned order release | | 110 | 0 | 0 | 140 | 0 | 0 | 0 | 0 |

The associated cost is

$$\text{Order cost:} \quad 2 \times 250 = 500$$

$$\text{Carrying cost:} \quad 100 \times 4 + 40 \times 4 + 20 \times 4 + 20 \times 4 = 720$$

for a total cost of 1220. This is the best cost obtained so far. The PPB method does tend to give lower cost than other methods if the demands from period to period have wild fluctuations.

### 11.4.5 Safety Stock Considerations

Use of MRP offers a number of benefits. They include low in-process inventory and ability to keep track of the quantities of products or items necessary in each period.

Knowing what is required, we can develop a good production schedule to satisfy these needs. MRP assumes a known demand for items. However, a variation in either supply or demand may occur as a result of increase in demand or longer-than-expected production and assembly times. Safety stock can be introduced to accommodate such variation. This is accomplished by modifying the requirement for a period where the safety stock is to be included by adding the safety stock to the net requirements. For example, we expect that the requirement for period 2 is not certain and we wish to have a safety stock of 5 units to accommodate this uncertainty. Then the net requirement for period 2 would be 20 + 5 = 25 units and the order-release policy would be based on this value. It is a common practice to reevaluate the MRP order-release strategy if the actual demand differs from the estimated demand. Thus, for example, the actual demand in period 1 was 30 units rather than 40, as was initially planned. With this new value for initial inventory, the order-release sequence for periods 2 on should be replanned. If the demand for a period was higher than what was expected, then the unfilled demand is added on to the next period as the back order for the previous period, and order-release policies are reevaluated.

### 11.4.6 Lead Time Considerations

It is interesting to note how the lead times are propagated within a structure tree. For example, consider the following simple tree structure with the demand for level 0 item, $A$, as shown.

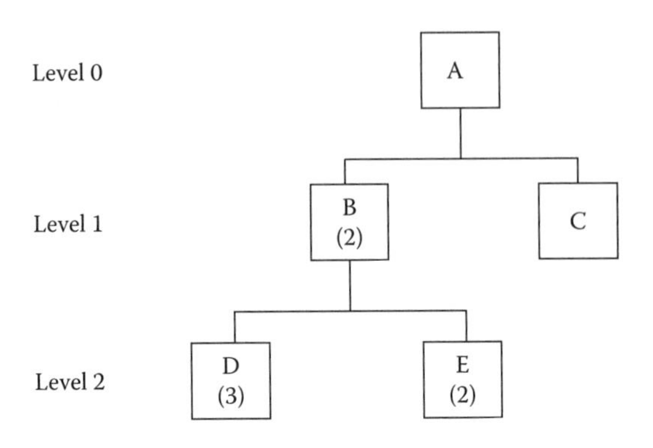


**Schedule for Item A**

| Period | 1 | 2 | 3 | 4 | 5 | 6 |
|---|---|---|---|---|---|---|
| Requirement | | | 15 | | 20 | 10 |

Suppose the lead time (or, here, production time) for A is one period (week). Because A is produced within our facility and not ordered from an outside supplier, the lead time is the same as production/assembly time; hence, the parts necessary to make A must be available 1 week early. Thus, the requirements for part B are:

**Schedule for Item B**

| Period | 1 | 2 | 3 | 4 | 5 | 6 |
|---|---|---|---|---|---|---|
| Requirement | | 30 | | 40 | 20 | |

Suppose lead time for B is also 1 week; because B is assembled in our plant, then the requirements for part E are:

**Schedule for Item E**

| Period | 1 | 2 | 3 | 4 | 5 | 6 |
|---|---|---|---|---|---|---|
| Requirement | 90 | | 120 | 60 | | |

The final order-release for item E from outside suppliers should be based on the preceding requirement schedule, taking into account the lead time associated with part E.

## 11.5 STORAGE/WAREHOUSE FUNCTIONS

In managing a storage and warehouse facility, one must perform many different activities related to the processing of raw materials, semifinished products, and finished goods. These tasks range from receiving, inspecting, and storing raw materials to packing, labeling, and shipping orders. The following is a brief description of common activities.

1. *Receiving.* The warehouse receives the material from outside suppliers and accepts responsibility for it. The operation consists unloading the goods from trucks and/or railroad cars and unpacking the containers.
2. *Identifying and sorting.* Material is identified and then recorded by using tags, codes, or other means. The items are sorted to find any breakage, and shortages are determined by checking receipts versus packing slips. Appropriate action is taken to inform the shippers and vendors of any discrepancies.
3. *Dispatching to storage.* The goods are transferred to appropriate areas for storage.
4. *Storing.* The units are held in inventory until needed.
5. *Picking the order.* Items needed for an order are retrieved from storage. Picking of the items for a particular order can be accomplished by one or more people depending on the number of items and their locations in the warehouse.
6. *Assembling the order.* All items in a single order are grouped together. Any shortage, breakage, or nonconforming item is recorded, and the item is replaced or the order is modified.
7. *Packaging.* All units in an order are packed together.
8. *Dispatching the shipment.* Appropriate shipping orders and documents are prepared, and the order is sent to the transport vehicles.
9. *Maintaining records.* Records such as the following are kept for each item: the amount received, on-hand inventory, orders received, and orders processed. These records are critical for good inventory management.

## 11.6 STORAGE AND WAREHOUSE OPERATIONS

Within a plant, management must decide whether to build a centralized warehouse or multiple storage facilities, each near its point of use (e.g., near each assembly line station). The latter approach reduces material handling and halting of production due to delays in delivery from a centralized warehouse. It also allows for tighter inventory control. Many times, such storage facilities can be built to use space that would otherwise not be utilized.

### 11.6.1 Storage Policies

Within a storage facility, several policies influence its layout, locations of storage cells, and assignment of items to these cells. These policies are briefly described next.

1. *Physical similarity.* Items with similar physical characteristics are grouped together in one area. For example, large items are stored in one area, and small items are located in another. This allows the use of similar

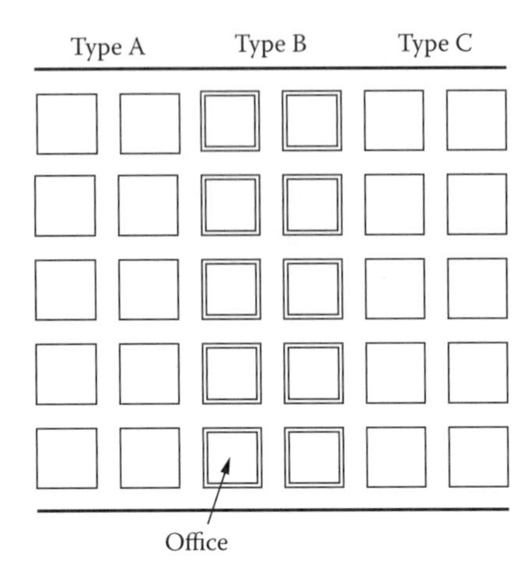


**FIGURE 11.4** Storage arrangements (layout pattern based on similarity, either physical or functional)

material-handling equipment and similar physical care for each area. Also, special environmental controls such as refrigeration, humidity, and fire safety can be concentrated in one area as the needs of the items dictate (Figure 11.4).

Aisles with physically similar materials (Courtesy of Jungheinrich AG)

2. *Functional similarity.* Functionally related items can be stored together. For example, electrically, hydraulically, and mechanically operated items are grouped in segregated storage areas. The system is especially convenient in manually operated storage facilities in which each warehouse worker becomes knowledgeable in a specific functional area (Figure 11.4).
3. *Popularity.* Every warehouse has items that are retrieved more often than others. In this system, these fast-moving items are stored close to receiving and shipping areas, and the slow-moving items are assigned to spaces that are farther away. This arrangement minimizes the distance traveled by warehouse workers in picking orders. Actual studies have shown that, on the average, 15% of the goods account for 85% of the work in a warehouse (Figure 11.5).
4. *Reserve stock separation.* It may be advantageous to separate reserve stocks from working stocks. All working stocks are kept together in a compact area from which picking is relatively easy. Reserve stocks from outlying areas replenish the working stocks as the need arises.
5. *Randomized storage.* Today, with modern information processing systems (computerized inventory control systems) it is no longer necessary to assign a fixed and unique location to an individual stock item. Changing from dedicated storage to randomized storage might result in considerable savings in the space requirement for the warehouse. The items are stored in spaces that are available when needed without reserving any space for items that are not currently in stock.
6. *High-security storage.* If there are items that are particularly valuable and subject to significant pilferage (e.g., gold, watches), an area might be needed that is under lock and key and/or other security measures.

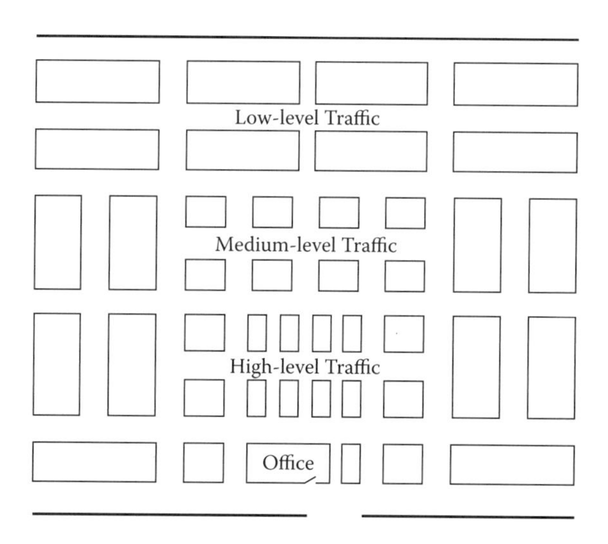


**FIGURE 11.5** General layout pattern based on popularity for one-entrance storage buildings

### 11.6.2 Order Picking Policies

Another important factor affecting the performance and layout of a warehouse is the policy followed in filling an order, called order picking. There are several such policies; we will limit our description to some of the more popular ones.

1. *Area system.* Items are stored in the warehouse in some logical manner. The warehouse personnel circulate through the area, picking the items required for an order until the entire order is filled.
2. *Modified area system.* This system is applicable where reserve stocks are separated from working stocks. Order picking follows the area system, while secondary personnel are utilized to replenish the working stock from the reserve stock.
3. *Zone system.* The warehouse is divided into zones, and the order is distributed among the order pickers, each picking units from his or her assigned zone.
4. *Sequential zone system.* Each order is divided into zones as in the zone system, but the order is passed from one zone to another as it is assembled. Many orders can be processed simultaneously as each proceeds from one zone to the next.
5. *Multiple-order or schedule system.* A group of orders is collected and analyzed to determine the total items needed from each zone. In a manner similar to the zone system, these items are picked by making one trip through each zone. The orders are assembled in a common area for further dispatching. A slight variation of this operation is scheduling simultaneous arrival of parts from each zone associated with each order, then putting them together for dispatching.

The area system is the simplest of all and is widely used when the average number of items in an order is not large. If this number increases, the order is either picked simultaneously (zone system) or sequentially (sequential zone system). The multiple-order system is beneficial only when there are large numbers of orders, each containing but a few items to be processed.

#### EXAMPLE—STORAGE ARRANGEMENT

Four different items are to be stored in the warehouse shown in Figure 11.6. Table 11.9 shows the number of pallets received each week, the number of trips from receiving to storage, the average size of each order shipped, and the number of trips from storage to shipping. Each of the 16 sections of the warehouse stores 100 pallets. The rectilinear distance from section to section is 10 units. Determine the most efficient storage arrangement for the warehouse.

#### Solution

First, determine the ratio of receiving trips to shipping trips, as shown in Table 11.10. The items with the higher ratios have more trips from receiving to storage than the reverse.

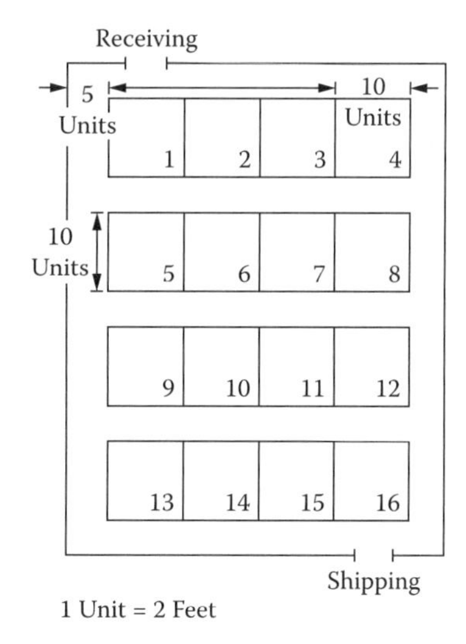


**FIGURE 11.6** Initial warehouse layout

**TABLE 11.9**
**Data on Load Movement**

| Item (1) | Pallets Received (2) | Average Receiving per Week (3) | Average Pallets per Shipment (4) | Shipping per Week (5) = (2)/(4) | Sections Needed (6) = (2)100 |
|---|---|---|---|---|---|
| A | 275 | 138 | 2.7 | 102 | 3 |
| B | 425 | 213 | 2.0 | 213 | 5 |
| C | 150 | 75 | 0.4 | 375 | 2 |
| D | 550 | 275 | 1.2 | 459 | 6 |

**TABLE 11.10**
**Calculation of Ratio of Receiving/Shipping for Each Item**

| Item | Receiving/Shipping |
|---|---|
| A | 138/102 = 1.35 |
| B | 213/213 = 1.00 |
| C | 75/375 = 0.20 |
| D | 275/459 = 0.60 |

**TABLE 11.11**
**Data Calculations for Each Section**

| Section | Rectilinear Distance to Receiving | Rectilinear Distance to Shipping |
|---|---|---|
| 1 | 10 | 5 |
| 2 | 20 | 85 |
| 3 | 30 | 75 |
| 4 | 40 | 65 |
| 5 | 35 | 80 |
| 6 | 45 | 70 |
| 7 | 55 | 60 |
| 8 | 65 | 50 |
| 9 | 50 | 65 |
| 10 | 60 | 55 |
| 11 | 70 | 45 |
| 12 | 80 | 35 |
| 13 | 65 | 40 |
| 14 | 75 | 30 |
| 15 | 85 | 20 |
| 16 | 95 | 10 |

Therefore, these items should be located as close to receiving as possible. Items with ratios less than 1 have more trips to shipping and should be located as close to shipping as possible; that is, item A should be close to receiving, and items C and D should be located close to shipping in that order. Item B, with a ratio of 1.0, may be placed in any available space.

This solution is based on an implied understanding of the workings of the warehouse. For instance, for item A for some order, two pallets might be carried from the warehouse and shipped; in another order, three pallets might be carried and shipped,

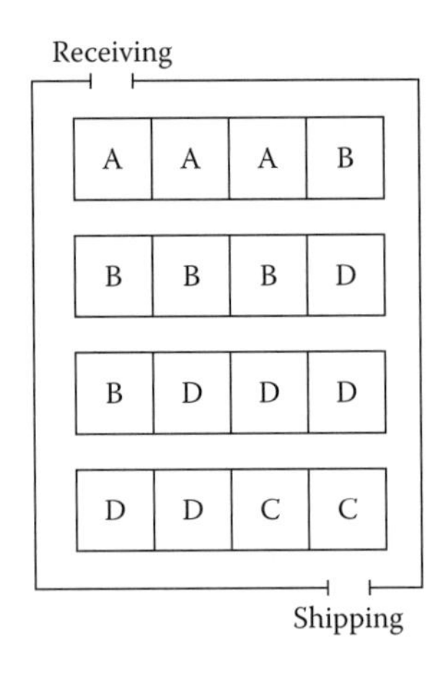


**FIGURE 11.7** Final warehouse layout

giving an average of 2.7 per trip. If for any reason only two pallets can be carried at a time, then the average shipping trips should be modified to 275/2 as 138, and a new solution to the problem develops.

Table 11.11 shows the distances calculated for rectilinear travel to each section from the receiving and shipping departments. The assignment of the items now can be made as follows. Item A requires the three sections closest to receiving: 1, 2, and 3. Item C requires the two sections closest to shipping: 15 and 16. Item D requires the six next closest sections to shipping: 8, 10, 11, 12, 13, and 14. Sections 4, 5, 6, 7, and 9 are left for item B. Figure 11.7 shows the final warehouse assignments.

## 11.7 ACCESSORIES

For storing individual and/or small items, various accessories are available. Almost all storage facilities and warehouses will use one or more of these to develop an orderly storage system.

1. *Bins*. There are many different sizes and shapes of bins that can be used when a huge variety of small parts is to be stored.
2. *Shelves*. In most cases, steel or wooden shelving is used for the storage of unpalletized loads or large items. Steel shelving is simply sheet metal that has been fastened to upright posts. The posts usually provide for flexibility in adjusting the height of the shelves and the vertical spaces between shelves.
3. *Racks*. The most commonly used items in storage are racks, which are described in detail in the next subsection.

Racks are the most commonly used items in storage, and are described in detail in the next subsection. (Courtesy of Ace World Companies)

4. *Stacking*. Unit loads on pallets or in boxes, bags, or sacks can be stacked on top of each other to better utilize vertical space.

5. *Conveyor storage.* Conveyer racks can make an effective storage accessory. These are a series of roller or skate wheel conveyors placed one above the other in adjacent stacks and slanted from input to output. The stored items should be contained in boxes or tote pans. Very compact storage can be achieved by using automatic loaders and eliminating excess aisles.
6. *Yard storage.* The high construction costs of enclosed warehouses have made outside storage more desirable for bulk items such as coal, sand, pulp wood, and scrap metal. The availability of containers and protective coatings has also contributed to greater use of yard storage.

## 11.7.1 Storage Racks

The purpose of storage racks is to facilitate storage and retrieval of loads in the warehouse. The racks are commonly made of steel frames with vertical posts and horizontal bars to support the loads with additional strength provided by diagonal or X braces. Different types of racks are available in the market, and the decision as to which one to use depends on the type of material that is to be stored. The following is a description of a few typical racks.

Height-adjustable bin rack stack

### 11.7.1.1 Selective Pallet Racks

The most commonly used storage rack is the selective pallet rack. This rack is formed with several pairs of supporting posts, the number depending on the length of the rack, several pairs of longitudinal beams placed to accommodate the height of the used pallet loads, and a number of horizontal braces at right angles to the beams in the rack to support the load. The depth of storage can be increased by placing two or more racks adjacent to each other. For example, two adjacent racks give storage room for two deep pallets, three adjacent racks can increase the room to stack pallets three deep, and so on. Owing to the simple design of this type of rack, it can easily be customized to provide maximum efficiency for almost any operation. A few common modifications are back-to-back ties (used to join two adjacent racks), drop-in skid supports, drop-in front-to-rear members, drum supports (used to support barrels), and deck surfaces.

#### 11.7.1.2 Movable-Shelf Racks

The movable-shelf rack is simply a selective storage pallet rack that has been modified to make it mobile. The modifications include diagonal rear bracing and a permanent top shelf that provides more stability to the rack during movement. This rack can be moved with machinery that is joined to the shelf by lugs that extend from the support beams of the shelf. The arms of a forklift or another vehicle are placed under these lugs to raise the rack. If there is a need to place these racks end-to-end, connectors will have been placed at both ends of each rack. In some cases, the rack may take the place of a pallet.

The pallet rack is designed in a way that the pallet trucks can go in and out easily

#### 11.7.1.3 Drive-In and Drive-Through Racks

Drive-in and drive-through racks are systems of racks arranged to form a central aisle. There are also several tunnels that are perpendicular to the central aisle. In each of these sections it is possible to stack a number of pallets. This configuration

Drive-in and drive-through racks are systems of racks arranged to form a central aisle.

makes it possible to drive a forklift down the central aisle for loading and unloading. This storage system is useful when a large number of the same items requires storage, for example, beer, tobacco, and frozen foods.

#### 11.7.1.4 Cantilever Racks

Cantilever racks are used for items that are extremely long (e.g., bar stock) and do not conform to other types of storage racks. These racks consist of arms that are fastened to two or more support posts, depending on the size of the item being stored. In some cases, it is convenient to have arms on both sides of the racks, called double-sided cantilever racks.

#### 11.7.1.5 Stacker Crane Racks

Stacker crane racks, used with AS/R systems (described in Section 11.9) and characterized by their extreme height, utilize the available vertical space in a warehouse. These racks usually range in height from 50 to 100 feet tall. Because of its extreme height, the basic structure of a stacker crane rack differs from that of most other racks. These racks are built with only one bay opening and the upright frame assemblies. To provide the needed stability, overhead ties and many braces are attached to the racks. The pallets can be supported by rails or arms such as those used on a cantilever rack. Stacker cranes are needed to load and unload the pallets in such a system. Other accessories that are found with these racks include overhead guide rail supports, building attachments, conveyor supports, mezzanine attachments, and loading/unloading stations.

#### 11.7.1.6 Portable Racks

Portable racks are designed to be mobile, with or without a load. Some of these racks also come equipped with knockdown or nesting features, allowing racks to be disassembled and stored compactly when not in use. The purpose of this is to provide the best possible utilization of the available space. Two of the more common portable racks are portable drum stacking racks and a portable rack for rolled strip or wire.

#### 11.7.1.7 Rack Buildings

In some cases, the racks support the roof and sides of a building. These rack buildings greatly reduce the cost of the storage facility. It is necessary to determine the roof live loads and wind loads to be sure that no building codes are violated.

These drive-in racks are one of the most fundamental inventory system components. The items inside the rack are transported by a vehicle with the rack.

#### 11.7.1.8 Fire Prevention

With many wooden pallets supporting cardboard boxes or some other combustible product, storage facilities are in danger of fires. Fire tests have shown that the best fire prevention method is a water sprinkler system. The type and location of the sprinkler head will depend on the items stored and their configuration.

**TABLE 11.12**
**Typical 1992 Costs for One Linear Foot, 8-Foot High × 36-Inch-Deep Sections With a 4000-Pound Capacity**

| Type of Rack | Cost |
|---|---|
| Pallet | $90/feet |
| Drive-in and drive-through | $135/feet |
| Cantilever | $105/feet |
| Portable (generally 3–6 feet) | $136/feet |

For further information about storage racks and systems, readers may contact Rack Manufacturer's Institute (1326 Freeport Road, Pittsburgh, PA 15238). See Table 11.12 for typical costs of some types of racks. Changes from 1992 data can be estimated by using the Industrial Commodities Index.

## 11.8 STOCK LOCATION

A system to identify the location of items of stock should be developed to permit quick and easy access to the right unit when needed. The significant location symbol system is one such coding system. It consists of a nine-digit number (e.g., 152012102); the first two numbers identify the building, the next one the floor, the next three the row, the next two the stack number, and the last digit the level:

| Building | Floor | Row | Stack | Level |
|---|---|---|---|---|
| 15 | 2 | 012 | 10 | 2 |

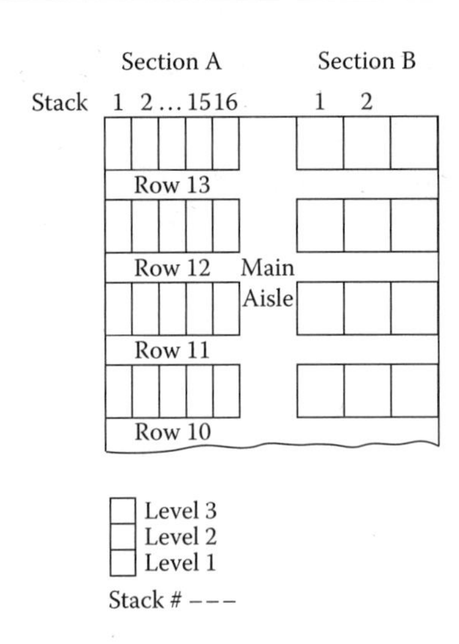


**FIGURE 11.8** Storage building layout

The code can be modified to suit the building and its layout. For example, in a building with one major aisle and different spacing between the stacks, as shown in Figure 11.8, the areas can be identified by the letters A and B, and a code such as A12153 would identify the location as being in section A, row 12, stack 15, third level.

## 11.9 AUTOMATED STORAGE AND RETRIEVAL

In recent years, the AS/R system has had a most dramatic impact on storage and warehousing operations. These high-rise storage units are becoming commonplace in companies that deal with numerous items in large volumes, companies that are unable to expand their present warehouses because of the limitation of available floor space, plants that operate in the environment of high labor cost and/or extended travel and search times in the warehouse, and those for whom delivering all orders accurately is critical. Unlike traditional warehouses, where the record of location and inventory levels may be kept manually, in AS/R such controls are maintained by computers within the system. In addition, the storage capacity of the warehouse is expanded two- to fivefold by using the same floor space by installing racks that allow high-density storage and by using specialized equipment that can work in narrow aisles. It is quite common to observe racks that are 80–90 feet high being served by computer-controlled machines (stacker cranes) carrying 3000- to 4000-pound loads. They can travel at speeds of 500 feet/minute in aisles that are only 6 inches wider than the narrow aisle cranes. Computer control of inventory can account for savings of up to 20% of the inventory cost. Such a fully mechanized system requires very little labor to operate; a single person can operate a warehouse containing thousands of parts. The system also minimizes the need for material-handling equipment and even material handling itself by reducing the average distance traveled and by being correct every time in identifying the location of an item. Pilferage and breakage, which are generally proportional to the amount of material handled, are also reduced. The AS/R system, however, requires a high initial investment, and a thorough economic analysis must be performed to determine the feasibility of using such a system in a particular warehouse.

The AS/R system has four major components: S/R machines, the storage structure, conveying devices, and controls.

Storage retrieval cranes form one of the most important parts of the system. These machines can carry heavy loads and can simultaneously move horizontally and vertically to reach the required location. They travel on floor-mounted rails guided by electrical signals and may be equipped to function in the single-command mode, allowing the machines to either store or retrieve in a trip, or they may have double-command mode capability, being able to perform both tasks in one trip.

The storage structure interfaces with S/R machines and, in doing so, requires very close tolerances in its construction. The guide rails within it must allow the S/R cranes to move in and out freely, stopping exactly at the required cubbyhole. The structure itself must be adequately protected against fire. These high-rise structures (heights of 90 feet are not unusual) can be freestanding within a building or can form part of the supporting frame of the building. In the overall design of the structure, as will be shown in the next section, many factors play important parts. They include load characteristics such as weight, shape, and size of the items to be stored, and environmental factors such as dimensions of the building, activity rate, and limitations imposed by the cost of construction.

Conveying devices are the auxiliary equipment that interface with the S/R machines and various departments within a plant such as shipping, receiving, and manufacturing. They include forklift trucks, various types of conveyors, towline and shuttle trolleys, guided vehicles, and other equipment. The items are transported by the conveying devices to and from the department, while they are further handled by the S/R machines in the storage structure. Many of the devices such as conveyors and guided vehicles can be made interactive with the AS/R system's computer to achieve maximum flexibility and rapid real-time response.

Computers and their software and controlling mechanisms that tie the computers to the S/R and summary machines are the key to the control of the AS/R system. They process the information and activate the necessary equipment. Most modern systems use several small computers, each controlling a separate device, rather than one large computer controlling the entire system. These small processors communicate with one large computer, which is in charge of inventory maintenance, cost calculation, and billing information. Such an arrangement is called a distributed system. It is flexible in design and use; despite the failure of one computer, the system can still operate, and the installation and maintenance of small computers are more easily accomplished than would be the case for one large computer. The distributed system also provides a faster response compared to a system consisting of a single computer.

### 11.9.1 Design of AS/R Systems

Designing an AS/R system means determining all three dimensions of the physical storage space. A vertical stack or storage, going from floor to ceiling, is referred to as a bay; a series of bays placed side by side are called rows, and the spaces between the rows form the aisles (Figure 11.9). The aisles are used for stacker cranes to move up and down between the rows. Each crane can serve both sides of the aisle and, as mentioned

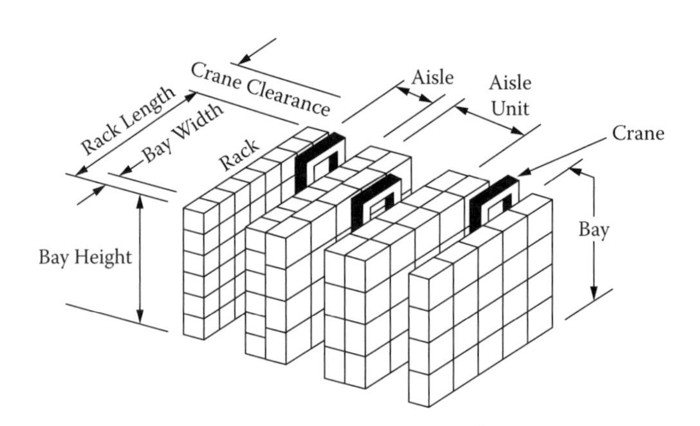


**FIGURE 11.9** AS/R system definitions

earlier, can be either a single- or dual-cycle crane. A dual-cycle crane can start from a pickup station, store an item, retrieve another item, and return to the pickup station in one command, while a single-cycle crane can work only one phase of the cycle at a time. For example, if the command is to store an item, the crane will go from the station to storage and perform the storing operation, but it must wait for a retrieve command before it can fetch another item from the storage and return to the pickup station.

Typically, one crane can perform 32 single operations per hour or 22 dual cycles per hour. Single operational mode is preferred for speed if the operating policy is to store all items at one time and retrieve all the necessary items at some other time. On the other hand, if storing and retrieving are to occur simultaneously, the dual-cycle crane is preferred. The efficiency of a crane system depends on several factors that go into making up the system. Generally, an 85% availability is considered as the average attainable. The steps involved in designing AS/R systems are

1. Determine the dimensions and weight of the load to be stored.
2. Determine the number of units to be stored.
3. Determine the throughput rate per hour.
4. Determine the number of cranes needed.
5. Determine the number of rows required.
6. Determine the building height and load height.
7. Determine the number of bays.
8. Determine the system length.
9. Determine the system width.

In determining the sizes of the loads involved, the storage will typically carry a variety of items in unit loads, many of which could be irregular in shape; therefore, the maximum required length, width, and height should be noted. Load orientation also plays an important part; loads that are to be stored parallel to the cubbyhole or at some angle will affect the width and depth of the required storage space. The weight of the loads influences the design of the structure; the maximum storage weight of an individual unit load should therefore also be noted.

The next step is to estimate how many storage spaces will be needed. In developing this number, one must consider not only the present requirements, but also the expected demands in the foreseeable future for the facility. About 2 years of lead time is required in building an AS/R system, and therefore the demand for at least 2 years in the future should be estimated so that the storage does not become inadequate soon after it goes into operation.

The third step is to determine the throughput. This value is the sum of the maximum number of loads in and loads out that will be carried per hour.

In developing the necessary number of cranes, one must first fix the mode of operations that will be most often used: single- or dual-crane systems. The number of cranes then can be determined as follows.

For single-cycle cranes:

$$\text{Number of cranes} = \frac{\text{throughput}}{32\ (\text{cycles/hour}) \times 0.85\ (\text{efficiency})}$$

For double-cycle cranes:

$$\text{Number of cranes} = \frac{\text{throughput}}{22\ (\text{cycles/hour}) \times 0.85\ (\text{efficiency})}$$

Because one crane can serve two rows, the number of rows in the system is obtained by multiplying the number of cranes by 2.

A note about operational cycles per hour for a crane: P & H, a major manufacturer and system installer of AS/R systems, suggests using 32 and 22 cycles/hour for single- and double-cycle cranes, respectively, for the initial design. The cycles per hour might change slightly depending on the length of the aisle and the length-to-height ratio. It is therefore important to contact the manufacturer after planning the initial design to obtain more accurate estimates.

The next step is to determine the height of the storage. This height may range from 30 to 90 feet; however, the most efficient height is between 50 and 70 feet. A value can be chosen within this range to determine how many loads can be stacked in such a structure. Light loads of less than 2500 pounds require a 6-inch clearance for rack support and crane entry, while heavy loads of more than 2500 pounds demand a slightly larger clearance of 9 inches. The total number of stacks is obtained by dividing the storage height by the sum of the load height and the needed clearance and then subtracting 1 to allow for floor and ceiling clearance.

The number of bays (a bay is one vertical stack of storage from floor to ceiling) needed can now be calculated by applying the following relationship:

$$\text{Number of bays} = \frac{\text{number of units to be stored}}{\text{number of rows} \times \text{number of stacks}}$$

The length of the storage facility depends on the bay width and the number of bays determined thus far. The length of a storage rack is equal to the width of the bay times the number of bays in a row. To this, 25 feet must be added for crane runout clearance and an additional allowance for any equipment.

Finally, we must determine the width of the system. To begin, we should establish an aisle and the two adjacent storage racks. A storage cell should be deep enough to accommodate the load, and the aisle should be wide enough to facilitate movement of the load. Thus the aisle unit, which includes 2 times the storage width plus the aisle width plus clearance, should be at least equal to 3 times the depth of a load plus 2 feet for clearance. The width of the storage system is then the aisle unit times number of aisles in the system.

For example, consider the following problem. We wish to store a unit load on $42 \times 48$-inch pallets that is 48 inches high and has a weight of 2000 pounds. Fifty dual-cycle transactions are anticipated per hour, and the total storage requirement is for 10,000 unit loads.

The load characteristics are the following: length, 48 inches; width, 42 inches; height, 48 inches; and load weight, 2000 pounds. Suppose we select the height of the storage building as 60 feet. Then the numbers of stacks that can be accommodated with a 4-foot height load is (taking only the integer portion of the quotient and allowing for 6-inch clearance between stacks)

$$\frac{60}{(4+0.5)} - 1 = 13.33 - 1 = 12 \text{ loads}$$

The number of dual-cycle cranes necessary to handle a throughput at 50 transactions per hour are 50/(22 × 0.85) = 2.67, or 3 (round off to the next higher integer value of division).

The number of bays necessary would be

$$\frac{10{,}000 \text{ storage units}}{2 \times 3 \text{ cranes } \times \text{ 12 loads nigh}} = 138.8, \quad \text{or} \quad 139 \text{ bays}$$

For each bay, the width is 42 inches for the load plus 6 inches of clearance, giving a total of 4 feet. The length of the storage is then 4 × 139 + 25 feet for crane clearance, or 581 feet.

To obtain the width of the system, multiply the aisle unit by the number of aisles (cranes). The depth (length) of a unit × 3 + 2 feet gives the aisle unit, which is 4 × 3 + 2 =14 feet. The width of the system is then 14 × 3 = 42 feet. Thus the storage dimensions are 42 × 581 × 60 feet.

It is obvious that, by choosing different heights for the building, it is possible to obtain different dimensions for the storage building.

### 11.9.2 Cost Estimate

P & H (1984) uses the cost parameters in Table 11.13, accurate to within 15%, as of 1984 (see the table note for information regarding 1993). Changes from that date can be approximated by using the Industrial Commodities Index or Iron Age Steel Index.

For the AS/R system configuration developed in the example, the estimated cost can be calculated as in Table 11.14.

The price will range somewhere between the low value and the high value. For budget purposes, one might consider the cost to be $3.78 million, the average of two values.

### 11.9.3 Order Picking in an AS/R System

Order picking can be either for a full unit load or for part loads. In a full unit load, two methods may be used: worker-ride or out-of-aisle. In the worker-ride system, the operator rides an S/R machine, picks the goods from storage, and retrieves the entire order. The operator is assisted by auxiliary lifting devices and by a computer terminal in the carrying platform, which displays information such as which item to pick, where it is in storage, and what quantity is required.

In out-of-aisle picking, the unit loads (boxes) are picked and brought to the ends of the aisles by machines. The operator then assembles the order, and the remaining items are returned to storage.

**TABLE 11.13**
**Estimated Costs for AS/R Components[a]**

| Equipment | Cost | |
|---|---|---|
| | Low | High |
| Storage spaces (drive-in design), each | $100 | $160 |
| Freestanding racks (building not included) | $150 | $220 |
| Or rack-supported building (roof, siding, slab, HVAC, sprinklers, lighting not included) | | |
| Floor-running, automated crane, each | $210,000 | $275,000 |
| Aisle transfer car (if required), each | $140,000 | $185,000 |
| Dedicated warehouse computer system | $80,000 | $150,000 |
| Basic control and inventory | | |
| Full function | $150,000 | $800,000 |
| Aisle hardware for crane or crane transfer car (included floor rail, top rail position sensors, electrical wiring, hydraulic end stops) | | |
| Cost per aisle | $15,000 | $20,000 |
| Plus cost per foot | $95 | $160 |
| Conveyors (powered chain or roller type) | | |
| Cost per aisle (input and output spurs) | $50,000 | $70,000 |
| Plus cost per foot (all other conveyors) | $500 | $1000 |
| Project services (includes project management, site supervision, and training) | $50,000 | $200,000 |

[a] 2008 Prices are approximately 3.5 times the values indicated in the table.

**TABLE 11.14**
**Cost Determination for AS/R System in 1984[a]**

| Equipment | Cost | |
|---|---|---|
| | Low | High |
| Storage space | | |
| Rack-supported building 12 loads high × 139 bays × 6 rows = 10,008 | $1,501,200 | $2,201,760 |
| Cranes, 3 aisles, automated | $630,000 | $825,000 |
| Dedicated warehouse computer, full Function | $150,000 | $800,000 |
| Aisle hardware | | |
| 3 aisles | $45,000 | $60,000 |
| 581 feet/aisle × 3 = 1743 feet | $165,585 | $278,880 |
| Conveyors, 3 aisles with I/O spurs | $150,000 | $210,000 |
| About 200 feet | $100,000 | $200,000 |
| Project services | $50,000 | $200,000 |
| Total cost | $2,791,785 | $4,775,640 |

[a] For 2008 costs, multiply the above values by 3.5.

## 11.10 LOADING DOCKS

Let us turn our attention to loading/unloading docks, another important feature in storage operations. Design of loading and unloading docks is a task that depends on the type of transportation used and the amount of material handled. Some warehouses load directly onto planes or trains, and require special docks to accommodate these situations. Most loading (and the type concentrated on here) is from warehouse loading dock to truck. The amount of material handled by weight at the busiest time of day divided by the productivity rate of 7500 pounds per worker-hour is used to determine the minimum number of docks needed. For example, suppose that trucks hold an average of 4290 pounds each; then it takes 0.572 worker-hours to unload each truck. If 10 minutes is allowed for the truck to leave the dock and the next one to pull up, then 44.3 minutes is needed to serve each truck. Therefore, one dock can serve 60/44.3 = 1.35 trucks per hour. If $n$ trucks are to be served per hour, the number of docks needed would be $[n/1.35] + 1$, where the quantity in brackets stands for the integer portion of the division. For example, serving three trucks per hour will require $[3.0/1.35] + 1 = 3$ docks.

One might consider the operation of docks as a queuing phenomenon, with trucks arriving randomly (following a Poisson distribution) and the time of service being exponentially distributed or following some other probability distributions. The standard formulas are noted in Appendix C.

In our example, suppose now that the arrival pattern of trucks follows the Poisson distribution (random) with a mean of 3/hour and that the time for service follows an exponential distribution with a mean of 1.35/hour, that is, $\lambda = 3$/hour and $\mu = 1.35$/hour, respectively. Clearly, because the service rate is less than the arrival rate, we need more than one dock.

Assume that we decide to provide three docks. The corresponding characteristic values are since $\lambda/\mu = 2.22$)

$$P_0 = \left[\frac{(2.22)^0}{0!} + \frac{(2.22)^1}{1!} + \frac{(2.22)^2}{2!} + \frac{(2.22)^3}{3!\left(1 - \frac{2.22}{3}\right)}\right]^{-1}$$

$$= 0.0788$$

$$P_1 = \frac{(2.22)^1}{1!} \times 0.0788 = 0.175$$

$$P_2 = \frac{(2.22)^2}{2!} \times 0.0788 = 0.194$$

$$P_3 = \frac{(2.22)^3}{3!} \times 0.0788 = 0.143$$

$P_0$, $P_1$, and $P_2$ represent probabilities of having zero, one, and two trucks in the system, respectively. With three docks in the warehouse, the sum of these probabilities indicates the probability that an arriving truck will find an empty dock and will not have to wait. The average number of trucks in the queue can now be determined:

$$L_q = \frac{(0.0788)(2.22)^3\left(\frac{2.22}{3}\right)}{3!\left(1-\frac{2.22}{3}\right)^2}$$

$$= 1.55 \text{ trucks}$$

$$L = \frac{\lambda}{\mu} + L_q = 2.22 + 1.55$$

$$= 3.77 \text{ trucks}$$

The average waiting time for the truck is $W_q = 1.55/3 = 0.5166$ hour, and the average time spent by a truck on the premises is

$$W = W_q + \frac{1}{\mu} = 0.5166 + \frac{1}{1.35}$$

$$= 1.257 \text{ hours}$$

Both quantities can be reduced by increasing the number of docks. We can determine the optimum number of docks by performing an economic analysis similar to ones shown below. Suppose the cost of operating one dock is $12/hour, which includes the cost of an operator, forklift truck, and construction and maintenance cost of the dock spread over an hourly basis. The cost of operation of the truck is $16/hour, including the driver's pay and truck expenses. Because each arriving truck spends 1.257 hours on the premises, it costs 1.257 × 16 = $20,112 per truck. On average, three trucks arrive per hour, giving an hourly cost of 3 × 20.112 = $60.34 for truck operations. We must add to this the hourly cost of operations of three docks, which is 3 × 12 = $36.00, giving a total hourly cost for the loading/unloading operation of $96.34.

Similar calculations for a different number of docks lead to the following results:

| Number of Docks | Cost of Loading/ Unloading per Hour |
|---|---|
| 3 | $96.34 |
| 4 | $88.19 |
| 5 | $96.66 |

Therefore, on the basis of present data, we should have four docks in the plant.

Important aspects of a dock include the length, width, and elevation. The elevation of the dock should be 48 inches for pickup trucks and 52 inches for larger trucks. If a variety of vehicles is used in loading and unloading operations, a scissor dock (table) can be used to vary the elevation. The width and length of the dock depend on the size of the load to be stored on it before being moved into the warehouse or into the truck. The space should be at least 12 feet wide to accommodate a truck. Other factors of importance are the volume of freight, the size of units handled, and the material-handling equipment used.

Suppose a warehouse will handle pallets with a maximum size of 48 square inches. The management expects to have no more than six pallets in storage on the dock at one time. At least 3 inches of clearance between the pallets is needed. To accommodate truck loading and unloading, the dock is already 12 feet wide. If 1 foot is added to the platform, three pallets can be stored across it with 3 inches on the sides and between the pallets; therefore, the platform should be made 13 feet wide. If three pallets are placed across the platform, only two rows will have to be placed down the length. That would be 8 feet 9 inches in length. A forklift truck will be used to handle the pallets. The dock width allows 13 feet for the truck and now another 12 feet must be added to the length for maneuvering space, because the forklift truck chosen is about 8 feet long and another 4 feet of clearance is needed for the truck to safely back up and swing around. The length of the dock is then 20 feet 9 inches. This means that the overall minimum dimensions should be 20 feet 9 inches × 13 feet.

After determining the number and size of the docks, approaches for the trucks must be designed. These approaches are to handle the traffic coming and going and to allow space for the trucks to turn around and back to the docks. Space should also be allowed for trucks waiting if all docks are occupied. Two points are to be kept in mind: first, the closer the turn from the main road is to 90°, the more difficult it is for the vehicles; and second, large trucks can make left turns more easily than right turns.

Two characteristics of the approach are the apron depth and the bay width. The apron depth is the amount of space needed to accommodate the truck length, and the bay width corresponds to the truck width. The angle of the dock can be varied to make the best use of the available space (Figure 11.10). Because truck lengths range from 40 to 70 feet, the apron depth should be sufficient to accommodate the largest vehicle delivering to or picking up from the plant.

Parking required for loading/unloading docks

1
2
3
4
5
6
THE DART
ADVANTAGE

Loading and unloading stations

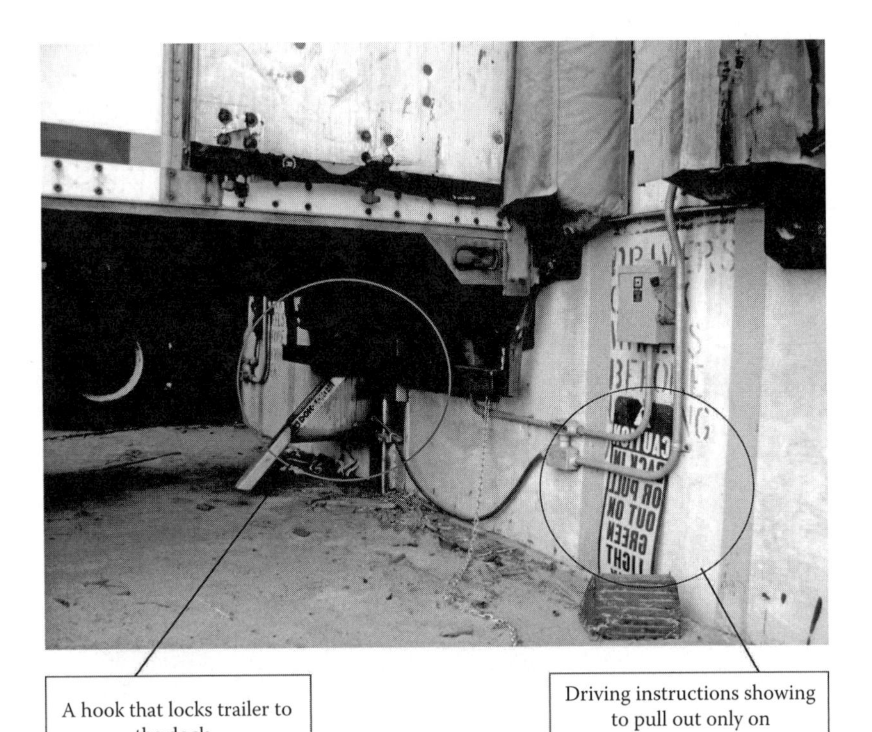
A hook that locks trailer to the dock.
Driving instructions showing to pull out only on green lights

Safety features on a dock

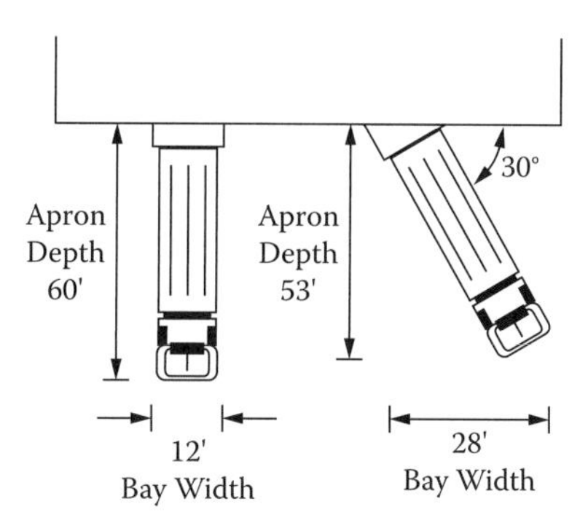


**FIGURE 11.10** Apron depth and bay width

The dock could be completely enclosed (Figure 11.11a) or simply covered by a roof, providing protection against the environment.

The parking surface for the trucks and trailers immediately adjacent to the dock should be almost level but with a slight incline to provide for adequate drainage. When a sawtooth dock arrangement is available (Figure 11.11b), it helps to reduce maneuvering problems.

## 11.11 DOCK DOORS

When docks are not in use, their entrances must be secured. The types of protection vary and can be selected according to the company's needs and the advantages of the particular door. Here is a list of door types and their characteristics:

1. *Air curtains* are used for high-traffic situations. High-velocity airstreams are blown down across the door opening. Wind, heat, cold, and insects are kept out by the airflow. Air curtains are not very effective for radical temperature differences or for controlling noise.
2. *Strip doors* are inexpensive, transparent plastic strips that part easily to accommodate traffic. Strip doors are not a good barrier to heat transfer (in or out) and insects, but they allow quick and easy entrances.

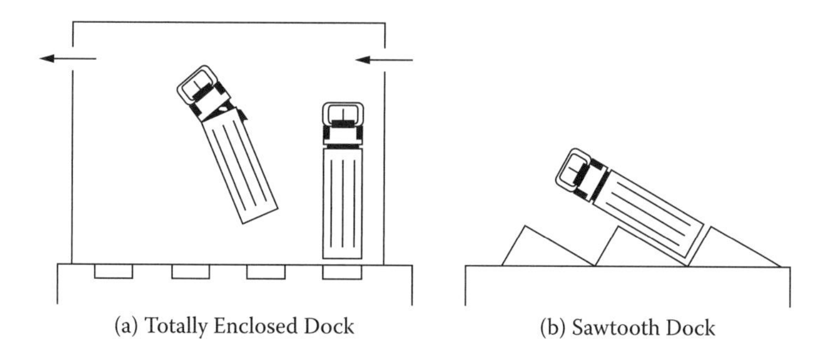


**FIGURE 11.11** Dock arrangements

3. *Hinged doors* can be made of wood or metal and may have windows. They offer maximum security, but it takes time to open and close the doors each time.
4. *Swinging doors* are also made of wood or metal with 180° swing radius. They are self-closing and can be secured.
5. *Impact doors* are swinging doors with impact bumpers so that trucks can push through and bump them open. They can be single or double doors and can be equipped with electronic controls that open the door upon impact.
6. *Sliders* are single or double doors on track-and-trolley hardware. They can be telescoping if there is not enough wall space to slide the door back. Sliders can be manual or power-operated.
7. *Overhead doors* are similar to garage doors. The doors travel on tracks up to the ceiling. They can be manual or power-operated, but they require time to open. Overhead doors provide the best use of space because the doors do not take up any vertical space.

Plastic strips and air curtain to maintain temperature and allow free fork-lift transportation

With the energy conservation and associated savings offered to a company, further sealing of the dock opening might be an important consideration. One way to achieve this is by using conventional insulation or weather stripping, which can be placed around the door. Air seals can also be used that can be inflated and sealed around the truck while it is being loaded or unloaded.

Another consideration in designing dock doors is the choice of controls used to open and close the doors. There are two types to consider, manual and power-operated, and there are some trade-offs in using each type. Power controls open and close the doors more quickly and require little labor; however, they are more expensive and require more maintenance. Manual controls require more time and labor, and they expose the building interior to outside temperatures for longer periods. The best choice for the type of control depends on the volume of activity, which should be analyzed carefully.

## 11.12 COMPUTER PROGRAM DESCRIPTION

### 11.12.1 MATERIAL REQUIREMENT PLANNING

The program "mrp" solves problems associated with MRP for a given BOM structure. The BOM will be referred to as BOM in the forthcoming discussion.

The user should type "mrp" at the DOS prompt to execute the program. The MAIN MENU is displayed and allows the user to choose from one of the five options displayed.

```
MAIN MENU
1. ENTER THE BOM FILE
2. PROCESS FILE
3. LOAD BOM FILE
4. SAVE BOM FILE
5. EXIT
ENTER YOUR SELECTION:
```

#### 11.12.1.1 Option 1 (Enter the BOM File)

The first option expects the user to enter the BOM structure of a product. The program allows the user to enter the BOM for multiple end products. This option is explained in detail with an example. Consider the following BOM structure for two end products.

**End Product 1**

| Part Number | Carrying Cost | Ordering Cost | Initial Inventory | Lead Time |
|---|---|---|---|---|
| 1 | 0.0[a] | 0.0[a] | 0 | 0 |
| 7 | 3.2 | 230.0 | 20 | 1 |
| 3 | 0.0[a] | 0.0[a] | 45 | 1 |
| 4 | 0.7 | 13.0 | 220 | 2 |
| 5 | 5.0 | 80.0 | 75 | 2 |

| Period | Demand |
|---|---|
| 1 | 10 |
| 2 | 20 |
| 3 | 10 |
| 4 | 15 |
| 5 | 0 |
| 6 | 10 |

[a]Carrying cost and ordering cost for assembled products are assumed zeros.

### End Product 2

| Part Number | Carrying Cost | Ordering Cost | Initial Inventory | Lead Time |
|---|---|---|---|---|
| 2 | 0.0[a] | 0.0[a] | 15 | 1 |
| 8 | 2.5 | 240.0 | 15 | 1 |
| 3 | 0.0[a] | 0.0[a] | 45 | 1 |
| 4 | 0.7 | 13.0 | 220 | 2 |
| 5 | 5.0 | 80.0 | 75 | 2 |

| Period | Demand |
|---|---|
| 1 | 15 |
| 2 | 12 |
| 3 | 0 |
| 4 | 5 |
| 5 | 10 |
| 6 | 0 |

[a] Carrying cost and ordering cost for assembled products are zero.

The output displayed by the computer is shown next; the values to be entered by the user are in boldface.

```
ENTER THE NO. OF END PRODUCTS (MAX. OF 10 ONLY):  2
ENTER THE NUMBER OF PERIODS (MAX. OF 20 ONLY):  6
ENTER DETAILS FOR END PRODUCT 1
PARENT NUMBER:  0
PART NUMBER:  1
CARRYING COST:  0.0
ORDERING COST:  0.0
INITIAL INVENTORY:  0
LEAD TIME:  0
CONFIRM ENTRIES (Y/N)?:  Y
DO YOU WANT TO CONTINUE WITH THIS END PRODUCT?:  Y
```

The user should note that the parent number for all end products is automatically assigned a 0 by the program. The user is not allowed to enter a parent number for an end product. The user has to check all entries before entering Y for the first question. In this case, because there is more than one component that makes the end product, the user should enter Y for the second question. In some cases, the end product is not made up of any subparts, in which case the user should enter N for the question "Do you want to continue?"

```
ENTER DEMAND FOR EACH    PERIOD
DEMAND FOR PERIOD 1:     10
DEMAND FOR PERIOD 2:     20
DEMAND FOR PERIOD 3:     10
DEMAND FOR PERIOD 4:     15
DEMAND FOR PERIOD 5:     0
DEMAND FOR PERIOD 6:     10
```

The screen is refreshed and the user is required to enter the inventory details of the parts that make the final end product. These values have to be carefully keyed in from the data tables given previously.

```
ENTER DETAILS FOR PARTS
PARENT NUMBER:      1
PART NUMBER:      7
CARRYING COST:      3.2
ORDERING COST:      230.0
INITIAL INVENTORY:      20
NO. OF UNITS REQUIRED:      1
LEVEL NUMBER:      1
LEAD TIME:
CONFIRM ALL ENTRIES (Y/N)?:      Y
DO YOU WANT TO CONTINUE WITH THIS END PRODUCT?:      Y
```

Once again, the user is expected to check all entries before answering Y to the first question, and to answer Y for the next question if there are more products required to make the end product. By referring to the BOM structure the user can enter the level numbers for each of the parts. The program can check for proper level-number sequencing but not for validity of the level number. The program gives erroneous results if the level number is not entered correctly. The program continues until all the parts of the end product have been keyed in. Once all the parts for the end product have been entered, the user should enter N to the second question to terminate further data entry for that end product. The entire process is repeated for the second end product.

Note: The user should enter a 0 for ordering cost and carrying cost for parts manufactured in the company. Valid costs should be entered for parts ordered from another vendor.

#### 11.12.1.2 Option 2 (Process the BOM File)

Once the data have been entered, they are ready for processing. The program gives the following output for the example data entered.

**PART 1**

| PERIOD | DEMAND | SCHEDULED RECEIPTS | PLANNED ORDER RELEASE |
|---|---|---|---|
| 1 | 10 | 10 | 10 |
| 2 | 20 | 20 | 20 |
| 3 | 10 | 10 | 10 |
| 4 | 15 | 15 | 15 |
| 5 | 0 | 0 | 0 |
| 6 | 10 | 10 | 10 |

**PART 2**

| PERIOD | DEMAND | SCHEDULED RECEIPTS | PLANNED ORDER RELEASE |
|---|---|---|---|
| 1 | 15 | 10 | 12 |
| 2 | 12 | 12 | 0 |
| 3 | 0 | 0 | 5 |
| 4 | 5 | 5 | 10 |
| 5 | 10 | 10 | 0 |
| 6 | 0 | 0 | 0 |

**PART 3**

| PERIOD | DEMAND | SCHEDULED RECEIPTS | PLANNED ORDER RELEASE |
|---|---|---|---|
| 1 | 44 | 0 | 39 |
| 2 | 40 | 39 | 30 |
| 3 | 30 | 30 | 50 |
| 4 | 50 | 50 | 0 |
| 5 | 0 | 0 | 20 |
| 6 | 20 | 20 | 0 |

**PART 4**

| PERIOD | DEMAND | SCHEDULED RECEIPTS | PLANNED ORDER RELEASE |
|---|---|---|---|
| 1 | 117 | 0 | 197 |
| 2 | 90 | 0 | 0 |
| 3 | 150 | 197 | 0 |
| 4 | 0 | 0 | 0 |
| 5 | 60 | 0 | 0 |
| 6 | 0 | 0 | 0 |

ORDERING COST: 13.00

CARRYING COST: 165.20

**PART 5**

| PERIOD | DEMAND | SCHEDULED RECEIPTS | PLANNED ORDER RELEASE |
|---|---|---|---|
| 1 | 39 | 0 | 64 |
| 2 | 30 | 0 | 0 |
| 3 | 50 | 64 | 0 |
| 4 | 0 | 0 | 0 |
| 5 | 20 | 0 | 0 |
| 6 | 0 | 0 | 0 |

ORDERING COST: 80.00

CARRYING COST: 410.00

PART 7

| PERIOD | DEMAND | SCHEDULED RECEIPTS | PLANNED ORDER RELEASE |
|---|---|---|---|
| 1 | 10 | 0 | 45 |
| 2 | 20 | 45 | 0 |
| 3 | 10 | 0 | 0 |
| 4 | 15 | 0 | 0 |
| 5 | 0 | 0 | 0 |
| 6 | 10 | 0 | 0 |

ORDERING COST: 230.00

CARRYING COST: 288.00

PART 8

| PERIOD | DEMAND | SCHEDULED RECEIPTS | PLANNED ORDER RELEASE |
|---|---|---|---|
| 1 | 12 | 0 | 0 |
| 2 | 0 | 0 | 12 |
| 3 | 5 | 12 | 0 |
| 4 | 10 | 0 | 0 |
| 5 | 0 | 0 | 0 |
| 6 | 0 | 0 | 0 |

ORDERI COST: 240.00

CARRYING COST: 40.00

To change the data: This example data is stored in the file mrp.DAT.

---

### 11.12.1.3 Option 3 (Load the BOM File)

This option helps the user retrieve any stored BOM data for further processing. Once this option has been chosen, the program prompts the following.

```
ENTER THE NAME OF THE FILE TO BE LOADED:  mrp.DAT
```

Upon refreshing the screen,

```
DO YOU WANT TO MAKE CHANGES IN THE PRODUCT DEMAND?:  Y
THE FOLLOWING END PRODUCT NUMBERS ARE AVAILABLE
FOR CHANGES
1
2
```

The user is given a list of end products available for changes in the demand, if there are any.

```
ENTER THE END PRODUCT NUMBER:  2

PERIOD    DEMAND
1           15
2           12
3            0
4            5
5           10
6            0
DO YOU WANT TO MAKE CHANGES IN THE DEMAND?:  Y
```

The user is expected to enter Y if and only if there is a necessity for change in the demand.

```
ENTER THE PERIOD: 2
ENTER THE NEW DEMAND: 77
```

The program gives the user an opportunity to change demands for more than one period. Once the user is through with the necessary changes to be made with the demand, the user can get back to the main menu and process this new BOM data file and view the results.

#### 11.12.1.4 Option 4 (Save the BOM File)

This option helps the user to save all the data entered in a file for future reference and processing. The program prompts the following to the user:

```
ENTER THE NAME OF THE FILE TO BE SAVED: mrp.DAT
```

#### 11.12.1.5 Option 5 (Exit)

This option exits from the program and takes the user to the DOS prompt.

### 11.12.2 Caveats

1. If the part numbers are repeated, the other parameters — initial inventory, carrying cost, ordering cost, and lead time of the part — should also be maintained. The program does not check for this feature.
2. If a subpart of one end product is another end product by itself, the BOM structure should be maintained throughout. Again, the program does not check the error if two different BOM structures are given the same end product.
3. The part number of any part should not exceed the value 20. The program will give unpredictable results if a greater number is entered.

## 11.13 SUMMARY

Application of the concept of JIT in manufacturing organizations requires that raw material and partially finished products are available where and when they are needed. Furthermore, the finished products should be shipped to their final destination immediately so as to avoid any finished goods inventory. Although desirable, such a system is very hard to achieve if one cannot predict with certainty all the demands and sales, or is at the mercy of subcontractors who are not reliable in their promised delivery dates or times. Most manufacturers therefore provide for storage and warehouse facilities to absorb variations in production and sales and also to take advantage of economic lot sizes and quantity discounts. Storage is almost always required in retail business, where a variety of customers desire immediate delivery of different products and also where display of the items is so very important to sales.

The basic concepts of the material-requirement-planning technique are presented in the chapter. MRP leads to planned order release and hence eliminates excessive and nonproductive storage of the items. Ordering and storage cost can play an important part in determining the order size.

To select a site for storage and warehousing, one should follow the site selection procedure described in Chapter 15 and apply the models from Chapters 17, 18, and

19. In general, the site should be accessible, appropriately zoned, and protected by the local authorities. Considerations for the building itself involve analysis of such factors as approach roads, size of the site, and inside layout of the building (aisle layout, column pattern, and storage rack arrangement).

The chapter contains two examples that show how the space requirement for each item in storage may be calculated. Such an analysis should be performed on all items stored in a warehouse. Obtaining a good estimate of the total space requirement is important in building a warehouse that is neither too large nor too small.

Many different activities are regularly conducted in storage and warehouse operations. They range from receiving, inspecting, and storing items to packaging, labeling, and shipping them. Physically arranging the warehouse so that these activities can take place efficiently is critical in keeping the cost of operation down. The items may be stacked based on physical similarity, functional similarity, or popularity, or by separating the reserve stock from the regular stock or storing items randomly but with computer control. Expensive and high-security items may need additional considerations. How the items are to be collected to fill an order also influences the layout of the storage structure. The area system, modified area system, zone system, sequential zone system, and multi-order system are some of the ways in which the order picking may be accomplished. The chapter provides an example to show how a storage arrangement may be developed.

Many accessories such as bins, shelves, and racks are available for storing small and individual items. Racks are a common feature in most warehouses, and they are available in many different styles and shapes. Some of the common types of racks are the selective pallet rack, movable-shelf rack, drive-in and drive-through racks, cantilever rack, stacker crane rack, and portable rack. Each type has certain properties that make it attractive in a particular situation. Storing and then retrieving items from a particular area in the storage structure requires development of a storage location system, and one such coding system is illustrated in the chapter.

AS/R is a nontraditional storage system that uses high-rise storage units, fast-moving stacker cranes, and controls and management systems that are under computer control. It has high initial cost but provides economy in operations by requiring smaller floor space, fewer workers for operation, and less time in order picking (which is also more accurate). It also reduces pilferage. An AS/R system has four major components: S/R machines, the storage structure, conveying devices, and controls. However, to obtain a good estimation of the initial cost, one must develop in detail two major components, S/R machines and the storage structure. The chapter presents the steps involved in the design of an AS/R system, and then illustrates them with a numerical example. By using cost data provided by P & H, a major AS/R system manufacturer, the cost estimate for the example problem is obtained.

Loading docks are another important feature in storage operations. The chapter shows how one may use queuing analysis to determine the number of docks necessary to support a certain level of loading and unloading activities. The chapter also discusses dock design, which involves determination of the physical dimensions and layout of docks.

When docks are not in use, their entrances must be secured by dock doors. A number of different types of doors are commercially available, with each having some distinct characteristics. Depending on the activity level and type of activities in a plant, one may use air curtains, swinging doors, impact doors, or overhead doors.

## PROBLEMS

1. What are the advantages and disadvantages of JIT?
2. What factors are to be considered in planning the construction of a warehouse?
3. Why are aisles important? What factors are to be considered in planning the aisle width?
4. A company has three rooms for storing the units on pallets (Figure 11.12). If a 6-foot aisle space is needed, along with clearances of 3 inches between stacks and 1 foot overhead, determine how many $4 \times 5 \times 3$-foot ($W \times L \times H$) pallets can be stored. The ceilings are 10 feet high. By what percentage will storage capacity increase if the two non-load-bearing walls are removed and 6-foot aisles can be rearranged either north-south or east-west?
5. Discuss which method of storage is used in each of these examples.
   a. A grocery store
   b. An auto parts store
   c. A department store such as Sears
   d. A catalog store
6. The demands per week for part 103 for the upcoming 10 weeks are:

| Week | 1 | 2 | 3 | 4 | 5 |
|---|---|---|---|---|---|
| Demand | 20 | 30 | 25 | 20 | 30 |
| Week | 6 | 7 | 8 | 9 | 10 |
| Demand | 15 | 40 | 15 | 10 | 45 |

   On-hand inventory for the part is 20 units. Lead time for receipt is 2 weeks, and the company is expecting scheduled receipts of 35 units in the second week. The order cost is $5/order and the carrying cost is $0.20/unit/week. Compare an MRP chart for each of the following procedures and calculate the cost of operation for each.
   a. If the company uses the lot-for-lot ordering procedure (i.e., order the exact quantity in each period that equals the net requirement for the period)
   b. If the lot size is fixed at 30 units
   c. If the ordering is done using EOQ
   d. If the ordering is done using PPB

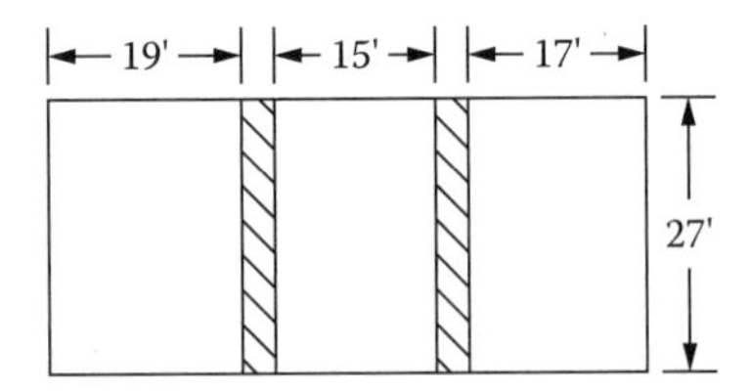


**FIGURE 11.12** Storage rooms

7. In problem 6, suppose the demand for period 3 is uncertain and a safety stock of 10 units is to be introduced for period 3. How will the answers to parts a, b, c, and d change?
8. In problem 6, if the actual demand in period 3 is 32 units, what is the ordering policy in each case?
   a. Using fixed lot size of 30 units for order
   b. Using the PPB method
9. In problem 8, if the actual demand in period 3 is 40 units, answer parts a and b.
10. The following information for end products is given. Determine the planned order releases and scheduled receipts for each of these products (Figure 11.13).

| Period | Demand |
|---|---|
| January | 12 |
| February | 20 |
| March | 30 |
| April | 17 |
| May | 22 |

**Keyboard**

| Period | Demand |
|---|---|
| January | 45 |
| February | 60 |
| March | 55 |
| April | 80 |
| May | 30 |

| Part | Carrying Cost | Ordering Cost | Initial Inventory | Lead Time |
|---|---|---|---|---|
| AT 386 | — | — | 10 | 1 |
| Hard disk | 0.4 | 35 | 5 | 2 |
| Monitor | 1.2 | 90 | 25 | 1 |
| Keyboard | — | — | 45 | 1 |
| Motherboard | — | — | 10 | 1 |
| Memory chips | 0.9 | 35 | 20 | 0 |
| Mouse | — | — | 5 | 1 |
| Floppy Drive | 0.3 | 25 | 0 | 0 |
| Controller | — | — | 10 | 0 |
| Frame (keyboard) | — | — | 0 | 0 |
| Cord | 0.01 | 2.2 | 20 | 0 |
| Frame (mouse) | 0.05 | 3.2 | 15 | 0 |
| Buttons | 0.01 | 1.2 | 45 | 0 |
| Keys | 1.3 | 12 | 300 | 0 |
| Screws | 0.4 | 5 | 250 | 0 |

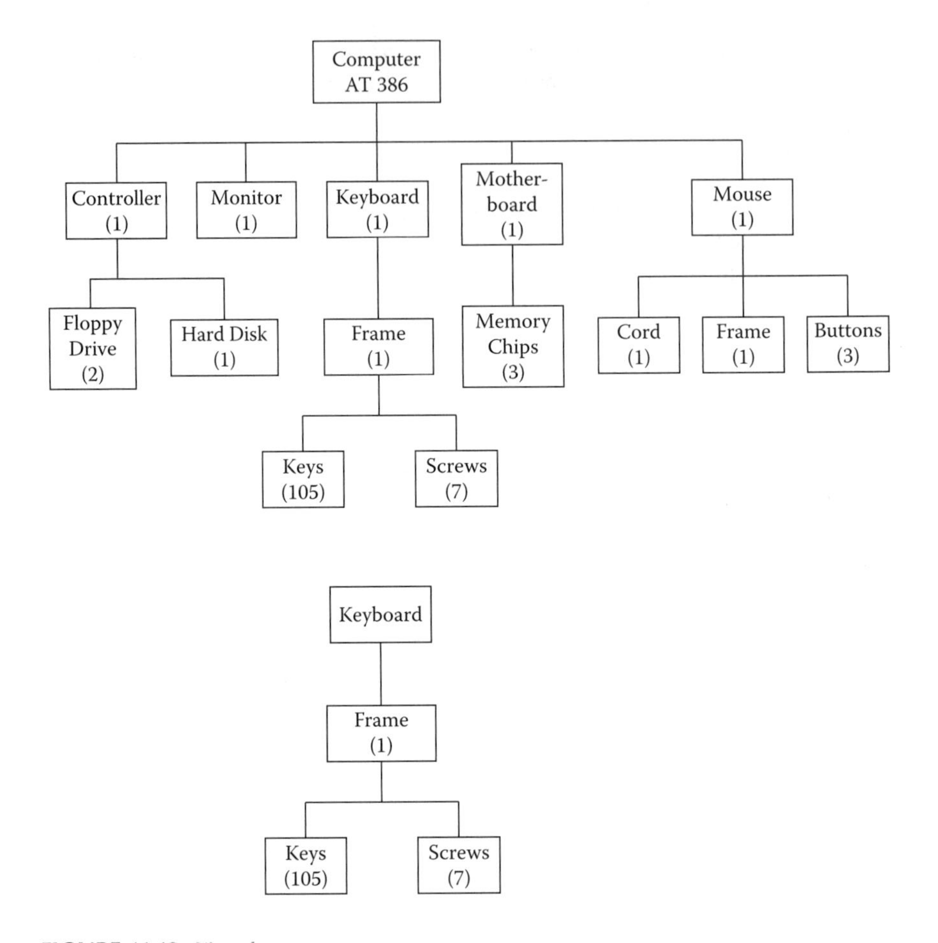


**FIGURE 11.13** Flowchart

11. Determine the best storage arrangement for the six items in the accompanying table, along with the number of sections required and the size of the building if it can have only four sections in a row with an arrangement as shown in the sketch in Figure 11.14.

    Other information is as follows: items B, E, and F have the same unit size.

| Item | Units Received | Receiving Trips | Average Units Shipped |
|---|---|---|---|
| A | 100 | 50 | 0.5 |
| B | 400 | 80 | 3 |
| C | 250 | 10 | 5 |
| D | 200 | 20 | 2 |
| E | 500 | 100 | 5 |
| F | 175 | 15 | 4 |

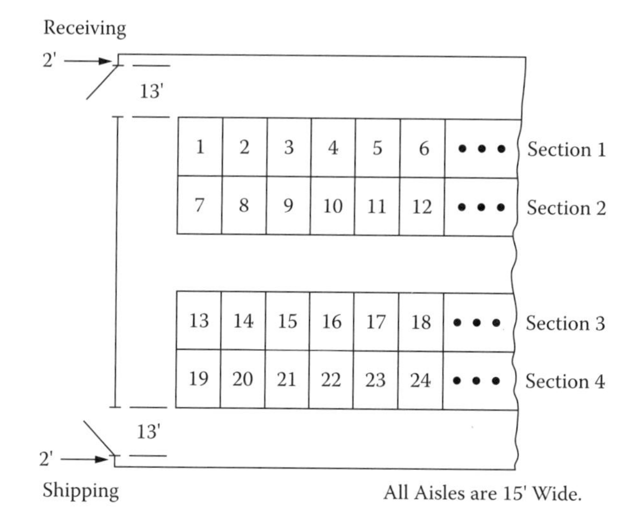


**FIGURE 11.14**

The unit size of items A and D is one-half the size of items B, E, and F, while C is one-third the size of items B, E, and F. One hundred units of item B fill one storage section.

12. Discuss the advantages of AS/R systems.
13. A plant produces 75 units/hour of an item with dimensions of 1 × 1 × 1 foot, storing 1 week's supplies in containers measuring 5 × 5 × 5 feet. A minimum of 3 inches of space is required between adjacent stacks. Determine the number of containers needed.
14. A 1-week inventory of the boxed telephones in problem 3 are to be stored by an AS/R system.
    a. Determine the number of cranes needed if single-cycle cranes are used.
    b. Determine the number of cranes needed if double-cycle cranes are used.
    c. Determine the system width and length for part a.
    d. Determine the system width and length for part b.
    e. Estimate the cost of the system.
15. A manufacturing company wishes to store a unit on a 36 × 48 × 24-inch pallet having a weight of 1400 pounds and 75 dual cycles per hour. The total storage is 18,000 unit loads. The height of the building is 80 feet, but clearances of 2 feet from the ceiling and 6 inches for the rack support are needed.
    a. Determine the number of stacks that can be accommodated with the height of the load.
    b. Determine the number of dual cranes needed.
    c. Determine the number of bays needed.

 d. Determine the storage dimensions.
 e. Determine the cost of the system.

16. A company receives materials delivered by trucks with a maximum weight of 3500 pounds each. If at most 10 trucks per hour will be delivering at the plant, determine how many docks are needed, using the standards presented in the chapter. Allow 10 minutes for one truck to leave and another to pull in.
17. The docks of Acme Plumbing Manufacturing receive trucks that hold 8000 pounds each. It takes 1.15 worker-hours to unload each truck, and 5 minutes is allowed for one truck to leave and another to pull in. If six trucks are to be serviced at the busiest time of the day, how many docks are needed?
18. Suppose the arrival pattern of the trucks follows a Poisson distribution with a mean of 2 trucks/hour, and the service rate follows an exponential distribution with a mean of 2 trucks/hour. Find the number of docks needed so that a truck does not spend more than ½ hour in the plant.
19. If in problem 18 the truck cost is $50/hour and the dock operator is paid $10/hour, what is the most economical dock size? (Ignore the constraint on the time spent by the truck.)
20. If in problem 18 it is decided that a truck should not wait more than 10 minutes (excluding time for loading and unloading), how many docks are required?
21. Discuss the types of dock doors and their strong and weak points. In one industrial plant in your area, observe the dock door and determine whether the choice was appropriate or whether you would recommend a change.

## SUGGESTED READINGS

### Storage and Warehousing

Binning, R.L., "New uses of traditional storage systems can minimize inventory and work-in-process," *Industrial Engineering*, 16(3), 81–83, March 1984.

Schonberger, R.J., "Just-in-time production systems: replacing complexity with simplicity in manufacturing management," *Industrial Engineering*, 16(10), 52–63, October 1984.

Taff, C.A., *Management of Physical Distribution and Transportation*, 7th edition, Richard D. Irwin, Inc., Homewood, IL, 1984.

### Operations Research

Hillier, F.S., and Lieberman, G.J., *Introduction to Operations Research*, 8th edition, McGraw-Hill, NY, 2004.

Phillips, D.T., Ravindran, A., and Solberg, J.J., *Operations Research — Principles and Practice*, John Wiley, New York, 1976.

Taha, H., *Operations Research — An Introduction*, 8th edition, Prentice Hall, Upper Saddle River, NJ, 2006.

## Automated Storage and Retrieval

Everything You Ever Wanted to Know About System Planning, Harnischfeger P & H, Milwaukee, WI, 1978.

9 Simple Steps to Determine the Layout, Design, and Estimated Cost of an Automated Storage/Retrieval System, Harnischfeger P & H, Milwaukee, WI, 1984.

## Material-Requirement Planning

Smith, S.B., *Computer-Based Production and Inventory Control*, Prentice Hall, Englewood Cliffs, NJ, 1989.

Vollmann, T.E., Berry, W.L., and Whybark, D.C., *Manufacturing Planning and Control Systems*, 5th edition, Richard D. Irwin, Inc., Homewood, IL, 1992.

# 12 Plant and Office Layout *Conventional Approach*

Since the beginning of organized manufacturing, considerable effort has been expended to make the manufacturing facility as efficient as possible. The locations and arrangements of the departments and work centers contribute in a large measure to the manner in which a facility is operating.

The "right" solution to plant layout problems is important for two reasons. First, material-handling costs range anywhere from 30% to 75% of total manufacturing costs. Any savings in material handling realized through a better arrangement of the departments is a direct contribution to the improvement of overall efficiency of the operation. Second, plant layout is a long-term, costly proposition, and any modification or rearrangement of the existing plant represents a large expense and cannot be easily accomplished.

Characteristics of plant layout problems and their data requirements will be discussed in detail in Chapter 13, which also illustrates the use of computers in solving such problems. The basic objective is to achieve an orderly and practical arrangement of departments and work centers to minimize the movement of material and/or personnel while allowing for sufficient working space and perhaps space for future expansion within an area that may be predefined. This goal is kept in mind in every phase of plant development from assignment of overall areas for each of the departments to the generating of a detailed layout of each individual department within its space. These applications are associated with, for example, production, warehousing, offices, toolrooms (TRs), food services, and maintenance. In this chapter, we discuss conventional methods of determining the plant layout.

## 12.1 PROCEDURE

In developing a plant layout, such as the sample shown in Figure 12.1, the procedure generally follows the steps indicated next.

First, determine the area required for each work center. Careful analysis must be performed to establish the desired or necessary content of each center and its associated area requirements. For example, an office center might include the offices for the president and the general manager; areas for sales, personnel, engineering, and accounting departments; conference rooms; reception areas; and space for administrative personnel. A production area could include an area to locate each machine; room to conduct maintenance on the machine; and space for the operator, in-process inventories, assorted tools, and auxiliary equipment. In addition, space must be reserved for quality control, general offices, maintenance benches, and storage areas as well as for water fountains, lavatories, ventilation ducting and equipment, and so forth. It is helpful to develop a simple table listing the contents and their associated

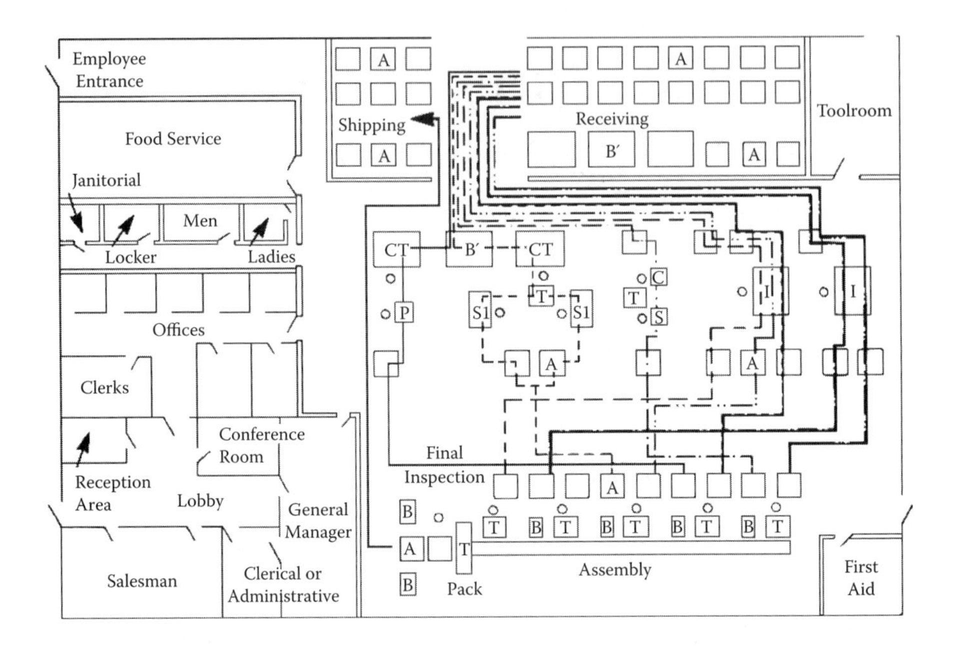


**FIGURE 12.1** Plant layout of Easy Light, Inc

area requirements for each work space to obtain an accurate estimate. Table 12.1 and the associated sketch in Figure 12.2 show how this might be done for the production area at a small manufacturing plant. Table 12.2 shows how this might be done for office space (for more on office area development, see Section 12.6).

**TABLE 12 .1**
**Space Requirements (Production Area)**

| Machine | Number Needed | Dimensions (feet) | Clearance[a] $5 \times l + 4 \times w$ (square feet) | Auxiliary Equipment | In Process Inventory | Singe Unit Area | Total |
|---|---|---|---|---|---|---|---|
| Lathe | 3 | $3 \times 8$ | 52 | $2 \times 4 = 8$ | 10 | 94 | 282 |
| Drill press | 2 | $3 \times 4$ | 32 | $2 \times 2 = 4$ | 15 | 63 | 126 |
| ⋮ | ⋮ | ⋮ | ⋮ | ⋮ | ⋮ | ⋮ | ⋮ |
| Quality control room | | | $20 \times 15$ feet | | | | 300 |
| Lavatory and showers | | | 3 at $3 \times 20$ feet | | | | 180 |
| Lockers and wash basins | | | 30 at $2 \times 6$ feet | | | | 360 |

[a] Estimated by the engineer on the project.

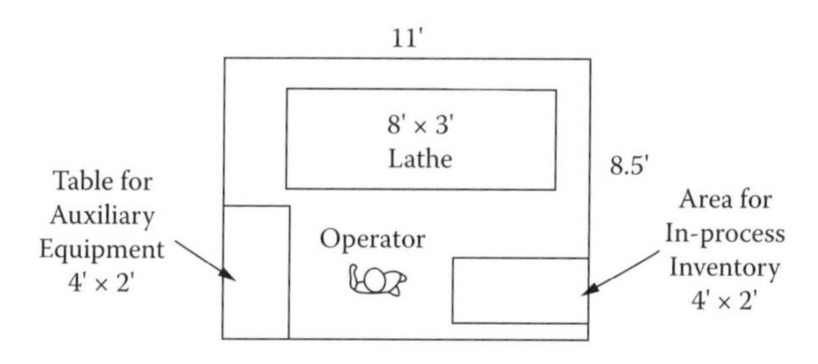


**FIGURE 12.2** Sketch of workstation for a lathe

It should be noted that allowances for aisles in the plant and hallways in offices must be added in their respective area calculations. This allowance is taken to be between 20% and 40% of the total area otherwise required.

The second step is to draw a rel-chart (relationship chart) or establish a from–to chart. The rel-chart qualitatively describes the degree of closeness that the analyst feels should exist between different work centers. This proximity might be dictated by the flow between the departments, convenience, necessity of using the same personnel, facilities being used in two or more departments, or need for communication. The standard codes used in describing this closeness, in descending order of priority, are shown in Table 12.3. In addition, there is code X to describe the undesirability of having two activities close together. For example, noisy punch press operation should be isolated from other manned areas, especially offices. Code X is assigned a negative value such as -1 (or even -10, based on severity) in future evaluations.

**TABLE 12.2**
**Office Space Requirements**

| **Office/Occupants** | **Area (square feet)** |
|---|---|
| President | 250 |
| General manager | 200 |
| Sales manager | 200 |
| Production manager | 200 |
| Accountants (4) | 800 |
| Engineers (6) | 775 |
| Sales representatives (6) | 600 |
| Administrative assistants (7) | 700 |
| Receptionist | 150 |
| Conference room | 250 |
| Copy room | 100 |
| Coffee room | 200 |
| Restrooms | 350 |
| Total | 4775 |

**TABLE 12.3**
**Rel-Chart Priority Codes**

| Code | Priority | Value |
|---|---|---|
| A | Absolutely necessary | 4 |
| E | Especially important | 3 |
| I | Important | 2 |
| O | Ordinary | 1 |
| U | Unimportant | 0 |
| X | Undesirable | −1 |

An example of a rel-chart is shown in Table 12.4. Only half of the entries are required because the table is symmetrical about the diagonal.

There is a close correlation between the from–to chart and the rel-chart.

The from–to chart describes the estimated trips or movements of unit loads between various departments, for example, 600 trips between production and warehouse per day. Thus, it forms a quantitative measurement of needed closeness; for instance, the work centers with the most trips between them should be as close together as possible to minimize travel. In transforming the quantitative measure into a qualitative measure — that is, transferring information from a from–to chart to a rel-chart — the desired degree of closeness between such departments is denoted by code A, E, or I, based on the number of trips between them. Similarly, the departments with almost no direct flow among them would be marked with a U or an O for closeness. In essence, the from–to chart supplies information similar to that from a rel-chart and the two can in fact easily be transformed into one.

For example, suppose the company estimates that its flow of goods between departments will be as that shown in Table 12.5.

**TABLE 12.4**
**Rel-Chart**

| Nodes | PR | WA | OF | TR | FS | MA | LR | SR |
|---|---|---|---|---|---|---|---|---|
| (1) Production (PR) | — | A | E | A | E | A | E | E |
| (2) Warehouse (WA) | | — | O | O | U | O | U | A |
| (3) Office (OF) | | | — | U | O | O | U | O |
| (4) TR | | | | — | O | A | U | U |
| (5) Food services (FS) | | | | | — | U | U | U |
| (6) Maintenance (MA) | | | | | | — | U | O |
| (7) Locker room (LR) | | | | | | | — | U |
| (8) Shipping/Receiving (SR) | | | | | | | | — |

**TABLE 12.5**
**Sample From–To Chart-Estimated Flow of Goods per Day**

| | To | | |
|---|---|---|---|
| **From** | **Production** | **Warehouse** | **Shipping/Receiving** |
| Production | | 600 | 100 |
| Warehouse | 2000 | | 600 |
| Shipping/Receiving | 300 | 2000 | |

The flow of incoming materials for production is from receiving to warehouse to production, while goods produced flow from production to warehouse to shipping. Sometimes, production materials received will go directly to production, while produced goods may go directly to shipping. The total flow between the departments is then given as shown in Table 12.6.

It is imperative, considering the large flow, that production and the warehouse be as close together as possible. The same holds true for the warehouse and shipping/receiving; therefore, in the rel-chart shown in Table 12.7, we assign A's between these departments. because there is some flow between receiving and production, these departments are assigned an E relationship.

Consider another illustration, which shows how to develop a from–to chart in a plant where multiple items are produced. Suppose three products are scheduled for production with a sequence of operations and with known weekly demands as shown in Table 12.8.

A flowchart is developed by noting the flow between departments as the products move from one department to another in their denned sequence. For example, there is a flow of 500 units of product 1 and 1000 units of product 2 between departments A and C, as shown in Table 12.9. The rest of the entries are similarly noted.

The flows between the departments can now be collected to produce a from–to chart, as in Table 12.10.

The entries show the total flow between two departments. The data could be converted to generate a rel-chart by noticing the strong flow between departments A

**TABLE 12.6**
**Total Flow Between the Departments**

| | To | | |
|---|---|---|---|
| **From** | **Production** | **Warehouse** | **Shipping/Receiving** |
| Production | 0 | 2600 | 400 |
| Warehouse | | 0 | 2600 |
| Shipping/Receiving | | | 0 |

**TABLE 12.7**
**Rel-Chart**

| | To | | |
|---|---|---|---|
| **From** | **Production** | **Warehouse** | **Shipping/Receiving** |
| Production | — | A | E |
| Warehouse | | — | |
| Shipping/Receiving | | | A |

**TABLE 12.8**
**Proposed Production Plan**

| Product | Sequence of Operations (Department) | Production per Week |
|---|---|---|
| 1 | A—C—D—F | 500 |
| 2 | B—A—C—D—F | 1000 |
| 3 | E—B—C—A—F—F | 300 |

**TABLE 12.9**
**Flowchart**

| | To | | | | | |
|---|---|---|---|---|---|---|
| **From** | **A** | **B** | **C** | **D** | **E** | **F** |
| A | | | 500<br>1000 | | | 300 |
| B | 1000 | | 300 | | | |
| C | 300 | | | 500<br>1000 | | |
| D | | | | | | 500 |
| E | | 300 | | | | 1000 |
| F | | | | | | |

**TABLE 12.10**
**From–To Chart**

| | A | B | C | D | E | F |
|---|---|---|---|---|---|---|
| A | | 1000 | 1800 | | | 300 |
| B | | | 300 | | 300 | |
| C | | | | 1500 | | |
| D | | | | | | 1500 |
| E | | | | | | |
| F | | | | | | |

and C, departments C and D, and departments D and F and assigning an A relationship between them. Departments A and B, with a flow of 1000 units, could be given an E relationship, and the departments with a flow of 300 units will be converted to an O relationship. The remaining relationships could be defined as unimportant and identified by the letter U. The result is shown in Table 12.11.

Although the flow may be a major reason for assigning the relationship code, it is not necessarily the only consideration. A number of factors may influence the relationship code, including (1) quantity of flow; (2) cost of material handling; (3) equipment used in material handling; (4) need for close communication; (5) need to share same personnel; (6) need to share some equipment; and (7) needed separation because of noise, danger, chemicals, fumes, or explosives. It is therefore necessary to analyze all the aspects, evaluating the importance of each and developing a consensus, before a code is finally assigned.

There is also an acceptable practice of displaying a rel-chart in a fashion similar to a mileage chart. Figure 12.3 shows the rel-chart from Table 12.11 in such a form. A number(s) may be entered below each relationship (the same can be done in a rel-chart) to give reason(s) for assigning the code. For example, besides a large quantity of flow between departments E and F, these two departments may be required to share a forklift truck. These reasons go into determining the A code between the departments; they are indicated below code A by 1, 3.

**TABLE 12.11**
**Rel Chart**

| | A | B | C | D | E | F |
|---|---|---|---|---|---|---|
| A | | E | A | U | U | O |
| B | | | O | U | O | U |
| C | | | | A | U | U |
| D | | | | | U | A |
| E | | | | | | U |
| F | | | | | | |

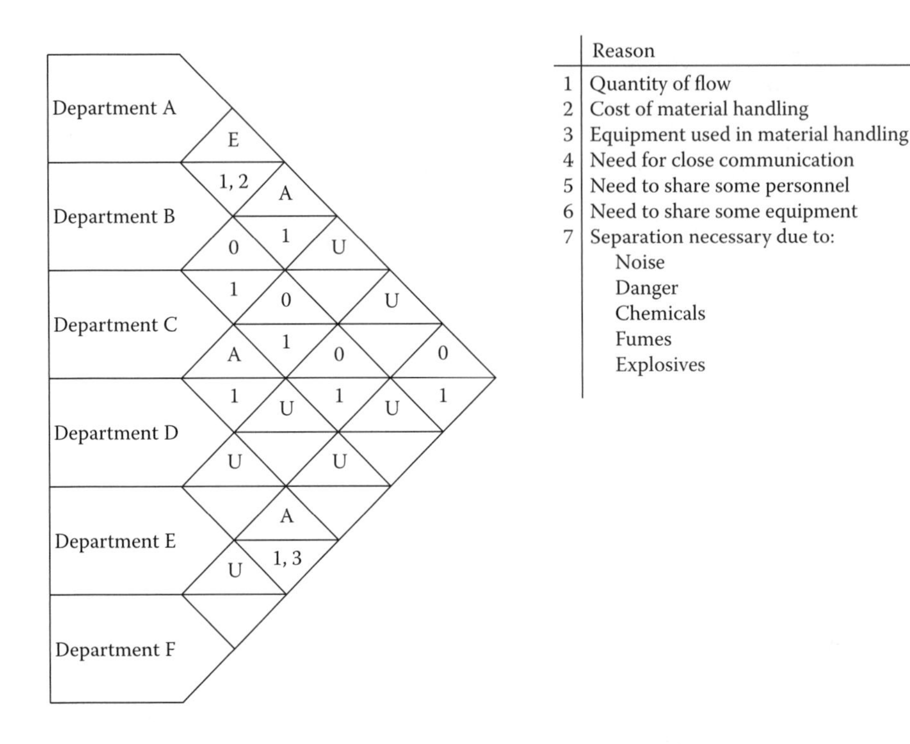


**FIGURE 12.3** Mileage chart format of rel-chart in Table 12.11

The remaining steps of the procedure are now first stated and then illustrated by applying them to an example problem.

The third step is to develop a graphical representation of the rel-chart. Muther (1973) suggested a way of transforming information from the rel-chart to a pictorial or graphical representation. The work centers are represented by nodes, and the number of lines between two nodes represents the closeness between the nodes (Figure 12.4).

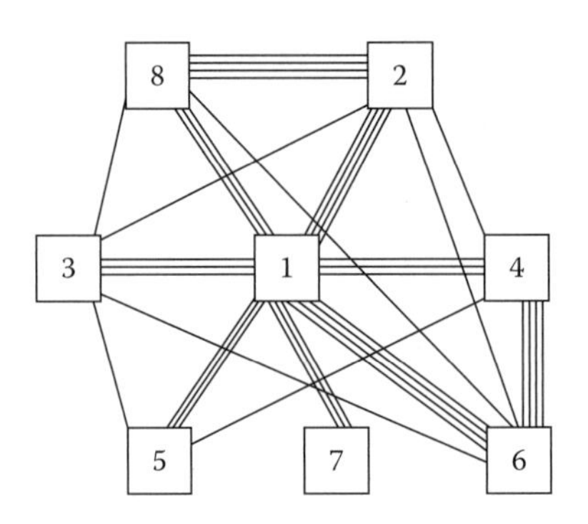


**FIGURE 12.4** Nodal representation

**TABLE 12.12**
**Rel-Chart Priority Codes**

| Code | Priority | Value |
|---|---|---|
| A | Absolutely necessary | 4 |
| E | Especially necessary | 3 |
| I | Important | 2 |
| O | Ordinary | 1 |
| U | Unimportant | 0 |
| X | Undesirable | −1 |

The decoding scheme is as follows: code A is shown by four lines, E by three lines, I by two lines, and O by one line. A wiggly line represents X.

The proper arrangement of the nodes and their relationships is critical in plant layout development, because it provides a basis for the starting arrangement of the departments. The objective is to arrange the nodes so that there is a minimum number of an area crossed when going from one department to another with the frequencies indicated by the decoded rel-chart. It might take a number of trials before one obtains a well-defined arrangement.

The procedure starts by conversion or decoding of the rel-chart to what we will denote as a value chart, using the values associated with the codes shown in Table 12.3, reproduced here as Table 12.12 for convenience. The measure of importance of each area, which ascertains the degree of closeness one department has with all other areas, is obtained by adding the row and column values for that department together. This is necessary because the rel-chart and the corresponding value chart are symmetrical about the diagonal, and only half of the table is normally displayed.

Select the department with the highest total and place it in the center of the nodal diagram. Locate around it any departments that have four relationships with it. Next, from the other departments in the diagram, select the one with the next highest total and place around it the departments with four relationships to it. Continue the procedure with each appropriate department in the diagram. After all four-relationship departments are exhausted, if some of the departments are still not in the diagram, continue with three-relationships, following the same sequence of steps as that used with four-relationships. Continue until all the areas are in the diagram, using two- and even one-relationships if necessary.

For the next step, go over the diagram again and adjust the positions of departments, switching them if necessary to satisfy closeness in between. For example, closeness of a department with three-relationships is more important than one with two-relationships. Again, the basic objective is to develop an arrangement in which departmental crossovers are at a minimum when travel between departments is made based on the value chart.

The fourth step is to develop an evaluation chart. The chart provides a measure of effectiveness of the nodal arrangement developed in step 3. Different arrangements can be evaluated by developing a chart for each, and the one with the lowest value is selected as the best arrangement to be used in step 5.

The evaluation chart is developed by first converting the nodal representation into a semiscaled grid representation. For each department, the necessary area is equated to the approximate number of blocks needed, using a convenient scale. (For example, 200 square feet could equal 1 block.) The total of all the blocks that are needed is determined, and an approximate square or rectangular shape is designed. For example, if a layout needs 50 blocks, the layout could be developed in a 10 × 5 or a 9 × 6 arrangement.

In developing such a grid, it might be beneficial to note the expected column span and accordingly fix one dimension, generally the width. For example, a 20-foot column span can be arranged as a 20 × 20-foot block of 400 square feet or could be divided as a 10 × 10-foot block of 100 square feet.

In the overall grid arrangement, we should make certain that the numbered blocks in one direction (e.g., width) are such that the resultant dimension is an integer multiple of the column span. Although ideally one might wish for a square layout (to minimize distance traveled), most plants are built in a rectangular form to accommodate the shape of the plot of land.

The next step is to place the individual departments within the grid arrangement. The departments are placed in the grid, represented by the necessary blocks for each, by using the arrangement of the nodal diagram in step 3 as a guide. The closeness measure can now be defined; it is equal to the shortest rectilinear distance between two areas multiplied by the value of the relationship between those two departments. An effectiveness evaluation chart, similar to a value chart, is useful in developing this measure of all departments. The grand total gives the measure of effectiveness of the nodal diagram; the grid chart with a minimum sum should be selected for the next step.

The fifth step is to develop templates to represent each area. Spacing of columns in the building — generally 20, 30, or 50 feet — gives a good first dimension. The other dimension is calculated on the basis of the area required for the center. For example, with a 20-foot span in a building, an area of 300 square feet needed for a quality control room is represented by 20 × 15 feet on a template.

The sixth and final step is to arrange the templates in the same fashion as the graphical representation of the rel-chart (step 3). The sizes and shapes of the templates might have to be changed to correct the resulting odd shape of the building and to minimize wasted space. (Constraints are sometimes imposed on changes in the shapes by the machines within the department.) Again, a few trials might be needed to obtain a desired layout.

The entire procedure is facilitated if some of the basic materials to be described in Section 12.3 are used in the development process. Base grid paper, for example, is excellent to use in determining the optimum shape of a department in step 4. At times a quick sketch might be better for adding clarity and understanding than working with templates.

## EXAMPLE LAYOUT FOR A SMALL MANUFACTURING FIRM

Steps 3–6 are now illustrated by applying them to the example associated with the rel-chart in Table 12.4. Suppose the area needed for each department (including aisle space) is as shown in Table 12.13. The letters in the rel-chart are converted to numbers using the codes given in Table 12.12 to begin the process, and the resulting value chart

**TABLE 12.13**
**Space Requirements**

| Department | Area (square feet) |
|---|---|
| (1) Production | 4800 |
| (2) Warehouse | 3050 |
| (3) Office | 2400 |
| (4) TR | 1150 |
| (5) Food service | 750 |
| (6) Maintenance | 1000 |
| (7) Locker room | 600 |
| (8) Shipping/Receiving | 1900 |

is shown in Table 12.14. This chart will be used as an aid in developing the nodal representation. To get the total measure of importance of a department, we add its relationship values to all other departments. This is easily done by adding all the numbers in the row and the column for an individual department. For example, the total for department 4 is obtained by adding the entries in row 4 (1 + 4 + 0 + 0) and column 4 (4 + 1 + 0) for a total of 10. Table 12.14 shows these values in the "Total" column.

Begin the nodal representation by placing the department (area) with the highest total (in this case, department 1). Place around it any departments with which it has a four-relationship (2, 4, 6, and 6), and place around it all the departments having a four-relationship with it. Here, we break the tie between departments 2 and 4 and departments 2 and 6 arbitrarily by selecting department 2 and placing department 8 beside it. Continue the process for all departments. Care must be taken to look ahead and notice other departmental relationships. For example, department 8 has a four-relationship with department 2, but it also has a three-relationship with department 1; therefore, it must be placed adjacent to both department 1 and department 2. After all these four-relationship

**TABLE 12.14**
**Value Chart**

| | Department | | | | | | | | |
|---|---|---|---|---|---|---|---|---|---|
| | 1 | 2 | 3 | 4 | 5 | 6 | 7 | 8 | Total |
| 1 | — | 4 | 3 | 4 | 3 | 4 | 3 | 3 | 24 |
| 2 | | — | 1 | 1 | 0 | 1 | 0 | 4 | 11 |
| 3 | | | — | 0 | 1 | 1 | 0 | 1 | 7 |
| 4 | | | | — | 1 | 4 | 0 | 0 | 10 |
| 5 | | | | | — | 0 | 0 | 0 | 5 |
| 6 | | | | | | — | 0 | 1 | 11 |
| 7 | | | | | | | — | 0 | 3 |
| 8 | | | | | | | | — | 9 |

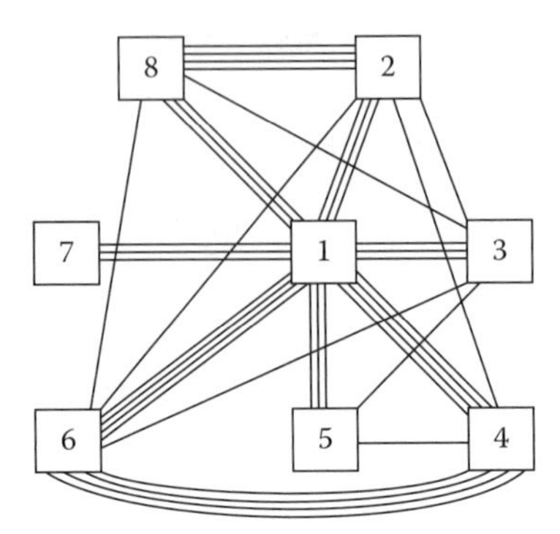


**FIGURE 12.5** Nodal representation with a potential placement problem

departments have been placed, go back to the department with the highest total and begin placing the three-relationships. After all three-relationships are exhausted, adjustments are normally made with the two- and one-relationships; however, in this case, all the departments are placed. The result is the diagram shown in Figure 12.4.

A good nodal representation has the fewest lines crossing the least number of departments. Depending on how the lines are placed and the number of departments involved in the layout, this can often be a tedious and difficult task to evaluate. Take, for example, the nodal representation of Figure 12.5. Departments 4 and 6 have four relationships with each other. The nodal representation shows the four lines crossing no department, but when the departments are actually placed, department 5 might fall directly between departments 4 and 6.

A grid formulation, as explained in step 4, can be quite helpful in evaluating potential layouts. Begin by reducing the area of each department to an approximate number of blocks. For our example, use a scale of 400 square feet equals one block (20 × 20 feet). For instance, the TR is 1150 square feet, which approximates three blocks. The numbers of blocks needed for each area are indicated in Table 12.15.

With the total blocks needed being 40, let us limit the dimensions of the grid to 5 × 8 blocks. The departments may be placed in the grid according to the nodal representation of Figure 12.4. The shape of the departments should remain as regular as possible, preferably rectangles, as shown in Figure 12.6.

**TABLE 12.15**
**Block Calculations**

| Department | Area | Blocks | Department | Area | Blocks |
|---|---|---|---|---|---|
| 1 | 4800 | 12 | 5 | 750 | 2 |
| 2 | 3050 | 8 | 6 | 1000 | 2 |
| 3 | 2400 | 6 | 7 | 600 | 2 |
| 4 | 1150 | 3 | 8 | 1900 | 5 |
| Total | | | | | 40 |

| | | | | |
|---|---|---|---|---|
| 8 | 8 | 2 | 2 | 2 |
| 8 | 8 | 2 | 2 | 2 |
| 3 | 8 | 2 | 2 | 4 |
| 3 | 1 | 1 | 1 | 4 |
| 3 | 1 | 1 | 1 | 4 |
| 3 | 1 | 1 | 1 | 6 |
| 3 | 1 | 1 | 1 | 6 |
| 3 | 5 | 5 | 7 | 7 |

**FIGURE 12.6** Grid representations

Now the grid can be evaluated with respect to distances and departmental crossings. The shortest rectilinear distance, the number of blocks crossed, between departments is multiplied by the value of the relationship between the corresponding departments. With the relationships from Table 12.4 and the rectilinear distances from Figure 12.6, the effectiveness of the departmental placement depicted in the latter is evaluated and the results are displayed in Table 12.16. For example, the shortest rectilinear distance between departments 2 and 6 is three blocks, and these departments have a one-relationship between them, giving $3 \times 1$ as the entry for row 2, column 6 in Table 12.16. The numbers are totaled for rows and then overall to get the effectiveness measure; in this case, it is 16.

To evaluate the alternative nodal representation of Figure 12.5, we shall apply the procedure followed with Figure 12.4 and compare the effectiveness measures for the

**TABLE 12.16**
**Effectiveness Calculation for Figure 12.4**

| | Department | | | | | | | | |
|---|---|---|---|---|---|---|---|---|---|
| | **1** | **2** | **3** | **4** | **5** | **6** | **7** | **8** | **Row Value** |
| 1 | — | 0 | 0 | 0 | 0 | 0 | 0 | 0 | 0 |
| 2 | | — | $1 \times 1$ | 0 | $4 \times 0$ | $3 \times 1$ | $4 \times 0$ | 0 | 4 |
| 3 | | | — | $3 \times 0$ | 0 | $3 \times 1$ | $2 \times 0$ | 0 | 3 |
| 4 | | | | — | $4 \times 1$ | 0 | $2 \times 0$ | $2 \times 0$ | 4 |
| 5 | | | | | — | $2 \times 0$ | 0 | $4 \times 0$ | 0 |
| 6 | | | | | | — | $2 \times 0$ | $5 \times 1$ | 5 |
| 7 | | | | | | | — | $6 \times 0$ | 0 |
| 8 | | | | | | | | — | 0 |
| Total | | | | | | | | | 16 |

| | | | | |
|---|---|---|---|---|
| 8 | 8 | 2 | 2 | 2 |
| 8 | 8 | 2 | 2 | 2 |
| 7 | 8 | 2 | 2 | 3 |
| 7 | 1 | 1 | 1 | 3 |
| 6 | 1 | 1 | 1 | 3 |
| 6 | 1 | 1 | 1 | 3 |
| 5 | 1 | 1 | 1 | 3 |
| 5 | 4 | 4 | 4 | 3 |

**FIGURE 12.7** Grid representation of Figure 12.5

two nodal diagrams. Figure 12.7 converts Figure 12.5 to a grid representation of the layout.

Although the nodal arrangement of Figure 12.5 appeared to have fewer lines crossing fewer departments, this will in fact not be the case in the actual layout. This observation is the result of comparing the effectiveness measure of Table 12.17 with that of Table 12.16 (25 vs. 16).

Small changes can be made in the grid without really affecting the final layout or the measure. In Figure 12.8, we see that all the departments have been shifted one block clockwise around department 1. The corresponding total in Table 12.18 has not changed.

## TABLE 12.17
## Effectiveness Calculation for Figure 12.5

| | Department | | | | | | | | |
|---|---|---|---|---|---|---|---|---|---|
| | **1** | **2** | **3** | **4** | **5** | **6** | **7** | **8** | **Row Value** |
| 1 | — | 0 | 0 | 0 | 0 | 0 | 0 | 0 | 0 |
| 2 | | — | 0 | $4 \times 1$ | $5 \times 0$ | $3 \times 1$ | $1 \times 0$ | 0 | 7 |
| 3 | | | — | 0 | $3 \times 1$ | $3 \times 1$ | $3 \times 0$ | $2 \times 1$ | 8 |
| 4 | | | | — | 0 | $2 \times 4$ | $4 \times 0$ | $4 \times 0$ | 8 |
| 5 | | | | | — | 0 | $2 \times 0$ | $4 \times 0$ | 0 |
| 6 | | | | | | — | 0 | $2 \times 1$ | 2 |
| 7 | | | | | | | — | 0 | 0 |
| 8 | | | | | | | | — | 0 |
| Total | | | | | | | | | 25 |

| 8 | 8 | 2 | 2 | 2 |
|---|---|---|---|---|
| 3 | 8 | 2 | 2 | 2 |
| 3 | 8 | 8 | 2 | 2 |
| 3 | 1 | 1 | 1 | 4 |
| 3 | 1 | 1 | 1 | 4 |
| 3 | 1 | 1 | 1 | 4 |
| 3 | 1 | 1 | 1 | 6 |
| 5 | 5 | 7 | 7 | 6 |

**FIGURE 12.8** Modified grid representation

Essentially, the layouts developed from Figures 12.4 and 12.8 would be equally effective and efficient. Figure 12.9 represents another possible arrangement with a score of 17.

In applying step 5 of the procedure, the grid should be converted to a scale model using templates such as those shown in Figure 12.10. The grid representation should be used as a starting point for developing the size and shape of the templates for the individual departments. Keep in mind that the grid is only an approximation of the required area; hence, it will frequently be necessary to enlarge or reduce the grid to meet exact specifications. These templates are laid out in Figure 12.11 according to the grid shown in Figure 12.4. The resulting layout is somewhat irregular and requires modification.

**TABLE 12.18**
**Effectiveness Calculation for Figure 12.8**

| | Department | | | | | | | | |
|---|---|---|---|---|---|---|---|---|---|
| | **1** | **2** | **3** | **4** | **5** | **6** | **7** | **8** | **Row Value** |
| 1 | — | 0 | 0 | 0 | 0 | 0 | 0 | 0 | 0 |
| 2 | | — | $1 \times 1$ | 0 | $6 \times 0$ | $3 \times 1$ | $4 \times 0$ | 0 | 4 |
| 3 | | | — | $3 \times 0$ | 0 | $3 \times 1$ | $2 \times 0$ | 0 | 3 |
| 4 | | | | — | $4 \times 1$ | 0 | $2 \times 0$ | $2 \times 0$ | 4 |
| 5 | | | | | — | $2 \times 0$ | 0 | $4 \times 0$ | 0 |
| 6 | | | | | | — | 0 | $5 \times 1$ | 5 |
| 7 | | | | | | | — | $4 \times 0$ | 0 |
| 8 | | | | | | | | — | 0 |
| Total | | | | | | | | | 16 |

**TABLE 12.19**
**Effectiveness Calculation for Figure 12.9**

| | Department | | | | | | | | |
|---|---|---|---|---|---|---|---|---|---|
| | **1** | **2** | **3** | **4** | **5** | **6** | **7** | **8** | **Row Value** |
| 1 | — | 0 | 0 | 0 | 0 | 0 | 0 | 0 | 0 |
| 2 | | — | $1 \times 1$ | 0 | $6 \times 0$ | $3 \times 1$ | $5 \times 0$ | 0 | 4 |
| 3 | | | — | $2 \times 0$ | 0 | $3 \times 1$ | $2 \times 0$ | 0 | 3 |
| 4 | | | | — | $3 \times 1$ | 0 | $2 \times 0$ | $4 \times 0$ | 3 |
| 5 | | | | | — | $2 \times 0$ | 0 | $5 \times 0$ | 0 |
| 6 | | | | | | — | 0 | $7 \times 1$ | 7 |
| 7 | | | | | | | — | $6 \times 0$ | 0 |
| 8 | | | | | | | | — | 0 |
| Total | | | | | | | | | 17 |

Small adjustments in departments 2 and 6 produced the smooth layout of Figure 12.12b. Modifying the layout is done basically by trial and error, but because the grid has already approximated the layout, the task should be a simple one, requiring only minor adjustments.

The layout planning for any plant can at first appear to be a tedious task. The steps given here are designed to simplify the job and produce an efficient layout.

| | | | |
|---|---|---|---|
| 8 | 8 | 2 | 2 |
| 8 | 8 | 2 | 2 |
| 3 | 8 | 2 | 2 |
| 3 | 1 | 1 | 2 |
| 3 | 1 | 1 | 2 |
| 3 | 1 | 1 | 4 |
| 3 | 1 | 1 | 4 |
| 3 | 1 | 1 | 4 |
| 5 | 1 | 1 | 6 |
| 5 | 7 | 7 | 6 |

**FIGURE 12.9** Elongated grid representation.

## 12.2 DETAILED LAYOUT

Once the positions of major centers in the overall layout have been established, the next step is to develop a detailed layout of each work center. There are three basic ways of representing these layouts: draw sketches, use two-dimensional templates, and use three-dimensional models.

A layout of the production area, for example, must indicate, to scale, the position of each machine and its operator. Adequate room, at least 18 inches of clear space all around the machine and at times more, must be provided to allow maintenance personnel to perform their tasks. The aisles, walls, exits, and columns must be clearly shown on the layout. Each machine should be labeled by name, and arrows should show the material flow — where the material is fed into and cleared from the machine. Space must be provided for in-process inventories, storage, auxiliary equipment, and maintenance benches.

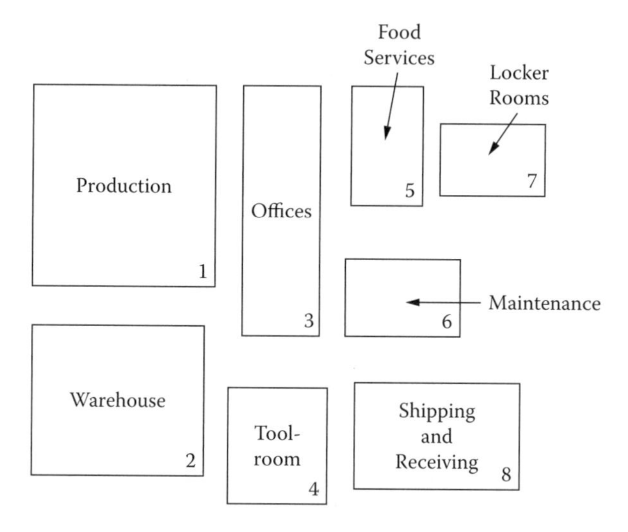


**FIGURE 12.10** Department templates

Overhead pipes and ducts are either indicated on the layout or, to avoid clutter, may be presented on an overlay. Outlets for both 240- and 120-volt electrical services, water, drains, compressed air, natural gas, bench lights, and other utilities should be shown and labeled. The flow of material can be indicated by using arrows directly on the layout or on a plastic overlay sheet.

Generally, one begins by selecting a scale to develop the plans; ¼ inch to 1 foot is the most commonly used scale by architects and contractors and therefore should be selected in facility development if at all possible. Draw a plan to show the outside walls; for an existing facility, draw columns, doors, windows, elevators, stairs, and

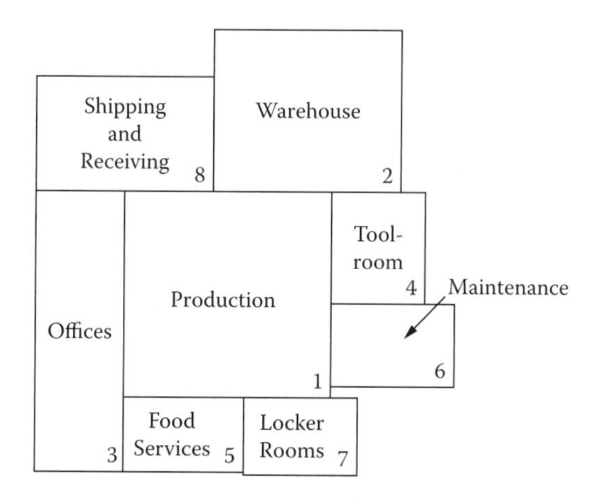


**FIGURE 12.11** Initial layout

(A)

(B)

**FIGURE 12.12** Final layout. (A) High-volume simplimatic conveyorized assembly line (courtesy of ALTRON, Inc., Anoka, MN): Space is assigned in a department for machines, people, and materials once flow pattern has been determined. The layout, such as the one shown here, should be planned as carefully as possible; after all the equipment had been placed, changes could prove too costly to be made. Here, the conveyor transports materials from one station to another for different assembly processes. (B) Job-Shop Layout (courtesy of Pdf Inc.): In a job shop layout, flexibility is critical to the design because the assemblies, processes, and material flow will vary. (C) In a layout base such as the one shown here, the base material should be sturdy and lightweight. The layout itself should be simple to understand and easily altered. (D) Here, transparent plastic overlays are used to show detailed department layouts and material flow paths. (E) Three-dimensional model of the office layout shown here; the use of model furniture and equipment helps the reader visualize the final layout (Photos by Daniel Zak, Krzysztof Plonka.)

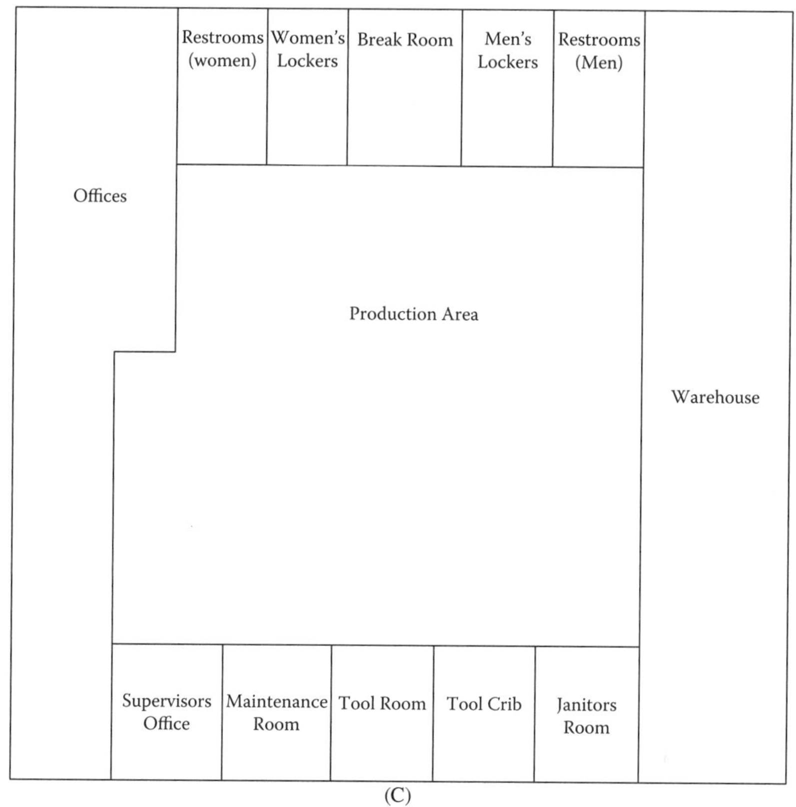


(C)

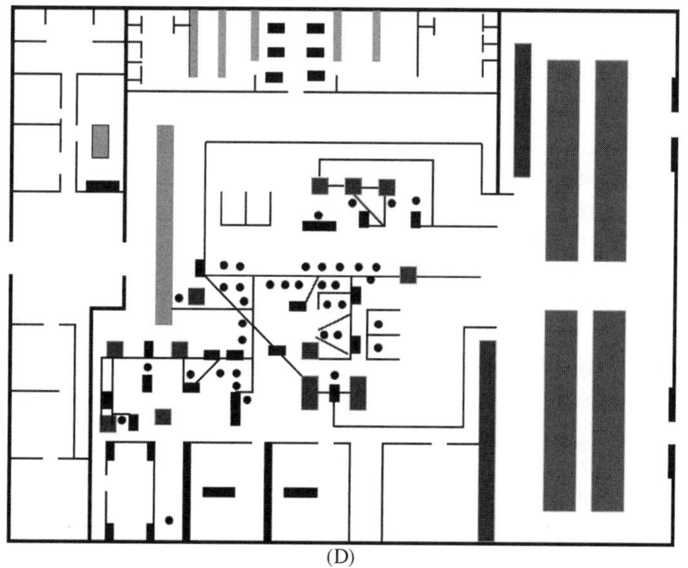

(D)

**FIGURE 12.12** (Continued)

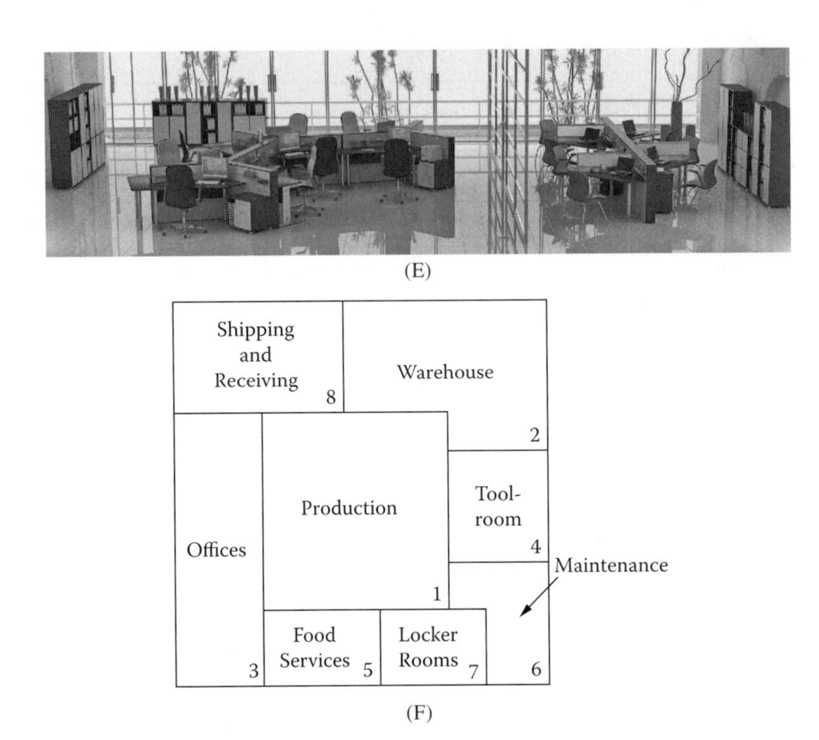


**FIGURE 12.12** (Continued)

other permanent features. Select the material flow pattern, thus fixing the receiving and shipping departments. Tentatively assign the major departments along the material flow line, making sure there is no interference. If a new building is being planned, the previous steps of placing such items as columns and doors and then defining the material flow pattern could be reversed to give more flexibility in design.

The next step is to define the positions of the aisles, which may or may not be the major material carrier routes. For example, if forklift trucks are used as the major material handler, then aisles also form the main artery of the material movement; however, even if conveyors are the material movers, aisles are still needed for people, maintenance carts, and truck movement. The aisles are also required for quick evacuation in case of an emergency. Table 12.20 shows the normal aisle widths needed for different types of equipment.

The major aisles are laid out to carry through traffic associated with all the departments. The department aisles branching off the major aisles are then established for each department.

The three characteristics of aisle are length, location, and width. The length of an aisle depends on the area of plant it is to cover and the location of the aisle. Two factors should be considered: All parts of the plant should be quickly accessible, but the more aisles are planned, the larger the plant must be to provide the same production area, increasing the cost of the building. Therefore, a minimum number of aisles should be placed to cover maximal space.

**TABLE 12.20**
**Suggested Aisle Widths for Various Flows**

| Flow | Aisle Width (feet) | Minimum Maneuvering Allowance (feet) |
|---|---|---|
| Tractor | 12 | 14 |
| 3-ton forklift | 11 | 12 |
| 2-ton forklift | 10 | 12 |
| 1-ton forklift | 9 | 11 |
| Narrow aisle truck | 6 | 12 |
| Manual platform truck | 5 | 10 |
| Personnel | 3 | |

The aisle's proposed use determines it width. Main aisles throughout the plant should be wide enough, 15–18 feet, to support two-way traffic of both vehicles and people without creating bottlenecks. Smaller aisles that are intended to support only one-way traffic may be 10–12 feet wide or less; if narrow aisle equipment is being used, aisles may be as narrow as 6 feet. Aisles that are to support only personnel traffic can be as small as 3 feet, depending on the amount of traffic they are to accommodate. The aisles should be located with uniform spacing to increase their effectiveness; they should be straight and not sinuous.

Machines are positioned within a given department once the flow pattern has been designated. Space is assigned for in-process inventory maintenance personnel, benches, operators, and others, as was planned in the initial space determination. The arrangements can be adjusted by moving the departmental aisles and material flow, which allows smooth movement of material and people to each machine. Some modifications to the shape of the department might also be necessary during this process.

The next step is to locate personnel and plant services such as washrooms, the cafeteria, vending machines, water fountains, maintenance facilities, and supervisory personnel. They must be placed to achieve utmost efficiency in operation. For example, the washrooms and the water fountains should ideally be within 60 feet of any person working in the plant. The cafeteria and the main tool crib should be placed such that they are accessible by each department without going through any other department. Checking in and out by personnel should be done at the main entrance gate to the plant building, thus allowing control of who may enter the building. The overall integration of a layout can be checked by the plant layout.

In an assembly-line layout (Figure 12.12A), space is assigned in a department for machines, people, and materials once the flow pattern has been determined. The layout, such as the one shown here, should be planned as carefully as possible; after all the equipment is in place, changes could prove too costly to be made. In a job-shop layout (Figure 12.12B), flexibility is critical to design because the assemblies, processes, and material flow will vary, following the activities of workers as they follow their daily activities of parking their cars; checking into the plant; working at machines and/or in their work centers; using personal facilities on breaks; requesting maintenance, quality control, and supervisory services; visiting a designated TR and tool crib; and finally leaving the plant at the end of their shift. All these activities should be completed without confusion or too much cross-traffic and congestion in aisles or service areas.

In detailing the plant layout, the first alternative of drawing sketches is easy and quickly completed. Usually, cross-section or grid paper is used as a base, and the details are drawn in. However, this is not a very flexible process; each change requires some redrawing.

Two-dimensional templates are more popular and relatively inexpensive to construct. Templates made of cardboard, plastic, metal, or some other material (usually to the scale of ¼ inch to 1 foot) are cut to the shapes of various machines. The templates are arranged and rearranged on the same scaled print of the floor layout until a satisfactory plan is obtained.

Three-dimensional models can be constructed and, in a process similar to that for the two-dimensional templates, located so as to achieve the optimum layout. The three-dimensional models are more realistic but are considerably more expensive. If overhead height is of concern, however, perhaps they would be more appropriate, and the models do make selling the plant layout easier.

## 12.3 MATERIALS USED IN PLANT LAYOUT ILLUSTRATIONS

To make the layout simpler to construct and less difficult to explain and follow, different materials can be used in its construction. Several are explained below.

The base material for the layout must be sturdy and strong, and it should be lightweight in case the display must be moved to different meetings during the course of its final evaluation. Cardboard, plastic, plywood, and metal are some of the more popular materials for the backboard.

The layout should be built or drawn on a background that makes its modification and understanding convenient. Tracing paper for drawing and transparent plastic sheets with grids for scales are good materials on which to develop the layout.

Plastic tape of different colors can be used to show defined lines representing walls, aisles, conveyors, railroad tracks, and material flow. Tape can also be used to show storage areas, pallets, benches, and other miscellaneous items.

Self-adhesive tape is available in a number of colors and widths. Material flow and aisle boundaries are generally designated by $^1/_{16}$-inch tape, conveyors are indicated by $^1/_8$-inch tape, and walls are shown by $^1/_4$- or $^3/_8$-inch tape.

To represent physical objects such as machines, benches, and stools, one can make templates. The material used in their construction should be easy to cut, paste, and form: paper, cardboard, plastic, sheet metal, and wood are popular choices.

Three-dimensional models can be built of materials similar to those used in making templates. Commercially made standard models are also available to represent different machines, office furniture, and other accessories.

All machines, material flows, storage areas, and other miscellaneous items should be labeled for clarity. Although hand labeling is permitted, a much neater and more professional-looking layout is obtained if a standard lettering set is used.

Clear plastic sheets with tapes representing flow lines can be used to overlay the basic plant layout to aid in determining the suitability of the plan while avoiding excessive cluttering.

In a layout base such as the one shown in Figure 12.12C, the base material should be sturdy and lightweight. The layout itself should be simple to understand and easily altered. In Figure 12.12D, transparent plastic overlays are used to show detailed department layouts and material flow paths. In the three-dimensional office model shown in Figure 12.12E, use of model furniture and equipment helps the reader visualize the final layout. The top view of the three-dimensional model shows how the grid of the plastic base guides accurate placement of office spaces and furniture.

## 12.4 DEVELOPING AND ANALYZING PLANT LAYOUTS

Let us now formally define steps that may be helpful in developing and analyzing alternative layouts.

1. Before beginning a layout, the designer should discuss with management the objectives, scope, available finances, and required time schedule for the project.
2. Develop the material flow requirements based on all the products that are scheduled or contemplated for production in the plant.
3. Discuss with management the appropriate layout types that may be used in the plant. Job shop, flow shop, assembly lines, and flexible manufacturing are some of the prominent types of layouts. The plant might have a combination of two or more of these types. Everyone concerned should review the data to make certain that all are in agreement with them. If there is disagreement, return to steps 1 and 2.
4. Develop overall departmental arrangements (layouts). There is seldom a case in which only one arrangement is possible. Provide at least two or three layouts for discussion.
5. For each overall layout, develop arrangements within each department. These might consist of placement of machines, storage areas, and departmental aisles along with major aisles and material-handling equipment such as conveyors and cranes. Review the alternatives with all departmental supervisory personnel and modify the layouts as appropriate.
6. A tentative solution to any problem should be evaluated with respect to defined objectives; plant layout is no exception. In assembly lines, the material and partial assemblies should flow smoothly and continuously from one station to the next in proper sequence following a well-defined flow pattern with no backtracking or congestion in the material flow

aisles. The same principles apply in the job-shop arrangement except that there might be multiple paths over which the products flow. Sufficient clear space should be available for each machine and its operator, and the layout should offer flexibility for future expansion or changes. The aisles should be wide enough to allow material-handling equipment to travel freely without obstruction. Noisy equipment such as punch presses should be located in an enclosed area. Space should be well utilized overall, and the employees should be satisfied with the layout; if any supervisor does not agree with the arrangement, operational problems may develop. The layout should be organized such that it integrates all the departments and their functions. Many mathematical techniques for plant layout construction and evaluation are now available in computer software. Several of these will be explained in detail in the next chapter.

7. Each alternative should now be evaluated with respect to factors such as capital requirements, operating costs, flexibility in terms of volume and product mix, unit cost of production, future expandability, ease of handling material, safety conditions, ease of supervision, inventory accumulation, maintenance and scheduling feasibility, employee satisfaction, utilization of floor space and volume, and effects of anticipated technological advancements.
8. The above comparisons could be made by using a tabular method of assigning weights. For example, each factor can be subjectively compared and assigned a weight, as shown in Table 12.21.

**TABLE 12.21**
**Tabular Method for Layout Evaluation**

| | | | Alternatives | | | |
|---|---|---|---|---|---|---|
| **Factors** | **Maximum Points** | **Minimum Points Needed** | **1** | **2** | **3** | **4** |
| **(1) Cost** | | | | | | |
| Initial capital investment | 100 | 50 | 80 | 70 | 90 | 95 |
| Operating cost | 80 | 50 | 75 | 79 | 73 | 70 |
| Maintenance cost | 80 | 30 | 50 | 60 | 40 | 50 |
| Unit cost of production | 100 | 60 | 85 | 80 | 95 | 100 |
| **(2) Flexibility in** | | | | | | |
| Future expansion | 50 | 30 | 40 | 10 | 45 | 40 |
| Volume change | 50 | 30 | 40 | — | 30 | 35 |
| Production mix change | 50 | 20 | 35 | | 40 | 40 |
| Technology change | 30 | 10 | 20 | — | 20 | 20 |
| Material handling ease | 60 | 40 | 50 | — | 55 | 45 |
| Total | | | 475 | — | 488 | 495 |

It is obvious that the lower the cost, the higher the weight that is associated with that alternative. In measuring flexibility, the higher the flexibility, the more weight the alternative would carry. Each factor has a maximum weight and a minimum weight assigned by the analyst. The minimum weight ensures that an alternative that is unsatisfactory in any one factor would be automatically rejected. The maximum weight for each factor reflects what the analyst believes the relative importance of the factor is. In the preceding example, alternative 2 does not offer flexibility for future expansion, a factor considered necessary by management, and therefore is not evaluated further. Alternative 4 is selected on the basis of the total point value.

Another method of evaluating the alternatives is to make financial comparisons only. Determine the cash flows for a planning period and then calculate the net present worth or return on investment. The alternative with the maximum net present worth (maximum return on investment) would then be selected.

There are some major problems with this method. Fixing the planning horizon is a task that cannot be taken lightly. Too short or too long a period can lead to selection of an alternative that indeed is not the best of those presented. The high costs of initial investment in building, material-handling equipment and systems, and modern processing equipment such as numerically controlled machines and robots make it important to be realistic in estimating the planning period. As with any engineering economic analysis, inflation, depreciation, tax considerations, and the minimum attractive rate of interest should be included in determining the net present worth. A thorough discussion with management in establishing the cash flows and other factors goes a long way in selling the ultimate solution to the management.

Selling a facility plan to management and the workers can indeed lead one to include features that are oriented more toward "satisfaction" than optimization. Managers are more concerned with long-term planning and with factors such as rate of return, flexibility in design, reliability of the system, and initial investment. Plant workers, on the other hand, are more interested in features such as inventory pile-ups, noise level, safety, light intensity, colors, heating and air conditioning, computer support, and other factors that will make them feel comfortable in their work surroundings and will help them perform their tasks. It is important when making the presentation to realize the makeup of the audience and try to resolve its concerns with one or more plant layout design alternatives.

## 12.5 PRESENTING THE LAYOUT

There are three phases in selling facility plans. Preparing a neat and organized facility layout and plot plan (Section 12.8) is the first step. A well-written report describing the benefits and high points of the new plan is an important follow-up step. Finally, the oral presentation integrating the written report and plans and answering questions from the audience is critical in selling the overall layout.

The written report should be designed specifically to sell the layout and should avoid any unnecessary sidetrack into detailed technical procedures. Most of the parts of the report are standard: a letter of transmittal, title page, index, introduction, body, conclusion, and appendices. The letter at the beginning of the report states the purpose of the report. The cover page gives the title and states to whom the report is presented, who is making the presentation, and the date. The body of the report should communicate to the reader the positive aspects of the proposed facility plan. It should be brief and accurate. The detailed justification might be illustrated by tables and charts. The description of the planned implementation should illustrate how the activities in the proposed facility plan could be achieved. The dates for starting and completing activities, along with possible necessary resources, are an important part of this segment. The conclusion should summarize the entire report in one or two pages. The appendices list any data or calculations that were used in formulating the report.

Probably the most important aspect of the project presentation is the oral report. It should be as brief as possible, asking the question of what is to be done and answering exactly how to do it. The information given orally should be directed at the interest of the group to whom it is to be presented. For example, if the vice-president of sales is present, emphasis should be on how this arrangement will help meet his or her forecast sales.

Use visual aids to hold attention and clarify points. Visual aid materials are numerous and may be chosen on the basis of cost and the attitude of the audience. Blueprints, plot plans, and three-dimensional models are popular items used in making the presentation. Other visual aids include flowcharts, cost sheets, templates, and computer graphics.

One should not become immersed in the details of how the plan was derived and rather should address oneself to the issues of what is to be done, why it is needed, and how it will be achieved. The presentation should last no more than 1 hour, even less if the ideas can be explained in a shorter time. Rehearse the presentation and do not put too much information on one slide. Dress appropriately and be confident. Listen carefully and address any questions that are raised. Be communicative without being aggressive. Be open-minded and answer questions truthfully. If you do not know the answer, admit it and ask whether the information may be supplied later; then follow up.

Selling changes to an existing plant presents some *special problems*. One should realize throughout the plant development and its presentation why people resist new ideas. Inertia, fear of the unknown, conflict of personalities, loss of authority, loss of job or job content, and tendency to defend an existing method of work because of fear that changes might mean criticism of one's work are some of the major causes of opposition. To work on these, one must stress the positive aspect of the new plan, and the changes should then be introduced gradually over an extended period. The planned implementation should illustrate this point. During development of the plan, letting others suggest ideas and/or letting them think that the necessary changes are really their ideas helps tremendously in obtaining their support for the overall plan. Convincing the boss and letting him or her sell the plan to subordinates is also a very effective method of selling. Ultimately, however, for the plan to succeed, all the people in the plant must be satisfied with the changes.

## 12.6 OFFICE LAYOUT

An office in a manufacturing facility is a communication center that is responsible for such activities as record keeping, accounting, production planning, employee management, inventory control, and sales. People who are responsible for these activities will work from the spaces provided within an office complex, and it is therefore appropriate in the overall plant layout analysis to also consider office layout.

The dimensions of individual offices are a function of the number of people using the office and the amount of furniture and equipment to be placed in the office. A desk might be 30 × 60 inches with a 24-inch chair. Filing cabinets and additional chairs are almost always necessary. In addition, at least 36 inches must be provided for chair clearance, and the room should include 36-inch walking spaces. After all necessary equipment had been decided upon, the dimensions of these items should be added to the necessary clearance and aisle dimensions to calculate the minimum office space required.

Consider an office that is to contain a desk that is 30 × 60 inches, a 24-inch armchair, a 24-inch side chair, and two file cabinets that are each 18 × 24 inches. The armchair requires clearance from the desk, while the side chair can sit against the wall. Figure 12.13 gives an example of a possible layout. The length of the office is 60 inches for the desk, 72 inches for two aisles, and an additional 24 inches for the cabinets and the side chair. Adding these together gives a length of 156 inches, or 13 feet. The width is 30 inches for the desk, 36 inches for the chair and clearance, and a 36-inch aisle for 102 inches, or 8.5 feet. Of course, all these are minimum dimensions and should be adjusted to suit the status or needs of the person occupying the office.

The following guidelines suggest typical average assignments; however, they can be made larger or smaller depending on needs and economy:

- Top executive: 250–500 square feet
- Executive: 200–400 square feet

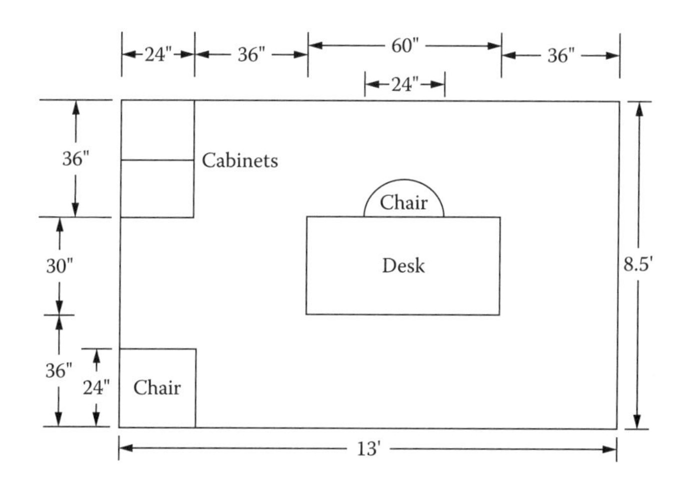


**FIGURE 12.13** An office layout

Office layout (courtesy of Integration New Media, Inc., Montreal, Canada; photographer: Vahe Kassardjian). In an open office plan such as this one, each person's work space is defined by partitions that can be rearranged according to the department's needs.

- Junior executive: 100–250 square feet
- Middle management (engineer, programmer): 80–150 square feet
- Clerical: 50–100 square feet
- Minimum workstation: 50 square feet

The steps involved in the office layout process are similar to those in a plant layout analysis. As mentioned before, the first step is to determine the departments and the number of people working and the facilities and furniture needed in each. This in turn leads to the estimation of the area needed for each department; for example, spaces must be provided for a conference room, a reception area, and the president's office. Table 12.2, reproduced here as Table 12.22, is a listing of space that might be needed and/or people who would be working in the various areas. The figures include a 20% allowance for hallways.

The space can be of the "open" type, in which a large number of desks are placed in one area, each person's working space being separated from the others by movable, 6- to 7-foot-high, sound-absorbing partitions. An office can be given more privacy by constructing walls to the ceiling, making it more of a permanent structure. In our example, major departments are given open space, although their managers might require private offices. The accounting department, for example, includes space for four accountants and two administrative assistants, and the accounting manager is provided with a private office. Table 12.23 shows the space distribution that is planned, including a 20% hallway allowance.

The next step is to construct a rel-chart between the work areas. The necessary communications and personal contacts between the departments are of major

**TABLE 12.22**
**Office Space Requirements**

| Office/Occupants | Area (square feet) |
|---|---|
| President | 250 |
| General manager | 200 |
| Sales manager | 200 |
| Production manager | 200 |
| Accountants (4) | 800 |
| Engineers (5) | 775 |
| Sales representatives (6) | 600 |
| Administrative personnel (7) | 700 |
| Receptionist | 150 |
| Conference room | 250 |
| Copy room | 100 |
| Coffee room | 200 |
| Restrooms | 350 |
| Total | 4775 |

consideration in developing the rel-chart. Managers should be close to the departments they are supervising, and the administrative assistants should be close to the departments for which they are responsible to maintain communication and records. The production department generally has more contacts with engineers than with accountants. The chain

**TABLE 12.23**
**Modified Space Allocation**

| Office | Area (square feet) |
|---|---|
| (1) President | 250 |
| (2) Executive assistant | 100 |
| (3) General manager | 200 |
| (4) Sales manager | 200 |
| (5) Production manager | 200 |
| (6) Managers' admin. | 100 |
| (7) Accountants (4) and administrative assistants (2) | 900 |
| (8) Engineers (5) and administrative assistants (1) | 875 |
| (9) Sales personnel (6) and administrative assistants (2) | 800 |
| (10) Receptionist | 150 |
| (11) Conference room | 250 |
| (12) Copy room | 100 |
| (13) Coffee room | 200 |
| (14) Restroom | 350 |
| Total | 4775 |

**TABLE 12.24**
**Closeness Relationships**

| | Office | | | | | | | | | | | | | |
|---|---|---|---|---|---|---|---|---|---|---|---|---|---|---|
| | **1** | **2** | **3** | **4** | **5** | **6** | **7** | **8** | **9** | **10** | **11** | **12** | **13** | **14** |
| 1 | — | A | A | I | I | U | O | U | U | U | U | U | O | O |
| 2 | | — | U | U | U | U | U | U | U | U | U | I | O | O |
| 3 | | | — | A | A | E | O | U | O | U | U | U | O | O |
| 4 | | | | — | U | E | E | O | A | U | U | U | O | O |
| 5 | | | | | — | E | E | A | O | U | U | U | O | O |
| 6 | | | | | | — | U | U | U | U | U | I | O | O |
| 7 | | | | | | | — | U | I | U | U | O | O | O |
| 8 | | | | | | | | — | U | U | U | O | O | O |
| 9 | | | | | | | | | — | U | U | O | O | O |
| 10 | | | | | | | | | | — | U | U | O | O |
| 11 | | | | | | | | | | | — | U | U | U |
| 12 | | | | | | | | | | | | — | U | U |
| 13 | | | | | | | | | | | | | — | E |
| 14 | | | | | | | | | | | | | | — |

of command and the organization chart also play important parts; the president's office is more inclined to have direct communication with departmental managers than with an individual within a department. Table 12.24 shows the closeness relationships.

Table 12.25 converts the latter representation to its value and calculates the total. Again, note that the entire table is not needed to calculate the total closeness. The sum of the row and column totals yields the same information. For example, for office 5, the row total is 13 and the column total is 6, giving an overall total of 19.

The closeness total from Table 12.25 is used to develop the nodal representation. The highest total is 19, associated with departments 3, 4, and 5; therefore, choose one of these arbitrarily, say department 3, and locate it in the center. Place all departments with four-relationships with it — departments 1, 4, and 5 — around it. Choosing the department already in the plot that has the next highest total — department 4 — place all four-relationship departments around it; that is, department 9. Since there are no four-relationships left, go on to the three-relationship departments, again beginning with the highest-total department. Following the procedure leads to the nodal representation shown in Figure 12.14. Departments 10, 11, and 12 can be placed essentially anywhere, and departments 13 and 14 need to only be adjacent to each other.

The nodal representation can now be checked by using an approximate area representation. Using the scale of 100 square feet per block, Figure 12.15 shows the approximate number of squares assigned to each department. The grid can be used further to evaluate the proposed layout.

In evaluating the grid, the shortest number of squares between two departments is multiplied by the relationship between those departments, all these numbers are added, and the arrangement with the smallest total should be the best. For the

**TABLE 12.25**
**Value Table**

| | Office | | | | | | | | | | | | | | |
|---|---|---|---|---|---|---|---|---|---|---|---|---|---|---|---|
| | **1** | **2** | **3** | **4** | **5** | **6** | **7** | **8** | **9** | **10** | **11** | **12** | **13** | **14** | **Total** |
| 1 | — | 4 | 4 | 2 | 2 | 0 | 1 | 0 | 0 | 0 | 0 | 0 | 1 | 1 | 15 |
| 2 | | — | 0 | 0 | 0 | 0 | 0 | 0 | 0 | 0 | 0 | 2 | 1 | 1 | 8 |
| 3 | | | — | 4 | 4 | 3 | 1 | 0 | 1 | 0 | 0 | 0 | 1 | 1 | 19 |
| 4 | | | | — | 0 | 3 | 3 | 1 | 4 | 0 | 0 | 0 | 1 | 1 | 19 |
| 5 | | | | | — | 3 | 3 | 4 | 1 | 0 | 0 | 0 | 1 | 1 | 19 |
| 6 | | | | | | — | 0 | 0 | 0 | 0 | 0 | 2 | 1 | 1 | 13 |
| 7 | | | | | | | — | 0 | 2 | 0 | 0 | 1 | 1 | 1 | 13 |
| 8 | | | | | | | | — | 0 | 0 | 0 | 1 | 1 | 1 | 8 |
| 9 | | | | | | | | | — | 0 | 0 | 1 | 1 | 1 | 11 |
| 10 | | | | | | | | | | — | 0 | 0 | 1 | 1 | 2 |
| 11 | | | | | | | | | | | — | 0 | 0 | 0 | 0 |
| 12 | | | | | | | | | | | | — | 0 | 0 | 7 |
| 13 | | | | | | | | | | | | | — | 3 | 13 |
| 14 | | | | | | | | | | | | | | — | 13 |

example, evaluating the effectiveness between departments 9 and 13 requires going straight down from department 9 to department 13. Three squares are crossed in the trip; because there is a one-relationship between the two departments, the effectiveness measure is $3 \times 1$, or 3. For brevity, only the product is shown in Table 12.26.

Accurately scaled sets of templates such as the ones shown in Figure 12.16 must now be drawn.

These templates are used to obtain a true picture of the layout. Converting the grid of Figure 12.14 to a definite layout results in the layout presented in Figure 12.17. This layout can be altered to correct the irregular shape of the design. The modified layout shown in Figure 12.18 is accurate and efficient and will be used to build the office.

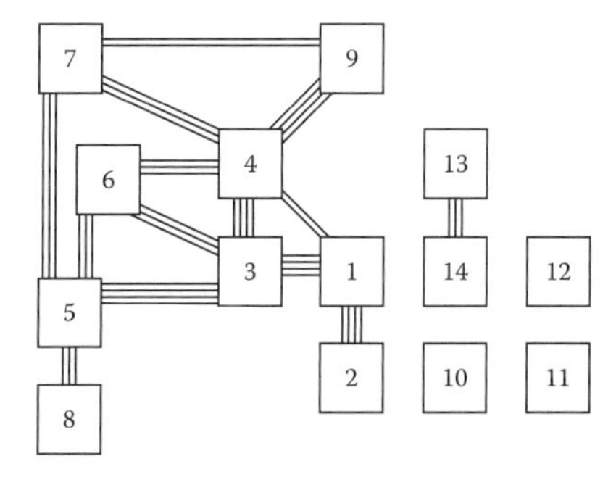


**FIGURE 12.14** Nodal representation for offices

| 7 | 7 | 7 | 11 | 11 | 11 |
|---|---|---|---|---|---|
| 7 | 7 | 7 | 9 | 9 | 9 |
| 7 | 7 | 7 | 9 | 9 | 9 |
| 5 | 6 | 4 | 4 | 9 | 9 |
| 5 | 3 | 3 | 1 | 1 | 12 |
| 8 | 8 | 8 | 2 | 14 | 14 |
| 8 | 8 | 8 | 13 | 14 | 4 |
| 8 | 8 | 8 | 13 | 10 | 10 |

**FIGURE 12.15** Grid representation for offices

## TABLE 12.26
## Effectiveness Calculation for Figure 12.14

| | Office | | | | | | | | | | | | | | |
|---|---|---|---|---|---|---|---|---|---|---|---|---|---|---|---|
| | **1** | **2** | **3** | **4** | **5** | **6** | **7** | **8** | **9** | **10** | **11** | **12** | **13** | **14** | **Total** |
| 1 | — | 0 | 0 | 0 | 4 | 0 | 2 | 0 | 0 | 0 | 0 | 0 | 1 | 0 | 7 |
| 2 | | — | 0 | 0 | 0 | 0 | 0 | 0 | 0 | 0 | 0 | 4 | 0 | 0 | 4 |
| 3 | | | — | 0 | 0 | 0 | 0 | 0 | 2 | 0 | 0 | 0 | 2 | 2 | 7 |
| 4 | | | | — | 0 | 0 | 0 | 0 | 0 | 0 | 0 | 0 | 2 | 2 | 4 |
| 5 | | | | | — | 0 | 0 | 0 | 3 | 0 | 0 | 0 | 4 | 4 | 11 |
| 6 | | | | | | — | 0 | 0 | 0 | 0 | 0 | 8 | 4 | 4 | 16 |
| 7 | | | | | | | — | 0 | 0 | 0 | 0 | 4 | 4 | 4 | 12 |
| 8 | | | | | | | | — | 0 | 0 | 0 | 1 | 0 | 1 | 2 |
| 9 | | | | | | | | | — | 0 | 0 | 0 | 3 | 1 | 4 |
| 10 | | | | | | | | | | — | 0 | 0 | 0 | 0 | 0 |
| 11 | | | | | | | | | | | — | 0 | 0 | 0 | 0 |
| 12 | | | | | | | | | | | | — | 0 | 0 | 0 |
| 13 | | | | | | | | | | | | | — | 0 | 0 |
| 14 | | | | | | | | | | | | | | — | 0 |
| Total | | | | | | | | | | | | | | | 67 |

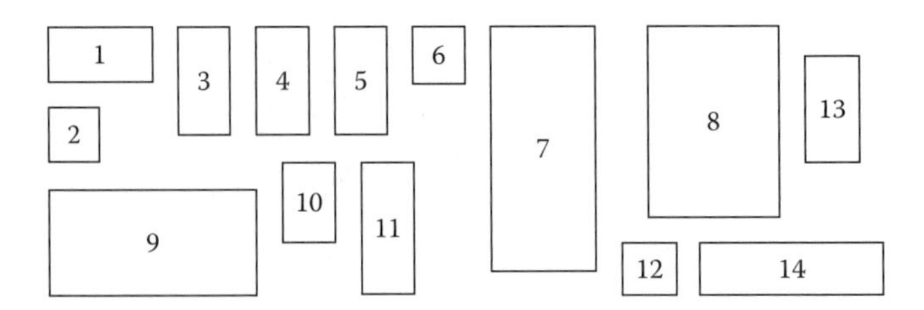


**FIGURE 12.16** Office templates

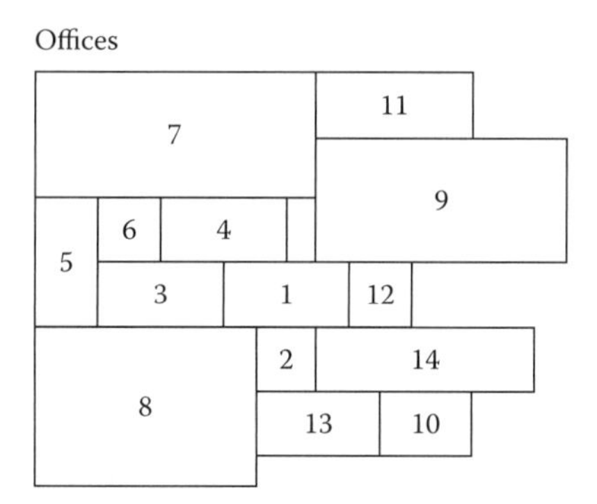


**FIGURE 12.17** Initial layout of offices

It should be apparent by now that the office layout problem is a smaller, but no less complex, version of the plant layout problem. Space requirements and departmental relationships must be carefully considered to produce the best office design resulting in working conditions that help promote efficiency.

## 12.7 RECENT TRENDS IN OFFICE LAYOUT

Use of either the private area or a combination of private and general office areas in planning an office layout needs to be carefully planned. It must provide maximum use of areas available. The disadvantages of conventional private offices are overcome by the general office areas using the open office concept based on the nature of the relationship between the employee and his or her job duties. In contrast, the private office layouts are based on the hierarchal structure of the organization.

In the design process of open office planning, the information flows and processes along with the cybernetics of the organization are considered. The information flows pertain to the paper flow, telephone communications, and face-to-face interaction.

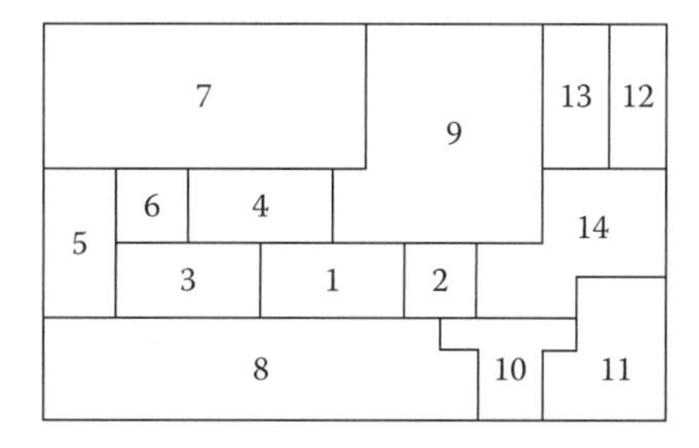


**FIGURE 12.18** Final layout of offices

The three alternatives in designing space around the open office concept include the modular workstation approach, the cluster workstation approach, and the landscape approach; in all these approaches, work areas are composed of panels and furniture components. The panels are even obtained in a variety of colors and finishes, including wood, metal, plastic, glass, carpet, and fabric.

### 12.7.1 Modular Workstation Approach

This approach creates individual work areas using panel-hung furniture components with storage components, and files of adjustable height are placed adjacent to desks or tables. It enables employees to have a complete office in terms of desk space, file space, storage space, and work-area lighting. Modular workstations are designed according to the specific job duties of their occupants.

Cluster workstations

The modular workstation approach is mainly preferred in situations that require considerable storage space; in addition, the work area can be specifically designed around the specific needs of the user. Even the changes in layout can be made easily and quickly.

### 12.7.2 Cluster Workstation Approach

In this approach, the employee work areas are clustered around a common core, such as a set of panels that extend from a hub, each employee's work area being defined by panels. Cluster workstations are not as elaborate as the other two models and are suitable for situations in which employees spend a portion of their workday away from their work area.

Cluster workstations
Courtesy of Freedom Corporation, Memphis, TN.

The advantages of a cluster workstation include economics, less expenses, and ease of making changes in the layout.

### 12.7.3 Office Landscaping Approach

Office landscaping, an alternative to conventional office layouts, is a concept that originated in East Germany in the 1950s. The idea is to make the office an open, airy, attractive, and pleasant place to work in, while being flexible enough to change as need dictates. Irregular placement of furniture and the use of partitions are characteristic of office landscaping.

The rows and rows of perfectly lined-up desks of the conventional office are eliminated in landscaping. At first glance, a typical landscaped office appears to be haphazardly arranged. Actually, personnel are grouped according to their need for communication. When privacy is required, partitions that are easily moved at any time are utilized. An example of a landscaped layout is shown in Figure 12.19.

The single major problem that must be addressed in landscaping is the level of noise. With many people in one large, open room, noise could become unbearable. However, noise can be reduced by reducing the smooth surfaces in the office that reflect sound. Using carpet instead of tiles on the floor and having textured walls and ceilings help in sound absorption. Live or artificial plants and attractive pictures on the wall give the office a homey feeling, which studies have shown to contribute to increased productivity.

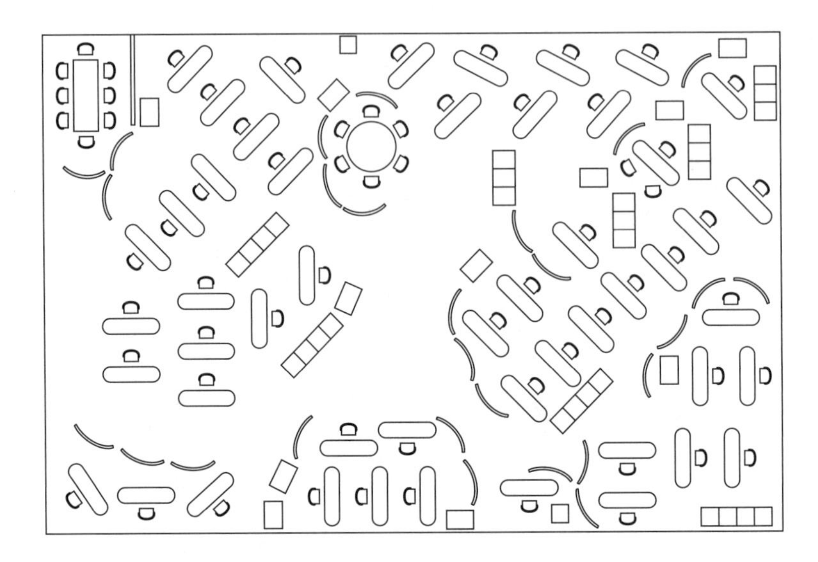

**FIGURE 12.19** A landscaped office

## 12.8 PLOT PLANNING

Once the production and office layout have been determined, the plant and other auxiliary building dimensions have been established, and a piece of land has been purchased or leased for the location of a new plant, complex steps must be taken to decide how the facilities are to be placed and arranged on the piece of land. This is known as plot planning. The basic functions of plot planning are

- Planning the best use of space available
- Planning the flow of materials throughout the facilities
- Planning for future needs

Plot planning involves more than just the location of the basic plant on a piece of property. For instance, if external warehouses are planned, they must be located in relation to the plant. Locations of driveways, entrances, and loading docks, as well as their relationships to roads and the location of parking relative to entrances must be considered. Other important aspects are locations of sidewalks, steps, and utility lines. Finally, space should be allocated now for future expansions. The plot plan is completed with landscaping and provision for installation of traffic control and other information signs.

The plot plan always begins with constant features, which include property lines, railways, waterways, and roadways. The main building is placed; then entrances, exits, roads for loading docks, sidewalks, and parking lots are located in relation to the building and the constant features. Keep in mind that at this point the plant layout has already been designed and is simply being placed in the best position. If the number of factors to be considered is relatively small, the task is not a difficult

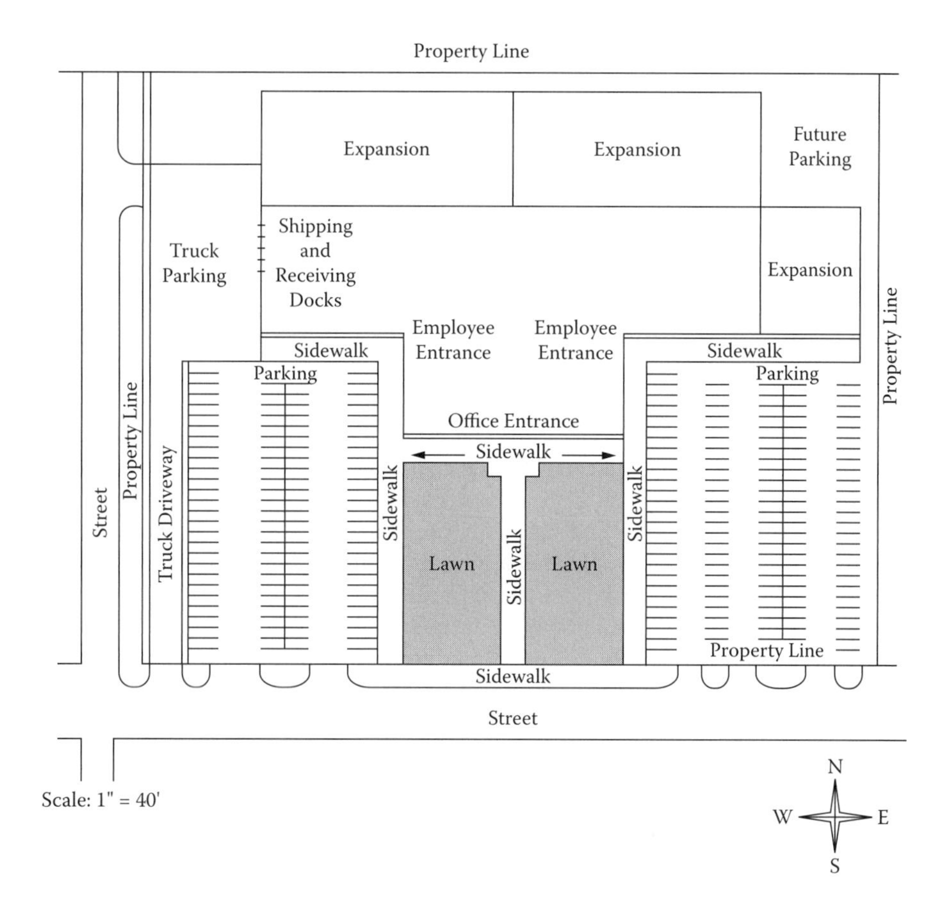


**FIGURE 12.20** A plot plan

one. If the complex consists of several buildings and a large piece of land, the task becomes more difficult. In such a case, the problem can be approached much like the plant layout problem. Identify all points of consideration and determine adjacency requirements. Then place the buildings, maintaining the necessary adjacencies to the extent possible.

The plot plan should always be drawn to scale with dimensions shown. To improve its attractiveness, a few minor changes can be made in the final drawing as well as adding landscaping. Figure 12.20 is an example of a plot plan.

## 12.9 COMPUTER PROGRAM DESCRIPTION

### 12.9.1 Computer Layout Design II

Computer Layout Design II (COMLAD II) is an interactive and user-friendly program. The program was written by Dr. Reza Ziai. Any first-time user should be able to run the program with a brief overview of the user's manual. The code for the program is written in QuickBASIC™. It allows the user to input data from a previously run example. The maximum size problem is 20 departments. COMLAD II uses a construction-type algorithm. It selects a department on the basis of its relationship with others. The placement is performed using the minimum rectilinear distance as the criterion. After the areas have been input, the generated layout is evaluated by computing a score that represents a closeness value.

The user may start a problem in two ways. He or she may give the number of products, their sequence of operations, and the volume of each through the designated departments. The product may be a physical item or a piece of data to be transferred. A rel-chart is developed, which may be modified. The user is given a chance to correct any mistakes while inputting the data. A second form of input data allows the user to input the relationship between departments directly.

Throughout the program, any value that is out of range or invalid is rejected and the user is prompted to input the right values. The numerical value of the relationship that each department has with all the others is summed to obtain a total departmental score. A square matrix is set up based on the number of departments to represent what is called a nodal diagram. The department with the highest total relationship score is placed into the matrix first. Departments that must be placed (code A) next to the first selected department are placed adjacent to it in descending order of their total relationship scores. This process of placement is continued for all other codes (E, I, O, U, and X) until all the departments have been entered.

Using the nodal diagram as a guide, the block layout is created by assigning areas to each department. The area may be given in terms of either the number of square feet or the number of blocks. In the former case, the column span of the building is also taken into account. Each department is represented as a combination of single blocks. The layout is evaluated by computing a layout score for it. The layout score is determined by summing the product of the shortest rectilinear distance and the relationship weighting for each pair of departments. The layout and its score are displayed on the screen.

After the initial layout is developed, the user can make and evaluate any alternative arrangements. The options available are

1. Interchange two departments that may or may not be of the same size.
2. Fix one or more department(s) in specific position(s).
3. Change the shape of a department by rearranging blocks.
4. Interchange any two or more of the single building blocks. This will help "shrink" (i.e., minimize or eliminate the empty blocks) or elongate the layout to fit the available area of the site.

In each case, the score is presented for the resultant layout, and the user may save it on the disk. It should be remembered that the score is not only a function of the relative position of each department but also influenced by its shape. The program allows further flexibility by permitting different width and length combinations for the overall layout. This may change the shape of an individual department, although the program tries to maintain a square or a rectangular shape for each department. Each of these options may override effects of the previous change, so several revisions should be made to obtain the desired final layout.

Finally, the user is able to query the relative distance between departments by placing the cursor at the desired blocks of the two departments under consideration. Thus, more realistic distances are computed. In most cases, products are received at one location and sent out from another within the same department. These distances are used in analyzing alternative material-handling equipment such as trucks, automated guided vehicles, and conveyors. A version of the program shown in Chapter 14 features such analysis and decision making.

The example described in the next several pages illustrates the prompts and screen layouts after each change.

How many departments? **12**
How many items will you be processing (MAX = 20)? **4**

*Note:* The sequence of the operations for each item is now requested.

Please give the sequence of operations (i.e., departments) for each product and its production rate (e.g., **→3→2→4→7**)
(Assign negative rate to the departments that should not be adjacent to each other.)
Sequence (terminate by pressing E/e, change by C/c) for item: **1**

**→1→3→4→6→8→10→11→e**

Do you wish to change the above sequence (Y/N)? **N**
Production rate for the above route? **800**
Sequence (terminate by pressing E/e, change by C/c) for item: **2**

**→1→2→5→7→8→11→12→e**

Do you wish to change the above sequence (Y/N) ? **N**
Production rate for the above route? **3200**
Sequence (terminate by pressing E/e, change by C/c) for item: **3**

**→3→4→7→8→9→12→e**

Do you wish to change the above sequence (Y/N)? **N**
Production rate for the above route? **5000**
Sequence (terminate by pressing E/e, change by C/c) for item: 4

**→1→4→5→7→11→e**

Do you wish to change the above sequence (Y/N)? **N**
Production rate for the above route? **2000**

| | 1 | 2 | 3 | 4 | 5 | 6 | 7 | 8 | 9 | 10 | 11 | 12 |
|---|---|---|---|---|---|---|---|---|---|---|---|---|
| 1 | S | I | U | O | U | U | U | U | U | U | U | U |
| 2 | I | S | U | U | I | U | U | U | U | U | U | U |
| 3 | U | U | S | E | U | U | U | U | U | U | U | U |
| 4 | O | U | E | S | O | U | I | U | U | U | U | U |
| 5 | U | I | U | O | S | U | I | U | U | U | U | U |
| 6 | U | U | U | U | U | S | U | U | U | U | U | U |
| 7 | U | U | U | I | I | U | S | A | U | U | O | U |
| 8 | U | U | U | U | U | U | A | S | I | U | I | U |
| 9 | U | U | U | U | U | U | U | I | S | U | U | I |
| 10 | U | U | U | U | U | U | U | U | U | S | U | U |
| 11 | U | U | U | U | U | U | U | O | I | U | S | I |
| 12 | U | U | U | U | U | U | U | U | I | U | I | S |

**FIGURE 12.21** Rel-chart

The rel-chart in Figure 12.21 is generated as a result of the preceding daily flow rate.

The user may change any of the preceding codes by moving the cursor around using the arrow keys. For this example, three codes in the third row are changed, as highlighted in Figure 12.22. The corresponding relationship totals are displayed in Figure 12.23.

The resultant nodal diagram is given in Figure 12.24. Areas for each department in this example are entered by giving the number of blocks for each, as in Figure 12.25.

Figure 12.26 shows the generated layout using the preceding information.

For any subsequent layout, once the "Enter" key is pressed, the bottom of the screen is cleared (starting with the line displaying the score) and the user is asked whether he or she wants to save the layout. After answering the question, a message is displayed at the bottom of the screen:

| | 1 | 2 | 3 | 4 | 5 | 6 | 7 | 8 | 9 | 10 | 11 | 12 |
|---|---|---|---|---|---|---|---|---|---|---|---|---|
| 1 | S | I | U | O | U | U | U | U | U | U | U | U |
| 2 | I | S | U | U | I | U | U | U | U | U | U | U |
| 3 | U | U | S | E | U | U | **A** | U | **E** | U | **O** | U |
| 4 | O | U | E | S | O | U | I | U | U | U | U | U |
| 5 | U | I | U | O | S | U | I | U | U | U | U | U |
| 6 | U | U | U | U | U | S | U | U | U | U | U | U |
| 7 | U | U | U | I | I | U | S | A | U | U | O | U |
| 8 | U | U | U | U | U | U | A | S | I | U | I | U |
| 9 | U | U | U | U | U | U | U | I | S | U | U | I |
| 10 | U | U | U | U | U | U | U | U | U | S | U | U |
| 11 | U | U | U | U | U | U | U | O | I | U | S | I |
| 12 | U | U | U | U | U | U | U | U | I | U | I | S |

**FIGURE 12.22** Modified rel-chart

```
TOTAL FOR DEPT 1  = 3
TOTAL FOR DEPT 2  = 4
TOTAL FOR DEPT 3  = 11
TOTAL FOR DEPT 4  = 7
TOTAL FOR DEPT 5  = 5
TOTAL FOR DEPT 6  = 0
TOTAL FOR DEPT 7  = 13
TOTAL FOR DEPT 8  = 8
TOTAL FOR DEPT 9  = 7
TOTAL FOR DEPT 10 = 0
TOTAL FOR DEPT 11 = 6
TOTAL FOR DEPT 12 = 4
```

**FIGURE 12.23** Departmental totals

Please select your option:

```
1) INTERCHANGE TWO DEPTS
2) FIX DEPTS
3) INTERCHANGE BLOCKS
4) FIX A DIMENSION
5) NO MORE MODIFICATIONS?
```

|  |  |  |  |
|---|---|---|---|
| 6 | 11 | 12 | 9 |
| 10 | 5 | 7 | 3 |
| 1 | 2 | 8 | 4 |

**FIGURE 12.24** Nodal diagram

The first option is selected in order to swap the positions of departments 4 and 6. Figure 12.27 shows the resultant layout.

The increase in the score indicates that this change was not a favorable one. For the next option, the length of the layout (left to right) is reduced to 12 columns. The first reduction will be at least as small as requested. The layout in Figure 12.28 shows the result.

The score has still remained high. As part of the previous option (four), the width (top to bottom) may also be fixed. For this example, it is set to 12 blocks. Figure 12.29 shows the resultant layout.

Number of blocks for department 1 = **8**
Number of blocks for department 2 = **12**
Number of blocks for department 3 = **10**
Number of blocks for department 4 = **6**
Number of blocks for department 5 = **11**
Number of blocks for department 6 = **8**
Number of blocks for department 7 = **14**
Number of blocks for department 8 = **9**
Number of blocks for department 9 = **5**
Number of blocks for department 10 = **4**
Number of blocks for department 11 = **12**
Number of blocks for department 12 = **9**

**FIGURE 12.25** Departmental areas

```
               A                   B                   C
    \1 2 3 4 5 6 7 8 9 0\1 2 3 4 5 6 7 8 9 0\1 2 3 4 5 . . .

  1              11111111
  2           6 61111111111
  3         6 6 611111112121212 9 9 9
  4         6 6 6         12121212 9 9
X 5          1010 5 5 5 7 7 7 7 3 3 3
  6          1010 5 5 5 7 7 7 7 3 3 3
  7   1 1 2 2 2 5 5 5 7 7 7 7 3 3 3
  8   1 1 2 2 2 5 5 8 8 8 7 7 4 4 3
  9   1 1 2 2 2     8 8 8   4 4 4 4
  0   1 1 2 2 2     8 8 8
> 1
  2
  3
  4
Y 5
  6
```

The SCORE for the above LAYOUT = 41

(if HARD COPY is desired, press the 'Shift' AND the 'Prt Sc' Keys AT THE SAME TIME)

HIT 'Return' to CONTINUE

**FIGURE 12.26** Initial layout

```
               A                   B                   C
    \1 2 3 4 5 6 7 8 9 0\1 2 3 4 5 6 7 8 9 0\1 2 3 4 5 . . .

  1              11111111
  2           4 41111111111
  3           4 411111112121212 9 9 9
  4           4 4         12121212 9 9
X 5          1010 5 5 5 7 7 7 7 3 3 3
  6          1010 5 5 5 7 7 7 7 3 3 3
  7   1 1 2 2 2 5 5 5 7 7 7 7 3 3 3
  8   1 1 2 2 2 5 5 8 8 8 7 7 6 6 3
  9   1 1 2 2 2     8 8 8   6 6 6 6
  0   1 1 2 2 2     8 8 8       6 6
> 1
  2
  3
  4
Y 5
  6
```

The SCORE for the above LAYOUT = 64

(if HARD COPY is desired, press the 'Shift' AND the 'Prt Sc' Keys AT THE SAME TIME)

HIT 'Return' to CONTINUE

**FIGURE 12.27** Modified layout showing two departments interchanged

```
                 A                             B                             C
        \1 2 3 4 5 6 7 8 9 0\1 2 3 4 5 6 7 8 9 0\1 2 3 4 5 . . .

     1        11111111
     2   4 41111111111
     3   4 411111112121212 9 9 9
     4   4 4        1212121212 9 9
   X 5        5 5 5 7 7 7 7 3 3 3
     6   2 2 5 5 5 7 7 7 7 3 3 3
     7   2 2 5 5 5 7 7 7 7 3 3 3
     8   2 2 5 5 8 8 8 7 7 6 6 3
     9   2 2 1010 8 8 8       6 6
     0   2 2 1010
   > 1   2 2 1 1
     2   1   1   1
     3   1   1   1
     4
   Y 5
     6
The SCORE for the above LAYOUT = 67

(if HARD COPY is desired, press the 'Shift' AND the 'Prt
Sc' Keys AT THE SAME TIME)

HIT 'Return' to CONTINUE
```

**FIGURE 12.28** Modified layout after the length had been reduced to 12 columns

```
                 A                             B                             C
        \1 2 3 4 5 6 7 8 9 0\1 2 3 4 5 6 7 8 9 0\1 2 3 4 5 . . .

     1           1 1 1 1 1
     2           1121212 1
     3   101011 1121212
     4   101011111212121212
   X 5    5 511111112121212 9 9
     6    5 5111111121212 9 9 9
     7    5 5111111 7 7 7 7 3 3 3
     8    5 5 4 4 7 7 7 7 7 3 3 3
     9    5 5 4 4 7 7 7 7 7 3 3 3
     0    2 5 4 4 2 8 8 8 8 6 6 3
   > 1    2 2 2 2 2 8 8 8 8 6 6 6
     2    2 2 2 2 2        8 6 6 6
     3
     4
   Y 5
     6
The SCORE for the above LAYOUT = 59

(if HARD COPY is desired, press the 'Shift' AND the 'Prt
Sc' Keys AT THE SAME TIME)

HIT 'Return' TO CONTINUE
```

**FIGURE 12.29** Modified layout after the width had been set to 12 rows

```
                A                   B                   C
          \1 2 3 4 5 6 7 8 9 0\1 2 3 4 5 6 7 8 9 0\1 2 3 4 5 ...

  1   12
  2   1212121212
  3   121212111111
  4       9 91111111111
X 5     9 9 9 711111111    101010
  6       8 8 7 7 7 7 5 5 6 610
  7       8 8 7 7 7 5 5 5 6 6 6
  8       8 8 7 7 7 5 5 5 6 6 6
  9     8 8 8 7 7 7 5 5 5 1 1 1 1
  0       4 4 3 3 3 2 2 2 1 1 1 1
> 1       4 4 3 3 3 2 2 2
  2       4 4 3 3 3 2 2 2
  3               3 2 2 2
  4
Y 5
  6
The SCORE for the above LAYOUT = 35

(if HARD COPY is desired, press the 'Shift' AND the 'Prt
Sc' Keys AT THE SAME TIME)

HIT 'Return' to CONTINUE
```

**FIGURE 12.30** Modified layout after department 12 had been positioned in a fixed location

For the next option, department 12 is positioned (fixed) in the upper left-hand corner, as Figure 12.30 shows.

The score has been reduced, indicating that the last change is also a good one; that is, it will reduce the material-handling cost. Option 3 is used to interchange blocks to make minor or detailed changes to the layout. The following prompt is displayed at the bottom of the screen

PLEASE PICK YOUR OPTION:

```
1. to interchange any TWO blocks
2. to interchange two groups of blocks, EACH ON ONE ROW
3. to Exit
4. YOUR CHOICE
```

Option 1 is picked, for which the coordinates of the two blocks need to be given, as the prompt shows:

```
ROW and COLUMN for the FIRST block -----(e.g., X4, A2)?  X1, A1
ROW and COLUMN for the SECOND block----(e.g., Y2, A7)?  X2, A6
```

Figure 12.31 shows the layout after the two blocks have been swapped. Note the highlighted corresponding block.

The swap has not changed the score. The user may now interchange two groups of blocks by picking option 2. The coordinates for the beginnings and ends of the rows need to be given as shown here.

```
                A                        B                        C
          \1 2 3 4 5 6 7 8 9 0\1 2 3 4 5 6 7 8 9 0\1 2 3 4 5 ...

      1   :::
      2   121212121212
      3   121212111111
      4     9 91111111111
    X 5   9 9 9 711111111   101010
      6     8 8 7 7 7 7 5 5 6 610
      7     8 8 7 7 7 5 5 5 6 6 6
      8     8 8 7 7 7 5 5 5 6 6 6
      9   8 8 8 7 7 7 5 5 5 1 1 1 1
      0     4 4 3 3 3 2 2 2 1 1 1 1
    > 1     4 4 3 3 3 2 2 2
      2     4 4 3 3 3 2 2 2
      3             3 2 2 2
      4
    Y 5
      6
```

The SCORE for the above LAYOUT = 35

(if HARD COPY is desired, press the 'Shift' AND the 'Prt Sc' Keys AT THE SAME TIME)

HIT 'Return' TO CONTINUE

beginnings and ends of the rows need to be given as shown here.

ROW and COLUMN for the first element of the FIRST group --,-- (e.g., X0, B1)? **Y3, A7**
ROW and COLUMN for the last element of the FIRST group -- (e.g., B1)? **A9**
ROW and COLUMN for the first element of the SECOND group --,-- (e.g., X3, A5)? **Y1, A0**
ROW and COLUMN for the last element of the SECOND group -- (e.g., A%)? **B2**

**FIGURE 12.31** Modified layout after two blocks had been swapped

```
ROW and COLUMN for the first element of the FIRST group - - -
  (e.g., X0, B1)?  Y3, A7
ROW and COLUMN for the last element of the FIRST group - - - -
  (e.g., B1)?  A9
ROW and COLUMN for the first element of the SECOND group- - - -
  (e.g., X3, A5)?  Y1, A0
ROW and COLUMN for the last element of the SECOND group - - - -
  (e.g., A%)?  B2
```

Figure 12.32 shows the layout after the above change has been made. Note the two highlighted blocks of department 3.

```
                    A                             B                         C
        \1 2 3 4 5 6 7 8 9 0\1 2 3 4 5 6 7 8 9 0\1 2 3 4 5 ...

    1
    2  121212121212
    3  121212111111
    4     9 91111111111
  X 5   9 9 9 711111111    101010
    6     8 8 7 7 7 7 5 5 6 610
    7     8 8 7 7 7 5 5 5 6 6 6
    8     8 8 7 7 7 5 5 5 6 6 6
    9   8 8 8 7 7 7 5 5 5 1 1 1 1
    0     4 4 3 3 3 2 2 2 1 1 1 1
  > 1     4 4 3 3 3 2 2 2 2 2 2
    2     4 4 3 3 3 2 2 2
    3             3
    4
  Y 5
    6
The SCORE for the about LAYOUT = 35

(if HARD COPY is desired, press the 'Shift' AND the 'Prt
Sc' Keys AT THE SAME TIME)

HIT 'Return' TO CONTINUE
```

**FIGURE 12.32** Modified layout after two rows had been swapped

If this last layout is satisfactory to the user, he or she may continue to determine the interdepartmental distances. The user may position the cursor at the desired location of each pair of departments to have the rectilinear distances calculated. Each block may represent a certain number of feet. Five feet per block has been selected here. Figure 12.33 shows the resultant screen. Only distances between departments 1 and 2 have been selected (by cursor).

| | 1 | 2 | 3 | 4 | 5 | 6 | 7 | 8 | 9 | 10 | 11 | 12 |
|---|---|---|---|---|---|---|---|---|---|---|---|---|
| 1 | 0 | 0 | 25 | 40 | 20 | 5 | 30 | 55 | 65 | 15 | 45 | 55 |
| 2 | 0 | 0 | 0 | 0 | 0 | 0 | 0 | 0 | 0 | 0 | 0 | 0 |
| 3 | 25 | 0 | 0 | 0 | 0 | 0 | 0 | 0 | 0 | 0 | 0 | 0 |
| 4 | 40 | 0 | 0 | 0 | 0 | 0 | 0 | 0 | 0 | 0 | 0 | 0 |
| 5 | 20 | 0 | 0 | 0 | 0 | 0 | 0 | 0 | 0 | 0 | 0 | 0 |
| 6 | 5 | 0 | 0 | 0 | 0 | 0 | 0 | 0 | 0 | 0 | 0 | 0 |
| 7 | 30 | 0 | 0 | 0 | 0 | 0 | 0 | 0 | 0 | 0 | 0 | 0 |
| 8 | 55 | 0 | 0 | 0 | 0 | 0 | 0 | 0 | 0 | 0 | 0 | 0 |
| 9 | 65 | 0 | 0 | 0 | 0 | 0 | 0 | 0 | 0 | 0 | 0 | 0 |
| 10 | 15 | 0 | 0 | 0 | 0 | 0 | 0 | 0 | 0 | 0 | 0 | 0 |
| 11 | 45 | 0 | 0 | 0 | 0 | 0 | 0 | 0 | 0 | 0 | 0 | 0 |
| 12 | 55 | 0 | 0 | 0 | 0 | 0 | 0 | 0 | 0 | 0 | 0 | 0 |

**FIGURE 12.33** Cursor-selected rectilinear distances

The distances serve in only an informational capacity at this point. However, they will be important in the material-handling function.

## 12.10 SUMMARY

A well-designed plant aids in reducing material-handling costs and contributes to the overall efficiency of operations. There are six basic steps that must be followed in obtaining a good plant layout. The procedure starts by determining the area requirement for each work center and is completed when templates of all the work centers have been prepared to obtain a regularly shaped building. Where the templates are to be placed, how large the building should be, and how you know that a good decision has been obtained are the types of questions answered by the remaining steps in the procedure. The procedure is efficient and effective and is illustrated in this chapter by applying it to the development of a layout for a small manufacturing firm. Effectiveness calculations measure the relative value of a design, and the one with the smallest effective value (this value can be negative for an X-type relationship) is generally the most efficient.

Once the positions of the departments are established, the next step is to develop a detailed layout for each department. All items such as aisles, columns, locations of machines, material flows, spaces for in-process inventories, and spaces for maintenance and auxiliary equipment must be clearly shown. Special attention should be given to aisle locations and their widths because they form the connecting paths between the various departments. Commonly used facilities such as washrooms, cafeterias, water fountains, and TRs should be placed so as to obtain utmost efficiency in operations. This chapter gives some guidelines suggesting how this can be achieved.

Various materials may be used to construct a layout. Two- and three-dimensional templates representing physical objects placed on a grid and plastic sheets laid on a sturdy base material are a good start. Plastic tapes of different colors and widths may be used to show aisles, walls, conveyors, material flow, storage areas, and other interesting features and may be further identified by professionally lettering them. Clutter can be reduced by developing different overlay sheets, with each overlay showing a different aspect of the layout.

Analysis of a good solution requires evaluating it both quantitatively and qualitatively. Solutions should offer smooth and continuous material flow, no congestion in aisles, sufficient clearance for machines and operators, and flexibility to change and expand. Quantitatively, a weighted analysis and/or rate-of-return analysis can be performed. A well-written report, a good oral presentation, and the question-and-answer session are the three critical steps in selling a design. At times, features oriented more toward satisfaction than optimization may have to be chosen to obtain consent from both management and workers, because each may have a different prospective on the design.

Development of an office layout follows the same six-step procedure as that of a plant layout. An example illustrating an office layout is presented in this chapter. Offices can be made more informal, airy, and pleasant to work in by following an office landscaping program. Office landscaping can also reduce noise levels.

Plot planning is the next step in development. It involves placing a newly designed plant with its warehouses, streets, driveways, parking lots, loading docks, and lawns on the plot of land that had been acquired for it. Other features such as sidewalks, utility lines, and security fences must now be considered. The plan should offer a degree of flexibility for change and future expansion.

In time, revaluation of the existing plant layout may be warranted if inefficiencies develop. It becomes helpful if the model representing the current layout is available. It is therefore advisable to maintain the records and layouts for current development for possible future use. One should also understand that there is a time lapse between the design phase and the construction phase and that situations that demand reanalysis might arise.

## PROBLEMS

1. Why is the plant layout problem so important?
2. Discuss reasons for placing the following departments and state what other areas they should be close to and what departments they should be removed from.
   a. Restrooms
   b. Paint department
   c. Receiving
3. Convert the following rel-chart into a nodal representation.

| | 1 | 2 | 3 | 4 | 5 |
|---|---|---|---|---|---|
| 1 | — | 1 | A | O | O |
| 2 | | — | U | O | E |
| 3 | | | — | A | X |
| 4 | | | | — | E |
| 5 | | | | | — |

4. List the steps for developing a layout.
5. Using the rel-chart in Problem 3, determine a layout for the five departments. Each department requires 1000 square feet, with each block representing 250 square feet.
6. Using the following information, determine a layout for the six departments. Assuming code X = −1 and building width = 4 blocks, with a block size of 20 × 20 feet, calculate the effectiveness (the lower the number, the better the arrangement, including negative values). Develop the final layout.

| Department | Size |
|---|---|
| (1) Painting | 2100 |
| (2) Welding | 1800 |
| (3) Maintenance | 700 |
| (4) Machine shop | 1500 |
| (5) Office | 1000 |
| (6) Storage and shipping | 2500 |

| | 1 | 2 | 3 | 4 | 5 | 6 |
|---|---|---|---|---|---|---|
| 1 | — | E | I | U | X | A |
| 2 | | — | E | E | U | U |
| 3 | | | — | I | U | U |
| 4 | | | | — | U | U |
| 5 | | | | | — | E |
| 6 | | | | | | — |

7. The production area of the XYZ Machine Company is a process-type layout. The present layout is as follows:

| | |
|---|---|
| A | C |
| D | E |
| B | F |

In the layout above, A, B, C, D, E, and F are the six processing areas. For safety reasons, it is undesirable to place areas A and F adjacent to each other. A study of the manufacturing activities of the XYZ Machine Company indicated that four major products accounted for 85% of the total manufacturing activity in the above production area last year. The production sequence of these products and the number of units produced per year can be found in the following table.

| Product | Production Sequence | Number of Units Produced per Week | Number of Units per Trip |
|---|---|---|---|
| 1 | B—E—F | 8000 | 50 |
| 2 | A—D—E—F | 500 | 25 |
| 3 | C—B—E—A | 5000 | 50 |
| 4 | A—C—B—E—F | 15,000 | 100 |

Standard 36 × 36-inch unitized pallets and a conventional narrow aisle for a lift truck are the material-handling equipment involved in the production of the four products. The average numbers of units carried per trip by the lift truck are also indicated in the above table (assume one unit of area for each department).

a. Develop an activity rel-chart.
b. Develop a layout of the processing areas.
c. Discuss the present layout versus the proposed layout.
d. If the areas for A, B, and C are twice as large as those for D, E, and F, discuss how the layout will change in terms of the relative positions of the departments and the shape of the overall layout.

8. How does the office layout problem differ from the plant layout problem?

9. Using the following information, determine a layout for a plant.

a. Office

| Department | Size |
|---|---|
| (1) Executive office | 500 |
| (2) Bookkeeping | 800 |
| (3) Sales | 800 |
| (4) Engineering | 500 |
| (5) Administrative assistant or clerical | 1000 |

| | 1 | 2 | 3 | 4 | 5 |
|---|---|---|---|---|---|
| 1 | — | E | E | E | U |
| 2 | | — | E | O | A |
| 3 | | | — | U | A |
| 4 | | | | — | O |
| 5 | | | | | — |

b. Manufacturing

| Department | Size |
|---|---|
| (1) Receiving raw materials | 1500 |
| (2) Storage of raw materials | 3000 |
| (3) Receiving finished product | 1800 |
| (4) Storage of finished product | 2000 |
| (5) Shipping | 250 |

| | 1 | 2 | 3 | 4 | 5 |
|---|---|---|---|---|---|
| 1 | — | A | U | U | O |
| 2 | | — | A | U | O |
| 3 | | | — | A | O |
| 4 | | | | — | O |
| 5 | | | | | — |

c. Storage

| Department | Size |
|---|---|
| (1) Receiving raw materials | 900 |
| (2) Storage of raw materials | 3000 |
| (3) Receiving finished product | 900 |
| (4) Storage of finished product | 5000 |
| (5) Shipping | 900 |

| | 1 | 2 | 3 | 4 | 5 |
|---|---|---|---|---|---|
| 1 | — | A | U | U | U |
| 2 | | — | U | U | U |
| 3 | | | — | A | U |
| 4 | | | | — | A |
| 5 | | | | | — |

d. Auxiliary

| Department | Size |
|---|---|
| (1) Restrooms | 900 |
| (2) Lockers | 3000 |
| (3) Cafeteria | 900 |
| (4) Maintenance | 5000 |

| | 1 | 2 | 3 | 4 |
|---|---|---|---|---|
| 1 | — | E | E | O |
| 2 | | — | U | U |
| 3 | | | — | U |
| 4 | | | | — |

| | A | B | C | D |
|---|---|---|---|---|
| A | — | O | E | O |
| B | | — | A | A |
| C | | | — | O |
| D | | | | |

Specifically, the bookkeeping department has an E relationship with receiving raw materials and with shipping, painting has an A relationship with receiving the finished product, and maintenance has an A relationship with the manufacturing department.

## SUGGESTED READINGS

### Plant and Office Layout

Conway, H.M., and Liston, L., *Industrial Facilities*, Planning Conway Publishing, Atlanta, 1976.

Francis, R.L., and White, J.A., *Facility Layout and Location—An Analytical Approach*, 2nd edition, Prentice Hall, Englewood Cliffs, 1991.

Muther, R., *Systematic Layout Planning*, Cahners Books, Boston, 1973.

Pile, J., *Open Office Planning,* Whitney Library of Design, New York, 1986.
Tompkins, J.A., White, J.A., Bozer, Y.A., Tanchoco, J.M.A., *Facilities Planning*, 3rd edition, 2002, John Wiley, New York.

## Operations Research

Phillips, D.T., Ravindran, A., and Solberg, J.J., *Operations Research — Principles and Practice,* John Wiley, New York, 1987.
Taha, H., *Operations Research — An Introduction*, 8th edition, Prentice Hall, /Upper Saddle River, NJ, 2006.

# 13 Computer-Aided Plant Layout

Traditionally, the development and evaluation of plant layouts was accomplished subjectively by designers who resorted to graphical techniques and template manipulation. With the emergence of operations research and the use of digital computers, however, more analytical-based procedures were applied in the layout developments. This trend culminated in what is now known as computer-aided plant layout. The approach involves using computer programs that assist in generating layouts quickly and comparing them on an objective basis. It is more suitable for the job shop than for product layouts, which by their very nature are simpler to develop. The term "computer" is used mainly to distinguish between the recent analytical approach and the traditional manual approach. Thus, although some of the analytical procedures do not produce block layouts in the final computer output, as others do, they can be included in the discussion of computer-aided plant layout, because the computer is still necessary to perform the required calculations.

Using the computer in solving the layout problem has several advantages over the traditional manual approach. First, the computer can perform the necessary calculations and generate several solutions much more quickly than manual procedures. Second, the computer can solve large-sized problems, which usually involve huge amounts of data. Third, because the computer can develop solutions quickly, it is more economical to use a computer as an aid in the design process than to depend on human planners and designers alone. And fourth, by using the computer, solutions will be developed on the basis of mathematical expressions and operations that can be evaluated objectively. These solutions are often better than those produced solely by subjective judgment.

It should be emphasized that these computer programs are by no means a substitute for human involvement in the design process; they are merely an aid to facility planners. These programs have several drawbacks, and designers must usually modify the solutions obtained from them to guarantee a realistic and useful design. In addition, these programs do not produce a detailed layout of the facility, but rather a block layout of the arrangement of work centers or departments with respect to each other. Hence, the output produced by these programs should not be considered the final step in the design process.

## 13.1 CHARACTERISTICS OF THE PROBLEMS

The plant layout problem has some characteristics that make it difficult to formulate and solve completely by analytical means. For one thing, because each location is a candidate for each department, all combinations of departments and locations have to be evaluated. Hence if there are $n$ departments and $n$ locations, the maximum number of combinations that has to be considered is $n!$ Evaluation of all possible combinations could take an inordinate amount of time for medium- to large-sized problems. Even if the numbers of departments and locations are not the same — say, $n$ and $m$, respectively, where $n < m$, the number of combinations that require evaluation is $m!(m - n)!$, which, although less than $n!$, is still large.

Minimizing the material-handling cost is usually considered the most important objective of the problem. However, several other critical objectives exist, such as achieving maximum adjacencies (closeness) between the departments, flexibility of arrangement and operation, and effective utilization of the available space. When considered simultaneously in an analytical model, these objectives complicate the formulation. Furthermore, it is difficult to translate qualitative information into quantitative objectives and constraints.

The layout problem (particularly in industry) involves locating areas, not points. These areas affect the shapes and dimensions of the departments, which in turn affect the distances between these departments. The number of configurations a department can assume is infinite, and the distance between the departments changes accordingly, thereby affecting the material-handling cost. The interrelation between departmental shapes and distances is difficult to control or predict.

When a new plant is to be constructed, the data required for solving the problem, such as the material flow between departments and the operating cost of material-handling equipment, are usually not available and must be estimated, which is not an easy task.

## 13.2 DATA REQUIREMENTS

The basic desired input for working with an analytical model varies from method to method. The following list gives the different types of data that might be required. Among these, departmental relationships and areas are the minimum necessary in every computerized method of analysis.

1. The area of each department expressed in square feet or as a number of unit squares
2. The rectilinear distances between candidate locations or between departments, usually measured between their centers
3. Departmental relationship measures that can be expressed either quantitatively in a from–to chart or qualitatively in a relationship chart (rel-chart).
4. A scale for plotting the layout by the computer, required in some of the procedures.

In addition, some have several options regarding departmental shapes, location, and layout outline.

## 13.3 APPROACHES AND TYPES OF PROCEDURES

In practice, plant layout problems can be solved by any one or more of the following approaches:

- Exact mathematical programming procedures, mainly branch and bound
- Heuristic
- Probabilistic approaches
- The application of graph theory

Notice, however, that whatever the approach used, it attempts to solve an optimization problem. The classification is mainly based on the dominant theoretical basis of the procedure. Thus, although the probabilistic approach is completely dependent on heuristics, it is not classified as such, because what distinguishes it from the other heuristics of computer-aided plant layout is the inclusion of probability in its solution procedures. Similarly, graph theory-based procedures use both heuristics and the branch-and-bound techniques in developing layouts; however, because they use the concepts and ideas of graph theory, they are classified as such.

The objective of these procedures depends on the manner in which departmental relationships are expressed. When qualitative relationships are used, as expressed in a rel-chart, the objective is to maximize the adjacencies (closeness) between departments. On the other hand, when departmental relationships are expressed quantitatively by a from–to chart, the objective is to minimize the material-handling cost — that is, the product of departmental distances, flow, and unit-handling cost. In many cases, especially for a new plant, the handling system will not be known until the layout has been established. A handling cost of 1 is assumed in such cases, which results in the cost to be minimized being equal to the flow multiplied by the distance only. (Henceforth, this cost will be referred to as the movement cost.)

The solution procedures for any of these approaches are of either the construction or the improvement type. Construction procedures, which are used when a layout is developed for the first time, involve three steps. First, a criterion for selecting the departments to enter the layout is established on the basis of departmental relationships. Second, the placement of each department in a suitable location is attempted such that the inherent objective is achieved. Finally, steps toward development of departmental configurations are undertaken. Construction procedures differ among themselves by the rules used to execute each step. The solution obtained from a construction procedure is almost always suboptimal, which is a result of the combined effect of the three steps.

Improvement procedures attempt to reduce the movement cost of an initial layout, which could be an existing layout or one developed by a construction procedure. The improvement is usually achieved by interchanging the locations of departments until no further reduction in movement cost is possible. Thus, the solution obtained will most likely be better than that of a construction procedure, but there is still no guarantee of optimality.

## 13.4 MATHEMATICAL PROGRAMMING

In this section, only two mathematical formulations are illustrated without going into the details of the solution procedures. The interested reader can explore this further by consulting the references, while those who prefer simpler heuristic approaches can go directly to the next section.

The plant layout problem is most often formulated as a quadratic assignment problem (Francis and White, 1974; Pierce and Crowston, 1971), where $m$ locations are available to which $n$ departments are to be assigned. All the assignments must be done simultaneously, and the arrangement that minimizes the movement cost is then selected. If $n \leq m$, let

$c_{ikjh}$ = cost of placing departments $i$ and $j$ at locations $k$ and $h$, respectively

$$x_{ik} = \begin{cases} 1 & \text{if department is locked at } k \\ 0 & \text{otherwise} \end{cases}$$

Then we

$$\text{Minimize } Z = \frac{1}{2}\sum_{i=1}^{n}\sum_{k=1}^{m}\sum_{j=1}^{n}\sum_{h=1}^{m} c_{ikjh} x_{ik} x_{jh}$$

$$\text{Subject to } \sum_{i=1}^{n} x_{ik} = 1 \quad k = 1,\ldots,m$$

$$\sum_{j=1}^{m} x_{ik} = 1 \quad i = 1,\ldots,n$$

$$x_{ik} = \{0, 1\} \quad \text{for all } i,k$$

Several branch-and-bound algorithms have been developed for solving this structure; however, they are feasible only for small-sized problems. The difficulty increases rapidly as the number of departments exceeds 15 because of the increase in the number of solution combinations. Moreover, unless some constraints are established to prohibit a large number of adjacencies of departments from being achieved or certain departments from being placed in certain locations, every branch in the solution tree is a feasible solution that might have to be evaluated completely. However, branch-and-bound procedures guarantee an optimal solution and can provide alternative layouts in addition to the optimal one.

An alternative mathematical formulation has been suggested (Bazaraa, 1975) as a quadratic set-covering problem and is solved by branch-and-bound procedures. The areas of the layout and the individual departments are expressed as unit squares. Several candidate locations are made available for each department to be located, which prevents each location from being available for every department as in the quadratic assignment problem. Restricting the availability of locations to departments helps in reducing the number of combinations that must be evaluated.

The procedure is capable of producing desirable shapes for the departments as well as for the layout outline by specifying the shapes for these as part of the input data. However, specifying these shapes implies additional constraints to the problem, which in turn might affect the objective function adversely. The model has two drawbacks concerning some of its assumptions. First, candidate locations for each department must be established in advance, which may not be an easy task. Second, the specification of departmental shapes might not be practical, because it is usually not known why a particular shape would be preferred over another or how a good department shape should look. The problem formulation is shown below. Let

$d(k_i, h_j)$ = rectilinear distance between the centers of the $k$th and the $h$th locations for departments $i$ and $j$

$f_{ij}$ = flow between departments $i$ and $j$

$F_{ik}$ = fixed cost of locating department $i$ at $k$

$I(i) = I$ = number of candidate locations for department $i$

$I(j) = J$ = number of candidate locations for department $j$

$J(k)$ = set of squares occupied by department $i$ if it is located at $k$

$$a_{ikt} = \begin{cases} i & \text{if block } t \in j(k) \\ 0 & \text{otherwise} \end{cases}$$

$$x_{ik} = \begin{cases} 1 & \text{if department } i \text{ is located at } k \\ 0 & \text{otherwise} \end{cases}$$

The problem requires that we

$$\text{Minimize } Z = \sum_{i=1}^{m}\sum_{j=1}^{m}\sum_{k=1}^{I}\sum_{h=1}^{j} f_{ij}x_{ik}x_{jh}d(x_i, h_j) + \sum_{i=1}^{m}\sum_{k=1}^{\text{I}} F_{ik}x_{ik}$$

Subject to

$$\sum_{k=1}^{I} x_{ik} = 1 \quad i = 1, 2, \quad , m$$

$$\sum_{i=1}^{m}\sum_{k=1}^{I} a_{ikt} \qquad x_{ik} \text{ for all } t$$

$$x_{ik} = \{0, 1\} \; k = 1, 2, \quad , I(i)$$

$$i = 1, 2, \&, m$$

A third formulation appeared a few years ago (Hassan, 1983) in which the layout problem was formulated as a generalized assignment problem. However, the solution procedure is heuristic, and therefore optimality is not guaranteed. Practical application of the above procedure is limited owing to its inherent complexity. Therefore, little would be gained in pursuing it through a numerical example or further discussion.

## 13.5 HEURISTICS

As mentioned previously, applying an exact procedure that produces an optimal solution is feasible only for small problems because of the number of calculations involved in the process; hence, the exact procedures are not very widely used. Heuristics, on the other hand, enjoy a large following. The main advantages of heuristics are their ability to produce good suboptimal solutions and their capacity for handling large problems with a reasonable amount of computational effort. Following are basic plant layout techniques that were developed and computerized some time back in programming languages that were available then. Though the methods are old, they illustrate different logics on which various algorithms are developed. Each type of thinking has certain advantages and some disadvantages. All the recent techniques follow, in some measure, similar ideas in development of new algorithms. It may therefore be beneficial to look at these layout programs as a starter.

### 13.5.1 Computerized Relationship Layout Planning

Computerized Relationship Layout Planning (CORELAP; Lee and Moore, 1967) is a construction program that uses a rel-chart and attempts to develop a layout whose objective is to achieve maximum adjacencies between departments. In CORELAP's selection method, each element of the rel-chart is assigned a numerical value, and departments are ranked in nonincreasing order of the sum of their relations with one another. The first department in the list is selected to enter the layout. Next departments are selected by determining which of them has the strongest relation with one of the previously selected departments, starting with the first selected department, then the second selected, and so on. When ties arise, the department with the larger sum of relations is selected first. CORELAP places the first department in the middle of the layout, and the rest of the departments grow around it. The program attempts to place each department adjacent to the one with which it has the strongest relation, or as close as possible. Departmental shapes are constructed to be as close to a square as possible. An example of the layout produced by the computer for CORELAP is shown in Figure 13.1

As an example of the selection method of CORELAP, consider the rel-chart shown in Table 13.1. Furthermore, assign the following values to the elements of the chart:

$$A = 6 \qquad O = 3$$
$$E = 5 \qquad U = 2$$
$$I = 4 \qquad X = 1$$

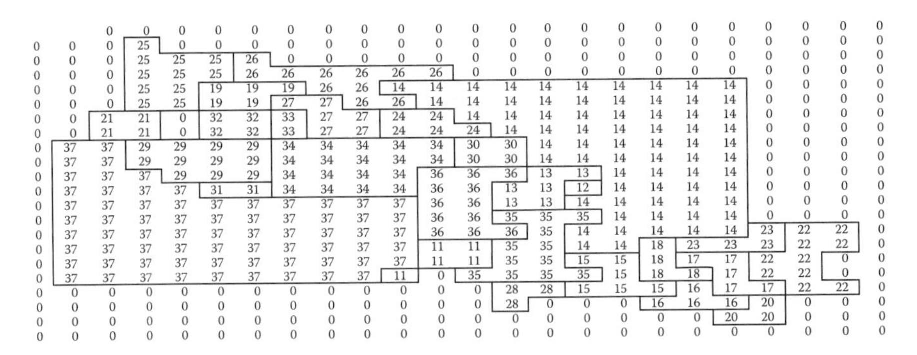

**FIGURE 13.1** Computer output produced by CORELAP (from Lee and Moore, 1967; with permission from the Institute of Industrial Engineers)

**TABLE 13.1**
**Rel-Chart for a CORELAP Illustration**

| Department | 1 | 2 | 3 | 4 | 5 |
|---|---|---|---|---|---|
| 1 | — | U | E | I | U |
| 2 | | — | A | O | O |
| 3 | | | — | E | I |
| 4 | | | | — | O |
| 5 | | | | | — |

Note that these values are different from those we used in Chapter 12. These values are the measures of closeness between the departments, and the way they are defined is one possible variation when using heuristic methods.

By substituting these values, the rel-chart now becomes that shown in Table 13.2.

The sum of the relations between a department and all other departments is obtained by adding entries in corresponding departments, rows, and columns. For example, the relationship value for the third department is obtained by adding the third row and the third column values; that is,

$$5 + 4 + 6 + 5 = 20$$

Similarly, if we perform the calculations for other departments, these sums are 13, 14, 20, 15, and 12. Arranging the departments according to these sums will result in the following ranking: 3, 4, 2, 1, 5.

Department 3 is selected first, because it has the largest sum. In looking for a department with an A relation with department 3, the program finds and selects department 2. No other department having an A relationship with department 3 exists. CORELAP next looks for an A relationship with department 2 but finds none. Next, the program hunts for an E relationship with department 3, and it finds both departments 1 and 4. The tie is resolved according to the sum of relationships; because department 4 has a larger sum of relationships compared with department 1, department 4 is chosen first, followed by department 1. The final order of selection is 3, 2, 4, 1, 5.

**TABLE 13.2**
**Value Chart for CORELAP Illustration**

| Department | 1 | 2 | 3 | 4 | 5 | Sum of Relations |
|---|---|---|---|---|---|---|
| 1 | — | 2 | 5 | 4 | 2 | 13 |
| 2 | | — | 6 | 3 | 3 | 14 |
| 3 | | | — | 5 | 4 | 20 |
| 4 | | | | — | 3 | 15 |
| 5 | | | | | — | 12 |

### 13.5.2 Plant Layout Analysis and Evaluation Technique

Another construction program is Plant Layout Analysis and Evaluation Technique (PLANET) (Apple and Deisenroth, 1972), which uses flow data expressed by the from–to chart or the rel-chart. Three methods are available for selecting departments after the latter are divided into groups according to a user-specified priority. The first method (A) selects from among all the departments the two having the strongest flow relationship with each other. The department having the strongest relationship with any of the previously selected departments is selected next. (In case of a tie, arbitrarily select either.) The second method (B) selects the first two departments in the same way as is done by method A, but the next department selected is the department having the strongest sum of relationships with all the previously selected departments. The third method (C) ranks departments in a decreasing order according to the sums of their flow values with each other and selects them according to this ranking. PLANET's objective is to minimize the total movement cost.

PLANET locates the first selected department in the middle of the layout area, and the rest of the departments cluster around it. The best location for each department is determined by examining several points around the periphery of the partial layout formed by the previously selected and placed departments. These points represent the candidate centers of the entering department. The location that results in minimum movement cost between the entering and placed departments is then selected. This search scheme is repeated for each selected department until the layout is complete. Once a location has been determined for a department, its shape is made as close to a square as possible. This is accomplished by placing the unit squares that constitute a department side by side in a spiral fashion. An example of the layout produced by PLANET is shown in Figure 13.2.

To demonstrate the selection methods of PLANET, consider the symmetric from–to chart in Table 13.3 and assume that all departments have the same priority.

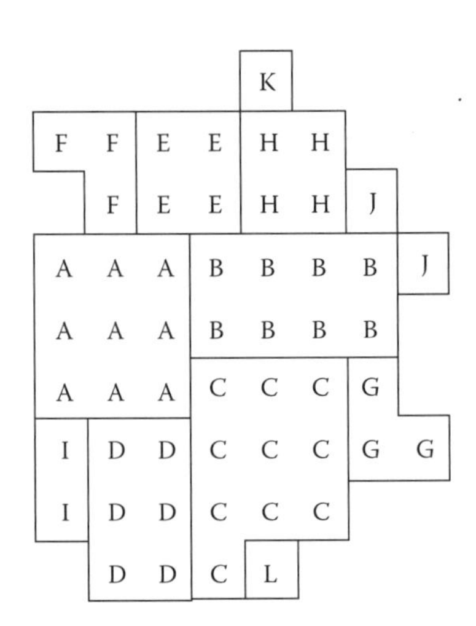


**FIGURE 13.2** Computer output of a layout by PLANET (the program does not plot the lines representing department borders)

**TABLE 13.3**
**From–To Chart for a PLANET Illustration**

| | A | B | C | D | E |
|---|---|---|---|---|---|
| A | — | 0 | 36 | 18 | 0 |
| B | | — | 9 | 10 | 0 |
| C | | | — | 12 | 8 |
| D | | | | — | 5 |
| E | | | | | — |

#### 13.5.2.1 Method A

Departments A and C are selected first, because they have the strongest relationship, with a flow of 36 between them. The program looks for the department with the next strongest relationship with either A or C, and department D is found to have the strongest relationship with A, with a flow of 18 between the two. Department D is selected, and the program then looks for the department with the strongest relationship with A, C, or D. The flow between B and D is 10, which is larger than that between E and any of A, C, or D; hence, B enters next and then E, because it is the remaining department. The order of selection is then A, C, D, B, E.

#### 13.5.2.2 Method B

Departments A and C are selected first as in method A. Next, the sum of the flow between B and both A and C is evaluated as 0 + 9 = 9, that between D and both A and C as 18 + 12 = 30, and that between E and both A and C as 0 + 8 = 8. Therefore, department D is selected because it has the largest sum with departments A and C.

Departments B and E are now compared by finding the largest sum with the previously selected departments. Departments A, B, C, and D yield 0 + 10 + 9 = 19, and department E yields 0 + 8 + 5 = 13. Accordingly, department B is selected next and then E; the order of selection is A, C, D, B, E.

#### 13.5.2.3 Method C

The sums of the relationships between the indicated departments and all others are:

$$\text{A:}\quad 0+36+18+0=54$$

$$\text{B:}\quad 0+9+10+0=19$$

$$\text{C:}\quad 36+9+12+8=65$$

$$\text{D:}\quad 18+10+12+15=45$$

$$\text{E:}\quad 0+0+8+5=13$$

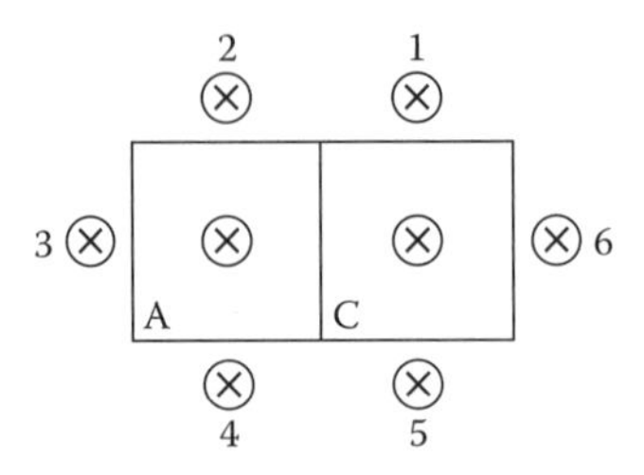


**FIGURE 13.3** Candidate locations for department D

If we arrange the values in decreasing order, the sequence of the departments becomes C, A, D, B, E.

Although all three selection methods produce the same order of selection in this example, this would not necessarily be the case with other data.

To demonstrate the placement procedure of PLANET, assume that all departments have the same area of 1 square foot. The first two departments (A, C) enter the layout and are placed as shown in Figure 13.3. The movement cost between the two departments is $36 \times 1 = 36$. The candidate location centers for the third entering department are numbered 1 through 6.

The distances are calculated as a rectangular measurement. For example, the distance between location 4 and the center of C (marked with ⊕) is two units, and the distance between location 5 and C is one unit.

The movement costs between entering department D and the placed departments at these locations are:

| Flow | | Department | Distance Location 1 | 2 | 3 | 4 | 5 | 6 |
|---|---|---|---|---|---|---|---|---|
| AD | CD | A | 2 | 1 | 1 | 1 | 2 | 2 |
| (18 | 12) × | C | 1 | 2 | 2 | 2 | 1 | 1 |
| | | | = (48, 42, 42, 42, 48, 48) | | | | | |

The minimum movement cost is 42 given by locations 2, 3, and 4, and any of them can be selected. Arbitrarily select location 4. The candidate locations for the next selected department, department B, are shown in Figure 13.4.

The movement costs at these locations are:

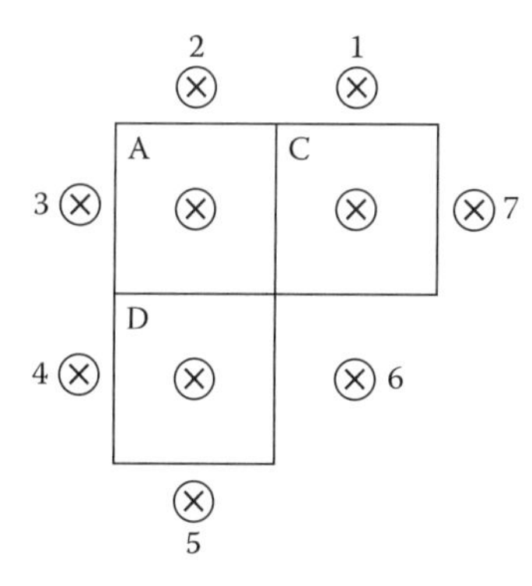


**FIGURE 13.4** Partial layout and candidate location for department B

| Flow | | Department | Distance Location | | | | | | |
|---|---|---|---|---|---|---|---|---|---|
| | | | 1 | 2 | 3 | 4 | 5 | 6 | 7 |
| AD | CD | A | 2 | 1 | 1 | 2 | 2 | 2 | 2 |
| (18 | 12) × | C | 1 | 2 | 2 | 3 | 3 | 1 | 1 |
| | | D | 3 | 2 | 2 | 1 | 1 | 1 | 3 |
| | | | = (39, 38, 38, 37, 37, 19, 39) | | | | | | |

The minimum movement cost is 19 and is given by location 6.

Similarly, the location of department E can be determined and is found to have a cost of 23. The final layout is shown in Figure 13.5. The total movement cost of this layout is 36 + 42 + 19 + 23 = 120.

### 13.5.3 Modular Allocation Technique

Another heuristic construction procedure, perhaps less popular than CORELAP and PLANET, is the Modular Allocation Technique (MAT) (Edwards et al., 1970), which attempts to solve the quadratic assignment problem. MAT is built on the theory that the sum of the products of two sets of numbers arranged in different order, decreasing and increasing, is minimum. Flow values between pairs of departments are arranged in decreasing order, while the distances between each pair of the candidate locations are arranged in increasing order. Each pair in the first set is matched against a pair in the second set, and departments are assigned to locations accordingly. Although in theory this process should

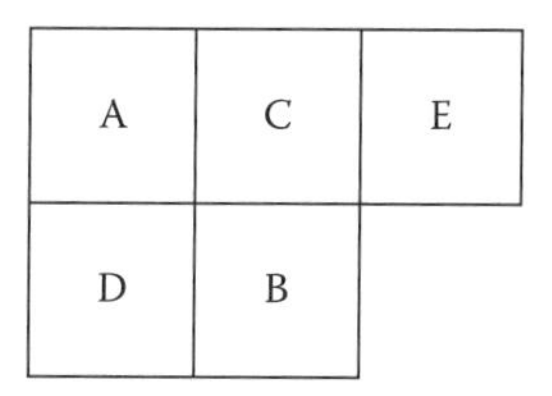


**FIGURE 13.5** Layout produced by PLANET for the example problem

lead to the optimal solution, in practice this does not usually occur owing to the conflicts that arise when departmental distances and flow relations are matched together. For example, in the assignment of a pair of departments to locations, one of the departments might have been assigned previously. In this case, the other department of the pair has to be assigned to the remaining location in the corresponding location pair (in which the first department has been assigned a location) if it is vacant or to some other location. Neither of these locations might be the best for that department.

### 13.5.4 Computerized Relative Allocation of Facilities Technique

Now let us study a layout method based on improvement procedures, Computerized Relative Allocation of Facilities Technique (CRAFT) (Buffa et al., 1964). It also relies heavily on heuristics because of the magnitude of the computations involved. The program requires an initial layout as an input. The iterative steps within the procedure attempt to interchange the locations of departments that have common borders or equal areas (most often pairwise, but other interchanges are also possible). The interchange that results in the greatest reduction in the movement cost is executed, and the procedure is repeated until no further reduction in the cost is possible. The maximum number of pairwise interchanges that can be performed in an iteration is less than $n(n-1)/2$, where $n$ is the number of departments.

As an example of the calculations of CRAFT, consider the initial layout and the symmetric from–to chart shown in Figures 13.6a and b, respectively. The distance between the departments in the initial layout is shown in the from–to chart in Figure 13.6c.

The movement cost of the initial layout is calculated by multiplying the elements of both charts in Figures 13.6b and c as follows:

$$\text{Movement cost} = 1(3) + 1(1) + 2(4) + 2(4) + 1(1) + 1(2) = 23$$

The possible pairwise interchanges for the initial layout are A-B, A-C, A-D, B-C, B-D, and C-D. If departments A and B are interchanged, the distance matrix of the resulting layout will be as shown in Figure 13.7a. The movement cost of the resulting layout will be 17, which results in a savings of 23 – 17 = 6. Next, departments A and C in the initial layout are used in an attempted interchange. The movement cost of the resulting layout is 20, with a cost saving of 3. The other interchanges are attempted, and the cost savings are calculated. The largest saving is obtained by interchanging departments A and B; in this case, the interchange is actually made. The resulting layout at the completion of this iteration is shown in Figure 13.7a. The distance between the departments is shown in Figure 13.7b. The movement cost of this current layout is 17.

With the layout of Figure 13.7a, new pairwise interchanges are attempted between A-C, A-D, B-C, B-D, and C-D. The interchange that results in the most savings in movement cost is performed, and a new layout results. The procedure stops when no cost savings is produced. Notice that the program does not actually interchange the location of departments when the cost savings are compared; only when the cost savings of all interchanges in an iteration have been calculated and the greatest saving has been found does the actual interchange take place. Figure 13.8 shows the computer output of the initial and final layouts when CRAFT is used.

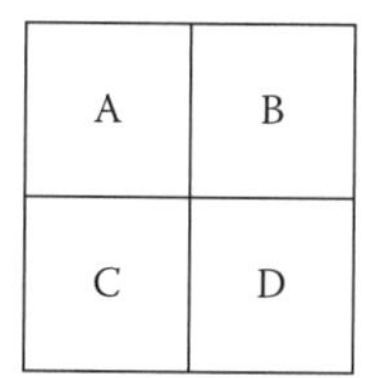


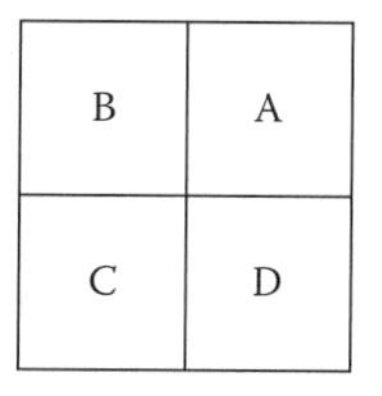


**FIGURE 13.6** CRAFT initial layout and data

**a. Initial Layout Used With CRAFT**

| From/To | A | B | C | D |
|---|---|---|---|---|
| A | — | 3 | 1 | 4 |
| B | | — | 4 | 1 |
| C | | | — | 2 |
| D | | | | — |

**b. Flow data**

| From/To | A | B | C | D |
|---|---|---|---|---|
| A | — | 1 | 1 | 2 |
| B | | — | 2 | 1 |
| C | | | — | 1 |
| D | | | | — |

**c. Distance between the departments**

**FIGURE 13.7** Results of interchanging departments A and B

**a. Current layout after interchanging departments A and Bt**

| From/To | A | B | C | D |
|---|---|---|---|---|
| A | — | 1 | 1 | 2 |
| B | | — | 2 | 1 |
| C | | | — | 1 |
| D | | | | — |

**b. Distance matrix if departments A and B are interchanged**

Other improvement procedures differ in the manner in which each carries out the interchange. Some of these procedures consider departmental areas, while others do not.

Examples of these procedures are the following.

1. A procedure by Hillier (1963) attempts to move a department to a location to the right of, left of, below, or above its current location. This procedure has been extended to allow exchanging nonadjacent departments.
2. Another procedure (Hillier and Connors, 1966) interchanges at each iteration the two departments having the highest movement costs with other departments.
3. CRAFT-M (Hicks and Cowan, 1976) is an extension of CRAFT that compares the savings in material-handling cost that result from interchanging

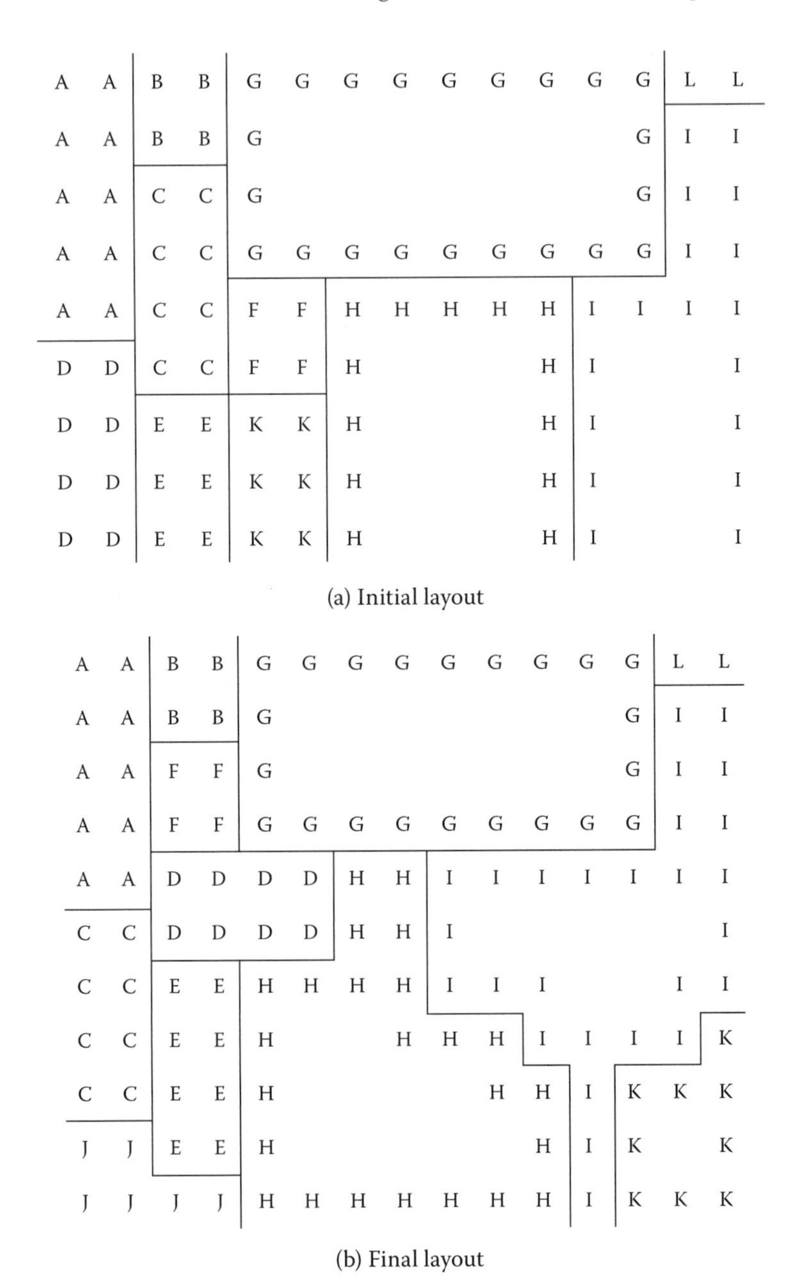


**FIGURE 13.8** Computer output of initial and final layouts produced by CRAFT (the program does not plot the lines separating the departments (a) Initial layout (b) Final layout

departments with the cost of reconstructing the departments and decides whether to perform the interchange or not.

4. SPACECRAFT (Johnson, 1982) is another extension of CRAFT that considers multiple floors and nonlinear material-handling costs.
5. Plant Layout Optimization Procedure (PLOP) (Lewis and Block, 1980) examines all the possible exchanges and performs the one that achieves the greatest cost reduction.

## 13.6 PROBABILISTIC APPROACHES

Probabilistic approaches, such as heuristics and exact methods, use known data. However, the probabilistic aspects stem from the manner in which departments are selected to enter the layout in construction procedures or to interchange locations in improvement ones. Three probabilistic approaches exist: random selection, biased sampling, and simulation.

### 13.6.1 Random Selection

Automated Layout Design Program (ALDEP) (Seehof and Evans, 1967) was the first program to use a probabilistic aspect in its solution methodology. It is a construction program that attempts to maximize the closeness between departments. The first department to enter the layout is selected randomly. The rest of the departments are selected according to whether their relationship with a placed department is higher than a user-specified value. If no department can be found, or if there are ties, the selection is accomplished randomly. The program places the first selected department in the left-hand side of the layout, starting from the top of the layout outline. The next department is placed adjacent to the most recently placed department and begins where that one ended. Departments are constructed as consisting of rectangular strips that are placed beside one another. The width of the strip is user specified.

### 13.6.2 Biased Sampling

The biased sampling approach is a probabilistic version of improvement procedures. The procedure depends on the fact that each problem has different solutions with different costs. These solutions are considered to be the population from which some solutions are sampled. The sampling is not performed randomly but in a biased manner to select only good solutions. Thus, for a given layout, for each exchange of department locations that reduces the movement cost, a probability of executing the exchange is calculated. The value of this probability is a function of the cost reduction anticipated by the exchange. Different samples are taken, and different solutions are obtained, each having an objective function value that is better than the initial one. Notice that the probability of performing the best cost-saving exchange is equal to 1 in a program such as CRAFT, while in the biased sampling approach, the probability of selecting exchanges is less than 1; therefore, any exchange that reduces cost may be executed, perhaps contributing to savings in computer time.

**TABLE 13.4**
**Classification of Plant Layout Approaches**

| Approach/ Classification | Mathematical Programming | Heuristics | Probabilistic Approaches | Graph Theory |
|---|---|---|---|---|
| Construction | | CORELAP<br>MAT<br>PLANET<br>FATE<br>COFAD<br>CRAFT<br>FRAT<br>Hillier | Simulation | Foulds (1983), Foulds and Robinson (1976, 1978), Seppanen and Moore (1970, 1975) |
| Improvement | | Procedure PLOP Vollman et al. (1968) | Biased sampling | |
| Both construction and improvement | Bazaraa procedure works in 1975 | | ALDEP | |

### 13.6.3 Simulation

One of the existing procedures of plant layout uses simulation in developing layouts. The procedure depends on considering all feasible layouts that can be produced (using the same set of data) as a statistical population. The simulator generates the sequence by which departments enter the layout in the biased sampling technique to increase the chances of obtaining good solutions. Each run generates a layout, and the simulation is terminated when no considerable improvement can be achieved over previous runs. Simulation usually requires large amounts of computation time compared with other computer-aided plant layout procedures.

## 13.7 RECENT PLANT LAYOUT SOFTWARE

Here are some of the recent plant layout software programs that are available commercially. We had no way of checking each program, but we are listing all their brochure information and website addresses, should the reader be interested in further contacts with the vendors.

### 13.7.1 CATIA Plant Layout 1

CATIA Plant Layout 1 enables organizations to optimize their manufacturing plant layout. This product is one piece of Dassault's integrated Digital Manufacturing Solutions. It deals specifically with the "spatial organization" and components of the plant, allowing quick easy layout and downstream evolution of the layout design. Through the CATIA V5 integrated product environment, you have a seamless solution to address all your manufacturing environment needs. It has the tools necessary to optimize production facility layout, leading to optimized factory production and output.

CATIA Plant Layout 1 provides an accessible solution for departments of small and medium enterprises. Its friendliness, intuitiveness, and ease of use make it possible for inexperienced users to use the system with minimal cost of implementation. The complete layout of the facility can be driven almost completely with the mouse, and the product has a user interface that helps you make the transition from the traditional 2D layout to a 3D layout. With the intelligence behind the plant layout model, plant designers and systems layout design teams can identify and solve problems with the layout or production process long before equipment is installed or moved inside the plant. The software enables you to not only complete designs faster but also significantly improve the quality of your designs. Today, every manufacturing company is looking for new methods to reduce the time it takes to ramp up, reduce the number of problems on the production floor (including downtime), and get products to the market faster. Dassault's Digital Manufacturing Solutions, including the CATIA Plant Layout 1 product line, assist you in achieving these objectives. CATIA Plant Layout 1, together with Dassault Data Management offerings, gives you the power to manage your plant layouts and associated components from initial design to plant operations in a way that is easily adapted to how you work.

#### 13.7.1.1 Features

- Includes an optimized user interface for easy maneuverability and layout design iteration.
- Provides an easy tool to define the pathway route for preliminary space claim.
- Provides optimized tools for quick and easy layout of complex configurations.
- Enables quick design changes through an extensive set of modification tools with intuitive "snap to 3D grids" function and direct manipulations.
- Parametric catalogs (design tables) enabling sizing of equipment at the time of placement.
- Provides specialized interference detection and equipment change-out capabilities.
- Analyzes area and footprint to determine resources used in each region of a plant.
- Features advanced 2D-to-3D capabilities that allow the user to reuse legacy 2D data and drawings.
- Generates customized area bills of materials.

#### 13.7.1.2 Key Customer Benefits

**Tools specifically for plant layout design and maintenance** CATIA Plant Layout 1 (PLO) enables you to optimize the "spatial organization" of a plant, allowing quick and easy layout and downstream evolution of the design. CATIA Plant Layout 1 provides an accessible solution for small and medium enterprises. Its friendliness, intuitiveness and ease of use make it possible for new users to benefit from the system with minimal cost of implementation. CATIA Plant Layout 1 helps you identify and solve layout of production process problems long before equipment is installed or moved inside a plant. Companies thus have the power to manage the plant components from initial design to plant operations and to be flexible in the way they work.

**Optimized user interface** CATIA Plant Layout 1 has an easy-to-use, interface, optimized to the needs of a plant designer. It offers familiar techniques, such as cut-and-paste and drag-and-drop. The organization of space in user-defined grids, simplifies the placement of plant features and equipment.

**Advanced plant design capability** CATIA Plant Layout 1 enables you to create layouts of plant sites by defining the buildings, the major areas, all the way down to the individual cells and stations. This is done by simply creating walls, fences, or boundaries with an extruded mode for enhanced visualization. Sub areas for facilities, work cells, stations, and lines can then be created within the plant. The system allows a hierarchical approach including true partition of space with shared boundaries, areas with multi patches, and so on. You can use the "snap to 3D grid" to easily position equipment. CATIA Plant Layout 1 enables you to perform space allocation and reservation for products and resources (equipment, tooling, robots, etc.). This can even be done for resources not yet designed.

**Plant-specific capabilities accelerate design** Beyond standard Windows commands, such as copy-and-paste, CATIA Plant Layout 1 provides advanced capabilities such as, direct manipulations, move, boundary align, mirror and offset from object and snap to drafting element. These enable you to quickly do the initial design and then to rapidly change the design as the plant layout evolves.

**Parametric catalog facilitates equipment sizing** CATIA Plant Layout 1 comes complete with a starter catalog of common equipment. The parameterized nature of these parts let you size each component as it is placed in the design. The parameters in the catalog can be imported, exported and edited as design tables in Microsoft Excel. You can modify any of the equipment included in the starter catalog and can even define their own catalog (using the CATIA Part Design product). In addition, a product can be placed from the catalog on any plane of an object. When used with the CATIA Systems Routing 2 product, CATIA Plant Layout 1 allows you to evolve the layout from space reservation to the actual 3D equipment and from routing pathways to the actual system lines and parts.

**Area and footprint analysis** Reports made up of Excel-format graphical charts can be easily customized to provide space utilization analysis based on equipment footprints.

**Interference detection and equipment change-out capabilities** When used in collaboration with the CATIA Space Analysis product, CATIA Plant Layout 1 lets you manage equipment within the factory without damage. Using CATIA Plant Layout 1 with the CATIA DMU Fitting Simulator

2 product, the change out of large equipment within the plant can be simulated off-line. This helps you to understand, in advance, the procedures that will be required to clear permanent fixtures in a facility during a complex move.

**Advanced 2D-to-3D capabilities** Engineers can reuse existing 2D plant and equipment drawings (including legacy data and DXF) to define facility areas/footprints. This functionality enables you to evolve from 2D to 3D design, reusing previous work, and benefiting from 3D space in solving factory layout problems.

**Customized bills of materials** You can generate customized bills of materials, even down to the areas level of detail, to facilitate material needs definitions.

(Material taken entirely from http://www-306.ibm.com/software/applications/plm/catiav5/prods/plo/).

### 13.7.2 PlantWise

PlantWise is a unique suite of knowledge-centric software solutions that enable rapid creation of fully piped 3D plant concept model for review and optimization. Used for creating bid packages, front-end design, and early detail design, PlantWise has demonstrated large reductions in total installed cost and project schedule. Used as a sales tool PlantWise provides customers with early visibility of the proposed plant solutions. With ease of collaborative "what-if" change implementation and feedback, PlantWise users converge on optimal plant layouts faster than with other design packages.

PlantWise includes three tightly integrated, knowledge-centric solution modules: PlantBuilder, AutoRouter, and PlantDrafter . With these tools, PlantWise users have demonstrated up to 10% reduction in total plant installed cost through layout optimization. Likewise users have recorded up to 20% reduction in detailed piping design hours by using PlantWise as a piping layout planning tool. Accurate MTOs and significantly reduced schedules are additional benefits documented by PlantWise users.

PlantWise users have several choices for transferring data into other plant design software . The highly flexible PlantWise Report Writer allows users to easily extract and format data suitable for importing into other software. Also, the AutoRouter Pipe Route Export provides a complete set of piping and piping component data from which users have written their own export routines for products such as PDS, and PDMS. To complement those generic interface capabilities, Design Power provides several interface solutions for Intergraph plant design products (Source: http://www.dp.com/plantwise/).

### 13.7.3 CADWorx Plant and Plant Professional by COADE, Inc.

The CADWorx Plant suites supply the most complete range of tools for hassle-free plant design. Piping, equipment, steel, HVAC and cable trays and database links are all included. CADWorx Plant Professional also features CADWorx Equipment, ISOGEN, NavisWorks® Roamer and live database links.

#### 13.7.3.1 2D and 3D Piping

CADWorx Plant allows you to work in 2D or 3D with full intelligence. Quick 2D plans can be converted into 3D models at anytime. Working in 3D with CADWorx Plant is easy, and, because CADWorx Plant uses standard AutoCAD solids, it delivers some of the best performance, smallest models and greatest compatibility in its class. No special enablers are required for viewing, plotting or even editing in native AutoCAD.

#### 13.7.3.2 Structural Steel

Each seat of CADWorx Plant is shipped with CADWorx Steel, giving CADWorx Plant the steel modeling capabilities of one of the most powerful and intuitive steel packages on the market. See the CADWorx Steel section for more information.

#### 13.7.3.3 Equipment

Equipment capabilities built into CADWorx Plant allow you to create any 2D or 3D equipment shape. Vessels, pumps, tanks and exchangers can be quickly added for review and interference checking.

#### 13.7.3.4 Ducting/Cable Trays

HVAC ducting and cable tray shapes are easily placed using routines built into CADWorx Plant. Select square, rectangular, round and oval shapes with transitions from a single intuitive dialog box. Key dimensional information is retained to allow you to quickly and easily design duct and tray runs with the minimum of effort.

#### 13.7.3.5 Automatic Isometrics

CADWorx Plant provides the industry's best looking automatic isometrics and does so with ease. Isometrics can be generated from individual piping layouts or project databases. Intelligent stop marks allow you to specify ideal break points for consistent drawings from revision to revision.

Settings for user borders, automatic dimensioning, tagging and bills of material are easily defined. CADWorx Plant makes automatic isometric generation a real cost saver by dramatically reducing errors and improving productivity.

#### 13.7.3.6 Piping Specifications

CADWorx Plant includes ready-to-use specifications with access to over 60,000 parametric -driven components. In keeping with the global character of CADWorx

Plant, specifications and data files are provided in imperial, metric and mixed metric formats.

The powerful specification editor lets you create and modify specifications on-the-fly or outside a drawing session, thus ensuring consistency and accuracy in your designs.

#### 13.7.3.7 Bills of Material

Because CADWorx components placed in the design are intelligent, you can create accurate user-configurable bills of material in the most popular database formats at any point in the design process.

#### 13.7.3.8 Export to External Database

CADWorx Plant allows you to export design and component information to an external database in a variety of database formats for easy reporting and global model updates.

#### 13.7.3.9 Bidirectional Links to Stress Analysis

You can output any part of a CADWorx Plant piping system to stress analysis in native CAESAR II file format. Routing, pipe support and system modifications made in CAESAR II can be read directly into the design by CADWorx Plant. CADWorx Plant's Bidirectional link with CAESAR II reduces rework and ensures that designers and engineers work with the latest and most accurate information available.

#### 13.7.3.10 Automated Stress Isometrics

CADWorx Plant can use CAESAR II output to automatically create stress isometrics. You can select the analysis results that you want to see on the stress isometrics.

The modules and capabilities that follow are included with CADWorx Plant Professional.

#### 13.7.3.11 Live Database Links

CADWorx Plant Professional lets any number of users connect to a common external database during the design. Model additions and deletions from any designer are constantly monitored to ensure that the model and external project databases are always in sync.

#### 13.7.3.12 CADWorx Equipment

COADE's parametric equipment modeler includes all the features needed to easily model 3D equipment for plant design. The package lets you quickly and effortlessly build vessels, exchangers, tanks, pumps and other components.

#### 13.7.3.13 Personal ISOGEN

Alias' full-featured ISOGEN package provides complete ISOGEN functionality to produce fully automatic isometrics in the world's most popular isometric format.

#### 13.7.3.14 Model Review and Walkthrough

NavisWorks Roamer allows real-time design review of even the largest 3D models. Intuitive tools allow the navigation, collaboration, presentation and coordination of your 3D models, making it easy to create movies and to review comments with other CADWorx Plant users.

(http://www.softscout.com/software/Engineering/Plant-Layout-and-Design/CADWorx-Plant-and-Plant-Profesional-2005.html).

### 13.7.4 Flow Planner

The Flow Planner takes the work out of diagramming material flow through manufacturing facilities and calculating the distance, time and cost of these moves. Working within AutoCAD, the Flow Planner automatically generates material flow diagrams and calculates material handling travel distances, time and cost. With variable-width flow lines color-coded by product, part or material handling method, users quickly see how layouts should be arranged and where excessive material handling should be eliminated from the manufacturing process (www.proplanner.com).

### 13.7.5 FactoryFLOW

For manufacturers worldwide, more efficient factory layouts directly result in reduced material handling, improved product quality, and increased profits. Tecnomatix FactoryFlow's layout evaluation tools model the effect of layout changes before undertaking the risk and expense of physically reworking inefficient layouts. And optimized factory designs bring factories online faster, compress time to launch, and improve production efficiency. Customer testimonials show that users have recovered their investment in software and training in the first year, and often in the first study.

FactoryFLOW is a graphical material handling system that enables engineers to optimize layouts based on material flow distances, frequency, and costs. Factory layouts are analyzed by using part routing information, material storage needs, material handling equipment specifications, and part packaging (containerization) information.

#### 13.7.5.1 The Shortest Distance Between Two Points

FactoryFLOW uses aisle network information to find the shortest distance between any two points to identify the closest incoming dock and storage area to a part's point of use. Material flow studies are performed on alternate layout configurations and automatically compared to determine which layout is better. FactoryFLOW can also be used to compute material handling equipment requirements and optimized tugger (milkrun) routes. Users can also use the available container information to autopopulate containers and bins on storage areas and racks in order to create operator walk paths. Factory layout information is stored in a FactoryFLOW database. FactoryFLOW uses this information to help engineers develop layouts that facilitate the manufacturing process. FactoryFLOW generates Euclidean (point-to-point) material flow diagrams, actual path flow diagrams, aisle congestion diagrams, and quantitative reports so engineers can compare layout options and improve production efficiency.

#### 13.7.5.2 The Competitive Advantage

A typical factory layout or engineering effort includes layout considerations and capacity, utilization, throughput, and resource constraint analysis. FactoryFLOW can stand alone in situations where the layout is the focus of the project. In situations where there are capacity or process issues, FactoryFLOW adds significant value to the simulation effort and improves the quality of the overall engineering work. Here are the key benefits you can expect from FactoryFLOW:

- Create initial layouts easily
- Improve layout productivity by determining the best location of machines and departments
- Reduce material handling needs and storage requirements
- Design work cell layouts on the process plan
- Optimize layouts based on qualitative factors such as noise, dirt and supervision needs
- Diagram material-flow intensity
- Calculate material handling costs and requirements
- Flow charts: The flow chart feature allows you to develop material routings using standard process symbols. You can elect multiple activity points and move arrows in the flowchart for mass routing change. Also, there is a capability to cut, copy, and paste multiple activity points for rapid editing of the material routing file.
- Data templates and equations: FactoryFLOW provides data templates that contain standard information to enable you to compute and track micro-activities such as the amount of time spent on cutting open cardboard boxes or walking.
- Material flow calculations: FactoryFLOW checks the data to verify that the proper devices are being used, and notifies you when material handling devices are under- or overutilized, so that you can track the use of your operating assets.
- Material handling equipment utilization: FactoryFLOW provides tools to assess the requirements for material handling equipment such as fork-lifts and tuggers. The analysis can create a variety of reports including the type of equipment, number of trips by route and material, and the level of utilization. This information is a key to understanding where savings could be made in equipment requirements by adjusting aspects of the factory layout.

- Container packing: The container placement routines automatically place containers on the shop floor as well as on racks, using an optimum container packing routine.
- Activity points: Activity points allow FactoryFLOW to determine exact work center locations when material flow diagrams are created.
- Walk Path generation: FactoryFLOW also has intelligent walk path creation algorithms that allow you to see the effect of material placement in a work cell almost immediately.
- Reports: Besides color-coded flow diagrams and graphs, FactoryFLOW allows you to create many types of detailed reports on the layout, material flow, time, and cost saving comparisons.

#### 13.7.5.3 FactoryCAD

Tecnomatix FactoryCAD is a factory layout application that gives you everything you need to create detailed, intelligent factory models. Instead of having to draw lines, arcs, and circles, FactoryCAD allows you to work with "smart objects" that represent virtually all the resources used in a factory, from floor and overhead conveyors, mezzanines and cranes to material handling containers and operators. With these objects, you can "snap" together a layout model without wasting time drawing the equipment.

Because FactoryCAD makes layout creation, modification, and visualization easier and faster, design flaws and issues can be identified and eliminated earlier in the design process, before physically building or modifying the factory. Re-using the layout data in other related applications saves time, allows more design iterations to be assessed, and makes the layout information more valuable. Overall, getting factories into production sooner with fewer last minute modifications provides significant financial benefits. Here are the key benefits you can expect from FactoryCAD:

- Reduced interpretation errors
- Layout design problems discovered very early
- Expensive redesign problems avoided
- Usable with other analysis packages
- 90% faster than traditional 3D modeling
- Up to 95% reduction in file sizes
- Able to create 2D/3D models in less time and effort
- Data re-use makes the layout information more valuable

FactoryCAD enhances AutoCAD and the AutoDesk Architectural Desktop product to deliver a complete factory design solution by providing a library of smart objects that represent factory equipment and resources. Each object has both 2D and 3D views and incorporates key performance factors. This data, with the layout parameters, can then be extracted from the FactoryCAD layout for input to production simulation tools through FactoryCAD's simulation data exchange (SDX) format. Similarly, cost factors can be extracted from the FactoryCAD layout for estimating purposes. For equipment that is not already represented in FactoryCAD's library, the object builder enables users to create their own lightweight parametric object models. FactoryCAD requires AutoCAD or AutoDesk Architectural Desktop.

With FactoryCAD, 3D models can be created faster than 2D drawings with conventional CAD. With smart object technology, stored file sizes are smaller than 2D drawing files, thus avoiding the data size and performance problems normally associated with modeling complete factories. FactoryCAD enables engineers to create full 3D models that provide much more information than 2D drawings, helping them discover potential layout problems early in the design process. Because these layout models can be leveraged directly in visualization, material flow, and discrete event simulation programs, they offer considerable time savings.

#### 13.7.5.4 Major Capabilities

- View 3D models with non-CAD viewers: Technology has been embedded into the smart objects allowing simple viewers such as Factory View, a Tecnomatix product; Vis Mockup, a Teamcenter™; product; Volo View, VIZ 4 (3D StudioVis), and many others to view the factory models.
- Conveyor objects: All types of conveyors are available for use from package conveyors such as belt conveyors, V-belt-driven live roller conveyors, gravity roller conveyors, structure track conveyors, and palletized assembly conveyors to highly sophisticated conveyors such as automotive floor conveyors, cross transfers, lift systems, chain-on-edge, and power and free conveyors.
- Robot objects: Many highly detailed models of popular robots such as ABB, Fanuc, Kuka, and Kawasaki are available as objects. With built-in forward kinematics, you can articulate these robots in any position desired.
- Additional material handling objects: You will find a comprehensive list of material handling devices ranging from bridge cranes and jib cranes to lift tables and turn tables for containers.
- Create objects with object builder: With the object builder toolkit, you can build custom 3D factory equipment objects of your own. These objects then can be modified on the fly similar to the smart factory objects found in Factory CAD.
- Share objects with object enabler: The object enabler toolkit, which can be passed to and shared with non-factory CAD users, enables FactoryCAD models and drawings to be viewed in other AutoDesk programs.
- SDX-enabled: All objects within FactoryCAD have SDX parameters (e.g., cycle time, scrap rate, load time, unload time, breakdowns, setups, etc.) built into objects. FactoryCAD also has the ability to read an SDX file to update its objects. A number of major discrete event simulation tools can read SDX data.
- Block and symbol management tools: In addition to smart factory objects, FactoryCAD includes hundreds of traditional symbols and blocks. FactoryCAD enables you to move freely in the libraries to select, add, copy, move, and delete blocks.
- Clearance Detection: As the Factory is being designed, you can constantly use Factory CAD's clearance detection features to prove your designs early in the design process.

- BOM generation: You can also generate intelligent bill of materials (BOM) reports of equipment in your plant. This is especially important for complex, multiple-segmented equipment such as cable trays and fences.
- CAD data import: Tooling and product CAD data from NX, Parasolid, VRML, or JT formats can be imported as a smart factory object.

(Source: ttp://www.plm.automation.siemens.com/en_us/products/tecnomatix/plant_design/factorycad/index.shtml)

## 13.8 SUMMARY

This chapter has presented an overview of the application of analytically based computerized procedures to the plant layout problem. The emphasis in this chapter has been on the conceptual bases of these procedures rather than on how to use the computer to run them. Some methods have been presented in more detail than others, owing to the ease of understanding them and because of their pioneering nature. The capabilities of the available procedures, as well as their similarities and differences, have been discussed in depth in the evaluation of procedures, particularly construction ones, in an attempt to show their hidden limitations.

Despite these limitations and some criticisms that are directed to the approach, its popularity is increasing because of the development of new and more powerful procedures and the increased use of computers. An important point to notice is that the layouts produced by these procedures serve as good starting points for planners, and their use is sure to grow.

Current advances in computer-aided plant layout have achieved one or more of the following enhancements in the procedures:

- Development of detailed layouts
- Use of the graphic and interactive capabilities of the computers
- Development of procedures capable of generating multifloor layouts
- An increase in using graph theory
- More powerful exact mathematical programming procedures
- Development of procedures capable of designing layouts for the group technology method of manufacturing
- Developing of procedures suitable for generating flexible layouts; that is, ones that account for future changes in production volumes or process sequence

## PROBLEMS

1. What is a computer-aided plant layout and what are its advantages over traditional approaches?
2. Differentiate between construction and improvement programs.
3. What is the relationship of computer-aided plant layout to systematic layout planning?
4. Should we implement the layout produced by the computer as it is? Justify your answer.

5. How do improvement programs agree with and differ from one another?
6. What should be the characteristics of a good plant layout program?
7. Do we need any one of the standard computer programs to find the optimal layout for the following examples?
   a. Given five locations and the following data:

| | A | B | C | D | E |
|---|---|---|---|---|---|
| A | — | 20 | 20 | 20 | 20 |
| B | | — | 20 | 20 | 20 |
| C | | | — | 20 | 20 |
| D | | | | — | 20 |
| E | | | | | — |

| Department | Area |
|---|---|
| A | 5 |
| B | 5 |
| C | 5 |
| D | 5 |
| E | 5 |

   b. Given five locations and the following data:

| | A | B | C | D | E |
|---|---|---|---|---|---|
| A | — | 25 | | | 5 |
| B | | — | 30 | | |
| C | | | — | 20 | 10 |
| D | | | | — | 5 |
| E | | | | | — |

8. For the data in the accompanying tables, run PLANET using scales of one unit square equals 10, 20, 50, and 100 square feet. Are there any changes in the resulting layouts or the objective function values?

| Department | Area (square feet) |
|---|---|
| A | 100 |
| B | 600 |
| C | 200 |
| D | 400 |
| E | 500 |
| F | 300 |

| | A | B | C | D | E | F |
|---|---|---|---|---|---|---|
| A | — | 35 | 19 | 48 | 10 | 25 |
| B | | — | 7 | 53 | 40 | 15 |
| C | | | — | 34 | 23 | 19 |
| D | | | | — | 7 | 9 |
| E | | | | | — | 16 |
| F | | | | | | — |

9. For the layout produced by PLANET in problem 8, run CRAFT. Is the improvement significant? Justify your answer.
10. For the data in the accompanying tables, run ALDEP using scales of 2, 5, and 10 and observe the changes in the resulting layout and objective function value.

| Department | Area (square feet) |
|---|---|
| A | 100 |
| B | 600 |
| C | 200 |
| D | 400 |
| E | 500 |
| F | 300 |

| | A | B | C | D | E | F |
|---|---|---|---|---|---|---|
| A | — | | | | | |
| B | A | — | | | | |
| C | I | E | — | | | |
| D | U | U | A | | | |
| E | X | O | I | | | |
| F | U | U | U | U | U | — |

## SUGGESTED READINGS

Apple, J.M., and Deisenroth, M.P., "A Computerized Plant Layout Analysis and Evaluation Technique (PLANET)," *AIIE Technical Papers*, Twenty-Third Conference, Anaheim, 1972.

Armour, G.C., and Buffa, E.S., "A heuristic algorithm and simulation approach to relative location of facilities," *Management Science*, 9(2), 215–219, 1963.

Bazaraa, M.S., "Computerized layout design: a branch and bound approach," *AIIE Transactions*, 7(4), 432–437, 1975.

Block, T.E., "On the complexity of facilities layout problems," *Management Science*, 25, 280–285, 1979.

BLOCPLAN, Moore Productivity Software, Blacksburg, 1986.

Brennan, B.M., "Computerized facilities planning, design, and management options for today's offices," *Industrial Engineering*, 17(5), 70–74, 1985.

Buffa, E.S., "Sequence analysis for functional layouts," *Journal of Industrial Engineering*, 6(2), 12-13, 25, 1955.

Buffa, E.S., Armour, G.C., and Vollmann, T.E., "Allocating facilities with CRAFT," *Harvard Business Review*, 42(2), 136–158, 1964.

Carrie, A.S., "Computer-aided layout planning — the way ahead," *International Journal of Production Research*, 18(3), 283–294, 1980.

Carrie, A.S., Moore, J.M., Roczniak, M., and Seppanen, J.J., "Graph theory and computer-aided facilities design," *OMEGA, The International Journal of Management Science*, 6(4), 353–361, 1978.

Dutta, K.N., and Sahu, S., "A multigoal heuristic for facilities design problems: MUGHAL," *International Journal of Production Research*, 20(2), 147–154, 1982.

Edwards, H.K., Gilbert, B.E., and Hale, M.E., "Modular Allocation Technique (MAT)," *Management Science*, 17(3), 161–169, 1970.

El-Rayah, T.E., and Hollier, R.H., "A review of plant design techniques," *International Journal of Production Research*, 8(3), 263–279, 1970.

Filley, R.D., "The emerging computer technologies boost value of, respect for facilities function," *Industrial Engineering*, 17(5), 27–29, 1985.

Foulds, L.R., "Techniques for facilities layout: deciding which pairs of activities should be adjacent," *Management Science*, 29(12), 1414–1426, 1983.

Foulds, L.R., and Griffin, J.W., "A graph theoretic heuristic for multi-floor building layout," *Proceedings of HE Spring Conference*, Chicago, IL, 1984.

Foulds, L.R., and Robinson, D.F., "A strategy for solving the plant layout problem," *Operational Research Quart*erly, 27(4), 845–855, 1976.

Foulds, L.R., and Robinson, D.F., "Graph theoretic heuristics for the plant layout problem," *International Journal of Production Research*, 16(1), 27–37, 1978.

Francis, R.L., and White, J.A., *Facility Layout and Location — An Analytical Approach*, 2nd edition, Prentice Hall, Englewood Cliffs, 1991.

Hassan, M.D., "A computerized model for the Selection of Materials Handling Equipment and Area Placement Evaluation (SHAPE)," unpublished Ph.D. dissertation, Texas A&M University, College Station, 1983.

Hassan, M.D., and Hogg, G., "Review of graph theory applications to the facilities layout problem," *OMEGA, The International Journal of Management Science.*, 15(4), 219–300, 1987.

Heider, C.H., "An *n*-step, 2-variable search algorithm for the component placement problem," *Naval Research in Logistics Quarterly*, 20(4), 699–724, 1973.

Hicks, P.E., and Cowan, T.E., "CRAFT-M for layout rearrangement," *Industrial Engineering*, 8(5), 30–35,1976.

Hillier, F.S., "Quantitative tools for plant layout analysis," *Journal of Industrial Engineering*, 14(1), 33–40, 1963.

Hillier, F.S., and Connors, M.M., "Quadratic assignment problem algorithms and the location of indivisible facilities," *Management Science*, 13(1), 42–57,1966.

Johnson, R.V., "SPACECRAFT for multi-floor layout planning," *Management Science*, 28(4), 407–417, 1982.

Khalil, T.M., "Facilities Relative Allocation Technique (FRAT)," *International Journal of Production Research*, 2(2), 174–183, 1973.

Krejcirik, M., "Computer-aided plant layout," *Computer-Aided Design*, 2(1), 7–19, 1969.

Lee, R.C., and Moore, J.M., "CORELAP — Computerized Relationship Layout Planning," *Journal of Industrial Engineering*, 18(3), 195–200, 1967.

Lewis, W.P., and Block, T.E., "On the application of computer aids to plant layout," *International Journal of Production Research*, 18(1), 11–20, 1980.

Moore, J.M., "Computer-aided facilities design: an international survey," *International Journal of Production Research*, 12(1), 21–40, 1974.
Moore, J.M., "The zone of compromise for evaluating layout arrangements," *International Journal of Production Research*, 18(1), 1–10, 1980.
Muther, R., *Systematic Layout Planning*, Cahners, Boston, 1973.
Muther, R., and McPherson, K., "Four approaches to computerized layout planning," *Industrial Engineering*, 2(2), 39–42, 1970.
Nugent, C.E., Vollman, T.E., and Ruml, J., "An experimental comparison of techniques for the assignment of facilities to locations," *Operations Research*, 16(1), 150–173, 1968.
Pierce, J.F., and Crowston, W.B., "Tree-search algorithm for quadratic assignment problems," *Naval Research in Logistics* Quarterly, 18(1), 1–36, 1971.
*Plant Layout*, Industrial Engineering and Management Press, Institute of Industrial Engineers, Norcross, GA, 1986.
Rosenblatt, M.J., "The facilities layout problem: a multi-goal approach," *International Journal of Production Research*, 17(4), 323–332, 1979.
Seehof, J.M., and Evans, W.O., "Automated layout design program," *Journal of Industrial Engineering*, 18(12), 690–695, 1967.
Seppanen, J., and Moore, J.M., "Facilities planning with graph theory," *Management Science*, 17(4), B-242–B-253, 1970.
Seppanen, J., and Moore, J.M., "String processing algorithms for plant layout problems," *International Journal of Production Research*, 13(3), 239–254, 1975.
Shore, R.H., and Tompkins, J.A., "Flexible facilities design," *AIIE Transactions*, 12(2), 200–205, 1980.
Tompkins, J.A., *Facilities Design*, North Carolina State University, Raleigh, NC, 1975.
Tompkins, J.A., "Safety and facilities design," *Industrial Engineering*, 8(1), 38–42, 1976.
Tompkins, J.A., and Reed, R., Jr., "Computerized facilities design," *AIIE Technical Papers*, Twenty-Fifth Conference, Chicago, IL, 1973.
Trybus, T.W., and Hopkins, L.D., "Humans vs. computer algorithms for the plant layout problem," *Management Science*, 26(6), 570–574, 1980.
Vollmann, T.E., and Buffa, E.S., "The facilities layout problem in perspective," *Management Science*, 13(10), B-450–B-468, 1966.
Vollmann, T.E., Nugent, C.E., and Zartler, R.L., "A computerized model for office layout," *Journal of Industrial Engineering*, 19(7), 321–327, 1968.
Zoller, K., and Adendorff, K., "Layout planning by computer simulation," *AIIE Transactions*, 4(2), 116–125,1972.

# 14 Simultaneous Development of Plant Layout and Material Handling*

A number of researchers have suggested that development of both plant layout and the associated selection of material-handling (MH) equipment should be evaluated simultaneously. Only then can we obtain a truly optimum combination. The optimum plant layout is highly dependent on the equipment available for MH. The distances between departments as well as modes and frequencies of MH equipment used contribute to the total cost. In general, a layout should have smooth and efficient material flow with the least MH cost. In most cases, this cost is not necessarily directly proportional to the distances between the departments. For example, the fixed cost of the equipment is not linear, because the addition of one forklift truck adds a fixed sum to the total capital expenditure. Furthermore, no two departments can be presumed independent of all other departments within a plant. For instance, placing departments A and B next to each other may prevent other departments from being in proximity to them. Thus, such placement may reduce the MH cost between departments A and B but may increase the same cost between other departments, causing an overall increase. In addition, a change in the distance between two departments may also cause a shift in the mode of transportation (conveyor to forklift truck or vice versa, conveyor of different length, or forklift truck of different capacity), modifying the cost.

In this chapter, we present two algorithms as an illustration of the approach to build both material handling and plant layout simultaneously. The first algorithm optimizes the assignment of MH equipment to a given plant layout. The second uses the first algorithm in an iterative fashion to develop a plant layout that would result in a minimum cost. A computer program, COMLAD3, written by Reza Ziai, performs both of these operations and can be found at http://www.crcpress.com/e_products/downloads/download.asp?cat_no=44222. In describing the procedure, we often refer to this program to illustrate the assumptions involved in building the program. However, the assumptions are flexible, because they influence only the calculations and not the procedure. Therefore, the same procedure could be applied even if some of the assumptions were not valid, except that the calculations would have to be done either by hand or with a computer program written by the reader. For the clarity of presentation, the algorithm for selecting MH equipment is first presented and

* The work presented in this chapter is mainly from Reza Ziai (1991).

illustrated with an example. Following that, the iterative process of improving plant layout and developing the associated optimum MH is explained.

## 14.1 AN ALGORITHM FOR MH EQUIPMENT SELECTION

The algorithm is developed in two phases. The initial solution phase consists of developing independent MH systems using suitable conveyors only and another MH system using is developed appropriate trucks only. The second phase consists of making judicial selection of either a conveyor of a specified type or a truck of a given capacity for each route. The procedure is based on assumptions:

1. Only one conveyor system is considered practical between any two departments — that is, on a particular route.
2. Different types of conveyors, such as belt and chain, with different widths are available.
3. Only an integral number of trucks are permitted for the entire plant.
4. Different-capacity trucks may be available, with different first costs and operating costs.
5. The truck type with the highest capacity can take care of the MH requirement of any route. For purposes of discussion, we have assumed that there are four types of trucks available; type 4 is the largest, so it can handle all MH requirements.

The assumptions within the COMLAD3 computer program are:

1. A constant speed ($S$) of 60 feet/minute is used for the conveyors based on industry practice.
2. Efficiencies of 85% for conveyors and 70% for trucks are assumed based on industry practice. Therefore, the operating time for the conveyor is $480 \times 0.85 = 408$ minutes per day and that of the trucks is $480 \times 0.7 = 336$ minutes per day.
3. The larger of the two dimensions of the item (i.e., length) is used to determine the belt width of the conveyor. Thus, the width of the item is used to determine the number of items that can be carried per given length of the conveyor. For widths greater than 1 foot, a clearance of 1 foot is used between two successive items, but for widths less than 1 foot, a clearance of 6 inches is sufficient.
4. Two types of conveyors are considered in the analysis: (a) belt-driven and (b) chain-driven. Based on industry practice, if the weight per foot is more than 100 pounds or the total weight per 100 feet of conveyor is greater than 3200 pounds, then a chain-driven conveyor is recommended. If an item's weight exceeds either of the two set values, conveyors are infeasible and trucks are used instead. As far as the analyst is concerned, only price differentiates the two types of conveyors. Chain-driven conveyors are more expensive than the belt-driven type because of the material and operating requirements, such as a larger horsepower motor.
5. Four types of trucks are available for MH:
   Type 1: Light duty, can lift up to 1000 pounds
   Type 2: Light/medium, can lift up to 2000 pounds

**TABLE 14.1**
**Conveyor Cost per Foot Used in COMLAD3**

| | Width (inches) | | | | | | | | | | | |
|---|---|---|---|---|---|---|---|---|---|---|---|---|
| | 6 | 8 | 10 | 12 | ... | 20 | 22 | 24 | ... | 32 | 34 | 36 |
| Belt | $39 | $42 | $45 | $48 | ... | $60 | $63 | $66 | ... | $78 | $81 | $84 |
| Chain | $54 | $58 | $62 | $66 | ... | $82 | $86 | $90 | ... | $106 | $110 | $114 |

Type 3: Medium, can lift up to 3000 pounds
Type 4: Heavy duty, can lift more than 3000 pounds

All the calculations are done with the assumption that truck type 4 can perform all MH; therefore, the data for type 4 must be real. If the available number of types is less than four, a large initial cost should be assigned to the lower-type trucks. For example, if there are only two types of truck available, type 1 and type 2 should be assigned a high initial cost, but the actual costs should be assigned to type 3 and type 4.

Some initial data must be collected on both conveyors and trucks. They include:

1. Determine the cost per foot for different widths for both a belt conveyor and a chain conveyor. COMLAD3 uses the cost data in Table 14.1. If the actual costs are different, the cost of any specific conveyor can be changed by the analyst during the execution of the program as a response to a question.
2. For trucks, perform the following calculations: Determine the daily fixed cost (capitalized cost) and operating cost for each truck type. The operating cost is calculated in units of dollars per foot to later account for different costs on different routes based on the distances between the departments.

The crux of the algorithm is to determine on which paths the conveyor system is efficient and on which paths the truck system is efficient. Furthermore, the capacities of trucks needed on different routes must be determined. For simplicity of illustration, we restrict the final solution to a maximum of two types of trucks.

A number of alternatives can be developed as solutions to the problem. For example, we may have a belt or chain conveyor with different widths on some routes and trucks with different capacities on other routes. Thus, the number of combinations can be very large.

The problem is further complicated because trucks with excess capacities may be used on some routes. The greater cost of this can be offset by sharing the trucks and thus eliminating conveyors and/or reducing the use of lower-capacity trucks on other routes. Therefore, an overall integrated approach is required; the following procedure incorporates such thinking.

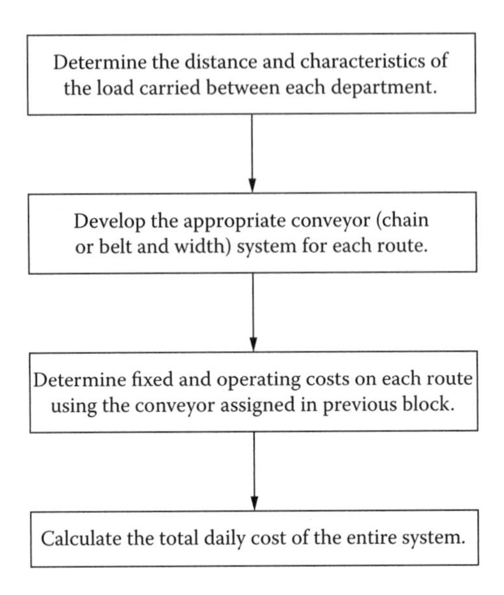


**FIGURE 14.1** Conveyor system

### 14.1.1 Phase I: Construction of an Initial Feasible Solution

#### 14.1.1.1 Calculating the Rectilinear Distances between the Departments

For the computer program supplied, this calculation may be done in one of two ways. In the layout generated by COMLADII (COMLAD3 includes COMLADII), these distances may be either from designated blocks within the two departments or from the centroids of the departments.

#### 14.1.1.2 Determining a Conveyor System for the Entire MH Task

This system is found by checking the feasibility of the type of conveyor between departments based on the load characteristics. The least expensive feasible alternative should be selected. Based on the conveyor assignments, calculate the total daily costs of the conveyor for the entire system. This cost is the sum of the daily operating and fixed costs of each conveyor between two departments. These steps are illustrated in Figure 14.1.

#### 14.1.1.3 Designing a Truck System for the Entire MH Task

Designing a truck system is an involved process requiring the following steps (Figure 14.2):

1. Determine the percentage of utilization, $\beta_i$, of each truck of type $i$ for each route. This value is achieved by finding the ratio of the required daily flow for each item to the daily capacity of each feasible truck type for that item (capacity required/capacity available). If more than one item is to be transferred by the truck on a given route, $\beta_i$ values for each item should be computed and

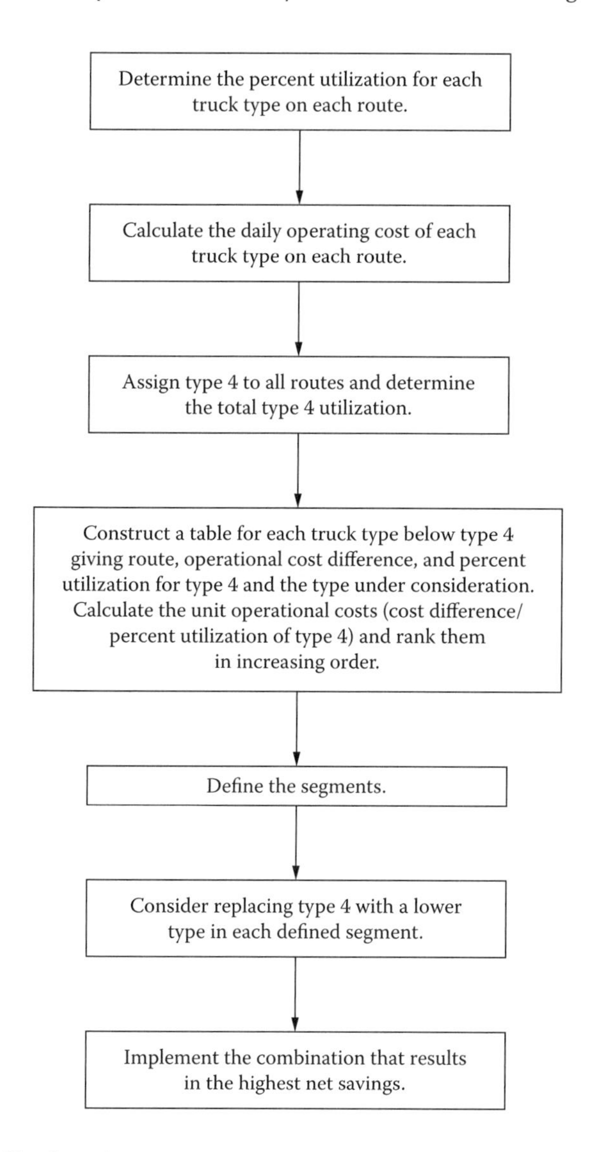


**FIGURE 14.2** Truck routine

summed up for that route. To calculate the percentage of utilization for each truck type, the analyst should know (default options for the computer program are specified in parentheses) the efficiency of the truck (0.70), pallet length and width (72 × 48 inches), number of pallet stacks or height (preset height of 24 inches), loading or unloading time (3 minutes), and average speed of the truck (5 miles/hour, or 440 feet/minute [fpm]). For an 8-hour operation, the time the truck is available per day is equal to 8 × 60 × efficiency of the truck.

Knowing the distance between two departments, the necessary time to travel in both directions is determined. If the loading and unloading times per trip are added to this, the value gives the total time necessary for one trip on the route.

If the size of an item is too large for a selected pallet size, the next larger-sized pallet should be used to accommodate the item. An interlocking-pattern pallet-loading program may be adopted (Kulick, 1982) to place the maximum number of items on a pallet. Thus, for each item, the number of pieces per pallet load and total weight of the pallet load (pounds) can be determined. Knowing the total number of pieces that must be carried between the two departments, we can establish the number of necessary trips. The percentage of utilization of a given type of truck on a route is then equal to the number of trips required times the time per trip divided by time the truck is available per day. If the weight of the pallet exceeds the load capacity of that particular truck type, the next feasible truck with higher capacity should be used.

2. Calculate the daily operating cost of use of each truck type on each route. This cost for a specific route is equal to the operating cost of truck type $i$ per day divided by the efficiency of operation and multiplied by the percentage of utilization of the truck on that specific route.
3. Construct a table for each truck type lower than type 4, including the following information: the two departments, associated route number, operational cost difference (increase) between the truck of type 4 and the truck under consideration on this route, the percentage of utilization for the truck of type 4 when used on this route, the percentage of utilization of the truck under consideration on the same route, and the unit operational cost and ranking (as explained next).

   The unit operational cost is obtained by dividing the cost differences by the percentage of utilization of the type 4 truck ($\beta_4$) on the route. Each ratio represents a normalized cost, or the cost differential for one whole unit of a type 4 truck on that route. It is important to use the normalized values so the costs can be compared on the same basis — that is, 1 unit of type 4.

   The last column should consist of a ranking of the unit costs in ascending order. The route with rank 1 is the one with the least increase in operating cost if the proposed lower-type truck is to operate on that route.

   It should be noted that because the type 4 truck is capable of carrying all items, it is initially assumed to be doing all the MH tasks on all routes unless the analyzer forces some other type to be the required truck on some route(s). This may happen, for example, if a narrow-aisle truck is the only one that can be used on a route and the truck is of a type other than 4.

   The procedure compares type 4 with all the lower types, namely, 3, 2, and 1. It replaces type 4 by a type that has given the maximum net savings, if such savings are possible. Thus, the net savings, as explained in step 5, is used only to find the most economical number of units of a lower-type truck to replace type 4 trucks. For instance, if no net savings is obtained on any route for any of the lower types, type 4 will be the most economical. There is, therefore, no need to compare any of the lower types against one another. By the same reasoning, for example, when type 3 is found to be more economical than type 4 on one or more routes, it will inevitably be more economical than type 2 or type 1 because type 4 has already been compared against those two.

4. Calculate the total of type 4 utilizations by adding the percentage of utilization on all routes for type 4. The smallest integer greater than or equal to this value is the minimum number of trucks required if all the routes are to be serviced by type 4 trucks only.
5. Consider replacing type 4 trucks with the next lower type under analysis in defined segments of the total utilization. These segments are obtained by subtracting the sum of type 4 utilizations from the next lower integer. For instance, if the sum in step 4 is 4.5, then the first intended segment (remainder) will be (4.5 – 4.0 =) 0.5. This means that at least 0.5 units of type 4 will be replaced by one or more units of a lower type on a route or routes if there is a net savings. The next segments are 1.5, 2.5, 3.5, and 4.5. Check the $\beta_4$ of the route with the lowest unit cost (rank 1). If it is equal or greater than the segment, go to step 6; otherwise, continue adding the $\beta_4$ of the next higher-unit cost in the ordered list to equal or exceed the segment. This is called a *combination*. Go to step 6.
6. Perform the following cost calculations for the combination just obtained. Add the operating cost difference(s) corresponding to the routes identified in the combination. This sum represents the increase in operating cost if the type 4 trucks, whose number is represented by the smallest integer greater than or equal to the sum of the associated $\beta_4$, are to be replaced with the lower type.

   The total daily operating cost of a lower-type truck is likely to be greater than that of a type 4 because of its lower carrying capacity, which results in a greater number of trips. But some savings may be realized because one or more units of type 4 are replaced by a lower type, which has a smaller fixed cost. To find out how many units of type 4 are being replaced, subtract the smallest integer that is greater than or equal to the present value of the sum of $\beta_4$ values from the number of trucks of type 4 already in service. The number of units of the lower type $i$ under consideration is the smallest integer greater than or equal to the sum of the $\beta_4$ values. For instance, suppose we are comparing truck type 4 with truck type 3, and for one combination the sum of $\beta_4$ values still in use is 2.3; the number of type 4 trucks still in use is 3. Also, if the total of $\beta_3$ values on the routes under consideration is 2.7, 3 units of type 3 will be needed. The initial requirement of 4.5 type 4 trucks, when translated to an integer value, yields 5 trucks. Therefore, (5 – 3 =) 2 units of type 4 may be replaced by 3 units of type 3. If the daily fixed cost of 1 unit of type 4 is \$50 and that of type 3 is \$25, the savings will be 2 × \$50 – 3 × \$25 = \$25. If the sum of operating cost differences on the routes for which the combination was obtained is less than \$25, then type 3 trucks will replace type 4 trucks on these routes; if the sum of the operating cost differences is greater than or equal to \$25, then the assignment will not change.

7. Repeat steps 5 and 6 for all the remaining combination segments.
8. Repeat steps 5, 6, and 7 for all the lower-type trucks.
9. Implement the replacement for the combination that renders the highest net savings of all. This will be the MH system consisting of trucks only. To check for the possibility of any improvement by combining trucks and conveyors, go to phase II.

### 14.1.2 Phase II: Improvement Procedure

In the improvement phase, the overall cost is lowered and truck utilization is increased, if possible. This is achieved by replacing trucks with conveyors, where economical. However, such replacement may further suggest a different truck combination that will further modify the number and type of trucks, thus additionally lowering the total cost. The steps that follow explain the procedure (Figure 14.3).

1. Replace the recommended truck type from phase I with a conveyor on every route for which the daily total cost of the conveyor is lower than the truck's. If the cost of the conveyor and the operating cost of the truck are the same, choose the conveyor to avoid the possible fixed cost of the truck. To get the new MH cost, add the costs on all the routes and the $\beta_i$ values for each truck type $i$ present. The smallest integer equal to or greater than the sum of $\beta_i$ values gives the number of trucks of type $i$ that is necessary. Multiply each by the corresponding daily fixed cost of truck type $i$ and add all these fixed costs to the total of all the other costs. This is the overall cost of the MH system.

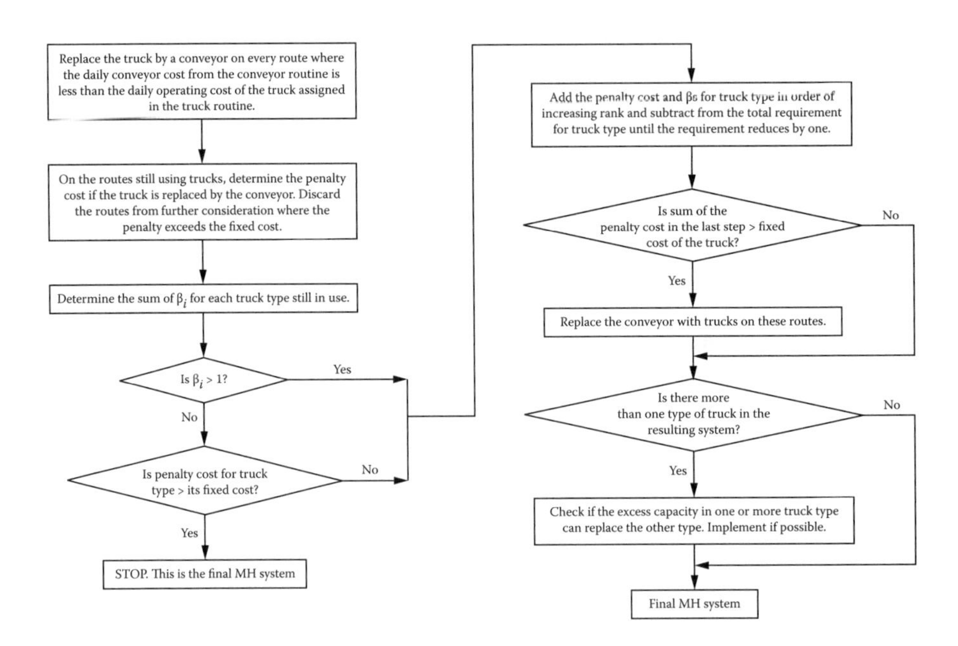


**FIGURE 14.3** Improvement procedure

It should be noted that converting the total requirement of $\beta_i$ values to a higher integer value at this step may create an excess capacity for the truck of type $i$. The following steps investigate if this excess capacity can be used to reduce MH cost even further.

2. Deduct the operating cost of the truck type assigned to a route (if any) from the proposed conveyor system (which was not selected due to its higher cost). Repeat this calculation for all the routes that have trucks assigned to them. The cost difference is a penalty that will be incurred if the conveyor replaces the truck. Discard any penalty that is equal to or greater than the daily fixed cost of truck type $i$ under consideration.
3. Add the $\beta_i$ values for each truck type in the present MH system. Go to step 4 for any truck type whose total $\beta$ is greater than 1. For the remaining truck types with total $\beta$ less than 1 (if any), check the penalty costs. If the sum of penalty costs is greater than the fixed cost of that truck type, discontinue any further analysis, because replacement of the truck would cost more than the associated fixed cost savings. It should be noted that the solution obtained in step 1 (or in step 6, which follows) will be the final one if the sum of $\beta$ values for all the truck types has been found to be less than 1 and no reduction in the cost is possible. Otherwise, continue with step 4.
4. Rank the penalty costs for each truck type in ascending order. Starting from the lowest rank, for a given truck type, subtract the corresponding $\beta_i$ from the total $\beta_i$ values to check whether it reduces the required number of type $i$ trucks by 1. If it does, go to step 5; otherwise, add the next $\beta_i$ values in the order, while adding their penalty costs until the required number of type $i$ trucks is reduced by 1. Go to step 5.
5. Replace the trucks on the routes under consideration with conveyors if the savings due to the reduction of truck units is more than the total penalty costs. The overall MH cost is reduced by this net savings. If no net savings is realized, the MH assignment found in step 1 will be the final one. Terminate phase II unless there is more than one type of truck in the system, in which case step 6 should be applied.
6. There are two ways of reducing the cost even further. Check whether the cost may be decreased by reducing the number of types of trucks or the number of truck units of a particular type. Because conveyors have replaced trucks on one or more routes, the capacity planning performed in phase I has been disturbed. Any new excess capacity on any one or more truck types may be enough to replace the other type.

Start with the highest type; subtract the sum of its $\beta$ values from the smallest integer value (representing the number of trucks needed) that is greater than the sum. The remainder is the excess capacity for that type of truck, which can be used to eliminate the lower-capacity trucks, if possible.

Add the $\beta$ values of the higher-capacity type under consideration on the routes where the lower type is presently assigned. If the latter sum equals its available excess capacity, replace the lower type on those routes by the higher type if a net saving is realized.

In addition, check the feasibility of introducing conveyors on the routes where the lower-type trucks are assigned. Start with the route that may have the lowest-cost conveyor. If inclusion of this or more conveyors while maintaining only the higher-type truck produces a savings, substitute the conveyor(s) on the appropriate route(s) (this step is illustrated in the following example). The result is the recommended and improved MH system for the given layout and input cost data.

## 14.2 ILLUSTRATIVE EXAMPLE 1

Two examples illustrate key elements of the algorithm.

### 14.2.1 Initial Data Preparation

The layout for a manufacturing facility that has six departments is developed, through which five items are processed. The items, along with their physical characteristics and production rates, are shown in Table 14.2. Based on the production sequence of the items, the interdepartmental transportation requirements are tabulated in Table 14.3. For example, items 3, 4, and 5 are transported between departments 1 and 3 using the procedure described in Chapter 12. The between-centroid distances are calculated (by COMLAD3) using square building blocks of 40 feet per side. The values are shown in Table 14.4.

Note that at this point the user may assign or impose in COMLAD3 either a conveyor or a truck between any two departments. This assignment overrides the system recommended by the algorithm. In other words, the algorithm considers only the *unassigned routes*. A higher cost will be incurred if the systems assigned by the user on the routes are not the same as those that would have been recommended by the algorithm.

Next, the type of conveyor, belt or chain, to use on each route is determined. In COMLAD3, the conveyor module is invoked. The user furnishes the maximum size of the available belt width (for this example, 44 inches), the maximum weight per foot allowed (here, 88 pounds), and the maximum weight per 100 feet of conveyor (here, 8888 pounds). Other system-related data are the efficiency of the conveyor (default = 0.85), the economical life (default = 5 years), the salvage value (default = $0), the annual compound interest rate (default = 12%), and the operating cost per hour (default = $2). All the default values are used in this example.

**TABLE 14.2**
**Item Characteristics and Daily Rates**

| Item Name | Length (inches) | Width (inches) | Width (pounds) | Required Rate/Day |
|---|---|---|---|---|
| 1 | 36 | 24 | 70 | 100 |
| 2 | 42 | 30 | 80 | 200 |
| 3 | 12 | 6 | 80 | 500 |
| 4 | 24 | 12 | 15 | 400 |
| 5 | 30 | 12 | 20 | 300 |

**TABLE 14.3**
**Items Transported Between Departments**

| Department | 1 | 2 | 3 | 4 | 5 | 6 |
|---|---|---|---|---|---|---|
| 1 | — | (1, 2) | (3, 4, 5) | (3) | (4) | (1, 2) |
| 2 | | — | (1) | (2) | (4, 5) | (4) |
| 3 | | | — | (3, 5) | (1) | (4) |
| 4 | | | | — | (5) | (2) |
| 5 | | | | | — | (1) |
| 6 | | | | | | — |

A belt-driven conveyor is recommended between departments 1 and 2. This recommendation is made because items 1 and 2, with corresponding weights of 70 and 80 pounds, are conveyed between these departments. This results in a maximum of 35 (70 pounds/2 feet) and 32.5 (80 pounds/2.5 feet) pounds per foot, respectively; with the clearance of 1 foot between each unit, the weight per 100-foot length is $100/(2+1) \times 70 = 2333.33$ pounds for item 1 and is $100/(2.5+1) \times 80 = 2285.7$ pounds for item 2. Neither item exceeds the loading limits set for the belt conveyor. On the other hand, for departments 1 and 4, a chain-driven conveyor is recommended because the weight per foot exceeds 100 pounds (80 pounds/0.5 foot = 160 pounds/foot). Table 14.5 shows the type of conveyor required for each pair of departments.

The standard conveyor cost table used in COMLAD3 is displayed in Table 14.1. For each 2-inch increment of the belt width, the belt-conveyor cost increases by $3/foot and the chain-conveyor cost increases by $4/foot.

The sum of the daily operating cost and the fixed cost per day for the conveyor system are displayed in Table 14.6. One cost calculation is illustrated next for the belt conveyor between departments 1 and 6.

The distance between these two departments is 60 feet, requiring a conveyor having a width of 42 inches (width of item 2), which has a default cost of $93 per foot (Table 14.1). Therefore, the initial fixed cost will be 60 feet × $93 = $5580. No salvage value has been assumed for this equipment. Hourly operating cost is $2.

**TABLE 14.4**
**Calculated Distances Between Departments**

| Department | 1 | 2 | 3 | 4 | 5 | 6 |
|---|---|---|---|---|---|---|
| 1 | — | 120 | 100 | 200 | 180 | 60 |
| 2 | | — | 220 | 120 | 140 | 140 |
| 3 | | | — | 140 | 280 | 120 |
| 4 | | | | — | 260 | 260 |
| 5 | | | | | — | 200 |
| 6 | | | | | | — |

**TABLE 14.5**
**Assigned Conveyor Type for Each Route**

| Department | 1 | 2 | 3 | 4 | 5 | 6 |
|---|---|---|---|---|---|---|
| 1 | — | Belt | Chain | Chain | Belt | Belt |
| 2 | | — | Belt | Belt | Belt | Belt |
| 3 | | | — | Chain | Belt | Belt |
| 4 | | | | — | Belt | Belt |
| 5 | | | | | — | Belt |
| 6 | | | | | | — |

The conveyor operates all day to avoid any interruption in production. There are 250 working days per year (5 days/week × 50 weeks/year). The capital recovery factor for 12% compounded annually is 0.2774. Thus, we have

$$\text{Salvage value} = 0\% \times \$5580 = \$0.00$$

$$\text{Net cost of capital/year} = (\$5580 - \$0) \times 0.2774 = \$1547.89$$

$$\text{Fixed cost/day} = (\$1547.89)/250 = \$6.19$$

$$\text{Operating cost/day} = \$2 \times 8\ \text{hr/day} = \$16.00$$

$$\text{Total cost/day on route } 5 = \$16 + \$6.19 = \$22.19, \text{ or } \$22$$

Note that a path between departments is designated as a route, for convenience. For example, departments 1 and 2 are connected by route 1. These routes are displayed in Table 14.7.

The next step is to determine which truck type to use on each route if all routes are to be serviced by trucks. COMLAD3 invokes the truck routine for this purpose. Once all the general information has been given, the program determines the feasibility and required daily utilization of each of the four types of trucks. Using this information, phase 1 of the heuristic algorithm assigns one of the four truck types to each route (between pairs of departments) and calculates the total number of trucks of each type necessary to achieve the minimum cost. Thus, there can be more than one type of truck in the solution.

**TABLE 14.6**
**Daily Conveyor Costs ($)**

| Department | 1 | 2 | 3 | 4 | 5 | 6 |
|---|---|---|---|---|---|---|
| 1 | — | 28 | 27 | 31 | 29 | 22 |
| 2 | | — | 37 | 28 | 28 | 26 |
| 3 | | | — | 32 | 42 | 25 |
| 4 | | | | — | 38 | 41 |
| 5 | | | | | — | 35 |
| 6 | | | | | | — |

**TABLE 14.7**
**Percentage of Utilization of the Four Truck Types on Each Route**

| Departments | Route No. | Type 4 | Type 3 | Type 2 | Type 1 |
|---|---|---|---|---|---|
| 1–2 | 1 | 49.01 | 92.06 | 122.55 | 220.59 |
| 1–3 | 2 | 28.33 | 0 | 0 | 0 |
| 1–4 | 3 | 3.83 | 0 | 0 | 0 |
| 1–5 | 4 | 12.25 | 18.38 | 32.68 | 73.53 |
| 1–6 | 5 | 49.01 | 92.06 | 122.55 | 220.59 |
| 2–3 | 6 | 12.25 | 18.38 | 24.51 | 73.53 |
| 2–4 | 7 | 36.76 | 73.53 | 98.04 | 147.06 |
| 2–5 | 8 | 24.50 | 40.44 | 69.44 | 147.06 |
| 2–6 | 9 | 12.25 | 18.38 | 32.68 | 73.53 |
| 3–4 | 10 | 16.08 | 0 | 0 | 0 |
| 3–5 | 11 | 12.25 | 18.38 | 24.51 | 73.53 |
| 3–6 | 12 | 12.25 | 18.38 | 24.51 | 73.53 |
| 4–5 | 13 | 12.25 | 22.06 | 36.76 | 73.53 |
| 4–6 | 14 | 36.76 | 73.53 | 98.04 | 147.06 |
| 5–6 | 15 | 12.25 | 18.38 | 24.51 | 73.53 |

To illustrate the calculations, consider item 1 transported between departments 1 and 2. The size of this item is 36 × 24 inches, using the default pallet size of 72 × 48 inches; for a type 4 truck, three items are placed on each pallet. Although four items may fit on this pallet size, the interlocking pattern used by the program permits only three in order to maintain a good structural support for the items on the pallet when they are stacked (remember, the units are not strapped together). We have decided to load four stacks because of the height consideration of the item. (The default number of stacks for truck type 4 is 5, and that for types 3, 2, and 1 is 4, 3, and 2, respectively. The analyzer can change these values during execution.) With four stacks, the total number of pieces for each load is 12 (3 × 4 = 12). Item 1 weighs 70 pounds, producing a total load of 840 pounds (12 × 70 = 840) to be carried. The number of trips per day may be determined by computing the average travel time and dividing it by the available daily time. The distance between departments 1 and 2 is 120 feet, and the default average speed is 440 fpm; therefore, the average travel time is 1 minute (i.e., next higher integer of 120 feet/440 fpm). This value is multiplied by 2 to account for the return trip. Adding the 3 minutes (default value) for loading and unloading per trip gives a total of 5 minutes (2 × 1 + 3 = 5). Using an efficiency of 70%, there will be 336 minutes available per day (0.70 × 480 = 336), resulting in 68 possible trips per day (the next greatest integer value for the ratio of 336 minutes per day to 5 minutes per trip). We noted earlier that truck type 4 is carrying 12 items per load (trip), which means that 816 items (12 × 68 = 816 per truck) may be transferred per day. One hundred units of item 1 must be carried (Table 14.2) between the two departments; because each truck type 4 is capable of carrying 816 units of item 1 per day (its total capacity), this truck will be occupied 12.25% of the time due only to

**TABLE 14.8**
**Data for Trucks**

| Truck Type | Fixed Cost | Operating Cost per Hour | Fixed Cost per Day (Calculated) |
|---|---|---|---|
| 1 | \$14,000 | \$12 | \$12.43 |
| 2 | \$15,000 | \$13 | \$13.32 |
| 3 | \$16,000 | \$14 | \$14.20 |
| 4 | \$26,000 | \$19 | \$23.08 |

this item (100/816 = 0.1225). The same analysis is performed for item 2, resulting in 36.76% utilization. Therefore, truck type 4 is being utilized by (12.25% + 36.76% = 49.01%) to satisfy the MH requirements between departments 1 and 2. This procedure is applied to other truck types and for all other departmental pairs. Table 14.7 shows the results of the calculations.

The zero entries in Table 14.7 indicate infeasibility. Here, truck types 1, 2, and 3 are not feasible on routes 2, 3, and 10 due to the heavy load of item 3. For example, when using the default pallet size and stack heights for loading, the total weight is greater than what is allowed for each truck type. Although the user may specify his or her own stack height to make a route feasible, in this example, we will continue using the default stack height options for truck types 1, 2, and 3.

The parameters in Table 14.8 are used to calculate the total daily costs for all four types of trucks on the 15 routes. COMLAD3 asks for the economical life of the equipment (default = 5 years), the initial cost of the truck (default = \$14,000), the salvage value (default = 20% of the initial value), the interest rate compounded annually (default = 12%), and the operating cost per hour (default = \$12). Fixed cost per hour is determined by calculations similar to the following:

For truck type 2,

$$\text{Salvage value} = 0.20 \times 15,000 = \$3000.00$$

$$\text{Net cost of capital/year} = (\$15,000 - \$3000) \times 0.2774 = \$3328.80$$

Assuming 250 working days per year,

$$\text{Fixed cost/day} = \frac{\$3328.80}{250} = \$13.32$$

The next step is to calculate the expected daily operating cost for each truck type on each route. The procedure for calculating this cost is illustrated for truck type 2 on route 15. Its hourly operating cost is \$13 (given in Table 14.8). Because the truck is used by 24.5% on route 15, its daily operating cost on route 15 is $\$13 \times 8 \times 0.245 =$ \$25.49, which is rounded off to \$25.

Table 14.9 depicts the *daily* fixed and operating costs for all four truck types.

**TABLE 14.9**
**Expected Daily Operating Costs of the Four Truck Types per Route**

| Departments | Route No. | Type 4 ($) | Type 3 ($) | Type 2 ($) | Type 1 ($) |
|---|---|---|---|---|---|
| 1–2 | 1 | 75 | 103 | 127 | 212 |
| 1–3 | 2 | 43 | 2800 | 2600 | 2400 |
| 1–4 | 3 | 6 | 2800 | 2600 | 2400 |
| 1–5 | 4 | 19 | 21 | 34 | 71 |
| 1–6 | 5 | 75 | 103 | 127 | 212 |
| 2–3 | 6 | 19 | 21 | 25 | 71 |
| 2–4 | 7 | 56 | 82 | 102 | 141 |
| 2–5 | 8 | 37 | 45 | 72 | 141 |
| 2–6 | 9 | 19 | 21 | 34 | 71 |
| 3–4 | 10 | 24 | 2800 | 2600 | 2400 |
| 3–5 | 11 | 19 | 21 | 25 | 71 |
| 3–6 | 12 | 19 | 21 | 34 | 71 |
| 4–5 | 13 | 19 | 25 | 38 | 71 |
| 4–6 | 14 | 56 | 82 | 102 | 141 |
| 5–6 | 15 | 19 | 21 | 25 | 71 |

### 14.2.2 Application of the Heuristic Procedure

All the basic data have now been collected, so we can start applying the heuristic procedure for MH equipment selection.

#### 14.2.2.1 Phase I: Construction of an Initial Feasible Solution

Table 14.9 shows the expected daily operating costs for all four truck types on the 15 routes. The zero entries of Table 14.7 are translated as a large operating cost entry to make this choice unattainable. In the program, this is done by assuming the value of the daily utilization ($\beta$) on those routes to be 25. The product of this number and the operating cost makes these trucks on the given routes uneconomical. This number (25) is not included in any further considerations.

Entries in column three of Table 14.10 are the differences between the operating costs of truck type 4 and those of truck type 3 (step 3 of phase I, designing a truck system). Operating cost differences between type 4 and both type 2 and type 1 are shown in Tables 14.11 and 14.12, respectively. The *X*'s indicate that the cost difference is not applicable to routes 2, 3, and 10, because truck types 3, 2, and 1 are not feasible for them, as noted earlier. In all three tables, column 6 contains the unit costs, which are obtained by dividing the entries in column 3 by their corresponding entries in column 4 $\beta_4$, step 2 of phase I). Unit costs are ranked in ascending order in column 7.

The next step is to develop the initial MH solution comprising only trucks. Because truck type 4 is feasible on any route where a truck is allowed, it is used as a reference for possible replacement. This replacement is done in several steps. At each step, one or more units of type 4 are considered to be replaced by the three

**TABLE 14.10**
**Differences Between Type 4 and Type 3 in Terms of Operating Costs, $\beta_4$, $\beta_3$, Unit Costs, and Their Corresponding Ranks**

| (1) Departments | (2) Route No. | (3) Operating Cost Difference ($) | (4) $\beta_4$ | (5) $\beta_3$ | (6) Unit Operating Cost ($) | (7) Rank |
|---|---|---|---|---|---|---|
| 1–2 | 1 | 28 | 0.4901 | 0.9206 | 57 | 10 |
| 1–3 | 2 | X | 0.2833 | 0.0 | 9999 | 14 |
| 1–4 | 3 | X | 0.0383 | 0.0 | 9999 | 15 |
| 1–5 | 4 | 2 | 0.1225 | 0.1838 | 16 | 1 |
| 1–6 | 5 | 28 | 0.4901 | 0.9206 | 57 | 9 |
| 2–3 | 6 | 2 | 0.1225 | 0.1838 | 16 | 2 |
| 2–4 | 7 | 26 | 0.3676 | 0.7353 | 71 | 12 |
| 2–5 | 8 | 8 | 0.2450 | 0.4044 | 33 | 7 |
| 2–6 | 9 | 2 | 0.1225 | 0.1838 | 16 | 3 |
| 3–4 | 10 | X | 0.1608 | 0.0 | 9999 | 13 |
| 3–5 | 11 | 2 | 0.1225 | 0.1838 | 16 | 4 |
| 3–6 | 12 | 2 | 0.1225 | 0.1838 | 16 | 5 |
| 4–5 | 13 | 6 | 0.1225 | 0.2206 | 49 | 8 |
| 4–6 | 14 | 26 | 0.3676 | 0.7353 | 71 | 11 |
| 5–6 | 15 | 2 | 0.1225 | 0.1838 | 16 | 6 |

**TABLE 14.11**
**Differences Between Type 4 and Type 2 in Terms of Operating Costs, $\beta_4$, $\beta_2$, Unit Costs, and Their Corresponding Ranks**

| (1) Departments | (2) Route No. | (3) Operating Cost Difference ($) | (4) $\beta_4$ | (5) $\beta_2$ | (6) Unit Operating Cost ($) | (7) Rank |
|---|---|---|---|---|---|---|
| 1–2 | 1 | 52 | 0.4901 | 1.2255 | 106 | 5 |
| 1–3 | 2 | X | 0.2883 | 0.0 | 9999 | 14 |
| 1–4 | 3 | X | 0.0383 | 0.0 | 9999 | 15 |
| 1–5 | 4 | 15 | 0.1225 | 0.3268 | 122 | 6 |
| 1–6 | 5 | 52 | 0.4901 | 1.2255 | 106 | 4 |
| 2–3 | 6 | 6 | 0.1225 | 0.2451 | 49 | 1 |
| 2–4 | 7 | 46 | 0.3676 | 0.9804 | 125 | 10 |
| 2–5 | 8 | 35 | 0.2450 | 0.6944 | 143 | 11 |
| 2–6 | 9 | 15 | 0.1225 | 0.3268 | 122 | 8 |
| 3–4 | 10 | X | 0.1608 | 0.0 | 9999 | 13 |
| 3–5 | 11 | 6 | 0.1225 | 0.2451 | 49 | 2 |
| 3–6 | 12 | 15 | 0.1225 | 0.2451 | 122 | 7 |
| 4–5 | 13 | 19 | 0.1225 | 0.3676 | 155 | 12 |
| 4–6 | 14 | 46 | 0.3676 | 0.9804 | 125 | 9 |
| 5–6 | 15 | 6 | 0.1225 | 0.2451 | 49 | 4 |

**TABLE 14.12**
**Differences Between Type 4 and Type 1 in Terms of Operating Costs, Unit Costs, $\beta_4$, $\beta_1$, and Their Corresponding Ranks**

| (1) Departments | (2) Route No. | (3) Operating Cost Difference ($) | (4) $\beta_4$ | (5) $\beta_1$ | (6) Unit Operating Cost ($) | (7) Rank |
|---|---|---|---|---|---|---|
| 1–2 | 1 | 137 | 0.4901 | 2.2059 | 279 | 3 |
| 1–3 | 2 | X | 0.2833 | 0.0 | 9999 | 14 |
| 1–4 | 3 | X | 0.0383 | 0.0 | 9999 | 15 |
| 1–5 | 4 | 52 | 0.1225 | 0.7353 | 424 | 5 |
| 1–6 | 5 | 137 | 0.4901 | 2.2059 | 279 | 4 |
| 2–3 | 6 | 52 | 0.1225 | 0.7353 | 424 | 11 |
| 2–4 | 7 | 85 | 0.3676 | 1.4706 | 231 | 1 |
| 2–5 | 8 | 104 | 0.2450 | 1.4706 | 424 | 6 |
| 2–6 | 9 | 52 | 0.1225 | 0.7353 | 424 | 7 |
| 3–4 | 10 | X | 0.1608 | 0.0 | 9999 | 13 |
| 3–5 | 11 | 52 | 0.1225 | 0.7353 | 424 | 8 |
| 3–6 | 12 | 52 | 0.1225 | 0.7353 | 424 | 9 |
| 4–5 | 13 | 52 | 0.1225 | 0.7353 | 424 | 10 |
| 4–6 | 14 | 85 | 0.3676 | 1.4706 | 231 | 2 |
| 5–6 | 15 | 52 | 0.1225 | 0.7353 | 424 | 12 |

lower-type trucks. The percentages of utilization of type 4 ($\beta_4$ values) for all the routes are summed (step 4 of process I) to start the procedure. The total is 3.30 for this example. Therefore, the first segment (__) of analysis is (3.30 – 3.00 =) 0.30.

Starting with type 3 and rank 1 in Table 14.10 (on route 4), the corresponding $\beta_4$ is examined (0.1225). Because it is not equal to the segment under consideration (i.e., 0.30), the $\beta_4$ for rank 2 (0.1225) is added to the previous one to get 0.2450. This sum is still less than 0.30, so the third rank's $\beta_4$ (0.1225) is added to the sum, resulting in a total of 0.3675. This sum is the first value larger than 0.30, which means that one unit of truck type 3 (the largest integer equal to 0.1838 + 0.1838 + 0.1838 = 0.5514) may replace one unit of type 4 (the largest integer equal to 0.3675) on routes 4, 6, and 9 *if* a net saving is realized. The following cost calculations are performed (step 5 of phase I) to determine the savings. The differences in operating costs on routes 4, 6, and 9 are summed to get a total of $6 ($2 + $2 + $2 = $6). Thus, an additional operating cost of $6 will be incurred in a day if type 3 is operated on these routes instead of type 4. Savings may occur through a reduction in fixed costs. One unit of type 4 is being replaced by one unit of type 3, so a daily savings of about $9 is realized due to the difference in their daily fixed costs ($23.08 – $14.20 = $8.88, rounded off to $9). This gives a daily net savings of $3 ($9 – $6 = $3), thus confirming the replacement of type 4 with type 3 for routes 4, 6, and 9. The calculations are summarized next.

| Route | Rank | Difference in Operating Costs ($) | $\beta_4$ | $\beta_3$ |
|---|---|---|---|---|
| 4 | 1 | 2 | 0.1225 | 0.1838 |
| 6 | 2 | 2 | 0.1225 | 0.1838 |
| 9 | 3 | 2 | 0.1225 | 0.1838 |
| Total | | 6 | 0.3675 | 0.5514 |

$$\text{Fixed cost difference (see Table 14.8)} = 1 \times \$23.08 - 1 \times 14.20$$

$$= \$8.88, \text{ or about } \$9$$

$$\text{Net saving} = \$9 - \$6 = \$3$$

These calculations are performed for the next segment (i.e., 1.30). Again, starting with rank 1, $\beta_4$ values are summed until they equal or exceed the present segment. Ranks 1 through 9, corresponding to routes 4, 6, 9, 11, 12, 15, 8, 13, and 5 (Table 14.10), have a total of 1.5926 for their $\beta$ values and operating cost increase. Replacement of 1.59, or 2, units of type 4 would require 2.648, or 3, units of type 3 trucks. The fixed cost saving between 2 units of type 4 and the 3 units that would be required of type 3 becomes $4, producing a net savings of -$50. This value is negative; therefore, it will cost $50 more to operate type 3 on the routes instead of type 4. The calculations are shown here:

| Route | Rank | Difference in Operating Costs ($) | $\beta_4$ | $\beta_3$ |
|---|---|---|---|---|
| 4 | 1 | 2 | 0.1225 | 0.1838 |
| 6 | 2 | 2 | 0.1225 | 0.1838 |
| 9 | 3 | 2 | 0.1225 | 0.1838 |
| 11 | 4 | 2 | 0.1225 | 0.1838 |
| 12 | 5 | 2 | 0.1225 | 0.1838 |
| 15 | 6 | 2 | 0.1225 | 0.1838 |
| 8 | 7 | 8 | 0.2450 | 0.4044 |
| 13 | 8 | 6 | 0.1225 | 0.2206 |
| 5 | 9 | 28 | 0.4901 | 0.9206 |
| Total | | 54 | 1.5926 | 2.6484 |

$$\text{Fixed cost difference} = 2 \times \$23.08 - 3 \times 14.20 = \$3.56, \text{ or about } \$4$$

$$\text{Net savings} = \$4 - \$54 = -\$50$$

The next segments (i.e., 2.30 and 3.30) are checked, but no savings are realized in both cases. Therefore, only the changes suggested by the first segment were economical.

Steps 4 and 5 are repeated for truck types 2 and 1. Because no savings are realized, the solution suggested by truck types 3 and 4 forms the initial solution. Table 14.13 shows the initial feasible solution with corresponding costs.

**TABLE 14.13**
**Initial Feasible Solution**

| Departments | Route No. | $\beta_4$ | Operating Cost ($) | $\beta_3$ | Operating Cost ($) |
|---|---|---|---|---|---|
| 1–2 | 1 | 0.4901 | 75 | — | — |
| 1–3 | 2 | 0.2833 | 43 | — | — |
| 1–4 | 3 | 0.0383 | 6 | — | — |
| 1–5 | 4 | — | — | 0.1838 | 21 |
| 1–6 | 5 | 0.4901 | 75 | — | — |
| 2–3 | 6 | — | — | 0.1838 | 21 |
| 2–4 | 7 | 0.3676 | 56 | — | — |
| 2–5 | 8 | 0.2450 | 37 | — | — |
| 2–6 | 9 | — | — | 0.1838 | 21 |
| 3–4 | 10 | 0.1608 | 24 | — | — |
| 3–5 | 11 | 0.1225 | 19 | — | — |
| 3–6 | 12 | 0.1225 | 19 | — | — |
| 4–5 | 13 | 0.1225 | 19 | — | — |
| 4–6 | 14 | 0.3676 | 56 | — | — |
| 5–6 | 15 | 0.1225 | 19 | — | — |
| Total | | 2.9328 | $448 | 0.5514 | $63 |

In the initial solution, 3 units of type 4 (the largest integer equal to 2.9328) and 1 unit of type 3 are needed. The corresponding total daily fixed costs are (3 × $23.08 + 1 × $14.20) = $83.44, and the total daily operating costs are ($448 + $63) = $511. Therefore, the total daily MH cost is about ($83.44 + $511) = $594.44. This will be greatly reduced in the improvement phase when conveyors replace the trucks.

#### 14.2.2.2 Phase II: Improvement Procedure

In step 1 of phase II, trucks are replaced by conveyors on routes 1, 2, 5, 7, 8, and 14 because their costs, as found in Table 14.6, are lower than the operating costs of the trucks (Table 14.14 depicts the costs of conveyors and trucks on all the routes.) The total MH cost for this improved solution (Table 14.15) becomes $380. This is a reduction of $214 ($594.44 – $380 = $214), or 36%. This savings comes from not only the lower conveyor costs but also the elimination of two units of truck type 4, which saves $46.16 (2 × $23.08 = $46.16), or $46.

The next step (step 2 of phase II) is to investigate whether any one or more units of the existing trucks may be replaced by conveyor(s) to lower the cost, mainly through a reduction in a fixed cost.

The operating costs of the two truck types on routes 3, 4, 6, 9, 10, 11, 12, 13, and 15 are subtracted from the proposed conveyor costs on each route (see Table 14.16 under penalty column) to start the process in step 2. Only those costs that are less than the daily fixed cost of the associated truck are considered (otherwise, no savings may be realized). This cost is an excess cost, or a penalty, if the conveyor is to be replaced by a truck. Because the sum of $\beta$ values for truck type 3, used on routes 9, 4, and 6, is less than 1 and the total of penalty costs on the corresponding routes is

**TABLE 14.14**
**Comparison of Conveyor Costs and Truck Operating Costs**

| Departments | Route No. | Conveyor Cost ($) | Truck 4 Operating Cost ($) | Truck 3 Operating Cost ($) |
|---|---|---|---|---|
| 1–2 | 1 | 28 | 75 | — |
| 1–3 | 2 | 27 | 43 | — |
| 1–4 | 3 | 31 | 6 | — |
| 1–5 | 4 | 29 | — | 21 |
| 1–6 | 5 | 22 | 75 | — |
| 2–3 | 6 | 37 | — | 21 |
| 2–4 | 7 | 28 | 56 | — |
| 2–5 | 8 | 28 | 37 | — |
| 2–6 | 9 | 26 | — | 21 |
| 3–4 | 10 | 32 | 24 | — |
| 3–5 | 11 | 42 | 19 | — |
| 3–6 | 12 | 25 | 19 | — |
| 4–5 | 13 | 38 | 19 | — |
| 4–6 | 14 | 41 | 56 | — |
| 5–6 | 15 | 35 | 19 | — |

**TABLE 14.15**
**First Improved Solution**

| | | | Trucks | | | |
|---|---|---|---|---|---|---|
| Departments | Route No. | Conveyor Cost ($) | $\beta_4$ | Operating Cost ($) | $\beta_3$ | Operating Cost ($) |
| 1–2 | 1 | 28 | — | — | — | — |
| 1–3 | 2 | 27 | — | — | — | — |
| 1–4 | 3 | — | 0.0383 | 6 | — | — |
| 1–5 | 4 | — | — | — | 0.1838 | 21 |
| 1–6 | 5 | 22 | — | — | — | — |
| 2–3 | 6 | — | — | — | 0.1838 | 21 |
| 2–4 | 7 | 28 | — | — | — | — |
| 2–5 | 8 | 28 | — | — | — | — |
| 2–6 | 9 | — | — | — | 0.1838 | 21 |
| 3–4 | 10 | — | 0.1608 | 24 | — | — |
| 3–5 | 11 | — | 0.1225 | 19 | — | — |
| 3–6 | 12 | — | 0.1225 | 19 | — | — |
| 4–5 | 13 | — | 0.1225 | 19 | — | — |
| 4–6 | 14 | 41 | — | — | — | — |
| 5–6 | 15 | — | 0.1225 | 19 | — | — |
| Total | | $174 | 0.6891 | $106 | 0.5514 | $63 |

Total cost = operating cost + fixed cost of trucks = 174 + 106 + 63 + 23.08 + 14.20 = 380

**TABLE 14.16**
**Conveyors, Trucks, and Penalty Costs for Illustrative Example 1**

| Departments | Route No. | Conveyor Cost ($) | System | Present System Costs ($) | $\beta_{i}$, ($i$ = 3 or 4) | Penalty | Type 4 Rank | Type 3 Rank |
|---|---|---|---|---|---|---|---|---|
| 1–2 | 1 | 28 | Conveyor | 28 | | | | |
| 1–3 | 2 | 27 | Conveyor | 27 | | | | |
| 1–4 | 3 | 31 | T-4[a] | 6 | 0.0383[b] | 25 | 6 | |
| 1–5 | 4 | 29 | T-3[c] | 21 | 0.1838[d] | 8 | | 2 |
| 1–6 | 5 | 22 | Conveyor | 22 | | | | |
| 2–3 | 6 | 37 | T-3 | 21 | 0.1838[d] | 16 | | 3 |
| 2–4 | 7 | 28 | Conveyor | 28 | | | | |
| 2–5 | 8 | 28 | Conveyor | 28 | | | | |
| 2–6 | 9 | 26 | T-3 | 21 | 0.1838[d] | 5 | | 1 |
| 3–4 | 10 | 32 | T-4 | 24 | 0.1608[c] | 8 | 2 | |
| 3–5 | 11 | 42 | T-4 | 19 | 0.1225[c] | 23 | 5 | |
| 3–6 | 12 | 25 | T-4 | 19 | 0.1225[c] | 6 | 1 | |
| 4–5 | 13 | 38 | T-4 | 19 | 0.1225[c] | 19 | 4 | |
| 4–6 | 14 | 41 | Conveyor | 41 | | | | |
| 5–6 | 15 | 35 | T-4 | 19 | 0.1225[c] | 16 | 3 | |

[a] T-4 = truck type 4.

[b] $i = 4$.

[c] T-3 = truck type 3.

[d] $i = 3$.

greater than its fixed cost ($29 > $14.20), type 3 trucks will obviously not be replaced by a conveyor and are not considered any further.

This procedure is repeated for truck type 4. The sum of $\beta_4$ values on routes 12, 10, 15, 13, 11, and 3 is 0.6891, which is less than 1; the associated total penalties are greater than the cost of 1 unit of truck type 4 ($97 < $23.08). Therefore, no savings are realized by replacing this truck with a conveyor, and the final solution will be the first solution, with a cost of 343 + 23 × 1 + 14 × 1= $380. The cost includes the operating cost and the fixed cost for 1 unit each of type 3 and 4 trucks.

The sum of $\beta_4$ is 0.6891, producing a remainder of 0.3109 when subtracted from the next largest integer (1 - 0.6891 = 0.3109). This is the excess capacity of type 4 available in the solution. The type 3 truck is assigned on routes 4, 6, and 9. The corresponding $\beta_4$ values (from Table 14.7 or 14.10) are 0.1225 each on the three routes, giving a sum of 0.3675. However, this total is greater than the excess capacity of type 4 available (0.3675 > 0.3109). This means that more than 1 unit of type 4 is needed to replace 1 unit of type 3. This extra unit costs $23.08 (from Table 14.8), and the added operating cost (from Table 14.9) is 19 + 19 + 19 = $57. The total additional cost is then ($23.08 + $57.00 =) $80.08. On the other hand, because truck type 3 would be replaced, the total cost may be reduced by its fixed cost and operating cost, $14.20 + $21 + $21 + $21 = $77.20. Therefore, this replacement costs $2.88 more ($80.08 – $72.20 = $2.88).

Before abandoning this idea, inclusion of a conveyor on one of these three routes is also considered. The corresponding conveyor costs on the three routes of 4, 6, and 9 are $29, $37, and $26, respectively. The lowest cost is $26, on route 9. If a conveyor is introduced on this route and routes 4 and 6 are served by type 4, the total $\beta_4$ will be less than 1, or 1.0566 – 0.1225 = 0.9341. Thus, the same unit of type 4 can handle all routes and no new type 4 will be needed. The MH cost for this new assignment is now calculated. If a savings is realized, it will be implemented.

The added cost due to the extra operating time of type 4 and the conveyor will be $26 + $19 + $19 = $64.00, but it will be reduced by $14.20 + $21 + $21 + $21 = $77.20, because a type 3 truck is eliminated. Thus, a savings of $13.20 is realized ($77.20 – $64.00 = $13.20). Because of this savings, a new MH assignment will be implemented, which includes only one truck of type 4 and one more conveyor (route 9), giving a final cost of $367 (Table 14.17).

The mathematical model (illustrated in Appendix C) representing this example required 118 variables and 55 constraints without the 122 nonnegative and integrality constraints (or 177 constraints in all).

The solution obtained using the heuristic approach was found to be very close to that determined by the mathematical model (Table 14.18). The latter result was $366.06, as compared with $367 from the heuristic approach. The minor difference in the total value is due to the effect of rounding in the computer program. Many additional examples were tested; the answers were found to differ by 3% or less.

**TABLE 14.17**
**Final Solution by the Heuristic Method for Illustrative Example 1**

| Departments | Route No. | System | Cost ($) | $\beta_i$ |
|---|---|---|---|---|
| 1–2 | 1 | Conveyor | 28 | s |
| 1–3 | 2 | Conveyor | 37 | — |
| 1–4 | 3 | T-4 | 6 | 0.0383 |
| 1–5 | 4 | T-4 | 19 | 0.1225 |
| 1–6 | 5 | Conveyor | 22 | — |
| 2–3 | 6 | T-4 | 19 | 0.1225 |
| 2–4 | 7 | Conveyor | 28 | — |
| 2–5 | 8 | Conveyor | 28 | — |
| 2–6 | 9 | Conveyor | 26 | — |
| 3–4 | 10 | T-4 | 24 | 0.1608 |
| 3–5 | 11 | T-4 | 19 | 0.1225 |
| 3–6 | 12 | T-4 | 19 | 0.1225 |
| 4–5 | 13 | T-4 | 19 | 0.1225 |
| 4–6 | 14 | Conveyor | 41 | — |
| 5-6 | 15 | T-4 | 19 | 0.1225 |
| Total | | | 344 | |

*Note:* All numbers in the cost column are rounded off.

Total cost = operating cost + fixed cost = 344 + 23.08 = $367

**TABLE 14.18**
**Final Solution by the Mathematical Model for Illustrative Example 1**

| Departments | Route No. | System | Operating Cost ($) | $\beta_i$ |
|---|---|---|---|---|
| 1–2 | 1 | Conveyor | 28 | — |
| 1–3 | 2 | Conveyor | 27 | — |
| 1–4 | 3 | T-4 | 5.82 | 0.0383 |
| 1–5 | 4 | T-4 | 18.62 | 0.1225 |
| 1–6 | 5 | Conveyor | 22 | — |
| 2–3 | 6 | T-4 | 18.62 | 0.1225 |
| 2–4 | 7 | Conveyor | 28 | — |
| 2–5 | 8 | Conveyor | 28 | — |
| 2–6 | 9 | T-4 | 18.62 | 0.1225 |
| 3–4 | 10 | T-4 | 24.44 | 0.1608 |
| 3–5 | 11 | T-A | 18.62 | 0.1225 |
| 3–6 | 12 | T-4 | 18.62 | 01225 |
| 4–5 | 13 | T-4 | 18.62 | 0.1225 |
| 4–6 | 14 | Conveyor | 41 | — |
| 5–6 | 15 | T-4 | 18.62 | 0.1225 |
| Total | | | 342.98 | |

Total cost = operating cost + fixed cost = 342.98 + 23.08 = $366

## 14.3 AN ITERATIVE PROCESS FOR LAYOUT AND OPTIMUM MH DEVELOPMENT

So far, one layout (the first layout) and the accompanying MH system have been produced. The optimizing algorithm has reduced the MH cost to a minimum value. The following iterative process is invoked to investigate whether this cost can be reduced even further by changing the layout:

1. A high MH cost between two departments may be reduced by bringing them closer together. To achieve this, develop a new rel-chart using MH cost between the departments as the basis for the relation ship. One method for assigning the codes is to determine the range of the route costs and divide the range using the following percentages: $A$ = top 15%, $E$ = next 20%, $I$ = next 30%, $O$ = next 20%, and $U$ = last 15%.
2. Use the modified rel-chart from step 1 to develop a new plant layout.
3. Using the new plant layout from step 2, develop the optimum MH and associated cost. This can be done by applying the MH selection algorithm.
4. The procedure is terminated if one of the following conditions is met:
   a. No change is noted in the resultant MH relationship matrix.
   b. After one or more iterations, the process repeats itself.
   c The number of iterations reaches the value set by the analyst. If the process has not terminated, go to step 1.

The first iteration is the result obtained with the initial layout and the associated MH system. To assign codes *A, E, I, O,* and *U* within the COMLAD3 program, the difference between the maximum and minimum route costs is determined. A range for each code is developed using the following values:

$A = 0.15$
$E = 0.20$
$I = 0.3$
$O = 0.2$
$U = 0.15$

For example, in our illustration, the maximum cost of \$41 is associated with route 14 between departments 4 and 6. The minimum cost of \$6 is associated with route 3 between departments 1 and 4. The difference in cost is \$35. All departments having an MH cost greater than or equal to \$41 – 0.15 × 35.00 = \$35.75 would be coded as an *A* relationship. Between \$35.75 and \$28.69 (i.e., \$35.75 – 0.20 × 35.00 = \$28.75) would be given an *E* relationship, and so on. In COMLAD3, the user may change these percentages if desired. Iteration 1 results are displayed in Figure 14.4.

The minimum cost and the corresponding MH assignment are reported to the analyst for every iteration. The results of the successive application of the procedure are displayed in Table 14.19. The COMLAD3 program is very efficient in performing these calculations. It should be noted that although each route connects two specific departments, the relative locations of these departments in the layout may be shifting from cycle to cycle. For example, route 1, which connects departments 1 and 2 and uses conveyors in both iterations 1 and 2, has a different associated cost because departments 1 and 2 are in different positions, resulting in different distances

```
                A                   B                   C
      1 2 3 4 5 6 7 8 9 0\1 2 3 4 5 6 7 8 9 0\1 2 3 4 5 6

  1         5
  2         5 5 5
  3   6 6 6 6 2 2 2
  4   6 6 1 1 2 2
X 5   6 6 1 1 2 2
  6     3 3 3 4 4 4
  7     3 3 3 4 4 4
  8               4
  9
  0
> 1
  2
  3
  4
Y 5
  6
```

**FIGURE 14.4** Layout generated by the computer program

**TABLE 14.19**
**Layout Design Solution for Illustrative Example 1**

| Departments | Route No. | Iteration 1 | | Iteration 2 | | Iteration 3 | | Iteration 4 | |
|---|---|---|---|---|---|---|---|---|---|
| | | System | Cost | System | Cost | System | Cost | System | Cost |
| 1–2 | 1 | Conveyor | 28 | Conveyor | 36 | Conveyor | 28 | Conveyor | 36 |
| 1–3 | 2 | Conveyor | 27 | Conveyor | 27 | T-4 | 34 | Conveyor | 27 |
| 1–4 | 3 | T-4 | 6 | T-4 | 5 | T-4 | 5 | T-4 | 5 |
| 1–5 | 4 | T-4 | 19 | T-4 | 15 | T-4 | 15 | T-4 | 15 |
| 1–6 | 5 | Conveyor | 22 | Conveyor | 26 | Conveyor | 28 | Conveyor | 28 |
| 2–3 | 6 | T-4 | 19 | T-4 | 15 | T-4 | 15 | T-4 | 15 |
| 2–4 | 7 | Conveyor | 28 | Conveyor | 32 | Conveyor | 28 | Conveyor | 39 |
| 2–5 | 8 | Conveyor | 28 | Conveyor | 36 | Conveyor | 26 | Conveyor | 26 |
| 2–6 | 9 | Conveyor | 26 | T-4 | 15 | T-4 | 15 | T-4 | 15 |
| 3–4 | 10 | T-4 | 24 | T-4 | 20 | Conveyor | 25 | T-4 | 20 |
| 3–5 | 11 | T-4 | 19 | T-4 | 15 | T-4 | 15 | T-4 | 15 |
| 3–6 | 12 | T-4 | 19 | T-4 | 15 | T-4 | 15 | T-4 | 15 |
| 4–5 | 13 | T-4 | 19 | T-4 | 15 | T-4 | 15 | T-4 | 28 |
| 4–6 | 14 | Conveyor | 41 | Conveyor | 30 | Conveyor | 32 | Conveyor | 28 |
| 5–6 | 15 | T-4 | 19 | T-4 | 15 | T-4 | 15 | T-4 | 15 |
| Total Cost including Fixed Cost | | | 367 | | 340 | | 334 | | 337 |

| Departments | Route No. | Iteration 5 | | Iteration 6 | | Iteration 7 | | Iteration 8 | |
|---|---|---|---|---|---|---|---|---|---|
| | | System | Cost | System | Cost | System | Cost | System | Cost |
| 1–2 | 1 | Conveyor | 28 | Conveyor | 28 | Conveyor | 34 | Conveyor | 28 |
| 1–3 | 2 | T-4 | 34 | Conveyor | 27 | Conveyor | 25 | Conveyor | 36 |
| 1–4 | 3 | T-4 | 5 | T-4 | 5 | T-4 | 5 | T-4 | 5 |
| 1–5 | 4 | T-4 | 15 | T-4 | 15 | T-4 | 15 | T-4 | 15 |
| 1–6 | 5 | Conveyor | 28 | Conveyor | 22 | Conveyor | 41 | Conveyor | 22 |
| 2–3 | 6 | T-4 | 15 | T-4 | 15 | T-4 | 15 | T-4 | 15 |
| 2–4 | 7 | Conveyor | 28 | Conveyor | 28 | Conveyor | 28 | Conveyor | 24 |
| 2–5 | 8 | Conveyor | 26 | T-4 | 30 | Conveyor | 26 | Conveyor | 33 |
| 2–6 | 9 | T-4 | 15 | T-4 | 15 | T-4 | 15 | T-4 | 15 |
| 3–4 | 10 | T-4 | 20 | T-4 | 20 | T-4 | 20 | T-4 | 20 |
| 3–5 | 11 | Conveyor | 25 | Conveyor | 25 | T-4 | 15 | T-4 | 15 |
| 3–6 | 12 | T-4 | 15 | T-4 | 15 | T-4 | 15 | T-4 | 15 |
| 4–5 | 13 | T-4 | 15 | T-4 | 15 | T-4 | 15 | T-4 | 15 |
| 4–6 | 14 | Conveyor | 28 | Conveyor | 41 | Conveyor | 32 | Conveyor | 26 |
| 5–6 | 15 | T-4 | 15 | T-4 | 15 | T-4 | 15 | T-4 | 15 |
| Total cost including fixed cost | | | 340 | | 339 | | 339 | | 322 |

*(continued)*

**TABLE 14.19 (CONTINUED)**

**Layout Design Solution for Illustrative Example 1**

| Departments | Route No. | Iteration 9 | | Iteration 10 | | Iteration 11 | |
|---|---|---|---|---|---|---|---|
| | | System | Cost | System | Cost | System | Cost |
| 1–2 | 1 | Conveyor | 28 | Conveyor | 28 | Conveyor | 28 |
| 1–3 | 2 | Conveyor | 27 | T-4 | 34 | Conveyor | 27 |
| 1–4 | 3 | T-4 | 5 | T-4 | 5 | T-4 | 5 |
| 1–5 | 4 | T-4 | 15 | T-4 | 15 | T-4 | 15 |
| 1–6 | 5 | Conveyor | 36 | Conveyor | 28 | Conveyor | 22 |
| 2–3 | 6 | T-4 | 15 | T-4 | 15 | T-4 | 15 |
| 2–4 | 7 | Conveyor | 36 | Conveyor | 28 | Conveyor | 28 |
| 2–5 | 8 | Conveyor | 26 | Conveyor | 26 | T-4 | 30 |
| 2–6 | 9 | T-4 | 15 | T-4 | 15 | T-4 | 15 |
| 3–4 | 10 | T-4 | 20 | T-4 | 20 | T-4 | 20 |
| 3–5 | 11 | T-4 | 15 | Conveyor | 25 | Conveyor | 25 |
| 3–6 | 12 | T-4 | 15 | T-4 | 15 | T-4 | 15 |
| 4–5 | 13 | T-4 | 15 | T-4 | 15 | T-4 | 15 |
| 4–6 | 14 | Conveyor | 28 | Conveyor | 28 | Conveyor | 41 |
| 5–6 | 15 | T-4 | 15 | T-4 | 15 | T-4 | 15 |
| Total Cost including Fixed Cost | | | 334 | | 335 | | 339 |

between them in these iterations. The simulation stops at iteration 11, because the solution for iteration 11 is the same as that for iteration 6. Any further analysis would only repeat the cycle from iteration 6 through iteration 11. The minimum cost of $322 is given in iteration 8. The final layout is shown in Figure 14.5.

The score for the above layout = 20
Distances for the routes:

| Route | Distance (feet) |
|---|---|
| Route = 1 | 120 |
| Route = 2 | 180 |
| Route = 3 | 160 |
| Route = 4 | 120 |
| Route = 5 | 60 |
| Route = 6 | 300 |
| Route = 7 | 80 |
| Route = 8 | 200 |

MH system is truck type 4 on route 4 with daily operating cost of $15
MH system is conveyor on route 5 with daily total cost of $22
MH system is truck type 4 on route 6 with daily operating cost of $15

```
                A                   B                   C                   D
     \1 2 3 4 5 6 7 8 9 0\1 2 3 4 5 6 7 8 9 0\1 2 3 4 5 6 7 8 9 0\1 2 3 4 5 6 7

  1     5 5       2 2 2
  2     5 5 1 1 2 2
  3   3 3 6 1 1 2 2 4 4
  4   3 3 6 6 6 6 4 4
X 5   3 3 6 6 6 4 4 4
  6
  7
  8
  9
  0
> 1
  2
  3
  4
Y 5
  6
```

The SCORE for the above LAYOUT = 20

**FIGURE 14.5** Optimum layout and associated MH

MH system is conveyor on route 7 with daily total cost of $24
MH system is conveyor on route 8 with daily total cost of $33
MH system is truck type 4 on route 9 with daily operating cost of $15
MH system is truck type 4 on route 10 with daily operating cost of $20
MH system is truck type 4 on route 11 with daily operating cost of $15
MH system is truck type 4 on route 12 with daily operating cost of $15
MH system is truck type 4 on route 13 with daily operating cost of $15
MH system is conveyor on route 14 with daily total cost of $26
MH system is truck type 4 on route 6 with daily operating cost of $15

Total conveyor cost = $69

Total variable truck cost = $130

Total fixed cost of trucks = $23

Total reduced number of trucks of type 4 = 1 at unit price of $23.08

Total MH cost = $322

## 14.4 ILLUSTRATIVE EXAMPLE 2

Example 2 involves minor changes to the first example to show step 6 of phase II of the algorithm further. The fixed and operating costs of the four truck types are changed, and the number of stacks per pallet for type 4 is increased to five. Table 14.20 shows

**TABLE 14.20**
**Initial Fixed and Operating Costs of the Four Truck Types for Illustrative Example 2**

| Truck Type | Initial Fixed Cost ($) | Operating Cost/Hour ($) |
|---|---|---|
| 1 | 13,000 | 13 |
| 2 | 15,000 | 14 |
| 3 | 16,000 | 15 |
| 4 | 18,000 | 16 |

the operating and fixed costs, while Table 14.21 shows the corresponding daily values. Table 14.22 is similar to Table 14.16, depicting the solution and penalty costs of truck type 4. As can be noted, the total $\beta$ values exceed 1.

Using the values in Table 14.22, the total cost for this MH system is equal to the total operating cost of $294 plus the cost of 2 units of truck type 4, or $2 \times \$16 = \$32$, for a total of $331. Implementing step 6 in phase 2, this cost may be reduced even further.

The eligible penalty costs (those less than the daily fixed cost of $16 of truck type 4) are ranked in ascending order of their value. Because there is only one truck size in the solution, the $\beta_4$ values are referred to as $\beta$ for convenience. Starting with rank 1 on route 7, the corresponding $\beta$(0.2941) is subtracted from the total $\beta$(1.6299), leaving 1.3358. The value of penalty 1 is less than the fixed cost of truck 4; therefore, the process continues. Route 14 has rank 2, and its $\beta$ (0.2941) is subtracted from the last total (1.3358), giving a remainder of 1.4170 and a total penalty of 2. Because the total $\beta$ is still above 1 and the penalty cost is less than 16, $\beta$ for the route ranked third, route 4, is subtracted from the total, giving a remainder of 0.8458; the penalty is increased to $6. The total of $\beta$ values is now below 1, which means that only 1 unit of truck type 4 may be used. This is 1 unit less than the original solution, and because the associated savings of $16 – $10 = $6 can be realized, the new arrangement is implemented. Thus, conveyors replace the truck on routes 7, 14, and 8.

The replacement of one more truck of type 4 would mean replacing all trucks from all routes by conveyors. The penalty cost of such replacement far exceeds $16, the maximum savings that could be realized. Hence, no further replacement is sought. Table 14.23 shows the final solution for Example 2.

**TABLE 14.21**
**Daily Fixed and Operating Truck Costs for Illustrative Example 2**

| | Truck Type | | | |
|---|---|---|---|---|
| | 1 | 2 | 3 | 4 |
| Daily fixed cost ($) | 11.54 | 13.32 | 14.20 | 15.98 |
| Daily operating cost ($) | 104.00 | 112.00 | 120.00 | 128.00 |

**TABLE 14.22**
**Conveyors, Trucks, and Penalty Costs for Illustrative Example 2**

| Departments | Route No. | Conveyor Cost ($) | System | Operating Cost of Present System ($) | $\beta$ (or $\beta_4$) | Penalty | Rank Type |
|---|---|---|---|---|---|---|---|
| 1–2 | 1 | 36 | Conveyor | 36 | | | |
| 1–3 | 2 | 25 | Conveyor | 27 | | | |
| 1–4 | 3 | 28 | T-4 | 4 | 0.0309 | 24 | |
| 1–5 | 4 | 22 | T-4 | 13 | 0.0980 | 9 | 4 |
| 1–6 | 5 | 24 | Conveyor | 24 | | | |
| 2–3 | 6 | 46 | T-4 | 13 | 0.0980 | 33 | |
| 2–4 | 7 | 39 | T-4 | 38 | 0.2941 | 1 | 1 |
| 2–5 | 8 | 29 | T-4 | 25 | 0.1961 | 4 | 3 |
| 2–6 | 9 | 25 | T-4 | 13 | 0.0980 | 12 | 6 |
| 3–4 | 10 | 30 | T-4 | 16 | 0.1287 | 14 | 7 |
| 3–5 | 11 | 31 | T-4 | 13 | 0.0980 | 18 | |
| 3–6 | 12 | 28 | T-4 | 13 | 0.0980 | 15 | 8 |
| 4–5 | 13 | 23 | T-4 | 13 | 0.0980 | 10 | 5 |
| 4–6 | 14 | 39 | T-4 | 38 | 0.2941 | 1 | 2 |
| 5–6 | 15 | 31 | T-4 | 13 | 0.0980 | 18 | |
| Total | | | | 299 | 1.6299 | | |

**TABLE 14.23**
**Final Solution by the Heuristic Method for Illustrative Example 2**

| Departments | Route No. | System | Cost ($) | $\beta_4$ |
|---|---|---|---|---|
| 1–2 | 1 | Conveyor | 36 | — |
| 1–3 | 2 | Conveyor | 27 | — |
| 1–4 | 3 | T-4 | 4 | 0.0306 |
| 1–5 | 4 | T-4 | 13 | 0.0980 |
| 1–6 | 5 | Conveyor | 24 | — |
| 2–3 | 6 | T-4 | 13 | 0.0980 |
| 2–4 | 7 | Conveyor | 43 | — |
| 2–5 | 8 | Conveyor | 29 | — |
| 2–6 | 9 | T-4 | 13 | 0.0980 |
| 3–4 | 10 | T-4 | 16 | 0.1286 |
| 3–5 | 11 | T-4 | 13 | 0.0980 |
| 3–6 | 12 | T-4 | 13 | 0.0980 |
| 4–5 | 13 | T-4 | 13 | 0.0980 |
| 4–6 | 14 | Conveyor | 43 | — |
| 5–6 | 15 | T-4 | 13 | 0.0980 |
| Total | 313 +16 (T-4) = 329 | | 329 | |

The mathematical solution for Illustrative Example 2 required 118 variables and 177 constraints, including nonnegative and integrality, the same as with Example 1.

## 14.5 COMPARISON OF THE ANALYTICAL AND HEURISTIC APPROACHES FOR MH SYSTEM DEVELOPMENT

Twenty examples were created and the results of the heuristic and mathematical methods were compared to test the validity of the heuristic method on the MH system development. Table 14.24 shows the percentage of differences in the final results of the two approaches for all 20 examples.

The deviations from the analytical results range from 0% to slightly more than 3%, with a mean of 1.015% and a standard deviation of 0.822%. These figures show that the heuristic procedure is performing very well.

**TABLE 14.24**
**Comparison of the Results From the Analytical and Heuristic Approaches**

| Example | Mathematical Model ($) | Heuristic Model ($) | Percentage of Difference |
|---|---|---|---|
| 1 | 1902 | 1926 | 1.3 |
| 2 | 2132 | 2168 | 1.7 |
| 3 | 3147 | 3168 | 0.7 |
| 4 | 4053 | 4094 | 1.0 |
| 5 | 3147 | 3168 | 0.7 |
| 6 | 3286 | 3286 | 0.0 |
| 7 | 2574 | 2613 | 1.5 |
| 8 | 2708 | 2745 | 1.4 |
| 9 | 2821 | 2868 | 1.7 |
| 10 | 2516 | 2552 | 1.4 |
| 11 | 2432 | 2443 | 0.4 |
| 12 | 2347 | 2375 | 1.2 |
| 13 | 1470 | 1470 | 0.0 |
| 14 | 1476 | 1476 | 0.0 |
| 15 | 1846 | 1894 | 2.6 |
| 16 | 1954 | 1957 | 0.2 |
| 17 | 2076 | 2080 | 0.3 |
| 18 | 1937 | 1954 | 0.9 |
| 19 | 2151 | 2219 | 3.2 |
| 20 | 1994 | 2019 | 1.2 |

A few examples are given to realize how fast the size of the mathematical model grows with respect to the number of departments and items.

If there are $N$ departments, the number of routes will be $N(N - 1)/2$. Thus, a facility with mere 12 departments will have 66 routes [i.e., $12(12 - 1)/2 = 66$]. The number of constraints (excluding nonnegativity and integrality ones) is the same as the total number of routes plus the product of the number of items per route and number of routes. If, in the preceding example, there are three items per route on the average, the number of constraints will jump to ($3 \times 66 =$) 198. This increase, in turn, raises the computational time exponentially. In addition, it should be noted that the computational time is also greatly affected by the values of the coefficients and the right-hand sides. In the 20 examples that were run, some of the problems required less than 1 hour, while others needed more than 24 hours. This variance occurs because the values determine, in the branch-and-bound solution procedure [the reader not familiar with this method can find it in any standard operations research textbook], the number of branches and how soon a limit (bound) is reached in the solution to the mathematical model.

## 14.6 COMPUTER PROGRAM DESCRIPTION

COMLAD3 can be found at http://www.crcpress.com/e_products/downloads/download.asp?cat_no=44222. As mentioned, COMLAD3 is built on the output of COMLADII; hence, the initial data input for COMLAD3 is the same as that for COMLADII. In the MH phase, the program is interactive and the input to the program is given in the form of answers to the questions on the screen. The output is displayed on the screen; the PRINT SCREEN command can be used to get a hard copy.

## 14.7 SUMMARY

This chapter presents a heuristic procedure to develop an MH system for an entire plant by using two types of conveyors (chain and belt) and four types of trucks, each with different capacity and cost. Although the heuristic procedure may appear a little complex, its application has constantly provided results that match the mathematical outcomes, which are more difficult to get and indeed may not be feasible at all for large problems. The main objective of the chapter is to show how MH and plant layout problems are interrelated and should be considered simultaneously to achieve a truly integrated system with minimum cost. The illustrative examples should provide some insights into the intricacies involved.

## PROBLEMS

1. Given four departments and four items: The economic life of the equipment is 5 years. The interest rate is 12% compounded annually. The work day is 8 hours. Use straight-line depreciation.

| Item | Route | Length (feet) | Width (feet) | Weight | Daily Rate |
|---|---|---|---|---|---|
| 1 | 1-2-4 | 3 | 2 | 100 | 300 |
| 2 | 3-1-4 | 1 | 1 | 30 | 5000 |
| 3 | 2-3-4 | 1 | ½ | 10 | 800 |
| 4 | 1-2-3-4 | ½ | ½ | 5 | 2000 |

| Route | 1-2 | 1-3 | 1-4 | 2-3 | 2-4 | 3-4 |
|---|---|---|---|---|---|---|
| Distance (feet) | 100 | 200 | 40 | 80 | 30 | 50 |

| | Truck Type | | | |
|---|---|---|---|---|
| | 1 | 2 | 3 | 4 |
| Fixed cost ($) | 13,000 | 14,000 | 15,000 | 16,000 |
| Operating cost per hour ($) | 13 | 14 | 15 | 16 |

| | Truck Type | | | |
|---|---|---|---|---|
| | 1 | 2 | 3 | 4 |
| Pallet size (inches) | 36 × 36 | 40 × 48 | 48 × 48 | 48 × 72 |
| No. of stacks/pallet | 2 | 3 | 4 | 5 |

Determine:

a. the length of time conveyors and trucks will be occupied per day per route

b. the daily cost of the trucks and conveyors per route

2. Given five departments and five items: The size of each square block for calculating the distances is 30 feet. The operating cost per hour for conveyors is $2. All other parameters use the default values.

| Item | Route | Length (feet) | Width (feet) | Weight | Daily Rate |
|---|---|---|---|---|---|
| 1 | 2-4-5-2 | 10 | 5 | 10 | 800 |
| 2 | 1-3-4-5 | 25 | 11 | 20 | 600 |
| 3 | 1-2-3-4 | 36 | 6 | 30 | 500 |
| 4 | 1-4-5-3-1 | 50 | 24 | 60 | 300 |
| 5 | 2-3-4-5-1 | 18 | 12 | 45 | 700 |

| Department | 1 | 2 | 3 | 4 | 5 |
|---|---|---|---|---|---|
| No. of blocks | 15 | 10 | 20 | 30 | 17 |

| | Truck Type | | | |
|---|---|---|---|---|
| | 1 | 2 | 3 | 4 |
| Fixed cost ($) | 16,000 | 20,000 | 22,000 | 24,000 |
| Operating cost per hour ($) | 11 | 12 | 13 | 15 |

Develop a plant layout and the associated MH for the first iteration.

a. Find the daily cost of trucks and conveyors per route.
b. Find the length of time conveyors and trucks will be occupied per day per route.

3. Given six departments and six items: The size of each square block used for calculating the distances is 4 feet. The operating cost per hour for conveyors is $6. Four stacks per pallet are used for type 4 trucks. All other parameters use the default values.

| Item | Route | Length (feet) | Width (feet) | Weight | Daily Rate |
|---|---|---|---|---|---|
| 1 | 1-3-4-5-6-1 | 40 | 4 | 30 | 1500 |
| 2 | 1-2-4-6-2 | 12 | 6 | 50 | 3500 |
| 3 | 2-3-4-5-2 | 32 | 15 | 20 | 1000 |
| 4 | 1-4-5-1 | 42 | 20 | 10 | 500 |
| 5 | 1-3-5-6-3-1 | 25 | 10 | 15 | 800 |
| 6 | 2-4-5-6-2 | 14 | 10 | 22 | 2000 |

| Department | 1 | 2 | 3 | 4 | 5 | 6 |
|---|---|---|---|---|---|---|
| Area (square feet) | 200 | 300 | 100 | 400 | 200 | 300 |

Using a column space of 4 feet, the following numbers of blocks will be calculated (see the layout):

| No. of blocks | 13 | 19 | 7 | 25 | 13 | 19 |
|---|---|---|---|---|---|---|

| | Truck Type | | | |
|---|---|---|---|---|
| | 1 | 2 | 3 | 4 |
| Fixed Cost ($) | 30,000 | 30,000 | 12,000 | 20,000 |
| Operating cost ($) | 30 | 30 | 14 | 16 |

Using COMLAD3, determine the optimum layout and associated MH assignments. How many iterations were required before the iterative process of developing a plant layout, the associated MH requirement, and then a plant layout again, etc., is completed?

a. Find the daily cost of trucks and conveyors per route.
b. Find the length of time conveyors and trucks will be occupied per day per route.

4. Given 18 departments and seven items: The size of each square block used for calculating the distances is 5 feet. The operating cost per hour for conveyors is $3. All other parameters use the default values.

| Item | Route | Length (feet) | Width (feet) | Weight | Daily Rate |
|---|---|---|---|---|---|
| 1 | 1-2-3-4-5-6-7-9-12-14-15-18-1 | 36 | 10 | 15 | 700 |
| 2 | 2-5-6-8-10-11-13-15-16-17-2 | 25 | 20 | 50 | 1000 |
| 3 | 1-3-4-5-7-8-9-10-11-13-16-17-18-1 | 15 | 10 | 12 | 800 |
| 4 | 1-4-7-9-11-12-13-14-16-18-1 | 45 | 20 | 40 | 500 |
| 5 | 2-4-6-8-9-10-15-16-18-2 | 24 | 12 | 30 | 400 |
| 6 | 1-2-5-8-10-12-13-17-18-1 | 14 | 10 | 40 | 1600 |
| 7 | 1-5-7-9-13-14-15-17-18-1 | 20 | 12 | 30 | 1200 |

| Department | 1 | 2 | 3 | 4 | 5 | 6 | 7 | 8 | 9 |
|---|---|---|---|---|---|---|---|---|---|
| No. of blocks | 12 | 9 | 8 | 10 | 11 | 12 | 4 | 5 | 8 |

| Department | 10 | 11 | 12 | 13 | 14 | 15 | 16 | 17 | 18 |
|---|---|---|---|---|---|---|---|---|---|
| No. of blocks | 14 | 8 | 8 | 16 | 4 | 5 | 8 | 9 | 14 |

| | Truck Type | | | |
|---|---|---|---|---|
| | 1 | 2 | 3 | 4 |
| Fixed Cost ($) | 30,000 | 30,000 | 12,000 | 20,000 |
| Operating cost ($) | 30 | 30 | 11 | 14 |

Note that larger fixed and operating costs have been assigned to types 1 and 2 trucks to prevent them from entering the solution.

Use COMLAD3 to determine the optimum layout and associated MH. Also try to make the layout rectangular and see if the MH has changed.

a. Find the daily cost of trucks and conveyors per route.
b. Find the length of time the conveyors and trucks will be occupied per day per route.

## SUGGESTED READINGS


Apple, James M., *Material Handling Systems Design*, 2nd edition, John Wiley, New York, 1985.

Buffa, E.S. Armour, G.C., and Vollmann, T.E., "Allocating facilities with CRAFT," *Harvard Business Review*, 42, 136–158, 1964.

Dutta, Kedar N., and Sahu, S., "A multigoal heuristic for facilities design problems: MUGHAL," *International Journal of Production Research*, 20(2), pp. 147 and 154, 1982.

Kulick, Alexis, "Interlocking pallet pattern simulation program," *Industrial Engineering*, 14(9), 22–24, 1982.

Tompkins, James A., and Reed, R., "An applied model for the facilities design problem," *International Journal of Production Research*, 14(5), 583–595, 1976.

Ziai, M. Reza, "Optimizing computerized plant layout design and material handling equipment selection," Dissertation, Louisiana Tech University, 1991.

# 15 Plant Site Selection and Service (Support) Considerations

This chapter briefly discusses the topics that relate the financial well-being of a manufacturing firm. In the first section, the traditional, nonquantitative method of selecting a plant site is examined. This analysis could serve as a screening process for the quantitative methods described in the location analysis segment in Part I of the book. Section 15.2 deals with utilities. No plant can function without utilities, which can be a major expense. Every effort should be made to reduce the cost, and Section 15.2 examines sources and alternatives for providing utilities. Section 15.3 deals with insurance. To remain in business, almost every firm must have insurance or be self-insured. This section examines types of insurance policies that a manufacturing organization must have and provides examples of total insurance programs. A safe operating environment goes a long way toward reducing insurance cost, and Section 15.4 discusses safety practices and gives an example of a program that encourages safety in a plant. All manufacturing organizations must comply with the Americans with Disabilities Act (ADA) of 1990. Section 15.5 discusses the act as it relates to manufacturing facilities. Taxes are a major financial burden that is unavoidable. The projected sales cost of the product must reflect the tax liabilities, and Section 15.5 describes some of the taxes that frequently must be considered. Section 15.6 focuses on financial statements; it briefly describes how a firm might evaluate total cost and determine the unit cost of production. If the sale price is unknown, it might be possible to develop a balance sheet from the information generated thus far.

## 15.1 PLANT SITE SELECTION

The geographical location of a new plant often has an important effect on the ultimate profitability of a venture. Frequently, the selection of the site for a company's first plant is made in a less than scientific manner; sometimes the city chosen is simply the hometown of the person beginning the business. The location for an additional plant usually receives more attention because industrial organizations are rather immobile and, once settled in an area, usually remain there. The decision to locate a new plant away from the area of the company's roots is usually made only after completing a thorough analysis. Even the selection of the country in which the plant is to be located is important, as evidenced by the numerous multinational companies that have located their assembly facilities nearer the final product market or have taken advantage of lower labor costs in different countries.

### 15.1.1 Factors Influencing Site Selection

Several factors are presented here that may be considered in site selection; the detailed listing that is actually used will depend on the type of business being located.

- Transportation facilities
- Labor supply
- Availability of land
- Nearness to markets
- Availability of suitable utilities
- Proximity to raw materials
- Geographical and weather characteristics
- Taxes and other laws
- Community attitudes
- National security
- Proximity to the company's existing plants

Some of these factors are interrelated, and one could easily produce a list with a slightly different breakdown. The number of alternative sites can usually be reduced by eliminating the locations that do not conform to the minimum requirements as defined by the evaluator in certain critical factors. Owing to the varied nature of different industries, however, these factors can vary from case to case. In each instance, however, the critical criteria should be determined and used in preliminary screening. The other factors will be considered as the site selection process narrows the field of alternative locations.

#### 15.1.1.1 Transportation Facilities

Suitable transportation facilities must be available to move personnel, equipment raw materials, and products to and from the plant. Waterways, railroads, and highways are the usual choices for shipping, but air transport might also be appropriate. The volume and types of raw materials and products will often determine the mode of transportation best suited to the plant.

Current transportation systems have made possible the decentralization of U.S. industry. Except in large, congested urban areas, people now travel more by personal cars than by railroads. The interstate highway system has made the suburbs as easily accessible as most cities, and so a plant can draw its work force from an area easily as large as 75 miles in diameter. Although large trucks are hauling increasing amounts of raw materials and finished products, many industries still choose railroads, especially when heavy or bulky material must be transported long distances. An innovation combining the better characteristics of trucks and railroads is the "piggyback" system, in which loaded truck trailers are driven to the rail depot, transported by rail, and then again driven to their ultimate destination. Shipping by barge is a low-cost form of transportation for manufacturing plants that are located on navigable waterways. Companies that provide or require bulk commodities such as coal, petroleum and petroleum products, iron ore, and bulk chemicals find barge transportation especially advantageous. Suitable airports in the vicinity of the plant permit the expeditious transportation of personnel or equipment as required. When the cost of shipping products or raw materials is a significant portion of the finished product cost, transportation can be a major consideration; however, when transportation is only a small part of the final product cost, it may not be as important a factor.

#### 15.1.1.2 Adequate Labor Supply

Even in the age of robotics, no company is able to operate without employees. The plant location study must ensure that the types and numbers of employees that will be needed will be available, even though most highly trained personnel are very mobile and can be recruited from other areas. The following factors are important: prevailing wages, workweek restrictions, existence of competing companies that can cause high turnover or labor unrest, productivity level, labor problems, and the education and experience of available potential employees. Many semiskilled factory positions can be filled by trainable unskilled workers. Some light industries have found that locating plants in smaller communities allows them to use lower-cost, more productive country labor. Other companies have chosen to locate new plants in rural areas to avoid the labor difficulties generally associated with areas that are staunchly union-oriented.

#### 15.1.1.3 Availability of Land

Communities that are attempting to attract new industry often provide land at a low cost, but the company must make certain that the site does not have problems that are too serious to be overcome. The soil characteristics and topography of the location must be evaluated because they can seriously affect building costs. The site should be almost level, and the soil load-bearing value should be greater than 2000 pounds per square foot. Additional space for future expansion required by unforeseen market conditions should be readily available and accessible.

#### 15.1.1.4 Nearness to Markets

It is highly desirable that a plant be located in or near the market area for the product, because costs for timely product distribution are a function of location. This is especially true for bulky items for which the cost of shipping is significant in comparison to the cost of materials and labor, as with fertilizer and building materials. For products for which the cost of labor and material is high, such as jewelry, computers, and watches, market proximity is not so critical. Consideration should be given to nearness to the market for any by-products that may result; for instance, sawdust from a lumber mill might be sold to a paper mill or a particle board manufacturer.

#### 15.1.1.5 Suitable Utilities

Many industrial plants require a great deal of electrical power and steam. Some plants burn fuel (oil, gas, wood, coal) to produce their own power and steam. For these plants, the availability of an inexpensive fuel supply is very important. For other plants, many areas of the country have ample supplies of low-cost electricity.

The availability of an adequate water supply is often important. Plants that consume small volumes of water often purchase it from public utilities, but plants that use large quantities might need to locate near a large river or lake or in an area where a deepwater well could be drilled. The state geological survey should be reviewed, and the U.S. Army Corps of Engineers should be contacted for available information on water tables, seasonal fluctuations of lake and river levels, and future plans for water sources.

The proposed location should have adequate facilities for waste disposal, not only of solids, but perhaps of gases and liquids as well. Care must be taken to ascertain that there are no unreasonable state or municipal regulations that will severely restrict the future plant's ability to dispose of waste economically. The permissible limits for alternative methods of disposal should be determined, and attention should be given to the possibility of providing additional waste disposal facilities.

If the plant site under evaluation is exposed to potential flood damage, the availability of flood protection facilities, such as drainage pumps, levees, canals, and causeways, must be reviewed. The quality and availability of a local fire department should also be determined. The exposure to fire damage because of the proximity of hazardous operations to the potential site should likewise be considered.

#### 15.1.1.6 Proximity of Raw Materials

The cost of shipping raw materials and fuel to the plant site should be considered along with the cost of transporting the products to market so as to minimize the total transportation cost as much as possible while balancing that cost against other operating expenses. Plants whose raw materials are perishable or bulky tend to locate near their source of raw materials. Plants whose raw materials lose much of their weight during the manufacturing process, such as ore refineries, steel plants, and paper mills, also often locate as near their raw material sources as is practical.

#### 15.1.1.7 Geographical and Weather Considerations

The geographical characteristics of the site can greatly affect the cost of the buildings to be constructed there as well as the plant operating costs. A severely cold climate will necessitate the building of additional shelters for equipment. A very hot climate could require the plant to have additional air-conditioning for personnel comfort and additional cooling towers for process equipment. The geographical and weather factors to be considered are altitude, temperature, humidity, average wind speed, annual rainfall, and terrain.

#### 15.1.1.8 Taxes and Legal Considerations

Because taxes will be an operating cost, the types, bases, and rates of taxes charged by state and local governments must be considered seriously. Many cities offer tax incentives such as exemptions to encourage companies to locate new plants in their communities. Some taxes to be evaluated are property, income, and sales taxes. Unemployment compensation taxes should also be determined. Local regulations concerning real estate, health and safety codes, truck transportation, roads, acquisition of easements and rights-of-way, zoning, building codes, and labor codes should all be evaluated.

#### 15.1.1.9 Community Considerations

The community (both local authorities and the people) under consideration should be pleased to have the plant located in its area. The community should be able to provide essential services such as police and fire protection, street maintenance, and trash and garbage disposal. Good community living conditions will be needed to attract and maintain motivated workers. Cultural facilities, churches, libraries, parks, good schools, community theaters, community symphonies, and recreational facilities are indications of a community's character and its growth potential. If something essential is lacking, the company should determine the cost of providing for the need.

#### 15.1.1.10 National Security

The U.S. government sometimes encourages companies that supply strategic materials to locate such that the sources for the products are widely dispersed. A company can increase its probability of being awarded a government contract if it locates its manufacturing facilities where the government wishes. In other cases, the government might want to locate a very security-sensitive manufacturing facility within the limits of a government installation.

#### 15.1.1.11 Proximity to an Existing Plant

Many companies locate a new satellite facility in the general area of another major plant. Doing so facilitates upper-level management supervision of the new plant. Executives and consultants can minimize travel time between plants if they are near each other. Care must be taken, however, to locate the plants far enough apart that they do not have to compete with each other for labor.

### 15.1.2 Procedures for Site Selection

From the preceding discussion it is evident that a large amount of information must be accumulated to aid in the decision-making process. Federal and state geological surveys can provide maps and other geographical information. The Department of Labor, the Federal Communications Commission, the Federal Power Commission, and the Federal Trade Commission are other good sources of indispensable information. Local agencies such as industrial commissions and chambers of commerce can provide details more specific to a particular location. Railroads frequently provide information regarding available shipping and the associated costs for any of numerous locations.

Once the necessary information about alternative locations has been gathered, the advantages and disadvantages of one location over another must be evaluated. One location might have the lowest raw material cost, another the lowest utility cost,

**TABLE 15.1**
**Sample Location Rating Procedure**

| Considerations | Maximum Weight | Location | | |
|---|---|---|---|---|
| | | *X* | *Y* | *Z* |
| Transportation facilities | 100 | 90 | 80 | 90 |
| Labor supply | 100 | 70 | 80 | 90 |
| Land | 100 | 80 | 70 | 70 |
| Markets | 100 | 40 | 70 | 80 |
| Utilities | 75 | 70 | 70 | 65 |
| Raw materials | 75 | 50 | 75 | 60 |
| Geographical/weather | 50 | 40 | 35 | 50 |
| Taxes and legal factors | 50 | 30 | 20 | 40 |
| Community | 40 | 20 | 40 | 35 |
| National security | 30 | 10 | 5 | 15 |
| Proximity to existing plant | 30 | 30 | 10 | 20 |
| Total | 750 | 530 | 555 | 615 |

and yet another offers the best labor supply. A commonly used aid in selecting from alternative sites is a rating procedure. Each of the major factors is rated from 0 to 100 with regard to its importance. Each individual location is then rated from 0 to the maximum for each factor. The scores for the locations determine the final ranking. An example of this procedure is shown in Table 15.1.

It must be understood that the ranking obtained by this method is, to a large extent, subjective. When the factors are evaluated, they express an analyst's feelings, which are measured in terms of assigned weights. It is quite possible that different analysts might choose varying weights for the same physical conditions, leading to entirely different site selections. Such variations in judgment can be eliminated if all the locations that meet certain minimum requirements are made candidates and the procedures developed in Chapters 17 through 19 are then applied on the basis of the cost data that can be collected for each site. Because the ultimate objective of any business is to provide goods and/or services at a profit, the site that can minimize the operational cost would be a natural choice. The models developed in these later chapters would evaluate each site in a quantitative manner.

The minimum requirements can be set in many different ways. For example, for each factor a minimum necessary value could be set, and the site that fails to meet the minimum is then eliminated from any future consideration; or a minimum could be set just for the important factors such as the first four considered in Table 15.1; or an overall minimum could be set, for instance at 550, eliminating location X from further consideration.

Another approach is to allow all the sites to be considered for objective evaluation by first applying the methods from Chapters 17 to 19 and then checking to see whether the best site selected also meets the minimum requirements of the subjective factors mentioned above. If it does not and an alternative site must be chosen, management will at least understand the "cost" associated with that decision.

### 15.1.3 Industrial Parks

A large tract of land designed and maintained for the use of several industries is referred to as an industrial park. The industrial park is usually located on a major highway or near an airport. The industries located in the park are usually quite varied and most often have no relationship to each other except for the location.

Industrial parks have become popular because of their several advantages. Successful industrial parks tend to draw more industry to the area. The locations provide access to transportation, are industrially zoned, and already provide utilities. Big buildings are standard, and because the entire area is industry, the neighbors are compatible.

The advantages of industrial parks are not limited to industry; decided advantages for the community also exist. These parks provide for efficient land usage and management, which is quite important in some areas. The centralized location of the area industry also helps to improve the appearance of the community. By drawing the localizing industry, parks help to offer new employment opportunities.

Difficulties certainly do exist that hinder the development of parks. To begin with, a large plot of land is required. This can be difficult to obtain, especially in large, developed cities. Careful planning is required to ensure a minimum of traffic congestion. Also, building costs can be high when buildings are constructed for adaptability instead of specialization so that if a plant is vacated, it can easily be used by a new company. Thus in considering developing an industrial park, the benefits to the community and the businesses must be weighed against the disadvantages. At the same time, an industry that can locate in an industrial park should examine such opportunities in its evaluation for the site.

## 15.2 UTILITIES SPECIFICATIONS

Specifications for utilities should be carefully developed because they can make up a significant portion of the variable operating cost of a manufacturing facility. Some items that are typically considered are

- water — drinking, cooling, deionized, and process
- electricity
- refrigeration
- steam
- compressed air — instrument, control, and power
- waste disposal and treatment
- air-conditioning/heating
- telephone

### 15.2.1 Water

Potable water might be provided by a public utility or by an in-plant source such as a well or a waterway. The best source is usually the public utility because its water is tested, treated, and certified as being safe for human consumption. If plant water is to be used, some testing and treatment might be required to ensure that it is satisfactory for drinking.

The piping and pumps for cooling water must be large enough to handle the peak demand even if the system has some scale buildup, which can occur as the result of long-term mineral deposits in the pipes effectively reducing their inside diameters. If practical, lines should be large enough to allow for future additions of equipment requiring cooling without necessitating significant piping changes.

Water used in air-conditioning and other equipment cooling can be obtained from the same source as drinking water; however, because it need not be of the same quality as the drinking water, well water is frequently less expensive to use for this purpose. Cooling water is usually recycled to a central cooling tower, where its temperature is lowered through partial evaporation; water that is lost to the atmosphere must then be replaced.

Provisions must be made in the design of the system to incorporate the necessary pH, algae, and scale control facilities. Instrumentation should be included to measure the total cooling water flow and the individual flow to each piece of equipment. This will allow proper monitoring of the flows and accurate allocation of the cooling water cost if desired.

Process water is that which is used directly in, or is consumed by the operation of the facility. In some cases, the process water need not be potable; in others, such as pharmaceutical manufacturing, the water might have to be mineral- and germ-free. The facilities necessary to treat the water must be carefully specified.

### 15.2.2 Electricity

The cost of electricity is often a significant operating expense for a manufacturing facility. Most electrical power is provided by a public utility to the plant transformer(s), and distribution of power within the plant is the user's responsibility. Typical contracts for purchasers of electrical power contain provisions for a demand charge, energy usage charge, and fuel adjustment charge.

The maximum power used by a facility during some specified period (e.g., 30 minutes) within the billing period determines the demand charge portion of the bill. The total energy consumed, in kilowatt-hours, during the billing period (normally a month) is the basis for the usage charge. Typical contracts specify a lower usage charge with increased energy consumption. A facility's usage and demand charges per unit will typically diminish with increased electrical energy requirements. Most contracts are also structured so that more savings are realized by improving the load factor (the ratio of the average use to demand) than by increasing the electrical load. The fuel adjustment charge is used to "pass on" to the user any increase in fuel cost experienced by the utility.

Private electrical power generation (often called "cogeneration") is seldom economical. In isolated instances, usually when a plant has a large demand for low-pressure steam, the possibility of producing high-pressure steam to generate electricity should be investigated. The low-pressure steam, the by-product of electrical power generation, could then be utilized in the plant.

### 15.2.3 Lighting

Lighting should be considered in the plant-planning stages. This is important because such things as the materials used and the colors chosen can greatly affect the lighting situation. Rough surfaces absorb more light than do smooth, shiny surfaces, and rooms decorated in dark colors require more lighting than those with bright colors. It is also important to plan the lighting for the specific tasks being performed in the various portions of the plant.

#### 15.2.3.1 Development of the Light Arrangement in the Work Area

There are a number of different lamp types that can be used to light a work area. However, their costs of operation can differ considerably. For example, incandescent light costs as much as 20 times more to operate than a fluorescent lamp, while sodium or mercury vapor lamp operating costs are twice that for fluorescent lamps. A complete light fixture consists of lamps, reflectors, ballast, and other parts, and is commonly called a luminary. To obtain the necessary uniform intensity of light in a work area, decisions must be made as to the number of luminaries necessary and their arrangement in the work area. The following is a procedure that may be applied to help in these decisions.

The procedure consists of determining how much light must be emitted from the light fixtures. The light reaching the work area depends on factors such as the type of light fixtures and their characteristics, working environment, ceiling and wall reflectivity, and the volume of space between lights and the work area.

##### *15.2.3.1.1 Factors Associated With Luminaries*

Each lamp manufacturer has individual specifications for the type of luminaries it supplies. For example, a supplier of 48- and 96-inch fluorescent tubes gives the product characteristics shown in Table 15.2.

In addition, the lamp color influences the output of a fluorescent lamp. The initial lumens (IL) should be multiplied by the following factors to account for lamp color: cool white, 1.00; warm white, 1.03; daylight, 0.086; living white, 0.76; and supermarket white, 0.75.

**TABLE 15.2**
**Fluorescent Tube Characteristics**

| | Initial Lumens (foot-candles) | Wattage | LLD | Rated Life (hours) |
|---|---|---|---|---|
| **48-inch tube** | | | | |
| Basic | 3000 | 40 | 0.80 | 9000 |
| High performance | 4200 | 60 | 0.80 | 12,000 |
| Super performance | 6900 | 110 | 0.79 | 12,000 |
| **96-inch tube** | | | | |
| Basic | 6300 | 75 | 0.90 | 12,000 |
| High performance | 9000 | 100 | 0.86 | 12,000 |
| Super performance | 15500 | 215 | 0.80 | 12,000 |

##### *15.2.3.1.2 Lamp Lumen Depreciation Factor*

The lamp lumen depreciation (LLD) factor gives the percentage of IL that the lamps will emit after the initial aging effect sets in (or the lamps are broken in). The work area is planned to operate at this depreciated value, because the majority of the lamp's useful life is at this level. After 70% of useful life, a rapid aging can cause considerable depreciation in a lamp's output, and it should then be replaced.

##### *15.2.3.1.3 Ballast Performance Efficiency*

Ballast performance efficiency (BPE) refers to the loss due to ballast performance in fluorescent lights. This loss is generally taken as 5% of the lamp's foot-candles; therefore, BPE is taken at a fixed value of $1 - 0.05 = 0.95$.

##### *15.2.3.1.4 Voltage Fluctuation Efficiency*

The voltage fluctuation efficiency (VFE) refers to changes due to voltage fluctuation. Each percentage of voltage fluctuation in the line can cause up to a 0.4% change in a lamp's output. Thus, a drop of 5% voltage at peak load results in a VFE of $1 - 5 \times 0.004 = 0.98$.

Manufacturers also specify a maximum ratio of spacing between lights and distance (height) between the work surface and luminaries. This ratio must be considered in the final lighting design. For fluorescent lights, this ratio is 1.3.

#### 15.2.3.2 Working Environment

The amount of light reaching a work area is affected by the cleanliness of the environment and how often the lamps are cleaned. A relative value can be assigned to each factor, for example: clean environment, 0; medium clean, 1; and dirty, 2. Similarly, if the lights are serviced (cleaned) every year, the value is 0; every 2 years, 1; and every 3 years, 2. Based on whether the lamps are enclosed or not, the lumen dirt depreciation (LDD) factor for fluorescent lamps can be estimated by using one of the following expressions.

For enclosed lights:

$$\text{LDD} = 0.88 - 0.05 \times (\text{environmental value} + \text{lamp service frequency value}) \quad (1)$$

For open lights:

$$\text{LDD} = 0.94 - 0.05 \times (\text{environmental value} + \text{lamp service frequency value}) \quad (2)$$

#### 15.2.3.3 Ceiling and Wall Reflectivity

Reflection of light from the ceiling and walls contributes to the total illumination in the workplace. The typical values of reflectivity, based on normal working surface, are:

Ceiling: Commercial building, 80%
Industrial buildings, 50%
Walls: 0%

These values may be modified based on surface colors. For example, the reflectivity (in almost linearly descending order) for light colors such as white, cream, and ivory is between 80% and 70%; for medium colors such as buff, green, gray, it is between 63% and 56%; and for dark colors such as tan, gray, olive green, dark oak, natural cement, and red brick, it is between 45% and 15%.

The volume of space between ceiling and luminaries also contributes to the final value of ceiling reflectivity. If luminaries are flush to the ceiling, then the net reflectivity is equal to the base ceiling reflectivity (BCR). If not, light manufacturers provide detail tables to evaluate this value. We will estimate it here by using the following expression, which is within ±10% accuracy within normal working ranges.

Net ceiling reflectivity:

$$\text{NCR} = 0.0858(1+\text{BCR})^{2.563} \times (1+\text{WR})^{0.9426} \times \text{CCR}^{-0.1593} \tag{3}$$

where

BCR = base ceiling color reflectivity
WR = wall reflectivity
CCR = ceiling cavity ratio

CCR is calculated as follows: If $L$ and $W$ are the length and width of the floor area, then first determine the ratio, LWR as

$$\text{LWR} = \frac{L+W}{L \times W}$$

Then

$$\text{CCR} = 5 \times (\text{height between luminaries and ceiling}) \times \text{LWR}$$

Note that it is not possible for NCR to have a higher value than BCR.

#### 15.2.3.4 Volume of Space Between Light Fixtures and Work Area

The effect of the volume of space between a luminary and a work area is given by the factor similar to the ceiling cavity ratio. The factor, called room cavity ratio (RCR), is determined as

$$\text{RCR} = 5 \times (\text{height between work platform area and luminaries}) \times \text{LWR}$$

##### *15.2.3.4.1 Coefficient of Utilization*

Coefficient of utilization (CU) is the ratio of light reaching the work area to the light emitted from the luminaries. Again, a manufacturer can supply a detail table, but the value of CU is dependent on net ceiling reflection, wall reflection, and RCR. For fluorescent luminaries with two lamps, it can be determined by the following expression to within ±10% accuracy within normal working ranges.

$$\text{CU} = 0.631(1+\text{NCR})^{0.353} \times (1+\text{WR})^{0.531} \times (\text{RCR})^{-0.454} \tag{4}$$

The range of CU value is from 88% to 15%.

#### 15.2.3.5 Minimum Illumination Level (MIL)

Minimum illumination level (MIL) is the minimum light intensity in foot candles (fc) per square foot necessary at the work area to perform normal activity. The value depends on the type of work. For example, for stock room and warehouse, 30 fc is acceptable, and for a typical office, 50 fc is sufficient. As the work becomes more precise and tedious, the level must increase. For example, a medium-difficult workplace such as a technical drawing room requires 150 fc, while a very precise work area such as an inspection department may need between 400 and 500 fc.

##### *15.2.3.5.1 Determination of the Number of Luminaries*

To determine the number of necessary luminaries, we can simply equate the total required illumination to the net total illumination supplied at the workplace by all luminaries.

Total required illumination of work place

$$\begin{aligned} &= \text{MIL} \times \text{area} \\ &= \text{MIL} \times L \times W \end{aligned} \tag{5}$$

If $N$ is the number of luminaries, with $M$ lamps in each, then the total illumination supplied at the workplace equals the number of luminaries times the number of lamps in each luminary times the LLD factor times the ballast power efficiency times the VFE times the LDD factor times the CU, or

$$\text{Total illumination} = N \times M \times \text{IL} \times \text{LLD} \times \text{BPE} \times \text{VFE} \times \text{LDD} \times \text{CU} \tag{6}$$

Equating (3) and (4) and solving for $N$, we get

$$N = \frac{\text{MIL} \times L \times W}{\text{M} \times \text{IL} \times \text{LLD} \times \text{BPE} \times \text{VFE} \times \text{LDD} \times \text{CU}} \tag{7}$$

##### *15.2.3.5.2 Determination of the Spacing*

Luminaries are arranged in full rows. Knowing the dimension (generally the width) of the workspace over which the luminaries are strung in a row and the length of each luminary, we can determine the number of luminaries that can be accommodated in one row, RL. The number of rows is obtained by dividing $N$ by RL. Some adjustment in the value of $N$ may be necessary at this stage. It is suggested that the spacing between side walls and the immediate rows should be one-third of the spacing between the interior rows.

Thus, if the spacing between two successive interior rows in $S$ and there are $R$ rows in all, then the total dimension covered by the rows is equal to $S(R + 1 + 2/3)$. There are $R - 1$ interior spaces in this arrangement. This dimension should be equated to the other dimension (generally the length) of the room to determine the value for $S$.

The maximum spacing may not exceed the ratio of height to spacing suggested by the lamp manufacturer. The entire procedure is illustrated in the following example.

## EXAMPLE

A light machine shop 40 × 80 feet and 20 feet high is to be illuminated. There are two alternatives presented: (1) to use 48-inch 60-watt light tubes with a light fixture that can hold two tubes and is 54 inches long; or (2) to use 96-inch super-performance tubes with a fixture that can hold two tubes and is 102 inches long. The luminaries are to be suspended 5 feet from the ceiling and are not to be enclosed. Voltage drop during peak load is 4%. The ceiling is painted white, and the walls are buff color. Develop the lighting layout for each alternative.

First, let us develop the layout for the 48-inch tube.

1. The following data are obtained from Table 15.2 and other calculations.

   For 48-inch, 60-watt tube: IL = 4200 fc; LLD = 0.8.
   Assume cool white lamps; hence the color multiplying factor is 1.
   Ballast performance efficiency: BPE = 0.95.
   VFE = 1 − 4 × 0.004 = 0.984.

2. Working environment in a light machine shop is medium clean with an environmental factor of 1. Assuming lamps would be cleaned every 2 years, the lamp cleaning frequency value is 1. Because the lights are not enclosed, the LDD factor — from equation (2) — is

$$LDD = 0.94 - 0.05(1 + 1) = 0.84$$

3. The BCR for white is 80% and WR for buff color is 63%. We need to establish the net ceiling reflectivity; for that we must first determine ceiling cavity ratio, CCR.

   For this room, L = 80 feet and W = 40 feet, giving LWR = (80 + 40)/(80 × 40) = 0.0375. The height between luminaries and ceiling is 5 feet. Hence CCR = 5 × 5 × 0.0375 = 0.9375. Using expression (3), the net ceiling reflectivity is

$$NCR = 0.0858 \times (1 + 0.8)^{2.563} \times (1 + 0.63)^{0.9426} \times (0.9375)^{-0.1593} = 0.6197$$

4. The next step is to determine the CU. Assuming the working surface is 3 feet from the floor, the height of the room cavity (from working surface to luminaries) is equal to 20 − 5 − 3 = 12 feet. Hence the RCR is 5 × 12 × 0.0375 = 1.35.

   Using expression (4),

$$CU = 0.631 \times (1 + 0.6197)^{0.353} \times (1 + 0.63)^{0.531} \times (1.35)^{-0.454} = 0.844$$

5. In a light-machine shop, 110 fc/square foot would be sufficient. A specific station may need additional light, which can be supplied with a localized light fixture.

   To determine the number of luminaries necessary, calculate the total illumination required and the total illumination supplied. From expression (5), the total illumination necessary is

$$110 \times 40 \times 80 = 352{,}000 \text{ fc}$$

   From expression (6), the illumination supplied by N luminaries with 2 tubes is

$$N \times 2 \times 4200 \times 0.8 \times 0.95 \times 0.984 \times 0.84 \times 0.844 = 4453.58N$$

Equating the two and solving for $N$, we get

$$N = \frac{352{,}000}{4453.58} = 79.03, \text{ or } 79$$

Next we should decide the layout. Each luminary is 54 inches, or 4.5 feet, long. Placing them across the width of the machine shop, we can have 40/4.5 = 8.88, or 8, luminaries in a row. Notice that because the width dimension is fixed, the number has to be rounded down.

The total number of rows required to set 79 luminaries is 79/8 = 9.875, or 10. The spacing S is determined by equating total spacing to length:

$$\left(10 - 1 + \frac{2}{3}\right)S = 80$$

or $S = 8.28$ feet.

This spacing is less than 1.3 times the mounting height (room cavity height), which is 12 feet in this example. Hence, the design is acceptable (Figure 15.1).

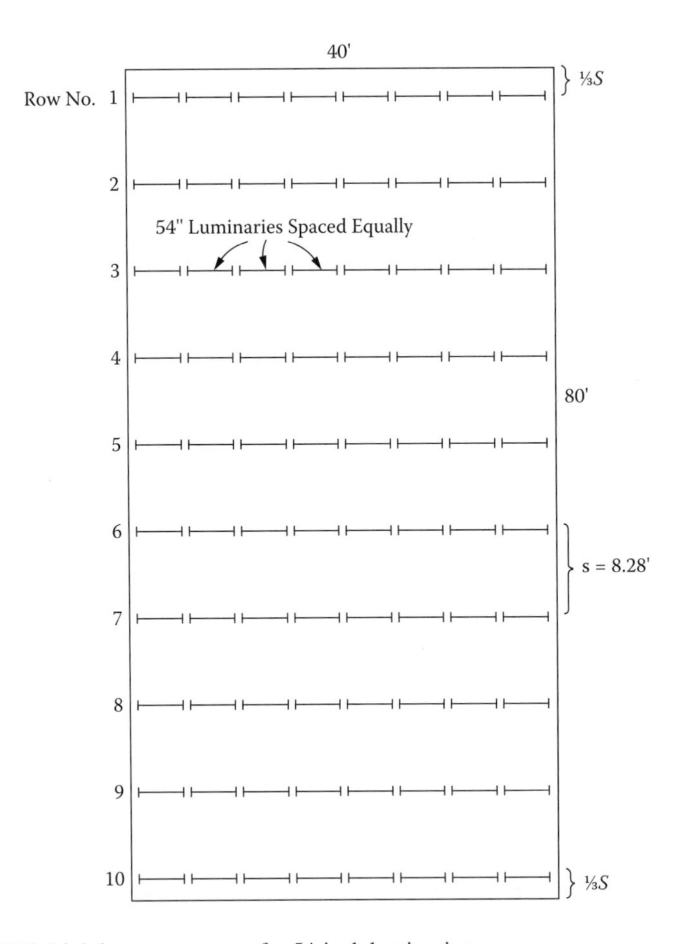


**FIGURE 15.1** Lighting arrangement for 54-inch luminaries

In checking the second alternative, installing 96-inch super-performance tubes, very few data are changed. The initial number of lumens is 15,500 per tube, with LLD = 0.8. Because the fixture is 102 inches long, the number of units that can be strung in a row is also changed to (40 × 12)/102 = 4.7, or 4. All other data are the same; hence, applying equation (6) and equating it to 352,000 fc, the total illumination needed is

$$N \times 2 \times 15{,}500 \times 0.8 \times 0.95 \times 0.984 \times 0.84 \times 0.844 = 352{,}000$$

or

$$N = 214, \text{ or } 22$$

With four fixtures in each row, we will need five rows. The spacing between the rows would be

$$(6 - 1 + 2/3)S = 80, \quad \text{or} \quad S = 14.13$$

This spacing is within limits of 12 × 1.3 = 15.6 feet allowed by the manufacturer, and hence this design is also acceptable.

### 15.2.4 Compressed Air

Many plants use compressed air for actuating valves and operating controls and instruments. The air for instruments must be thoroughly filtered so that it will be as free of particulates as possible. Compressed air that is used to drive equipment usually does not require such special care, except for moisture removal.

### 15.2.5 Process Waste Products

It is not uncommon for a by-product of a process that is considered waste to be recyclable. The by-product might be usable in another process in the plant or by another company. With chemical treatment, some substances can be converted into useful products. Flue gases can be used to preheat materials needing oven treatment. The economic advantages of reusing or selling process waste products make the usefulness of such substances worth investigating.

### 15.2.6 Steam

Steam can be furnished by low-pressure (15 psi or less) boilers for building heat and by power boilers (greater than 15 psi) for steam-driven air compressors, turbine pumps, and many other processes. Generating steam at a high pressure is more efficient than at a low pressure, even if it must be reduced in pressure downstream from the boiler. Initial investment and a large, continuing operating cost make steam an expensive utility; but in most cases there is no substitute when it is to be used in the process.

### 15.2.7 Ventilation

Often dusts, mists, and chemical vapors are given off in plant processes. These by-products are not only irritants but can be health as well as fire hazards. Proper ventilation is necessary to provide breathable air for employees. The type of ventilation provided depends on the type and concentration of the contaminant and should be planned early in the plant planning stage.

The type of ventilation used is based on one of two principles: dilution or removal of the contaminant. If the concentration is low or the contaminant is not particularly hazardous, fresh air can be pumped into the room to dilute the concentration. This method requires little maintenance. In particular, hazardous substances must be removed from the air by drawing the contaminated air out, filtering it, and replacing it with decontaminated air. Filters must be constantly maintained. One should be absolutely certain that contaminants are not passed through to other areas through the ventilation system.

### 15.2.8 Waste Disposal and Treatment

Although waste treatment and disposal are not often thought of as utilities, they can be considered as such. The cost for treatment depends on the facility served, and the cost factors included in this element of plant operation include those for depreciation, chemical usage, labor, and maintenance. Engineering and cost studies can be conducted to determine the feasibility of disposing of wastes by in-plant treatment, discharging into municipal facilities, incineration, and/or transporting to a landfill.

### 15.2.9 Air-Conditioning and Heating

To keep the interior environments of plants and offices comfortable, the temperature within should be maintained at a fairly constant level: in the range of 68–78°F with a relative humidity of about 50%. Depending on the seasonal variation, the ambient air temperature might necessitate the use of heating or air-conditioning systems or both. Such units can be either electric or gas-operated, and their efficiency ratings need to be considered in making purchase recommendations.

To determine the capacity of cooling and heating units for a facility, it is necessary to determine BTU (British thermal unit) loss or gain for the building during summer and winter months. For an office or home, these factors are established based on average conditions, but for a manufacturing plant the expected severe conditions form the base. In either case, some initial data must be established. These include data related to the physical characteristics of the building and data related to heat-generating sources within the building. In noting the physical characteristics, we must observe the length, width, and height of the building, along with the material used in construction and insulation of outside walls, roof, and floor. The glass area and type (single-panel, double-panel), door and window frame construction (metal, wood), and direction in which doors and windows face also contribute in BTU calculations. The temperature differences during summer and winter seasons between the outside and inside of the building and the minimum required fresh-air flow by building codes or design should also be noted. Within the building, there are heat-generating sources such as lights, equipment, and people. The data should be collected on total wattage for lights, total horsepower of all the motors within the building, and number of people expected to occupy the building. As an estimate, 1 ton of air-conditioning is required for every 30 people, 3.5 kilowatts (kW) of light, or 4.7-hp-motor use.

The blueprints and architectural models provide a good starting point for generating the data on physical layout for a facility yet to be built. A discussion with management should also lead to establishment of inside building factors. For an existing facility we can observe and measure the necessary data without much difficulty. In short, the following data are collected :

1. Walls, ceilings, and roof: dimensions, construction material, thickness, insulation, shade factor, and building orientation
2. Windows, doors: area, type of glass (single-paned or double-paned), construction
3. Interior load: total wattage of lighting and appliances, horsepower of motors in use, number of people for which the facility is to be designed
4. Ventilation and infiltration: fresh-air design requirements based on either floor area or per person.

## ILLUSTRATIVE EXAMPLE FOR DETERMINING A/C AND HEATING REQUIREMENTS

The calculations are quite straightforward, especially with the help of the forms provided by the TRANE company (Figure 15.2). We illustrate their use by determining the air-conditioning and heating requirements for a house with the following specifications.

The building is 2400 square feet, with dimensions of 60 × 40 feet. Inside, height to the ceiling is 8 feet. The front entrance is facing south and has a wooden entrance door 6 × 4 feet. As you enter, it has a living room with a glass panel 16 feet long and 8 feet high facing south. There are 10 windows in the building, five facing north and five facing south; each has a 3 × 4-foot window pane.

There is a playroom with two 8 × 4 foot sliding doors facing east. There are two side doors, each 6 × 3 feet. All glass windows are double-paned insulated windows. The building is constructed of masonry wall with R-11 insulation. The ceiling has 3 feet of R-11 insulation. The floor is made of concrete slab with 1-inch edge insulation. The ducts are in the attic with R-6 insulation. The building, on average, would be occupied by four people. Design for temperature difference of 25°F for cooling and 50°F for heating.

The completed evaluation form is shown in Figure 15.3. Summary calculations are:

| Units | Direction | Size (feet) | Area per Unit (square feet) | Total Area (square feet) |
|---|---|---|---|---|
| **Windows, Double Paned** | | | | |
| 1 | South | 16 × 8 | 128 | 128 |
| 5 | South | 3 × 4 | 12 | 60 |
| 5 | North | 3 × 4 | 12 | 60 |
| 2 | East | 8 × 4 | 32 | 64 |
| | | | Sum | 312 |
| **Doors, Wooden (no glass)** | | | | |
| 1 | Front | 6 × 4 | 24 | 24 |
| 2 | Side | 6 × 3 | 18 | 36 |
| | | | Total | 60 |

Grand sum of windows and door areas: 312 + 60 = 372

Whole House Worksheet
New Construction

Customer's Name ______________________ Address ______________________

City ______________ State ____________ Zip ____________ Telephone Number ______________

**WINTER: Inside Design Temp** ________ **°F – Outside Design Temp** ________ **°F = Heating Temp Difference** ________ **°F**

**SUMMER: Outside Design Temp** ________ **°F – Inside Design Temp** ________ **°F = Cooling Temp Difference** ________ **°F**

| HEATING | | COMMON DATA SECTION | | COOLING | |
|---|---|---|---|---|---|
| BTUH LOSS | HEATING FACTOR | SUBJECT | SQ. FT. | COOLING FACTOR | BTUH GAIN |
| | FROM | GROSS WALL | | FROM | |
| | TABLE E | DOORS & WINDOWS (Table A or B) | | TABLE E | |
| | | NET WALL | | | |
| | | | | | |
| | | CEILING | | | |
| | | | | | |
| | | FLOORS | | | |
| Infiltration Btu/hr = | Heating Table D | x 10 x 1.1/60 x Volume (Cu.Ft) | Volume (Cu. Ft) x | 1.1/60 x ΔT x | Cooling Table D = Infiltration Btu/hr |
| = | | x 0.18333 x | x | 0.01833 x x | = |
| | | SUB-TOTAL BTUH LOSS (per 10°F) | | | |
| x | | ADJUSTMENT FACTOR (Table C) | | | |
| | | TOTAL BTUH LOSS | | | |
| | | PEOPLE ______ x 300 BTUH GAIN (Assume 2 persons per bedroom) | | | |
| | | APPLIANCES BTUH | | | 1200 |
| | | SUB-TOTAL BTUH GAIN (room sensible only) | | | |
| x | | DUCT LOSS/GAIN FACTOR (Table F) | | | x |
| | | SUB-TOTAL BTUH (Sensible Gain) | | | |
| | | MOISTURE REMOVAL (sub total x 1.3) | | | x 1.3 |
| | | **TOTAL BTUH LOSS/GAIN** | | | |

**TABLE A – HEATING – DOORS & WOOD FRAME WINDOWS (PER 10°F)**

For sliding glass doors - use factors for the same type window construction.

| Window & Door Types | Frames Wood | TIM | Metal | x Area | = Btuh Loss |
|---|---|---|---|---|---|
| Single Pane Clear | 9.90 | 10.45 | 11.55 | | |
| With Storm | 4.75 | 5.25 | 6.50 | | |
| Double Pane Clear | 5.51 | 6.09 | 7.25 | | |
| With Storm | 3.41 | 3.85 | 4.90 | | |
| Triple Pane Clear | 3.80 | 4.39 | 5.46 | | |
| Jalousie Single | – | – | 11.00 | | |
| Single w/storm | – | – | 5.0 | | |
| Skylights Single | 11.07 | 11.69 | 12.92 | | |
| Double | 6.65 | 7.35 | 8.75 | | |
| Door Wood Only | 4.60 | – | – | | |
| Wood w/storm | 3.20 | – | – | | |
| Urethane Core (R-5) | – | – | 1.90 | | |
| Urethane Core (R-5) w/storm | – | – | 1.70 | | |
| | | | TOTALS | | |

**TABLE B – COOLING – DOORS & WINDOWS**

Factors assume windows have inside shading by draperies or venetian blinds and sliding doors are treated as windows.

| | SINGLE GLASS TEMP. DIFF. | | | DOUBLE GLASS TEMP. DIFF. | | | TRIPLE GLASS TEMP. DIFF. | | | x Area | = BTUH GAIN |
|---|---|---|---|---|---|---|---|---|---|---|---|
| Direction | 15° | 20° | 25° | 15° | 20° | 25° | 15° | 20° | 25° | | |
| N | 18 | 22 | 26 | 14 | 16 | 18 | 11 | 12 | 13 | | |
| NE & NW | 37 | 41 | 45 | 31 | 33 | 35 | 26 | 27 | 28 | | |
| E & W | 52 | 56 | 60 | 44 | 46 | 48 | 38 | 39 | 40 | | |
| SE & SW | 45 | 49 | 53 | 39 | 41 | 43 | 33 | 34 | 35 | | |
| S | 28 | 32 | 36 | 23 | 25 | 27 | 19 | 20 | 21 | | |
| Skylights | 164 | 168 | 172 | 141 | 143 | 145 | 132 | 136 | 140 | | |
| Wood ① | 8.6 | 10.9 | 13.2 | 8.5 | 10.9 | 13.2 | 8.6 | 10.9 | 13.2 | | |
| Metal ② | 3.5 | 4.5 | 5.4 | 3.5 | 4.5 | 5.4 | 3.5 | 4.5 | 5.4 | | |
| | | | | | | | | | TOTALS | | |

① For wood doors and polystyrene core metal doors
② For urethane core metal doors

**TABLE D – INFILTRATION MULTIPLIERS**

**Winter Air Changes Per Hour**

| Floor Area | 900 or less | 900-1500 | 1500-2100 | over 2100 |
|---|---|---|---|---|
| Best | 0.4 | 0.4 | 0.3 | 0.3 |
| Average | 1.2 | 1.0 | 0.8 | 0.7 |
| Poor | 2.2 | 1.6 | 1.2 | 1.0 |
| For each fireplace add: | | Best 0.1 | Average 0.2 | Poor 0.6 |

**Summer Air Changes Per Hour**

| Floor Area | 900 or less | 900-1500 | 1500-2100 | over 2100 |
|---|---|---|---|---|
| Best | 0.2 | 0.2 | 0.2 | 0.2 |
| Average | 0.5 | 0.5 | 0.4 | 0.4 |
| Poor | 0.8 | 0.7 | 0.6 | 0.5 |

**TABLE C – ADJUSTMENT FACTORS – (HEATING)**

| °F Temperature Diff. | 30 | 40 | 50 | 60 | 70 | 80 | 90 |
|---|---|---|---|---|---|---|---|
| Adjustment Factor | 3 | 4 | 5 | 6 | 7 | 8 | 9 |

Courtesy of Trane Company, Tyler, Texas

**FIGURE 15.2** New construction worksheet (courtesy of Trane Company, Tyler, TX)

**TABLE E**
**CONSTRUCTION FACTORS HEATING & COOLING**

| Heating Factor (1) | TYPE OF CONSTRUCTION | Cooling Factor (°F. Temp. Diff.) 15° | 20° | 25° |
|---|---|---|---|---|
| **WALLS (Use Sq. Ft.)** | | | | |
| | **Walls – wood frame w/sheeting & siding, veneer or other finish** | | | |
| 2.71 | A) No insulation, 1/2" Gypsum Board | 5.0 | 6.4 | 7.8 |
| 0.90 | B) R-11 Cavity insulation + 1/2" Gypsum Board | 1.7 | 2.1 | 2.6 |
| 0.80 | C) R-13 Cavity insulation + 1/2" Gypsum Board | 1.5 | 1.9 | 2.3 |
| 0.70 | D) R-13 Cavity insulation + 3/4" Bead Board (R-2.7) | 1.1 | 1.4 | 1.7 |
| 0.60 | E) R-19 Cavity insulation + 1/2" Gypsum Board | 1.1 | 1.4 | 1.7 |
| 0.50 | F) R-19 Cavity insulation + 3/4" Extruded Poly | 0.9 | 1.2 | 1.4 |
| | **Masonry Walls** | | | |
| 5.10 | A) Above grade No insulation | 5.8 | 8.3 | 10.9 |
| 1.44 | B) Above grade + R-5 | 1.6 | 2.3 | 3.1 |
| 0.77 | C) Above grade + R-11 | 0.9 | 1.3 | 1.6 |
| 1.25 | D) Below grade No insulation | 0.0 | 0.0 | 0.0 |
| 0.74 | E) Below grade + R-5 | 0.0 | 0.0 | 0.0 |
| 0.51 | F) Below grade + R-11 | 0.0 | 0.0 | 0.0 |
| | **CEILINGS (Use Sq. Ft.)** | | | |
| 5.99 | A) No insulation | 17.0 | 19.2 | 21.4 |
| 1.20 | B) 2"-2½" insulation R-7 | 4.4 | 4.9 | 5.5 |
| 0.88 | C) 3"-3½" insulation R-11 | 3.2 | 3.7 | 4.1 |
| 0.53 | D) 5½"-6½" insulation R-19 | 2.1 | 2.3 | 2.6 |
| 0.48 | E) 6"-7" insulation R-22 | 1.9 | 2.1 | 2.4 |
| 0.33 | F) 8"-9½" insulation R-30 | 1.3 | 1.5. | 1.6 |
| 0.26 | G) 10"-12" insulation R-38 | 1.0 | 1.1 | 1.3 |
| 0.23 | H) 12"-13" insulation R-44 | 0.9 | 1.0 | 1.1 |
| 3.08 | I) Cathedral type No insulation (roof/ceiling combination) | 11.2 | 12.6 | 14.1 |
| 0.72 | J) Cathedral type R-11 (roof/ceiling combination) | 2.8 | 3.2 | 3.5 |
| 0.49 | K) Cathedral type R-19 (roof/ceiling combination) | 1.9 | 2.2 | 2.4 |
| 0.45 | L) Cathedral type R-22 (roof/ceiling combination) | 1.8 | 2.0 | 2.2 |
| 0.40 | M) Cathedral type R-26 (roof ceiling combination) | 1.6 | 1.8 | 2.0 |
| | **FLOORS (Use Sq. Ft. OR Linear Ft.)** | | | |
| 1.56 | **Floors over unconditioned space (use sq.ft)** A) Over basement or enclosed crawl space (not vented) | 0.0 | 0.0 | 0.0 |
| 0.40 | B) Same as "A" + R-11 insulation | 0.0 | 0.0 | 0.0 |
| 0.26 | C) Same as "A" + R-19 insulation | 0.0 | 0.0 | 0.0 |
| 3.12 | D) Over vented space or garage | 3.9 | 5.8 | 7.7 |
| 0.60 | E) Over vented space or garage + R-11 insulation | 0.8 | 1.3 | 1.7 |
| 0.52 | F) Over vented space or garage + R-19 insulation | 0.5 | 0.8 | 1.1 |
| 0.24 | **Basement Floors (use sq. ft.)** | 0.0 | 0.0 | 0.0 |
| 8.10 | **Concrete slab floor unheated (use linear ft.)** A) No edge insulation | 0.0 | 0.0 | 0.0 |
| 4.10 | B) 1" edge insulation R-5 | 0.0 | 0.0 | 0.0 |
| 2.10 | C) 2" edge insulation R-9 | 0.0 | 0.0 | 0.0 |
| 19.00 | **Concrete slab floor duct in slab (use linear ft.)** A) No edge insulation | 0.0 | 0.0 | 0.0 |
| 11.40 | B) 1" edge insulation R-5 | 0.0 | 0.0 | 0.0 |
| 9.30 | C) 2" edge insulation R-9 | 0.0 | 0.0 | 0.0 |

**FIGURE 15.2** (Continued)

| Case I - Supply Air Temperature Below 120°F | Duct Loss Multipliers | |
|---|---|---|
| Duct Location and Insulation Value | Winter Design Below 15°F | Winter Design Above 15°F |
| Exposed to Outdoor Ambient | | |
| Attic, Garage, Exterior Wall, Open Crawl Space - None | 1.30 | 1.25 |
| Attic, Garage, Exterior Wall, Open Crawl Space - R2 | 1.20 | 1.15 |
| Attic, Garage, Exterior Wall, Open Crawl Space - R4 | 1.15 | 1.10 |
| Attic, Garage, Exterior Wall, Open Crawl Space - R6 | 1.10 | 1.05 |
| Enclosed in Unheated Space | | |
| Vented or Unvented Crawl Space or Basement - None | 1.20 | 1.15 |
| Vented or Unvented Crawl Space or Basement - R2 | 1.15 | 1.10 |
| Vented or Unvented Crawl Space or Basement - R4 | 1.10 | 1.05 |
| Vented or Unvented Crawl Space or Basement - R6 | 1.05 | 1.00 |
| Duct Buried In or Under Concrete Slab | | |
| No Edge Insulation | 1.25 | 1.20 |
| Edge Insulation R Value = 3 to 4 | 1.15 | 1.10 |
| Edge Insulation R Value = 5 to 7 | 1.10 | 1.05 |
| Edge Insulation R Value = 7 to 9 | 1.05 | 1.00 |
| **Case II - Supply Air Temperature Above 120°F** | | |
| Duct Location and Insulation Value | Winter Design Below 15°F | Winter Design Above 15°F |
| Exposed to Outdoor Ambient | | |
| Attic, Garage, Exterior Wall, Open Crawl Space - None | 1.35 | 1.30 |
| Attic, Garage, Exterior Wall, Open Crawl Space - R2 | 1.25 | 1.20 |
| Attic, Garage, Exterior Wall, Open Crawl Space - R4 | 1.20 | 1.15 |
| Attic, Garage, Exterior Wall, Open Crawl Space - R6 | 1.15 | 1.10 |
| Enclosed in Unheated Space | | |
| Vented or Unvented Crawl Space or Basement - None | 1.25 | 1.20 |
| Vented or Unvented Crawl Space or Basement - R2 | 1.20 | 1.15 |
| Vented or Unvented Crawl Space or Basement - R4 | 1.15 | 1.10 |
| Vented or Unvented Crawl Space or Basement - R6 | 1.10 | 1.05 |
| Duct Buried In or Under Concrete Slab | | |
| No Edge Insulation | 1.30 | 1.25 |
| Edge Insulation R Value = 3 to 4 | 1.20 | 1.15 |
| Edge Insulation R Value = 5 to 7 | 1.15 | 1.10 |
| Edge Insulation R Value = 7 to 9 | 1.10 | 1.05 |

**DUCT GAIN MULTIPLIERS**

| Duct Location and Insulation Value | Duct Gain Multiplier |
|---|---|
| Exposed to Outdoor Ambient | |
| Attic, Garage, Exterior Wall, Open Crawl Space - None | 1.30 |
| Attic, Garage, Exterior Wall, Open Crawl Space - R2 | 1.20 |
| Attic, Garage, Exterior Wall, Open Crawl Space - R4 | 1.15 |
| Attic, Garage, Exterior Wall, Open Crawl Space - R6 | 1.10 |
| Enclosed in Unconditioned Space | |
| Vented or Unvented Crawl Space or Basement - None | 1.15 |
| Vented or Unvented Crawl Space or Basement - R2 | 1.10 |
| Vented or Unvented Crawl Space or Basement - R4 | 1.05 |
| Vented or Unvented Crawl Space or Basement - R6 | 1.00 |
| Duct Buried In or Under Concrete Slab | |
| No Edge Insulation | 1.10 |
| Edge Insulation R Value = 3 to 4 | 1.05 |
| Edge Insulation R Value = 5 to 7 | 1.00 |
| Edge Insulation R Value = 7 to 9 | 1.00 |

**FIGURE 15.2** (Continued)

*Add-On & Replacement*
## *Whole House Worksheet*

Customer's Name ________ Address ________

City ________ State ________ Zip ________ Telephone Number ________

**WINTER: Inside Design Temp** 20 °F – **Outside Design Temp** 70 °F = **Heating Temp Difference** 50 °F

**SUMMER: Outside Design Temp** 100 °F – **Inside Design Temp** 75 °F = **Cooling Temp Difference** 25 °F

| HEATING | | COMMON DATA SECTION | | COOLING | |
|---|---|---|---|---|---|
| BTUH LOSS | HEATING FACTOR | SUBJECT | SQ. FT. | COOLING FACTOR | BTUH GAIN |
| | FROM TABLE E | GROSS WALL | 1600 | FROM TABLE E | |
| 2538 | | DOORS & WINDOWS (Table A or B) | 372 | | 9228 |
| 946 | | NET WALL | 1228 | | 1965 |
| | | | | | |
| 2112 | | CEILING | 2400 | | 9840 |
| | | | | | |
| 9840 | 2400 | FLOORS | 2400 | | |
| Infiltration Btu/hr = | Heating Table D | x 10 x 1.1/60 x Volume (Cu.Ft) | Volume (Cu. Ft) x | 1.1/60 x ΔT x | Cooling Table D = Infiltration Btu/hr |
| 1056 = | 3 | x 0.18333 x 19200 | 19200 x | 0.01833 x 25 x | .2 = 1760 |
| 15436 | | SUB-TOTAL BTUH LOSS (per 10°F) | | | |
| 4 x 5 | | ADJUSTMENT FACTOR (Table C) | | | |
| 81039 | | TOTAL BTUH LOSS | | | |
| | | PEOPLE 4 x 300 BTUH GAIN (Assume 2 persons per bedroom) | | | 1200 |
| | | APPLIANCES BTUH | | | 1200 |
| | | SUB-TOTAL BTUH GAIN (room sensible only) | | | 25193 |
| x 1.05 | | DUCT LOSS/GAIN FACTOR (Table F) | | | x |
| | | SUB-TOTAL BTUH (Sensible Gain) | | | 25193 |
| | | MOISTURE REMOVAL (sub total x 1.3) | | | x 1.3 |
| 79590 | | **TOTAL BTUH LOSS/GAIN** | | | 32750 |

**TABLE A – HEATING – DOORS & WOOD FRAME WINDOWS (PER 10°F)**

For sliding glass doors - use factors for the same type window construction.

| Window & Door Types | Frames Wood | TIM | Metal | x Area | = Btuh Loss |
|---|---|---|---|---|---|
| Single Pane Clear | 9.90 | 10.45 | 11.55 | | |
| With Storm | 4.75 | 5.25 | 6.50 | | |
| Double Pane Clear | 5.51 | 6.09 | 7.25 | 312 | 2262 |
| With Storm | 3.41 | 3.85 | 4.90 | | |
| Triple Pane Clear | 3.80 | 4.39 | 5.46 | | |
| Jalousie Single | – | – | 11.00 | | |
| Single w/storm | – | – | 5.0 | | |
| Skylights Single | 11.07 | 11.69 | 12.92 | | |
| Double | 6.65 | 7.35 | 8.75 | | |
| Door Wood Only | 4.60 | – | – | 60 | 0 |
| Wood w/storm | 3.20 | – | – | | |
| Urethane Core (R-5) | – | – | 1.90 | | |
| Urethane Core (R-5) w/storm | – | – | 1.70 | | |
| | | | TOTALS | | 2262 |

**TABLE B – COOLING – DOORS & WINDOWS**

Factors assume windows have inside shading by draperies or venetian blinds and sliding doors are treated as windows.

| | SINGLE GLASS TEMP. DIFF. | | | DOUBLE GLASS TEMP. DIFF. | | | TRIPLE GLASS TEMP. DIFF. | | | x Area | = BTUH GAIN |
|---|---|---|---|---|---|---|---|---|---|---|---|
| Direction | 15° | 20° | 25° | 15° | 20° | 25° | 15° | 20° | 25° | | |
| N | 18 | 22 | 26 | 14 | 16 | 18 | 11 | 12 | 13 | 60 | 1080 |
| NE & NW | 37 | 41 | 45 | 31 | 33 | 35 | 26 | 27 | 28 | | |
| E & W | 52 | 56 | 60 | 44 | 46 | 48 | 38 | 39 | 40 | 64 | 3072 |
| SE & SW | 45 | 49 | 53 | 39 | 41 | 43 | 33 | 34 | 35 | | |
| S | 28 | 32 | 36 | 23 | 25 | 27 | 19 | 20 | 21 | 188 | 5076 |
| Skylights | 164 | 168 | 172 | 141 | 143 | 145 | 132 | 138 | 143 | | |
| Wood ① | 8.6 | 10.9 | 13.2 | 8.6 | 10.9 | 13.2 | 8.6 | 10.9 | 13.2 | | |
| Metal ② | 3.5 | 4.5 | 5.4 | 3.5 | 4.5 | 5.4 | 3.5 | 4.5 | 5.4 | | |
| | | | | | | | | | TOTALS | | 9228 |

① For wood doors and polystyrene core metal doors
② For urethane core metal doors

**TABLE C – ADJUSTMENT FACTORS – (HEATING)**

| °F Temperature Diff. | 30 | 40 | 50 | 60 | 70 | 80 | 90 |
|---|---|---|---|---|---|---|---|
| Adjustment Factor | 3 | 4 | 5 | 6 | 7 | 8 | 9 |

Courtesy of Trane Company, Tyler, Texas

**TABLE D – INFILTRATION MULTIPLIERS**

**Winter Air Changes Per Hour**

| Floor Area | 900 or less | 900-1500 | 1500-2100 | over 2100 |
|---|---|---|---|---|
| Best | 0.4 | 0.4 | 0.3 | 0.3 |
| Average | 1.2 | 1.0 | 0.8 | 0.7 |
| Poor | 2.2 | 1.6 | 1.2 | 1.0 |
| For each fireplace add: | | Best 0.1 | Average 0.2 | Poor 0.6 |

**Summer Air Changes Per Hour**

| Floor Area | 900 or less | 900-1500 | 1500-2100 | over 2100 |
|---|---|---|---|---|
| Best | 0.2 | 0.2 | 0.2 | 0.2 |
| Average | 0.5 | 0.5 | 0.4 | 0.4 |
| Poor | 0.8 | 0.7 | 0.6 | 0.5 |

**FIGURE 15.3** Completed whole house worksheet

The BTU loss/gain calculations are:

1. The gross area of the walls is $(2 \times 8)(60 + 40) = 1600$ square feet.
2. The heating BTU lost through doors and windows for 10°F is from Table A of Figure 15.2, as follows.

   For windows: double panel, metal frame $7.25 \times 312 = 2262$
   For doors: wooden door $4.6 \times 60 = 276$
   Total $= 2262 + 276 = 2538$
   Cooling BTU gained for 25°F, from Table B:

| | |
|---|---|
| North | $18 \times 60 = 1080$ |
| East and west | $48 \times 64 = 3072$ |
| South | $27 \times 188 = 5072$ |
| Sum | 9228 |

3. The total area of the doors and windows is 372 square feet; hence the net wall area is $1600 - 372 = 1228$. Masonry wall 3-inch insulation from Table E:

$$\text{Heating: } 0.77 \times 1228 = 946$$

$$\text{Cooling: } 1.6 \times 1228 = 1965$$

4. The ceiling has 3-inch R-11 insulation. The area is $40 \times 60$ feet $= 2400$ square feet. From Table E:

$$\text{Heating: } 0.88 \times 2400 = 2112$$

$$\text{Cooling: } 4.1 \times 2400 = 9840$$

5. The floor is a concrete slab with 1-inch edge insulation (R-5). The area is also 2400 square feet. From Table E:

$$\text{Heating: } 4.1 \times 2400 = 9840$$

$$\text{Cooling: } 0 \times 2400 = 0$$

6. Infiltration: From Table D with floor area greater than 2100 and best (not severe) winter air change: 0.3. Use the formula in Figure 15.3.

$$\text{Heat loss: } 0.3 \times 0.18333 \times 2400 \times 8 = 1056$$

$$\text{Cooling gain: } 2400 \times 0.01833 \times 25 \text{ (temperature difference)} \times 0.2 = 1760$$

7. Subtotal of BTU heat loss (per 10°F) = 15,160

Adjustment factor for heating (from Table C) for 50°F temperature difference: 5

$$\text{Total BTU heat loss } 5 \times 15{,}436 = 77{,}180$$

8. The cooling gain for people is $4 \times 300 = 1200$.
9. The subtotal for cooling gain is

$$9228 + 1965 + 9840 + 1760 + 1200 + 1200 = 25193.$$

10. The duct loss and gain factor for vented crawl space with R-6 insulation is, from Table F,

Winter design below 15°F factor: 1.05 Cooling gain: 1.0

11. The total heat loss in winter is 81,039 BTU; the total heat gain in summer is 32,750 BTU.
12. Unit specification is as follows:
For heating: 80,000 BTUs
For cooling, since 1 ton = 12,000 BTUs:
$\frac{32,750}{12,000} = 2.72$, or 3.0 ton unit

Similar calculations can be made in developing plant and office requirements. The variations are due mainly to the fact that in air-conditioning calculations, the heat gain through sources such as glass, air ventilation, lights, motors, and people should be based on the most adverse condition during the working period. The table values in house calculations are average values over a 24-hour period, including the use of standard appliances. Heat generated by people also depends on factors such as the distribution of male, female, children, and adults, and activities performed in the workplace, such as standing, sitting, and eating, and type of jobs, such as clerical, light work, or heavy manual work. Nevertheless, a good estimate for plant air-conditioning can be made using the tables provided and adding the heat gains by lights and motors people as:

Heat gains due to incandescent lighting:

$$3.4 \times (\text{installed wattage})$$

Heat gain due to fluorescent lighting:

$$4.1 \times (\text{installed wattage})$$

Heat gain from motors varies based on actual use of the motor, which may be working at a different (usually lower) horsepower than what is specified on the plate. The comparison of actual current drawn with the plate current at which the horsepower was rated gives the working horsepower, as illustrated in the example in Table 15.3. This table is also used to determine BTUs generated by the motors.

### 15.2.10 Telephone Services

A company's management has many choices in selecting a telephone system. They can contact the local telephone company or any of the private firms that install such systems. The equipment can be bought or rented. Local and long distance telephone service can be provided by any of a number of companies such as Verizon, Sprint, and AT&T. The charges in each case should be examined before contracting for the services; some companies charge a low transmission fee but set monthly fees, some charge no setup cost but have a higher transmission cost, and some companies give discounts for calling during certain periods, while others do not.

### 15.2.11 Conservation Measures

The high cost of utilities can be reduced by conducting and following up on careful energy audits and by following conservation measures. A large energy loss occurs if

**TABLE 15.3**
**Heat Gain From Electric Motors (Continuous Operation — BTU/hour) (Courtesy of Trane Company, Tyler, TX)**

| | | | Location of Motor and Driven Equipment with Respect to Conditioned Space or Air Stream | | |
|---|---|---|---|---|---|
| | | | A | B | C |
| Motor Nameplate or Rated Horsepower | Motor Type | Full Load Motor Efficiency (%) | Motor In, Driven Equipment In (BTU/hour) | Motor Out, Driven Equipment In (BTU/hour) | Motor In, Driven Equipment Out (BTU/hour) |
| 0.05 (1/20) | Shaded pole | 35 | 360 | 130 | 240 |
| 0.08 (1/12) | | 35 | 580 | 200 | 380 |
| 0.125 (1/8) | | 35 | 900 | 320 | 590 |
| 0.16 (1/6) | | 35 | 1160 | 400 | 760 |
| 0.25 (1/4) | Split phase | 54 | 1180 | 640 | 540 |
| 0.33 (1/33) | | 56 | 1500 | 840 | 660 |
| 0.50 (1/2) | | 60 | 2120 | 1270 | 850 |
| 0.75 (3/4) | 3 phase | 72 | 2650 | 1900 | 740 |
| 1 | | 75 | 3390 | 2550 | 850 |
| 1.5 | | 77 | 4960 | 3820 | 1140 |
| 2 | | 79 | 6440 | 5090 | 1350 |
| 3 | | 81 | 9430 | 7640 | 1790 |
| 5 | | 82 | 15,500 | 12,700 | 2790 |
| 7.5 | | 84 | 22,700 | 19,100 | 3640 |
| 10 | | 85 | 29,900 | 24,500 | 4490 |
| 15 | | 86 | 44,400 | 38,200 | 6210 |
| 20 | | 87 | 58,500 | 50,900 | 7610 |
| 25 | | 88 | 72,300 | 63,600 | 8680 |
| 30 | | 89 | 85,700 | 76,350 | 9440 |
| 40 | | 89 | 114,000 | 102,000 | 12,600 |
| 50 | | 89 | 143,000 | 127,000 | 15,700 |
| 60 | | 89 | 172,000 | 153,000 | 18,900 |
| 75 | | 90 | 212,000 | 191,000 | 21,200 |
| 100 | | 90 | 283,000 | 255,000 | 28,300 |
| 125 | | 90 | 353,000 | 318,000 | 35,300 |
| 150 | | 91 | 420,000 | 382,000 | 37,800 |
| 200 | | 91 | 559,000 | 509,000 | 50,300 |
| 250 | | 91 | 699,000 | 636,000 | 629,004 |

*Notes:* Normally motors do not operate at the nameplate power or run continuously. Appropriate load and usage factors should be used with this table.

the insulation in the building is poor, but the condition can be corrected by adding or replacing insulation in ceilings and walls.

Example:
Nameplate horsepower = 2 hp at 7 A Pro-rated load = 6440 × (3.5/8) × (5.4/7)
Measured amperes = 5.4
Run time = 3.5 hours per 8-hour shift Motor load = 2174 BTU/hour

Other means to conserve energy are making windows airtight, installing self-closing doors, and hanging heavy plastic strip (2–6 inches wide) curtains on large door openings such as those in receiving and shipping that are seldom closed. Turning off or setting back heating and air-conditioning units when not in use and setting the temperature a couple of degrees lower than standard during winter months and higher than standard during summer also contribute to energy savings. It is desirable that all natural gas-fired equipment such as boilers, hot water heaters, and furnaces be regularly inspected and adjusted if necessary.

In parking lots and warehouses, low-pressure sodium lights, which provide a yellow illumination, are very economical and are quite satisfactory. In the plant and offices, metal halide lights are preferred to fluorescent lights, both giving the same color of light but the latter being more expensive to operate than the former. In a plant with high ceilings, infrared or radiant heaters located close to the floor are more economical than heating the entire area. In some cases, ducts direct heat or ventilation to the immediate vicinity of workers to avoid the added expense of heating or cooling an entire room or building when the space is very big and largely unoccupied. If possible, heating with electric resistance heaters should be avoided because they cost almost 2½ times as much to operate compared to gas-fired heaters. Management can investigate the feasibility of using waste heat from manufacturing in heating or preheating. Combustible scrap such as wood, sawdust, and solvents can be used in heating whenever it is possible to safely do so.

Floating balls or wafers of polypropylene in vats used in dipping and painting operations can save as much as 60–70% of the heat loss due to radiation. Production processes with waste gases that need to be exhausted should be isolated and provided with their own air inlets so as not to use any treated air (heated or cooled) from the plant. Preventive maintenance programs in most cases eliminate air leaks in compressed air systems.

Management should become familiar with utility charge structures, especially the structure for electricity. If there is significant reduction in cost by having a high power factor, the possibility of using a power factor controller can be investigated. Running electric motors at low loads should be avoided. If there is an item of equipment with a large electrical demand, such as a smelting furnace or a foundry, consideration should be given to using a demand limiter. Whenever equipment such as an electric motor is to be replaced, it is desirable for the management to investigate the feasibility of buying a high-efficiency replacement. Using special drive belts in pulley-driven devices is more efficient than operating with regular V belts.

## EXAMPLE—UTILITY COSTS FOR ACE, INC.

The following is a typical set of utility specifications and associated cost estimates for a small manufacturing plant.

1. *Water.* It is estimated that the plant will require approximately 10 gallons of water per hour, or $^1/_6$ gallons per minute. A 6-inch-diameter water line will be installed. The annual cost for water is estimated at $3800.
2. *Sewage.* The sewage outflow is estimated to be three-fourths of the water intake. The plant will connect onto the city sewerage system at a cost of $2253 per year and dispose of waste through a 4-inch-diameter pipe. The annual cost of the water disposed is approximately $4800 listed above.
3. *Telephone.* Three lines will be required to service seven phones. Two lines will be connected to the front desk, and the third line will be located in the maintenance department. The costs are:
   Installation: $250
   Monthly charge: $130
   Total annual telephone cost: $1800
4. *Gas.* Not required
5. *Electricity requirements*:
   Offices: 120 volts, single-phase, 50 amperes
   Heating/air-conditioning: 460 volts, 3-phase, 100 amperes
   Production machines (each of three lathes): 230 volts, 3-phase, 20 amperes
   Drill press: 230 volts, 3-phase, 10 amperes
   Total annual electricity cost: $6500
   Total annual cost for utilities: $20,713

## EXAMPLE—SAMPLE CALCULATION FOR DETERMINING THE AMOUNT OF COOLING WATER REQUIRED

Given that a cooling tower must handle a load of 5 million BTU/hour, that the tower can reduce the water temperature by an average of 75°F during the year, and the following data:

$q$ = 5,000,000 BTU/hour
$c_p$ = 1 BTU/pound °F
$t_t$ = 115°F
$t_2$ = 75°F
$e_w$ = 8.33 pounds/gallon

where
$q$ = cooling load, BTU/hour
$c_p$ = heat capacity of cooling, water, BTU pound °F
$t_1$ = inlet cooling water temperature, °F
$t_2$ = outlet cooling water temperature, °F
$e_w$ = density of water, pounds/gallon

Determine the minimum cooling water flow rate required for the manufacturing plant.

### SOLUTION

It is usually not desirable to heat cooling water above 120°F because mineral compounds will begin to precipitate, causing fouling of equipment. Knowing that the maximum allowable cooling water inlet temperature will be 115°F, we have

$$f_w = \frac{q}{c_p(t_t - t_2)60}$$

where
$f_w$ = required cooling water flow rate, gallons/minutes (gpm).

$$= \frac{5,000,000}{1(115-75)(8.33)60 \text{ in}} \frac{\text{BTU}}{\text{hr}} \frac{\text{1b}}{\text{BTU}} \frac{1}{\text{YF}} \frac{1}{\text{1b}} \frac{galhr}{\text{min}} = 250 \text{ (gpm)}$$

## EXAMPLE—SAMPLE CALCULATION SHOWING THE PROFITABILITY OF A HIGH POWER FACTOR FOR ELECTRICAL DEMAND

Given that a manufacturer has an industrial electrical power rate schedule that specifies a demand charge of $3.00 per kilovolt-ampere (kVA) per month and that the plant's induction motors operate at an average power factor of 0.63 and draw 55 kW from the 440 V supply, determine (1) the capacitor rating required to raise the power factor to 0.85 and (2) the annual savings that would result from such an increase in the power factor. The symbols used are:

$P$ = power (sometimes called real power), measured in kW
$S$ = apparent power, measured in kVA
$Q$ = reactive power, measured in kVAR
kW = kilowatts; measure of power
kVA = kilovolt-amperes; measure of apparent power
kVAR = kilovolt-amperes, reactive; measure of reactive power
$P_f$ = power factor, cosine of the angle by which apparent power leads or lags real power

### Solution

Use the power triangle method (Figure 15.4).
First, for the 0.63 power factor:

$$\arccos 0.63 = \alpha_1 = 50.945$$

$$\tan 50.945 = 1.2325$$

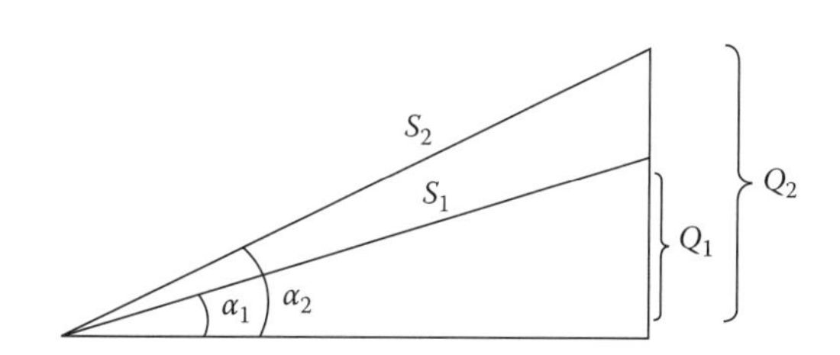


**FIGURE 15.4** Power factor correction

$$Q_1 = 55 \times 1.2325 = 67.79 \text{ kVAR}$$

$$S_1 = \sqrt{55^2 + 67.8^2} = 87.3 \text{ kVA}$$

For the 0.85 power factor:

$$\arccos 0.85 = \alpha_2 = 31.788$$

$$\tan 31.788 = 0.6197$$

$$Q_2 = 55 \times 0.6197 = 34.08 \text{ kVAR}$$

$$S_2 = \sqrt{55^2 + 34.1^2} = 64.71 \text{ kVA}$$

Reactive kVA at 63% power factor = 67.8 kVAR
Reactive kVA at 85% power factor = 34.1 kVAR
The reactive load rating of the capacitor required to raise the power factor from 63% to 85% is 67.8 − 34.1 = 33.7 kVAR.

Second, determine the annual savings:
kVA demand at 63% power factor = 55 kW/0.63 = 87.3 kVA
kVA demand at 85% power factor = 55 kW/0.85 = 64.7 kVA
The annual demand charge savings are:

$$\text{Month/year} \times (\text{kVA at 63\% power factor} - \text{kVA at 85\% power factor})$$

$$\times \text{ cost/kVA/month}$$

$$= 12 \times (87.3 - 64.7) \times 3.00$$

$$= \$813.60$$

The initial cost of the capacitor, then, can be economically evaluated against the energy savings.

## 15.3 INSURANCE

Potential loss to a firm due to fire, other accidents on the premises, or lawsuits resulting from allegedly defective products could be considerable, even catastrophic. Insurance is a contractual system wherein the risk of individual loss is spread over the large number of members who pay a fee and share in both the protection and the risk. The contract is an insurance policy, the fee is a premium, the member is the insured, and the administrator of the system is the insurance company.

The system under which all insurance companies operate is based in large measure on statistical principles. For example, actuarial tables have been developed (and are periodically reviewed and revised) that provide listings of life expectancies for people in a variety of categories. Reference to the appropriate table will advise how many more years the average 34-year-old white male in the United States, for example, may expect to continue living. Many such tables inject other variables such as level of education, principal occupation, and location of residence. It should be anticipated that, all other things being equal, an individual in a category of people expected to die within 5 years will pay a higher premium for a new life insurance policy than will someone in the group that statistically is expecting to live an additional 25 years.

### 15.3.1 Types of Insurance

Life insurance is but one of a great many different types of insurance coverage available. Some insurance companies specialize; some write only life insurance; others limit themselves to insuring carnivals and circuses; still others insure nothing but boilers and machinery.

A basic division for the larger, multiline companies is the division between personal lines and commercial lines. The former might include such coverages as private automobiles, life, homeowners, and personal liability, which are written for individuals. The commercial lines portion of the company handles business insurance ranging from that for the mom-and-pop delicatessen on the corner to coverage for the largest oil refinery or automobile assembly plant. Because our objective is to discuss insurance as it relates to business and industry, we will concentrate on the latter.

### 15.3.2 Commercial Lines

At times, the owner or risk manager of a business might feel that he or she is faced with an infinite number of different coverages from which to choose. Certainly no one wishes to be overinsured, but being underinsured in a particular area could prove devastating. Choosing insurance means making business decisions that are no less important than decisions in other facets of the operation.

Many insurance companies split the commercial lines into property and casualty; however, there are sometimes overlaps. Some arbitrary distinction must be made, and we shall consider property to include primarily fire, along with contractors' equipment and boiler and machinery protection. A fire policy is not nearly as limited as the name would suggest; it also includes windstorm, lightning, hail, explosion, and water and smoke damage. By endorsement (amendment), it can also provide coverage against glass breakage, fire sprinkler leakage, riot damage, damage by aircraft and other vehicles, and other perils. When the direct loss is insured, protection against the indirect loss due to business interruption is also available by endorsement.

Casualty insurance normally includes workers' compensation business automobiles (including trucks), the various types of liability, and burglary. When a package policy is written that contains the elements of both property and casualty — for example, fire and liability — it is usually the property side of the insurance company that issues it.

#### 15.3.2.1 Essential Coverages

It would be a poor management practice for a business to own its building and not have property insurance, or to open its doors to the public without premises liability insurance. In this litigation-prone society, a manufacturer without product liability insurance is in danger of bankruptcy.

Practically all states require annual certification that a company's boilers are safe to operate. If the boilers are insured, the insurer is obliged to conduct the inspections necessary to make the determination.

It would be extremely difficult to find many businesses that do not insure their vehicles. Just as with a personal automobile, the company truck can be insured against collision, theft, liability, and comprehensive loss. Some large companies are able to self-insure against everything except liability, but very few have the resources necessary to sustain the large liability loss that can occur at any time. Hence liability insurance is almost universal.

Elements that influence the amounts of the premiums for different lines of coverage are relatively simple in most cases. For example, the installation of an automatic sprinkler system will reduce the fire premium, while the use of flammable liquids in the manufacturing process will increase it. Company drivers who have successfully completed a defensive driving course might contribute to a reduction in the automobile premium; however, a number of moving violations and/or accidents by a driver will result in an increase.

#### *15.3.2.1.1 Workers' Compensation*

Probably the most important type of insurance for business and industry is workers' compensation (WC), owing mainly to the very high premiums and losses associated with this line of coverage. Because the factors affecting the WC premium determination are more complicated than those for other types of insurance, we will limit the remainder of our discussion to WC.

Every state regulates this protection to some degree because the welfare of so many people is involved; more than three of every four employees in the United States are covered by WC laws. A case could be made for a federal WC law because of the tremendous variance among the laws of the individual states. Some states prohibit self-insurance, and some permit it; a few require that a state insurance fund be used, some simply allow it, and others do not even have such a fund. Private insurance is either required or permitted in almost all states.

Self-insurance for WC is not as simple as saying that if an employee is injured, he or she will be compensated from the company's operating funds. States that permit self-insurance have strict rules that require approval by the state for each prospective plan. The employees are protected by requiring the employer to post a bond or securities, or set up a cash reserve to indemnify each worker who might be injured if the injury resulted from an accident arising from his or her employment and such accident occurred during the course of that employment. (These three elements for a legitimate claim are applicable to all WC coverage, not just self-insurance.)

The biggest difference found in regulations in the various states is in the schedules of payments for injuries. In some states, the loss of a thumb will bring the employee higher payments than the loss of an arm at the shoulder will in other states. In some states, the loss of a leg will not pay as much as will the loss of a foot in other states. The WC laws in all states require payment of medical benefits; but some do not pay the entire bill or may limit the time for which the bills are paid, or they may set an upper limit on the total amount that will be paid.

The premium for WC insurance can vary considerably on the basis of several factors. The first of these is the location of the business. Each state regulates the premium paid in that state, but sometimes the insurance company is permitted to reduce the premium below the maximum to be competitive.

A second factor is the categories and numbers of workers employed. The Insurance Services Office prepares a set of tables that, in effect, tell how hazardous all of the different job categories are. The underwriter finds the appropriate factor in the table, multiplies the payroll for the category by that factor, and thus determines the basic premium for that group of workers. The justification for the factor should be fairly obvious; for example, the likelihood of a compensable injury occurring in an office with only two workers is far less than it is for a company with 25 pulpwood cutters. Chain saws in the woods are more dangerous than typewriters in the office. Payroll enters the pictures not only to account for the number of employees, but to consider how expensive an injury might be. Compensation awards are based on percentages of salaries, so the more highly paid the worker, the higher the award. Thus the premium must be higher to statistically cover that predictable (within limits) cost.

A third factor is the loss experience of the particular business. The Bureau of Labor Statistics annually publishes a series of tables that list two important items: the rates of incidence (frequency) of all accidents and, more serious, the lost-time accidents for specific types of businesses (e.g., tire manufacturers) and for broad categories (e.g., all manufacturing). The insured firm (the account) can be compared to its particular category, and the underwriter can then form an opinion concerning the desirability of insuring that company. The loss ratio (value of claims paid divided by the premium) is an indicator of the probable profitability of the account to the insurance company. From this historical (experience) ratio, the underwriter can determine by what factor the premium previously determined must be multiplied so that the insurance company, or the state fund, will probably not suffer a net loss for this account for the coming policy period.

A fourth factor can be the size of the company to be insured. A major cost to an insurance company is that for its overhead and administration, usually in the range of 30–40 cents of every dollar of premium collected. The cost, however, to service a WC policy with a premium of $1,000,000 is nowhere near 100 times as high as that to service the $10,000 policy. Because of this and perhaps other favorable factors (instituting training and safety programs and installing newer, safer equipment), premiums may be "deviated" or discounted if the underwriter feels the reduction is justified.

A different system of rating is the retrospective rating plan. The premium as determined above is set as the target, and the account pays this amount at the beginning of the policy period (normally 1 year). At the end of the period the target is adjusted upward or downward depending on the loss experience of the account for the period. Thus the employer is rewarded for safe operation and penalized if its employees suffered more (or more costly) injuries than were anticipated.

#### *15.3.2.1.2 Insurance Planning*

In order for the reader to gain some appreciation of the cost of business insurance, an example is presented that itemizes the costs of insurance for one type of business.

**TABLE 15.4**
**Annual Insurance Premiums for the Folding Table Manufacturer**

| Insurance | Premium |
|---|---|
| Property ($16.38 per $1000 of value) | 7127 |
| Workers' compensation | |
| Office ($0.32 per $100 of salary) | 480 |
| Production ($6.84 per $100 of wages) | 24,624 |
| Hospital and medical ($430 per month per employee) | 103,200[a] |
| Premises liability ($230 per month) | 2760 |
| Product liability ($0.55 per $100 of sales) | 1320 |
| Automobile (1 pickup truck) | 1015 |
| Total | 140,526 |

[a]This premium is partially offset by workers' contributions.

### EXAMPLE—FOLDING TABLE MANUFACTURER

This small company is located in a town of some 10,000 persons. It operates out of a prefabricated metal building of approximately 7900 square feet valued at $435,000. There are five office workers (salaries totaling $12,500 per month) and 15 hourly production employees (salaries of $30,000 per month). Sales average $2,400,000 per year, and products are shipped by common carrier. The annual insurance premiums for this company are itemized in Table 15.4.

## 15.3 SAFETY

Accidents can seriously affect the financial resources of a company. Because of the possible high cost in terms of human suffering and lost production, a business should place particular emphasis on safety measures.

When an accident occurs, either machines, people, or both are involved. If equipment is damaged, the machinery must be repaired or replaced, and the unsafe conditions must be rectified. The cost of an accident in such cases includes the use of emergency equipment, loss of production, and the need for cleanup equipment and crews.

In addition to humanitarian concerns, the loss to the company when an employee is injured can be extreme. A trained employee is at least temporarily displaced, causing a slowdown in production and perhaps a degradation of quality, all of which can contribute to a reduction of income. There is also the additional cost in terms of payment of claims and legal fees, training of a new employee, and an increased insurance premium.

If an accident results in injury or damage that is minor, it should be taken as a warning to be heeded. The employee who gets up rubbing his or her lower posterior after slipping in spilled oil is lucky; if corrective action is not taken, the next person passing by might not be able to get up at all.

### 15.3.1 When to Plan

Safety considerations should be deliberated and implemented during the early stages of plant development, preferably when the building is being planned. Appropriate aisles and adequate storage space should be provided. Exits that are easily accessible should be properly located and prominently marked. Stairs, ramps, and ladders should be sturdy and coated with slip-resistant material.

After the building has been designed, the next step in setting up a plant is usually the selection and installation of equipment. The safety standards for the equipment are very important; for instance, there should be a minimum of exposed moving parts that could injure an operator. Also, when the equipment is set up, there should be no obstruction of movement of people and materials through aisles or loading spaces.

When hazardous materials are being handled, extra precautions must be taken to ensure the safety of the employees. These materials should be properly stored and handled, and ventilation should carry away dangerous vapors. Fire equipment should be readily available around combustibles, and eye and body showers should be installed to be used in case of chemical burns. Potential danger to the eyes, as in the case of electric arc welding, can be guarded against by the use of proper goggles or shields.

Once a safe plant has been set up, it is relatively easy to maintain the safe conditions; this job becomes the responsibility of every employee. When all the safety procedures have been defined, they should be followed religiously. Care should be taken never to leave equipment or materials protruding into or obstructing aisles or work spaces. Dangerous areas should never be entered without the proper equipment and instructions. Potentially hazardous situations such as slippery floors should be corrected quickly.

If the equipment in a plant is properly set up and all care has been taken to maintain a safe environment, then most resulting accidents will be caused by improper procedures or human error. This is why it is important to see that employees are properly trained and that appropriate safety methods are constantly being implemented. Employees should be aware of potential hazards such as worn machinery and should report such conditions immediately.

### 15.3.2 Regulation

Because the safety of all employees is so very important, the federal government has enacted laws to help ensure safe and healthful working conditions for all. The most important law in this regard, passed in 1970, is known as the Occupational Safety and Health Act, or OSHA. (The abbreviation also stands for Occupational Safety and Health Administration.)

One of the purposes of OSHA is encouraging efforts to reduce the number of occupational hazards. OSHA has authorized the Secretary of Labor to set mandatory safety and health standards and to divide the responsibilities between employers and employees. From the accident prevention standpoint the act has provided for the National Institute for Safety and Health to conduct research in occupational safety and health problems. A company that is found to be in violation of any of the standards set by OSHA may be punished by a fine, or its management may be imprisoned in extreme cases.

Government regulations are not limited to OSHA. Most states have their own safety standards that must be adhered to by industries within their jurisdictions. In many instances, OSHA has provided grants to states to assist in their developing occupational health and safety standards.

### 15.3.3 Responsibility within a Plant

The responsibility for safety in a given plant, as well as for almost everything else there, rests ultimately with the president of the corporation. The president may be able to assign the responsibility for devising and implementing safety procedures to the plant manager, but that just makes the manager accountable to the president, who remains accountable to the stockholders and the government. If the plant manager should hire a safety manager, the latter would be directly accountable to the plant manager, who in turn still reports to the president, and so on. Of course, if any individual in the chain were to violate the law (OSHA) with regard to safety, that person would be subject to prosecution just as he or she would be for breaking any other law.

Viewed from the positive side, industrial safety is good because it protects the health of workers, reduces injuries, helps prevent deaths, and is cost-effective: it increases productivity. The primary difficulty in implementing a safety program is apathy; very many people are victims of the it-can't-happen-to-me syndrome. Unfortunately, it can and does — maybe only once, but once is often enough.

Because constant attention is required to establish and maintain safe conditions and procedures in an industrial setting, the best alternative is generally to have a safety professional (safety manager or safety engineer) on the staff. We know intuitively that a plant with a two-person production department cannot support a staff including a full-time safety manager, but we should also realize that a metal-fabricating plant employing a production force of 1000 people would probably be negligent if it did not employ a safety manager.

In addition to top management and the safety professional, production supervisors bear a great deal of responsibility for the safe operation of a plant. These are the people who are in continual contact with the individual workers, and thus the supervisors are in a position to observe unsafe conditions of their facilities and equipment and unsafe procedures by their crews. The maintenance supervisor is, of course, responsible for ensuring that all items of equipment and machinery are kept in safe operating condition; and the workers, the people with the most to gain or lose, are responsible for following safe procedures and reporting unsafe conditions.

### 15.3.4 Safety Program

Every business should have a safety program, the details and complexity of which depend on the size and type of the operation. The plan should always include a policy statement and, in almost every case, safety rules. Most plants need written work procedures that should emphasize safety. Likewise, most will profit from safety meetings. Safety inspections should be held regularly, and the possibility of instituting a system of providing safety incentive awards should be investigated.

Like all policy statements, the safety policy statement should be in general terms, but it must be sincere, and it must be followed. Lip service to a principle is quickly recognized for what it is and can be more damaging than doing nothing. The statement should be signed by the most senior official at the plant or by the president in the home office if a company-wide policy is appropriate. There should be widespread dissemination of the policy statement; each employee should receive a copy when it is first published (in the pay envelope perhaps), each new employee should be have one upon being hired, and every bulletin board in a plant should display a copy prominently. (We include as an example the safety policy statement for JOROCO Electronics, Inc. See p. 654)

Safety rules vary widely with the needs of the organization, and it can prove dangerous to attempt to implement a "standard" set of rules that are probably not specific enough for a given set of circumstances. As in the case of the safety policy statement, safety rules should reach every employee. The vehicle might be an employee handbook, which would also contain information about such matters as insurance, taxes, vacations, holidays, promotion policies, and retirement. The ever-present bulletin board should supplement the handbook. (The safety rules for the JOROCO Electronics are furnished as an example.)

Safety meetings can be either an invaluable means of reinforcing the idea of safety in the minds of the workers or a boring, unproductive waste of time. Each meeting must be organized, and everyone must know that it is important. Sometimes the safety meetings are small, brief, informal gatherings held by the supervisor during working hours; sometimes they include a company or departmental dinner after work. In the latter case, the plant manager or the president might say a few words about the safety record, and the safety director or an outside consultant might present a formal program on a timely topic. Regardless of the format chosen, the employees should always be afforded the opportunity to ask questions and present opinions.

Incentive awards can be given to remind the employees about the importance of safety and to encourage them to work in a safe manner. There is almost no limit to what the award can be. If the business has several locations, the plant with the best safety record for the year could be awarded a bonus, either a fixed amount or a percentage of salary, as a tangible sign of appreciation by management. Pen-and-pencil desk sets, plaques, playing cards, certificates, calendars, jackets, T-shirts, sweatshirts, belt buckles, and ball caps are items that, when suitably engraved or imprinted, can be presented during safety meetings in recognition of accident-free periods (months or years) for individuals or groups. Some companies have a standing offer of an all-expense-paid vacation in some attractive locale for employees who complete a lengthy period (perhaps 10 years) without an accident. In general, the awards system should combine relatively frequent presentations of inexpensive gifts for minor achievements with the occasional awarding of more valuable items for significant accomplishments.

The safety manager, or the person with the collateral duty, usually operates in a staff function, reporting directly to the plant manager or the president. The responsibility for a properly functioning safety program is this person's, and all matters relating to safety must be routed to or via his or her desk. The safety manager should conduct inspections in addition to those made by department heads and area supervisors, and should maintain records of those inspections and check on efforts to correct deficiencies. One of the safety manager's more important tasks is analyzing accident reports and taking action to make certain that no one repeats the circumstances that caused an earlier accident. He or she is also responsible for the preparation and posting of all OSHA-required reports.

## EXAMPLE—SAFETY POLICY STATEMENT

The primary purpose of this statement is to establish throughout JOROCO Electronics, Inc., the concept that our employees and property are our most important assets. Protection of people from injury and the conservation of property are our top priorities and will receive support and participation by all levels of management.

It is the policy of JOROCO Electronics to ensure that this plant remains a safe place in which to work, one that is free from fire and accident. I seek your help in accomplishing this goal and will appreciate your efforts toward that end.

The safety of our employees and the prevention of loss are principal responsibilities of management. Efficient production cannot exist without accident prevention. Management and supervisory personnel must keep in mind that the safety and well-being of all employees are their primary responsibility. It can be met only by consistently striving to promote safe work practices and maintaining property and equipment in safe operating order.

The keystone in our safety program is the manager because the manager is continually involved with employees, and safety is people. Neither a foreman, a supervisor, nor an operating head may at any time delegate any portion of his or her responsibility for safety to a staff organization; it is an operating function.

Plant workers must always remember that safe practices are an integral part of all operations. If workers have not followed every safety rule and taken every precaution to protect themselves and their fellow workers, the job has not been completed efficiently. Production and safety are inseparable at JOROCO.

Accidents do not have to happen; they are preventable, and we must not believe otherwise. We all have our part, and it includes acting safely and talking safety at all times. A healthy attitude toward preventing accidents and improving our safety record can be achieved if each of us does his or her part.

Reducing accidents and their related insurance expenses will also enable us to be more competitive in our industry. As we improve as a team in this regard, we enhance our chances as individuals for greater tangible benefits.

**J.R. Coleman**
President

## EXAMPLE—SAFETY RULES

In keeping with the established policies of JOROCO Electronics, Inc. the following rules pertaining to the safety of personnel and property are made effective immediately. Anyone failing to comply with an applicable rule will be subject to disciplinary action including termination if appropriate.

1. Personal protective equipment must be worn as specified.
   a. Hard hats are to be worn in production and loading areas.
   b. Steel-toed safety shoes are to be worn by all employees in production and loading areas.
   c. Goggles are to be worn by all employees operating drills or working in painting or polyurethaning areas.
   d. Earplugs are to be worn in designated high-noise areas.
   e. Respirators are to be used in painting and polyurethane-dispensing areas.
2. Documented procedures, to be followed in an emergency situation, are to be made available to all employees. Evacuation procedures are posted along with emergency phone numbers throughout the plant. Exits are to be clearly marked and kept easily accessible.
3. Sprinklers are installed and fire extinguishers spaced according to regulations set forth by the fire marshal. These are to be tested by authorized personnel only and may not be used except in the event of an emergency.
4. Eyewash and shower equipment are placed in areas where hazardous or flammable materials are handled. The posted instructions regarding their use must be followed to prevent possible serious injury.
5. First aid kits are available in each department and are to be used for minor injuries suffered on the job. Department heads will ensure that the kits are inventoried monthly and replenished as necessary.
6. All accidents and injuries shall be reported to the appropriate supervisor as soon as practicable. Those persons at the scene where serious injury has occurred should take immediate action in accordance with those procedures referenced in paragraph 2 above. The supervisor is to complete an accident report and forward it to the safety manager as soon as possible following any accident.
7. Continuing safety is the responsibility of all plant employees. Each department head will publish safety procedures for operation of all machinery and equipment within his or her jurisdiction. The procedures will become a part of these safety rules and will be enforced as such. No employee is to operate any equipment with which he or she is not familiar.
8. Each area supervisor is required to complete a safety check sheet weekly. He or she will mark appropriate answers — yes, no, or not applicable — for each of the 22 statements and give an explanation for any unsatisfactory conditions (Figure 15.5).

### 15.3.5 Americans with Disabilities Act of 1990

The ADA, passed in 1990, guarantees a qualified individual with a disability access to employment, public accommodations, transportation, public services, and telecommunications. The law applies to all state and federal agencies and covers organizations in the private sector having 15 or more employees. A qualified individual with a disability is a person who, with reasonable accommodation, can perform the essential functions of the employment position that individual holds or desires. As long as it does not cause an undue hardship, an employer has an obligation to provide a reasonable accommodation, which is defined as modification to a job or work environment that would enable the individual with a disability to perform the job. The undue hardship to an employer may involve high cost or substantial work in redesigning the work place, disruption of business, or causing the nature or operation

DEPARTMENT________________ DATE ________________________

| | YES | NO | N.A. |
|---|---|---|---|
| 1. RECORD OF INJURIES IS UP TO DATE | | | |
| 2. WORK AREAS ARE CLEAN | | | |
| 3. AISLES ARE CLEAR | | | |
| 4. ALL FLOORS ARE CLEAN AND DRY | | | |
| 5. LOAD LIMITS FOR MACHINERY AND STORAGE AREAS ARE CLEARLY MARKED | | | |
| 6. ALL PLATFORMS 4 FEET ABOVE FLOOR LEVEL HAVE SECURE RAILS | | | |
| 7. EXITS ARE CLEARLY MARKED AND EASILY ACCESSIBLE | | | |
| 8. NOISE EXPOSURE LEVELS HAVE BEEN CHECKED FOR PERMISSIBLE LEVELS | | | |
| 9. HARD HATS ARE WORN BY WORKERS | | | |
| 10. STEEL-TOED SHOES ARE WORN BY WORKERS | | | |
| 11.GOGGLES ARE WORN | | | |
| 12. RESPIRATORS ARE USED | | | |
| 13. "NO SMOKING" SIGNS ARE POSTED | | | |
| 14. FIRST AID KIT IS COMPLETE AND EASILY ACCESSIBLE | | | |
| 15. FIRE EXTINGUISHERS ARE MOUNTED IN PROMINENT LOCATIONS | | | |
| 16. FIRE EXTINGUISHERS ARE PROPERLY CHARGED | | | |
| 17. EVACUATION PLAN IS CLEARLY VISIBLE | | | |
| 18. EMERGENCY PHONE NUMBERS ARE POSTED | | | |
| 19. EQUIPMENT HAS BEEN INSPECTED FOR HAZARDS | | | |
| 20. PREVENTIVE MAINTENANCE HAS BEEN PROPERLY PERFORMED | | | |
| 21. ALL OVERHEAD LIGHTS ARE IN WORKING ORDER | | | |
| 22. STEPS HAVE BEEN TAKEN TO MAKE NEEDED REPAIRS AND TO CORRECT ANY UNSAFE CONDITIONS | | | |

EXPLANATIONS AND COMMENTS:

**FIGURE 15.5** Safety check sheet

of business to change. Most often in a manufacturing plant small modifications to the work area are all that may be needed to satisfy the intent of the law and accommodate a large percentage of disabled workers. We shall review some of the most common recommendations suggested for individuals who require wheelchairs or visual and hearing assisted devices.

### 15.3.6 Assisting Individuals in Wheelchairs

#### 15.3.6.1 Flooring

1. Carpet in the office or laboratory area should be well secured with no more than ½-inch pile.
2. For steel floor used on walkways, the grating should not be greater than ½ inch wide, with the elongated (long) side perpendicular to the direction of travel to accommodate wheelchair travel.

#### 15.3.6.2 Parking

At least one handicap parking space, 96 inches wide, must be provided for every 25 total parking spaces for the first 100 parking spaces. Above 100 spaces, at least one handicap parking space must be provided for every 50 spaces for the next 100 spaces; then one handicap parking for every 100 parking spaces must be provided. A vertical clearance of 114 inches should be provided for passenger loading zones. Other reserved routes for wheelchair vans should have a minimum clearance of 98 inches.

#### 15.3.6.3 Accessible Paths

1. There should be at least one accessible path coming from handicap parking, sidewalks, and roads in the complex to the production facility.
2. There should be at least one accessible route that connects the entrance to the workplace to the handicapped worker's work area and areas he or she may be required to go, such as the quality control room, storeroom, assembly shop, locker room, rest room, and conference room.
3. The minimum clearance for single wheelchair passage should be 32 inches at the door entrance and 36 inches for moving in aisles. For two wheelchairs to pass, the aisle's width should be increased to 60 inches.
4. The surface should be slip-resistant, stable, and firm. Level change of less than ¼ inch should be vertical without an edge. For changes in levels between ¼ and ½ inch, a slope of no more than 1:2 should be used. For level change greater than ½ inch, a ramp should be built.

#### 15.3.6.4 Workstation

The minimum floor space needed to accommodate one immobile wheelchair should be 30 × 48 inches. For a worker in a wheelchair, the maximum front reach height is 48 inches, and the minimum is 15 inches. These specifications are useful in the event the worker needs to push operational buttons for devices such as lathes, milling machines, or control panels.

### 15.3.7 Assisting Visually and Hearing Impaired People

Accommodations for visually and hearing impaired require the following additional considerations.

#### 15.3.7.1 Headroom

All halls, walks, passageways, and moving space between the machines should have minimum headroom of 80 inches. If the clearance for any such area is less than 80 inches, an obstruction or barrier should be provided to warn people of changing clearance.

#### 15.3.7.2 Objects Protruding from Walls

In a production facility there are numerous areas where an object may protrude into a walkway. Such obstructions might include raw materials or finished products stacked near a machine, telephones, water fountains, and fire extinguishers and/or alarms in hallways and walkways. Any obstacles at the floor level should be avoided. Objects that must protrude, such as water fountains and wall telephones, should be installed in an enclosure on the wall such that the bottom edge is not more than 27 inches from the ground. This allows a walking cane to strike the enclosure, providing necessary warning to the person.

### 15.3.8 Other Considerations

#### 15.3.8.1 Emergencies

There should be both audible and visual alarms in a facility. The audible-alarm sound level should be 15 dBa more than the existing sound level in the facility unless the sound level is already too high in the production area due to heavy machinery use. In that case, the sound level of the alarm should be 5 dBa louder. Visual alarms should have flashing signals with a minimum intensity of 75 candlepower with a flash rate of between 1 and 3 Hz

There should be a designated, readily accessible room or area where disabled workers can be gathered in case of a fire. The room should be separated from the rest of the production facility by a smoke barrier and should have a minimum fire rating of 1 hour.

#### 15.3.8.2 Lavatories

The counter surface should be no higher than 34 inches and should provide a clearance of 29 inches above the finished floor to the bottom of the counter. Hot-water drain pipes under the lavatories should be insulated. If the faucets are self-closing, then they should remain open for at least 10 seconds.

#### 15.3.8.3 Toilet Stalls, Urinals, and Drinking Fountains

All facilities should be easily accessible. The toilet stall for a wheelchair should have minimum depth of 54 inches and should have wall-mounted water closets. Grab bars should be provided to assist the disabled in maintaining balance. Urinals should be wall-hung with elongated rim at 17 inches (maximum) above the floor space. A clear space of 30 × 48 inches should be provided in front of the urinal. A drinking fountain's spout should be in the front of the unit, no more than 30 inches from the floor. The control should be mounted in the front and clear space of 30 × 48 inches should be maintained to accommodate a wheel chair.

Presented in the Appendix E are typical checklists. Such checklists are commonly used to determine if facilities are in compliance with the ADA law.

## 15.4 TAXES

Just like individuals, corporations and other businesses are subject to federal, state, and local taxes. They range from an income tax on earnings to contributions to the unemployment compensation fund for the workers. Tax laws change frequently; one detailed source of information about current applicable laws is *Prentice Hall's Federal Taxation*. It is a complete listing of tax laws, rates, and information in an 18-volume set published annually by Prentice Hall. The first volume is an index of tax topics making reference to paragraph numbers in the other volumes.

The following is an overview of the major taxes paid by corporations.

### 15.4.1 Federal Income Tax

The largest single tax paid by corporations is federal income tax, which the company pays on the profits made each year. The 1986 tax laws, as implemented in 1992, give the tax rates shown in Table 15.5.

A corporation with a taxable income exceeding $100,000 must pay an additional tax equal to 5% of the amount in excess of $100,000 up to a maximum additional tax of $11,750.

This extra tax phases out the benefits of the graduated rates for corporations with taxable income between $100,000 and $335,000. A corporation having taxable income of $335,000 or more gets no benefits from the graduated rates and pays, in effect, a flat rate of 34% on all taxable income.

### 15.4.2 Social Security Act

In 1935, Congress passed the Social Security Act, which includes provisions for a system of federal-state unemployment insurance, insurance for old age, old age pensions for the needy, pensions for the blind and for relatives caring for orphans and other destitute children, and an appropriation for the Public Health Service. The entire program is officially named the Old Age, Survivors, and Disability Insurance. These programs are supported by taxes placed on employers and employees based on a percentage of the employee's income.

Payments to individuals qualifying under this act depend on the specific situation. Unemployment benefits, for example, differ from state to state. Old age pensions are based on the amount of taxes paid into the program, the number of years paid, and past and present income. Cost-of-living increases are included for all these programs.

In 1965, Medicare was added to the Social Security Act. Medicare is a federal insurance program for people age 65 and older and some disabled people. The cost of the hospital insurance is financed by social security taxes on employers and employees. Anyone eligible for social security is automatically eligible for this insurance, which covers basic hospital costs, post-hospital extended care, physicians' services, drugs, X-rays, skilled nursing care, and the like. The optional medical insurance is financed by monthly premiums paid by the government and the insured people. It pays for services not covered or not fully covered by the hospital insurance.

Contributions to Social Security are required for most employees by the Federal Insurance Contributions Act (FICA). To receive retirement benefits upon reaching the required age, a person must have paid FICA taxes for at least 40 quarters. That is because the calendar year is divided into four quarters (January through March, April through June, and so forth), the individual must have paid taxes during 40 of those periods to be eligible for retirement benefits. When the law was first passed, the minimum was just six quarters so that people who were near retirement age at that time would not be denied the opportunity to receive Social Security benefits. The tax rate has been raised several times and is imposed on both the employer and the employee, based on the employee's income. The present tax rate structure for FICA is shown in Table 15.6.

## TABLE 15.5
## Corporate Income Tax Rates–2008

| Taxable Income Over | Not Over | Tax Rate |
|---|---|---|
| $ 0 | $ 50,000 | 15% |
| 50,000 | 75,000 | 25% |
| 75,000 | 100,000 | 34% |
| 100,000 | 335,000 | 39% |
| 335,000 | 10,000,000 | 34% |
| 10,000,000 | 15,000,000 | 35% |
| 15,000,000 | 18,333,333 | 38% |
| 18,333,333 | | 35% |

Personal Service Corporations

Personal service corporations are subject to a flat tax of 35% regardless of their income.

Personal Holding Company

Personal holding companies are subject to an additional tax on any undistributed personal holding company income.

| Year | Rate |
|---|---|
| 2007 | 15.0% |
| 2006 | 15.0% |
| 2005 | 15.0% |
| 2004 | 15.0% |
| 2003 | 15.0% |
| 2002 | 38.6% |
| 2001 | 39.1% |
| 2000 and prior years | 39.6% |

Accumulated Earnings Tax

In addition to the regular tax, a corporation may be liable for an additional tax on accumulated taxable income in excess of $250,000 ($150,000 for personal service corporations).

| Year | Rate |
|---|---|
| 2008 | 15.0% |
| 2007 | 15.0% |
| 2006 | 15.0% |
| 2005 | 15.0% |
| 2004 | 15.0% |
| 2003 | 15.0% |
| 2002 | 38.6% |
| 2001 | 39.1% |
| 2000 and prior years | 39.6% |

## TABLE 15.5 (CONTINUED)
## Corporate Income Tax Rates–2008

Miscellaneous Business Data
Maximum Section 179 Expense Deduction

| 2008 | 2007 | 2006 | 2005 | 2004 | 2003 |
|---|---|---|---|---|---|
| $250,000 | $125,000 | $108,000 | $105,000 | $102,000 | $100,000 |

Phaseout–$800,000 for 2008; $500,000 for 2007; $430,000 for 2006; $420,000 for 2005; $410,000 for 2004; $400,000 for 2003; $200,000 for prior years.

Health Savings Accounts–2009
Maximum annual HSA contributions deductible–$3,000 for individual; $5,950 for family coverage

Catch-up contributions for individuals 55 or older (but less than 65) is $1,000 (pro rate for year).

Deductible Amounts that define High Deductible Plan:

| Amounts Exceed | Annual Deductible Not Less Than | Deductibles, Co-Payments and other but not Premiums not to |
|---|---|---|
| Self Only | $1,150 | $5,800 |
| Family Coverage | 2,300 | 11,600 |

### 15.4.3 Unemployment Compensation

The Federal Unemployment Tax Act (FUTA) imposes a tax on employers alone to provide financial aid for the unemployed. Employers have been paying the federal government 3.5% of the first $7000 of annual wages of each employee, with a reduction of up to 2.7% of the total for contributions made to state unemployment funds.

## TABLE 15.6
## The Following Chart Shows the Social Security and Medicare Rates for Each Employer and Employee

| | 2008 | 2007 | 2006 | 2005 |
|---|---|---|---|---|
| Social Security | 6.2% | 6.2% | 6.2% | 6.2% |
| Social Security Wage base | $102,000 | $97,500 | $94,200 | $90,000 |
| Medicare | 1.45% | 1.45% | 1.45% | 1.45% |
| Medicare Maximum | no ceiling | no ceiling | no ceiling | no ceiling |

Notes:

Students that are enrolled half time or more are exempt from Social Security and Medicare tax.

International employees holding F or J visas are exempt. Holders of H Visas are subject to both social security and medicare taxes.

As of January 1, 1985, the tax rate was increased to 6.2% with a credit of up to 5.4% for state taxes.

### 15.4.4 Sales Taxes

A tax imposed on the sale of a broad range of items is known as a sales tax, a major source of revenue for states and localities. These taxes are frequently raised as the needs increase. As might be anticipated, there is a great deal of variance in rates between states; for instance, only a few states allow the establishment of local sales taxes. Items such as alcohol, cigarettes, gasoline, and hotel room rental may be taxed differently, frequently at a higher rate, while food and other necessities may be exempt or have lower rates. Sales taxes are normally paid by consumers; but if a business buys a product or service for its own use, it is usually liable for the sales tax.

### 15.4.5 Property Taxes

Businesses and corporations are required to pay local taxes, the single largest one being property taxes. Property taxes are levied in every city across the United States, although the actual rate varies considerably from city to city.

It can be as low as $0.05/$100 value/year or as high as $3/$100 value/year. Most cities depend on this tax as their single major source of revenue.

Both individuals and businesses are subject to property taxes, which may also be known as real estate taxes. The real estate taxed includes any land owned, houses, and business buildings. In addition to these, businesses may be subject to taxes levied on heavy machinery and inventory.

### 15.4.5 Comments

The list of tax laws is seemingly endless, and many laws can be imposed for specific situations; there are taxes on inheritances, gifts, and accumulated earnings (intangible property), to name a few. But income, property, sales, FICA, and FUTA taxes constitute the major tax burden for any company and are a major source of income for federal, state, and local governments. For more detailed information, the *Prentice Hall's Federal Taxation* is recommended.

## 15.5 FINANCIAL STATEMENTS

The objective of financial statements is to clearly state the status of the work center that the report covers. Within a company, the reports are generally constructed by using the inverted pyramid approach. Detailed reports are generated for each work center (load center), and summary reports are compiled for each higher level of management for the specific area of responsibility. At the end of each year, a company's financial activities are presented to the stockholders in terms of a balance sheet that states the company's sales, operating expenses, and profit. The company's projected plans are very much dependent on its overall financial standing. Figure 15.6a–d

| | |
|---|---|
| INDIRECT SUPPLIES | $ 1,000 |
| INSURANCE | 51,444 |
| SALARIES: | |
| SUPERVISION | 49,820 |
| OFFICE AND CLERICAL | 49,964 |
| INDIRECT LABOR | 83,040 |
| UTILITIES | 198,036 |
| REPAIR AND MAINTENANCE | 1,000 |
| DEPRECIATION: | |
| BUILDING (DDB: 40 YEARS) | 21,571 |
| MACHINERY (DDB: 12 YEARS) | 41,920 |
| TOTAL | 497,795 |
| BURDEN THROUGH DIRECT LABOR COST: | |
| DIRECT LABOR COST/MONTH | 14,240 |
| I.M.E. 497,795/12 | 41,483 |

$$\text{OVERHEAD RATE} = \frac{41{,}483 \times 100}{14{,}240} = 290\% \text{ OF DIRECT COST}$$

(a)

| | |
|---|---|
| DEPARTMENT, WOODWORKING: | |
| MATERIAL | $20,625 |
| LABOR | 6,240 |
| OVERHEAD | 18,096 |
| DEPARTMENT, METALWORKING: | |
| MATERIAL | 12,767 |
| LABOR | 1,280 |
| OVERHEAD | 3,712 |
| DEPARTMENT, CLIP ASSEMBLY: | |
| MATERIAL | 5,000 |
| LABOR | 5,600 |
| OVERHEAD | 16,240 |
| DEPARTMENT, FINAL ASSEMBLY: | |
| MATERIAL | 1,217 |
| LABOR | 1,120 |
| OVERHEAD | 3,248 |
| DEPARTMENT, PACKAGING: | |
| MATERIAL | 2,222 |
| LABOR | 2,240 |
| OVERHEAD | 6,496 |
| TOTAL STANDARD COST | $106,103 |
| PRICE PER UNIT | $0.67 |
| SELLING PRICE PER UNIT | $1.20 |

(b)

**FIGURE 15.6** (a) Annual indirect manufacturing expense (b) Clip Corp. statement of cost of goods made, monthly costs — 166,667 units, 9 × 12-inch clipboard (c) Clip Corp. balance sheet for the period (d) Clip Corp. profit-and-loss statement (first year forecast)

| | |
|---|---|
| ASSETS: | |
| CURRENT ASSETS: | |
| INVENTORIES: | |
| MATERIALS | $33,588 |
| FIXED ASSETS: | |
| LAND | 6,000 |
| BUILDING | 435,600 |
| EQUIPMENT | 254,518 |
| TOTAL ASSETS | 727,706 |
| LIABILITIES AND OWNER'S EQUITY: | |
| LIABILITIES: | |
| ACCOUNTS PAYABLE | 33,588 |
| OWNER'S EQUITY: | |
| CAPITAL STOCK + RETAINED EARNINGS | 694,118 |
| TOTAL LIABILITIES | 727,706 |

(c)

| | |
|---|---|
| SALES | $2,400,000 |
| LESS: COST OF GOODS SOLD | 1,333,236 |
| LESS: SELLING AND ADMINISTRATIVE EXPENSE | 300,000 |
| LESS: INCOME TAXES | 260,699 |
| NET PROFIT | 506,065 |

$$\text{PAYBACK PERIOD} = \frac{\text{INVESTMENT}}{\text{UNIFORM ANNUAL CASH BENEFIT}} = \frac{\text{INVESTMENT}}{\text{NET PROFIT + DEPRECIATION CHARGED}}$$

INITIAL INVESTMENT: $727,706

$$\text{PAYBACK PERIOD} = \frac{727{,}206}{506{,}065 + 63{,}441} = 1.277 \text{ YEARS}$$

(d)

**FIGURE 15.6** (Continued)

shows examples of a manufacturing expense statement, a statement of costs of goods made, a balance sheet, and a profit-and-loss statement for a manufacturing firm.

## 15.6 COMPUTER PROGRAM DESCRIPTION

A program written by Kailash Bafna and Shamik Bafna (see Appendix G) is quite useful in forming the detailed financial analysis during the planning stages for a manufacturing plant. It permits the flexibility required to conduct repeated trials with "what if" questions and provides as an output well-formulated reports. The financial analysis is conducted for a 5-year planning period, with the following set of calculations for each year.

1. *Sales revenues.* Annual revenues are computed for varying volumes and price changes.
2. *Operating expenses.* Annual expenses are computed for various items considering proposed volumes and the inflation rate.
3. *Depreciation calculations.* The program computes the annual depreciation for various categories of assets using present tax laws.

1. NAME OF YOUR COMPANY: WARD-BAND, INC.
2. HOW MANY PRODUCTS DO YOU HAVE: 3

| INFORMATION NEEDED | PRODUCT #1 | PRODUCT #2 | PRODUCT #3 |
|---|---|---|---|
| PLANT CAPACITY (IN PCS.) | 100,000 | 30,000 | 60,000 |
| FIRST YEAR SELLING PRICE (IN $) | 15.00 | 38.50 | 22.25 |

| | YEAR 1 | YEAR 2 | YEAR 3 | YEAR 4 | YEAR 5 |
|---|---|---|---|---|---|
| PERCENT OF DESIGN CAPACITY | 75 | 100 | 110 | 120 | 125 |

AVERAGE RATE OF SELLING PRICE INCREASE (% PER YEAR): 6

3. COSTS FOR THE FIRST YEAR FOR THE FOLLOWING:

| | |
|---|---|
| RAW MATERIALS | $784,200 |
| PURCHASED PARTS | $570,500 |
| HOURLY EMPLOYEES | $727,400 |
| SALARIED EMPLOYEES | $403,700 |
| INSURANCE | $ 12,000 |
| MAINTENANCE | $ 48,000 |
| UTILITIES | $ 24,000 |
| MISCELLANEOUS | $ 60,000 |

AVERAGE RATE OF COST INCREASE (% PER YEAR): 8

4.

| ITEM | INITIAL COST | SALVAGE |
|---|---|---|
| PRODUCTION EQUIPMENT | 350,000 | 10,000 |
| TOOL ROOM EQUIPMENT | 50,000 | 5,000 |
| MECHANICAL HANDLING EQUIPMENT | 75,000 | 3,000 |
| AUXILIARY HANDLING EQUIPMENT | 150,000 | 5,000 |
| OFFICE EQUIPMENT | 80,000 | 4,000 |
| BUILDING COST | 550,000 | 75,000 |
| LAND | 120,000 | – |

5. WHAT PERCENT OF CAPITAL IS BORROWED? 60
WHAT IS THE INTEREST RATE OF BORROWED CAPITAL (%)? 14
IN HOW MANY YEARS IS THE LOAN TO BE PAID OFF? 10
WHAT PERCENT OF THE ANNUAL REVENUES IS THE WORKING CAPITAL? 17
WHAT IS THE INTEREST RATE OF WORKING CAPITAL (%)? 16

(a)

**FIGURE 15.7** (a) Financial analysis worksheet (b) Sample sales revenues (c) Sample operating expenses. (d) Sample depreciation calculations (e) Sample interest calculations (f) Sample tax computation (g) Sample income statement (h) Sample earnings on investment

4. *Interest calculations.* Given the debt ratio and other relevant information, annual interest payments are computed. Interest on working capital is also determined.
5. *Tax computations.* Using present tax laws, the annual tax liabilities are determined.
6. *Income statement.* This is generated for the 5 years, showing the annual net incomes.
7. *Earnings on investment.* The after-tax cash flow for the five years is shown. Rate-of-return and payback are also determined.

The data entry form and samples of reports generated by the program are shown in Figure 15.7a–h.

WARD-BAND, INC.

Sales Revenues

| Item | Year 1 | Year 2 | Year 3 | Year 4 | Year 5 |
|---|---|---|---|---|---|
| Percent of Capacity | 75 | 100 | 110 | 120 | 125 |
| Product 1: | | | | | |
| Volume (units) | 75000 | 100000 | 110000 | 120000 | 125000 |
| Selling Price ($) | 15 | 15.9 | 16.85 | 17.87 | 18.94 |
| Total Sales ($) | 1125000 | 1590000 | 1853940 | 2143829 | 2367144 |
| Product 2: | | | | | |
| Volume (units) | 22500 | 30000 | 33000 | 36000 | 37500 |
| Selling Price ($) | 38.5 | 40.81 | 43.26 | 45.85 | 48.61 |
| Total Sales ($) | 866250 | 1224300 | 1427534 | 1650748 | 1822701 |
| Product 3: | | | | | |
| Volume (units) | 45000 | 60000 | 66000 | 72000 | 75000 |
| Selling Price ($) | 22.25 | 23.59 | 25 | 26.5 | 28.09 |
| Total Sales ($) | 1001250 | 1415100 | 1650007 | 1908008 | 2106758 |
| Total Revenue ($) | 2992500 | 4229400 | 4931481 | 5702585 | 6296604 |

Note:
1. An annual selling price increase of 6% is assumed.
2. Installed capacity for Product 1 equals 100000 units per year.
3. Installed capacity for Product 2 equals 30000 units per year.
4. Installed capacity for Product 3 equals 60000 units per year.

(b)

**FIGURE 15.7** (Continued)

WARD-BAND, INC.

Operating Expenses

| Item | Year 1 | Year 2 | Year 3 | Year 4 | Year 5 |
|---|---|---|---|---|---|
| * Raw Materials | 784200 | 1129248 | 1341547 | 1580586 | 1778159 |
| * Purchased Parts | 570500 | 821520 | 975966 | 1149865 | 1293598 |
| * Hourly Employees | 727400 | 1047456 | 1244378 | 1466103 | 1649366 |
| - Salaried Employees | 403700 | 435996 | 470876 | 508546 | 549229 |
| - Insurance | 12000 | 12960 | 13997 | 15117 | 16326 |
| * Maintenance | 48000 | 69120 | 82115 | 96746 | 108839 |
| * Utilities | 24000 | 34560 | 41057 | 48373 | 54420 |
| - Miscellaneous | 60000 | 64800 | 69984 | 75583 | 81629 |
| Total Expenses | 2629800 | 3615660 | 4239919 | 4940919 | 5531568 |

Note: 1. An annual cost increase of 8% is assumed.
2. Asterisks (*) indicate directly variable costs. Dashes (-) indicate fixed costs.

(c)

WARD-BAND, INC.

Depreciation Calculations

| Item | First Cost | Salvage Value | Tax Life | Annual Deprec. | Book Value Sixth Year |
|---|---|---|---|---|---|
| Production Equip. | 350000 | 10000 | 7 | 48571 | 107143 |
| Tool Room Equip. | 50000 | 5000 | 7 | 6429 | 17857 |
| Mech. Handling Equip. | 75000 | 3000 | 7 | 10286 | 23571 |
| Aux. Handling Equip. | 150000 | 5000 | 12 | 12083 | 89583 |
| Office Equip. | 80000 | 4000 | 10 | 7600 | 42000 |
| Building | 550000 | 75000 | 30 | 15833 | 470833 |
| Land | 120000 | | | 0 | |
| Total | 1375000 | | | 100802 | 870988 |

Note: Straight-line depreciation is assumed in all cases.

(d)

**FIGURE 15.7** (Continued)

**WARD-BAND, INC.**

Interest Calculations

| Item | Year 1 | Year 2 | Year 3 | Year 4 | Year 5 |
|---|---|---|---|---|---|
| **Borrowed Capital:** | | | | | |
| Loan Balance | 825000 | 782336 | 733700 | 678254 | 615046 |
| Annual Payment | 158164 | 158164 | 158164 | 158164 | 158164 |
| Principal Paid | 42664 | 48637 | 55446 | 63208 | 72057 |
| Interest on B. C. | 115500 | 109527 | 102718 | 94956 | 86106 |
| **Working Capital:** | | | | | |
| Amount of W. C. | 508725 | 718998 | 838352 | 969439 | 1070423 |
| Interest on W. C. | 81396 | 115040 | 134136 | 155110 | 171268 |
| Total Interest | 196896 | 224567 | 236854 | 250066 | 257374 |

Note:
1. Loan is 60% of total capital.
2. Capital is borrowed at 14% per year.
3. Loan is paid off in equal annual installments in 10 years.
4. Working capital is assumed to be 17% of annual revenues.
5. Interest on working capital is charged at 16% per year.

(e)

**WARD-BAND, INC.**

Tax Computation

| Item | Year 1 | Year 2 | Year 3 | Year 4 | Year 5 |
|---|---|---|---|---|---|
| Gross Tax Income | 65002 | 288371 | 353905 | 410797 | 406860 |
| Loss Carry Forward | 0 | 0 | 0 | 0 | 0 |
| Net Tax Income | 65002 | 288371 | 353905 | 410797 | 406860 |
| Gross Tax | 16601 | 92546 | 120328 | 139671 | 138332 |
| Net Tax | 16601 | 92546 | 120328 | 139671 | 138332 |

(f)

**FIGURE 15.7** (Continued)

WARD-BAND, INC.

Income Statement

| Item | Year 1 | Year 2 | Year 3 | Year 4 | Year 5 |
|---|---|---|---|---|---|
| Sales Revenues | 2992500 | 4229400 | 4931481 | 5702585 | 6296604 |
| Operating Expenses | 2629800 | 3615660 | 4239919 | 4940919 | 5531568 |
| Gross Profit | 362700 | 613740 | 691562 | 761666 | 765036 |
| Depreciation | 100802 | 100802 | 100802 | 100802 | 100802 |
| Interest | 196896 | 224567 | 236854 | 250066 | 257374 |
| Gross Tax Income | 65002 | 288371 | 353905 | 410797 | 406860 |
| Net Tax | 16601 | 92546 | 120328 | 139671 | 138332 |
| Net Income | 48401 | 195825 | 233577 | 271126 | 268527 |

(g)

WARD – BAND, INC.

Earnings on Investment

| Item | Year 1 | Year 2 | Year 3 | Year 4 | Year 5 |
|---|---|---|---|---|---|
| Net Income | 48401 | 195825 | 233577 | 271126 | 268527 |
| Add Depreciation | 100802 | 100802 | 100802 | 100802 | 100802 |
| Net Cash Flow | 149203 | 296627 | 334380 | 371929 | 369330 |

Return on investment over a 5-year period = 16.01%

Payback on Investment = 4.6 years.

(h)

**FIGURE 15.7** (Continued)

## 15.7 SUMMARY

The chapter discusses six major topics: plant site selection, utilities, insurance, safety, the ADA of 1990, and taxes. Although the topics seem unrelated, they all have considerable influence on the financial well-being of a manufacturing facility. The financial statements at the end of the chapter illustrate how the financial information may be displayed so that it is easy to understand and interpret.

In evaluating the sites for a new plant, one must consider a multitude of factors. However, it is not necessary that each factor be given equal importance. The need changes from industry to industry, and a factor considered to be very important in one industry may not be as crucial in another. A tabular method is suggested in the chapter for selecting one site from among the possibilities. However, selecting a site based on this method may be quite subjective. One may be more objective by combining the tabular method with the location methods discussed in Chapters 17 through 19. Locating a plant in an industrial park has some definite advantages, and should an opportunity arise to do so, it may be beneficial to investigate it further.

Utilities are the lifelines of industry — without them, an industry cannot function. The expense of utilities can be controlled by conservation and planning. Typically, we include water, electricity, refrigeration, steam, compressed air, water disposal and treatment, air-conditioning and heating, and telephone under the category of utilities. For each, the chapter discusses what the requirements in a plant could be and proposes a few ways of increasing efficiency in utilization. A chart illustrating typical utility specifications for a small manufacturing firm shows how the total cost for utilities may be estimated.

Making contributions to insurance is an expense that cannot be avoided if one intends to stay in business for long. Various types of insurances are available such as fire, WC, business, automobile, product and general liability, burglary, and medical and dental care. Some coverages are optional but strongly recommended, while others such as WC are required by law. The benefits provided under WC are quite different in each state, and therefore associated premiums also vary from state to state. Similarly, one might choose different levels of protection for other types of insurances, thereby reducing or increasing the total cost.

Safety is extremely important in a place of business. Unsafe conditions can lead to an increase in the number of accidents, which results in either higher premiums for insurance or cancellation of the policy. In addition, accidents cause human suffering and loss of production. Safety requires planning — from the time a plant is built to every instance in its operation. OSHA, a regulatory arm of the U.S. government, issues regulations that must be followed. Within a plant, the major responsibility to develop a good safety program lies with the management, and an example of one such safety program is given in the chapter.

The ADA of 1990 tries to provide access to employment for all qualified individuals. In most cases, some minor modifications to an existing facility may be needed to comply with the act. The act also provides for exemption to an employer if he or she can prove undue hardship with compliance. Sample checklists are provided in Appendix E.

Taxes are unavoidable and should be taken into account in planning. A major tax for any business is the federal income tax. However, state tax, Social Security, property taxes, and sales tax also contribute to the total tax bill. Tax laws and rates change frequently, and it is a good idea to refer to current publications for up-to-date information on the subject.

A financial statement or chart shows the financial state of a control unit such as a department within an organization or of the whole organization, depending on the level at which the statement is developed. The details involved in the statement are level dependent; lower levels in the organization receive very detailed reports, while only summaries are presented to the top level of management. The chapter presents samples of various financial reports, along with information about a computer program that can be used to evaluate "what if" questions that occur in the overall planning of a manufacturing plant.

## PROBLEMS

1. Discuss the factors in selecting a site for:
   a. a hospital
   b. a fresh-produce processing plant
   c. a clothing manufacturer
   d. an electronics corporation
2. Discuss specific circumstances when it would be best to ship by:
   a. air transport
   b. highway
   c. rail
3. Specify interior colors and lighting arrangement for an assembly operation in a building that is 30 × 50 feet and 15 feet high.
4. In problem 3 the following additional data are available: The cost of operation per tube is as follows.

| | 48 Inches | 96 Inches |
|---|---|---|
| Basic | $0.02/hour | $0.03/hour |
| High performance | $0.03/hour | $0.05/hour |

   Assuming the cost of installation per row of lights is $2000/row and the plant would be in operation for 10 years, working 40 hours/week and 48 weeks/year, which lighting arrangement would you recommend?
5. Develop heating and air-conditioning requirements for an office building with the following data obtained from a blueprint. The building is 35 × 58 feet with an 8-foot ceiling, and R-13 insulation in the walls, and R-22 insulation (7-inch mineral wool) in the ceiling.

| | | | |
|---|---|---|---|
| Windows: | 1 | 36 × 60 | facing |
| | 1 | 36 × 60 | facing |
| | 2 | 36 × 60 | facing |
| | 1 | 48 × 60 | facing |
| | 1 | 24 × 60 | facing |
| Doors: | | | facing |
| Sliding (glass) | 1 | 60 × 60 | facing |
| Wood | 2 | 36 × 84 | facing |

All windows are double-paned glass windows and doors are wooden with quantity, dimensions (in inches), and orientation as above. The building will be occupied by 10 people and should be designed for the following temperatures:

| | Winter | Summer |
|---|---|---|
| Indoor | 70 | 73 |
| Outdoor | 20 | 102 |

6. In problem 3 the building is made of masonry walls with R-11 insulation. The ceiling is insulated with 6 inches of R-19 and has a concrete slab with no insulation. It has a wooden door 60 × 84 inches facing south for employee entrance and a 12 × 10 foot door on the north side for supply and dock facilities. This door may be treated as a single-pane clear door with a metal frame. There are five motors in the plant: two are rated at 10 hp at 8 amperes and three rated at 5 hp at 6 amperes. All the motors are drawing 80% of the current when in operation (i.e., measured horsepower is equal to 80% of the rated horsepower) and are generally in use for 5 hours in an 8-hour day. There are 20 people working in the plant. If the plant is located in the same area as the office building in problem 15.5, determine the heating and cooling requirements for the plant (you must consider the lighting arrangement you developed).
7. What are the advantages of company insurance? What types of insurance should be purchased?
8. Discuss the influences on WC insurance premiums.
9. Discuss the safety equipment that should be used for the following jobs:
   a. moving pallets of materials in shipping
   b. operating a drill press
   c. working at a nuclear power plant
10. What major impact has OSHA had on plant and employee safety?
11. Go to a business in your community and prepare the safety check sheet in reference to the company's operation.
12. An electronics corporation had a net profit before taxes of $150,000 last year. If the state tax is 10% of the federal tax, what is the overall income tax for the company?
13. What benefits does the Social Security Act of 1935 provide?

14. A company wishes to build a second cooling tower to accommodate additions to its plant. Its present tower handles 5 million BTU/hour but has the capacity to handle 280 gallons/minute. The outlet cooling water temperature is 78°F. The system needs to handle 6,500,000 BTU/hour. Determine the cooling water flow rate of the second tower.
15. A company wants to reduce its power expenses by $1000 per year. If the electrical power rate schedule specifies a demand charge of $2.87 per kVa per month, and the average power factor is 0.58, drawing 55 kW from the 440 V supply, determine how much the power factor would need to be increased to save this money.
16. What means of conserving energy are available for manufacturing plants? (Exclude reduction of operations.)

## SUGGESTED READINGS

### Industrial Site

Hamilton, F.E.I., *Spatial Perspectives on Industrial Organizations and Decision-Making*, John Wiley, New York, 1974.
Lochmoeller, D.C., Munch, D.A., Thorme, O.J., and Viets, M.A., *Industrial Development Handbook*, The Urban Land Institute, Washington, D.C., 1975.
Richardson, H.W., *Regional Economics*, Praeger Publishers, New York, 1971.
Smith, D.M., *Industrial Location*, 2nd revised edition, John Wiley, New York, 1981.

### Utilities Specifications

Bonbright, J.C., *Principles of Public Utility Rates*, Columbia University Press, New York, 1961.
Lewis, B.T., and Morrow, J.P., *Facilities*, McGraw-Hill, New York, 1973.
*Lighting Design and Application*, Illumination Engineering Society of North America, New York, 1985.
Lowther, J., *Industrial Energy Conservation*, Louisiana Tech University, Ruston, LA, 1985.
Load Calculation for Commercial Summer and Winter Air Conditioning, Air Conditioning Contractors of America, Washington, D.C., 1988.
Residential Load Calculation Methods, The Trane Company, Tyler, TX, 1990.
Yarbrough, R.B., *Electrical Engineering Reference Manual*, 6th edn, Professional Publications, Inc., Belmont, CA, 1997.

### Insurance

Green, M.R., and Swander, P., *Insurance Insights*, Southwestern Publishing Company, Cincinnati, 1974.
Mehr, R.I., and Cammack, E., *Principles of Insurance*, Richard D. Irwin, Homewood, IL, 1972.
Pfaffle, A.E., and Nicosia, S., *Risk Analysis Guide to Insurance and Employee Benefits*, AMACOM, New York, 1977.

### Taxes

*Corporation and Partnership Tax Return Guide*, Research Institute of America, Inc., New York, 1999.
*Prentice Hall's Federal Taxation*, Prentice Hall, Englewood Cliffs, NJ, 2008.
RIA Federal Tax Handbook, Research Institute of America Inc., New York, 2008.

## Safety

Hawdley, W., *Industrial Safety Handbook*, McGraw-Hill, London, 1977.
Lewis, B.T., and Marron, J.D., *Facilities*, McGraw-Hill, New York, 1973.
Marshall, G., *Safety Engineering*, 3rd sub edition, American Society of Safety Eng., Des Plaines, IL.
OSHA *Safety and Health Standards*, U.S. Department of Labor, Washington, D.C., 2000.

# 16 Computer-Integrated Manufacturing System

There has been a dramatic increase in the use of computers in manufacturing-related activities. In the 1950s, these applications were restricted to a few well-defined problems in administration and finance, in which they produced straightforward solutions along with good documentation, mainly handling data sequentially. With the development of high-speed processing and low-cost memory, storing and manipulating large sets of nonsequential data have been greatly facilitated and new applications in manufacturing have been developed. Inventory control, production scheduling, production requirements, and capacity planning are some of the well-known areas in which computers were successfully used in the 1970s. With further advancements in both hardware and software, it is possible to further unite all phases of manufacturing. A computer-integrated manufacturing system (CIMS) is a concept for integrating all components involved in the production of an item. It starts with the initial stages of planning and design and encompasses the final stages of manufacturing, packaging, and shipping. It is the combining of all existing technologies to manage and control business in its entirety, which has led to the term "automated factory" to describe the ultimately successful application of CIMS. CIMS includes components such as computer-aided design (CAD), computer-aided manufacturing [CAM; i.e., use of computer numerically controlled (CNC) and direct numerically controlled (DNC) machines], artificial intelligence (AI), computer process planning, database technology and management, expert systems, information flow, just-in-time concepts, material requirement planning (MRP), automated inspection methods, process and adaptive control, and robotics. All these elements work together using a common database. Data are entered from the operations of the entire plant — from each level of production on the shop floor, by receiving and shipping, marketing, engineering design, and designated individuals within other departments who are involved in the operation of the plant. The information is continuously processed, updated, and reported to people on an as-needed basis in an appropriate format to assist in decision making. The same data set is used by NC machines to manufacture products and by automated material handling (AMH) systems to move them from one workstation to another. This integration of mechanical, electrical, and informational systems is the backbone of CIMSs.

## 16.1 SYSTEMS AND FILES

The system is designed to offer great flexibility in production by using NC machines and increased speed and flexibility in material-handling systems by using computers and automated guided vehicles (AGVs). It can handle a production mix consisting

**TABLE 16.1**
**CIMS Classification Characters**

| System Classification | No. of Different Parts to Produce | No. of Parts Produced per Hour | Features |
|---|---|---|---|
| Transfer line | 1–5 | >15, normally in hundreds | High utilization of machines and plant space; minimum labor cost, mainly single-purpose machining with sequences controlled by programmable controller; no flexibility |
| Special system | 3–10 | 1–15 | Produces parts in batches; considerable time spent in changeover; may contain one or more machines dedicated to the product |
| FMS | 10–50 | Up to 10 | Series of flexible machines with mainly numerical control; equally flexible but highly automated material handling; allows random processing |
| Manufacturing cell or stand-alone system | >20 | About 3 or 4 | FMS with independent control of each machine and can process units in batch mode |

of up to 800 different parts with a production volume anywhere from a few units to 10,000–12,000 units. The terms "special systems," "flexible manufacturing system" (FMS), and "manufacturing cell" are often used to describe CIMS with a different combination of part variety and production volume. Table 16.1 shows a popular classification guideline suggesting the features of a system to be utilized. A special system offers less production flexibility and more material-handling automation, including fast-moving computerized conveyor systems, while a manufacturing cell offers more flexibility in manufacturing with a small production volume on the average for each part. Here, material handling can be obtained more independently by using AGV systems. As is true of any manufacturing system, the purpose of CIMS is to produce salable goods at a minimum cost in the shortest possible time. Unlike traditional manufacturing, however, CIMS employs the philosophy that management must work to optimize the whole business operation rather than individual functions or workstations.

CIMS requires a change in management style. It allows management personnel to make decisions on the basis of up-to-the-minute information that is made readily available to them through the use of the computer, which is essential for maintaining, manipulating, storing, and recalling the database. Such a database may consist of information on machine utilization in the plant; scheduling of parts based on machine availability, promised delivery dates, and the possibility of creating final assemblies without building large in-process inventories; personnel assignments as machine coupling becomes more common and nonproductive time decreases; and/or decisions on tools that are to be selected and loaded onto an NC machine from the tool drum, based on the information supplied to the machines by the computer concerning parts and operations. The CIMS database might also consist of

- A file for programs of parts to be used in NC machines
- A file on routing describing the workstations through which the part must be processed
- A file on production of parts indicating for each part the production rate for each machine, in-process stores, and quality control requirements, among others
- A file for each workstation indicating the work scheduled, the tools needed, and the operator assigned
- A file on all tools indicating tool use, its life, maintenance needed, and the cost for each item

The changes in management concepts are evident if one considers the cooperation the system demands from various departments. No longer can each department within a plant, such as design, manufacturing, marketing, shipping, receiving, personnel, or purchasing, consider itself a closed cell in its operations. All employees, even including operators on the shop floor, must work jointly using the same manufacturing database.

Each person is responsible for keeping the database current and accurate and for using the information from it when needed. This type of system requires total commitment from people in top management. In particular, they must (1) train everyone in the plant to work with CIMS; (2) employ people who will take responsibility within the sphere of their influence instead of waiting to be told what to do (they must also develop the reward system that will encourage such behavior); and (3) develop a matrix or similar type of organization that allows and encourages different departments to communicate with one another.

It should be clearly understood by all who are working in the plant that there is a person in charge of the CIMS database who is responsible for maintaining and protecting it; however, this database is not assigned exclusively to any single person. Everyone can use it when needed, and everyone is responsible for its upkeep.

## 16.2 COMPONENTS OF CIMS

Most of the components have already been described in other chapters specifically related to the activities discussed in those chapters. To make the description of CIMS more nearly complete, we briefly review those functions here.

### 16.2.1 Computer-Aided Design

A manufacturer must begin with the design, drawings, and part details for a product, and CAD helps in every phase of these operations. CAD is especially useful because, on the average, even a small change in one drawing typically affects about five other drawings. Using such a system improves productivity three to four times and also improves the quality of design and drawing details. CAD in design analysis is sometimes referred to as computer-aided engineering. Computer-aided engineering can, for example, simulate how a tool will perform on a workpiece; how a setup of fixtures will function; how a mold will operate under the effects of friction, heat, and mold pressure; and how these factors will affect the raw material and ultimately the final product. CAD helps form an interaction between designers and manufacturing engineers during the design stage of the product. Manufacturing engineers can develop production processes, tools, and dies for the proposed design and analyze its feasibility during the initial stages. If changes are needed, they can be made before the product goes to the production phase.

Vendors can also be made part of the team to evaluate supply requirements, materials, and tools that would be needed for the job. Again, initial suggestions by suppliers can be very useful in modifying the product design and lowering the unit cost (more can be found on CAD in Chapter 2).

### 16.2.2 Group Technology

A large quantity of detailed design data can be put in order by early application of group technology (GT), which systematically codes the data and design analysis. Coding can be done by using design attributes such as size, shape, weight, and material of the part, tools, and equipment needed in parts production. A computer can group similarly coded items together and store this information in its memory. Generally, in an existing production facility, a new product design may be able to use anywhere from 20% to 60% of the existing drawings with little or no change required. Without a computer and GT, however, a designer would most likely design and draw the parts again rather than go through thousands of detailed drawings to find the parts that will fit into the new design. GT also facilitates the formulation of machine cells for flexible manufacturing centers, as explained in Chapter 5, thus expanding and expediting the work of CAD by reducing product design time as well as setup and waiting times.

However, GT is time consuming. Although computer software can allocate the GT codes to a part, each item of design data must be entered manually into the computer.

### 16.2.3 Computer-Aided Manufacturing

Further integration of CAD data to production is achieved through CAM. It has three major components: machine tools, production machines, and electronic controls and relays with which to operate production machines.

NC machines (CNC, DNC) form the foundation of CAM. NC machines are driven by internal computers. CNC machines are NC machines controlled by their own external computer. Large numbers of tool programs can be stored in the computer memory and recalled as the need arises, thereby avoiding the necessity of storing all information on every NC machine. A DNC machine is a number of CNC machines connected to one large computer. The loading, unloading, and modification of programs in DNC machines can be accomplished more quickly and conveniently than in CNC machines.

NC machines have higher spindle speeds and more torque on tool tips than do conventional machines, increasing cutting speeds. They also have sensors mounted on the tool spindles and work surface for real-time adaptive tool control.

CNC and DNC machines can be programmed off-line, increasing productivity. The controls for machines are provided by the programmable logic controller (PLC), which replaces thousands of relays and contains active memory for input/output functions. A PLC is a solid-state device that is able to withstand high temperatures and dusty environments, the conditions almost always present in production facilities (see Chapter 4). If the design and production quantities of a workpiece require increased specialization, process-specialized machine tools can also be incorporated into a system. For example, these specialized machine tools could be used to perform critical tolerance and/or multiple-spindle drilling, tapping, or boring operations and, in more sophisticated systems, washing, assembly, and even inspection operations.

### 16.2.4 Robots

A robot is a "worker" within the CIMS. It is a unit consisting of sensors, computers, and controls, the use of which is obtained in terms of its arm movements (and sometimes leg or wheel movements). A robot by itself is not very useful unless it is integrated in AMH, adaptive processes, and/or multitask functions.

A robot can be used as a part feeder to a single machine or a group of machines, a material handler in a complex assembly, or a visual sensor in an inspection process. Robots are also used in manufacturing involving such tedious tasks as welding and spray painting.

More and more robots are installed in CIMSs as they become affordable. Flexibility and quality of work are two major traits that encourage their use in production. Because robots do not have human feelings, nor do they perspire from heat, shed hair, or breathe fresh air, they are especially suited for working on monotonous and hazardous jobs (e.g., handling toxic chemicals), or where a dirt-free, clean operation is absolutely necessary (e.g., making computer chips). Robots can be attached to the ceiling, freeing the floor area for other activities. With adjustable speeds and motions, robots work precisely and repeatedly, integrating themselves completely with the rhythm of the other equipment within the system.

A robot is programmed either by being physically led through the steps it is required to perform or by use of a specific programming language. Research is continuing to standardize programming languages and to make vision robots more affordable for any size of manufacturing firm. There are also attempts to increase the flexibility of robots by programming on-line — that is, without removing them from production.

### 16.2.5 Material Handling

The purpose of the material-handling system in CIMS is to automatically deliver and remove workpieces from the machine tools. This eliminates the manual handling of the workpieces, which in turn lowers the work-in-process time. The material-handling system under computer control also maintains a constant supply of workpieces to the machine tools.

With the increased use of computers, servomechanisms, and electronic controls, material handling has been transformed into AMH. For example, a modern forklift truck has a display monitor, computer keyboard, radio communications, and bar code scanners to work effectively with different data files. The just-in-time philosophy reduces material in inventory and thus the need for material handling. The MRP system maintains accurate data on what parts are needed and how many are available at what time on the basis of the production schedule, again reducing material handling. The automated storage and retrieval system automatically stores and dispenses items from storage. AGVs collect and deliver products on flexible paths defined by under-floor wires or an optical guidance system. They are provided with safety bumpers at both ends, which stop the vehicle upon contact with any object. Although controlled by a computer, they also allow an operator to override the system and drive the vehicle away from its guided path if needed. For high-volume transportation, conveyors still play major a part, even they are automated now. For example, powered roller conveyors are used to transport heavy workpieces. Standard machines are linked by a waist-high conveyor loop, allowing them to carry workpieces through stations for different machining jobs. Towline carts are run by an under-the-floor towline system. Chains located in slots in the floor pull the carts along permissible paths. A cart is towed to the appropriate machine under computer control; when it arrives, a plate rises from the floor and disconnects the cart's chain drive. A piston pushes the cart against a hard stop for accurate positioning before loading or unloading the pallet. In many instances, conveyors provide movement over a fixed path within the work cells, even acting as mobile assembly platforms, while AGVs provide interdepartmental transport. All units within AMH are becoming more flexible, increasing nonsynchronized handling; that is, each unit is able to move independently of the others. There are many additional paths over which AGVs can travel while making decisions automatically: picking up speed and slowing down when needed, positioning themselves precisely under a workstation, and initiating loading and unloading actions by conveyors and by robots. The productivity of modern forklift trucks is increased tremendously by computers and other equipment. In all, AMH eliminates material handling when possible and performs other necessary material handling efficiently to increase productivity.

### 16.2.6 Computer Systems

In most systems, machine tools, material handling, and control activities are under computer direction. The magnitude of computer control is based on the system's level of sophistication: the more complex the CIMS, the greater the computer involvement.

Such a computer requires both hardware and software. The computer hardware depends on considerations such as the number of NC machines in the system, the number of parts involved, the amount of data stored, and the physical locations of the data terminals. Examples of some hardware requirements are:

- A computer for overall control of the system.
- Equipment for control of the material-handling system.

- Mirrored disks, one for on-line containing system data files and programs and the other for providing backup.
- Video display terminals for entry of system commands and display of reports and messages.
- Shop terminals used for semiautomatic entry of commands.

### 16.2.7 Control Software

Large amounts of data generated when a product requires many items have created a need for software to efficiently control the system. Typically, a medium-sized firm has 4 million to 5 million pieces of information that require storage and manipulation. The computer software consists of two major databases — manufacturing and product. The manufacturing database collects data from the process computer and the process controller. The data are used to report shop activities and to plan and control future activities. The product database is used by programmable equipment to perform manufacturing activities such as machining, inspecting, and counting. The data banks might consist of bills of materials, customer orders and promised dates of deliveries, inventories, production schedules, personnel schedules, and tool and equipment requirements, among others. We briefly describe some of the available software in the following paragraphs.

Manufacturing planning and control and its components, manufacturing resource planning (MRPII) and material requirement planning (MRP) are the key software packages in the market today. The MRP system answers pertinent questions such as which products are scheduled to be made, the parts and quantities that are needed to make these products, how much stock of each part the plant has, and how much it must purchase and when. MRPII, which is a simulation program, schedules the operations of each workstation, each machine, and indeed the entire plant. In addition, MRPII provides the status of each machine and its productivity. It makes sure that the required parts arrive at the correct station at the right time in the necessary quantities. Manufacturing planning and control integrates MRP, MRPII, CAD, CAM, and other software.

Other computer programs include one on statistical quality control, which provides information on whether a process is in control, and a package for testing the accuracy and reliability of tools and equipment. Simulation packages that allow "what if" analysis of sales, finances, tooling, and overall operation of a plant are also available. It is certain that more and better software will be developed to increase automation in plants as time passes.

Computer software is an important element of CIMS. It is expensive and yet the only truly flexible part of the system. As mentioned above, there are some standard packages such as MRPII; however, whenever possible, modularity is essential so as not to tie down the entire system to a single program. Many tool programs, for example, are available from various vendors, and a manufacturer should be able to select the appropriate ones and integrate them into its own CIMS. As the products change over the years, such flexibility becomes critical in reducing software costs.

AI (artificial intelligence), or the expert system, is an additional computer software. It consists of two phases — software to deduce new information, called the "inference engine," and software to help modify decision-making rules based on the

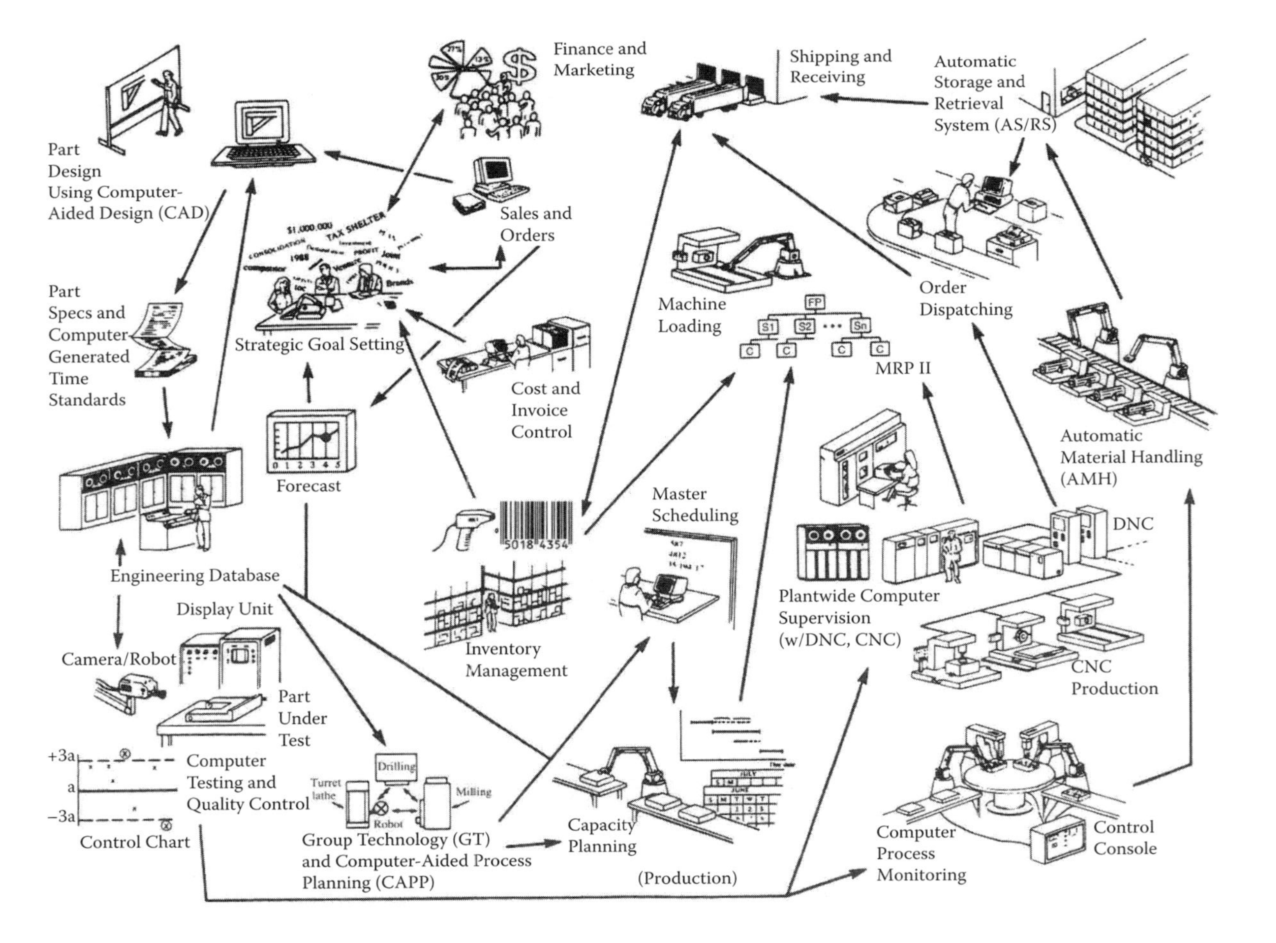


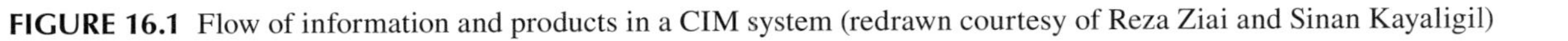
**FIGURE 16.1** Flow of information and products in a CIM system (redrawn courtesy of Reza Ziai and Sinan Kayaligil)

latest available information, called a "knowledge base." Observed AI applications in manufacturing are found in the industrial environment (production planning, maintenance, CAD use), the service environment (training, customer services, route selections in shipping), and the management environment (strategic planning, market planning, financial planning).

The flow of information and products in CIMS is illustrated in Figure 16.1.

## 16.3 BENEFITS AND DEFICIENCIES OF CIMS

CIMS is receptive to changes. In a CNC system consisting of 10 NC machines, it is possible to produce either 10 parts of the same kind or, at the other extreme, 10 different parts. Changes in designs and demands can easily be accommodated by CAD systems. Once the software has been developed, it is equally convenient to produce one part or a thousand parts. A manufacturer can therefore respond rapidly to changes in product design, demand, or product mix. The manufacturer can effectively serve both small customers asking for only a few parts and major customers looking for large quantities of many parts. Each can be made with consistent quality with a minimum of waste. Inventories and setup times are reduced considerably in comparison with those of traditional manufacturing, giving short turn-around times and lower costs. CIMS has the ability to handle parts with complex designs and to produce them continuously with a minimum of unscheduled downtimes. Each worker becomes a "micromanager," increasing the level of his or her contribution to the overall working of the plant. Increased productivity and improved quality of work are the results of a job-satisfied worker.

CIMS also has other benefits. According to a study sponsored by the American Society of Manufacturing Engineers, the use of CIMS has:

- reduced engineering design cost by 15%
- reduced overall lead time by 30–60%
- increased production by 40–70%
- reduced work-in-process inventory by 30–60%
- reduced personnel costs by 5–20%.

There are also some difficulties associated with the implementation of CIMS. Total CIMSs are expensive, costing anywhere from $1,000,000 to $100,000,000, depending on the size of the plant. Some companies can afford the capital investment to make the entire existing plant compatible with CIMS. Furthermore, there may be some reluctance toward computerization within a typical firm when people see their jobs being changed dramatically or even disappear as a result of CIMS. Most manufacturing plants currently have a combination of both batch and flow-through processing, which makes changeover and control difficult.

## 16.4 PLANNING FOR CIMS

Transferring a manufacturing system from the traditional surrounding to CIMS is indeed a major task. Two prime issues must be addressed before such implementation is possible. These are (1) what planning and design considerations are necessary for a given industrial application, and (2) how the large investment normally required for CIMS can be justified.

The outline below might prove to be beneficial in formulating and analyzing the requirements in such a changeover.

1. *Define objectives.* Analyze the engineering and manufacturing capacities and needs and envision how automation might be accomplished. Consider what traditional competitors have done in their efforts to modify and manufacture the product. Give special attention to leading users of factory automation. Such analysis might provide the company with a set of goals to achieve and might also indicate where the competitors are heading and what must be done to remain viable.
2. *Formulate a project team.* The team evaluates the alternatives and performs the necessary engineering and financial analysis. It is not uncommon to have a time span of 3 years or more before changeover to CIMS, from concept to production, is accomplished. There must be continuity during this time, and the project team provides it. The team might consist of one senior manager and a representative from each functional area that would be affected by the changeover, such as design, production, manufacturing, and finance. It might also be beneficial to include workers or union representation on the team; this inclusion can facilitate acceptance of the changeover by the shop floor personnel.
3. *Keep up to date with the technology.* CIMS might be a new field to many of the people on the team. To understand the technology, members can attend seminars and conferences and read technical articles. The team might ask vendors to arrange for a field trip to a plant where their product has been successfully implemented. The educational process should also be extended within the company. For example, before implementing a complete CIMS, it might be beneficial to obtain experience with some of the individual components of CIMS, such as NC machines and robots, as they are used in producing specific components of the company. Such knowledge is valuable in developing the useful integration of efforts.
4. *Evaluate the concepts.* The company should conduct an analysis to determine suitability of CIMS based on present and anticipated product mixes. GT should be applied wherever appropriate. The analysis should include current data on unit cost, lead time, quality, manpower, material handling, utilities, and any other appropriate items. A number of different manufacturing systems should be analyzed — for example, stand-alone cell versus integrated manufacturing. If CIMS is appropriate, several detailed possible configurations should be developed with different types of equipment.
5. *Justify the investment.* In making an economic analysis, both benefits and costs must be estimated. It is relatively easy to determine the investment needed in the purchase and installation of the hardware and software for CIMS. Less obvious is the cost of training and new personnel requirements for implementation of the system. On the benefit side, certain

improvements such as quality of the product and reduction of personnel can be easily quantified, but difficulty might arise in trying to estimate the benefit of better material handling and reduction of lead times. To develop a correct financial appraisal, one must consider:

a. the total CIMS picture to include both direct and indirect costs as well as benefits to the company in its future developments;
b. a long time span of 5 years or more to account for long-term impact of the investment; and
c. probability assignment in the evaluation to estimate uncertainties in the future.

We may ask questions such as:

a. What will be the result if we do not invest in CIMS?
b. Are there other alternatives that will keep the company competitive?
c. What is the competition doing worldwide and how will this affect the future?
d. Are there any new products that can be manufactured using CIMS that might open up new markets?

If the analysis leads to the decision to invest in CIMS, the next step is to ask for specific proposals from vendors.

6. *Evaluate and implement the system.* Information collected so far will lead one to develop a sound proposal to obtain firm bids from different vendors. Many discussions are required for vendors to understand and develop a system for any unique application that may be associated with the company's product. Proposals should be evaluated not only on the cost basis but also according to factors such as experience of vendors, quality of their products, and reliability and availability of technical assistance, should it be needed. Once the vendors have been selected, close cooperation between different suppliers and an understanding of the company's objective are essential to obtain an integrated system.

## 16.5 SYSTEM PROVIDERS

Many manufacturers, such as Allen-Bradley, Sperry, IBM, Tata Industries, Litton, and Easton, can develop an appropriate group among themselves, each performing the task for which it has the greatest expertise, to develop the entire system. For example, IBM computers might manage an FMS consisting of individual CNC machines using Allen-Bradley controllers for individual CNC tools. The computer systems might be developed by Tata, Hewlett-Packard, or IBM. Machine tools might come from Cincinnati Milacron. Material handling might be designed by Litton or Eaton-Kenway, while the robots might be supplied by GCA or GMF.

With experience and higher productivity in the computer industry, the prices of CIMS are sure to come down, making them more affordable to small- and medium-sized firms. Then again, one might build the system in modular fashion, reducing the initial expenditure.

A partial listing of CIMS suppliers is given in Table 16.2.

**TABLE 16.2**
**CIMS Suppliers**

| Supplier | Address | Supplier | Address |
|---|---|---|---|
| Harnischfeger Engineers Inc. | P.O. Box 554, Milwaukee, WI 53204 | Sperry Corporation | P.O. Box 500, Blue Bell, PA 19426-0024 |
| Xerox Computer Services | P.O. Box 66924, 5310 Beethoven St., Los Angeles, CA 90086 | Tandem Computer | 19191 Vallco Parkway, Cupertino, CA 95014 |
| | | Acco Babcock Material Handling | P.O. Box 460 Bailies Lane, Frederick, MD 21701 |
| Schlumberger | P.O. Box 986, 4251 Plymouth Road, Ann Arbor, MI 48106-0986 | | |
| Scientific Systems Services | 2000 Commerce Drive, Melbourne, FL 32901 | Tata | Mumbai, India. |
| Accu-Sort Systems | 511 School House Road, Telford, PA 18969 | Eaton-Kenway | 515 East 100 South Salt Lake City, UT 84102 |
| White Storage and Retrieval Systems | 30 Boright Ave., Kenilworth, NJ 07033 | Insta Read | 2400 Diversified Way, Orlando, FL 32804 |
| Advanced Systems, Inc. | 155E Algonquin Road, Arlington Heights, IL 60005 | Interlake Material Handling/Storage Products | 100 Tower Drive, Burr Ridge, IL 60521 |
| Allen-Bradley | 1201 South & Second St., Milwaukee, WI 53206 | GMF Robotics Corporation | 5600 New King St., Troy, MI 48098 |
| | | Litton UHS | 7900 Tanners Gate, Florence, KY 41042 |

## PROBLEMS

1. Discuss the philosophy behind CIMS.
2. Are there any additional subject areas that a person must learn to work in a CIMS environment?
3. Define the following:
   a. CAD
   b. GT
   c. CAM
   d. AMH
4. Distinguish between NC, CNC, and DNC.
5. What is a robot? Give examples of its use in industry.
6. What are some of the main advantages of using robots? What are the disadvantages?
7. Identify three methods of instituting AMH in place of material handling.
8. How is computer software used in industry today?
9. Savings in what areas will contribute toward justification of expenditures in CIMS?

10. Differentiate between two types of computer software available for control in industry.
11. Give two specific examples of use of AI in industry.
12. Discuss the advantages and disadvantages of CIMS.
13. Visit a local manufacturing facility. Identify specific ways in which CIMS could contribute to its processes.

## SUGGESTED READINGS

Asfahl, C.R., *Robots and Manufacturing Automation,* John Wiley, New York, 1985.

Fisher, E.L., "Expert systems can aid intelligent CIM decision making," *Industrial Engineering*, 17(3), 78–83, 1985.

Goldhar, J.D., and Jelinek, M., "Computer-integrated flexible manufacturing — organizational, economic, and strategic implications," *InterFaces*, 15(3), 94–105, 1985.

Gould, L.S., "Computer-integrated manufacturing systems," *Modern Material Handling*, 91–144, 1985.

"Measuring the effect of NC," *American Machinist*, 113–144, 1983.

Ogorek, M., "CNC Standard Formats," *Manufacturing Engineering*, 43–45, 1985.

de Toni, A., and Tonchia, S., *Manufacturing Flexibility: a literature review*, International Journal of Production Research, 36(6), 1587–1617, 1998.

Waldner, J-B., *Principles of Computer-Integrated Manufacturing*, John Wiley, New York, 1992.

# *PART 2*

## *Facility Location*

# 17 Basic Facility Location Problems*

Facility location problems deal with the question of where to place facilities and how to assign customers. We initiate our analysis for choosing locations by discussing very basic models in which there is only one type of cost to consider, that of transportation. We assume that each facility is sufficiently large that it can serve all its customers if necessary — that is, that there is no limitation on a facility's capacity. Sample problems are presented in the following discussion, which is divided into two parts: the single-facility placement problem and the multiple-facility placement problem.

## 17.1 SINGLE-FACILITY PLACEMENT PROBLEM

Suppose that a new copy machine is to be installed in an office building complex. The machine can be set up in any one of five possible locations, and six offices will be assigned to use it. The demand from each office (i.e., the expected number of trips made daily) and the cost (measured in terms of the time that a person spends in walking to and from each location) are given in Table 17.1.

The table shows that it takes 3 minutes for customer A to go to and return from location 1. We wish to determine where the machine should be installed to minimize the total cost — that is, the total time spent by all users walking to and from the copy machine.

### 17.1.1 Solution Procedure

When there is only one facility to place, the solution procedure consists of three very basic steps:

1. Calculate the total transportation cost matrix (also called the demand cost matrix). An element of this matrix is obtained by multiplying the cost of assigning a unit of demand, from a customer to a location, by the total demand from the customer. For example, the cost of assigning customer A to location 1 is $3 \times 15 = 45$ units of time.
2. Take the sum of each column. These values represent the total cost of assigning all customers to each location.
3. Select the location with the minimum total cost. This is the location in which the machine should be placed.

* Portions of this chapter have been adapted from the *Journal of Operations Management* (vol. 1, No. 4, 1981), with permission from the American Production and Inventory Control Society (Falls Church, VA).

**TABLE 17.1**
**Time and Demand**

| Customer (Office) | Location 1 | 2 | 3 | 4 | 5 | Demand |
|---|---|---|---|---|---|---|
| A | 3 | 5 | 4 | 1 | 6 | 15 |
| B | 5 | 2 | 4 | 4 | 2 | 20 |
| C | 3 | 6 | 3 | 4 | 5 | 10 |
| D | 5 | 3 | 8 | 5 | 6 | 12 |
| E | 3 | 9 | 5 | 2 | 8 | 10 |
| F | 9 | 10 | 7 | 9 | 2 | 7 |

### 17.1.2 Application to the Sample Problem

The total transportation cost matrix for this problem, generated in step 1, is given in Table 17.2. The last row, the total, is the sum of each column. Because the total cost for location 4 is the least, this site is selected, and all the customers are assigned to this machine. The procedure is quick and requires minimal effort.

## 17.2 MULTIPLE-FACILITY PLACEMENT PROBLEM

Let us now turn to an example of a more complex type of problem. A manager of automobile service stations has sufficient funds to buy two identical diagnostic machines. He currently operates five service stations within the city, and he has identified five areas from which demands will occur. The transportation time from the center of gravity of each area cluster to each location is given in Table 17.3.

**TABLE 17.2**

| Customer | Location 1 | 2 | 3 | 4 | 5 |
|---|---|---|---|---|---|
| A | 45 | 75 | 60 | 15 | 90 |
| B | 100 | 40 | 80 | 80 | 40 |
| C | 30 | 60 | 30 | 40 | 50 |
| D | 60 | 36 | 96 | 60 | 72 |
| E | 30 | 90 | 50 | 20 | 80 |
| F | 63 | 70 | 49 | 63 | 14 |
| Total | 328 | 371 | 565 | 278 | 346 |

**TABLE 17.3**
**Transportation Time and Demand Data**

| | Time for Location | | | | | |
|---|---|---|---|---|---|---|
| **Customer** | **1** | **2** | **3** | **4** | **5** | **Demand** |
| A | 5 | 3 | 2 | 8 | 5 | 100 |
| B | 3 | 5 | 2 | 6 | 7 | 50 |
| C | 5 | 2 | 0 | 1 | 0 | 150 |
| D | 2 | 1 | 8 | 2 | 3 | 200 |
| E | 3 | 2 | 4 | 0 | 4 | 300 |

Also shown is the demand from each customer (area) for such a machine. The requirement is to place the two new machines in locations that will minimize the total transportation time, because the manager considers this to be the measure of efficiency.

A unique feature of this type of problem is that since the capacity of each facility is unlimited, there is no need to divide a customer's demand and assign parts of it to the two facilities. Indeed, a customer should be assigned to the facility for which the cost of transportation is at a minimum. The problem is then twofold. First, choose from the available locations those in which the facilities should be placed, and then assign the customers to these facilities such that the cost of these assignments is at a minimum.

We present two approaches to solving this problem: first the brute force approach and then a heuristic approach. As the name suggests, the brute force approach tries out all possibilities. However, even here, some reductions in the required calculations are possible by using the basic idea that the customer should be assigned to a facility for which the assignment cost is at a minimum.

## 17.3 BRUTE FORCE APPROACH

Suppose we wish to place $K$ facilities in $M$ available locations. In how many ways can this be done? A theorem in statistics gives the answer.

There are $(M/K)$ ways of selecting $K$ distinct objects from a set of $M$ objects where

$$\left(\frac{M}{K}\right) = \frac{M!}{K!(M-K)!}$$

In our problem, because there are five locations and two facilities in place, we have $(M/K) = 5!/2!3! = 10$ possible ways of placing these facilities. If all combinations are examined and demand is assigned to each such that we obtain the minimum cost for the combination, then the overall minimum can be easily found. The following procedure achieves this objective.

**TABLE 17.4**
**Demand Cost**

| Customer | Location 1 | 2 | 3 | 4 | 5 |
|---|---|---|---|---|---|
| A | 500 | 300 | 200 | 800 | 500 |
| B | 150 | 250 | 100 | 300 | 350 |
| C | 750 | 300 | 0 | 150 | 0 |
| D | 400 | 200 | 1600 | 400 | 600 |
| E | 900 | 600 | 1200 | 0 | 1200 |

First, calculate the total cost of assigning all the demand from a customer to a location. The procedure is similar to that illustrated in Section 17.1. In Table 17.4, for example, the cost of assigning customer A to location 3 would be the product of the cost of assigning a unit of demand from customer A to location 3 (2 units of time) multiplied by that customer's demand (100 units) — that is, 200. All other like products are calculated in a similar fashion and entered in Table 17.4.

Let us consider one possible combination, say, locations 1 and 2. This means that one of the two available machines is placed in location 1 and the other is placed in location 2. If that is the case, then each customer should be assigned to the location for which the cost of transportation is at a minimum. Customer A is assigned to location 2; customer B, to location 1; customer C, to location 2; customer D, to location 2; and customer E, to location 2. This solution has the cost of 300 + 150 + 300 + 200 + 600 = 1550. Similar analyses for the other 10 combinations led to the data in Table 17.5.

The minimum cost is that associated with the combination of locations 3 and 4; hence, these locations are selected as the best in which to place the facilities, and the

**TABLE 17.5**
**Minimum Cost**

| Customer | Location Pairs (Combinations) 1–2 | 1–3 | 1–4 | 1–5 | 2–3 | 2–4 | 2–5 | 3–4 | 3–5 | 4–5 |
|---|---|---|---|---|---|---|---|---|---|---|
| A | 300 | 200 | 500 | 500 | 200 | 300 | 300 | 200 | 200 | 500 |
| B | 150 | 100 | 150 | 150 | 100 | 250 | 250 | 100 | 100 | 300 |
| C | 300 | 0 | 150 | 0 | 0 | 150 | 0 | 0 | 0 | 0 |
| D | 200 | 400 | 400 | 400 | 200 | 200 | 200 | 400 | 600 | 400 |
| E | 600 | 900 | 0 | 900 | 600 | 0 | 600 | 0 | 1200 | 0 |
| Total | 1550 | 1600 | 1200 | 1950 | 1100 | 900 | 1350 | 700 | 2100 | 1200 |

customers are assigned in the manner that led to the minimum cost. Formally, then, the steps of the procedure are:

1. Construct a demand cost table. An entry *ij* in the demand cost table represents the cost of allocating all the demand from source (customer) *j* to location *i*.
2. Formulate all $(M/K)$ possible combinations.
3. In the first combination, determine the demand assignment by assigning demand to the facility where the cost of the assignment is the minimum. To do this:
   a. Consider only the columns of the demand cost matrix associated with the particular combination.
   b. Assign each demand to the facility having the smallest demand cost entry in its row for those columns selected in step a.
4. Calculate the total cost for the combination by taking the sum of the costs of the assignments in step 3.
5. Repeat steps 3 and 4 for all remaining combinations.
6. Select the combination with the overall minimum cost as the optimum combination; the associated demand assignments are the optimum assignments.

Computationally, the method is fast and simple, and relatively small problems can be quickly worked out by hand. Table 17.6 shows that for $K > 1$, the number of possible combinations increases dramatically for $M > K + 1$. The decision of whether to use the brute force method should take into consideration the number of combinations that must be examined. To solve larger problems, we recommend using the heuristic method, which is presented in the next section.

**TABLE 17.6**
**Possible Combinations**

| | Facilities to Be Placed ($K$) | | | | | |
|---|---|---|---|---|---|---|
| Available locations ($M$) | 1 | 2 | 3 | 4 | 5 | 6 |
| 1 | 1 | NA | NA | NA | NA | NA |
| 2 | 2 | 1 | NA | NA | NA | NA |
| 3 | 3 | 3 | 1 | NA | NA | NA |
| 4 | 4 | 6 | 4 | 1 | NA | NA |
| 5 | 5 | 10 | 10 | 5 | 1 | NA |
| 6 | 6 | 15 | 20 | 15 | 6 | 1 |

## 17.4 HEURISTIC METHOD FOR PROBLEMS INVOLVING FACILITIES WITH UNLIMITED CAPACITIES

Most heuristic methods follow certain logical steps that generally lead to optimum or near-optimum results. For the problem at hand, the heuristic method suggested in this section works well. The procedure is based on the following logic.

If we are to place the facilities one at a time, then we should place each in a location where the maximum benefit is derived from it. To begin, if we have only one facility, it should be located where the cost of assigning all customers to that location is the minimum. If additional facilities are available, they should be placed one at a time in positions that contribute to the maximum savings. Thus, each subsequent facility that is placed in an additional location should successively improve upon each previous solution. If the solution does not improve, then we do not need the additional facility. Therefore, the method not only indicates where the facilities should be located and how the customers should be assigned, but also points out the optimum number of facilities we should have. The procedure continues until either all the facilities are located or there is no further improvement in the solution.

As illustrated in the flowchart in Figure 17.1, the steps involved in the procedure are:

1. Formulate the total cost table. An entry $ij$ in the total cost table represents the cost of allocating all demands from customer $j$ to location $i$.
2. Total each column. The sum represents the total cost if demands from all customers are assigned to that location.
3. Assign the first facility to the location with the minimum total cost. We will denote a location with a facility as an assigned location and denote one without a facility as an unassigned location.
4. If no additional facility is available for assignment, go to step 8; otherwise, continue.
5. Determine the savings to be derived by moving each customer from the assigned location(s) to an unassigned location(s). The savings may occur because of a change in transportation cost, but if there are no savings,* mark "—" in the appropriate column.
6. Total each unassigned column. The sum represents the savings that could be achieved if an assignment is made in that location.
7. Make an assignment in the location that indicates the maximum savings. Transfer the customers who contributed to these savings to the new location, which now becomes an assigned location. Return to step 4.
8. All the assignments are made. Calculate the minimum cost and schedule.

### 17.4.1 Application

To demonstrate the heuristic procedure, we now apply it to the auto shop manager's problem in Section 17.2. The cost data, in terms of time, are the same as those in Table 17.3; however, suppose now that the manager has three machines to place.

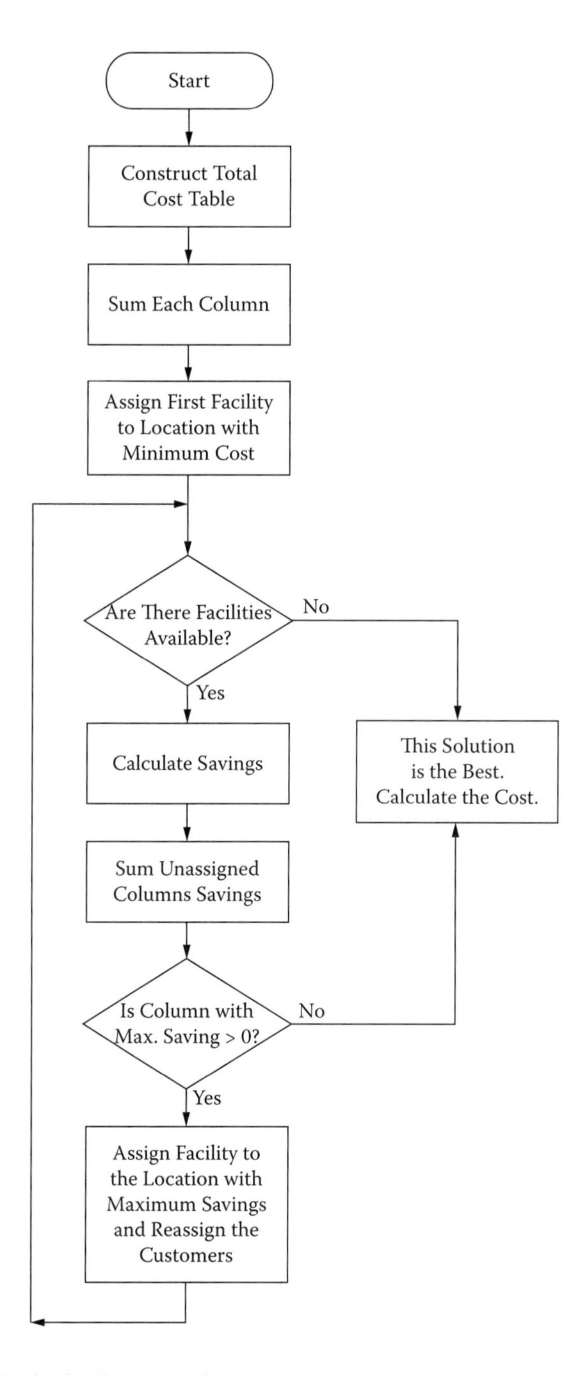


**FIGURE 17.1** Basic facility relocation problem flowchart for the heuristic procedure in Section 17.4

**TABLE 17.7**
**Demand Cost (Time)**

| | Location | | | | | |
|---|---|---|---|---|---|---|
| **Customer** | **1** | **2** | **3** | **4** | **5** | **Demand** |
| A | 500 | 300 | 200 | 800 | 500 | 100 |
| B | 150 | 250 | 100 | 300 | 350 | 50 |
| C | 750 | 300 | 0 | 150 | 0 | 150 |
| D | 400 | 200 | 1600 | 400 | 600 | 200 |
| E | 900 | 600 | 1200 | 0 | 1200 | 300 |
| Total | 2700 | 1650 | 3100 | 1650 | 2650 | 800 |

Table 17.4, a demand cost table, is reproduced for convenience as Table 17.7 and is expanded to include the customers' demands and the totals of the columns.

Because both columns 2 and 4 have the minimum total cost of 1650, the first machine could be assigned to either location 2 or location 4. We arbitrarily select location 4. Now, to decide where to place the second machine, we must calculate the savings (if any) to be gained by moving customers who are currently assigned to location 4 to any other unassigned location. For example, consider customer A. If customer A is moved from location 4 to location 1, the move will save 800 – 500 = 300. Similarly, for location 2, the saving is 800 – 300 = 500. Other savings are calculated in the same manner, and Table 17.8 provides the results. It should be noted that if moving to a different location results in negative savings, then that location is marked by "—." For example, there are negative savings when customer B is moved from location 4 to location 5 (300 – 350 = –50), and accordingly, the "—" notation is entered in the appropriate position. As shown in Table 17.8, the savings are calculated for all the customers and for each of the unassigned locations.

**TABLE 17.8**
**Savings Table I for Moving From Location 4**

| | Location | | | | |
|---|---|---|---|---|---|
| **Customer** | **1** | **2** | **3** | **4** | **5** |
| A | 300 | 500 | (600) | | 300 |
| B | 150 | 50 | (200) | | — |
| C | — | — | (150) | | 150 |
| D | — | 200 | — | o | — |
| E | — | — | — | o | — |
| Total | 450 | 750 | 950 | | 450 |

**TABLE 17.9**
**Savings Table II for Moving From Location 3 or 4**

| | Location | | | | |
|---|---|---|---|---|---|
| **Customer** | **1** | **2** | **3** | **4** | **5** |
| A | — | — | o | | — |
| B | — | — | o | | — |
| C | — | — | o | | o |
| D | o | 200 | — | | — |
| E | — | | — | o | — |
| Total | — | 200 | — | | — |

The total of the savings for each location is obtained by adding the column entries. For example, the total of the savings for location 1 is only 300 + 150 = 450. The second machine should be placed in a location that has the maximum savings, in this case, location 3. The customers who contributed to these savings, customers A, B, and C, are now reassigned to location 3. The others, customers D and E, are retained at location 4. The assignments at this stage are shown by circles in Table 17.8.

To decide where to place the third machine, we must recalculate the savings to be realized by moving the customers from their currently assigned locations to the unassigned locations. Again, if the amount saved is negative, we will mark the unassigned location by "—." Customer A is presently attached to location 3, and if we move this customer to location 1, the savings is 200 – 500 = –300. This –300 is a cost and is therefore marked by "—." Similarly, for location 2, the savings is 100 – 150 = –50, and for location 5, it is 200 – 500 = –300. For customer D, currently assigned to location 4, the corresponding calculations are as follows: For location 1, the amount saved is 400 – 400 = 0; for location 2, 400 – 200 = 200; and for location 5, 400 – 600 = –200, which is marked by "—." All results are shown in Table 17.9.

Because location 2 has the only savings, a total of $200, we will place the third machine in this location. The customer contributing to these savings, customer D, is now assigned to location 2. All the assignments are shown by circles in Table 17.9.

Thus, our final assignment is:

| **Customer** | **A** | **B** | **C** | **D** | **E** |
|---|---|---|---|---|---|
| Location cost | 3 | 3 | 3 | 2 | 4 |
| Cost | 200 | 100 | 0 | 200 | 0 |

The total cost for this solution is 500. Incidentally, we would have reached the identical solution had we chosen location 2 instead of location 4 for assigning the first machine.

Now suppose that a fourth machine is available. Could we profitably use it? We decide by determining whether any additional saving is possible; the results of our calculations are displayed in Table 17.10.

**TABLE 17.10**
**Savings Table III for Moving From Location 2, 3, or 4**

| | Location | | | | |
|---|---|---|---|---|---|
| **Customer** | **1** | **2** | **3** | **4** | **5** |
| A | — | | | | |
| B | — | | | | — |
| C | — | | | | o |
| D | — | | | | — |
| E | — | | | | — |

The only unassigned locations are 1 and 5. We find no savings in either of these two locations, so there is no need for the fourth machine. Our heuristic approach has identified the optimum number of machines as three.

Note that Table 17.10 indicates that customer C could be assigned to a fourth machine at location 5 at no increase in transportation cost; however, because there would likewise be no savings, any additional installation, operating, and maintenance costs for a fourth machine could never be recovered.

### 17.4.2 Validity

How good is the method? We have found by experimentation that this procedure provides the exact result about 90% of the time. For the remaining 10%, when it deviates from the theoretical minimum, it differs by an average of 5% from the optimum solution. (In one example, it did deviate by as much as 18%.) In most applications, the data are estimated to start with, and the 5% variation is negligible. This is especially important when one observes the simplicity of the procedure and the speed with which it can be applied.

The input to the computer program for the method is outlined in Section 17.6; the program itself is given in Appendix G.

## 17.5 OTHER METHODS

We will now briefly review the mathematical formulation of the model for readers who are interested in the exact mathematical solution. Let us define:

$X_{ij}$ = 1 if the demand from source $i$ is assigned to location $j$ and 0 otherwise
$C_{ij}$ = cost of assigning a unit of demand from source $i$ to location $j$, where $i = 1, 2, \ldots, n$ and $j = 1, 2, \ldots, m$
$D_i$ = demand from source $i$
$K$ = number of facilities available for placement.

Next, we wish to minimize total cost (TC), where

$$TC = \sum_{j=1}^{m} \sum_{j=1}^{n} C_{ij} \times d_i \times X_{ij}$$

subject to the constraints that are developed as follows.

Since the demand for each source must be assigned to at least one location, we have

$$\sum_{j=1}^{m} X_{ij} \geq 1 \quad \text{for each } i$$

Similarly, because only $K$ locations are to be selected and only $X_{tj}$ variables assigning demands to these locations can take positive values, we have

$$\sum_{i=1}^{n} X_{ij} \leq n\, I_j \quad \text{for each } j$$

$$\sum_{j=1}^{m} I_j = K$$

$$I_j = \begin{cases} 1 & \text{if the facility is assigned to location } j \\ 0 & \text{otherwise} \end{cases}$$

These formulations can be solved by integer programming.

The other methods suggested are branch and bound, implicit enumerations, and decomposition. We will not dwell on any of these methods because the purpose of this chapter was to present a method that requires hardly more than basic algebraic skills.

## 17.6 COMPUTER PROGRAM DESCRIPTION

### 17.6.1 Heuristic Method

1. Description of parameters
   IROW = number of customers
   ICOL = number of possible locations
   IMC = number of facilities to be located
   COST(I,J) = total cost of assigning customer $i$ to location $j$ (i.e., product or demand from customer $i$ and unit cost of assigning customer $i$ to location $j$)
2. Input lines
   Line type 1: IROW, ICOL
   IMC Line type 2: COST(*I*,*J*) (in row)
3. Printed output
   a. Listings of input data
   b. Final assignment for each customer (i.e., each customer number assigned to a location number)

## 17.7 SUMMARY

This chapter has covered two types of facility location problems in which the only factor considered is transportation cost: the single-facility placement problem and the multiple-facility placement problem. In the former case, the minimum cost is determined simply by multiplying each customer's demand by each location's unit cost. The facility is then placed in the location that is found to have the lowest overall cost when all the customers' demand costs are totaled by location.

The multiple-facility problem is slightly more complicated, and we discussed two methods for its solution: the brute force method and the heuristic approach. Brute force attacks the problem directly by setting up all possible combinations of locations and calculating the minimum transportation cost for each customer to be served by each location combination. The best combination of locations is that which has the overall minimum cost when the totals for all customers at each of the combinations are compared. This procedure becomes unwieldy when the number of possible combinations becomes large.

The heuristic method assigns the facilities to one location at a time. The first is assigned to service all customers and is placed in the location having the lowest total demand cost for all customers. Then, by working with one location at a time, the savings that would be gained if each customer were to be served from that location are determined. Only positive savings are of interest, and the location is selected if there is a net gain. Redistribution of the customers follows.

## PROBLEMS

1. A warehouse is to be built in one of two possible locations to store merchandise to be shipped to three stores.

| | Location | | |
|---|---|---|---|
| Store | A | B | Demand |
| 1 | 15 | 24 | 50 |
| 2 | 21 | 13 | 35 |
| 3 | 17 | 8 | 60 |

   a. Given the distances from the stores to the possible locations and the demands per week, as shown in the accompanying table, use the heuristic method to determine the best location for the warehouse.
   b. Solve the problem by mathematical formulation and compare the results with the solution to part a.

2. A water fountain can be located in one of three locations to accommodate four offices. The company would like to minimize the time spent away from the offices.
   a. Given the distances from the offices to the possible locations and the number of employees in each location, as shown in the accompanying table, use the brute force method to determine the best location.

| Office | Location 1 | Location 2 | Location 3 | Employees |
|---|---|---|---|---|
| 1 | 15 | 45 | 30 | 2 |
| 2 | 38 | 27 | 25 | 4 |
| 3 | 70 | 63 | 58 | 5 |
| 4 | 40 | 90 | 50 | 3 |

b. Solve by heuristic methods and compare the results with those for part a.
c. Determine the best locations if two water fountains can be installed.

3. Four locations in a department are being considered for the installation of a new drill to be operated by five workers who will have to travel to and from their stations in order to use it. The time required for each worker to travel to each location and the number of trips to be made by each are shown in the accompanying table.

| Worker | Location A | Location B | Location C | Location D | Demand |
|---|---|---|---|---|---|
| 1 | 0.50 | 1.00 | 1.20 | 0.80 | 8 |
| 2 | 0.70 | 0.30 | 2.00 | 0.75 | 15 |
| 3 | 1.60 | 2.20 | 1.30 | 0.90 | 22 |
| 4 | 0.40 | 0.60 | 1.40 | 2.10 | 17 |
| 5 | 1.30 | 0.30 | 0.50 | 0.80 | 11 |

Determine the best location for the drill. If another drill machine could be purchased, what would be the two best locations for these machines?

4. Three computers were purchased for use by six departments in a small appliance manufacturing factory. The transportation time from the center of each department to each of the possible locations for the computers is shown in the accompanying table, along with each department's daily demands.

| Department | Location 1 | Location 2 | Location 3 | Location 4 | Location 5 | Location 6 | Demand |
|---|---|---|---|---|---|---|---|
| A | 7.0 | 4.0 | 2.0 | 6.0 | 3.5 | 1.0 | 8 |
| B | 4.0 | 2.5 | 3.0 | 0.0 | 5.0 | 2.0 | 16 |
| C | 2.0 | 3.0 | 1.0 | 4.0 | 4.0 | 3.0 | 24 |
| D | 1.0 | 6.0 | 7.0 | 5.0 | 2.0 | 0.0 | 10 |
| E | 3.0 | 8.0 | 2.0 | 7.5 | 6.0 | 8.0 | 4 |
| F | 5.0 | 0.0 | 9.0 | 4.0 | 4.0 | 5.0 | 15 |

a. Use the brute force method to determine the best locations for the computers.

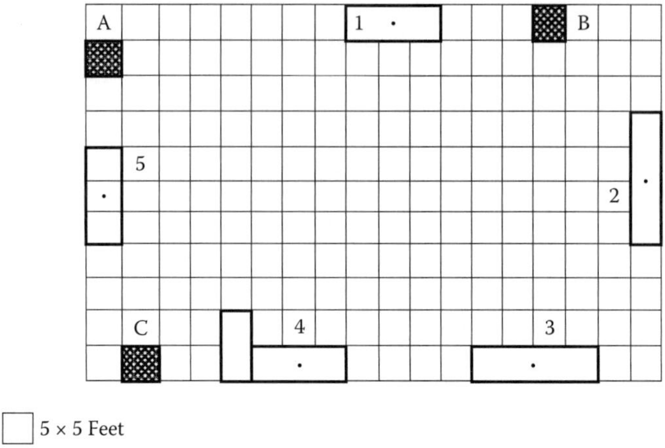

A 1 B 5 2 C 4 3

5 × 5 Feet

| Station | 1 | 2 | 3 | 4 | 5 |
|---|---|---|---|---|---|
| Demand | 10 | 13 | 9 | 21 | 15 |

**FIGURE 17.2**

b. Solve by using the heuristic method.
c. Determine whether more computers can be profitably added.
Note: Problem 4 can be solved using the computer.

5. Discuss possible instances in which it would become difficult or impractical to use the brute force method.
6. a. The diagram in Figure 17.2 shows a department with five stations (1, 2, 3, 4, and 5) and three possible locations for a new piece of machinery (A, B, and C). Using rectilinear distances from the points in the stations to the possible locations, determine the best location. Assume that the machines and demands are located at the centers of their respective squares, as marked. The demands are as shown in the accompanying table.
   b. Repeat part 17.6a if two new machines are to be located.
7. Conduct a 10 × 10 problem (10 customers and 10 possible locations) and solve it by the brute force and heuristic methods. Compare the results when placing
   a. two facilities
   b. three facilities
   Note that problem 7 can be solved using the computer.

## SUGGESTED READINGS

Sule, D.R., *Logistics of Facility Location and Allocation*, Marcel Dekker, Inc., New York, 2001.

# 18 Location Analysis with Fixed Costs

Under most circumstances, the site where a facility is to be located is not in a suitable condition to accept the facility on an "as-is" basis, and site preparation or modification work is required. This involves an additional expenditure; however, it is a one-time, initial cost known as a fixed cost. The numerical values of the fixed costs for installing a particular facility may vary from location to location. For example, in choosing a site for a plant, locations might differ in available conveniences. In some locations, land might have to be purchased and a building constructed; in others, suitable existing buildings might be available that could be converted to the desired plant. In other locations, modifications to transportation facilities might be needed to use the existing buildings as production facilities. Thus the one-time costs to build the warehouses (or other facilities) in each location are generally different.

In some other instances, location (or relocation) of a piece of equipment may be planned. Examples include relocation of a lathe in a machine shop, placement of a new X-ray machine in a hospital, and installation of a new copy machine in an office building. In each case, a number of suitable locations might be available; however, the cost of placing a machine might be different in each location. It could depend on such factors as the present condition of the site and/or the modifications required to the foundation as well as the surroundings. In some areas, preparation of the site would demand no additional expenditure, but in other areas it might require extensive work such as the installation of plumbing fixtures, sewage systems, and electric outlets; redesigning the area; and even perhaps the installation of air-conditioning systems.

Environmental regulations might also dictate an additional fixed cost for a site. For example, in locating a chemical or power plant, regulations governing exhaust emissions may be important. Such regulations within a city might require a special piece of equipment, such as a scrubber for a coal-burning power plant, while in outlying areas these restrictions might not exist.

The objective for the problems to be studied in this chapter is to select the location(s) and assign the demand(s) in a manner that minimizes the total cost, that is, the sum of transportation and fixed costs. Before proceeding, however, it is important to understand the time units for which these costs are defined. The fixed cost is a one-time cost and is therefore to be recovered over the entire life of the facility. On the other hand, a transportation cost resulting from assigning a customer's demand(s) to a location depends on the period for which the demands are determined. A demand may be expressed for any convenient unit of time — a week, a month, or a year — provided that all the demands and all the fixed costs are for the same period.

If this is not already the case, they must be converted to a convenient unit. For fixed costs this can be easily done by estimating the life of each facility and then using an annuity factor to spread the fixed cost (depreciation and interest cost) for each facility over a unit period (see Appendix A.) The effect, if any, of income taxes is ignored in all illustrative examples throughout this chapter. If the tax rate is known, we suggest developing after-tax figures and then using the procedures, which are applicable in either case without loss of generality.

## 18.1 HEURISTIC METHOD FOR SOLVING PROBLEMS WITH FIXED COSTS

The following heuristic method has been found to be very effective in solving problems with fixed costs. As illustrated in the flowchart in Figure 18.1, the steps involved in the procedure are as follows:

1. Construct the demand cost table as before (Chapter 17). If the number of facilities available for placement is two or more, go to step 2. Otherwise, take the sum of each column, which will represent the total variable (transportation) cost if the demands from all sources are assigned to that location. Then add the total variable cost for each column to the corresponding fixed cost. Select the location with the minimum overall cost and assign all customers to that location. This is the optimum solution if only one facility is available, and so the solution process terminates.
2. A good initial solution when there is more than one facility is obtained by constructing a "minimum savings table" as follows. For each customer, calculate the difference between the cost of assigning the customer to the minimum cost location and the next higher cost location. Let us call this difference the "minimum increment." This represents the minimum amount saved if the customer is assigned to the location with the minimum cost. All other locations to which this customer could be assigned would only increase the cost of assignment. In constructing the minimum savings table, assign each customer's minimum increment to its minimum cost location.
3. Calculate the total savings for each column in the minimum savings table, and subtract the fixed cost for each location from the corresponding column total. This gives the net savings for each location. Because we must have at least one facility to satisfy the demands, the first facility should be located where the cost is lowest. Assign all customers to this facility. This now becomes an assigned location.
4. Determine the savings in moving each customer from the assigned location(s) to the nonassigned location(s). If there is no saving for a particular location, mark it by "—" in the appropriate column.
5. Compute the total for each unassigned column. Each sum represents the savings that could be achieved if the customers were moved to that column. From each sum, subtract the fixed cost associated with that

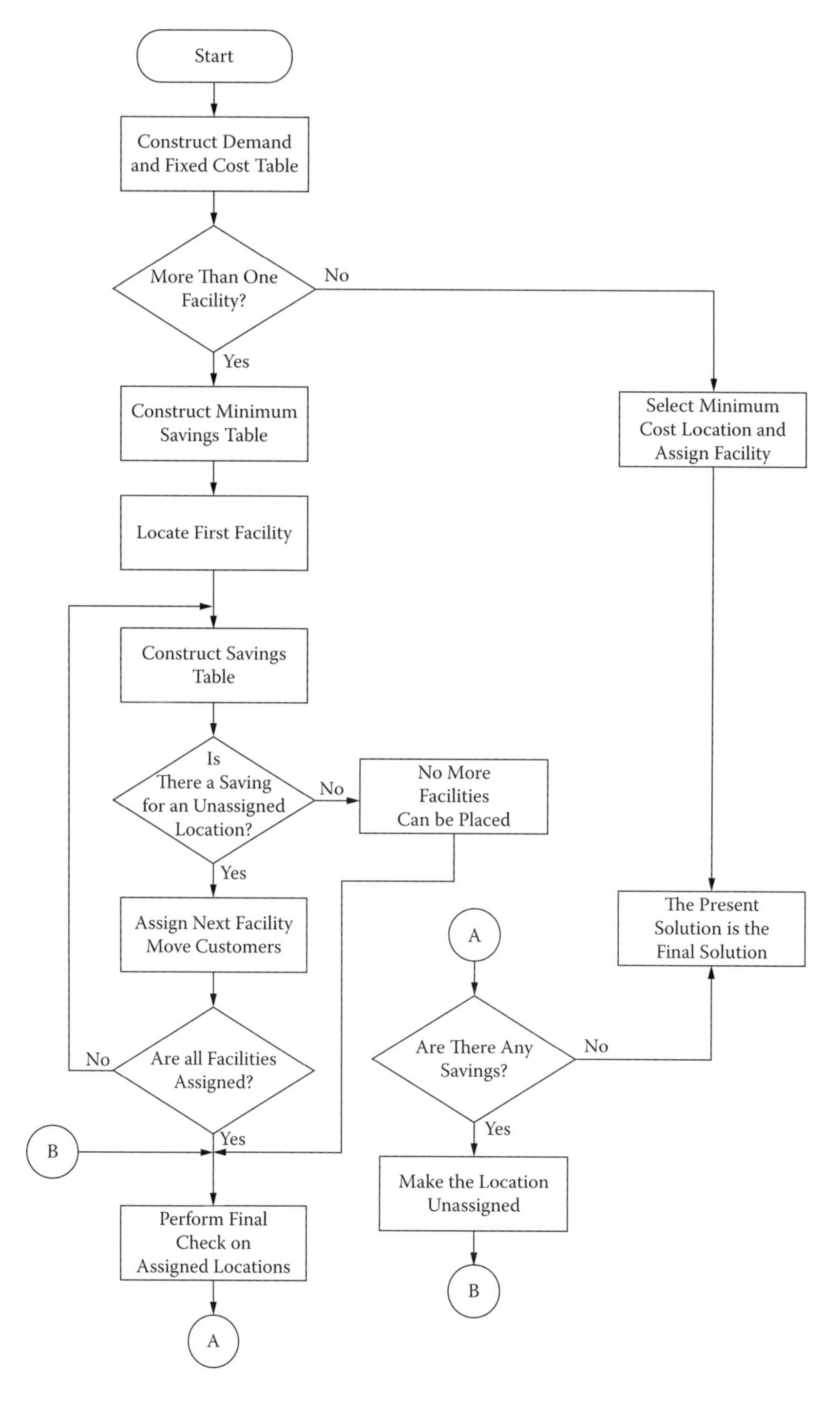


**FIGURE 18.1** Location analysis with fixed cost schematic diagram for the procedure in Section 18.1

location and make the next assignments in the location that has the maximum net savings. This is because we would use the additional facility only if it were going to reduce the total cost. The location just chosen now becomes an assigned location and a part of the "assigned group." Move the customers that contributed to the savings to this location. If all the facilities have been assigned, or if no column with savings can be found, proceed to step 6. Otherwise, return to step 4.

6. This step is the final check analysis. For each assigned location, determine the cost that would be incurred if each of the currently assigned customers for that location were moved to the assigned location with the next higher transportation cost. (Note that this may result in the customers' being moved to different assigned locations.) Total these costs. If this sum is less than the fixed cost for the assigned location under consideration, then additional savings could be achieved by not selecting the present location for placing a facility, that is, by making this location an unassigned location. Perform this check on all the assigned locations, and select the assigned location with the maximum savings if it exists. Make this location an unassigned location by moving each of the customers currently assigned to this location to the assigned location with the next higher transportation cost. Repeat step 6 until no further savings can be obtained. We then have the optimum solution.

## 18.2 APPLICATION

Consider the auto shop problem analyzed in Chapter 17. Recall that we had five customers and five possible locations in which the facilities could be placed. In Chapter 17 we assumed that the costs were in units of time; in this chapter, we will measure them in terms of dollars.

Table 18.1 provides the transportation costs and demands and leads directly to Table 18.2, the demand cost table for the problem.

**TABLE 18.1**
**Transportation Cost and Demand**

| | Location 3 | | | | | |
|---|---|---|---|---|---|---|
| **Customer** | **1** | **2** | **3** | **4** | **5** | **Demand** |
| A | 5 | 3 | 2 | 8 | 5 | 100 |
| B | 3 | 5 | 2 | 6 | 7 | 50 |
| C | 5 | 2 | 0 | 1 | 0 | 150 |
| D | 2 | 1 | 8 | 2 | 3 | 200 |
| E | 3 | 2 | 4 | 0 | 4 | 300 |

**TABLE 18.2**
**Demand Cost**

| | Location 3 | | | | |
|---|---|---|---|---|---|
| **Customer** | **1** | **2** | **3** | **4** | **5** |
| A | 500 | 300 | 200 | 800 | 500 |
| B | 150 | 250 | 100 | 300 | 350 |
| C | 750 | 300 | 0 | 150 | 0 |
| D | 400 | 200 | 1600 | 400 | 600 |
| E | 900 | 600 | 1200 | 0 | 1200 |

Suppose now that the fixed cost associated with each location, if a facility is placed there, is as follows:

| Location | 1 | 2 | 3 | 4 | 5 |
|---|---|---|---|---|---|
| Fixed Cost | 5000 | 4000 | 4500 | 6000 | 4200 |

In addition, assume that the life of the equipment is affected by the location in which it is placed. This may be attributable to such things as climate, available maintenance, or operating procedures. The estimated life expectancy for a facility in each location is:

| Location | 1 | 2 | 3 | 4 | 5 |
|---|---|---|---|---|---|
| Expected life (years) | 8 | 7 | 10 | 10 | 9 |

We wish to place the facilities and assign the customers in a manner that will minimize the total cost of operation per year.

### 18.2.1 Solution

Because the economic life of the equipment is location-dependent, the first step is to convert all costs to the same basic time units. The demand cost matrix is based on annual demand; therefore, it is convenient to convert the fixed costs to an annual basis. If the interest rate is known, we can use the capital recovery factor (Appendix A) to obtain the annual equivalent fixed cost. If the interest rate is not known or is considered insignificant, then an approximate estimate of the annual cost is obtained simply by dividing the fixed cost by the expected useful life.

Assume that the interest rate is 10%. Then by using the capital recovery factor, $A_i$, the fixed cost per year for location $i$ is:

| | | | |
|---|---|---|---|
| $A_1$ | = \$5000 (A/P, 10%, 8) | = 5000(0.1874) | = 937 |
| $A_2$ | = \$4000 (A/P, 10%, 7) | = 4000(0.2046) | = 818 |
| $A_3$ | = \$4500 (A/P, 10%, 10) | = 45,000(0.1627) | = 732 |
| $A_4$ | = \$6000 (A/P, 10%, 10) | = 6000(0.1627) | = 976 |
| $A_5$ | = \$4200 (A/P, 10%, 9) | = 4200(0.1736) | = 729 |

Because we were given the total cost table in the data and there is more than one facility to place, we will proceed directly to step 2 of the procedure. For customer A, location 3 has the minimum transportation cost of \$200, and the next larger amount is \$300 for location 2. The difference of \$100 is the minimum savings if this customer were serviced by location 3. This amount for customer A is entered in location 3 of the minimum savings table (Table 18.3). Similar analyses are performed for all other customers, and the results are entered in the table. Finally, the fixed cost for each location is entered as a negative savings.

The net savings for each location are determined by taking the sum of each column. In this case, each one has a net negative savings that actually indicates additional cost rather than a savings. Because the equipment must be placed and customers assigned, the first machine is placed where the cost is minimum, that is, at location 4. All customers are assigned to this location, as shown in Table 18.4.

The next step is to determine whether a second machine is needed and, if so, where it should be placed. To do so, we calculate the savings resulting from moving a customer from its currently assigned location to an unassigned location. Let us do this for customer A. The savings that results from moving customer A from location 4 to location 1 is 800 – 500 = \$300, which is entered in Table 18.5 for customer A in location 1. Similar calculations are performed for all customers for each unassigned location. Table 18.5 shows the results.

Again, a negative savings is indicated by a dash. Taking the total of the savings and the fixed cost for each location makes it clear that the maximum net savings are at location 3. The facility is thus placed in location 3, and the customers that contributed

**TABLE 18.3**
**Initial Assignment Minimum Savings Table**

| | Location | | | | |
|---|---|---|---|---|---|
| **Customer** | **1** | **2** | **3** | **4** | **5** |
| A | | | 100 | | |
| B | | | 50 | | |
| C | | | 0 | | |
| D | | 200 | | | |
| E | | | | 600 | |
| Fixed cost | –937 | –818 | –732 | –976 | –729 |
| Net savings | –937 | –618 | –582 | –376 | –729 |

**TABLE 18.4**
**Demand Assignments I**

| Customer | Location 1 | 2 | 3 | 4 | 5 |
|---|---|---|---|---|---|
| A | | | | × | |
| B | | | | × | |
| C | | | | × | |
| D | | | | × | |
| E | | | | × | |

to these savings — A, B, and C — are assigned there. The present demand assignments are shown in Table 18.6.

Table 18.7 details the process of selecting a location for the third machine (facility). The transportation cost savings is calculated on the basis of moving a customer from the present assignment. For example, moving customer A from location 3 to location 2 will result in 200 – 300 = –$100 savings (Table 18.2) and is marked by a dash in location 2, while moving customer D (presently assigned to location 4) to location 2 will result in 400 – 200 = $200 savings, as indicated for customer D in the second column. The results of similar calculations for all remaining combinations of customers and currently unassigned locations are shown.

After including the fixed cost, the net savings are found to be negative for all unassigned locations; therefore, the third machine is not profitable, and the process is stopped here. Thus, the previous solution is still the best at this stage. Table 18.8 presents the costs for those assignments.

**TABLE 18.5**
**Savings Table I for Moving From Location 4**

| Customer | Location 1 | 2 | 3 | 4 | 5 |
|---|---|---|---|---|---|
| A | 300 | 500 | 600 | | 300 |
| B | 150 | 50 | 200 | | — |
| C | — | — | 150 | | 150 |
| D | — | 200 | — | | — |
| E | — | — | — | | — |
| Total savings | 450 | 750 | 950 | | 450 |
| Fixed cost | –973 | –818 | –732 | | –729 |
| Net savings | –487 | –68 | 218 | | –279 |

**TABLE 18.6**
**Demand Assignments II**

| | Location | | | | |
|---|---|---|---|---|---|
| **Customer** | **1** | **2** | **3** | **4** | **5** |
| A | | | × | | |
| B | | | × | | |
| C | | | × | | |
| D | | | | × | |
| E | | | | × | |

**TABLE 18.7**
**Savings Table II**

| | Location | | | | |
|---|---|---|---|---|---|
| **Customer** | **1** | **2** | **3** | **4** | **5** |
| A | — | — | | | — |
| B | — | — | | | — |
| C | — | — | | | — |
| D | — | 200 | | | — |
| E | — | — | | | — |
| Total | 0 | 200 | | | 0 |
| Fixed cost | –937 | –818 | | | –729 |
| Net savings | –937 | –618 | | | –729 |

**TABLE 18.8**
**Total Cost for the Final Solution**

| | Location | | | | | |
|---|---|---|---|---|---|---|
| **Customer** | **1** | **2** | **3** | **4** | **5** | **Total** |
| A | | | 200 | | | 200 |
| B | | | 100 | | | 100 |
| C | | | 0 | | | 0 |
| D | | | | 400 | | 400 |
| E | | | | 0 | | 0 |
| Demand cost | | | 300 | 400 | | 700 |
| Fixed cost | | | 723 | 976 | | 1668 |
| Total cost | | | 1032 | 1376 | | 2368 |

The next step is to perform a final check analysis. Because there are only two assigned locations, the final check is rather simple. Customers A, B, and C are currently assigned to location 3. If all of them are moved to location 4, because the next minimum transportation cost assigned location for each customer happens to be the same in this example, the transportation cost is increased by 600 + 200 + 150 = $950. The fixed cost savings associated with making location 3 unassigned is $732 (Table 18.3). Because the savings is smaller than the cost, this move is not made.

Similarly, if customers D and E are moved from location 4 to location 3, the transportation cost is increased by 1200 + 1200 = $2400. The fixed cost savings for location 4 is $976. Consequently, this move is also rejected. Thus, the present solution is indeed the best solution.

Incidentally, one may also start the solution procedure by locating the first facility in the minimum cost location obtained in step 1 and ignoring steps 2 and 3. At times this starting point has given a better result. The reader should examine each problem using both starting assignments and then select a solution that gives the best results.

## 18.3 UNASSIGNABLE LOCATION

Quite often we come across a problem in which one or more customers cannot be assigned to certain locations. For example, consider the placement of a fire station in a community. Fire codes might dictate the maximum allowable response time of the fire department to an alarm, and travel time is obviously a major component of that time. This restriction on the maximum value for response time might not permit the selection of certain locations as suitable sites for placement of fire stations to serve the outlying areas. To solve this type of problem, we modify the fixed cost procedure by converting the "*" wherever it appears in the demand cost table (which indicates that designated customers cannot be assigned to that location) to some large cost. However, assigning an infinite cost ($\infty$) will make it difficult to determine the net savings (they will be indistinguishable from one location to another) as we move the demands from assigned locations to unassigned locations. As a general rule, this cost should be more than the largest fixed cost plus the two largest transportation costs. With this modification to the demand cost table, we can then apply the method described in Section 18.1 to solve the resulting problem.

### FIRE STATION PROBLEM

Consider the case of eight customers that are to be assigned to facilities placed in five locations. The demand cost data (i.e., demand multiplied by cost per unit demand) and the fixed cost data are given in Table 18.9.

Remember, an asterisk (*) for a cost entry means that the indicated customer cannot be assigned to the associated location. The first task is to determine the appropriate value for the "*" based on the data in Table 18.9. The largest fixed cost is 400; the largest transportation cost is 240, and the next largest is 210. With these data,

**TABLE 18.9**
**Demand Cost and Fixed Cost**

| | Location | | | | |
|---|---|---|---|---|---|
| **Customer** | **1** | **2** | **3** | **4** | **5** |
| A | 120 | 210 | 180 | 210 | 170 |
| B | 180 | * | 190 | 190 | 150 |
| C | 100 | 150 | 110 | 150 | 110 |
| D | * | 240 | 195 | 180 | 150 |
| E | 60 | 55 | 50 | 65 | 70 |
| F | * | 210 | * | 120 | 195 |
| G | 180 | 110 | * | 160 | 200 |
| H | * | 165 | 195 | 120 | * |
| Fixed cost | 200 | 200 | 200 | 400 | 300 |

the "*" could be replaced by 400 + 240 + 210 = 850 or, say, $900. This is shown in Table 18.10.

The next step is to determine the values for the minimum savings table, Table 18.11. We construct the table by taking the difference between the minimum and the next higher cost of assignment for each customer and entering that amount for the location having the minimum cost for that customer.

Because location 1 has the greatest net savings (lowest net cost), the first facility is placed there, and all the customers are assigned to that location. To determine the placement of the second facility, we continue the procedure by calculating the values for the first savings table, Table 18.12.

**TABLE 18.10**
**Modified Demand Cost and Fixed Cost**

| | Location | | | | |
|---|---|---|---|---|---|
| **Customer** | **1** | **2** | **3** | **4** | **5** |
| A | 120 | 210 | 180 | 210 | 170 |
| B | 180 | 900 | 190 | 190 | 150 |
| C | 100 | 150 | 110 | 150 | 110 |
| D | 900 | 240 | 195 | 180 | 150 |
| E | 60 | 55 | 50 | 65 | 70 |
| F | 900 | 210 | 900 | 120 | 195 |
| G | 180 | 110 | 900 | 160 | 200 |
| H | 900 | 165 | 195 | 120 | 900 |
| Fixed cost | 200 | 200 | 200 | 400 | 300 |

**TABLE 18.11**
**Minimum Savings for the Initial Placement**

| Customer | Location 1 | 2 | 3 | 4 | 5 |
|---|---|---|---|---|---|
| A | 50 | | | | |
| B | | | | | 30 |
| C | 10 | | | | |
| D | | | | | 30 |
| E | | | 5 | | |
| F | | | | 75 | |
| G | | 50 | | | |
| H | | | | 45 | |
| Total | 60 | 50 | 5 | 120 | 60 |
| Fixed cost | −200 | −200 | −200 | −400 | −300 |
| Net savings | −140 | −150 | −195 | −280 | −240 |

Because location 2 has the maximum savings, place the second machine there, and assign customers D, E, F, G, and H to it. The revised assignment is shown in Table 18.13.

To check for placement of a third machine with additional savings, let us calculate the second savings table, Table 18.14.

Because there are no net savings in any unassigned location, the demand assignment in Table 18.13 provides the present best solution as shown in the total cost table, Table 18.15.

**TABLE 18.12**
**Savings Table I for Moving From Location 1**

| Customer | Location 1 | 2 | 3 | 4 | 5 |
|---|---|---|---|---|---|
| A | | — | — | — | — |
| B | | — | — | — | 30 |
| C | | — | — | — | — |
| D | | 660 | 705 | 720 | 750 |
| E | | 5 | 10 | — | — |
| F | | 690 | — | 780 | 705 |
| G | | 70 | — | 20 | — |
| H | | 735 | 705 | 780 | — |
| Total | | 2160 | 1420 | 2300 | 1485 |
| Fixed cost | | −200 | −200 | −400 | −300 |
| Net savings | | 1960 | 1220 | 1900 | 1185 |

**TABLE 18.13**
**Demand Assignment**

| Customer | Location 1 | 2 | 3 | 4 | 5 |
|---|---|---|---|---|---|
| A | × | | | | |
| B | × | | | | |
| C | × | | | | |
| D | | × | | | |
| E | | × | | | |
| F | | × | | | |
| G | | × | | | |
| H | | × | | | |

The next step is to perform the final check. If we move the customers assigned to location 1 — A, B, and C — to location 2, we will increase the transportation cost by 90 + 720 + 50 = $860 (Table 18.10). The corresponding savings by making location 1 an unassigned location is its fixed cost, $200. Because the savings is less than the cost, this move is not made. Similarly, if the customers currently assigned to location 2 are moved to location 1, it will increase the transportation cost by 660 + 5 + 690 + 70 + 735 = $2160. The fixed cost savings for location 2 is $200. Again, the move is rejected. The solution in Table 18.15 is the optimum solution.

It may seem fortuitous that we have customers assigned only to locations where the assignments are possible; however, this is not a matter of luck. As long as we are

**TABLE 18.14**
**Savings Table II for Moving From Location 1 or 2**

| Customer | Location 1 | 2 | 3 | 4 | 5 |
|---|---|---|---|---|---|
| A | | | — | — | — |
| B | | | — | — | 30 |
| C | | | — | — | — |
| D | | | 45 | 60 | 90 |
| E | | | 5 | — | — |
| F | | | | 90 | 15 |
| G | | | — | — | — |
| H | | | — | 45 | — |
| Total | | | 50 | 195 | 135 |
| Fixed cost | | | –200 | –400 | –300 |
| Net savings | | | –150 | –205 | –165 |

**TABLE 18.15**
**Total Cost for the Final Solution**

| | Location | | | | | |
|---|---|---|---|---|---|---|
| **Customer** | **1** | **2** | **3** | **4** | **5** | **Total** |
| A | 120 | | | | | 120 |
| B | 180 | *[a] | | | | 180 |
| C | 100 | | | | | 100 |
| D | *[a] | 240 | | | | 240 |
| E | | 55 | | | | 55 |
| F | *[a] | 210 | *[a] | | | 210 |
| G | | 110 | *[a] | | | 110 |
| H | *[a] | 165 | | | * | 165 |
| Demand Cost | 400 | 780 | | | | 1180 |
| Fixed Cost | 200 | 200 | | | | 400 |
| Total Cost | 600 | 980 | | | | 1580 |

[a]Indicates unpermitted customer/location combination.

free to choose the number of facilities, and as long as there is a location to which a customer can be assigned, then a valid assignment for each customer is always possible.

The problem may become infeasible, that is, one without solution, if there is only a limited number of facilities. For instance, if we had only one facility in the previous example, all customers would have been assigned to location 1 because it has the lowest total cost. But customers D, E, and H cannot be assigned to this location, and so the solution is infeasible. One way to resolve this is to drop from the data the location where a customer with an artificially large cost is assigned in the final solution and solve the problem again with the remaining data. If a solution exists, continued application of this modification process will obtain the minimum cost solution. If there is no solution, the process will so indicate.

Let us return to the example. It is obvious that if there is only one facility to locate, it will not be in location 1. By using the modification process, we find that it will be placed in location 4 because this is the only location where it is possible to assign all customers.

## 18.4 COMPUTER PROGRAM DESCRIPTION

### 18.4.1 Purpose

The computer program for the fixed cost facility location problem using heuristics finds the optimum facility location/allocation for both single-facility and multiple-facility problems. It will also determine the optimum number of facilities, provided that the number of facilities is not restricted.

### 18.4.2 Program Description

1. *Usage*. The program consists of a main program and a subroutine. All array sizes are specified in DIMENSION or COMMON statements, which may be altered for the particular problem. The program is currently set to handle up to 50 locations and 60 customers. If the number of locations and/or the number of customers are/is to be increased, the DIMENSION and COMMON statements must be modified. The program will calculate the total variable cost per unit of demand, convert fixed cost values to the same period as the variable cost values, and handle unassignable locations.
2. *Description of parameters, including dimensions of arrays*

| | |
|---|---|
| ASIGN (60) | = assignment matrix, customer to location used for the final output |
| ASIGNB (60) | = assignment matrix, customer to location, used in the final check table calculation |
| COUNT | = counter that holds the number of locations that have been assigned; used in the final check table calculation |
| FINFC(50) | = final check fixed cost table |
| FINMAT(60, 50) | = final check variable cost table |
| FLAG | = flag used to control the NET subroutine operation |
| FXCM(50) | = fixed cost table |
| FXCMB(50) | = fixed cost table B used to maintain the original fixed cost values |
| IDMND | = customer demand |
| K | = used for location assignment from subroutine NET; also used as a DO loop counter |
| LOCST | = cost of assigning all demand at the location (demand x unit cost) |
| MINSAV | = used to hold minimum saving for (customer, location) assignment in minimum savings table |
| MSAV | = savings table counter |
| NOCUST | = number of customers |
| NOFAC | = maximum number of facilities available for placement in locations |
| NOLOC | = number of locations |
| EXPLIF | = expected life of the facility |
| PC | = print code, where 0 is print input tables and the solution and 1 is print input tables, intermediate tables, and the solution |
| PIDU(50) | = period in demand units (demand/period) |
| PRNTR | = set to installation device code for printer (i.e., 6) |
| RATE | = interest rate |
| RDR | = set to installation device code for reader (i.e., 5) |
| SAVMAT(60, 50) | = savings table |
| SAVNET(50) | = net savings table |
| STK(60) | = list of assigned locations, used in final check table calculation |
| VARCM(60, 50) | = variable cost table |

```
                                INPUT

                               COLUMN

12345678901234567890123456789012345678901234567890123456789012345678901234567890123456
                              1ST LINE
                              --------

   5   5   0   3   1
                              2ND LINE
                              --------

 100  50 150 200 300
                              3RD LINE
                              --------

    5      3      2      8      5
    3      5      2      6      7
    5      2      0      1      0
    2      1      8      2      3
    3      2      4      0      4
                              4TH LINE
                              --------

   5000    4000    4500    6000    4200
                              5TH LINE
                              --------

0807101009
                              6TH LINE
                              --------

0000010
```

**FIGURE 18.2** Sample example input for facility location with fixed cost program

3. *Input formats* (Figure 18.2).
   Line type 1: NOCUST, NOLOC, FOFAC, PC
   Line type 2: IDMND, one per customer
   Line type 3: LOCST, one per location
   Line type 4: FXCMB, fixed cost per location
   Line type 5: EXPLIF, expected life for each facility
   Line type 6: RATE
   If the rate is greater than 0, a type 7 line is required.
   Line type 7: PIDU, per location
4. *Output* (Figure 18.3). This program prints, as appropriate, a single facility table, a minimum savings table, savings tables, final check tables, a customer allocation table, and the cost of the solution.

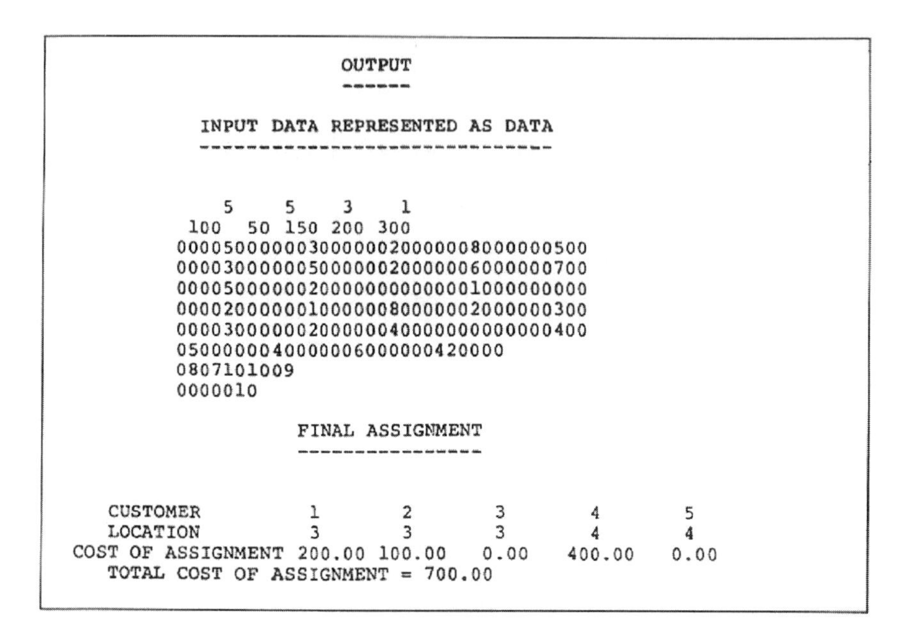

```
                    OUTPUT
                    ------

      INPUT DATA REPRESENTED AS DATA
      ------------------------------

        5     5    3    1
      100   50 150 200 300
     0000500000030000000200000008000000500
     0000300000005000000200000006000000700
     0000500000002000000000000001000000000
     0000200000001000000800000002000000300
     0000300000002000000400000000000000400
     05000000400000006000000420000
     0807101009
     0000010

                  FINAL ASSIGNMENT
                  ----------------

   CUSTOMER             1       2       3       4       5
   LOCATION             3       3       3       4       4
COST OF ASSIGNMENT 200.00 100.00    0.00  400.00    0.00
   TOTAL COST OF ASSIGNMENT = 700.00
```

**FIGURE 18.3** Output of the program from the sample example

5. Summary of user requirements
   a. Specify values of NOCUST, NOLOC, NOFAC, and PC. If the program is to optimize the number of facilities to be allocated, then enter NOFAC = NOLOC.
   b. Enter values for FXCMB.
   c. Enter values of EXPLIF.
   d. If fixed costs are for different periods, enter the current rate (e.g., 10% = 0.10) and PIDU for each location.

## 18.5 SUMMARY

New concepts in this chapter are the fixed cost, the savings table, and the minimum savings table. Also, we showed that problems with unassignable locations could be readily solved by modifying the demand cost table to show a high cost for unassignable locations.

Both sample problems that were presented assumed that a product (or service) was already complete (production costs were not considered) and required only being transported to the customer. The total cost table showed the solution as the total of the fixed cost (site preparation, construction, etc.) and the transportation cost. Articles that expand the cost structure are published frequently in management science and industrial engineering journals; however, such analysis is beyond the scope of this book.

## PROBLEMS

1. Define fixed cost and give three examples in which such cost is significant and may not be ignored.
2. A farmer wishes to locate a tobacco-drying barn in one of three locations to receive tobacco from his three fields. His fixed cost is associated with building the facilities to accommodate his trucks and house the equipment. These costs vary from location to location and are \$1200, \$1050, and \$1500 per year for locations A, B, and C, respectively. The cost of truck travel is estimated as \$0.70/mile. The distance from fields to locations and truckloads of tobacco that are expected from each field are given in the accompanying table. Determine the best location for the barn.

| | Location | | | |
|---|---|---|---|---|
| **Field** | **A** | **B** | **C** | **Loads** |
| 1 | 20 | 2 | 11 | 200 |
| 2 | 7 | 14 | 6 | 175 |
| 3 | 1 | 28 | 21 | 160 |

3. In problem 2 of Chapter 17, the cost of placement of the water fountain may vary from location to location because of the cost of installation and plumbing.

| | Location | | | |
|---|---|---|---|---|
| **Office** | **1** | **2** | **3** | **Employees** |
| 1 | 15 | 45 | 30 | 2 |
| 2 | 38 | 27 | 25 | 4 |
| 3 | 70 | 63 | 58 | 5 |
| 4 | 40 | 90 | 50 | 3 |
| Cost | 110 | 125 | 102 | |

   a. Given the cost per trip between each office and a location, the number of employees making trips from each office, and the cost per location for installing the water fountain (in the same units as trips), as shown in the accompanying table, determine by the heuristic method the best location(s) for the fountain.
   b. Solve by any other method and compare the results.
4. A plant wishes to install two pieces of machinery for use by three departments. Using the travel cost from each department to each possible new location (in dollars per trip), the demand per month for the new machinery by each department, the building cost for each location (in dollars), and the life expectancy of the machinery at each location, given in the

accompanying table, choose the two best of the five locations being considered. Assume 10% interest. Note that this problem can be solved by using a computer.

| | Cost Per Trip | | | | | |
|---|---|---|---|---|---|---|
| | Location | | | | | |
| Department | A | B | C | D | E | Demand |
| 1 | 15 | 8 | 2 | 5 | 0.5 | 2000 |
| 2 | 4 | 3 | 7 | 2 | 1 | 5000 |
| 3 | 5 | 1 | 2 | 2.5 | 3 | 7500 |

| Location | Building Cost | Life Expectancy |
|---|---|---|
| A | 100,000 | 15 |
| B | 150,000 | 12 |
| C | 136,000 | 17 |
| D | 95,000 | 10 |
| E | 125,000 | 8 |

5. Three locations are available for two machines to be operated by five people. Each of the five people is located at a different operating station. The accompanying table gives the fixed cost per week for each of the three possible locations, the travel cost per trip from each operating station to each possible location, and the number of trips per day each operator will make to the new machine. Determine the best possible locations for the two machines and identify the customers each machine should serve.

| | Location Cost | | | |
|---|---|---|---|---|
| Customer | 1 | 2 | 3 | Demand |
| A | 2.0 | 0.5 | 0.75 | 10 |
| B | 1.0 | 2.1 | 1.4 | 17 |
| C | 0.8 | 1.2 | 1.9 | 25 |
| D | 0.6 | 0.4 | 1.1 | 8 |
| E | 1.75 | 2.2 | 2.4 | 14 |
| Fixed cost | 200 | 140 | 175 | |

6. In the diagram in Figure 18.4, 1, 2, 3, 4, and 5 are possible locations, and A, B, C, D, and E are the departments that will be using the machines to be located. Using rectilinear distances, determine the optimum number of machines to be placed and the best locations. Assume that machines and demands are located at the centers of the squares. The operating cost for

| | | | | | | | | | |
|---|---|---|---|---|---|---|---|---|---|
| | A | | | | 5 | | | | |
| | | | | | | | | | |
| 1 | | | | | | 4 | | | |
| | | | | | | | E | | |
| B | | | | | | | | | 3 |
| | | | | | | | | | |
| | 2 | | C | | | | | D | |

5 × 5 Feet

| Location | I | 2 | 3 | 4 | 5 |
|---|---|---|---|---|---|
| Operating Cost/Month | 400 | 350 | 375 | 420 | 315 |
| Department | A | B | C | D | E |
| Demand/Mon | 40 | 25 | 15 | 36 | 20 |

**FIGURE 18.4**

the machine at each possible site and the demand for use of the machine by each department are given in the accompanying table. The travel cost is \$0.01 per foot. This problem can be solved by using a computer.

7. Discuss when it would be better or worse to use a computer program to solve a location problem instead of doing the computations by hand.
8. Four mailboxes are to be erected in a certain city, which is divided into 12 sections (Figure 18.5). Seven possible locations are shown. Dark lines indicate sections of the city. Use the demand for the use of mailboxes for each section, the fixed cost for using each location, and the life expectancy of each of the mailboxes at each of the locations provided in the accompanying table. To determine the best locations, assume that the customers are evenly dispersed throughout the city. Use rectilinear distances from the center of an area to the location as the average distance, but if the location is in the middle of the section in question, use 2.5 miles as the average distance. Note that this problem can be solved by using a computer.

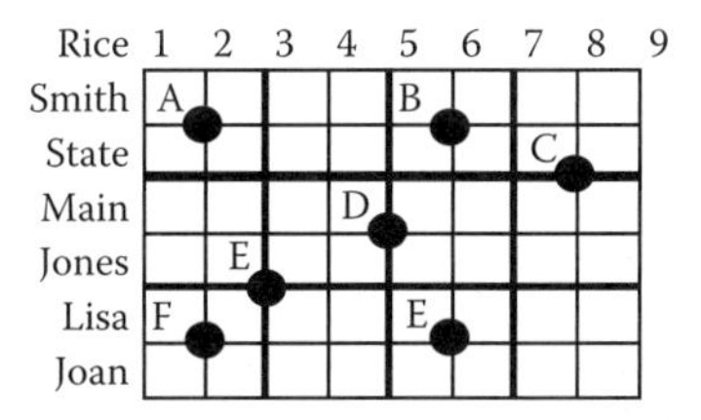


**FIGURE 18.5**

| | Location | | | | | | |
|---|---|---|---|---|---|---|---|
| | A | B | C | D | E | F | G |
| Fixed cost | 500 | 600 | 750 | 640 | 475 | 550 | 580 |
| Life | 10 | 8 | 12 | 14 | 9 | 11 | 13 |

| Section | Demand (in Thousands) |
|---|---|
| State to 3rd | 10 |
| State, 3rd 5th | 17 |
| State, 5th, 7th | 12 |
| State, 7th, 9th | 15 |
| State, Jones, 1st, 3rd | 9 |
| State, Jones, 3rd, 5th | 8 |
| State, Jones, 5th, 7th | 11 |
| State, Jones, 7th, 9th | 13 |
| Jones, Joan, 1st, 3rd | 16 |
| Jones, Joan, 3rd, 5th | 14 |
| Jones, Joan, 5th, 7th | 11 |
| Jones, Joan, 7th, 9th | 7 |

## SUGGESTED READINGS

Atkins, R.J., and Shriver, R.H., "New approach to facilities location," *Harvard Business Review*, May-June 1985.

Barcelo, J., and Casanovas, J., "Heuristic Lagrangian algorithm for capacitated plant location problem," *European Journal of Operations Research*, 15(2), 212 226, Feb. 1984.

Davis, P.S., and Ray, T.L., "A branch-bound algorithm for the capacitated facilities location problem," *Naval Logistic Quarterly Review*, 16, Oct. 1969.

Francis, R.L., and White, J.A., *Facility Layout and Location — An Analytical Approach*, Prentice Hall, Englewood Cliffs, NJ, 1974.

Khumawala, B.M., "An efficient branch and bound algorithm for the warehouse location problem," Krannert School of Industrial Administration Institute Paper Series No. 294, Purdue University, Dec. 1980.

Phillips, D.T., Ravindran, A., and Solberg, J.J., *Operations Research — Principles and Practice*, John Wiley, New York, 1976.

Plastria, F., "Localization in single facility location," *European Journal of Operations Research*, 18(2), 208–214, Nov. 1984.

Revelle, C., Marks, D., and Libman, J.C., "An analysis of private and public sector location models," *Management Science*, 16, July 1970.

Sa, G., "Branch and bound and approximate solutions to the capacitated plant location problem," *Operations Research*, Nov.-Dec. 1969.

Portions of this chapter have been adapted form the journal of operations Management vol. No. 4, May 1981 with permission of the American Production and Inventory Control Society, Falls Church, VA.

# 19 Continuous Facility Location

The facility location models that we have studied so far are based on the assumption that there are a few preselected locations that are suitable for the new facilities and that the problem is to select from among these locations one or more that would minimize a particular cost function. Preselection of sites may follow some qualitative considerations; for example, in selecting a site for a plant, one might evaluate factors such as availability of labor, land, transportation facilities, and market proximity. (Details can be found in Chapter 15.) Many locations may satisfy the basic requirements, but we then must select from these the one that would minimize a cost function. The solution of most industrial problems follows the above-mentioned two-step procedure.

There is another class of problems, however, for which one is free to choose any site and place the new facility there; it is called the continuous or universal facility location problem. This method can be used to determine the ideal location only when the sole consideration involved is transportation cost. The term "ideal" is used to stress the fact that on such a site, construction or placement of a facility may or may not be physically possible. It does, however, guide the analyst to the location(s) that would result in the minimum transportation costs; and if it should prove infeasible to utilize such a location, the "penalty" incurred by selecting an alternative location is better understood.

In this chapter, we shall study the following variations of the problem: (1) single-facility location, (2) one-at-a-time facility location, and (3) multiple-facility location of the same type. In each case, the demand points or the customers and their $x$-$y$ coordinates, and the necessary number of trips from each customer to a facility are known. The cost is measured in terms of the number of trips multiplied by the distance that a customer has to travel to reach the facility. In manufacturing plants the travel is in aisles, and so the distance is measured in rectilinear fashion. When direct travel by the shortest path is possible, the cost can be measured as either quadratically proportional to the distance or linearly proportional, called Euclidean distance (Figure 19.1).

Quadratic cost implicitly states that customers that are farther away will contribute much more to the cost than those nearby. This is appropriate when one also wants to implicitly include time as another parameter in the cost. For example, in placing a public facility such as a hospital within a community, it might be important that the hospital not be very far from any center of population that it is expected to serve, timely treatment in case of emergency being a critical criterion.

Euclidean cost, on the other hand, maintains the direct proportionality of the cost with distance and is appropriate when only the travel distance, and not time, is worthy of attention.

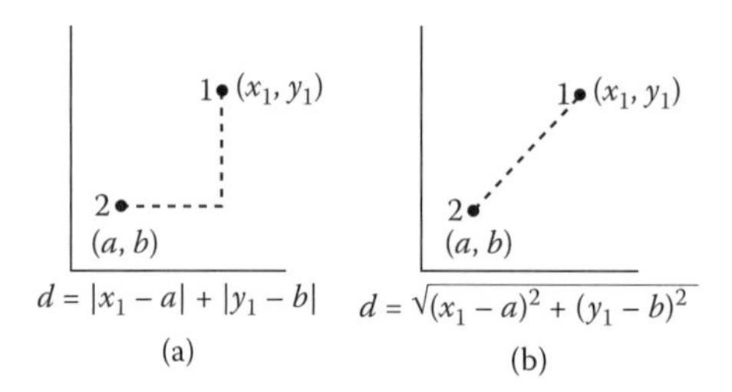


**FIGURE 19.1** (a) Rectilinear distance and (b) Euclidean distance

## 19.1 SINGLE-FACILITY LOCATIONS

By means of examples, we will explore the various possible conditions that might be encountered in single-facility location problems considering only the cost of transportation.

### 19.1.1 RECTILINEAR DISTANCE COST

Suppose the introduction of a new product will require parts from a new molding machine to be transported to six existing machines in the plant. The coordinates of the existing machines $(x, y)$ and the number of trips that would be made each day from the molding machine to these machines $(t)$ are given in Table 19.1. Figure 19.2 shows the plotted coordinates. The problem is to determine the appropriate location for the new molding machine.

The cost is assumed to be directly proportional to the distance traveled. Therefore, if the location of the molding machine is given by $(a, b)$, the objective is to minimize

$$d = \sum_{i=1}^{n} t_i \left[ | x_i - a | + | (y_i - b) | \right]$$

where $n$ is the number of existing machines.

**TABLE 19.1**
**Coordinates of Existing Machines**

| Machine | Coordinates $(x, y)$ | Trips/Day $(t)$ |
|---|---|---|
| 1 | 20, 46 | 20 |
| 2 | 15, 28 | 15 |
| 3 | 26, 35 | 30 |
| 4 | 50, 20 | 18 |
| 5 | 45, 15 | 20 |
| 6 | 1, 6 | 15 |

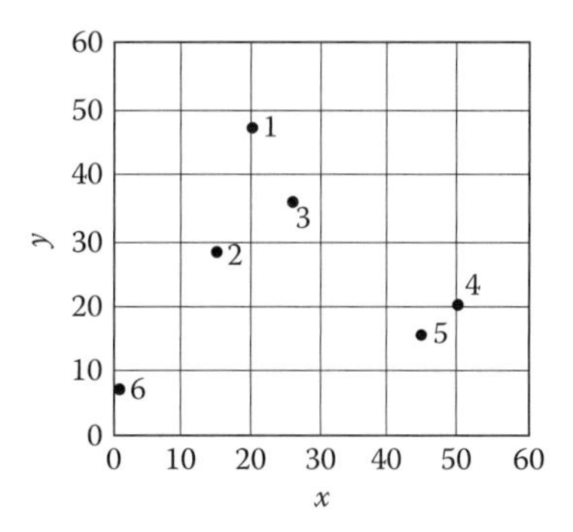


**FIGURE 19.2** Coordinates of existing machines

The function is linear, however; hence it is not possible to obtain the optimum by the usual method of taking the first derivative, setting it equal to zero, and solving the resulting expression.

To demonstrate a basic concept of how such a function can be optimized, suppose we take another example in which there are only two customers, who are located at $(x_1, y_1)$ and $(x_2, y_2)$, respectively, and who are making $t_1$ and $t_2$ trips to a facility. The cost of travel is given by

$$t_1 \,|(x_1 - a)| + t_1 \,|(y_1 - b)| + t_2 \,|(x_2 - a)| + t_2 \,|(y_2 - b)|$$

the value of which is to be minimized. The expression can be separated into two parts, min: $t_1 \,|(x_1 - a)| + t_2 \,|(x_2 - a)|$ and min: $t_1 \,|(y_1 - b)| + t_2 \,|(y_2 - b)|$.

In minimizing on the $x$-axis, since the function is linear, the optimum is on the boundary; that is, $a$ is equal to either $x_1$ or $x_2$. Similarly, for $y$ the minimum cost is obtained when $b$ is either $y_1$ or $y_2$. In general, the new facility should be located at the $x$ and $y$ coordinates of one of the existing customers (Figure 19.3).

The details of the procedure are as follows. To determine the best abscissa $x$ for the new machine, list the existing machines (facilities) in their order of increasing

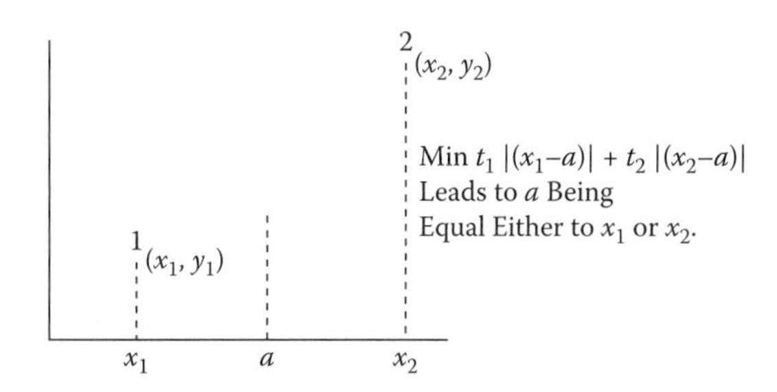


**FIGURE 19.3** Cost analysis for $x$-axis

**TABLE 19.2**
**Data for Machine Location Problem**

| Customer | Abscissa | Number of Trips | Cumulative Trips |
|---|---|---|---|
| 1 | 13 | 10 | 10 |
| 3 | 21 | 15 | 25 |
| 2 | 24 | 35 | 60 |

values of abscissae and the corresponding trips. The optimum $x$ coordinate is one associated with the 50th percentile, or the median value of the trips. It should be understood that the trips originate from discrete points, and therefore the median is associated with one of those points.

For example, if the cumulative trips are as shown in Table 19.2, the median is the 30th trip, which is made by customer 2. Repeat the same process for the $y$ ordinates of the existing machines to obtain the optimum $y$ coordinate of the new facility.

To illustrate the procedure, Table 19.3 lists the machine arrangement based on the abscissa values for the data in Table 19.1.

The 50th percentile would be the 59th trip (118/2). That trip occurs at machine 3, with an abscissa of 26, which thus is the optimum $x$-axis for the new molding machine.

To obtain the optimum ordinate, arrange the machines in ascending order of ordinate values as shown in Table 19.4. The 50th percentile trip, that is, the 59th trip, occurs with machine 2, whose ordinate is 28. This becomes the $y$-axis value of the new facility. Thus the location for the new molding machine is (26, 28).

### 19.1.2 Single-Facility Location — Quadratic Cost

Now, suppose that the cost is proportional to the square of the distance traveled rather than being linearly proportional as in the previous case. The procedure would be to

**TABLE 19.3**
**The *x*-Axis (Abscissa) Determination**

| Machine | Abscissa | Number of Trips | Cumulative Trips |
|---|---|---|---|
| 6 | 1 | 15 | 15 |
| 2 | 15 | 15 | 30 |
| 1 | 20 | 20 | 50 |
| 3 | 26 | 30 | 80 |
| 5 | 45 | 20 | 100 |
| 4 | 50 | 18 | 118 |

**TABLE 19.4**
**The *y*-Axis (Abscissa) Determination**

| Machine | Ordinate | Number of Trips | Cumulative Trips |
|---|---|---|---|
| 6 | 6 | 15 | 15 |
| 5 | 15 | 20 | 35 |
| 4 | 20 | 18 | 53 |
| 3 | 28 | 15 | 68 |
| 2 | 35 | 30 | 98 |
| 1 | 46 | 20 | 118 |

minimize the objective function

$$d = \sum_{i=1}^{n} t_i \left( (x_i - a)^2 + (y = b)^2 \right)$$

Taking partial derivatives with respect to $a$ and $b$, equating them to zero, and solving the resultant expressions, we obtain this solution:

$$a = \frac{\sum_{i=1}^{n} t_i x_i}{\sum_{i=1}^{n} t_i}$$

and

$$b = \frac{\sum_{i=1}^{n} t_i y_i}{\sum_{i=1}^{n} t_i}$$

The solution is referred to as a centroid, or the center-of-gravity, solution. Referring to the example with data in Table 19.1, the centroid solution is

$$a = [(20 \times 20) + (15 \times 15) + (26 \times 30) + (50 \times 18) + (45 \times 20) + (1 \times 15)]$$

$$\div (20 + 15 + 30 + 18 + 20 + 15) = 27.3$$

$$b = [(20 \times 46) + (15 \times 28) + (30 \times 35) + (18 \times 20) + (20 \times 15) + (15 \times 6)]$$

$$\div (20 + 15 + 30 + 18 + 20 + 15) = 26.6$$

Thus the location for the new molding machine is (27.3, 26.6).

### 19.1.3 Euclidean Distance Cost

The Euclidean distance between two points, $T$ and $S$, with coordinates $(x, y)$ and (a, $b$), respectively, is defined as

$$d = [(x - a)^2 + (y - b)^2]^{1/2}$$

When the cost is proportional to Euclidean distance, the objective is to minimize the resulting cost expression:

$$d = \sum^{n} t_i \left[ (x_i - a)^2 + (y_i - b)^2 \right]^{1/2}$$

Following a procedure similar to that in the previous subsection — that is, taking partial derivatives with respect to $a$ and $b$ and setting them equal to zero — leads to

$$\frac{\partial}{\partial a} = 0 = \sum_{i=1}^{n} \frac{t_i (x_i - a)}{\left[ (x_i - a)^2 + (y_i - b)^2 \right]^{1/2}} \tag{1}$$

$$\frac{\partial}{\partial a} = 0 = \sum_{i=1}^{n} \frac{t_i (y_i - a)}{\left[ (x_i - a)^2 + (y_i - b)^2 \right]^{1/2}} \tag{2}$$

The optimum location should be obvious if we could solve the resulting expressions. However, if the new machine could be placed, at least mathematically, at the same coordinates as one of the old machines — that is, for some $i$, $x = a$ and $y = b$ — both expressions (1) and (2) are undefined, and there is great difficulty in trying to obtain a unique solution.

An alternative is to engage in an iterative solution procedure. It can be proved mathematically that the objective function is convex as shown in Figure 19.4.

One can begin with a centroid solution and then perform a search with respect to each dimension until the optimum solution is obtained.

The procedure is easy to follow, and for the small problems most often encountered in practice it provides sufficient accuracy and speed. There are a few alternative rules that permit a reduction in the computational effort (Feldman et al., 1966; Scriabin and Vergin, 1985). If a problem is very large or computational time is expensive, the reader is advised to review these rules. Because of the level of mathematics involved, however, they will not be discussed here.

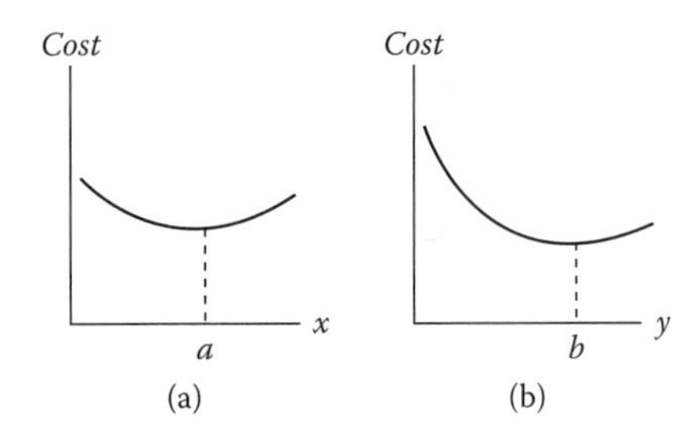


**FIGURE 19.4** (a) Cost as a function of $x$ dimension and (b) cost as a function of $y$ dimension

## 19.2 MULTIPLE FACILITIES

The two prominent variations in multiple-facility problems are:

1. *Multiple facilities of different types*. Each type is independently demanded by the existing customers. Furthermore, there may also be demand between the new facilities.
2. *Multiple facilities of the same type*. Customers are divided among them so that any one customer is served by only one new facility.

### 19.2.1 Multiple Facilities of Different Types

Even though alternative methods have been developed for obtaining the solution to each type of cost function, the simplest procedure remains the iterative technique in which the new facilities are located one at a time.

The procedure for the rectilinear type of cost calculation is presented in the following subsection, and quadratic and Euclidean types are discussed in the subsection after that.

#### 19.2.1.1 Rectilinear Cost

Consider a manufacturing plant within which four machines (customers) will require the services of two new facilities that are to be installed in the plant. The present location of each machine and the number of trips between them and each of the new facilities are given in Table 19.5. There are also nine trips expected between the two new facilities. The problem is to determine the optimum locations for the new facilities.

We will begin the process, called "one-at-a-time," by determining the abscissas for the new facilities. As a first step, ignore the trips between the new facilities and determine the $x$-axis value for facility 1 [$F_1$] with respect to the present machines (customers)]. Recall from Section 19.1 that the procedure calls for listing the present machines in order of increasing abscissas and selecting the $x$ value associated with the 50th percentile of the cumulative number of trips (Table 19.6).

**TABLE 19.5**
**Multiple Facilities of Different Types**

| | | Number of Trips | |
|---|---|---|---|
| Machine | Coordinates $(x, y)$ | Facility 1 | Facility 2 |
| 1 | 10, 15 | 10 | 8 |
| 2 | 8,20 | 4 | 6 |
| 3 | 4,7 | 6 | 12 |
| 4 | 15,9 | 0 | 30 |
| $F_1$ | $a_1, b_1$ | — | 9 |
| $F_2$ | $a_2, b_2$ | 9 | — |

**TABLE 19.6**
**The x-Axis (Abscissa) Data for Facility 1 (First Iterations)**

| Machine | Abscissa | Number of Trips | Cumulative Trips |
|---|---|---|---|
| 3 | 4 | 6 | 6 |
| 2 | 8 | 4 | 10 |
| 1 | 10 | 10 | 20 |
| 4 | 15 | 0 | 20 |

The 50th percentile is 10 trips (20/2), and the corresponding abscissa is 8. Therefore, as a first try the abscissa for the first facility is fixed at 8.

We now repeat the process for the second facility ($F_2$), except that we will also consider the trips between the facilities, because the first facility is at least temporarily located on the $x$-axis. Table 19.7 lists the data in proper sequence to make this determination.

The 50th percentile is 32.5, and the associated $x$ value, which must be the same as that of one of the existing customers, is 10. Therefore, the temporary abscissa for the second facility is 10.

Now go back and determine the abscissa for facility 1 knowing that the temporary $x$ location for facility 2 is 10. Table 19.8 shows the necessary update of Table 19.6 to find the new $F_t$ abscissa.

The 50th percentile, 14.5, is associated with $x = 10$, and the new temporary abscissa for $F_1$ is different from the previous value for the same facility.

The process must be repeated with $x = 10$ as the new temporary abscissa for facility 1. The details are shown in Table 19.9.

The abscissa associated with the 50th percentile of 32.5 trips is 10; that is, $x = 10$, which is the same as before for $F_2$. Therefore, the search for the abscissa is complete, and the latest temporary value becomes permanent.

Similarly, the search process for the $F_1$ and $F_2$ ordinates leads to Tables 19.10 and 19.11. Keep in mind that the nine trips between the two new facilities are ignored during the first iteration.

**TABLE 19.7**
**The x-Axis (Abscissa) Data for Facility 2 (First Iteration)**

| Machine | Abscissa | Number of Trips | Cumulative Trips |
|---|---|---|---|
| 3 | 4 | 12 | 12 |
| 2 | 8 | 6 | 18 |
| $F_1$ | 8 | 9 | 27 |
| 1 | 10 | 8 | 35 |
| 4 | 15 | 30 | 65 |

**TABLE 19.8**
**The *x*-Axis (Abscissa) Data for Facility 1 (Second Iteration)**

| Machine | Abscissa | Number of Trips | Cumulative Trips |
|---|---|---|---|
| 3 | 4 | 6 | 6 |
| 2 | 8 | 4 | 10 |
| 1 | 10 | 10 | 20 |
| $F_2$ | 10 | 9 | 29 |
| 4 | 15 | 0 | 29 |

**TABLE 19.9**
**The *x*-Axis (Abscissa) Data for Facility 2 (Second Iteration)**

| Machine | Abscissa | Number of Trips | Cumulative Trips |
|---|---|---|---|
| 3 | 4 | 12 | 12 |
| 2 | 8 | 6 | 18 |
| 1 | 10 | 8 | 26 |
| $F_1$ | 10 | 9 | 35 |
| 4 | 15 | 30 | 65 |

**TABLE 19.10**
**The *y*-Axis (Ordinate) Data for Facility 1 (First Iteration)**

| Machine | Ordinate | Number of Trips | Cumulative Trips |
|---|---|---|---|
| 3 | 7 | 6 | 6 |
| 4 | 9 | 0 | 6 |
| 1 | 15 | 10 | 16 |
| 2 | 20 | 4 | 20 |

**TABLE 19.11**
**The *y*-Axis (Ordinate) Data for Facility 2 (First Iteration)**

| Machine | Ordinate | Number of Trips | Cumulative Trips |
|---|---|---|---|
| 3 | 7 | 12 | 12 |
| 4 | 9 | 30 | 42 |
| 1 | 9 | 8 | 50 |
| $F_1$ | 15 | 9 | 59 |
| 2 | 20 | 6 | 65 |

**TABLE 19.12**
**The *y*-Axis (Ordinate) Data for Facility 1 (Second Iteration)**

| Machine | Ordinate | Number of Trips | Cumulative Trips |
|---|---|---|---|
| 3 | 7 | 6 | 6 |
| 4 | 9 | 0 | 6 |
| $F_2$ | 9 | 9 | 15 |
| 1 | 15 | 10 | 25 |
| 2 | 20 | 4 | 29 |

The median cumulative trip value is 20/2; therefore, the temporary ordinate is 15 (machine 1). Table 19.11 presents the data in the appropriate order to determine the first temporary ordinate for $F_2$.

The median cumulative trip value is 32.5, and this corresponds with an ordinate of 9, determined by machine 4. Table 19.12 continues the procedure, providing the second iteration data for $F_1$.

The median cumulative trip value is 14.5, which equates to an ordinate of 9, determined by facility 2. Because this is a different value from that found during the first iteration (Table 19.10), we must perform at least one more iteration as shown in Table 19.13.

The median is 32.5, and this leads to an ordinate of 9, determined by machine 4. This *y* value remains unchanged from that of the first iteration; hence, the procedure terminates.

The optimum locations for the new facilities are:

Facility 1: (10, 9)

Facility 2: (10, 9)

The solution yields the same location for both facilities, which, of course, could present a problem. It does, however, point out a drawback of the procedure: it might try

**TABLE 19.13**
**The *y*-Axis (Ordinate) Data for Facility 2 (Second Iteration)**

| Machine | Ordinate | Number of Trips | Cumulative Trips |
|---|---|---|---|
| 3 | 7 | 12 | 12 |
| 4 | 9 | 30 | 42 |
| $F_1$ | 9 | 9 | 51 |
| 1 | 15 | 8 | 59 |
| 2 | 20 | 6 | 65 |

to place multiple facilities at the same point. In practice, the facilities would be set side-by-side within existing practical limitations including the shape and size of the facilities, space required for operating and maintenance personnel, and the area necessary for material flow and inventory storage. The other techniques might provide closer-to-optimum solutions, but the required mathematics is beyond the scope of this book. For most practical purposes, the "one-at-a-time method" works quite satisfactorily.

#### 19.2.1.2 Quadratic and Euclidean Distance Cost

The one-at-a-time method is also applicable when the cost function is either quadratic or Euclidean. The only modification needed is using the appropriate placement method as described in Section 19.1.

## 19.3 MULTIPLE FACILITIES OF THE SAME TYPE

With multiple facilities that are of the same type, the problem of facility placement becomes that of determining the demand points (customers) that should be assigned to each facility and then deciding the locations of the facilities to minimize the cost of travel from those assigned customers. The solution procedure is based on recognizing such a cluster of customers that may be assigned to a single facility, and then applying the procedures of single-facility placement to obtain the optimum location for each facility. We will illustrate the procedure, called the combinational approach, by means of an example.

Suppose there are four customers with the data listed in Table 19.14, and two facilities are to be located.

We begin by scaling the demand from each customer using the smallest demand (50) as the scaling factor. For customer 1, for example, the scaled demand is 150/50, or 3.0. The complete results are shown in Table 19.15.

As the next step, using each customer's $x$ and $y$ coordinates from Table 19.14, we plot the locations of the customers with the same scale on both $x$ and $y$ axes on graph paper, as shown in Figure 19.5a, to visualize the present layout.

Now we are ready to begin the combinational process. The cost of travel for a customer is given by the Euclidean distance between the facility and the customer, multiplied by the demand from the customer. If we fix a cost value and divide it by

**TABLE 19.14**
**Multiple Facilities of the Same Type**

| Customer | Demand | Coordinates $(x, y)$ |
|---|---|---|
| 1 | 150 | 60, 20 |
| 2 | 100 | 20, 100 |
| 3 | 50 | 10, 20 |
| 4 | 60 | 30, 60 |

**TABLE 19.15**
**Customer Demand**
**(Scale Factor of 50)**

| Customer | Scaled Demand |
|---|---|
| 1 | 3.0 |
| 2 | 2.0 |
| 3 | 1.0 |
| 4 | 1.20 |

the individual demands, we will obtain the distances that each customer may move and still have the same costs for travel. The value of the constant can be kept small by using scaled demands rather than the actual demands. Using the distance calculated for each customer, we can draw circles centered at each customer to represent the identical cost of travel between the customer and a facility anywhere on the periphery of that circle. The customers are combined on the basis of the principle of encompassment; that is, when the circle from one customer encompasses or encloses the location of another customer, those two customers should be combined and represented by a single equivalent customer with its total demand equal to the sum of the demands of the two customers. Its equivalent location is calculated on the basis of the appropriate cost function. In this case, we will use the centroid rule (Section 19.1).

To determine the cost constant, use the following procedure: Calculate the distance from each customer to the nearest neighboring customer, and multiply by the

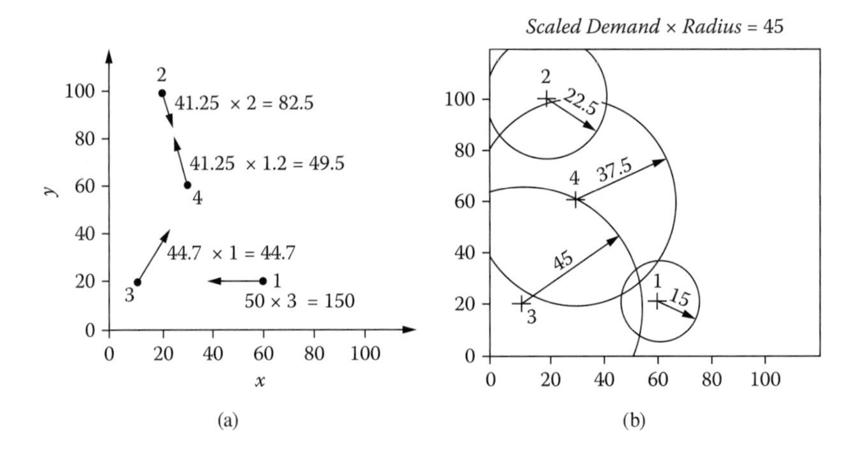


**FIGURE 19.5** (a) Determination of cost constant and (b) the initial layout

scaled demand for the customer. Choose the minimum value as the cost constant for this setup.

Once two customers are combined, the procedure is repeated, the new layout being considered as an initial layout. The procedure continues until we have exactly as many remaining customers as the number of facilities needing placement. The facilities are placed in the positions identified by the remaining customers in the final graph.

In our example, as shown in Figure 19.5b, with a cost constant of 45 from Figure 19.5a, customers 3 and 4 are first combined. Their equivalent facility coordinates are

$$x = \frac{(1)\times 10 + (1.2)\times 30}{2.2} = 20.9$$

$$y = \frac{(1)\times 20 + (1.2)\times 60}{2.2} = 41.8$$

with a total equivalent scaled demand of 2.2.

With a cost constant of 96.8, obtained by the method explained earlier, as shown in Figure 19.6, the radius from the equivalent facility encompasses customer 1. Combining these two leads to a subsequent equivalent customer with a scaled demand of 5.2 (2.2 + 3) and coordinates of

$$x = \frac{(202)\times 20.9 + (3)\times 60}{5.2} = 43.45$$

$$y = \frac{(2.2)\times 20.9 + (3)\times 60}{5.2} = 29.22$$

Now only two customers remain, and because there are two facilities to place, the procedure terminates. The final solution is given in Table 19.16 and includes the assignment of specific customers to the individual facilities (Figure 19.7).

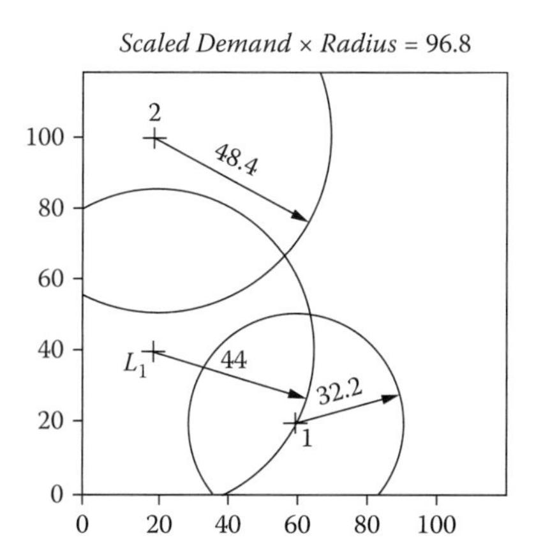


**FIGURE 19.6** The second layout

**TABLE 19.16**
**Combinational Approach (Final Solution)**

| Facility | Coordinates | Assigned Customers |
|---|---|---|
| 1 | (43.45, 29.22) | 1, 3, 4 |
| 2 | (20, 100) | |

It should be noted that the combinational procedure is heuristic in nature and provides a nearly optimum solution. Furthermore, because it is assumed that the facility capacities are unlimited, this procedure results in the customer demands never being split between any two or more facilities.

## 19.4 COMPUTER PROGRAM DESCRIPTION

Three programs are listed in Appendix G that represent computerization of some of the methods from this chapter. Two can be used for solving single-facility location problems; one, for multiple facilities of the same type problems.

### 19.4.1 Single-Facility Problems

#### 19.4.1.1 Rectilinear Distance Cost Method

The program is in interactive mode with the information to be supplied based on:

1. *How many machines are there?* (Enter number of facilities.)
2. *What is the x coordinate for the machine; the y coordinate; the number of trips?* (Enter $x$ and $y$ coordinates and number of trips for each machine.)

This program outputs the $x$-$y$ coordinates of the location for the new machine (facility).

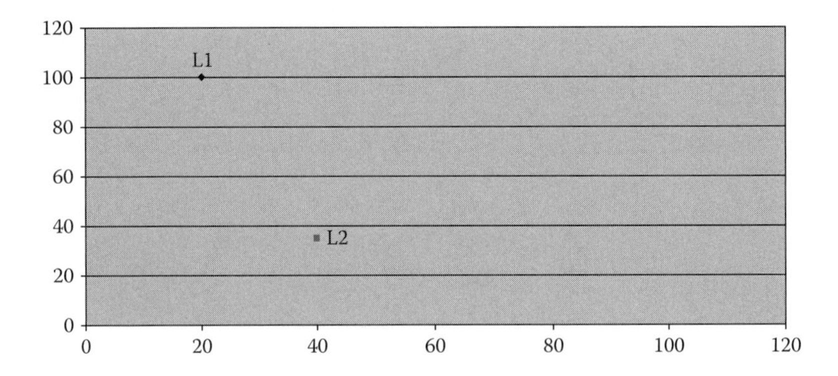


**FIGURE 19.7** The final solution

#### 19.4.1.2 Quadratic Cost Model

The program is in interactive mode with information supplied with the same set of questions as in the rectilinear distance cost method. This program outputs the $x$-$y$ coordinates of the location for each new machine (facility).

### 19.4.2 Multiple Facilities of the Same Type Problems

The program is in interactive mode with data supplied as answers to:

*How many facilities need placement?* (Enter number of new facilities to be located.)
*How many customers are there?* (Enter the number of customers.)
*What is the x coordinate of the customer; the y coordinate; the demand?* (Enter $x$ and $y$ coordinates and demand for each customer.)

This program outputs (1) the coordinates of the new facilities and (2) customer assignments to the new facilities.

## 19.5 SUMMARY

Three models for placing facilities are presented when one is free to choose any location within the planning area for their placement. A major component of the cost, which is minimized, is the distance between facilities. This distance is measured either in a rectilinear or a Euclidean manner.

In a single-facility placement problem, the unit processed on the existing facilities also requires services of the new facility so that the total cost of demand times distance traveled is minimized.

Two prominent variations for multiple-facility location problems exist. In the first, each new facility is independently demanded by the unit processed on the existing facilities. Furthermore, there may also be some flow between the new facilities. In the second variation, all new facilities are of the same type. The problem is to determine the best location for each of the new facilities and also to assign existing customers to them, so that one customer is served by one new facility only.

The procedures illustrated here are simple and efficient. Some of the procedures are also computerized and are presented in Appendix G.

## PROBLEMS

1. Define the continuous facility location problem.
2. Distinguish between rectilinear and Euclidean distances.
3. a. Using Figure 19.8, calculate the distance from 1 to the center of each of the other locations assuming you are traveling by forklift truck.
   b. Calculate these distances assuming you are traveling by overhead monorail.
4. a. Using Figure 19.8, determine the best location for a new piece of equipment. Use the rectilinear distance from the center of each location.

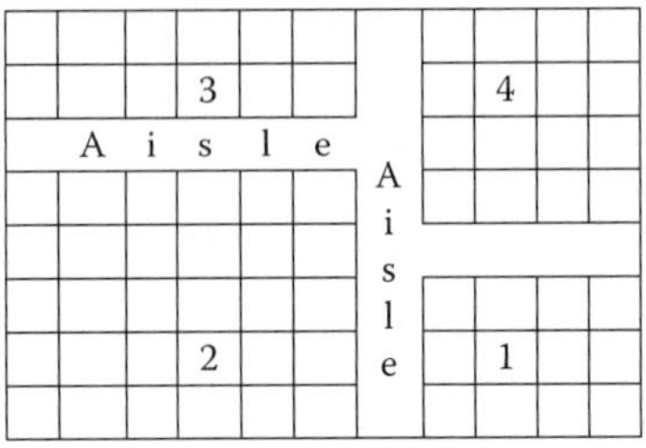

**FIGURE 19.8** Distance from 1 to the center of each of the other locations traveling by fork-lift truck.

Assume that demand from each section is proportional to the area of the section.

b. Solve again using Euclidean distance.

5. On the 15th floor of an office building, a coffeemaker is to be placed. There is no limit to the number of trips that may be made as long as work is done. A study was conducted to see how many trips per day on the average each person makes. Given the data in the accompanying table, determine where the coffeemaker should be placed.

| Customer | Coordinates | Trips/Day |
|---|---|---|
| 1 | 5, 5 | 3 |
| 2 | 10, 40 | 5 |
| 3 | 70, 45 | 2 |
| 4 | 65, 20 | 1 |
| 5 | 25, 25 | 3 |

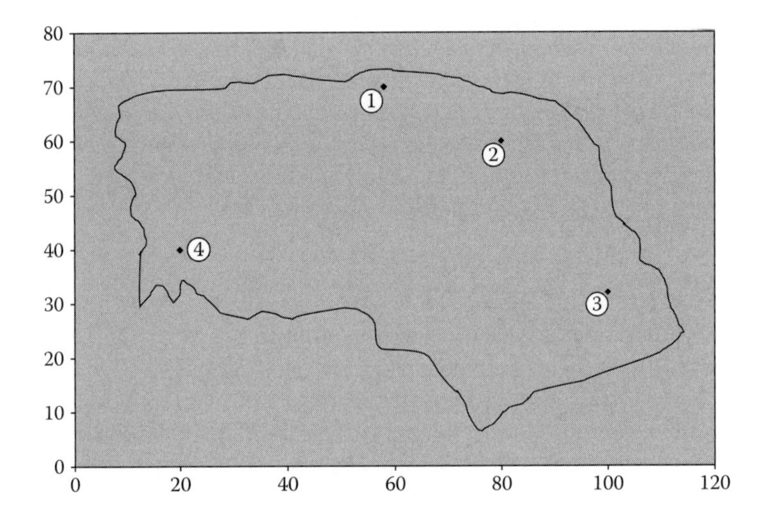


**FIGURE 19.9**

6. a. In the Gulf of Mexico, Tell Oil has four rigs. The location of each is given in Figure 19.9. Food for all the rigs is brought out once a month by ships to a central warehouse. From here, the food is distributed according to the data in the accompanying table. Where should the warehouse be located?

| Rig | 1 | 2 | 3 | 4 |
|---|---|---|---|---|
| Trips/Month | 7 | 10 | 8 | 12 |

b. Rework part a, assuming there are two warehouses to place.

7. The island of Margarita is famous for its delicious lobsters. They are caught at sea and brought to three main harbors around the island. A refrigerated warehouse is needed to which the lobsters will be shipped by airplane from the harbors, as shown in the accompanying table. Because the product's freshness is so important, where should the warehouse be located to minimize the cost of losing market due to lack of freshness?

| Harbors | Coordinates | Trips/Day |
|---|---|---|
| 1 | 4, 2 | 5 |
| 2 | 4, 7 | 7 |
| 3 | 10, 5 | 11 |

8. A manufacturing company is to be located along the east-to-west route in the southern section of the United States. The company will serve the three states of Louisiana, Mississippi, and Arkansas. The accompanying table lists the coordinates and the number of trips to be made to each city where the products will be sold. Determine the optimum location for the plant site.

| City | Coordinates | Trips/Month |
|---|---|---|
| 1 | 100,440 | 6 |
| 2 | 20,300 | 8 |
| 3 | 200,300 | 9 |
| 4 | 400,400 | 11 |

9. After being cut at saws 1 and 2, wood is sent either to the sander at $F_1$ or to the lathe at $F_2$. Determine the best location for the sander and the lathe so as to minimize the distance traveled. The number of trips per hour between each existing machine and the new machine, along with the coordinates of the existing machine, are given in the accompanying table.

| Machine | Coordinates (x, y) | Number of Trips Facility 1 | Facility 2 |
|---|---|---|---|
| 1 | 1, 1 | 10 | 15 |
| 2 | 1, 3 | 8 | 7 |
| $F_1$ | 3, 1 | 12 | 10 |
| $F_2$ | 5, 1 | 7 | 6 |

10.a. A corporation wishes to add on a new wing for its president and vice president. Administrative assistants are located at 1, engineering at 2, accounting at 3, and sales at 4. Locate these two offices so as to minimize the distance traveled. The average number of trips per day that the president and vice president make to these offices is given in the accompanying table.

| Machine | Coordinates (x, y) | Number of Trips Facility 1 | Facility 2 |
|---|---|---|---|
| 1 | 1, 1 | 10 | 15 |
| 2 | 1, 3 | 8 | 7 |
| 3 | 3, 1 | 12 | 10 |
| 4 | 5, 1 | 7 | 6 |
| $F_1$ | | — | 6 |
| $F_2$ | | 5 | — |

b. Rework part a using the quadratic cost function.

11. Describe the variations of the multiple-facility problem.
12. Solve problem 5 for the placement of two coffee dispensers.
13. Solve problem 6b for the placement of two identical warehouses.

## SUGGESTED READINGS


Blair, E.L., and Miller, S., "Interactive approach to facilities design using microcomputers," *Computers in Industrial Engineering*, 9(1), 91–102, 1985.

Brady, S.D., Rosenthal, R.E., and Young, D., "Interactive graphical minimax location of multiple facilities with general constraints," *HE Transactions*, 15(3), Sept. 1983.

Feldman, E., Lehrer, F.A., and Ray, T.L., "Warehouse locations under continuous economies of scale," *Management Science*, 2, May 1966.

Francis, R.L., and White, J.A., *Facility Layout and Location — An Analytical Approach*, Prentice Hall, Englewood Cliffs, NJ, 1974.

Khumawala, B.M., "An efficient branch and bound algorithm for the warehouse location problem," Krannert School of Industrial Administration Institute Paper Series No. 294, Purdue University, Dec. 1980.

Phillips, D.T., Ravindran, A., and Solberg, J.J., *Operations Research — Principle and Practice*, John Wiley, New York, 1976.

Scriabin, M., and Vergin, R.C., "A cluster-analytic approach to facility layout," *Management Science*, 1, 33–49, Jan. 1985.

# Appendix A: Engineering Economy Formulas

All of us realize the time value of money. One thousand dollars in hand today is worth more than a promised \$1000 one year from now; the present sum could be invested, for example, at an interest rate of 10% to earn \$100, making a total year-end capital of \$1100. When we buy an automobile and agree to pay monthly installments, the sum of the installment payments is more than the amount owed on the car, again because of the interest payment on the owed capital, or the time value of money.

It is possible to determine the effect of the time value of money by using mathematical expressions. We will not go into details of derivations, but only state the final results. Most standard books on engineering economy also tabulate the values resulting from using the formulas for specific interest rates and time periods; it might be easier for the reader to use these tabulated values. The following notations are used in Table A.I.

$i$ = interest rate per interest period in decimals (e.g., 10% per year interest is 0.10; if the compounding is done semiannually, $i = 0.1 \times \% = 0.05$)

$n$ = number of interest periods (e.g., if compounding is done semiannually, $n = \text{years} \times 2$)

$P$ = present value of the money

$F$ = future value of the money; that is, the value of $P$, $n$ periods from now at an interest rate of $i$ (e.g., \$1100 at the end of one year, $P = 1000$, and $i = 10\%$)

$A$ = annuity, or equal payments (receipts) at the end of each period, for $n$ periods, the present worth of which is $P$ at an interest rate of $i$% (e.g., monthly car payments)

$G$ = a uniform arithmetic gradient in payments, starting with zero and increasing at a constant amount $G$, from period to period for $n$ periods.

## TABLE A.I
## Time/Money Relationships

| Given | To Find | Multiply the Given by the Factor | Name of the Factor | Mathematical Expression of the Factor | Time/Money Display |
|---|---|---|---|---|---|
| $P$ | $F$ | $(F/P, i\%, n)$ | Compound amount | $\frac{(1+i)^a}{(1+i)^{-a}}$ | 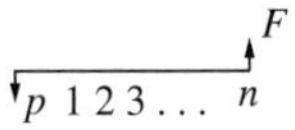 $F$; $p$ 1 2 3 . . . $n$ |
| $F$ | $P$ | $(P/F, i\%, n)$ | Present worth | | |
| $P$ | A | $(A/P, i\%, n)$ | Capital recovery series | $\frac{\frac{i(i+i)^a}{(1i)^a-1}}{\frac{(1+i)^a-1}{i(1+i)^a}}$ | 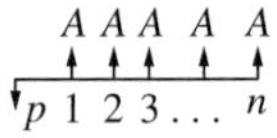 $A$ $A$ $A$ $A$ $A$; $p$ 1 2 3 . . . $n$ |
| $A$ | $P$ | $(P/A, i\%, n)$ | Present worth | | |
| $F$ | A | $(A/F, i\%, ri)$ | Sinking fund series | $\frac{\frac{i}{(1+i)^a}}{\frac{(1+i)^a}{i}}$ | 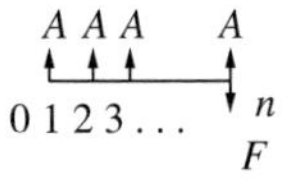 $A$ $A$ $A$ $A$; 0 1 2 3 . . . $n$; $F$ |
| $A$ | $F$ | $(F/A, i\%, n)$ | Compound amount | | |
| $G$ | $P$ | $(P/G, i\%, ri)$ | Gradient present worth | $\frac{1}{i}\left[\frac{(1+i)^a-1}{i}\right] \times \left[\frac{i}{(1+i)^a}\right]$ | 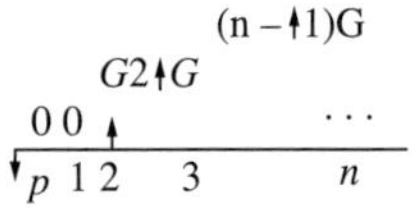 (n – 1)G; $G$2 $G$; 0 0 . . .; $p$ 1 2 3 $n$ |

# Appendix B: Quality Assurance Manual, Spring Controls, Inc.

## B.1.1 SCOPE

The purpose of this manual is to establish methods and procedures to be followed at this facility to assure compliance with quality standards required by contractual agreements with customers, particularly those stipulating military and government specifications (see MIL-I-45208).

Methods and procedures specified in this manual are applicable to the inspection of the end product and products in process in manufacture or components thereof.

This manual also is to convey to suppliers of Spring Controls, Inc., their responsibility in the control of the quality of their products. All suppliers shall be provided with drawings, documents, or specifications which shall be the acceptable standards.

### B.1.1.1 Revision Control

| Date | Section | Page | Revised Data |
|---|---|---|---|
| 1/2/.. | 3.0 | 1 | Added "per MIL-STD-105D" |
| 1/2/.. | 6.0 | 1 | Added "per MIL-STD-105D" |
| 7/5/.. | 7.0 | 3 | Added list of Process Vendors |

### B.1.2.1 Quality Control Organization Chart

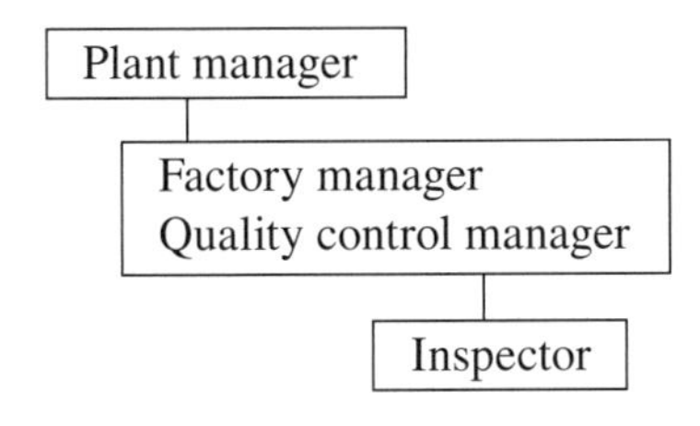

## B.2.1 BLUEPRINT CHANGE CONTROL

1. Blueprint changes are controlled by the Production Control Department.
2. When a blueprint change is received, the master file is to be updated with the new changes.
3. Obsolete blueprints are to be marked void and filed in the master file for reference.

## B.3.1 INCOMING MATERIAL INSPECTION

1. All raw materials, parts, and assemblies purchased for use in connection with the production of orders for the government or its subcontractors are procured against the supplier's certifications. All such items shall be subjected to inspection after receipt per MIL-STD-105D to assure conformance to purchase order requirements.
2. Certified test reports covering all raw materials and their specifications are to be retained at our plant subject to inspection upon request.
3. Spring wire and flat spring stock are to be checked for dimensional accuracy. Material that does not conform to ordered specifications is to be rejected and returned to suppliers accompanied by a rejection report. A corrective action report is required from the suppliers. Raw stocks that are accepted are to be segregated and tagged upon receipt and stocked in their respective places provided in our warehouse.

## B.4.1 MATERIAL CONTROL

1. Raw material in stock must be carefully controlled at all times. All raw stock shall be tagged and stored according to material, type, size, and weight.
2. Raw stock is to be checked out of Material Stores to Production as required for work and processed by a shop traveler. The shop traveler is released by Inventory Control for Production only after verification that the material meets all requirements.
3. After completion by Production of a specific requirement, all raw material not used must be returned to stock and replaced in its designated place, properly tagged. Appropriate notations shall be made on the shop traveler and returned to Inventory Control for the purpose of controlling the expended amount of material.

## B.5.1 FIRST ARTICLE INSPECTION

1. Tooling shall be checked for accuracy in conformance with the latest change of the appropriate drawing by the production supervisor prior to setup.
2. A number of the first parts produced shall be inspected, and production shall not begin until these parts are approved by the inspector. Parts not acceptable must be scrapped, and new samples must be submitted until accepted by the Inspection Department.

3. On the first production article, a comprehensive inspection and tests shall be performed of that article to assure its conformance with all blueprints and specification requirements. Approval shall be stamped on the shop traveler by the inspector.

## B.6.1 FINAL INSPECTION

1. Final inspection will be made by the quality control inspector to establish conformance with the customer's purchase order and drawing requirements.
2. Final inspection includes checking every blueprint requirement. Inspection is both visual and mechanical, and covers such items as physical dimensions, load, and torsional testing.
3. Inspection shall include sampling in accordance with approved statistical sampling methods per MIL-STD-105D or as specified by contract.

### B.6.1.1 Inspection Report

1. Customer:
2. Date:
3. Part Number:
4. Purchase Order Number:
5. Applicable Specifications:
6. Blueprint Dimensions:
7. Actual Dimensions:

### B.6.1.2 Instructions for Inspection Report

In filling out the inspection report, it should be noticed that each title has a number. These instructions "key" on that number, and begin with the items at the top of the form.

| Number | Instruction |
|---|---|
| 1. Customer | Enter name of customer |
| 2. Date | Enter date of inspection |
| 3. Part Number | Enter part number |
| 4. Purchase Order Number | Enter purchase order number |
| 5. Applicable Specifications | Enter specifications for material finish, heat treatment, etc. |

## B.7.1 OUTSIDE VENDOR QUALITY AND EVALUATION CONTROL

1. A Quality Control Evaluation Report will be made on each vendor doing any special process or subcontract work for Spring Controls, Inc. This report is intended to furnish data relative to the capability of the vendor to control the quality and conformance of applicable supplies, parts, and services to our requirements.
2. This report shall be prepared by the Quality Control Department at least once each year and as otherwise deemed necessary. The report will be

required for any new vendor prior to release of work to that vendor and must be approved by Quality Control.

3. The Quality Control Department shall maintain a list of approved sources for all processes, subcontractors, and other facilities as may be necessary to control outside sources. All outside processing of parts requiring government specifications shall be certified by the processor that the processing complies with these specifications. In addition, parts shall be inspected by the Inspection Department in accordance with approved statistical sampling methods or as specified by contract.
4. Copies of outside certifications shall be kept on file in the office of the Quality Control Department and shall be available for review as required.
5. All processes to be completed by outside vendors shall be controlled in a manner pertaining to the specification. All outside vendors must be approved by the prime customer and/or the government agency which may have control over process approval.
6. All outside vendors, in addition to approval by the prime customer, must be approved by the Quality Control Department of Spring Controls, Inc., and must maintain such records as requested in addition to specification records.

### B.7.1.1 Listing of Approved Processors

| Vendor | Process |
|---|---|
| | All types electroplating |
| | Glass bead peen |
| | Passivation |
| | Phosphate |
| | Parkerize and oil |
| | All types electroplating |
| | All types heat treating |

## B.8.1 INSPECTION MEASURING AND TESTING EQUIPMENT

1. All inspection equipment including personally owned calipers and micrometers shall be calibrated quarterly or more frequently as deemed necessary.
2. All inspection equipment shall be listed in a handbook and held in the Inspection Department to indicate when each item should be checked and recorded. This list shall be reviewed by the chief inspector on the first of such month for the purpose of calling in each test piece of equipment for inspection and review. All tests shall be completed within thirty (30) days prior to the expiration date.
3. The inspection equipment used in the Inspection Department shall receive periodic comparisons as outlined below and shall be recorded reflecting the date, type of equipment inspected, and any remarks. This is performed

by a certified outside source, and certification shall be maintained in the Quality Control Department.

4. Spring testers shall be tested and recorded for accuracy every three (3) months and any other time as deemed necessary. All testers shall be tagged showing the actual date of test and when the next test is due.
5. All micrometers and calipers used by the Inspection Department shall be checked with gage blocks traceable to the National Bureau of Standards and recorded twice monthly. They shall be labeled and recorded in the Inspection and Testing Book showing the date and identification of the person performing the test.
6. All furnaces and control panels shall be calibrated for uniform heating by an outside certified furnace inspection service every six (6) months. Certifications will be required and shall be maintained in the Quality Control Department.
7. Gage blocks shall be tested once each year, and a certification of calibration source will contain a statement to the effect that the calibration of equipment is traceable to the National Bureau of Standards.
8. A label as designated by the Quality Control Department will be established to use on small gages and other equipment to indicate the period of time said equipment was checked and recorded.

## B.9.1 SHIPPING AND PACKAGING

1. All packaging and shipping is to be in accordance with the customer's requirements. Our normal procedures ensure that the packing list and identification tags are inside the package. The package is further identified by placing the customer's name, part number, and purchase order number on the outside of the package.
2. Any deviation from the above procedure must be in accordance with the customer's specification.
3. The packing list further states, "Test reports covering all the material itemized as requiring tests are on file subject to examination and indicate conformance with applicable specification requirements."
4. Conformance certifications and/or physical and chemical test reports when required will be attached to the packing list. All certifications will show the customer order number, part number, and quantity shipped.
5. Adequate packaging is to be employed to ensure that all parts receive protection from damage and corrosion during shipment and handling.
6. Final packaging shall be approved by the Inspection Department.
7. The above paragraphs are to be posted in the packaging and shipping area.

## B.10.1 INSPECTION STAMPS AND TAGS

1. All inspectors shall have a stamp that will be numbered to identify them on all articles inspected and stamped. These stamps shall not be used by anyone other than the designated personnel assigned a numbered stamp. The inspector should use diligence in controlling his assigned stamp.

2. Inspection Department stamps shall be used on all first article production parts and final inspections.
3. A list of inspection stamp numbers will be maintained in the Inspection Department with the name of the inspector having control and the date of issuance. New stamps shall be issued only with the approval of Quality Control.

## B.11.1 DISCREPANCY REPORT FORM

1. Customer:
2. Part Number:
4. Purchase Order Number:
3. Date Received:
5. Quantity Received:
6. Discrepancies:
7. Responsible Department:
8. Inspector:
9. Preventive Action:
10. Final Disposition: Rework Scrap
11. Approval:
12. Quality Control Manager:

### B.11.1.1 Procedure for Reporting Discrepancies

This procedure is to be used to report discrepancies on rejected parts. In filling out the discrepancy report, it should be noticed that each title has a number. These instructions "key" on that number, and begin at the top of the form. All entries must be typed or printed.

| Number | Instructions |
|---|---|
| 1. Customer | Enter name of customer |
| 2. Part number | Enter part number |
| 3. Purchase order number | Enter purchase order number |
| 4. Date received | Enter date parts were received |
| 5. Quantity received | Enter quantity received |
| 6. Discrepancies | Indicate what is wrong with the part; be as detailed as necessary to show the location of the discrepancy on the part; quote drawings, specification, and condition of the part, etc. |
| 7. Responsible department | Enter name of department responsible for the discrepancy |
| 8. Inspector | Enter name and number of inspector who checked parts |
| 9. Preventive action | Enter a statement of the action to prevent recurrence. Include the cause of the discrepancy and the positive corrective action taken. |
| 10. Final disposition | Indicate whether parts are to be reworked or scrapped |
| 11. Approval | Signature of a company official approving final disposition |
| 12. Quality Control Manager | Signature of the Quality Control Manager |

## B.12.1 QUALITY CONTROL

1. It shall be the responsibilty of the Quality Control Department to see that articles are produced in accordance with the requirements of the contract and/or purchase order and the amendments or supplements thereto.
2. The responsibility of producing and maintaining all specifications required by purchase orders, drawing, and contracts shall rest within the Production Control Department.
3. The quality control requirements of Spring Controls, Inc., are believed be consistent with the requirements established by MIL-I-45028.
4. All changes in the Quality Assurance Manual will be incorporated into all existing manuals that are in the hands of prime customers or subcontractors. It shall be the Quality Control Department's responsibility that all outstanding manuals be maintained to the latest revision.

## B.13.1 SHOP TRAVELER

Shop travelers are required to include:

1. Customer name
2. Purchase order number
3. Part number and revision
4. Date of order entry
5. Date order is due
6. Order quantity
7. Type and size of material
8. Manufacturing procedures
9. Finish and heat treatment specifications
10. Name of vendor on outside process

## B.14.1 TYPES OF RECORDS AND NUMBER YEARS MAINTAINED

| Item | Period (years) |
|---|---|
| (1) Purchase orders from customers | 5 |
| (2) Raw material purchase orders | 5 |
| (3) Material certifications | 5 |
| (4) Special process certifications: Heat treatment, plating, etc. | 5 |
| (5) Inspection reports | 5 |

## B.15.1 SPECIAL PROCESS CONTROL

1. Special processing such as heat treating, Magnaflux, plating, etc., is to be performed by outside vendors.
2. All special processing must be certified to specifications by the vendor performing the process. In addition, parts must be inspected by our Inspection Department in accordance with approved statistical sampling methods or as specified by the customer. This manual is reproduced courtesy of John Cochran.

# Appendix C: Queuing Results

## NOTATION:

$\lambda$ = arrival rate in the system
$\mu$ = service rate of each server
$n$ = number of customers in queuing system
$P_n$ = probability that exactly $n$ customers are in queuing system
$L$ = expected number of customers in queuing system
$L_q$ = expected number of customers in queue
$W$ = expected waiting time in the system (including service time) = $L/\lambda$
$W_q$ = expected waiting time in queue = $L_q/\lambda$
$\rho$ = utilization ratio
$\lambda_e$ = effective arrival rate (finite queue models) = $\lambda P_n$, if there are only $n$ spaces in the system

Single server (Poisson arrival, exponential service, M/M/l):

$$\rho = \frac{\lambda}{\mu}$$

$$P_o = 1 - \rho$$

$$P_n = \rho^n P_o$$

$$L = \frac{\rho}{1-\rho}$$

$$L_q = \frac{\lambda^2}{\mu(\mu - \lambda)}$$

Multiserver model (Poisson arrival, exponential service):

1. Unlimited queue (M/M/C):

$$p_0 = \frac{1}{\left[\sum_{n=0}^{C-1} \frac{(\lambda/\mu)^n}{n!} + \frac{(\lambda/\mu)^C}{C!} \cdot \frac{1}{1-(\lambda/C_\mu)}\right]}$$

$$p_n = \frac{\left(\frac{\lambda}{\mu}\right)^n p_o}{n!} \quad \text{if } 0 \le n \le c$$

$$= \frac{\left(\frac{\lambda}{\mu}\right)^n P_o}{C!C^{n-c}} \quad \text{if } n \geq c$$

$$L_q = \frac{P_o(\lambda/\mu)^c \rho}{C!(1-\rho)^2}$$

where

$$\rho = \lambda/C_\mu < 1$$

$$L = L_q + \frac{\lambda}{\mu}$$

2. Queue size limited to $K$ (M/M/C/K):

$$P_o = \frac{1}{\left[\sum_{n=o}^{c-1} \frac{(\lambda/\mu)^n}{n!} + \frac{(\lambda/\mu)^c}{C!} \cdot \left(\frac{1-\rho^{k-c+1}}{1-\rho}\right)\right]}$$

where

$$\rho = \lambda/C_\mu \leq 1$$

$$P_n = \left\{ \frac{(\lambda/\mu)n}{n!} P_o \right.$$

$$L_q = \frac{(\lambda\mu)^C \rho P_o}{C!(1-\rho)^2}\{1-[(K-C)(1-\rho)+1]\rho_{K-C}\}$$

$$L = L_q + \frac{\lambda}{\mu}(1-P_K)$$

# Appendix D: Mathematical Formulation for Material Equipment Selection (Chapter 14)

## A MATHEMATICAL MODEL FOR EQUIPMENT SELECTION

A mixed-integer linear programming model is developed to obtain the optimum selection of conveyors and trucks (or some other two classes of material-handling systems). The cost coefficients are calculated as discussed previously. In effect, different types of conveyors and trucks considered here are represented only by their corresponding costs. Automated guided vehicles may also be classified like the trucks to obtain similar cost breakdowns to include in the mathematical model.

Minimize

$$Z = \sum_{j=1}^{n} \left[ g_i y_i + \sum_{i=1}^{4} D_i \left( \sum_{k=1}^{p} (X_{ijk}) \right) \right] + \sum_{i=1}^{4} F_i T_i$$

Subject to

$$\sum_{i=1}^{4} A_{ik} X_{ijk} + B_k y_i = B_k \quad \text{for all } k \text{ for each}$$

$$\sum_{k=1}^{p} \sum_{i=1}^{4} X_{ijk} \leq h_j, \quad M \text{ for all } j$$

$$y_j = (1 - h_j) \quad \text{for all } j$$

$$\sum_{j=1}^{n} \sum_{k=1}^{p} x_{ijk} \leq T_j \quad \text{for all } i$$

where:

$g_j$ = total daily conveyor cost on route $j$
$D_i$ = daily operating cost of truck type $i$
$F_i$ = daily fixed cost of truck type $i$
$X_{iJk}$ = percent capacity needed for truck type $i$ on route $j$ to carry item $k$

$T_i$ = an integer number of truck type $i$
$A_{ik}$ = daily capacity of truck type $i$ for item $k$
$B_k$ = daily flow rate (demand) for item $k$

$$y_j = \begin{cases} 1 & \text{conveyor present on route } j \\ 0 & \text{otherwise} \end{cases}$$

$h_j$ = auxiliary binary variable to introduce either conveyor or truck on route $j$
$M$ = a large positive number (e.g., 99999)

$i$: truck type $i$; $i = 1, 2, 3, 4$
$j$: route $j$; $j = 1, 2, \ldots, n$
$k$: item type $k$; $k = 1, 2, \ldots, p$
$p$: number of items (products)
$n$: number of items

This model assigns conveyors or trucks to routes. One or more truck types may be shared on a given route. Although one obtains optimum (minimum) cost, the computations prove to be prohibitive for even smaller problems. For example, with 5 departments, having 15 feasible routes and only 5 items, 126 variables and 56 constraints were required.

# Appendix E: Samples of Americans with Disabilities Act of 1990 Checklists

Facility ID: ________________ Level: __________________

## Part C: Interior Route

Use this interior route checklist for each level (story) of the facility. Identify the story level in upper right-hand corner.

Interior accessible route means: a barrier-free path of travel from the facility entrance along halls, corridors, aisles, skywalks, tunnels, and other spaces that connect with programs, activities, and services offered and lead to restrooms, water fountains, telephones, etc. (ADAAG 4.3)

### Doors and Doorways

| | | |
|---|---|---|
| 1. Do doorways on this floor have clear opening of at least 32″ when open? | YES | NO |
| 2. Is the maximum opening force 5 lb on interior hinged, sliding, or folding doors? | YES | NO |
| 3. Are doors operable by a single effort? | YES | NO |
| 4. Is the floor on the inside and outside of the doorways level for a distance of 5 ft from the door in the direction the door swings? | YES | NO |
| 5. Are all thresholds no higher than 1/2″ with beveled edge? | YES | NO |
| 6. Do doors with closers allow enough closing time for use of the doors by physically disabled persons? | YES | NO |
| 7. a. Are doors identified with either raised or indented letters/numerals that identify the area? | YES | NO |
| b. If doors are marked as indicated in 7a, are the markings at a level of between 4′6″– 5′6″ above the floor? | YES | NO |
| 8. Are there signs to denote accessible route if the distance is long and difficult to determine? | YES | NO |
| Review NO responses. Are interior doors accessible on the pathway to program/activity/service offered? | YES | NO |

Comments:

Courtesy of Environmental Safety Office, Louisiana Tech University

## PART C: EXTERIOR ROUTE

### Walks/Walkways

*Need*: People who walk with difficulty or use wheelchairs, crutches, canes, or walkers need a wide, smooth, level, firm surface. Sight-impaired people need a path free of hazards such as low-hanging or protruding objects undetectable by a cane. (ADAAG 4.3–4.3.10)

Exterior accessible route means: A barrier-free path of travel from the parking lot, public transportation debarkation area, or passenger-loading zone along walks/walkways of a continuing common surface not interrupted by steps or abrupt changes in levels that leads to an accessible entrance(s).

| | | |
|---|---|---|
| 1. Do walks have a minimum of 36″ clear width except at doors? | YES | NO |
| 2. Is there at least a 60″ × 60″ passing space every 200 ft? | YES | NO |
| 3. Is there a minimum of 80″ clear headroom to avoid over head hazards? | YES | NO |
| 4. Surface: nonslip, firm, and stable? | YES | NO |
| 5. Slope does not exceed 1″ of rise to 20″ of incline? | YES | NO |
| 6. Are routes not interrupted by changes in level of no more than 1/2″ or steps? | YES | NO |
| 7. Are grates set in the direction of the route no more than 1/2″ wide? | YES | NO |
| 8. Is route clear of benches, water fountains, etc., with leading edges at or below 27″ that reduce the width of route space to less than 36″? | YES | NO |
| 9. Is there at least one (1) accessible route from transportation stops, parking lot, or street to facility? | YES | NO |
| 10. a. Are there signs to indicate accessible route? | YES | NO |
| b. To denote direction to accessible entrance? | YES | NO |
| Review NO responses. Is there an exterior route(s) to the facility that is accessible? | YES | NO |

Comments:

## PART C: EXTERIOR ROUTE

### Curb Ramps

*Need*: Curb ramps or curb cuts allow people who are mobility impaired access to raised walkways. (ADAAG 4.7)

| | | |
|---|---|---|
| 1. Are there curb ramp(s): | | |
| a. Located whenever accessible route crosses a curb and where cars do not park? | YES | NO |
| b. Slope does not exceed 1″ of rise to 12″ of incline (1:12)? | YES | NO |
| c. At least 36″ wide, excluding flared sides? | YES | NO |
| d. Surface is firm, stable, and nonslip? | YES | NO |
| e. If ramp is 72″ long (6″ rise), are there hand rails? | YES | NO |
| f. Do flared sides have a slope of no more than 1:10? | YES | NO |
| g. If at intersection, is curb ramp located within and to one side of marked crossing? | YES | NO |
| h. Is there flush/smooth transition to street level? | YES | NO |

| | | |
|---|---|---|
| 2. a. Is there a sign to denote location of curb ramp? | YES | NO |
| b. Is it painted blue or yellow or other color to prevent illegal parking that would block it? | YES | NO |

Review NO responses. Are curb ramps (cuts) accessible? YES NO

Comments:

---

## Part C: Exterior Route

### Parking

*Need*: Mobility-impaired people need parking spaces wide enough to open car doors fully or lower a lift from a van in order to get out with a wheelchair or mobility aid, that are close to the building or facility, and that are on an accessible route from parking lot to building. The numbers presented in this table are required *minimum standards*. Where there is need or accommodations are requested, the numbers should exceed the table. [ADAAG 4.1.2(5) and 4.6]

### Suggested Guideline

| Total Parking | Required Minimum |
|---|---|
| 1–25 | 1 |
| 26–50 | 2 |
| 51–75 | 3 |
| 76–100 | 4 |
| 101–150 | 5 |
| 151–200 | 6 |
| 201–300 | 7 |
| 301–400 | 8 |

| | | |
|---|---|---|
| 1. If any visitor, participant, or employee parking is provided, are spaces reserved for handicap parking? | YES | NO |
| 2. Reserved space(s) located closest to accessible entrance; on accessible route? | YES | NO |
| 3. Is the space(s) at least 96″ wide? | YES | NO |
| 4. At least one (1) of eight (8) handicap parking spaces provided at least 96″ wide with access aisle and labeled "van accessible"? | YES | NO |
| 5. Access aisle next to other handicap spaces at least 60″ wide? | YES | NO |
| 6. Slope of space/access aisle no more than 1:50? | YES | NO |
| 7. a. Sign with accessible symbol designating space? | YES | NO |
| b. Mounted at a height not obscured by parked vehicle? | YES | NO |
| 8. Surface: nonslip, firm, and stable? | YES | NO |
| 9. Are the spaces/access aisles painted blue or other color to denote reserved accessible area? | YES | NO |

Review NO responses. Are parking spaces accommodating? YES NO

Comments:

---

## PART C: EXTERIOR ROUTE

### Ramp

*Need*: People in wheelchairs need gently sloped ramps with handrails, no drop-offs, and a smooth, stable surface with level top and bottom platforms for resting and turning. [ADAAG 4.8]

---

| | | |
|---|---|---|
| 1. Is slope as gentle as possible and no more than 1:12? | YES | NO |
| 2. Cross slope (perpendicular to direction of travel): no more than 1:50? | YES | NO |
| 3. Surface: nonslip, firm, and stable? | YES | NO |
| 4. Walls, railings, or edge protection curbs at least 2″ high to prevent slipping off ramp? | YES | NO |
| 5. a. Level landing is as wide as ramp and at least 60″ long at top of ramp and each turn of ramp? | YES | NO |
| b. And 72″ long at the bottom? | YES | NO |
| 6. Outside ramp is at least 60″ wide and rises no more than 30″ without an intervening fiat area? | YES | NO |
| 7. Handrails: | | |
| a. Provided on both sides? | YES | NO |
| b. Diameter of gripping surface 1 1⁄4? to 11⁄2?? | YES | NO |
| c. If on or next to wall, wall and handrail are 11⁄2″ and wall surfaces smooth? | YES | NO |
| d. (1) If ramp rise is more than 6″ and length is more than 72″, are there handrails between 30″ and 34″ high? | YES | NO |
| (2) Rails extend 1 ft beyond top and bottom of ramp? | YES | NO |
| e. Are handrail ends and edges rounded or returned to the floor, wall, or post? | YES | NO |
| f. Are handrails solidly anchored with fittings that do not rotate? | YES | NO |
| g. Are handrails parallel to slope of ground surface ramp? | YES | NO |
| 8. Are there signs to direct persons to ramp? | YES | NO |
| Review NO responses. Are ramps accessible? | YES | NO |

Note: If access into the facility is provided by means of a lift, describe in detail the method used to access the lift.

Comments:

---

Facility ID: ______________ Level: _______________

## PART C: INTERIOR ROUTE

### Elevators

*Need*: All persons with disabilities benefit from using elevators. Each elevator must provide adequate maneuvering space and time to get to and enter the car, must be conveniently located, and must have marked controls. Blind persons need audible indications on direction of travel and floors and tactile information to be visual. [ADAAG 4.10]

| | | |
|---|---|---|
| 1. How many elevators are there in this building?<br>How many passenger elevators?<br>How many freight elevators? | | |
| 2. At least one serves each level on accessible route in a multistory facility? | YES | NO |
| 3. Is it an automatic self-leveling elevator with reopening devices? | YES | NO |
| 4. Car dimensions: | | |
| a. If door opens in the center, floor at least 51″ × 80″? | YES | NO |
| Or | | |
| b. If doors open on one side floor at least 51″ × 80″? | YES | NO |
| 5. Hall call buttons: centered 42″ or less from floor and lighted? | YES | NO |
| 6. Car controls: highest control 48″? | YES | NO |
| a. Buttons at least 3/4″ in their smallest dimension? | YES | NO |
| b. Marked with raised and/or Braille characters? | YES | NO |
| 7. Door remains open minimum of 3 seconds? | YES | NO |
| 8. Visual and audible floor indicators provided? | YES | NO |
| 9. If emergency information systems provided, audible alarms (bells or audible instructions) and visual signals (flashing alarms or written instructions) used? | YES | NO |
| 10. Automatically corrects over- or undertravel within 1/2″ when stopping at floor? | YES | NO |
| 11. Door width at least 36″? | YES | NO |
| 12. Floor is firm, stable, and nonslip? | YES | NO |
| 13. No more than 11/4″ gap between car and landing platform? | YES | NO |
| Review NO responses. Are elevator(s) accessible? | YES | NO |
| Note: If only a freight elevator is in this facility, is it available for use by mobility-impaired persons? | YES | NO |

Comments:

Facility ID: _____________ Level: _____________

## Part C: Interior Route

### Stairs

*Need*: Sight-impaired people need stairs of uniform tread and width with handrails that guide them and indicate landings. [ADAAG 4.9]

| | | |
|---|---|---|
| 1. a. Stairstep heights are uniform? | YES | NO |
| b. Step depths are at least 11″ and uniform? | YES | NO |
| 2. a. No overhangs on steps greater than 1 1/2″? | YES | NO |
| b. Are overhangs curved? | YES | NO |
| 3. Handrails meet requirement? (see Ramps) | YES | NO |
| Review NO responses. Are stairs accessible? | YES | NO |

Comments:

### Floors and Interior Ramps

*Need*: Floor surfaces should be as slip-proof as possible with attention to any rugs that could be a safety hazard. Any changes in floor levels should have access either by ramp or lift.

---

1. Do floors have nonslip surface? YES NO
2. a. Are floors on this level (story) at a common level? YES NO
   b. Is there a ramp to connect levels? YES NO
      (1) Is the ramp at least 44″ wide? YES NO
      (2) Is the slope no greater than 1:12? YES NO
      (3) If the rise is more than 30″, are there intervening flat areas? YES NO
3. If a lift is used, describe how it is accessed.

Review NO responses. Does this level have an accessible route? YES NO

Comments:

---

Facility ID: ___________ Level: __________

## Part C: Interior Route

### Restrooms

*Need*: Mobility-impaired people need restrooms that they can get into and use easily and safely. Fixtures need adequate clear floor space for close approach and turning, and some require sturdily mounted grab bars for support or transfer. Controls and hardware must be within reach and easily operable. Hot, sharp, abrasive, or protruding objects are hazards. [ADAAG 4.16–4.24]

---

1. a. Are restrooms provided for the general public? YES NO
   b. Are restrooms provided for employees only? YES NO
   c. How many restrooms are on this floor?
   Men, Women, Unisex
2. If there are restrooms provided for the general public, is at least one set provided on an accessible route? YES NO
3. a. Entrance door has at least 32″ clear opening? YES NO
   b. Identified by sign with accessibility symbol? YES NO
4. Unobstructed 60″ × 60″ space inside restroom to allow for turning wheelchair? YES NO
5. Toilet stall door at least 32″ clear opening width? YES NO
6. a. In stall, 59″ × 60″ floor space for floor-mounted toilet; YES NO
   or
   b. 56″ × 60″ space for wall-hung toilet? YES NO
7. 48″ clear distance from front of toilet bowl to opposite wall or floor? YES NO
8. In stall, front partition and at least one side partition provide toe clearance of at least 9″ above the floor? (If depth of stall is more than 60″, toe clearance not needed.) YES NO

| | | |
|---|---|---|
| 9. a. Grab bars are 33–36″ high? | YES | NO |
| b. Located on back and side of stall? | YES | NO |
| c. Are 1 1⁄4″ to 1 1⁄2″ in diameter? | YES | NO |
| d. Are no more than 1 1⁄2? from the wall? | YES | NO |
| e. Sharp edges and protrusions eliminated? | YES | NO |
| 10. Toilet is 17″–20″ high and located 18″ from center of toilet to closest wall? | YES | NO |
| 11. For wall-mounted urinal, basin opening is no more than | | |
| a. 17″ from floor? | YES | NO |
| b. Has elongated rim? | YES | NO |
| c. Clear floor space 30″ × 48″ in front of urinals? | YES | NO |
| 12. a. Toilet paper dispenser 5″ above grab bar? | YES | NO |
| And | | |
| b. 36″ from rear wall? | YES | NO |
| 13. Sinks: | | |
| a. Height maximum 34″? | YES | NO |
| b. Drain and hot water pipes insulated? | YES | NO |
| c. Minimum 29″ clearance below apron of sink? | YES | NO |
| d. Clear floor space 30″ × 48″ in front of sink? | YES | NO |
| 14. Faucets: | | |
| a. Controls mounted no more than 48″ above the floor? | YES | NO |
| b. Hand-operated or automatic but do not require tight gripping, pinching, or twisting of wrist? | YES | NO |
| 15. Where there are mirrors, is the bottom at least 40″ above the floor and the top at least 72″ above the floor? | YES | NO |
| 16. Towel dispenser and disposal units: operable part at least 40″ above floor? | YES | NO |

Review NO responses. How many restrooms are accessible? Men, Women, Unisex

Comments:

---

Facility ID: ___________ Level: __________

## Part C: Interior Route

### Drinking Fountains

*Need*: Persons in wheelchairs need drinking fountains mounted low so they can reach the spout. They need to be able to pull up under the fountain or along its side. Persons who have difficulty using their hands need controls that can be easily operated. [ADAAG 4.15]

---

| | | |
|---|---|---|
| 1. Are water fountains located on this floor? | YES | NO |
| 2. a. If fountains are available, 50% accessible on each floor? | YES | NO |
| b. If only one is available, is it on an accessible route? | YES | NO |
| 3. Spout mounted 34″ above floor in front of unit with water flow at least 4″ high and parallel to front of unit? | YES | NO |
| 3 Controls operable with one hand without grasping or twisting? | YES | NO |

| | | |
|---|---|---|
| 4. a. Wall mounted: bottom of apron to floor at least 27″? | YES | NO |
| b. At least 30″ × 48″ clear space in front of fountain? | YES | NO |
| Review NO responses. Are drinking fountains accessible? | YES | NO |

Comments:

---

## Hazardous Areas and Warning Signals

*Need*: Visually impaired people need audible emergency warning systems and need to be alerted by touch to hazardous areas. Persons with hearing impairments need visual alarms. [ADAAG 4.28]

---

| | | |
|---|---|---|
| 1. If warning systems are provided, both visual (flashing) and audible provided? | YES | NO |
| 2. a. Door knobs to hazardous areas roughened (knurled knobs)? | YES | NO |
| b. Doors labeled in raised or indented letters? | YES | NO |
| c. Identification labels at a height between 4′6″ and 5′6″? | YES | NO |
| Are hazardous areas identified? | YES | NO |
| Are warning signals both visual and audible? | YES | NO |
| Review NO responses. | YES | NO |

Comments:

---

Facility ID: ___________ Level: __________

## Part C: Interior Route

### Assembly, Meeting, and Conference Areas

*Need*: People in wheelchairs need a level area from which they can view the performance area. Both the seating area and the performance area must be on an accessible route (including raised portable stages). Persons with hearing impairments need an auxiliary listening system. [ADAAG 4.32 and 4.33]

### Minimum Guidelines

| Total Capacity | W/C Seating |
|---|---|
| 50–75 | 3 |
| 76–100 | 4 |
| 101–150 | 5 |
| 151–200 | 6 |
| 201–300 | 7 |
| 301–400 | 8 |

| | | |
|---|---|---|
| 1. Are wheelchair spaces provided? | YES | NO |
| 2. Are wheelchair locations adjacent to accessible route and whenever possible, ramped to different seating levels? | YES | NO |
| 3. Are seating positions dispersed? | YES | NO |
| 4. Is the performing area on an accessible route? | YES | NO |
| 5. For large areas, is an amplification system available (volume controls, wireless headphones, infrared-audio loops, and radio frequency acceptable)? | YES | NO |
| Review NO responses. Do these areas accommodate persons with disabilities? | YES | NO |

Comments:

Facility ID: ___________ Level: __________

## Part C: Interior Route

### Public Telephones

*Need*: Persons in wheelchairs need adequate clear floor space to pull up to the telephone and a low mounting height so they can reach all operable parts. Persons with hearing impairments need volume controls. [ADAAG 4.31]

| | | |
|---|---|---|
| 1. If public telephones are provided, is at least one accessible per floor? | YES | NO |
| 2. Located on an accessible route with clear floor space 30″ × 48″ in front of phone? | YES | NO |
| 3. a. Highest operable control 48″ high for front approach? | YES | NO |
| b. 54″ for parallel approach? | YES | NO |
| 4. Push-button controls? | YES | NO |
| 5. Any provision for persons with hearing impairments? | YES | NO |
| 6. Is there signage to denote handicap access? | YES | NO |
| Review NO responses. Are public telephone(s) accessible? | YES | NO |
| Note: Is there a TTY/TDD installed in this facility for use by deaf/hearing-impaired persons? | YES | NO |

Comments:

Facility ID: ___________ Level: __________

## Part C: Interior Route

### Exhibits, Signs, and Information Displays

*Need*: Persons with disabilities need exhibits, signs, and information displays adequately lighted, in high-contrast colors, in large, easy-to-read print, and at levels where the materials may be read by short people or by persons in wheelchairs. Tactile objects allow visually impaired persons to enjoy exhibits and displays. Audio information should be available for hearing-impaired persons in some other format. The services available that provide accessibility, as well as general information about the building or site, should inform persons on the extent of the building's or site's accessibility.

| | | |
|---|---|---|
| 1. TTY/TDD or other capability for deaf persons available? | YES | NO |
| 2. Availability of services (e.g., sign language, captioned films) publicized? | YES | NO |
| 3. Exhibits, signs, and labels between 54″ and 65″ high? | YES | NO |
| 4. If there are publicity materials on facility and programs, are accessible features listed? | YES | NO |
| 5. Signs adequately lighted, in high-contrast colors and under nonglare glass? | | |
| 6. Do direction, warning, and emergency exit signs have raised lettering? | YES | NO |
| Review NO responses. Are these items accessible? | YES | NO |
| Comments: | | |

## Seating, Tables, and Work Areas

*Need*: Persons in wheelchairs need seating with flat, clear floor space in front of tables, counters, and work areas, as well as sufficient knee clearance.

| | | |
|---|---|---|
| 1. Located on an accessible route? | YES | NO |
| 2. a. If frontal approach, 30″ × 30″ plus 19″ under table? | YES | NO |
| b. If L-shaped approach, 36″ × 30″ plus 19″ under table? | YES | NO |
| 3. Tops of tables and work surfaces 28″–34″ from floor? | YES | NO |
| 4. Knee clearance at least 27″ high, 30″ wide, and 19″ deep? | YES | NO |
| Review NO responses. Are these items in work area accessible? | YES | NO |
| Comments: | | |

# Appendix F: Case Studies

## CASE STUDY 1: FACILITIES PLANNING DESIGN PROJECT

### Conduct of Study

This case study has been used in a senior-level facilities planning course. It is meant to be performed over a 3- to 5-week period. The case study is also intended to be performed by a group of three students. The case study provides information at various levels of abstraction in a wide variety of situations in which decisions are made.

### Atlantic Connectors Inc. Plans Expansion

You are the facilities planning manager for Atlantic Connectors Inc. (ACI). ACI produces a variety of connectors according to various customer specifications. ACI's upper management is planning to expand production by the construction of three new facilities. You have been asked to find locations for the new facility structures and to provide a layout for the new facilities. Assume for initial planning purposes that each of the facilities will have the same layout and operation.

There are five distribution points that ACI would like to service through the new facility. Seven suitable locations have been identified and are considered to be of equal merit on all factors except transportation costs. The trailer-truck transportation costs from our distributors to the candidate facilities are shown in Table F.1. The expected number of trailer-truck trips per month to each distribution center is also shown. The expected construction costs at each site are shown at the bottom of the table. Also note that distribution center 3 cannot be serviced if the facility is located at sites 5 or 6 and that distribution center 2 cannot be serviced if the facility is located at site 7.

The new ACI facility will be composed of five main functional units: offices, production, warehousing, shipping and receiving, and maintenance.

The office area will be composed of four functional units: food and other services, design and engineering, administration (management, sales, personnel, purchasing, etc.), and production floor supervisors. Food services will need to serve and seat a maximum of 200 people at any particular time. Administration comprises 80 people. Space planning can be based on an average enclosed office space for one person (desk, files, accessories, phone, extra chair, etc.) plus a 15% allowance for storage, filing, and supplies space. Hallways and walkways should add another 20% allowance to the average size office space. One large (40 people maximum) and two small (10 people maximum) meeting rooms will also be necessary for use by everyone, even the shop-floor workers. Design and engineering personnel work closely with the sales staff and production personnel to provide the customer with a quality product. The design and engineering area consists of approximately 20 people.

**TABLE F.1**

| Distribution Center | Site 1 | Site 2 | Site 3 | Site 4 | Site 5 | Site 6 | Site 7 | Trips |
|---|---|---|---|---|---|---|---|---|
| 1 | 64 | 102 | 58 | 198 | 224 | 105 | 90 | 44 |
| 2 | 12 | 76 | 94 | 154 | 166 | 94 | 74 | 38 |
| 3 | 168 | 212 | 88 | 38 | 66 | 121 | 68 | 52 |
| 4 | 112 | 148 | 124 | 82 | 20 | 77 | 55 | 32 |
| 5 | 94 | 119 | 111 | 88 | 49 | 40 | 210 | 49 |
| Fixed cost per year | 38,340 | 34,210 | 39,450 | 33,940 | 29,220 | 36,654 | 32,119 | |

Production floor supervision for ACI is typically one-fourth the number of people in design and engineering.

The main production floor area will be composed of eight functional units: toolroom, injection molding, assembly, machining, finishing, heat treating, forming, and testing.

After speaking with the design and process engineers, you have learned that the six basic products offered by ACI must follow the steps shown next:

| Product | Units/Day | Units/Load | Sequence of Operations |
|---|---|---|---|
| 1 | 358 | 4 | Storage, form, machine, treat, store, machine, finish, test, storage |
| 2 | 310 | 50 | Storage, assemble, storage |
| 3 | 1120 | 15 | Storage, machine, assemble, machine, finish, Storage |
| 4 | 18 | 1 | Storage, form, machine, assemble, storage |
| 5 | 150 | 12 | Storage, heat treat, form, finish, storage |
| 6 | 3100 | 50 | Storage, injection molding, finish, storage |

Assume that a forklift carries each of the unit loads through the sequence of operations. In addition, you have learned that all incoming materials are sorted, inspected, and placed in the warehouse before reaching the production floor. Also, as products are produced, they will first be stored in the warehouse before being picked, packaged, and shipped.

The machining area is composed of six types of machines, whose space requirements are shown next. The area measurements include only the machine itself and no other working area. Realistic estimates must be made for the other elements, which may need space allowances in the machining area.

| Machine Type | Area (feet) | Number of Machines |
|---|---|---|
| 1 | $5 \times 11$ | 2 |
| 2 | $4 \times 4$ | 4 |
| 3 | $7 \times 4$ | 1 |
| 4 | $13 \times 9$ | 1 |
| 5 | $4 \times 3$ | 3 |
| 6 | $7 \times 6$ | 1 |

For initial planning purposes, any unspecified areas will be 20% smaller than the assembly area. The assembly area will be 50% larger than the machining area. The toolroom is typically 8% of the size of the entire production area. Each of the production areas that produce or modify parts must use the toolroom, except testing, which has its own local tool storage. Trips per day to the toolroom from each of the other production areas are shown next. Each of the trips is made by a forklift.

| Production Area | Trips |
|---|---|
| Forming | 11 |
| Machining | 41 |
| Heat Treating | 15 |
| Finishing | 19 |
| Assembly | 27 |
| Injection Molding | 14 |
| Testing | 6 |

The assembly operations for products 2, 3, and 4 require additional materials from a storage area within warehousing. A forklift is used for transportation of the parts.

| Product | Units/Assembly | Units/Load |
|---|---|---|
| 2 | 2 | 200 |
| 3 | 5 | 100 |
| 4 | 1 | 20 |

The heat-treating area receives unit loads according to a Poisson distribution with a mean value of approximately 20 minutes between loads. The heat-treating area is able to perform heat-treat operations according to an exponential distribution with a mean value of 12 minutes. You wish to determine the loads you can expect to be waiting for heat treatment. Understanding the queues that may build up in front of the heat-treat area will help you plan for the storage that will be necessary for work-in-process inventory.

The maintenance personnel are responsible not only for production equipment, but also for any office equipment that may need preventive maintenance or repair. Typically, the total number of trips from maintenance to production is 10 times the number of trips from maintenance to the office area. The maintenance area is normally one-eighth the size of the entire production area.

You wish to know how many forklifts will be needed to service the trips between production departments and between the production and warehousing area. Estimate total travel distance based on the distance between the central points of the individual production and warehousing areas. Assume that a forklift may travel 150 feet/minute, must be refueled near the maintenance area once for every 3 hours of operation, and must operate with a congestion factor of 0.8. Also, the forklifts must be taken to maintenance twice per day for a 30- and a 10-minute inspection and minor preventive maintenance.

The warehousing area will need to maintain certain levels of stock in order to best serve the customers. The minimum storage needs are:

| Product | Day's Supply | Unit Size | Dimensions (feet) |
|---|---|---|---|
| 1 | 5 | Crate of 40 | $3 \times 3 \times 3$ |
| 2 | 2 | Boxes of 30 | $2 \times 3 \times 2$ |
| 3 | 2 | Crate of 10 | $2 \times 4 \times 2$ |
| 4 | 3 | Individually | $4 \times 5 \times 3$ |
| 5 | 1 | 24 per pallet | $4 \times 4 \times 5$ |
| 6 | 3 | 100 per box | $4 \times 4 \times 3$ |

Analysis of past orders has determined the following observations: Product 2 is usually shipped with product 4 and product 1. Products 4 and 6 are always shipped alone. Product 3 is sometimes shipped with product 1; otherwise it is shipped alone.

You need to design an internal layout for the shipping/receiving area but not for the maintenance area. Consider maintenance to be treated as a "black box," focusing only on the flow of material and information between that area and other functional areas.

## CASE STUDY 2: FACILITIES PLANNING DESIGN PROJECT

### Conduct of Study

This case study is intended for use in a senior-level facilities planning course. It is meant to be performed by a group of two students over a period of 2–3 weeks. This case study provides information at various levels of abstraction in a wide variety of situations in which decisions are made.

### Atlantic Cycles Inc. Plans New Product Line

You are the facilities planning manager for Atlantic Cycles Inc., a producer of bicycles. As part of its diversification plans, the management of ACI plans to introduce a new model of exercise bicycle. Presently, ACI manufactures approximately 210 bicycles in a 450-minute shift, and there are two such shifts in one working day. The present production system is of the job-shop type. Management is considering replacing the existing job shop unit with a cellular manufacturing unit. As ACI's facilities planning manager, you have been asked to report on the impact that will be made by the addition of the exercise bicycle to ACI's current production capabilities.

You have the following data from the existing plant:

1. The present production rate is 210 regular bicycles in a 450-minute shift.
2. The following is the list of all the existing equipment within the company.

| Equipment Type | Quantity | Operating Time |
|---|---|---|
| Electric saw | 2 | 1 cut/15 seconds |
| Tube bender | 1 | 1 cut/30 seconds |
| Welding | 6 | 1 weld/60 seconds |
| Forging | 2 | 60 seconds/large sprocket |
| | | 30 seconds/small sprocket |
| Molding | 1 | 2 parts/90 seconds |

A list of all the parts with their numbers is given in Table F.2. Table F.3 shows a detailed bill of materials of all the parts to be manufactured by the company. The rest of the parts are either purchased from the market or subcontracted to vendors.

The machining requirements for both the models are:

| Regular Bicycle | Exercise Bicycle |
|---|---|
| Frame | Frame and Handle Bar |
| Cutting: 1, 2, 3, 4, 5, 6, 7 | Cutting: 1, 2, 3, 4, 5, 6 |
| Bending: 5, 7 | Bending: 2 (no. 2 bends) |
| Casting: 10, 11, 12, 13 | Welding: (a) 10-2 (b) 9-2 (c) 6-2 (d) 3 4[a] |
| Welding: 8, 9, 10, 11, 12, 13 | Saddle and Seat Post |
| Handle Bar and Stem | Cutting: 7, 8, 9, 10 |
| Cutting: 18, 20 | Punch press: 10 |
| Bending: 18 | Molding: 12 (no. 2) |
| Molding: 19 | Drill press: 8, 9 |
| Casting: 20 | |
| Saddle Post | |
| Cutting: 26 | |
| Molding: 25A (plastic saddle) | |
| Drive Chain | |
| Forging: 29, 30, 31 | |

[a]When a number follows a dash, it signifies the number of times an operation is performed.

**TABLE F.2**
**Part Name and Part Number**

| Part Name | Part Number |
|---|---|
| Top tube | 1 |
| Seat tube | 2 |
| Down tube | 3 |
| Head tube | 4 |
| Fork blade | 5 |
| Chainstay | 6 |
| Seatstay | 7 |
| Rear fork tip | 8 |
| Front fork tip | 9 |
| Top tube lug | 10 |
| Down tube lug | 11 |
| Seat Lug | 12 |
| Bottom bracket | 13 |
| Side pull brake | 14 |
| Brake shoes | 15 |
| Brake lever | 16 |
| Brake cable | 17 |
| Handlebars | 18 |
| Handlebar plugs | 19 |
| Handlebar stem | 20 |
| Expander bolt | 21 |
| Binder bit | 22 |
| Wedge | 23 |
| Grippers | 24 |
| Saddle | 25 |
| Seat Post | 26 |
| Screw (¼ inch diam., ¾ inch long) | 27 |
| Screw (¼ inch diam., 1 inch long) | 28 |
| Crank spider | 29 |
| Large sprocket | 30 |
| Small sprocket | 31 |
| Crankarms | 32 |
| Crank cover plate | 33 |
| Crank attaching nut | 34 |
| Washer | 35 |
| Ball bearing | 36 |
| Fixed cup | 37 |
| Screw (¼ inch diam., ½ inch long) | 38 |
| Chain | 39 |
| Back wheel | 40 |
| Front wheel | 41 |
| Front derailleur | 42 |
| Back derailleur | 43 |
| Pedal | 44 |
| Bicycle tube | 45 |
| Tire | 46 |
| Shift layers | 47 |

**TABLE F.3**
**Exercise Bicycle**

| Part | Part Number |
|---|---|
| Hub | 1 |
| Frame legs | 2 |
| Handlebar tube | 3 |
| Saddle post tube | 4 |
| Handlebar | 5 |
| Balance bar | 6 |
| Handlebar post | 7 |
| Saddle post | 8 |
| Mount brackets | 9 |
| Axle mount | 10 |
| Tension strap | 11 |
| Chain guard | 12 |

The rest of the parts of the regular and exercise bicycles are either purchased from the market or are subcontracted to the vendors. Figure F.1 and Figure F.2 show flowcharts of the parts for the two bicycles. Figure F.3 contains a layout of the present job-shop production system.

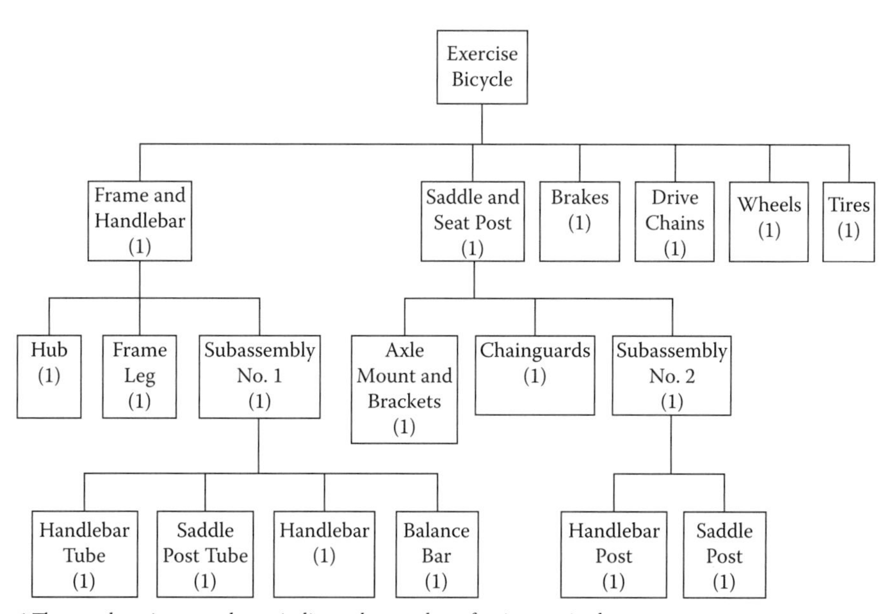


**FIGURE F1** Flowchart of parts for the exercise bicycle

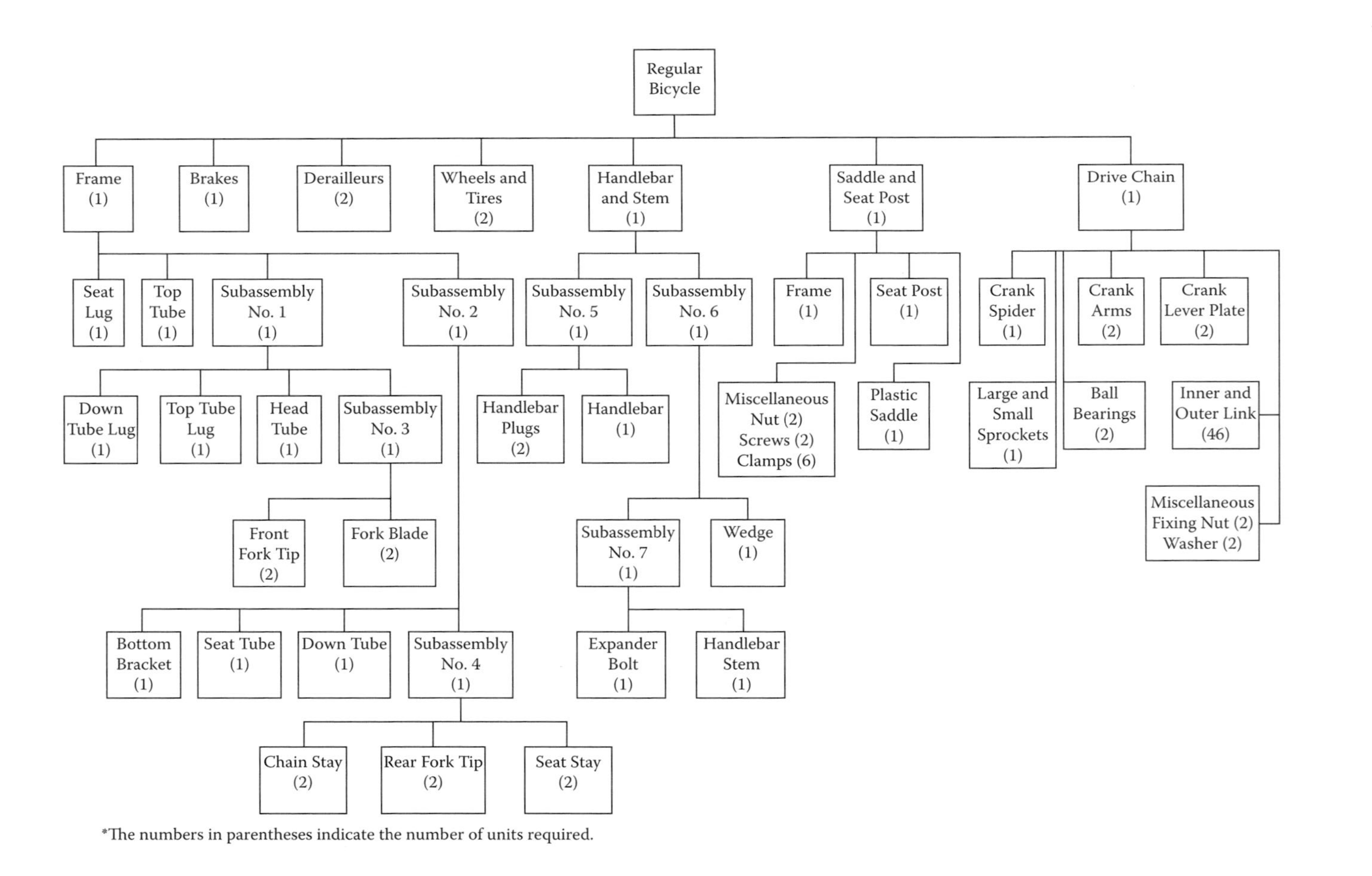


*The numbers in parentheses indicate the number of units required.

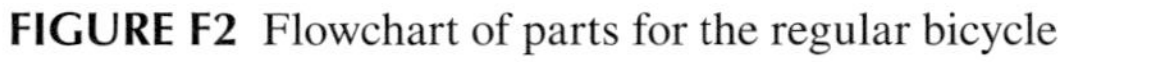

**FIGURE F2** Flowchart of parts for the regular bicycle

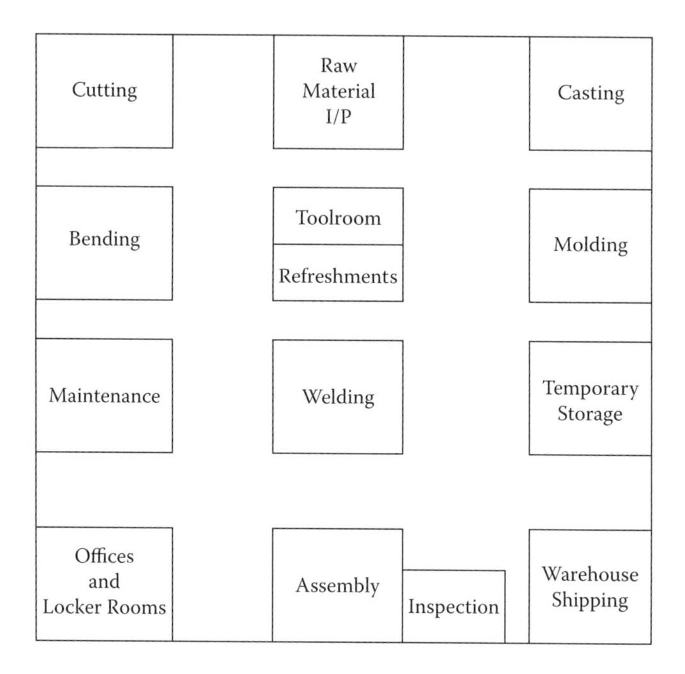


**FIGURE F.3** Job-shop floor plan of the existing facility

## Conduct of Impact Study

Design a cell layout for the manufacturing unit, incorporating group technology principles. Also assume the number of additional machines that need to be purchased to achieve a total production rate of approximately 600 bicycles per shift. There are two such shifts in one day. Also suggest the possible material-handling system for the same. Identify the areas where automation could be used and give an economic justification for its use. In the end, determine the total costs the company will incur in this diversification effort.

Assume the total market demand for both types of bicycles is evenly divided. The figures for the last 4 years of the combined total market demand are:

| Year | Demand |
|---|---|
| 1989 | 700,000 |
| 1990 | 769,000 |
| 1991 | 829,000 |
| 1992 | 912,000 |

At present the shortage is met by importing the bicycles. But this is a costly option; hence indigenously manufactured bicycles of comparable quality would be in great demand.

# Appendix G: Manufacturing Facilities: Computer Programs

These programs, written in BASIC, FORTRAN, and C, can be found at http://www.crcpress.com/e_products/downloads/download.asp?cat_no=44222. Follow the instructions given here to copy the programs on your hard disk.

1. Make a local directory named COMLADII. COMLADII is the directory on which plant layout program COMLADII and material-handling program COMLAD3 store the data. To make the directory, type the following at the DOS prompt:

```
    cd\
md COMLADII
```

2. The contents of the floppy occupy approximately 0.70 megabytes of disk space. Make sure that the hard disk has enough space to accommodate the content.
3. Copy the contents of the floppy to the COMLADII directory by typing the following. If b: is not assigned to the 3½-inch floppy drive in your machine, use a.

```
copy b:*.* c:\COMLADII
      cd COMLADII
```

4. To display the programs available, run MENU.
5. The following qbasic files should also be available to run programs written in Quick-BASIC:

```
qbasic.EXE
qbasic.HLP
qbasic.INI
```

6. To run qbasic files, type qbasic at the DOS prompt, open MENU_1.bas, and run the program. (To run the program, press ALT-F to pull down the file MENU_1.B and choose the option Open to open MENU_1.bas. After MENU_1.bas is loaded, press ALT-R to run the program.)

## QUICK-BASIC PROGRAMS

The following programs are available in Quick-BASIC. To run a program, simply type either the given name — that is, HEUR #1 — or the program number — for instance, 1.

### (1) HEUR # I

The program follows the location procedure described in Section 17.4. There are a number of customers (ICUST), each requiring services from one of the machines (IMC). The machines can be located in any one of a number of locations (ICOL), where ICOL > IMC. The cost of serving a customer $i$ from location $j$ is $C_{ij}$. The program develops the best locations for placing the machines and assigns customers to machines to minimize the cost.

### (3) SNGL-RDL AND (4) SNGL-QAD

These programs are associated with the continuous or universal facility location problem and follow the procedure described in Section 19.2. We are free to choose any site in the area to place new facilities. The facilities are of *different* types and may even have transportation among them. Thus, a new facility is a facility when it is serving customers, whereas it is a customer when it requires the services from other facilities. The cost is measured in terms of number of trips times either the rectilinear distance (SNGL-RDL) or a quadratic or Euclidean distance (SNGL-QAD) between a facility and a customer. The objective is to place the facilities to minimize the total cost.

### (5) MULT-FAC

In continuous or universal location problems, multiple facilities of the same type are to be placed to serve, with minimum cost, a number of customers with different service demands. The combinational approach described in Section 19.3 is followed.

### (6) RATE-IRR

The program calculates the internal rate of return given the life of the project, the initial investment, and expected yearly income.

### (8) ASMB-LCR AND (9) ASMB-RPW

These two programs are for the assembly line balancing described in Section 5.4. The input to the programs are required cycle time, precedence relationships of the tasks, and the task times. The outputs are the number of stations required in the assembly line, the task assignments in each station, and the efficiency of the assembly line. Program 8 is based on the largest-candidate rule, whereas program 9 is based on the ranked-position-weight method.

## (10) Q-SINGLE

The program analyzes the single-server M/M/1 queuing model. Given arrival rate and service rate, it calculates probability of $i$ units being in the system, mean number of units in the system and in the queue, and average waiting times in the system and in the queue.

## (11) Q-MULTI

The program analyzes the multiserver model M/M/C/K, with $C$ servers and a finite queue length of $K$ units. Given the arrival rate, service rate, number of servers, and maximum allowed queue length, the program calculates the probability of $i$ units being in the system, the mean number of units in the system, and queue and average waiting times in the system and in the queue.

## (13) DATAGUIDE

The program gives a condensed manual describing the data input formats for commercially available popular plant-layout programs such as CRAFT, CORELAP, ALDEP, and PLANET. See Chapter 13 for the description of these programs and associated procedures.

## (14) FIN ANAL

The program displays yearly financial analysis over a 5-year period, given the initial investment and other cost data and expected sales and yearly trend, for each product manufactured. The program produces the following statements: sales, operating expenses, depreciation calculations, interest calculations, tax calculations, income statements, and earnings on investment. Refer to Section 15.7 for an example.

# FORTRAN PROGRAMS

## (2) HEUR #2

Because the program is written in FORTRAN, it runs with a FORTRAN or WATFIV compiler. The program follows the procedure in Section 18.2. It extends the problem described in HEUR #1 by allowing site preparation cost or fixed cost for each location. The cost may vary from location to location. The program places facilities in locations and assigns customers to the facilities to minimize the sum of transportation and fixed costs.

# EXECUTABLE FILES PROGRAMS

## (I) MRP.EXE

The program is associated with material-requirement planning, described in Section 11.4. The program develops, for each component, the order release schedule using the part-period balancing technique. For each product, the input consists of its bill of materials, initial inventory, and order and carrying costs for each component. Multiple products can be processed simultaneously. The output for each component is its planned order release that minimizes the total of order and carrying costs.

### (II) GROUP.EXE AND LARGEGROUP.EXE

These two programs are associated with group technology. They follow the tabular method described in Section 5.3 to form machine grouping and assign components to each group. For the same machine-component input matrix, it may be possible to develop different machine grouping based on the user-specified value of the percent, *P* (see Section 5.3 for details). The resulting groups may be further evaluated by additional objectives, as described in Section 5.4.

The programs can also be used to form machine grouping to reduce material handling when the production rates of the products differ considerably, as described in Section 10.2.

The program "GROUP" displays the results on the screen, whereas the program "LARGEGROUP" stores the output onto a file that can be printed later. Depending on the size of the machine-component matrix, either program, GROUP or LARGEGROUP, may be used.

### (III) SML.EXE

The program determines the optimum or near-optimum solution for placement of *N* machines in *N* locations in a job shop or cellular manufacturing.

The problem is referred to as a quadratic assignment problem in the literature. The program follows the procedure described in Section 10.3. The program also allows for a fixed cost such as site preparation cost.

### (IV) COMLADII.EXE*

The program develops a plant layout based on the information on products such as volume and necessary sequence of operations. Also necessary are the data on the area requirement for each department. The procedure followed in developing the plant layout is described in Chapter 12. The program allows a number of options, such as fixing the width or the length of the plant, fixing a department in certain position, the flexibility of interchanging two departments, and the flexibility of changing the shape of a department.

### (V) COMLAD3.EXE*

The program develops a plant layout and associated material handling by alternately optimizing each so that at the end, the best layout and material handling combination is obtained. The procedure followed is described in Chapter 14. Chain or belt conveyors of different widths and trucks with different capacities and costs may be used. In the iterative process, for a given layout the program selects the best MH combination, which may involve sharing of the certain trucks and/or use of conveyors on certain routes. The layout is then changed to reduce distances between departments that have the most material handling between them.

# Index

## A

## D

## E

## G



## H



## I

## J



## K



## L

## M

## N



## O

## P

## Q

## R



## S

## T

## U



## V



## W

## Y



## Z